Climate: Present, Past and Future

First published in 1972, this first volume of Professor Lamb's study of our changing climate deals with the fundamentals of climate and climatology, as well as providing global data on the contemporary climates of the twentieth century.

'Professor Lamb has indeed served us well, in a complex field of study of the utmost significance to mankind' – *The Times Higher Education Supplement*

'Destined to become one of the classics of climatology' – *Geographical Magazine*

'Professor Lamb is to be congratulated on producing such an informative book, one which is almost certainly destined to become a classic synthesis of our present understanding of physical and dynamical climatology.' – *Times Literary Supplement*

Climate: Present, Past and Future

Volume 1

Fundamentals and Climate Now

H. H. Lamb

First published in 1972
by Methuen

This edition first published in 2011 by Routledge
2 Park Square, Milton Park, Abingdon, Oxon, OX14 4RN

Simultaneously published in the USA and Canada
by Routledge
711 Third Avenue, New York, NY 10017

Routledge is an imprint of the Taylor & Francis Group, an informa business

Publisher's Note
The publisher has gone to great lengths to ensure the quality of this reprint but points out that some imperfections in the original copies may be apparent.

Disclaimer
The publisher has made every effort to trace copyright holders and welcomes correspondence from those they have been unable to contact.

A Library of Congress record exists under ISBN: 0416115306

ISBN 13: 978-0-415-67959-6 (set)
ISBN 13: 978-0-415-67950-3 (hbk)
ISBN 13: 978-0-203-80431-5 (ebk)
ISBN 13: 978-0-415-68222-0 (pbk)

Clima
Present, Past and Futu

CLIMATE:
Present, Past and Future

VOLUME 1

FUNDAMENTALS AND CLIMATE NOW

H. H. LAMB

First published in 1972
by Routledge,
2 Park Square, Milton Park, Abingdon, Oxon, OX14 4RN

270 Madison Ave, New York NY 10016
Reprinted 2001 by Routledge

Transferred to Digital Printing 2006

Routledge is an imprint of the Taylor & Francis Group

Filmset in Photon Times 11 on 13 pt. by
Richard Clay (The Chaucer Press) Ltd,
Bungay, Suffolk

ISBN 0 416 11530 6 (hbk)

Publisher's Note
The publisher has gone to great lengths to ensure the quality of this reprint but points out that some imperfections in the original may be apparent

Printed and bound by CPI Antony Rowe, Eastbourne

Contents

Contents

Part II Climate now

Tables

Part I

Tables

Figures

Part I

Figures

Figures

Figures

Figures

Part II

Figures

Acknowledgments

The author acknowledges his indebtedness to two successive Directors-General of the Meteorological Office, Sir Graham Sutton, F.R.S., and Dr B. J. Mason, F.R.S., who arranged the facilities for research and study over the past fifteen years through which he built up the knowledge to write this book. Much of the ground covered is either new or has never been accessible in a reference book before. There are still very few meteorologists engaged in research on changes of climate, especially the climatic record of the past and the probable physical causes and mechanisms underlying the changes observed. Until lately all this was left by meteorologists, apart from a few honoured exceptions such as the late Sir George Simpson and Dr C. E. P. Brooks, to be the province of geologists, botanists and others who uncovered the evidence but who could only hazard interpretations in an uncharted field. There are still far too few meteorologists, or others, with the basic knowledge of past climates and of the processes that continually bring about changes of weather and climate, when one considers the importance of the latter to the practical problems of the present day. The subject is, however, beginning to develop. Only now has the fund of available knowledge reached the stage at which a wide, and balanced, survey such as is here attempted can be made. The first volume of this work has taken seven years to write. For this very reason, the author extends his thanks to all those (probably including colleagues and employers) whose patience has been tried by the effects of his exertions or by the long waiting for their completion; especial thanks are due to his wife and family whose devotion has been equal to putting up with late hours and much forbearance in whatever direction was required.

Several colleagues have helped by reading and commenting on parts of the text. Mr C. Hawson's advice helped greatly in Chapter 5; Mr R. P. W. Lewis, also of the Meteorological Office, and Dr G. Brier of the United States Weather Bureau, looked over the statistical methods section of Chapter 6; and the author here repeats his thanks to them. The contributions of others are acknowledged at appropriate points in the book.

The author gratefully acknowledges the kindness of the copyright-holders in allowing him to reproduce their diagrams and maps. Specifically, he expresses his thanks to Professor C. W. ALLEN and the Royal Meteorological Society for the use of fig. 2.11; to Dr R. Y. ANDERSON of the University of New Mexico for fig. 6.9 (*a*); to Professor FRANZ BAUR of Bad Homburg for figs. 2.1 and 10.26 which are taken with little adaptation from his work; to Professor E. A. BERNARD of Louvain for the design of fig. App. III.1, which essentially follows his presentation of MILANKOVITCH's *mathematische Klimalehre*; to Professor J. BJERKNES for figs. 10.4, 10.5 and 10.10; to Dr J. BLÜTHGEN of the University of Münster, Westphalia, and his publishers WALTER DE GRUYTER, Berlin, for allowing the adaptation of his maps used in figs. 4.6 (*a*) and (*b*); to Dr G. BRIER for fig. 6.4 (*a*); to Professor T. J. CHANDLER and the Royal Geographical Society for figs. 11.1 (*a*) and (*b*); to Professor H. FLOHN of the University of Bonn for figs. 3.20, 3.21, 4.15 and 4.21; to Dr A. H. GORDON for figs. 4.16 and 4.17; to Professor F. K. HARE for figs. 5.4 and 5.5; to Professor H. E. LANDSBERG for fig. 6.4 (*b*); to Professor G. H. LILJEQUIST and *Geografiska Annaler* for fig. 6.1.; to Professor F. LINK and *Planetary and Space Science* for fig. App. I.1; to Dr H. VAN LOON for figs. 4.4 and 4.5; to the late Dr M. MILANKOVITCH and his publishers GEBRÜDER BORNTRAEGER, Berlin, for fig. 2.5; to Dr V. MIRONOVITCH and the Meteorological Institute of the Free University of Berlin for fig. 10.25; to Dr J. M. MITCHELL for figs. 6.2 and 6.5; to Dr J. NAMIAS for fig. 10.3; to Mr B. N. PARKER for figs. App. IV.1 and 2; to Professor H. E. SUESS for figs. 2.4 (*a*) and (*b*), 10.24 and fig. App. I.2; to Dr A. J. J. VAN WOERKOM for figs. 2.6 and 2.7; to various Soviet research workers who have sent him their published papers, the worth of which is acknowledged in this book; to the American Meteorological Society for fig. 10.24; to the New York Academy of Sciences for figs. 6.4 (*a*) and (*b*) and 6.9; and to the Comptroller of Her Majesty's Stationery Office, London for permission to reproduce figs. 3.3, 3.16 (*a*)–(*d*), 3.17 (*a*) and (*b*), 4.1, 5.1, 7.14 and 11.2–11.8 inclusive.

The author acknowledges his indebtedness to the Royal Greenwich Observatory for the data in Table App. I.2; to Dr J. R. BRAY for the data in Tables 10.4–6 and to Dr D. J. SCHOVE for the data in Table App. I.4.

The author has especial cause to be grateful for the Russian translations service of the Meteorological Office librarians, and workers in many countries who sent him their publications.

Foreword

The study of climate necessarily involves the collection and processing of large amounts of data gathered from all parts of the globe, but for many years meteorologists could do little more than to identify and describe the main climatic zones of the Earth. We are still a long way from a complete dynamical and physical explanation of climatic features, but it is now recognized that a proper examination of the long-term and large-area aspects of atmospheric variations must be as deeply rooted in physics as is the study of transient weather systems. The author of this work is a professional meteorologist who has spent his life among mathematicians and physicists, and it is natural that his account of climate begins with its fundamental physical aspects. But climatology is more than a branch of physics and it is in the wider aspects of its study that the unique nature of this book lies.

The climatologist has to consider not only what is happening today, as revealed by the vast amount of information collected every day by the national weather services, but also what happened long before man invented instruments to measure and record the atmospheric elements. He has to call on many fields of knowledge for his facts—on history, geography, archaeology and geophysics, to name but a few. The essential clues to past variations of climate may have to be sought in the story of the wanderings of primitive peoples in search of food, or equally, by the analysis of ice deposited centuries ago in Greenland.

Mr Lamb is recognized, not only in Great Britain, but in all countries of the world as a leader (if not the leader) in this work. He is a man of wide learning and boundless energy who has devoted his life to the study of climate by all possible means. His eminence in this work was recognized some years ago when he was awarded a Special Merit Promotion in the Meteorological Office to enable him to engage in his researches without the trammels of routine duties. He is now to enter academic life as the Head of the newly formed Climatic Research Unit of the University of East Anglia.

This is the book that I always hoped Mr Lamb would write. It is a treasure house of information gathered in a lifetime of dedicated work. I know of no other work in this field that approaches it in scope and reliability. I have no doubt that what I have been reading are the proofsheets of a classic of meteorology, and that here, if anywhere, climatology really enters into its own.

Sir Graham Sutton,

C.B.E., D.SC., F.R.S.,

formerly Director-General of the Meteorological Office

tropical warmth[1] in England, continental Europe and North America, but the general belief was that for at least the last two thousand years the climate had ceased to undergo any important change. Two things may have contributed a good deal to this attitude. Firstly, Classical studies revealed that there was much, if not quite perfect, similarity between the climate of the Mediterranean two thousand years ago and in modern times. (We shall come later in this work to regard the similarity of these two epochs two thousand years apart as to some degree fortuitous. And the fact that Britain impressed Tacitus and other Italians in various centuries as a misty isle with a mild climate only means that *relative to Italy or other parts of the continent* Britain's climate has maintained the same sort of differences of frequency of fogs and of extreme temperatures: it does not necessarily imply that the frequencies in either place have stayed constant – they could have undergone more or less parallel variations. Indeed, we shall come across many indications that something like this happened.) Secondly, it did happen that the European climate in the 1880s and 1890s had reverted to figures close to those ruling between 1760 and 1820.

One must probably also make some allowance for people becoming mesmerized with the importance of averages taken over long periods of time in an era when averages were the only statistic that it was practicable to compute – and even they represented a fearful labour.[2]

Modern computing equipment and statistical methods have made it possible to compute a variety of statistics that tell one much more about a series of observations than mere averages can do. Moreover, from 1900 onwards for thirty to fifty years climates in almost all parts of the world underwent an unmistakable warming, significant in many of its aspects, both in the statistical sense and as regards effects upon the natural environment, agriculture and the human economy. And latterly there have been signs of reversion and renewed instability of climate that have produced in England since 1950 several of the wettest summers in a 200-year-long record, and in the winters an increased incidence of snow, as well as two that were respectively the coldest (1962–3) and the driest (1963–4) winters for over two hundred years. Few people nowadays therefore – and certainly no

---

by I. Venetz in 1821 and published in 1833. A similar conclusion about the glaciers in Norway was published by J. Esmark in Christiania in 1824. The concept of a general ice age, with a great ice sheet over all northern Europe, made widely known by J. L. R. Agassiz between 1837 and 1847, seems to have been first put forward by R. Bernhardi in 1832. (See R. F. Flint 1957 and M. Schwarzbach 1961 for further history of the development of the idea.)

1. Deduced in 1686 by Robert Hooke, F.R.S., from fossil ammonite and turtle shells found at Portland on the south coast of England.

2. Arago, writing in 1858 about a series of daily observations made in Florence in the 1650s and 1660s with some of the earliest thermometers, the calibration of which had been determined by Libri twenty-six years earlier in 1832, was able to quote the extreme values of warmth and cold represented by the Florentine series but had to remark that a better comparison with contemporary experience would be possible when the averages had been worked out: at that date they were still not ready. Comparison of extremes and ranges was common practice even till the early part of the present century, because averages took so long to compute.

Introduction

This book is a study of the development and history of climate. It is intended to meet a variety of needs. Students of meteorology and climatology and workers in other disciplines concerned with the impact of climate upon human affairs, upon the animal and plant kingdoms and upon the surface of the Earth itself, will find in it an account of what makes climates what they are, and of how they vary and have varied. Practical men – farmers, engineers, industrialists and especially those concerned with long-term economic development and planning – may gain from it a new appreciation of things to be allowed for where climate affects, or might come to affect, their schemes.

The question of whether climate is effectively constant is one on which prevalent opinion – even expert opinion – has changed several times. It was alarm about 'the sudden variations in the behaviour of the seasons', to which the climate of Europe seemed in the late eighteenth and early nineteenth centuries to have become 'more and more subject', that apparently led to the institution of the first country-wide networks of weather observation posts where temperature, barometric pressure, rainfall and so on were to be recorded daily. This was attempted by the government of France in 1775 and done by Prussia in 1817, and instruments were supplied for the purpose. An international network of observation points scattered between Greenland and eastern Europe was arranged in 1781 by the first meteorological society, in the Palatinate of the Rhine, the *Societas Meteorologica Palatina*, which owed much to the interest of the ruling prince. By the end of the nineteenth century, however, leading climatologists, including HANN in Vienna and MOSSMAN in Edinburgh, who had studied the accumulating observation records, were of the opinion that climate was so nearly constant that one had only to average the figures over a long enough period of time to acquire results of lasting validity, mean values to which the climate would always return. Climate was commonly defined just as 'average weather' and climatology as the statistical (and, it was thought, dullest) branch of meteorology.

It was admitted, to be sure, that there had been changes of climate in the geological past. It was known that many thousands of years ago there had been ice ages[1] and also eras of

1. As far back as the eighteenth century Swiss country-folk familiar with the scored and glacier-smoothed rock-faces and boulders to be found far down the valleys and in the forests were aware that the Alpine glaciers had had a far greater extent in some former time. The evidence was put before the *Societas Helvetica* in Luzern

tropical warmth[1] in England, continental Europe and North America, but the general belief was that for at least the last two thousand years the climate had ceased to undergo any important change. Two things may have contributed a good deal to this attitude. Firstly, Classical studies revealed that there was much, if not quite perfect, similarity between the climate of the Mediterranean two thousand years ago and in modern times. (We shall come later in this work to regard the similarity of these two epochs two thousand years apart as to some degree fortuitous. And the fact that Britain impressed Tacitus and other Italians in various centuries as a misty isle with a mild climate only means that *relative to Italy or other parts of the continent* Britain's climate has maintained the same sort of differences of frequency of fogs and of extreme temperatures: it does not necessarily imply that the frequencies in either place have stayed constant – they could have undergone more or less parallel variations. Indeed, we shall come across many indications that something like this happened.) Secondly, it did happen that the European climate in the 1880s and 1890s had reverted to figures close to those ruling between 1760 and 1820.

One must probably also make some allowance for people becoming mesmerized with the importance of averages taken over long periods of time in an era when averages were the only statistic that it was practicable to compute – and even they represented a fearful labour.[2]

Modern computing equipment and statistical methods have made it possible to compute a variety of statistics that tell one much more about a series of observations than mere averages can do. Moreover, from 1900 onwards for thirty to fifty years climates in almost all parts of the world underwent an unmistakable warming, significant in many of its aspects, both in the statistical sense and as regards effects upon the natural environment, agriculture and the human economy. And latterly there have been signs of reversion and renewed instability of climate that have produced in England since 1950 several of the wettest summers in a 200-year-long record, and in the winters an increased incidence of snow, as well as two that were respectively the coldest (1962–3) and the driest (1963–4) winters for over two hundred years. Few people nowadays therefore – and certainly no

---

by I. VENETZ in 1821 and published in 1833. A similar conclusion about the glaciers in Norway was published by J. ESMARK in Christiania in 1824. The concept of a general ice age, with a great ice sheet over all northern Europe, made widely known by J. L. R. AGASSIZ between 1837 and 1847, seems to have been first put forward by R. BERNHARDI in 1832. (See R. F. FLINT 1957 and M. SCHWARZBACH 1961 for further history of the development of the idea.)

1. Deduced in 1686 by ROBERT HOOKE, F.R.S., from fossil ammonite and turtle shells found at Portland on the south coast of England.

2. ARAGO, writing in 1858 about a series of daily observations made in Florence in the 1650s and 1660s with some of the earliest thermometers, the calibration of which had been determined by LIBRI twenty-six years earlier in 1832, was able to quote the extreme values of warmth and cold represented by the Florentine series but had to remark that a better comparison with contemporary experience would be possible when the averages had been worked out: at that date they were still not ready. Comparison of extremes and ranges was common practice even till the early part of the present century, because averages took so long to compute.

informed person – would still place much reliance on the easy assumption of climatic constancy so prevalent at the beginning of the century. But it is high time to get away from these repeated changes in the fashion of thought and take stock of the situation,

(*a*) to see the problem in that perspective which can only be gained from careful numerical assessments and comparisons extending over the longest possible period of time and over a world-wide survey; and
(*b*) to seek to understand the matter in terms of the processes going on in the atmosphere and oceans which must produce climatic changes; and to take note of any changes that may be relevant in the terrestrial and extraterrestrial environment.

These are the things this book is about. Only so can the erratic course of opinion be checked and substituted by the sure foundation of knowledge. Moreover, in recent years climatology, developing as a branch of physical and dynamical meteorology, and statistics, have both become a great deal livelier subjects.

The author is a meteorologist, and the book sets out to present the meteorology of climate and its changes. It contains the results of the first strictly meteorological investigation of the climatic sequence of the last thousand years besides giving a digest of the known facts, recent investigations and tentative conclusions regarding the climatic history of much longer periods of time, involving evidence contributed from many other branches of learning. In the nature of the case, these other fields, such as botany and geology, from which so far most of our knowledge of ancient climates has come, receive very cursory treatment. The author is conscious of the long years, indeed generations, of patient research that have gone to produce these results, and is very appreciative of the help and guidance he has received from original workers too numerous to mention save where their published works are cited. He has tried to do justice to all, while showing how the sum total of their evidence at present appears to the meteorologist; yet some unintended changes of emphasis or representation of evidence in fields that are far from meteorology are likely to have arisen. Those readers requiring deeper knowledge in fields other than meteorology should supplement this book by consulting the works cited.

Meteorology must, however, in the end provide the central viewpoint and unified vision of what is essentially a meteorological phenomenon. It sees the variations in different aspects of the weather in different parts of the world as manifestations of deviations in a single chain of events – the supply of heat from the sun, which, despite much that is wasted, warms the Earth and drives the winds that redistribute the heat to all parts of the world. Only some of the heat is conveyed by the winds; much is stored in the ocean and carried along with the ocean currents that the winds drive and which release heat to the winds in high latitudes. In the great circulation of winds and oceans, moisture is put into the air from the sea surface and transported to every part of the globe and to great heights in the atmosphere. The moisture produces what we call weather; its condensation produces clouds, rain and snow and in the process converts large amounts of latent heat into sensible heat (i.e. into warmth that can be felt) in places thousands of miles away from the tropical oceans where most of the evaporation occurs.

This way of looking at climate presents it as an organic whole, in which the different elements are intelligibly related to one another and combine to tell the same story. It becomes possible to use quite diverse fragments of information about the climates of any age to confirm and fit together like the pieces of a jig-saw puzzle, indicating the outlines of a connected picture – a global pattern – and finally to glimpse the sequence of global climatic patterns. This overall view is needed if we are to see the development, variations and changes of climate as an intelligible physical process.

This presentation is likely also to be of use to research in the disciplines other than meteorology that encounter evidence of the course of the world's climatic history. Practically every branch of learning is potentially involved, since the behaviour of climate touches all aspects of our Earthly environment. Hints are not lacking that astronomical factors are also involved, at least in the longest-term changes, and that semi-regular variations of solar behaviour may play a part in some of the variations from year to year and century to century. The record of former hot climates and deserts, of ancient floods and masses of ice, is registered – not always clearly and unambiguously – in the rocks and soils. The present and past distribution of plant and animal species is ultimately a response to climate and the environment that climate conditions, or has conditioned. The fate of human populations living near the climatic limit, whether of cold as in Greenland and Iceland, or of drought as in central Asia and parts of Africa, America and Australia, may be determined – and has been determined – by climate. So geologists, biologists, medical men, historians, archaeologists, oceanographers and others may all find something here that concerns them. It seems a reasonable hope, moreover, that the meteorological treatment and arrangement of the book may make clearer the significance of some of the evidence turned up in other fields of study. This could lead to the advancement of knowledge, by encouraging the unearthing of further evidence.

The aim of this work is to base understanding as closely as possible upon observed facts and known physical or mechanical processes rather than to present elaborate theoretical ideas that have little sure foundation. Over much of the field of past climates, particularly in regard to the ice ages, there has long been, and still is, a surfeit of rival, and often unrelated, theories. One reason for this is certainly that the greatest overall changes of world climate can only be brought about by the working together of many contributory influences, in some cases including things of very diverse natures. In a word, the big changes are polycausal. The chapters in Part I attempt to identify the most important influences in the genesis of climate and climatic variations and to isolate the characteristic effects of each. In some cases quantitative estimates of the magnitudes of the effects can be given. But this is by no means always possible as yet, particularly when several factors operate simultaneously. Whenever the causation of a given climatic shift remains unresolved, or presents difficulties despite fairly confident identification of the major influences at work, this is admitted. The search for fuller understanding is most likely to be helped in this way.

In the pooling of knowledge of past climates we see that all learning is really one, and that any rigid separation of science and humane studies can only hinder our quest. There is need for common sense and the systematic methods of science in sorting out the significance and trustworthiness of evidence. Many would-be interpreters of climatic history have gone astray through uncritical acceptance of travellers' tales. To the visitor from Rome or Greece, Britain has always seemed remarkably mild (though misty) for its latitude and Russia has always appeared to have dreadful winters. We also need the verdict of some immortal 'traveller through time' who knew these countries two thousand years ago, visited them again two hundred or even fifty years back and came again today. In some respects he could notice differences. The quest to find out the facts of past climates – let alone the agencies and processes responsible for different climatic behaviour – is one of the most fascinating in all science. It is a never-ending detective story, though like all detective stories it requires disciplined thought and testing of conclusions at every stage. Advance in knowledge of climatic history is bound to be a co-operative effort.

In a subject of such diversity one book cannot go into all details. Nor do we attempt here to follow equally far along every line of inquiry. This book seeks to provide the essential framework of knowledge on which further studies aimed at fuller, or more detailed, understanding must be built. Essentially, what has to be done is to lay the foundations of a hitherto neglected subject, physical and dynamical climatology, which embraces the long-range processes of meteorology and other physical aspects of the Earth's environment. To further this aim an extensive bibliography is given. Another aim, however, has been to give within this and the forthcoming volume more of the known or ascertained facts of climatic history than have hitherto been available in a single work as well as adequate reference data on present-day climate.

Meteorology was in no position to help very much in the interpretation of past climates and climatic relationships until quite recent years. There was no sufficient understanding of the general circulation of the atmosphere. Such understanding could not be expected before what has turned out to be the mainstream of the wind circulation, the powerful flow of the wind in the upper troposphere, was within our means of observation and so had been recognized for what it is and submitted to continual, day-by-day survey of its behaviour. It was the daily sending up of radio-sonde balloons and radar directed at them to follow the upper winds, at a world-wide network of places developed between about 1940 and 1955, that first made this possible. By now, theory, though still incomplete, has provided a far-reaching understanding of atmospheric flow that has been applied to the improvement of daily weather forecasting, to the introduction of dynamical methods of numerical prediction of the weather map up to three to five days ahead, to the introduction of longer-range weather forecasting and to a new view of climatic variations and vicissitudes.

Before one can produce a reasonable interpretation or classification of climates, describe climatic regions and define their boundaries – let alone the shifts of these boundaries from one decade, or one epoch, to another – one must be acquainted with the fundamental processes: the manner of the heat supply, the budget of gain and loss, the winds and ocean currents, and the moisture cycle. These will be considered in the opening chapters. A genetic interpretation of climate leads directly to understanding of the geographical distribution of climates and their boundaries. Such a genetic basis must be part of any simple classification of climates that aids the memory and makes the details 'fall into place'. We can then begin also to grasp the vagaries of climate as physical phenomena – i.e. to recognize some of the processes at work in them too. We are no longer forced to treat them only as haphazard events or at best as the systematic, but still obscure, element that emerges from statistical analysis of a time series of observations.

This first volume of the book is devoted to explaining the present-day world distribution of climates and providing a skeleton reference of facts and figures for comparison. Fuller details can be found in the climatic atlases of different countries and regions, in climatic tables and in the year-books of official weather services. This understanding of the present day provides a basis from which one can proceed to examine and discuss the differences that are found to have characterized other climatic epochs. There will follow in volume two a general survey of the multifarious evidence of climatic differences in the recent and more remote past. The magnitudes of the differences and the probable maximum rates of change should both engage attention. Information under these headings relating to quite recent years throws new light on the proper and judicious use of climatic statistics in planning for the future. Any climatic table records what happened during some period (that should always be specified) in the past. Yet most users of such tables are concerned only with the future. A climatic table can never be a substitute for a forecast, and to use one as a forecast involves the – usually unwitting – assumption of climatic constancy. Nevertheless climatic forecasting must wait until the proper scientific foundation, a knowledge of the physical factors and processes entailed, exists.

In the meantime, however, practical decisions involving the future have to be made. The available climatic statistics of the past have to be used for as much as they are worth; meteorological services and individual users have, moreover, to choose the most appropriate range of past years to include in a climatic table. This may differ according to the purpose the user has in mind. The most relevant past period is not necessarily the most recent thirty to fifty years, though it should probably always include the last ten years and beyond that such decades as seem, on grounds of similar environment in terms of ocean temperatures and extent of ice, and similarity of prevailing wind circulation patterns and strength, to show the best analogy. When concerned with *long* periods in the future, it becomes important to survey (if possible) a very much longer period of the past.

The title of this book should not be taken to mean that it provides a forecast of the climatic trend over future ages. For this, no adequate scientific basis exists as yet. The investigations described do, however, point to certain indicators and symptoms of variation or change (particularly in the sun, the atmosphere and the oceans) which should be watched and appear likely to prove valuable. In so far as a genuine insight is gained into the physical processes at work in climatic changes, these indicators may constitute part of the framework on which a scientific system of climatic forecasting can be built. And in these ways the reader will be made aware of the problems involved in thinking of future climates and can make his plans accordingly.

Until climatic forecasting becomes possible – including prediction of the magnitudes and distribution of effects of various external influences, whether solar, terrestrial or induced by man – any large-scale attempt to manipulate or modify world climates would be extremely hazardous and foolhardy. It could lead to disaster for wide regions.

Volume two will summarize much of what is so far known of past climates era by era either in the form of regional climatic histories, partly tabulated, or under the different categories of evidence – in some cases, both. The meteorological treatment in the present volume reveals at least some of the physical and dynamical processes at work in the climatic variations of the recent past; these may be of use in seasonal weather forecasting and in a first approach to prediction of the climatic trend for a few years or decades ahead. It will be seen, however, that any such forecast that could be developed in the foreseeable future could only be an estimate of the 'natural' or 'undisturbed' trend, 'other things being equal'. In addition to its probably wide margin of error due to incompleteness of the scientific basis, the success of such a forecast would be threatened by the supervention of influences of a manifestly unpredictable nature – great eruptions of volcanic dust and other natural events or changes due to the works of man, intended or unintended.

Some of the possible external events that might supervene to change the course of climatic history could have effects of disastrous magnitude. Such would be the case, for example, if increasing carbon dioxide in the atmosphere were to warm the climates as much as some estimates predict, and therefore raise world sea level by melting the Antarctic ice sheet, or if certain other events were to induce significant cooling. Generally some regions would gain and others would lose by any changes. Further research and stock-taking of the quantities involved is urgently called for, if we are to be able to face these questions. The book provides an introductory guide to such problems.

For the present, the problem of planning comes down to estimating the prudent margins of error for variability of climate that should be allowed in connexion with any project. Exploitation of every favourable turn of the climate may be gainful only if it is done with awareness of the threat of climatic reverse that hangs over such ventures and that will affect first the exotic crop and the marginal land, and enterprise.

Part I · Fundamentals

CHAPTER 1

Concepts and definitions: the weather, the atmosphere and the sun

In surveying the Earth's climatic history, and putting in perspective the succession of different regimes that can be recognized, care is needed in the use of words to describe times of different duration. This book follows usages that are widely current; though some readers may be surprised by, or look askance at, the use of the word 'epoch' in this context. Certain other common usages have been avoided, because they are either ambiguous or downright misleading. There is a present need to define the terms used in the climatic time scale and establish a stricter practice.

The following points should be noted:

Era is best reserved to describe long periods of time, especially the main geological eras lasting millions to hundreds of millions of years.

Age is necessarily more flexible in its use than era, because it is in general use by the public to mean various, vaguely long, time spans. Because of its use in the terms 'ice age' and 'interglacial ages', it may be permissible in this context to use it chiefly to describe periods lasting some thousands to some tens or even hundreds of thousands of years.

Epoch is generally used in this connexion to mean a shorter time, lasting some tens to some hundreds of years, marked by some climatic character which distinguishes it from the rest of the age or era in which it occurs. (The Greek origins of this word suggest either a *fixed point in time* – often taken to mark the beginning of a new departure or regime – or *holding a position*: on the latter ground the current climatic and geological usage seems justified.)

Period is a word in such wide use by the lay public that it may be impossible, or improper, to seek to give it a restricted, specialized meaning. In so far as it already has an established specialized meaning this is the period-length of a recurring, usually sinusoidal, oscillation.

Cycle should be restricted to mean phenomena that are 'cyclic', i.e. periodic, being repeated at regular intervals, in the course of a series of usually sinusoidal oscillations. (This word is often wrongly used, regardless of regularity or of any longer perspective of recurrences, by ill-informed writers to describe any run of a few years distinguished by more or less warmth, or rainfall, than the years that immediately preceded and followed them.)

Definitions in the realm of weather

Weather is taken to mean the totality of atmospheric conditions at any particular place and time – the instantaneous state of the atmosphere and especially those elements of it which directly affect living things. Its meaning may be extended to include exposure to radiation from the sun and to the sky, as permitted by the state of the atmosphere.

The **elements of the weather** are such things as temperature, atmospheric pressure, humidity, cloudiness, rain, sunshine, wind, visibility, considered separately.

Spells of weather are recognized by the continuance of some type, or repetitive sequence, of weather over several days or weeks at a time – as when all the afternoons are more or less cloudy and the nights and mornings fine, the temperature mostly low or near the average for the time of year, and the winds from west and northwest – or some other type.

It is convenient to distinguish **short spells** (or 'runs of weather') lasting just a few days and **long spells** lasting several weeks (apart from trivial interruptions of not more than three days). Sometimes the use of the word 'spell' leaves uncertainty as to which is meant. (In German a separate word-form *Witterung* is recognized as describing a sequence of weather (*Wetter*) over a few days, or sometimes longer, associated with some definable wind and weather pattern and with the prevalence of some repetitive characteristics.)

Diurnal changes of wind and weather are changes that occur directly because of the changing radiation conditions as day follows night and night follows day – for instance, when convection clouds over sun-heated land decay in the evening, and the gusty wind-currents that fed the convection die away, only to reappear next day.

Outside the tropics the diurnal rhythm is not always well marked, but it is so at times and occurs within spells in which continuance of the same general weather type from day to day can readily be recognized.

Seasonal changes can sometimes be recognized in ways rather similar to the diurnal rhythm, the characteristic difference being the longer time-scale. Spells of weather are sometimes so long that they give their stamp to the character of a whole season, and their ending marks the end of a **natural season**, a concept that is strengthened by some approach to regularity of date in the middle and higher latitudes – as with the mid-June return of the west winds and the late October wind and rain in most years over much of Europe, followed by anticyclones in mid November – and by great regularity of occurrence within some broader range of dates practically every year in low latitudes – as with the Indian monsoon. It may happen even in middle and high latitudes that successive summers or winters are characterized by rather similar spells; though outside the tropics this is seldom the case for more than two or three years together, and there is some suggestion that it is more liable to occur in alternate years.

Climate is the sum total of the weather experienced at a place in the course of the year and over the years. It comprises not only those conditions that can obviously be described as 'near average' or 'normal' but also the extremes and all the variations.

As this definition indicates, it is often advantageous to decide what range of conditions it is reasonable or admissible to call **normal** or **near average**. This is best done by using some standard statistical measure in relation to the frequencies of different conditions – for instance, the 'middle tercile' (that third part of the whole array of observations which is clustered about or comes closest to the average value) or the broader range of those observations which fall within one standard deviation on either side of the mean value.

We must notice that, because the actual ranges and the average values and other statistics will all differ according to what years we consider, we can only define the climate in terms of some period of time – some chosen run of years, a particular decade or decades, some **epoch**.

The epoch or **period of reference** (or **datum period**) should always be specified, since the figures for different periods themselves differ somewhat. Moreover, the figures representing all elements or aspects of the climate under discussion at any one time should relate to the *same* period of years. Otherwise, there is no necessary physical connexion between them, and understanding becomes impossible.

Climatology, the study of climate, the long-term aspects and total effects of meteorological processes, is (like meteorology) a branch of Earth-physics (geophysics). It is concerned with the various conditions of the atmosphere that occur and with everything that, habitually or occasionally, influences the condition of the atmosphere, either locally or over great regions of the Earth. Like any other branch of physics, it is also concerned with measuring the effects of such influences, and seeking to discover laws or principles of general application governing their behaviour and interconnexions.

Changes of climate inevitably involve the slowest, and longest-lasting, and probably the largest-scale, processes that meteorology is concerned with. They appear as changes of the average level of temperature in different parts of the world of the order of 2–12°C as between ice ages and warm interglacial times, and changes from a fraction of a degree up to 2–4°C as between various post-glacial periods of much shorter duration. The extreme values occurring within these different periods, and the frequencies of conditions over-stepping this or that threshold, undergo corresponding changes. Rainfall variations are also involved; in certain cases, for instance where the arid zone undergoes some displacement, where mild moisture-bringing winds penetrate further or less far into the polar regions, or where the prevailing winds and lee effects at a mountain barrier undergo a shift, the average annual rainfall may be multiplied many times or reduced to a small fraction of its former value. Apart from these extreme cases, changes of average annual rainfall commonly amounting to 5–15% and, in regions threatened with aridity, occasionally 50–100% between periods of the order of one century and another, are found in many parts of the world.

Genesis of climate. The climate of any place is produced by:

(1) *Radiation:* the net gain or loss of heat in the form of radiation. This is determined by the amount of energy available as incoming solar radiation at the given latitude, combined with the clarity (transparency) of the atmosphere and height of the place above sea level, minus the amount of radiation emitted by the Earth's surface and atmosphere at the temperatures prevailing.

This radiation balance undergoes characteristic diurnal and seasonal changes. It is also dependent upon the weather, especially the cloudiness.

(2) *Transport of heat.* Winds and ocean currents carry heat to and from any place. This is convection on the largest scale. Some of the heat is brought along by the wind as latent heat (absorbed during evaporation at a water surface and released again in the air as sensible, or feelable, heat during the condensation of water droplets and ice crystals).

(3) *Transport of moisture.* This is carried by the winds, partly as invisible vapour and partly in its condensed forms as fog, clouds, rain, snow, hail, etc.

(4) *Transport of other matter* carried by the winds, e.g. solid matter picked up from the surface – dust, loose dry soil, sand, fine snow and ice grains, sea-salt (left in the air after evaporation of ocean spray), smoke particles and other forms of pollution, solid, liquid and gaseous.

(5) *Nature of the ground surface.*

(i) Slope of the ground, particularly its aspect to the sun and sky, since these determine the intake of radiation per unit area.

(ii) Nature of the surface as regards efficiency of absorption and reflection of radiation.

(iii) Specific heats and the thermal conductivity of the rocks and soil.

(iv) Porosity, aeration, drainage and prevailing dryness or wetness of the soil.

(v) Extent of water bodies, and prevailing temperature of the surface waters, in the region. (The water temperature is greatly affected by such things as the depth of the water body and convection within it, its volume and sources of supply, ranging from cold mountain rivers to warm ocean currents. But in all cases the water temperature is liable to less day-to-day, and month-to-month, variation than the surrounding land.)

(vi) Great mountain chains, and even small hills, act as obstacles to the free flow of the winds. Uplift and vertical motion of the air, due to this and other causes, produce expansion (on ascent) and compression (on descent), accompanied respectively by *adiabatic* decreases and increases of temperature of the air. In consequence, clouds are formed and evaporated. The clouds, if their development is sufficient, produce either showers and thunderstorms or gentle, steady rain – according to the type of vertical motion prevailing in them. And meanwhile the upper surface of the clouds reflects and loses much of the incoming radiation, somewhat like an expanse of snow.

(vii) Over some regions, such as forest, the vegetation cover effectively presents a raised surface to the winds and to radiation exchanges with the sun and sky. This raised surface or 'vegetation mat' is insulated from the ground by the stagnant air layer trapped within and beneath it, where a special climate with peculiar values of temperature, humidity and illumination prevails.

(viii) Built-up areas present a case in some ways analogous to (vii), though the climate within them is increasingly artificial. Not only are the quantities of artificial heat increasingly significant but rain (and often even snow) which falls on the streets and pavements is soon run into pipes and no longer available for evaporation, etc., or to exercise any further influence on the local weather. Towns and industrial areas are also sources of pollution.

Items (1), (2), (3), occasionally (4), 5 (ii), (5) (v), and occasionally (5) (vi) and (vii), control the climates of wide regions. Other items under (5) are essentially local. One may think of these controls chiefly as three: *radiation, wind (and ocean) circulation and nature of the surface.* These are dealt with in more detail in the next chapters. It is difficult, however, to deal with each in isolation, because each affects and reacts upon the others.

The atmosphere

The composition and structure of the Earth's atmosphere are dealt with in the next chapters and will be better understood after reading them. Nevertheless it will be convenient to introduce here the broad framework of the atmosphere's structure and the terms used to describe it.

Composition. Air is a gas consisting of a mixture mainly of nitrogen (about 78%) and oxygen (about 21%) and much smaller quantities of other gases, mostly in almost constant proportions according to modern observation; there is a variable admixture of water vapour.

Suspended matter consists of variable quantities of condensed water, in the forms of ice crystals (snowflakes, etc.) and drops (cloud droplets which may grow into drizzle or raindrops and freeze into hailstones), and miscellaneous (chiefly solid) impurities. Most of the *particle sizes* of the suspended matter are microscopic (under 0·01 mm diameter); *cloud and fog droplets* are for the most part under 0·1 mm in diameter; particles with much larger diameter fall out of the air at rates familiar in falling *drizzle* (droplet diameters 0·1–0·5 mm, terminal velocities of the order of 1 m/sec (metre per second)) and *rain* (drop diameters from 0·5 up to 5·5 mm, terminal velocities from 2 to 9 m/sec). Within the lower atmosphere (troposphere) impurities of all sizes are washed out by the drizzle and rain, etc.

Distribution with height. The density of the atmosphere is concentrated towards the surface of the Earth by the pressure of the overlying air layers. This, together with as much of the layer structure of the atmosphere as we shall have occasion to mention, is seen in fig. 1.1. As shown by the pressure scale at the left of the diagram, half the atmosphere's mass is contained within the bottom 5–6 km; over three-quarters of it is within the troposphere, the region within which all weather is produced. There is no weather, as the word is ordinarily meant, above that because only minute quantities of water vapour get carried up any higher. The higher layers of the atmosphere seen in fig. 1.1 will concern us very little and only so far as they may be capable of affecting the flow, or checking the vertical motion, and indirectly affecting cloud development or radiation conditions, within the lower atmosphere.

Layering. Vertical mixing of the atmosphere depends upon its thermal structure and occurs mainly by convection within those layers where temperature decreases fairly rapidly with increasing height – viz. above the heating surfaces represented by the Earth's surface and the upper surface of the ozone layer (see the temperature curve at the right-hand side of fig. 1.1). Heat is also produced by ionization processes above 80–100 km, but such remote, very sparse, regions of the atmosphere do not greatly concern us here. Vertical motion is inhibited in regions where the temperature increases, or changes little, with height. This tends to isolate the upper and lower layers from each other, as regards both composition (gases, water vapour, particulate matter) and general flow of the winds.

The terms used are:

Troposphere. The lower atmosphere, well mixed by vertical motions, which carry up water vapour, etc., from the Earth's surface and, in the process, produce the weather. Temperature generally decreases with increasing height. About 90% of the atmosphere's mass is within the troposphere over latitudes less than 30°; generally 70–80% over other latitudes.

Tropopause. The upper limit of the troposphere; it is normally defined by a sharp decrease (or cessation) of the rate of temperature lapse with increasing height. The tropopause is found at heights mainly about 11 km over middle and high latitudes, but 16 or 17 km over lower latitudes.

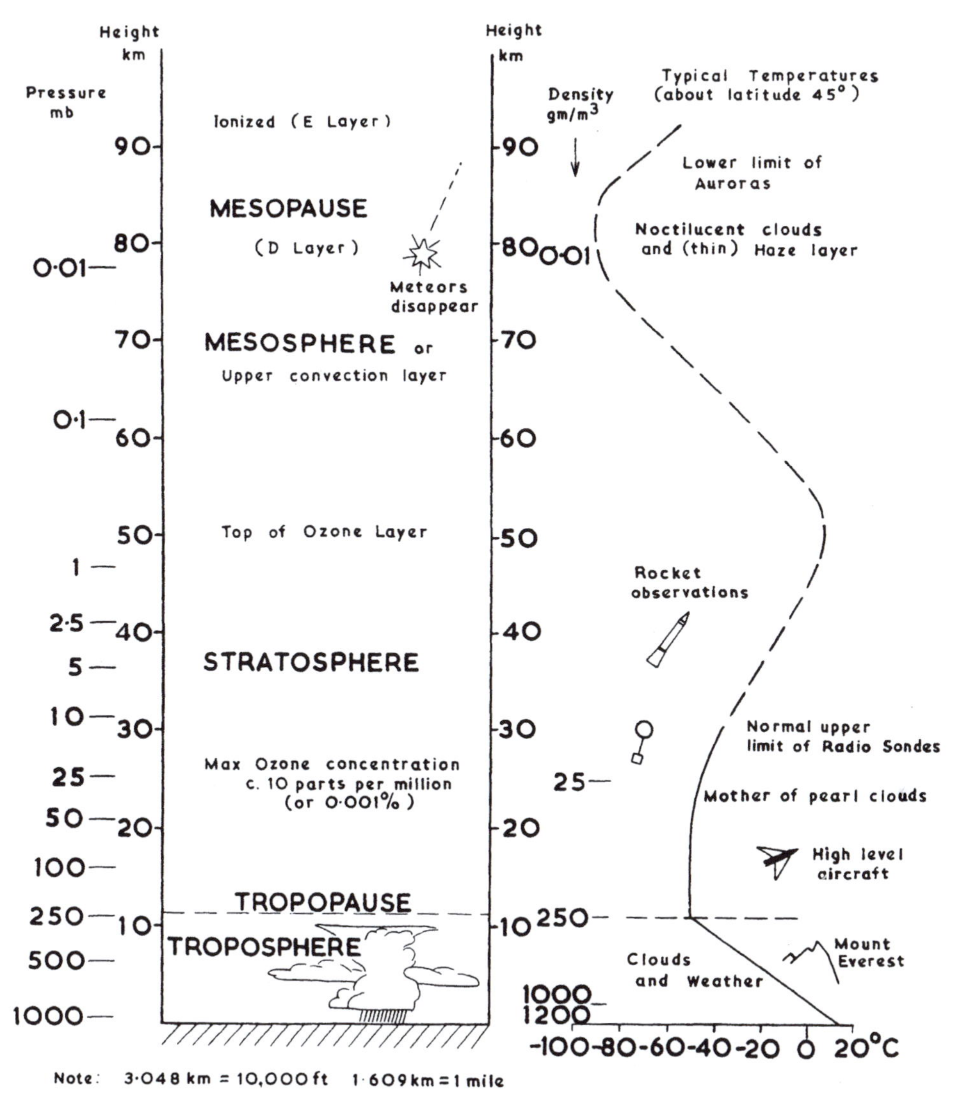

Fig. 1.1 Vertical divisions of the atmosphere.

Note: Since 1960 rocket observations have been pressed to greater heights, many to 65 km, some to 130 km.

The minimum heights (perigee) of artificial satellite (sputnik) orbits are commonly 150–200 km.

Stratosphere. The region above the tropopause as far as the level of maximum temperature (sometimes called the 'stratopause') about the top of the ozone layer (50–55 km). Temperature in the lower stratosphere often changes little with height, then generally increases to the top of the ozone layer. In the region of winter darkness over the poles, the polar winter night, the upper stratosphere is unheated and no ozone is produced; consequently the temperature then, on the whole, decreases with height all the way up to the D layer or mesopause, and the limits of the stratosphere become indefinite.

Mesosphere. Region of decreasing temperature with height above the levels warmed by absorption of solar radiation to produce ozone. The mesosphere is well mixed by convective motion, and any minute quantities of water vapour penetrating above 50 km may be carried up and occasionally condensed as tenuous clouds at the top of this layer – the luminous night clouds (*noctilucent clouds*) lit by the sun over high latitudes in summer. Thin haze which has also been reported at the top of the mesosphere may consist partly of volcanic and partly meteoric dust.

Mesopause. The top of the mesosphere, at 80–85 km. Temperature begins to increase strongly with height above this, owing to heat abstracted from the solar beam in ionizing the atmospheric gas molecules. At the mesopause atmospheric density is only about one hundred thousandth part of what it is at the Earth's surface.

Ionosphere. The regions above about 80 km in which the atmosphere is ionized and charged ions and free electrons are sufficiently abundant to reflect radio waves. Particular layers, known as the E, F_1 and F_2 layers, at about 110, 160 and 250 km, represent maxima of ionization, due to absorption of some of the shortest-wave solar radiation.

Thermosphere. The atmosphere above the mesopause to its outermost regions. The temperature increases with height, apparently throughout.

Exosphere. The outermost part of the atmosphere, around 700 km above the Earth's surface and beyond, where the atmosphere is so thin that collisions between atmospheric particles are rare and those with velocities great enough escape into space. The numbers of such escapes are presumably more or less balanced by the particles coming in, but it is probable that the composition and also the total mass of the atmosphere has changed in the long course of the Earth's history.[1]

For further definitions of the features of atmospheric structure and atmospheric phenomena the reader should refer to *The Meteorological Glossary*.

1. The age of the Earth is currently estimated as about 5000 million years.

The sun

Dimensions and other aspects of the sun's behaviour and structure that concern, or may concern, climates on the Earth are:

Diameter: about 100 times the Earth's diameter.

Mass: about one-third of a million times the Earth's mass.

Rotation period: the speed of rotation varies with the solar latitude (ϕ_s), being fastest at the solar equator. The *sidereal rotation* (i.e. the rotation relative to the stellar background) is $14°{\cdot}38 - 2{\cdot}7 \sin^2 \phi_s$ in degrees of solar longitude per day. The *synodic rotation* (i.e. the rotation relative to the Earth) is taken as $13°{\cdot}39 - 2{\cdot}7 \sin^2 \phi_s$. The sidereal *minus* the synodic rotation is equal to the Earth's orbital progress per day, but the figure varies because the Earth's orbit is elliptical. These figures give 25·0 days for the sidereal rotation period at the sun's equator. The rotation period relative to the Earth averages $26{\cdot}90 + 5{\cdot}2 \sin^2 \phi_s$ days according to solar latitude – i.e. it averages 26·9 days at the sun's equator, about 28·2 days at $\phi_s = 30°$ and 30·8 days at $\phi_s = 60°$.

Average distance of the sun from the Earth: approximately 149 500 000 km (92 900 000 miles).

Angle subtended by the sun's disk as seen from the Earth: $\frac{1}{2}°$.

Photosphere: Source of most of the radiation reaching the Earth, its surface therefore constitutes the brilliant orb normally seen from the Earth. Its temperature appears to be about 6000°C. The deeper, hidden layers of the sun are believed to be at much higher temperatures, rising to perhaps 20 million °C, at the centre.

Sunspots: Dark spots or areas on the surface of the sun, occasionally visible to the naked eye at sunset or when the sun is seen through a suitably thick haze or fog. They are, in fact, regions of only relatively less luminosity in the photosphere with a temperature about 4000°C. Their occurrence varies with an irregular period of average length about 11 years. The only systematic indices available of the history of solar disturbance over any long period of past years relate to sunspots, particularly the Zürich series of *relative sunspot numbers* (annual numbers from A.D. 1700, monthly values from 1749) computed by a formula due to R. Wolf of the Zürich Observatory, which takes some account of the grouping as well as total number of spots, and the Greenwich series of *sunspot group areas* (from the daily sun photographs made at the Royal Greenwich Observatory, starting

1874).[1] The series of daily observations of the sun are, in fact, completed by arrangements between observatories in widely separated parts of the world. The average, and the maximum, amount of solar disturbance within the 11-year sunspot cycles differs from cycle to cycle: various longer periodicities may be involved – e.g. about 89 and 178 years. (The greatest sunspot maxima so far observed were in 1778 and 1957.) The spots appear to be vortices within the photosphere and are centres of strong magnetic fields. The Earth's magnetic field is disturbed by the associated phenomena. The distribution of magnetic polarity in sunspots follows a pattern which reverses in successive 11-year cycles: in some connexions, therefore, a 22-year cycle may be recognized.

Sunspots are the central parts, or cores (VITINSKII 1965), of disturbed or 'active' regions of the sun which include a variety of kinds of disturbance. Around the sunspots bright **faculae**, in which the solar output of radiation is intensified, develop. Although the occurrence of faculae keeps more or less in step with the development of sunspots, the faculae fade away very much more slowly than the spots with which they were associated; faculae may remain visible for up to a few months after the disappearance of a large sunspot group, marking where the spots were. A small group of faculae is also commonly the first visible sign of development of a new group of sunspots, though only a few hours beforehand. Some faculae are not obviously related to spots.

Chromosphere: an outer region of the sun, above the photosphere, seen from the Earth only during total eclipses when it appears as a region of rosy colour. It is within the clouds of hydrogen and gaseous calcium in the chromosphere that the short-lived, brilliant **solar flares** occur either above sunspots or near the edges of the spots. Flares, too, are therefore associated with the same general course of variation of solar disturbance as sunspots. The flares are sources of intense bursts of *ultra-violet* and *long-wave* (*radio-wave*) *radiation*, which reach the Earth in about 8 minutes; they also eject streams of electrically charged particles (*corpuscles*) which take about 26 hours to reach the Earth, disturbing the Earth's magnetic field on arrival (*geomagnetic storms*).

Corona: the outermost regions of the sun that are ever visible, seen only during total eclipses of the main orb. Extensions out to 23 million miles have been photographed. The corona evidently consists of very sparse ions and electrons moving at such high speeds as to imply a temperature of about 1 000 000°C. There is no sharp boundary to this outer region

1. The formula for the Wolf relative sunspot number W is

$$W = k(10g + f)$$

where g is the number of groups of spots, f is the total number of spots present and k is a constant.

The sunspot group areas S, measured in millionth parts of the visible hemisphere of the sun from the Greenwich sun photographs, are found to be strongly correlated with the Wolf number (correlation coefficient + 0·85). The relationship may be conveniently expressed as

$$S = 16{\cdot}7\ W$$

of the sun; and, according to one modern view, the Earth itself may be regarded as within the tenuous outer parts of the sun's atmosphere though protected from the dangerous types of radiation (which would destroy life) by their absorption in the Earth's atmosphere.

Solar constant, i.e. the average intensity of the solar beam at the Earth's distance before atmospheric absorption affects it. Over the years 1926–50 this was found to be:

$$1{\cdot}99 \text{ g cal/cm}^2\text{/min (within } \pm 5\%\text{)}.$$

This is equivalent to 1·40 kW/m^2. Measurements in the 1960s have indicated 1·94 g cal/cm^2/min (within $\pm 1\%$) as the best value of the solar 'constant' in these years: the value is liable, however, to fall below this when sunspot activity is either notably great or little (see p. 18).

For more information about the sun, and our observational knowledge of it, the reader may usefully consult H. W. NEWTON's *The Face of the Sun*, London (Pelican books) 1958. Some further information is, however, given in Chapters 2 and 6 of this book and in Appendix I, where historical data are tabulated.

CHAPTER 2

Radiation and the Earth's heat supply

The source of all the energy that heats the air, the ground and the seas, and drives the winds and ocean currents, is the sun. By comparison, the flow of heat from the Earth's interior (*geothermal heat*) is quite negligible, and is generally supposed to have been so for at least the last 500 million years. Averaged over the Earth, and over the year, about 720 g cal of radiation for each square centimetre of surface (corresponding to 2 cal/cm^2/min falling on any surface normal to the solar beam) are received daily from the sun at the outer limit of the atmosphere; and between 300 and 350 cal reach and are absorbed at the Earth's surface. The geothermal heat flux from the Earth's interior averages 0·1–0·2 cal/cm^2/day (see footnote 1, p. 17); it is locally rather higher, 0·4–0·6 cal/cm^2/day, in the extensive Tertiary-era volcanic areas in Iceland, and only very locally in contemporary volcanic areas are average values ten times or more than this attained (though in a volcanic vent enormous heat fluxes occur during eruptions). Even so, it is useful to remember that the temperature in all parts of the world normally increases with depth in the ground: the world average rate is about 3°C/100 m depth in the uppermost layers of the crust – lower down the rate eases off, so that at 35 km depth the average temperature is thought to be about 600°C. In much of Iceland – the youngest volcanic land area of its size in the world – the rate of temperature increase with depth in the uppermost layers is as great as 10–15°C/100 m, rising in the immediate vicinity of hot springs and steam holes to over 10° per metre.

Heat reaches, and is lost from, the surface of the Earth and all points within the atmosphere, by radiation, convection and conduction. Some surfaces reflect back and waste much of the radiation falling on them. Conduction plays a minor part because neither the air nor most types of ground are good conductors.[1] Well aerated, dry soil or sands and (most of

1. The thermal conductivities in g cal/cm^2/sec for a temperature gradient of 1°C/cm depth of different substances and surfaces are given by ALISSOW *et al.* (1956, pp. 226–8) as:

Air	0·00005
Snow	Mostly 0·0003–0·0005, but less or more than these limits if the snow be either very loose (porous) or wet/hard-packed respectively
Water	0·00124
Dry sand or soils	0·0020
Turf	0·0021

all) loose, new-fallen snow are bad conductors of heat, so that the surface is exposed to the full impact of radiation, mitigated almost only by convective heat transfer in the air above. Water is not much better than some snow as a conductor, but it can also transport heat by convection. Moreover, the specific heat of water is big – i.e. water has a large heat capacity; it requires a greater quantity of heat per unit mass to change its temperature than any other common substance. Hence, water surfaces and wet ground are subject to much smaller ranges of temperature with the transient changes of season and weather, and from night to day, than either dry ground or snow or ice. In the case of water bodies, convection below, as well as in the air above, the surface commonly plays a large part in moderating temperature changes at the surface. And when a large body of water, such as a deep ocean, is heated up, say, by one degree, in the course of some long-continued climatic change, a great amount of heat is taken in and stored in the water in the process; this heat may at some later time be given off again and used to heat cold air passing over the water surface.

Since the Earth's main heat supply comes by radiation from the sun, radiation exchanges are the first stage in the production of the climates we observe, and should be considered first. The further stages are the circulation of the winds in the atmosphere and of the water currents in the oceans, which transport and redistribute (i.e. to some places advecting, from other places removing) both heat and available water vapour.

Any body whose temperature is not at absolute zero[1] radiates heat. The amount of radiation it gives out depends on the fourth power of its absolute temperature: in accordance with the STEFAN-BOLTZMANN law, the intensity I of the radiation (rate of energy flow) from unit surface area of a 'black body', or perfect radiating surface, is given by

$$I = kT^4$$

where k is a constant and T is the absolute temperature of the surface. Both the sun, with the temperature of its visible surface of the order of 6000°A, and the Earth, with its average surface temperature nearly 300°A, therefore radiate heat. The emission from the sun is enormously the more intense because of its size and high temperature, but the energy received per unit area of any surface offered at right angles to the beam falls off with

---

Wet sand or soils	0·0040
Ice	0·0051–0·0053
Rock	0·0097–0·0107

Notice that the conductivity of snow is only one-fifth to one-tenth of that of bare ground and only one-tenth to one-twentieth of that of solid ice.

Soils become much better conductors of heat when water replaces air in the interstices.

Note, however, the contrast with the conductivities of metals at typical atmospheric temperatures:

Bronze	0·10	Zinc	0·27
Steel	0·11	Copper	0·92
Cast iron	0·15	Silver	1·0

1. Absolute zero is −273°C. A temperature on the absolute or KELVIN scale (°A or °K) may be obtained from the Centigrade temperature simply by adding 273.

increasing distance according to the inverse square law. The result is that at the existing temperature of the Earth, and with the existing properties of the atmosphere in transmitting radiation, the heat given out by the Earth approximately balances the heat received. It is of critical importance for the very existence of life on this planet that this balance between incoming solar and outgoing terrestrial radiation occurs with temperatures at the surface of the Earth which permit the widespread occurrence of water in the liquid state. Indeed, surface air and water temperatures are mostly confined to the range between 0° and 30°C most generally tolerated by the forms of life developed on the Earth.

It is instructive to consider the overall radiation balance a little further at this stage.

The total energy emitted by the sun may be expressed as

$$4\pi R_s^2 \,.\, ke_s T_s^4$$

where R_s is the radius of the sun, k is the STEFAN-BOLTZMANN constant, e_s measures the effectiveness of the sun as a radiator of heat in comparison with a black body and T_s is the effective radiating surface temperature of the sun. This last must in reality be a complex entity to which the very different temperatures of various outer layers of the sun contribute.

At the distance of the Earth d_E this solar energy is spread over an imaginary spherical surface of area $4\pi d_E^2$.

So I the intensity of the solar beam per cm^2 at the distance of the Earth is given by

$$I = \frac{R_s^2}{d_E^2} \,.\, ke_s T_s^4$$

This intensity is effectively spread at any given time over that half of the Earth which is in sunlight, the total amount therefore being equal to the solar intensity I multiplied by the Earth's cross-sectional area πR_E^2 where R_E is the Earth's radius.. However, only some fraction α of this is actually absorbed and not reflected away. The amount absorbed by the Earth and its atmosphere therefore is

$$\alpha \,.\, \pi R_E^2 \left(\frac{R_s^2}{d_E^2}\right) \,.\, ke_s T_s^4$$

This must be approximately balanced by the total amount of heat which the Earth radiates in all directions to space, namely

$$4\pi R_E^2 \,.\, ke_E T_E^4$$

where e_E is the effective emissivity of the Earth and T_E is the effective temperature of the Earth's radiating surface.

Since $$\alpha\pi R_E^2 \left(\frac{R_s^2}{d_E^2}\right) ke_s T_s^4 = 4\pi R_E^2 ke_E T_E^4$$

$$T_E^4 = \frac{\alpha}{4} \cdot \frac{e_s}{e_E} \cdot \frac{R_s^2}{d_E^2} \cdot T_s^4$$

Putting observed values of α, R_s and d_E into this equation and taking reasonable values of the other quantities, we find that the effective temperature of the Earth's radiating surface T_E is of the order of 245–250°A (i.e. −28 to −23°C). This means that the effective radiating surface of the Earth must be at some height up in the atmosphere, and that radiation exchanges within the atmosphere which return much of the Earth's emitted radiation to the surface are responsible for keeping the average surface temperature of the Earth (at present about 288°A or +15°C) some 40°C higher than it would otherwise be.

Observed amount of energy supplied by the sun: constancy or variability

The intensity of the solar beam at the Earth's distance from the sun appeared from measurements by the observatories of the Smithsonian Institution over the years 1883–1913 to average 1·94 cal/cm²/min of surface presented at right angles to the beam: these units are also known as *langleys* (or *ly*) per minute. More recent calculations have suggested values varying from 1·89 to 2·07 ly/min, and Professor C. W. ALLEN (1958) on the basis of observations by the Australian Commonwealth Observatory from 1926 to about 1950 has recommended adoption of 1·98 or 1·99. For most practical purposes it may be taken as 2 cal/cm²/min. This is called the *Solar Constant*; it is the amount of radiation which in the absence of the atmosphere would be available to heat the surface.[1] Since its calculation may be in error by up to 5% on account of the allowances that have to be made for atmospheric effects, there is no certainty as to whether the value has been rising up to 1950 (as the figures quoted might suggest) or has been strictly constant over the years of observation. Nor have instruments mounted on artificial Earth-satellites so far supplied any comparative information about more recent years owing to the rapidity with which the calibration is reported to deteriorate when exposed to full radiation outside the atmosphere.

On the question of variability of the sun's energy output over various longer time-spans, it is in any case impossible to give a verdict from observations extending over only a few decades. Within those decades no variations of the total output have been detected which are great enough to exceed the margin of experimental error and uncertainty due to adjustment of the observed values to conditions outside the atmosphere. The long history of the development of life on the Earth over hundreds of millions of years[2] sets a limit to the range of any variations of solar output which can have occurred during that time.

Over the brief period for which good pyranometer observations then existed, between 1920 and 1955, ALLEN (1958) considered, apparently on the basis of the annual averages, that the 'solar constant' could be regarded as truly constant. Only within the extreme short-wave (ultra-violet) and long-wave (radio-wave) parts of the solar spectrum have large

1. The figure means that the average flux of heat from the Earth's interior (quoted on p. 14) is of the order of 10^{-4}–10^{-5} times the solar constant. Moonlight is weaker still – only at full moon about 10^{-5} times the solar constant.

2. The beginning of the Palaeozoic is dated about 560 million years ago; the Proterozoic stretches back much further (SCHWARZBACH 1961).

variations been demonstrated: for instance, monthly values of ultra-violet intensity may differ by 40% or more, and the smoothed curve shows a range of about 10% of the intensity on these wave lengths, varying in phase with the 11-year sunspot cycle. But, as these figures could be taken to imply discernible changes of the solar constant (and solar temperature) of which there was little sign, ALLEN believed that the measured variations must be due to changes of atmospheric transmission in the ultra-violet and infra-red. Such changes should affect the temperature of the upper stratosphere and the higher layers where the absorption occurs, about which far too little was then known. Whether and how such events affect the denser, lower layers of the atmosphere is as yet unproven, though some recent evidence will be given in a later chapter. An alternative suggestion is that the changes of transmission occur primarily in the solar atmosphere: this would mean that the solar constant should undergo some small variation which could not so far be established with certainty because it is presumably a good deal less than the 5% margin of error mentioned above.

Evidence of variations of the solar constant and of the 11-year cycle

The German meteorologist and former head of the first institute for long-range weather forecasting research (established by the Ministry of Agriculture for Prussia in 1929), Professor FRANZ BAUR (1964), drew attention at a joint meeting of astronomers and meteorologists at Bad Homburg in 1963 to the fact that the monthly solar constant values – the *preferred solar constant* – reported by C. G. ABBOT at the Smithsonian Institution, Washington from 1920 to 1955, appeared to show a systematic variation within the 11-year sunspot cycle over a range of about $\frac{1}{2}$% (see BAUR 1963). Unfortunately since 1955, following Dr ABBOT's retirement as Director of the Astrophysical Observatory, the observation routine has ceased; and it will probably be some years before artificial satellites above the atmosphere can begin to provide equivalent or better observation of solar radiation variations. BAUR noted (fig. 2.1) that the intensity of the solar beam appeared to undergo a double variation within the 11-year sunspot cycle, with maximum strength at about $\frac{3}{10}$–$\frac{4}{10}$ of the rising phase of sunspot activity and again between $\frac{2}{10}$ and $\frac{6}{10}$ of the declining phase of solar disturbance; a sharp minimum of the solar constant appeared just before sunspot minimum, and a lesser minimum of output at about $\frac{6}{10}$–$\frac{8}{10}$ of the rising phase.

The general nature of this finding is supported by subsequent observations (1961–8) using instruments mounted on balloons flown to heights above 30 km, from which KONDRATIEV and NIKOLSKY (1970) report that the solar 'constant' (measured values extrapolated to before first encounter of the solar beam with the Earth's atmosphere) appears to attain a maximum value (1·94 cal/cm^2/min) when the WOLF (Zürich) sunspot number is between 80 and 100 and decreases with both lower and higher sunspot numbers. The greatest decreases in these years (below 1·94 ly) did not appear likely to exceed 2–2·5%. The time-scale in terms of subdivisions of the mounting and declining phases of solar disturbance has been found convenient for comparisons of supposed sunspot-cycle

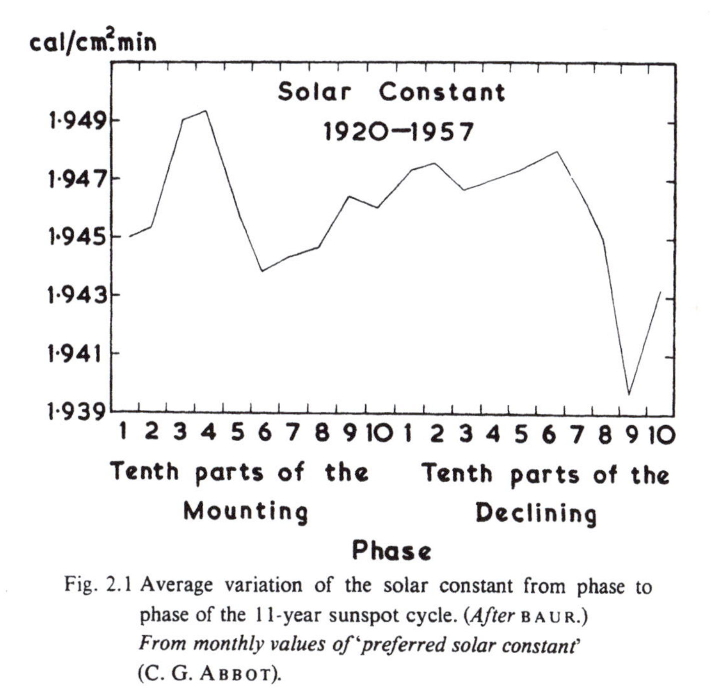

Fig. 2.1 Average variation of the solar constant from phase to phase of the 11-year sunspot cycle. (*After* BAUR.) *From monthly values of 'preferred solar constant'* (C. G. ABBOT).

effects because of the somewhat unequal length of these cycles. The history of sunspot activity since the year 1700 is seen in fig. 2.2. BAUR's contention led Dr ÖPIK of the Astronomical Observatory, Armagh, Northern Ireland, to comment as a solar physicist's view that, from what is known of processes going on in the sun (see, for instance, ÖPIK 1958), the so-called 'solar constant' certainly cannot be strictly constant; but the probable magnitudes and likely period-lengths of the variations cannot at present be determined.

It remains to be seen whether the detail of these variations within the 11-year sunspot cycles persists and proves statistically significant over longer periods of time, but BAUR (e.g. 1956, pp. 100, 118; 1959, p. 34) has been able to point to one or two climatic variables such as summer and winter character in Europe, that depend on the arrangement and vigour of the large-scale wind circulation, which appear to show something like this double oscillation within the 11-year sunspot cycle from the time of the earliest records at his disposal about 1750–1800.

The sunspot period averages 11·1 years. Since 1750 most cycles have been about 11 years, but the length has ranged from about 8 to 17 years.[1] There is some evidence that an

1. A strong relationship has been found (BRAY 1965) between the period length or (still more so) the period interval and the amount of sunspot activity, such that the shorter the time elapsed from the maximum of the previous sunspot cycle to the maximum of the current cycle (the period interval), the greater the sunspot numbers characterizing the current cycle (correlation coefficient for twenty-four cycles from the 1705 to the 1957 maximum −0·64, a figure that has a less than one in a thousand probability of occurring among random

approximately 11-year cycle has been a persistent characteristic of the sun's behaviour over geological ages. At least, effects presumed because of identity of period to be due to this have been recognized in tree-growth, as shown by annual ring widths, and in the yearly layers of varved sediments of various kinds and many ages.[1] Some work suggests that a 22-year periodicity may be more generally traceable in rock sediments than an 11-year cycle (ANDERSON 1961). Whether the production of such effects at the Earth's surface entails weather is not certain, because rather few weather phenomena have been found which show any appearance of a simple relationship to the 11-year sunspot cycle; but it is hard to imagine effects both upon the trees and the formation of sedimentary rocks without the medium of weather variations.

There are also longer-term variations of solar disturbance shown by differences of various kinds, including amplitude and period length, between different '11-year' cycles. Periods of 22–23, 80–89, 178, about 400 and 1700 years have been most frequently suggested as having some significance: periods of 567 or alternatively 737 years have been attributed to the disturbing gravitational effect on the sun of conjunctions of the planets in their orbits. In some cases, notably the 1700-year periodicity, effects on world weather, traceable, for instance, through variations of lake levels in arid regions, have been alleged; indeed, these effects may sometimes have been the origin of the suggestion about some solar cycle of the length in question as the supposed cause. Such examples of arguing in a circle bear witness not only to the uncontrolled enthusiasm which has inspired, and confounded, much work on sunspot cycles and weather, but to the real difficulty of establishing what, if any, long periodicities affecting weather and climate exist. The suggested 1700-year periodicity must be treated with reserve, at least as regards the nature of its connexion, if any, with the sun. A better perspective on these appearances of cyclic phenomena and some quite different suggestions about a 1700–2000-year periodicity may be gained from the work reported in Chapter 6.

Although the search for regular periods has attractions as a possible aid to prediction and simplicity of understanding, this approach quickly becomes very complicated and commonly inconclusive when all possible 'overtones' (sub-periods) of these periods, and all possible 'beat frequencies' ('beating' between periods of different length), are allowed for. In these ways, for instance, it should be possible to connect with the 11-year sunspot cycle

---

unconnected numbers). The average period interval between the 1907 and 1957 maxima, over cycles that were mostly vigorous and some very vigorous, was just 10 years. The same period length (approximately) is believed to have prevailed in various earlier times including the twelfth and mid-eighteenth centuries A.D., when there were, or are believed to have been, some vigorous cycles, contrasting with an average interval of 11·3–11·9 years in the seventeenth and nineteenth centuries.

1. Professor Dr G. RICHTER-BERNBURG of the Niedersachsisches Landesamt, Hannover and Professor T. S. WESTOLL of Durham University, reported this periodicity in varved evaporites of the Permian geological epoch in Germany and in various other samples back to the Pre-Cambrian, over 500 million years ago, at the NATO Advanced Studies Institute's Conference on Palaeoclimates at Newcastle, January 1963.

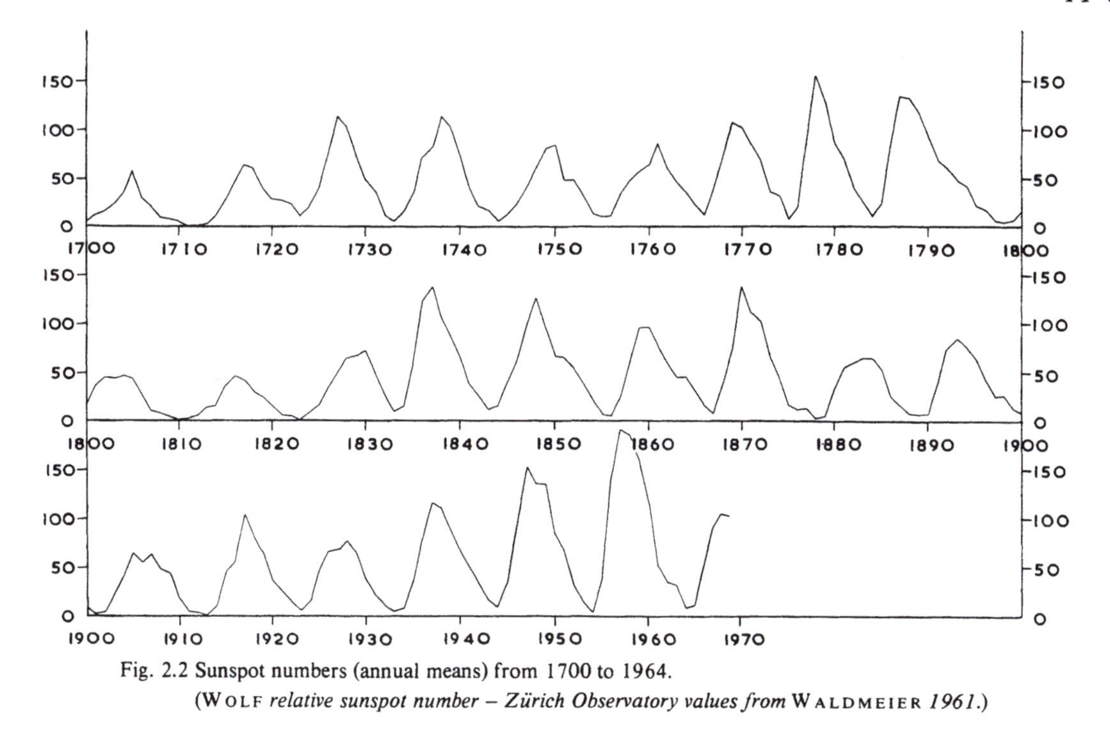

Fig. 2.2 Sunspot numbers (annual means) from 1700 to 1964.
(WOLF *relative sunspot number – Zürich Observatory values from* WALDMEIER *1961*.)

recurrences of weather phenomena after about 2·2,[1] 2·8, 5·5, 22–23, 33–35 and 55–57 years, and so many other intervals besides, over which recurrence tendencies have been suspected, that a periodicity of practically any length could be accommodated (see, for instance, the useful review by BERLAGE 1957). A glimpse of the main variations of sunspot number since A.D. 1700 is seen in fig. 2.2. Since 1750 the monthly mean values of WOLF's relative sunspot number[2] have ranged from 0 to the peak value of 254 in October 1957 (WALDMEIER 1961). The very big maxima of solar disturbance in 1778 and 1957 separated by an interval of 179 years stand out, as well as the occurrence of a group of big maxima about the middle of that interval, in 1837, 1848 and 1870. The other cycles mostly produced much less disturbance, though the maxima in 1727, 1738, 1788 and 1947 were high or fairly high. From the fragmentary records of eye observations in earlier times, and from the occurrence of associated phenomena, particularly aurorae, it is known that there were years of extraordinarily great sunspot activity between 1360 and 1380 (maxima 1362–5 and 1372), possibly exceeding the degree of solar disturbance in the 1957

1. LANDSBERG (1962) reports evidence from varved sediments and tree rings of the persistence of a 2·2-year recurrence period over many thousands of years. Samples of this evidence are given later in Chapter 6.

2. See Tables App. I. 5 and 6 for yearly values and highest and lowest (smoothed) monthly values of WOLF (Zürich) sunspot number.

maximum some 590 years later. LINK (1964) and SCHOVE (1955), on the basis of multifarious reports from Europe and China, also suggest great sunspot maxima late in the tenth century A.D., around 1100–1140 and around 1560 (in 1528, 1558 and 1572 according to SCHOVE), evidently representing further instances of a spacing of the order of 180–200 years between major maxima. Details are given in Appendix I. There seem to have been periods of relative quiescence of the sun between about 1290 and 1350 and marked quiescence early and late in the fifteenth century and from about 1590 onwards. There was very little activity on the sun (stressed also by BROOKS 1949, p. 366) between about 1645 and 1723, most of all between 1680 and 1712, in noticeable coincidence with a period – the so-called Little Ice Age – of extreme (generally cold) climatic conditions in many parts of the world.

Sunspots, in which the sun's emission of radiant energy is dimmed by 50–75% and the temperature of its apparent surface (the photosphere) is lowered by 1000–2000°C, are normally accompanied by other types of solar disturbance. (Convenient references on solar activity and solar disturbances are KUIPER 1953, NEWTON 1958, SEVERNY 1959.) These include (i) occasional flares or eruptions, in which high temperatures and bursts of both ultra-violet and long-wave radiation and particle emissions (*corpuscles*) are produced, and (ii) bright faculae, which are usually present around the spots. Some 10–30% of all faculae, however, show no close or obvious association with sunspots. The total areas of faculae therefore vary somewhat independently of the area of sunspots. In faculae the emission of radiation is enhanced; the temperature of the photospheric surface is slightly raised (by perhaps 100°C), while at the top of the faculae temperatures appear to be up to 2000°C above normal. Sunspot numbers, which measure only the areas of dimming of the sun, cannot when taken alone be expected to be a very good index of variation of solar output, even though the occurrences of other types of solar disturbance show some correlation with the abundance of sunspots. Logically, one should expect better results from an index which measures the total area of faculae (brighter sun) *minus* the total area of spots (dimmed sun). The fluctuations of such an index might very well give a more reliable measure of fluctuations of the solar 'constant' than has so far been obtainable from direct radiation measurements. BAUR (1949, p. 970) has proposed a solar index (S.I.) of this type, given by the expression

$$\text{S.I.} = 100\left(\frac{F}{\bar{F}} - \frac{D}{\bar{D}}\right)$$

where F is the total area of the sun's visible face occupied by faculae and $\bar{F}$ is the long-term average value of F, D is the total area of sunspots and $\bar{D}$ the long-term average value of D. Measurements have been made from the daily sun photograph series of the Royal Greenwich Observatory, taking the areas of that half of the sun's face that is turned towards the Earth and correcting for foreshortening – i.e. correcting to the spherical shape of the sun. Rather like the values of the 'preferred solar constant' (fig. 2.1) this index shows a double oscillation within the 11-year sunspot cycle, maximum values occurring during the rising and falling phases of the sunspot cycle and smaller values about the times of either sunspot extreme. No long series of monthly or yearly values of BAUR's solar index seems to have been published hitherto; values for 1874–1964 are given here in Appendix I. Yearly averages of sunspot number and faculae areas will also be found in the Appendix, as well as some estimates of the course of solar disturbance since 700 B.C. Monthly values of sunspot number and faculae since about 1750 and 1880 respectively are quite widely available – e.g. in WALDMEIER (1961) and in LINKE's *Meteorologisches Taschenbuch* (1962 volume, pp. 787–801). MIRONOVITCH (1960) shows the course of BAUR's solar index from 1874 to 1958 in terms of accumulating departures from the overall period average, a graph (at first steadily rising and then after the 1930s sharply falling) which appears highly correlated with the frequency of the zonal westerly type of the Earth's wind circulation and with various other indices of the atmosphere's circulation vigour.

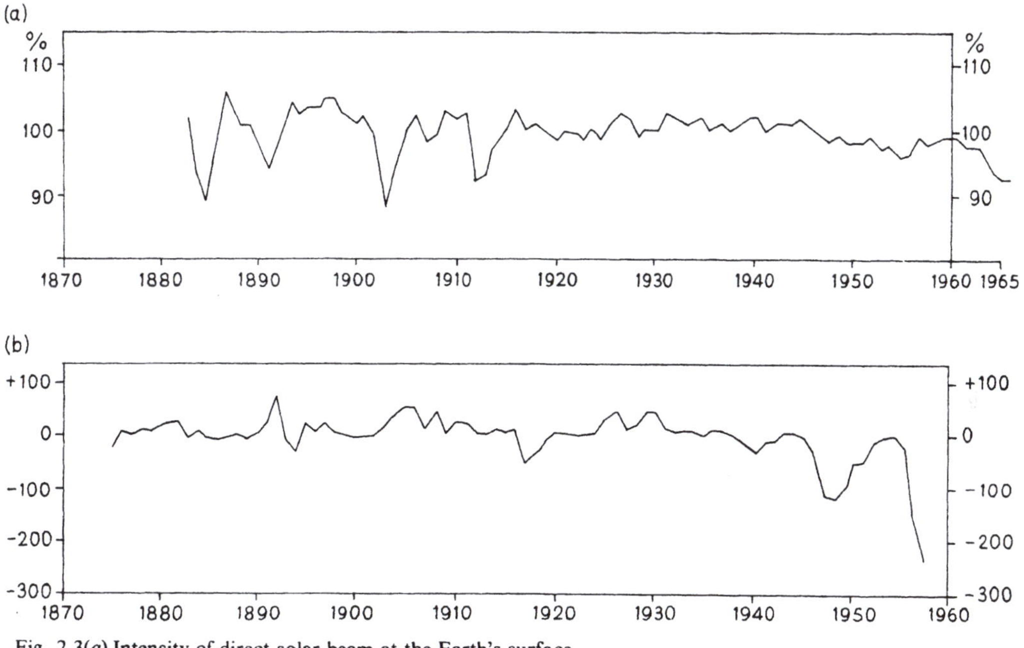

Fig. 2.3(*a*) Intensity of direct solar beam at the Earth's surface.

Yearly values 1883–1965 from pyrheliometric measurements at observatories mainly between 30 and 60°N.

(*Adapted from* PIVAROVA *1968*.)

Vertical scale; percentage of the overall mean.

Monthly values from the same records are plotted in fig. 2.12.

(*b*) BAUR's solar index $100\left(\frac{F}{\bar{F}} - \frac{D}{\bar{D}}\right)$.

Yearly values 1874–1957 derived from measurements of faculae on the Greenwich series of daily sun photographs and the WOLF relative sunspot numbers (Zürich data).

No means exist yet of testing whether the variations of BAUR's solar index correspond to real variations, doubtless minute in percentage terms, in the strength of the solar beam reaching the Earth. Yearly values of this index from 1874 to 1957 are given by the graph in the lower part (*b*) of fig. 2.3. One notices the increased fluctuations, and on the whole lower values, of the index in the cycles of exceptional solar disturbance which culminated with very great numbers of sunspots in 1947 and 1957. The best available runs of yearly mean values of the strength of the solar beam, plotted in the upper part (*a*) of fig. 2.3, illustrate the inadequacy of our knowledge from direct measurements of possible slight variations in the intensity of the incident radiation approaching the Earth. Comparisons with BAUR's solar index cannot be reliable because of:

(i) the wide error margin, amounting to several per cent, already mentioned, which applies to the measurements of the direct radiation, especially before 1893 when only one observatory was functioning. Moreover, the years of the insolation record after 1938 are from Japanese observatories and may have been imperfectly homogenized with the earlier part of the record, which rests mainly on American and European observatories. There is, however, rather wide observational evidence of a declining trend since the 1940s;
(ii) the deficits of measured insolation in 1884–6, 1888–92, 1902–5, 1907, 1912–15 caused by volcanic dust in the atmosphere after great eruptions. The sharply reduced values of the measured solar beam intensity in these years are, in fact, the only variations in these radiation curves which plainly exceed the margin of error.

However, there seems to be a certain gross parallelism between the overall course of the radiation and solar index curves since 1920; and, if the volcanic dust years be eliminated and isolated high values of the radiation measured in 1883 and 1887 and of the solar index in 1892 be treated as suspect, this parallelism could extend throughout the period covered. Comparisons of the values of the two parameters taken from the curves indicate no correlation in the individual years, but correlation coefficients about +0·4 connect the average values for the few 11-year sunspot cycles represented since 1880 (which cannot be regarded, however, as statistically significant). Of more interest is the coincidence of the groups of years of highest solar index in the early 1900s and around 1930 with the periods of maximum vigour of the general wind circulation (as indicated by measures of the prevailing westerlies in middle latitudes, etc.). Moreover, the big fluctuations of BAUR's solar index in the extreme sunspot cycles after about 1940 coincide with decades of increased variability of the wind circulation, as well as with generally falling values of indices of the strength of the zonal windstreams, and apparently increasing instability of climate – i.e. increased variance of the climatic elements in many parts of the world.

Radiocarbon evidence and its relevance to solar variation

Investigations in the field of radiocarbon dating have produced independent physical evidence, which indicates the probability of some variations of solar energy emission and effects in the neighbourhood of the Earth over the last thousand years. And already there is some similar information extending back over several thousand years. The application of radioactive carbon measurements to age determination depends on the fact that all living matter acquires a minute proportion of the radioactive isotope of carbon, ^{14}C, with the atmospheric carbon dioxide assimilated by the living vegetation. The proportion of the radioactive isotope should be the same as in the atmospheric CO_2. From the moment of death the radioactive carbon fixed in the vegetation during life decays at a known (exponential) rate, without replenishment. After about 5730 years the ^{14}C present has fallen to half the original amount; after a further lapse of 5730 years (the 'half-life' of ^{14}C) the amount remaining is halved again – and so on. The nuclear reactions concerned and the decay formulae are specified in Appendix II. The approximate date of death can therefore be estimated by accurate assay of the quantity of radioactivity still remaining in any sample. Age estimates obtained in this way depend on the assumption that the minute proportion of radioactive carbon atoms in the atmospheric carbon dioxide is always and everywhere the same, and that it has been so over the last 40 000–70 000 years – the period for which samples still contain enough radioactivity for measurement. Now, the radioactive ^{14}C atoms continually originate in the high atmosphere by neutrons colliding with atoms of ordinary atmospheric nitrogen, ^{14}N, some of which lose a proton and are thus converted into radioactive carbon. The neutrons responsible are themselves produced by cosmic ray particles (mostly protons) from outer space (i.e. from elsewhere in the galaxy) bombarding the outer atmosphere. The incidence of cosmic rays is reduced at times of solar disturbance, probably through their being deflected away from the Earth by powerful, though temporary, magnetic fields associated with streams of 'solar corpuscles' (which produce at the same time great disturbances of the Earth's magnetic field known as *geomagnetic storms*). Repeated careful radiocarbon measurements by WILLIS *et al.* (1960), and subsequently by many others (e.g. SUESS 1965), on objects of known ages have shown those originating between about A.D. 1300 and 1800 to be too rich in ^{14}C by 1–3% (fig. 2.4 (*a*)), apparently implying a weakened and rarer output of corpuscular streams from the sun during much of that time. (A deviation of 1% in the atmospheric ^{14}C would give a dating error of about 80 years.) That there may be a connexion between the periods of implied quiescence of solar flare activity – accompanying the known quiescence of sunspot activity and presumably of faculae – and the cold climatic period between about 1430 and 1850, is indicated by a correlation coefficient of $-0{\cdot}80$ between successive 50-year mean values of ^{14}C and (entirely independently derived) temperatures prevailing in England a century and a half later, both series extending over 1150 years since A.D. 650 and 800 respectively (LAMB 1965 *a*, *b*). This relationship is illustrated here by fig. 2.4 (*c*). SUESS (1968) reports an even stronger relationship between the indicated values of ^{14}C amount over this period and the LAMB Index of Winter Severity (LAMB 1963) in eastern Europe. A quicker and more straightforward response to any change in the radiation balance might be expected in continental interior regions and in the Arctic than near the great oceans. As regards the A.D.

era, statistical significance beyond the 5% level of the positive anomalies of ^{14}C in the atmosphere around the years 1450, 1550 and 1650, and of negative anomalies around the years 400 and 1200 – both the latter believed to be periods of rather warm climate in Europe – seems assured (RALPH and MICHAEL 1967).

We conclude that the sun's output of particle streams and of ultra-violet radiation – the latter increases and decreases more or less in parallel with the variations of sunspot activity in the 11-year cycles – was probably materially weakened during the periods of sharply cooling climates between about 1400 and 1700 or later. Furthermore, there is some indirect evidence, partly through the association mentioned above between the vigour of the general wind circulation (particularly the middle latitudes westerlies) and the areas of solar faculae, and partly through the degree of mutual association that appears to exist between the various types of solar disturbance, that there may have been a significant reduction of the solar 'constant' – though this is probably a matter of 1% at most.[1]

Regarding the observation that the quantity of ^{14}C in the atmosphere was enhanced during certain well-marked periods of cold climate, and particularly during those centuries when the cooling was setting in, the effect of the oceans must be considered. During phases of climatic cooling there must be less vertical stability in the oceans, owing to the reduced temperature (and therefore density) difference between the surface water and the colder (denser) layers beneath: hence the vertical circulation of the oceans should be increased. This must tend to bring to the surface more water from the ocean deep which has been long out of contact with the atmosphere and is therefore deficient in ^{14}C. For this reason the exchanges of carbon dioxide between air and sea at the ocean surface should militate against the observed increase of ^{14}C in the atmosphere, and there seems no question of this increase having come out of the oceans – further evidence that the increase of ^{14}C should be attributed to an extraterrestrial cause. On the other hand, with a weakened atmospheric circulation, as observed during some periods of colder climate, e.g. around A.D. 1800, and therewith lessened drag on the ocean surface, the (horizontal) ocean currents should be weaker, producing less upwelling and weaker convergence and divergence effects (which force vertical motion in the ocean). However, in the more extreme types of cold climate regime, e.g. ice ages, the general atmospheric circulation appears to be *strengthened*, though concentrated over the lower latitudes; and in the onset phases of climatic cooling the atmospheric circulation may be generally quite strong, though displaced towards lower latitudes – in both these cases tending to bring to the surface deep water with its characteristically low levels of radiocarbon activity.

There is no such certainty about the net effect of the oceans in some climatic stages like

1. Indications that the supply of solar radiation between 1780 and 1820, and for some time before that, was weakened by 1–2% compared to 1900–50 have been obtained (LAMB 1963) from a consideration of the ocean temperatures prevailing in the North and South Atlantic between 55°N and 40°S. Part of that reduction is, however, thought to have been attributable to abnormal incidence of veils of volcanic dust in the atmosphere.

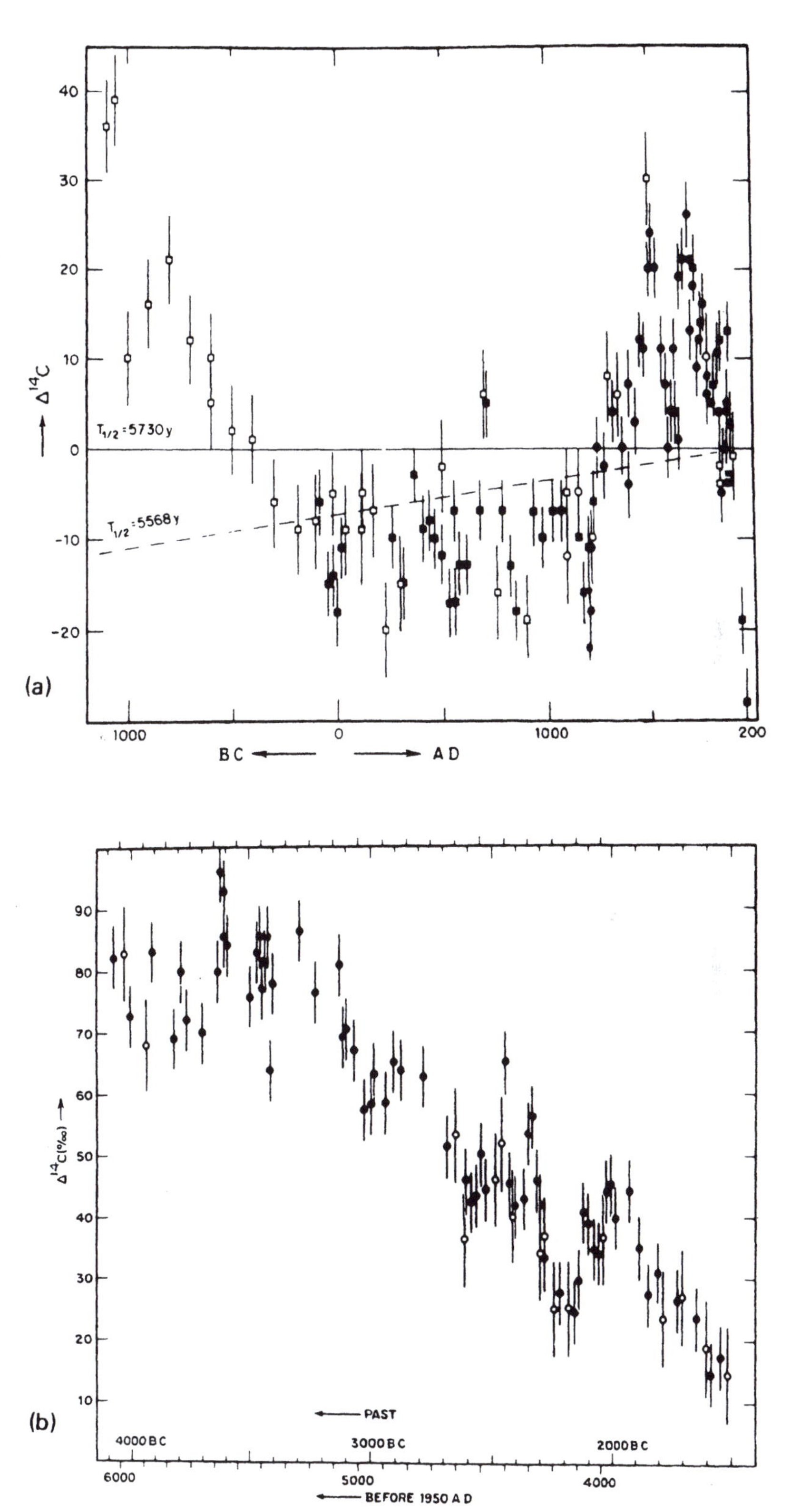

Fig. 2.4(*a*) Variations per thousandth of the quantity of radioactive carbon (^{14}C) in the atmosphere since 1100 B.C. (*After* H. E. SUESS *1965*.) Collection of observations derived from measurements of ^{14}C activity in wood of known ages performed in various laboratories. Deviations from atmospheric ^{14}C concentration as it was at A.D. 1850, before most of the pollution created in the modern industrial era. Calculated for a half-life of the ^{14}C atoms of 5730 years (distances above and below the sloping broken line measure the deviations which would be implied if the half-life were 5568 years, as formerly assumed). Circles denote wood from European trees, squares American trees. Open squares or circles indicate results of counting atomic disintegrations over 2 days, solid squares or circles over 4 days. This is not in itself a radiation diagram, though there is some ground for supposing that (inverted) it corresponds to a variation of solar activity. The periods of added radio-carbon concentration around 700 B.C. and A.D. 1400–1700 both appear to have been times of general fall of temperature over the world. (See also fig. App.I.2, p. 475.)

(*b*) Differences per thousandth of the atmospheric ^{14}C concentration indicated by wood of known ages between 1500 and 4100 B.C. from that of modern wood growing between A.D. 1870 and 1885. (*After* H. E. SUESS, *personal communication.*)

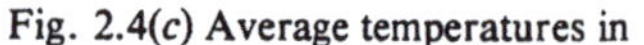

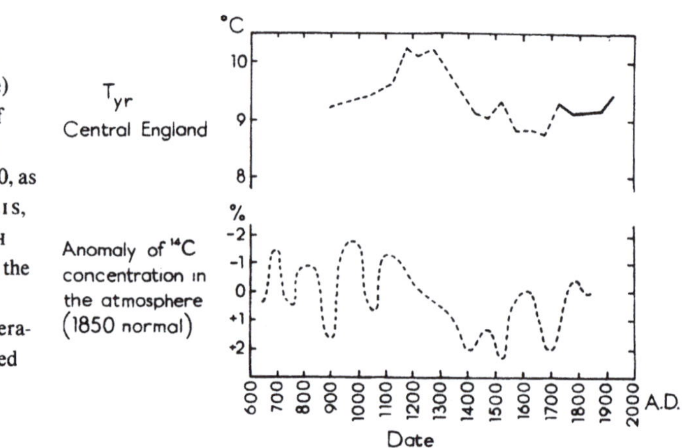

Fig. 2.4(*c*) Average temperatures in England and (lower curve) variation in the amount of radioactive carbon in the atmosphere since A.D. 600, as first established by WILLIS, TAUBER and MÜNNICH (1960), inverted to reveal the near parallelism with the course of prevailing temperatures in England, as derived by LAMB (1965*a*).

that around A.D. 1800 upon the amount of radiocarbon in the atmosphere as there is about the changes in the production rate that go with varying solar activity.

Radiocarbon dating studies on known tree chronologies back to 5000 B.C. from the *Sequoia gigantea* (Giant Redwoods) and still longer-living Bristlecone Pine in western North America, reported by SUESS and partly reproduced here in figs. 2.4 (*a*) and (*b*), indicate other marked increases of ^{14}C in the atmosphere for a few centuries around 1000 B.C. and at some earlier times (notably about 3300 B.C.) similar to that around A.D. 1500–1700 which may have been associated with the fluctuation towards sharply colder climates then developing. There is much evidence of a decline towards colder climates also from 1000 B.C. onwards till about 500 B.C.

Apart from these greatest deviations from the shape of a smooth curve shown by the ^{14}C measurements, there are two other prominent characteristics:

(i) An approximately 200-year oscillation of considerable amplitude in the amount of ^{14}C appears to be a constant feature throughout the record, though possibly replaced by a 400-year oscillation in the earlier millennia B.C. This continues through, and is superposed on, all the longer-term variations of the last 1000–3000 years already described, showing, for instance, strong maxima about A.D. 1500 and 1700. By its constant phase relationship in recent centuries to the great sunspot maxima (STUIVER 1961) and to the known maxima of the prevailing westerly winds in middle latitudes (cf. fig. 6.8), this ^{14}C oscillation may be regarded as evidence of the continuance of a solar variation of about this period length through many repeats.

(ii) A much longer-term variation from a maximum of ^{14}C about 4500 B.C. (general deviation of ^{14}C values about that time +9% from the A.D. 1850 standard) to a minimum level around 0 B.C. (deviation −2% from the same standard). This could be a sinusoidal oscillation related to that of estimated period 10 300 years in the Earth's magnetic field, which had a maximum field strength about double its present value in Roman times that may also have inhibited the arrival of cosmic rays from elsewhere in the galaxy. This would imply that this longest-term oscillation in the ^{14}C in the atmosphere has nothing to do with the sun. SUESS (1968) has, however, pointed out that if the ^{14}C had really been declining since the last ice age, this would imply that the ice age corresponded to a very long period of quiescent sun.

DAMON (1968) has derived tentative values of ^{14}C anomaly back to nearly 10 000 B.C. by comparing radiocarbon 'dates' on botanical material associated with late glacial and early post-glacial changes of climate and vegetation in northern Europe with the absolute dates obtained by counting the annual varves of Swedish lake-beds. DAMON finds that the average departures of ^{14}C amount in the atmosphere from modern values rise to +9 to +12% around 4000 B.C. but fall back to the modern level around 6500 B.C. (when the radiocarbon dates again match the true ages) and become −9% around 8000–9000 B.C. Before that the graph recurves to smaller departures once more, e.g. −5% near 10 000 B.C. This sequence appears to fit the hypothesis of a strong, indeed overriding, influence from the Earth's magnetic field strength variations and no great ^{14}C anomaly on the time-scale of the last glaciation.

The case for believing that at least the shorter-term regular variations discovered (by all the radiocarbon dating laboratories concerned) in the amount of radioactive carbon ^{14}C in the atmosphere over the last several thousand years represent variations in the amount of disturbance on the sun has been strengthened by computations by BRAY (1967). A crude solar disturbance index, which should be roughly equivalent to average sunspot number in the years of maxima of the 11-year cycles, was constructed from available knowledge of sunspots and aurorae from 527 B.C. to 1964 (see Appendix I). Throughout these centuries, runs of three or four 11-year cycles with high solar disturbance (maximum yearly sunspot numbers probably over 100) alternated with runs of three or four cycles with low solar disturbance (maximum yearly sunspot numbers below 100); and in all but two of the twenty-four change-overs for which adequate ^{14}C data before and after could be compared, the amount of ^{14}C in the atmosphere underwent a change in the opposite sense to that of the solar activity. (In one of the two exceptions there was no change of ^{14}C; in only one case did a ^{14}C change occur in the same direction as that of the solar activity.) There were also much longer periods of rather generally high or low solar activity, and these maintained the same inverse relationship to ^{14}C amount in the atmosphere. A correlation coefficient of −0·51, with a less than 1% probability of occurring by chance, was obtained for the thirty-one long periods of high or low solar activity since 129 B.C. for which adequate ^{14}C data existed. BRAY (1966, 1967) also observed an apparently significant in-phase relationship between these long periods of high or low solar activity and prevailing surface temperature levels as derived by LAMB (cf. fig. 2.4 (*c*) and p. 25 above) and MANLEY for central England or indicated by forest growth and glacier changes in North America and Europe (BRAY 1965). MITCHELL (1965) has drawn renewed attention to a corresponding parallelism, originally spotted by C. G. ABBOT, between the average sunspot number of each cycle from 1749 to 1923 and the summer temperatures prevailing at places in central and northern Europe. It appears, however, that cycles in which the sunspot numbers are exceptionally high (as around 1778, 1947 and 1957) may have some different effects.

Present knowledge of solar variability may be put briefly as follows. On the one hand, no significant variation of the total energy output has so far been firmly demonstrated by radiation measurements; indirect evidence is becoming impressive, but any variations that do occur must be a very small percentage. On the other hand, large percentage variations of the sun's output at the extreme short-wave end and in the long-wave part of the spectrum, and of the emission of streams of corpuscles, do occur. These variations go very roughly with the increases and decreases of sunspot number, though the latter cannot be regarded as a sufficient index for all these and other kinds of solar disturbance. Effects in the high atmosphere (ionosphere), the production of aurorae and magnetic storms, are well known. Some effect upon the stratosphere also seems likely through variations in the production of ozone, and there should be consequent changes of temperature in the top of the ozone layer: whether any effects upon this layer influence weather processes in the lower atmosphere remains a matter of investigation, about which we shall report in a later chapter. Corpuscular (i.e. atomic particle) streams shot out from the sun, but conveying amounts of energy several orders of magnitude less than the solar constant, are concentrated by the Earth's magnetic field towards a zone about the magnetic pole, and may from time to time be of importance in the stratosphere there. If coupling between stratosphere and troposphere does occur, or damping of developments in the troposphere, it seems likely that changes of tropopause height (possibly due to temperatures in the ozone layer) should be involved. So far, however, instruments and observation techniques have been inadequate to measure them.

Variations in the Earth's orbit, its distance from the sun and the tilt of its polar axis

Much greater variations of the Earth's annual radiation budget must occur through very long-term, periodic changes in the Earth's orbit, the tilt of its rotation axis and the seasonal variation of the Earth's distance from the sun characteristic of epochs defined by the orbital situation. Such epochs commonly change their character only slowly, over some thousands of years. But the changes are big, and some effects on climate appear inescapable, probably including the causation of the alternation of ice ages and warm interglacial periods during the Quaternary era (approximately the last million years), when the large-scale geography has been much as now.

The seasonal differences of radiation budget that occur every year within each epoch, including the present, due to changing length of day and altitude of the midday sun, and partly due to changing distance of the Earth from the sun, are also enormously greater than any secular variations of solar output such as we have been discussing.

The Earth's distance from the sun undergoes regular changes each year, on the way round its orbit, which is slightly elliptical with the sun at one focus of the ellipse. Consequently the strength of the solar beam reaching the Earth varies regularly about its mean value. At the present epoch, the Earth is in perihelion (nearest the sun) near the middle of the northern winter, on 2–3 January, and in aphelion (farthest) in the northern summer, on 5–6 July; this makes the solar beam near the Earth about 3½% stronger than the average solar constant in January and 3½% weaker than the average in July. Since the Earth moves faster round its orbit when it is near the sun, and the distance between the equinoctial positions is shorter, the northern winter measured from equinox to equinox is at present slightly shorter (7½ days less) than the northern summer. The net result should be to make the northern hemisphere winters warmer, and summers cooler, than the southern hemisphere equivalents; but the effect is more than overcome by the unlike distribution of land and water in the two hemispheres, and especially by the reflectivity of the wide area of winter snow cover, extending over half-way to the equator on the northern continents, and of the persistent ice and snow of Antarctica at all seasons.

When one considers the climatic difference between times stretching over many thousands of years, the astronomical 'constants' of the Earth's orbit must be treated as variables. There are three distinct variations going on simultaneously all the time. How they arise is explained with the aid of a diagram in Appendix III.

Firstly, the tilt of the Earth's axis of rotation relative to the plane of its orbit (the *obliquity of the ecliptic*) is believed to vary at least between 21·8° and 24·4° over a regular period of about 40 000 years. At present it is almost 23°27′ (23·44°) and decreasing by about half a second of arc (0·00013°) a year; the last maximum was about 10 000 years ago. This variation changes the latitudes of the tropics and polar circles, which are respectively the limiting latitudes reached by the zenith sun at noon in midsummer and by the day-long polar night in winter; it must affect the thermal range of the seasons.

Secondly, the *ellipticity* and 'eccentricity' of the Earth's orbit varies between extremes of about zero (circular) and 0·06[1] in the course of an oscillation of variable amplitude and period: the latter averages 96 600 years. The ellipticity has been 0·02 or less through most

1. The ellipticity, e_1, of the orbit is given by the expression $e_1 = \frac{1}{a}\sqrt{a^2 - b^2}$, where a is the major and b the minor axis of the ellipse. The eccentricity, e_c, the distance between the centre of the ellipse and the focus where the sun is, is given by $e_c = e_1 \,.\, a$. The extreme values of the Earth's distance from the sun on its way round the orbit are therefore $a(1 - e_1)$ and $a(1 + e_1)$.

The present value of e_1 is 0·0167, the last maximum having been a very low one (about 0·019) some 10 000 years ago; the previous minimum level was some 40 000 years ago (about 0·010). Much greater maxima (0·04–0·05 or rather over) occurred around 110 000, 200 000 and 300 000 years ago according to a graph by VAN WOERKOM (1953, p. 150) and the intervening minimum levels were about 0·02. There were other great maxima (0·05 or over) nearly 600 000, 700 000 and 960 000 years ago, whereas the maxima near 400 000, 500 000, 800 000 and 880 000 years ago were more modest (about 0·03).

of the last 100 000 years (VAN WOERKOM 1953). A time-span of the order of 100 000 years separating similar occurrences seems, however, to be the most pronounced feature of curves that reconstruct past climate sequences from field evidence (see, for instance, EMILIANI, 1961, figs. 6 and 9). The appearance is of a recurring glacial cycle, or sequence of cold climatic developments, the curves not sinusoidal in shape but developing towards a cold climax after which there is a swift throw-back to a peak of interglacial warmth, followed in turn by a slow decline into the next cold phase. The decline is manifestly overlaid by various smaller amplitude fluctuations on time-scales from a few tens of thousands of years downwards. When the orbit is most elliptical, the intensity of the solar beam reaching the Earth must undergo a seasonal range of about 30% between aphelion and perihelion; at present it is only 7%; when the orbit is circular, it becomes 0.

A third cyclic variation with period about 21 000 years is superimposed on these other two because of the rotation of the Earth's elliptical orbit and the slow precession of the equinoxes around the orbit. This means that the season at which the Earth is nearest the sun (*perihelion*) gradually changes, getting about one day later every 70 years: 10 000 years ago it was in the northern hemisphere summer.

The combined effects upon radiation receipt and world climates resulting from the three types of variation of the Earth's orbital arrangement can best be understood by studying the variations of total solar radiation available at this distance from the sun, latitude by latitude, in the summer and winter halves of the year taken separately, neglecting at this stage the effects of the atmosphere and the winds in it. These quantities have been calculated by MILANKOVITCH (1930), and his results for various latitudes from 25° to 75°N over the last 130 000 years are reproduced here in fig. 2.5. The radiation variations, which MILANKOVITCH expressed in various ways including equivalent changes of latitude, are here given in terms of his equivalent changes of prevailing temperature in °C (MILANKOVITCH 1930, 1938; see also a more extended account in English given by ZEUNER 1959). Within the time covered by fig. 2.5 the radiation falling on the Earth at 65°N in the summer half of the year varied from the modern amount at 60–62°N to only the equivalent of that at 74–80°N. The total radiation received during the year was, however, much more nearly constant. MILANKOVITCH's conversion of his results to corresponding temperatures made use of the fact that the average height of the permanent snow line latitude by latitude in the present epoch has a correlation coefficient of +0·83 with the radiation available in the summer half of the year. Between latitudes 40° and 90°N MILANKOVITCH found that the correlation coefficient approaches unity (+0·996). It therefore appears legitimate to regard the height of the snow line as broadly determined by the summer radiation. To this extent the neglect of atmospheric heat transport, for which the MILANKOVITCH theory is often assailed, appears justified. Changes of the total radiation available in summer can accordingly be expressed as equivalent to certain changes of latitude, and from the difference of height of the snow line that goes with that latitude change

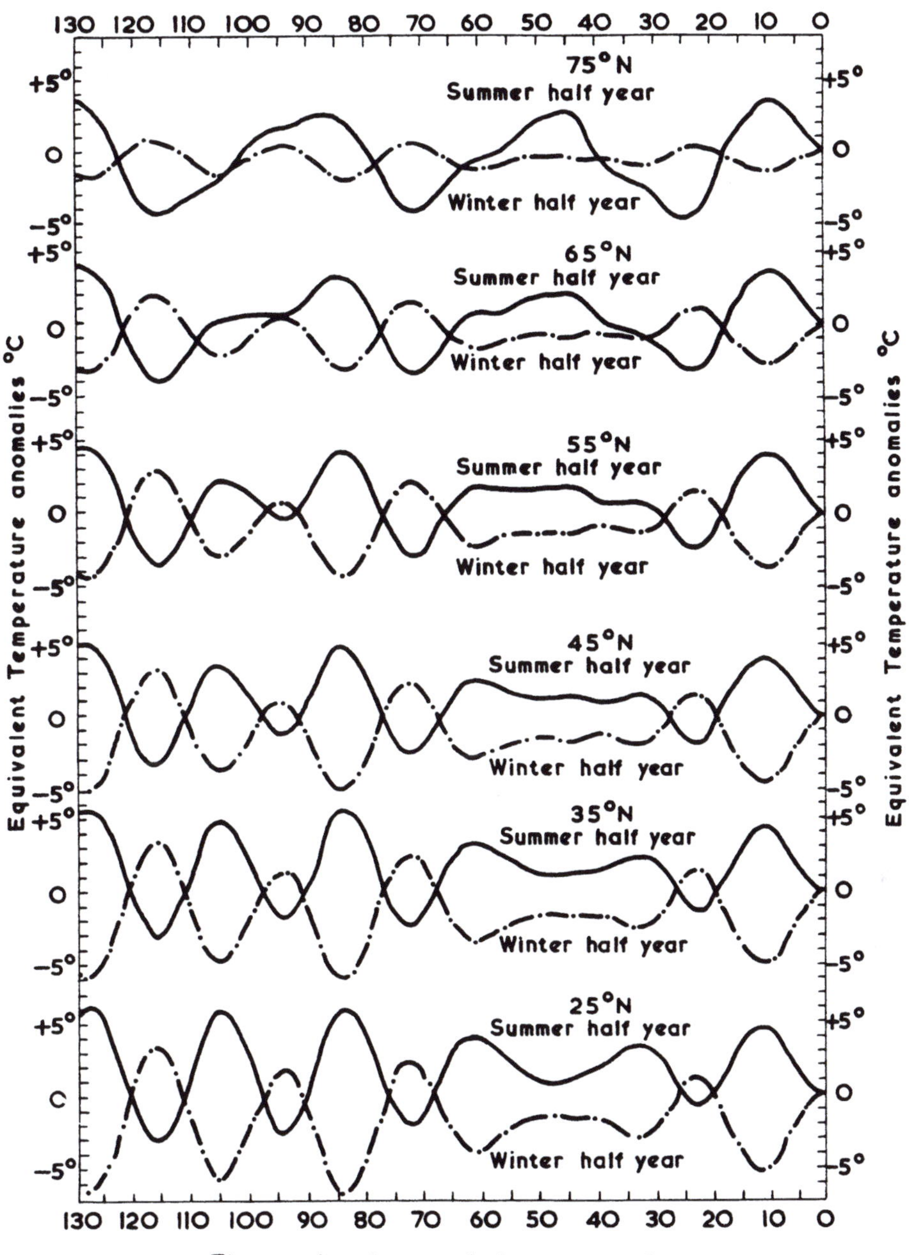

Fig. 2.5 Radiation available at various latitudes 25–75°N in the summer and winter half-years over the last 130 000 years according to MILANKOVITCH.

Vertical scale – equivalent departures from present average temperatures in °C: a 1° change of temperature is taken to correspond to about 150 m change of snow-line.

today, taken in consideration with the average fall of temperature with increasing height of 0·7°C/100 m, the equivalent change of temperature can be worked out. The same scale of equivalence between radiation and changes of prevailing temperature for each latitude has been used, with more doubtful validity, for the winter half-year.

MILANKOVITCH's conversion of his radiation results to equivalent temperature changes has been disputed by SIMPSON (1940), who considers the temperature anomalies indicated by MILANKOVITCH are generally three to five times too big. This criticism appears at first sight directly applicable to the warmest post-glacial climates around 5000 years ago, about which a good deal is now known. In the light of modern knowledge, MILANKOVITCH's temperature figures (seen in fig. 2.5) for the last maximum of the summer radiation in high northern latitudes indicate positive temperature anomalies in summer which are about twice as great as those 'observed' (i.e. derived from botanical and other evidence) in the post-glacial climatic optimum; and negative temperature anomalies are indicated for the winters of that epoch, when the 'observed' temperature anomalies were much smaller and still mostly positive. However, it should be remembered that MILANKOVITCH's data really refer to the radiation available outside the atmosphere and the effects that that would have if other things could be ignored. In fact, the post-glacial warm epoch occurred about 5000 years *after* the last maximum of summer radiation in northern latitudes, probably because it took that long to get rid of the accumulated ice and to warm the oceans – or, more precisely, to reach the minimum extent of ice of post-glacial times and the maximum warmth of the oceans; and by then, the radiation conditions indicated in fig. 2.5 were already more moderate. Both MILANKOVITCH's temperature conversion and SIMPSON's criticism of it ignore such lag effects; SIMPSON (1940) also largely ignores the effects of winter to summer and summer to winter persistence of temperature anomaly character introduced either by extensive ice caps or warm oceans. A more recent estimate (J. M. MITCHELL unpublished) puts the range of MILANKOVITCH variations of world surface temperature, having regard to the effects of the circulation of the atmosphere but not to the changes of albedo introduced by changes in the extent of snow cover, at 2°C; when the latter changes are taken into account a range of 5° or 6°C appears likely.

It seems reasonable to conclude, despite SIMPSON, that the effects of the orbital changes calculated by MILANKOVITCH are of the right order of magnitude to cause ice ages and warm epochs, and there is evidence now from radiocarbon dating that the timing indicated for the last glaciation and the warmest epoch since were about right, if a 5000-year lag imposed by the Earth's glaciers and water bodies (the 'hydrosphere') with their large heat capacity be accepted (FAIRBRIDGE 1961, LAMB *et al.* 1966). Radiocarbon dating of the rise of sea level that accompanied the melting of the ice sheets of the last (Würm III) ice age establishes that the melting took altogether about 15 000 years, from the ice maximum about 17 000 B.C. to the minimum extent of ice around 2000 B.C. MILANKOVITCH's equivalent temperature changes should only be taken therefore as representing the hypothetical situation that would arise if the lag and smoothing effects imposed by the atmosphere, ice and oceans did not operate.

Computed values of the obliquity of the ecliptic, of the eccentricity and of the position of perihelion in the Earth's orbit over the last 1 000 000 years have been presented in graphs by VAN WOERKOM (1953) and are the basis of the curves here shown in figs. 2.6 and 2.7.

The effect on the radiation budget of changing obliquity is insignificant near the equator, and the long-period changes that do occur in low latitudes are therefore dominated by the other two orbital variables. In high latitudes changes of obliquity have more effect on the quantity of heat coming in than the other two variables. The summer radiation in high latitudes increases, and the winter supply decreases, the greater the obliquity. At latitudes poleward of 43° the year's total incoming radiation varies up and down in the same sense as the radiation for the summer half-year; but the changes of summer radiation are greater – at 65° latitude four to five times greater – than the changes of the year's total, except at the pole where they are identical. Since it is cold summers rather than cold winters that must be most critical for the accumulation of snow and ice from one year to the next, and hence for the initiation of ice ages, we may usefully concentrate some attention on the variations of the supply of solar radiation (*insolation*) in summer, particularly near 65°N where the Earth is largely ringed with mountainous land with considerable winter snowfall.

The best values now to hand for the available radiation are probably those provided by the calculations of BROUWER and VAN WOERKOM using corrected figures for the planetary masses which affect the Earth's orbit, though the differences from MILANKOVITCH are rather slight, especially in the last few hundred thousand years. Fig. 2.6 indicates, from this source, the history of the summer radiation supply over the last million years – the 'Pleistocene' or 'Quaternary' era of geology – during which several ice ages and warm interglacial periods have occurred. MILANKOVITCH's tentative identification of features corresponding to nine glacial maxima in Europe has been entered on the diagram. Presumably the troughs of the radiation curve for 65°N should be taken as corresponding to the initiation and build-up of the great ice sheets and are not expected to coincide with their maximum extent. On this basis the understanding of the three main stages of the last (Würm) ice age and its warmer interstadials appears to agree rather well with the results of radioactive isotope methods of dating (see LAMB *et al.* 1966). The concordance of the results for greater ages is still doubtful. Dating methods and identification of the evidence are not yet good enough for times so long ago. Ages assigned by different methods to the beginning of the Pleistocene and its ice ages range from 350 000 to $1\frac{1}{2}$ million years ago. The radiation curve suggests that more numerous occurrences of ice ages, or of variations within ice ages and of partly glacial climates, may yet come to be recognized.

Fig. 2.7 shows the calculated variations (VAN WOERKOM 1953) of radiation available in the summer half-year at 65°N and 65°S compared. Because the obliquity of the Earth's axis is the most significant orbital variable for radiation available in high latitudes, there is some tendency apparent in the curves for deep minima of the summer insolation in high

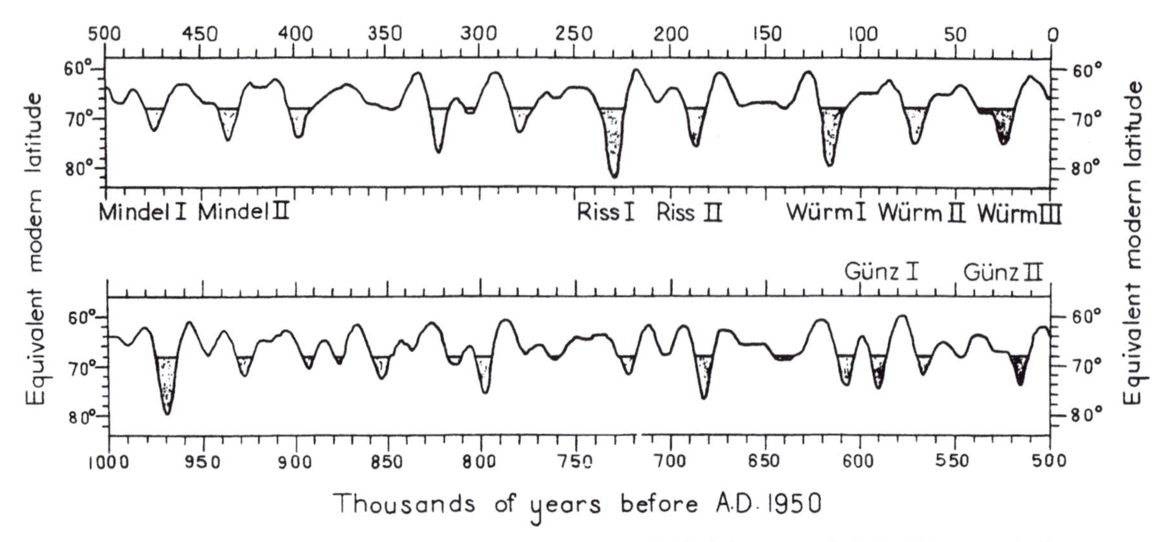

Fig. 2.6 Amounts of solar radiation available in the summer half of the year at latitude 65°N over the last million years, expressed as equivalent to the radiation available at various latitudes at the present day. (*After* MILANKOVITCH *(1930), following recalculation by* D. BROUWER *and* A. J. J. VAN WOERKOM *using newer figures for the masses of the planets that affect the Earth's orbit.*)
The latitude scale reads downwards because increasing latitude decreases the total of radiation available, whether over the year or over the summer half-year.
Areas where the curve goes beyond equivalence to the modern latitude 68° are shaded, as done by MILANKOVITCH (1930), since those periods have been supposed important in connexion with the initiation of ice ages (KÖPPEN and WEGENER 1924) and MILANKOVITCH's tentative identification of the association of features of this curve with major ice age stages is indicated on the diagram.

northern latitudes to be paired with deep minima in high southern latitudes about 10 000 years later. It is however the variations at 65 °N which give the best fit with the onset of the most securely dated glaciations. BROUWER and VAN WOERKOM, as well as MILANKOVITCH, indicate many periods in the last million years when radiation available in the Antarctic summer was less than today, separated by much shorter periods of only slightly greater radiation. If, therefore, present radiation conditions do not suffice to reduce the great ice sheet covering Antarctica materially, it is likely that Antarctica has been glaciated throughout the Quaternary era (and presumably should have been glaciated as long as that continent has existed in about its present latitude). Because most of the world's mountainous land in suitable latitudes to become glaciated by a lowering of summer temperatures lies in the northern hemisphere, it has to be expected that the changes of world temperature should follow most nearly the course of the radiation changes affecting the relevant latitudes between about 45° and 75 °N.

The variations of insolation here discussed appear capable of producing alternations between ice ages and warm interglacial periods separated by tens of thousands of years. This

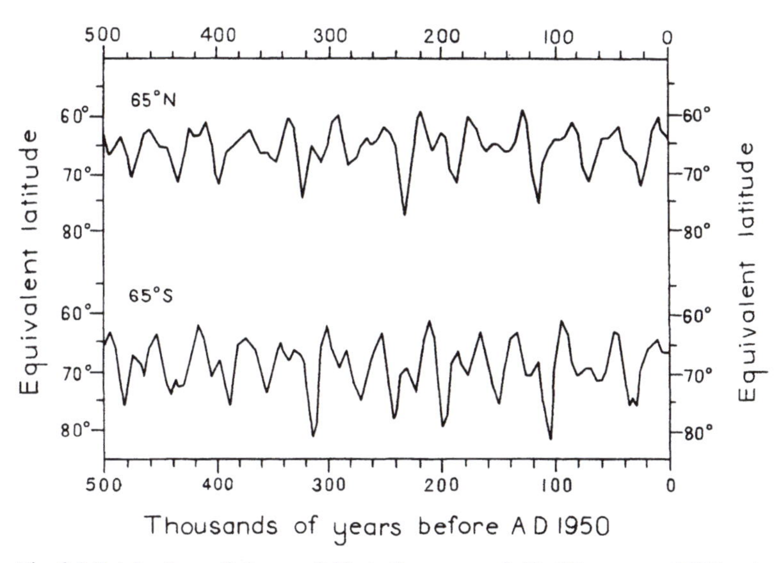

Fig. 2.7 Total solar radiation available in the summer half of the year at 65°N and 65°S over the last 500 000 years.
(*After* VAN WOERKOM *1953*.)
(*Reproduced here by kind permission of the author and Harvard University Press.*)

is, at least roughly speaking, the observed time-scale. At some stages of Pleistocene history, however, fig. 2.6 suggests that non-glacial, or almost non-glacial, climates might be expected to have lasted for 100 000 years or more, possibly explaining the occurrence of a long interglacial period. The implications of the MILANKOVITCH orbital variations as regards pluvial (rainy) and interpluvial (arid) periods in the African tropics during the Quaternary have been explored theoretically by Professor E. A. BERNARD (1962); in this matter, too, it is claimed that the theory is verified by the observations, though the latter can hardly be regarded as firmly enough dated to settle the point. Some extreme advocates of the MILANKOVITCH theory of climatic variations (e.g. ZEUNER 1958, 1959) have used it as a system of 'dating' all kinds of evidence of past climates during the Quaternary era.

Climatic changes over geological time

For the changes of world climate over the much longer eras of geological time, including warm eras with immunity to glaciation and little or no ice on the polar seas over many millions of years, changes of the major geography of land and ocean and mountain barriers, and the positions of equator and poles, may have to be considered. Of such changes there is increasing evidence from rock-magnetism and related studies. On this basis, it may be reasonable to assume that the Earth's liability to the periodic glaciations characteristic of the Quaternary era began when the North Pole some time within the Tertiary entered the nearly enclosed Arctic Ocean, which is ringed about by mountainous land occupying most of the zone 60–70°N, and the other end of the axis of rotation came within the mountainous Antarctic landmass.

E. J. ÖPIK (1958, 1964) has pointed out an apparent tendency for the dominance of warm ice-free eras to be interrupted by groups of ice ages at intervals of about 250–300 million years – viz. in the Quaternary, Permo-Carboniferous, Eo-Cambrian and several in the Pre-Cambrian eras – and believes that such a rhythm must be explained in terms of the physics of the sun's interior. He regards the individual glaciations within each group of ice ages – four in the Quaternary, five in the Permo-Carboniferous – as also due to solar variability, in this case on a time-scale of 10 000–100 000 years, which he calls 'flickering'.

Present distribution of available solar radiation by latitude over the year

The distribution over the year, in different latitudes, of the solar radiation supply at the present epoch is shown in more detail in fig. 2.8, as calculated by MILANKOVITCH. The most prominent features are

(i) the periods of polar winter night, during which no direct solar radiation is received in the area of darkness, and
(ii) the summer maximum daily totals, which are also over the polar regions because of the long daylight.

The effect of greater elevation of the midday sun, combined with the length of day, produces a secondary maximum of solar radiation available daily at the summer solstice near latitude $43\frac{1}{2}$°. Totalled over the year the radiation available at the poles at the present epoch is only about 42% of that supplied at the equator: decreasing obliquity of the ecliptic should decrease the proportion – it would fall to zero if there were no tilt of the polar axis – but any changes of this tilt appear to have been small, at any rate within the last million years. One notices in fig. 2.8 that the Antarctic summer maximum is rather stronger than the Arctic one, because the Earth is nearer the sun in January in the present millennium.

The energy supply to each unit area of surface is reduced where the rays of the sun fall at a slanting angle, as may be understood by watching how the area of shadow thrown by a round tree increases as the sun gets lower in the sky. Each unit of the energy supply is spread over a larger area as the solar elevation decreases: the intensity of the supply per square centimetre of surface varies as the sine of the angle of elevation. The daily total of energy available in high latitudes reaches high values in summer because of the length of

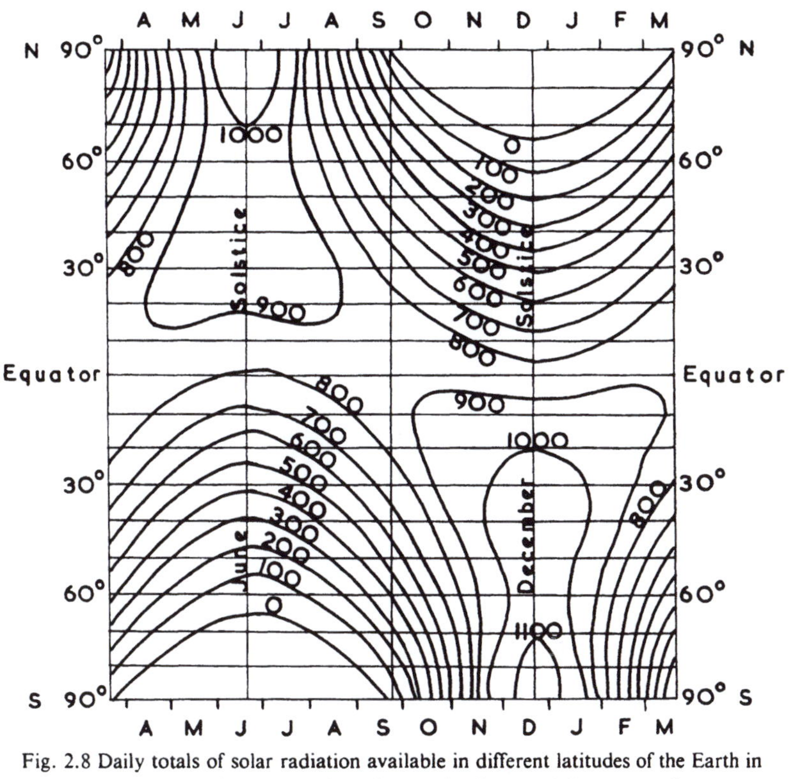

Fig. 2.8 Daily totals of solar radiation available in different latitudes of the Earth in the absence of the atmosphere, in g cal/cm², around the year, at the present epoch.

day.[1] The aspect of any place with regard to the midday sun is important. Surfaces sloping towards the sun get the radiation intensity (though not the length of day) appropriate to a lower latitude. Depletion of the energy in the solar beam by atmospheric absorption and scattering is, however, increased by low angles of elevation and corresponding increase of path length through the atmosphere.

To understand how much of the available energy supply is actually used to heat the Earth's surface and the air, and where and how the heating occurs, one must consider the spectral composition of the solar beam, and of the radiation returned by the Earth, as well as the absorbing and reflecting properties of the Earth's surface and the atmosphere.

1. The elevation of the midsummer sun at midnight is $23\frac{1}{2}°$ at the pole and $13\frac{1}{2}°$ at latitude 80°. The latter figure is the same as the height in the sky of the midday sun in December in central England (53°N), but the polar midnight sun gets more radiation through to the Earth's surface in clear weather because of the extreme transparency of the polar atmosphere – a quality that it owes to its cleanness and low moisture content. Heating rates somewhat over 0·4 cal/cm²/min are possible in places near 80°N.

Wave-length composition of solar and terrestrial radiation

The wave-length composition (so to speak, the 'colour') of the radiation given out by any body varies with its temperature. Sample curves (spectra) of energy radiated within different wave-length intervals by a 'black body' or perfect radiator at different temperatures are given in fig. 2.9. It is seen that, at the temperatures prevailing on the Earth and in the lower

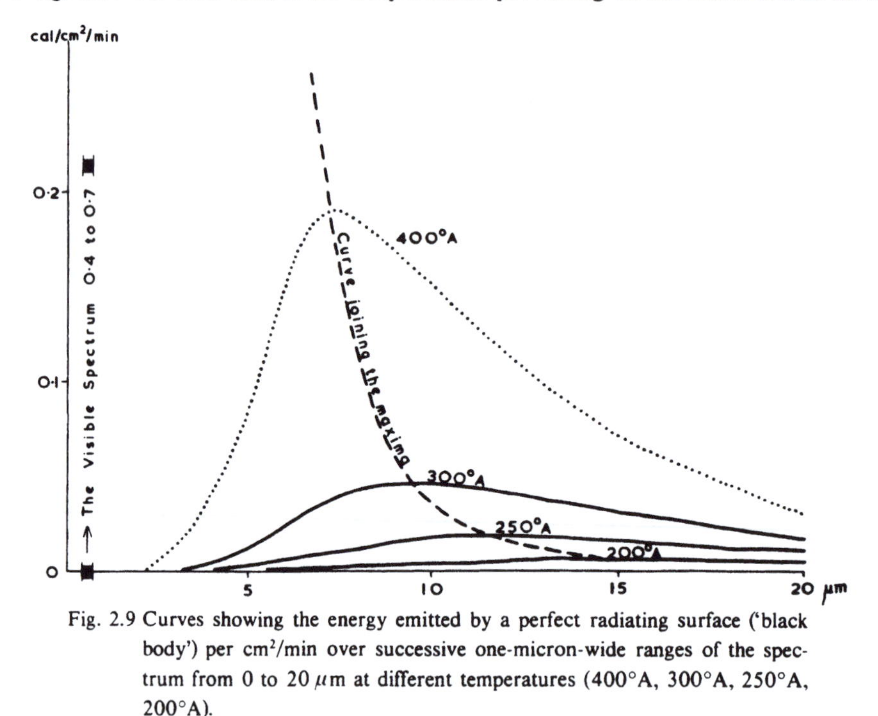

Fig. 2.9 Curves showing the energy emitted by a perfect radiating surface ('black body') per cm²/min over successive one-micron-wide ranges of the spectrum from 0 to 20 μm at different temperatures (400°A, 300°A, 250°A, 200°A).

atmosphere, the terrestrial radiation is long-wave, outside the visible spectrum. The rapid decrease of energy given out as the temperature falls is also obvious. The wave length of maximum emission (λ_{max}) lengthens as the temperature falls, in such a way that

$$T \times (\lambda_{max}) = \text{constant}$$

This is WIEN's law. T is the absolute temperature; and when λ is measured in microns (abbreviation $\mu m = 10^{-6}$ m), the value of the constant in the equation is about 2900. Thus, over 99% of the terrestrial radiation is at wave lengths between 3 and 100 μm, entirely within the infra-red. The greatest part of the sun's radiation is at much shorter wave lengths, 99% of it between 0·17 and 4 μm, and about half of it within the visible spectrum (0·40–0·74 μm);[1] maximum solar emission is in the blue-green part of the spectrum, about 0·5

1. According to S. FRITZ, the distribution of energy in the solar beam approaching the Earth is 45% in the visible spectrum, 46% in the infra-red and 9% in the ultra-violet.

μm. Nevertheless, the hotter a body, the more energy it emits at all wave lengths (cf. fig. 2.9).

The sun does not appear bluish, nor even bluish-white, partly because the sensitivity of the human eye is greater to wave lengths near the middle of the visible spectrum, but also because the atmosphere scatters the shorter wave lengths; these therefore reach the ground partly by indirect paths through the atmosphere, and are seen as the blue light from the sky. The sun itself appears yellowish or red according to the length of the beam's path through the atmosphere and the amount of mist, haze and impurities present.

Fig. 2.10 illustrates the general distribution of radiation intensity by wave length in the solar beam (I) outside the atmosphere, and (II–V) at the Earth's surface for different

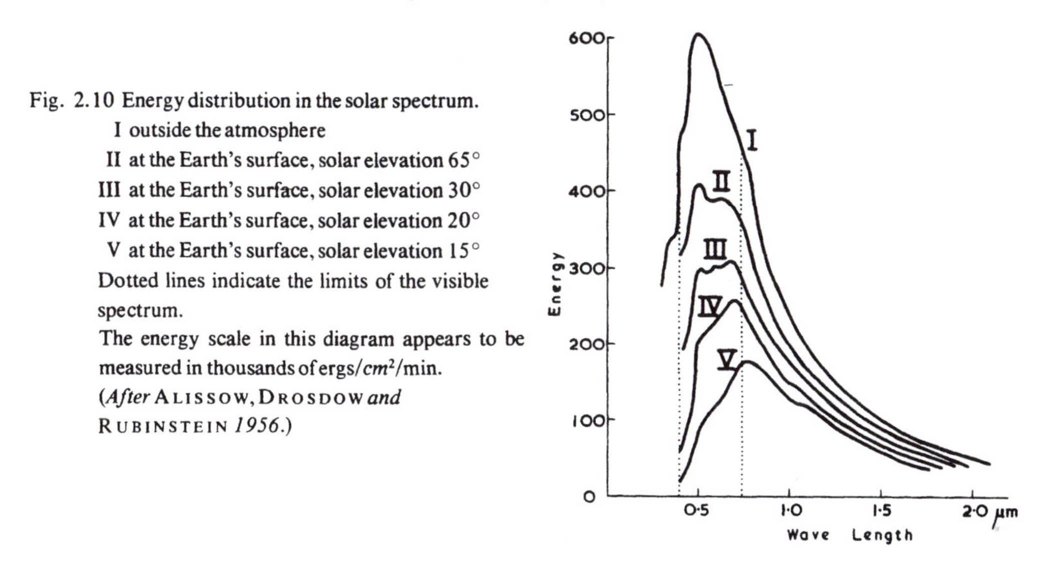

Fig. 2.10 Energy distribution in the solar spectrum.
I outside the atmosphere
II at the Earth's surface, solar elevation 65°
III at the Earth's surface, solar elevation 30°
IV at the Earth's surface, solar elevation 20°
V at the Earth's surface, solar elevation 15°
Dotted lines indicate the limits of the visible spectrum.
The energy scale in this diagram appears to be measured in thousands of ergs/cm^2/min.
(*After* ALISSOW, DROSDOW *and* RUBINSTEIN *1956*.)

elevations of the sun, such as correspond to different latitudes and seasons at midday or different times of the day at one place. The shift of maximum intensity of the energy in the beam from 0·49 μm with little or no distance passed through the atmosphere to the red end of the spectrum, about 0·7 μm, with greater path lengths when the sun is low, is accompanied by a reduction of the total energy getting through.

Absorption in the atmosphere

The spectrum of energy received from the sun is found to depart from the ideal form for a perfect radiator at any temperature. This is partly due to absorption by certain chemical elements (hydrogen, calcium, iron) in the cooler parts of the sun's own atmosphere (as revealed by the *Fraunhofer lines*) but is largely due to absorption in the Earth's atmosphere. Fig. 2.11 illustrates, in more detail than fig. 2.10, the normal spectral composition of the solar beam on arrival at the Earth's surface. The 'gashes' and 'dents' which make it depart from a smooth curve are the marks of energy intercepted and absorbed in the atmosphere.

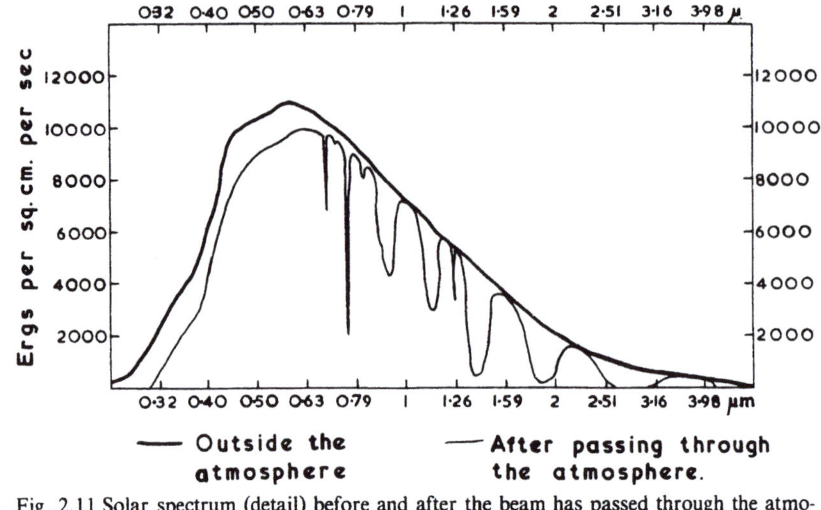

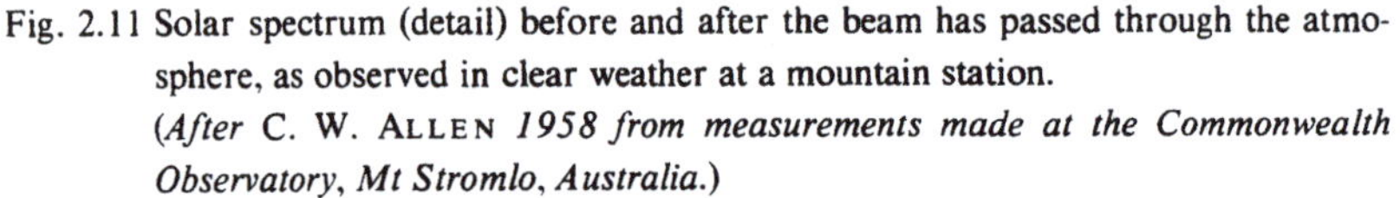

Fig. 2.11 Solar spectrum (detail) before and after the beam has passed through the atmosphere, as observed in clear weather at a mountain station.
(*After* C. W. ALLEN *1958 from measurements made at the Commonwealth Observatory, Mt Stromlo, Australia.*)

The main constituents of the atmosphere and the wave lengths which each substance characteristically absorbs are listed in Table 2.1.

The absorptive power of various substances is commonly defined by their *absorption coefficients*. The absorption coefficient is k, the constant, in the expression

$$I = I_0 e^{-kn}.$$

I_0 is the intensity of the initial radiation supply, reduced to I after passing through a thickness n, defined by the mass present in a column of unit cross-section, of the absorbing substance. Since the factor e^{-kn} is constant for each unit mass of the substance passed through, the radiation intensity is reduced by the same proportion on passing through each successive unitary layer.

The factor e^{-k} is called the *transmission coefficient* and commonly quoted as a percentage.

The expression may also be written $I = I_0\ (10^{-\alpha n})$ to eliminate the Naperian logarithm base. $\alpha = 0{\cdot}4343k$. α is called the *decimal coefficient of absorption*, to distinguish it from k, the *Naperian coefficient.*

The wave-length bands within which the substances named are efficient absorbers are shown in bold print in Table 2.1. Some further comments are necessary here; but readers who need a fuller account should consult the textbooks of physics and original research papers (particularly SIMPSON 1928, 1929).

TABLE 2.1

Constituents of the atmosphere, their distribution and principal absorption bands

Substance	*Molecular weight*	*Proportion by volume in the atmosphere*	*Height distribution*	*Main wave lengths absorbed (high percentage absorption indicated by* **bold** *print)*
Nitrogen (N_2)	28	78·1% of dry air	Uniform in troposphere and stratosphere	No important absorption
Oxygen (O_2)	32	20·95% of dry air	Uniform in troposphere and stratosphere	0·69 and 0·76 μm
Argon (Ar)	40	0·93% of dry air	Uniform in troposphere and stratosphere	No important absorption
Carbon dioxide (CO_2)	44	0·03% of dry air	Uniform in troposphere and stratosphere	4·3 μm and **12–18 μm** (max. at 14·7 μm)
plus smaller quantities of other gases, principally Neon (Ne), Helium (He), Krypton (Kr), Hydrogen (H_2) and Xenon (Xe), not important absorbers in the Earth's atmosphere.				
Ozone (O_3)	48	1×10^{-6} % in the troposphere. Never more than about 10–13 parts per million in the layer of maximum concentration.	Practically confined to layers between about 10–15 and 50 km, with peak concentration at 20–25 km.	**0·23–0·32 μm,** 0·61 μm, 4·8 μm, 9·6 μm and 13–15 μm.
Water vapour H_2O)	18	Mostly 0·5–1·5%, but up to 3% near the ground in the tropics. The amount of water vapour in terms of precipitable water in the total air column is commonly of the order of 30 mm depth of water in temperate and lower latitudes. A typical figure for the stratosphere appears to be 0·3 mm.	Concentration decreases greatly with increasing height and with decrease of temperature.	0·72, 0·81, 0·93, 1·13, 1·4, 1·8–1·9 and 2·0–2·1 μm (maxima at 2·01 and 2·05 μm), 2·2–3·2 μm (maximum at 2·7 μm), **5–8 μm** (max. at 6·7 μm), **11–80 μm.**
Liquid water drops or droplets (H_2O)	18	Clouds are estimated to contain the equivalent of a 1 mm-thick layer of water within a vertical thickness of cloud that ranges from 10 km down to 200 m, depending on cloud density and type. Total liquid water suspended in an air column therefore ranges from 0 to 5 or 6 cm.		Feeble band about 0·3 μm. Absorption effectively begins at wave lengths over 0·9 μm, where about 10% is absorbed in passing through 10 mm of water. **3 μm, 6 μm, 12–18 μm.**

Understanding of how the missing portions of the observed solar spectrum (fig. 2.11) should be interpreted has been arrived at by laboratory experiments on each substance, gas or liquid, responsible for the absorption. There are (in most cases, minor) differences between the results from different experiments and different laboratories: this is partly due to the difficulty of the experiments, but also because with some substances the absorption depends on the pressure and temperature. Behaviour in the conditions prevailing in the atmosphere may therefore differ somewhat from the laboratory results, but can be estimated. It is, in any case, abundantly clear which substances in the atmosphere operate chiefly to absorb part of the incoming solar radiation (e.g. oxygen and ozone) and which have their chief absorption within the spectrum of outgoing terrestrial radiation, on which they therefore have by far the greatest proportionate effect (e.g. carbon dioxide and water in vapour or liquid form).

The percentage of incoming solar energy absorbed by the oxygen in the atmosphere is small; it is represented by lines (fig. 2.11) at about 0·69 and 0·76 μm.

The ozone (O_3) formed from oxygen (O_2) molecules and free oxygen atoms (O) in the upper atmosphere by photochemical action is responsible for absorbing, in the process of its formation, the entire energy in the short-wave end of the solar beam at wave lengths less than about 0·29 μm. Most of this absorption occurs in the top of the ozone layer, near the 50-km level, which is the layer of maximum temperature in the stratosphere in consequence. Ozone is chemically unstable, the dissociation of O_3 molecules also being brought about by solar ultra-violet radiation; when the temperature gets too high, the rate of dissociation exceeds the rate of production, till a balance is attained. The total amount of ozone present in the atmosphere, and its effect on temperatures in the stratosphere, must vary with any variations in the amount of energy received from the sun at the short-wave end of the solar spectrum; it must also depend on the temperatures prevailing in the ozone layer, and on the rate at which ozone is transported by the atmosphere to lower levels, to the area of polar winter darkness, and to regions of lower temperature – to regions in which it is not exposed to the radiation that excites dissociation. Though the total intensity of the solar beam appears nearly invariant, the intensity on wave lengths that affect ozone production and decay does vary by a large percentage (cf. p. 30), apparently in association with sunspots and solar flares. This is, so far, the only physical process certainly known[1] by which solar disturbances affect the Earth's atmosphere below the ionosphere, though the questions of whether, and how, the primary changes at the top of the ozone layer in the stratosphere operate upon the much greater mass of the lower atmosphere (to affect weather processes) remain open. Several suggestions have been made, and will be referred to in the appropriate sections of this book, but up to now none have been convincingly demonstrated.

1. The small variations of the solar constant, for which there is some evidence, discussed earlier in this chapter, should, if they are real, affect the heating of the Earth's surface and the absorption of solar radiation at all levels in the atmosphere.

Ozone also absorbs energy on some wave lengths in the infra-red within the spectrum of terrestrial emission. Concentrations of ozone such as have been measured over Europe, equivalent to a layer of pure ozone 2–4 mm thick at surface temperatures and pressures, would absorb from 7 to 30% of the radiation falling on them of wave lengths 4·5–5 μm and 9·4–9·8 μm and up to 5% of that between 12·5 and 15·5 μm. But terrestrial radiation in these wave lengths is likely to have been completely absorbed by water vapour and carbon dioxide at levels well below the ozone layer.

The carbon dioxide (CO_2) content of the atmosphere is believed, from comparison with such measurements as were made around 1890–1900, to have increased from about 290 ppm (parts per million) about that time to about 315 ppm by 1960, due to the increased burning of fossil fuels, chiefly coal and oil. The increase was 5% by 1944, 10% by 1967. The CO_2 is apparently evenly mixed throughout the atmosphere (constant 'mixing ratio'), except near local sources and close to vegetation. The concentrations quoted mean that the total carbon dioxide present is equivalent to a layer of this gas at the surface 2·5–2·8 cm thick. This is more than enough to absorb all radiation between 4·0 and 4·8 μm, and enough to absorb about 40% of the radiation at wave lengths between 14 and 16 μm. In the absence of water vapour, the carbon dioxide at present in the atmosphere would absorb 15–20% of the total energy emitted from the Earth. Water vapour is, however, a still more effective absorber of radiation on most of the same wave lengths as CO_2, and indeed together they appear to absorb all the terrestrial radiation on wave lengths greater than 14 μm. So it seems that changes in the amount of carbon dioxide in the atmosphere can only affect wave lengths in the CO_2 band about 4·3 μm and between about 12 and 14 μm, for which wave length absorption by water vapour is incomplete. In the stratosphere, where the quantity of water vapour is very small, absorption by carbon dioxide just over a narrow band about its maximum at 14·7 μm exceeds that by water vapour.

Absorption curves reproduced by SHAW (1930) and PLASS (1954) indicate that, in the absence of water vapour, the carbon dioxide in the atmosphere would absorb all the radiation between wave lengths 12·5 and 17·5 μm emitted by the Earth's surface (taken as having an effective, or overall mean, temperature of 288°A), this main absorption band alone accounting for 14% of the Earth's emission. In fact, however, the water vapour present absorbs in the lower atmosphere almost all the radiation between those wave lengths emitted by the surface wherever temperature is over 283°A (CALLENDAR 1960) and its re-radiation returns much of this to the Earth; so the influence of CO_2 must be limited to the colder, drier climates and seasons, and to higher levels in the atmosphere. Some authors have held that long-term variations of the amount of CO_2 could make a difference of up to 1% more or less of the Earth's surface emission of long-wave radiation being absorbed in the atmosphere; others rate the effect much lower. PLASS (1956) estimated that doubling the present CO_2 content of the atmosphere should raise world temperatures by 3·6°C and believed that the observed 1°C rise from the industrial revolution to 1950 could be accounted for by the undoubted increase of CO_2; others (e.g. MATTHEWS 1959) believe that the effect could, at most, have been a quarter of this. MANABE and WETHERALD (1967) put it at just half PLASS's figure. Probably over most of the Earth the water vapour and carbon dioxide together already more than suffice to absorb all the radiation emitted on the wave lengths concerned. This point, and any future changes, might be settled by observations of the spectrum of Earth radiation from satellites. The observed decline of global temperature since 1945 implies some other factor exercising about three times as strong an effect (in the opposite direction) as the CO_2 increase.

Man's estimated present output of 9×10^9 t/yr (metric tons per year) of CO_2 from burning fossil fuels is not, however, all retained in the atmosphere. The oceans are a great reservoir of CO_2, holding about twenty-seven times as much of it as the atmosphere. Moreover, as the concentration in the oceans increases, more CO_2 becomes fixed as calcium bicarbonate in the deposits of the ocean bed. The vertical circulation of the oceans is slow and when there is a change in the amount of CO_2 in the atmosphere, it probably takes 10 000–50 000 years before a new equilibrium is attained between these three reservoirs. Moreover, the exchanges are complicated by the difference in the solubility of CO_2 at different temperatures: the oceans tend to give off CO_2 when they are warm and dissolve more when they are cold. J. MURRAY MITCHELL (unpublished lecture at the American Association for the Advancement of Science symposium on environmental pollution 1969) estimates that the amount of CO_2 in the atmosphere by 1980 will be 15%, by 1990 20%, above the 1890 concentration.

Water vapour is by far the most effective constituent of the atmosphere for absorbing radiation. The total quantity present in an air column between the surface and the greatest heights reached by water vapour is of the order of a few centimetres depth of precipitable water. Even the air over the deserts contains enough water vapour to absorb practically all radiation at wave lengths between about 5·5 and 7 μm and from 14 to 80 μm, plus smaller proportions of adjacent wave lengths. There is also some absorption in bands near the peak wave lengths of the solar spectrum. Moist air absorbs 10% or more of all the radiation at wave lengths emitted at temperatures generally prevailing near the Earth's surface, but dry stratospheric air at temperatures far below the freezing point is almost completely transparent to wave lengths between 8·5 and 11 μm.

The height at which radiation is absorbed by the water vapour in the atmosphere is determined by the high value of its absorption coefficient. Depletion of the outgoing infrared radiation due to absorption by water vapour is rapid in the immediate neighbourhood of the source of emission, whether that be the Earth's surface or some point in the atmosphere. Absorption is strongest in the layers near the ground, where the moisture content is greatest. Variations of absorption due to changing humidity chiefly allow those wave lengths to which water vapour is semi-transparent to be transmitted a greater distance through the atmosphere when the air is dry.

Liquid water drops suspended in the air in clouds and fogs absorb radiation over much the same wave lengths as water vapour, except that in addition they absorb all (mainly solar) radiation between about 2·5 and 3·3 μm. (The most important loss of incoming solar radiation for points beneath a cloud is, however, due to reflection at the cloud top.)

Solid particles, dust and haze, suspended as impurities in the atmosphere, also play a part in intercepting, absorbing, scattering and reflecting radiation. The effect is more variable in amount than those associated with ozone, carbon dioxide or water vapour. The most extensive and persistent dust veils are those consisting of the finest particles occasionally thrown up to heights of 20–30 km and more in the stratosphere by great volcanic explosions. The dimensions of the particles that reach these heights, and have a long residence time in the atmosphere, are about 1 μm across, probably for the most part in the form of shell-like fragments of shattered, solidified lava bubbles, so that their fall-speeds are even less than the terminal velocities of minute spherical globules of the same diameter. Particles of this size, starting to settle from heights of the order of 30 km, may remain from 2 to 12 years in the stratosphere before falling into the lower atmosphere where they are soon washed to Earth by the drizzle, rain and snow. The greatest dimension of these particles is smaller than the wave lengths of most of the terrestrial radiation, which they are therefore unable to absorb. Their absorption is concentrated at wave lengths less than 1–2 μm, and tends therefore to deplete the incoming solar radiation while not materially affecting the escape of terrestrial radiation to space. The depletion of solar radiation by volcanic dust veils is believed to be of the order of thirty times as great as the depletion of the outgoing terrestrial emission.

An indication of the magnitude, and the duration, of the effect after great volcanic eruptions in 1883, 1888, 1902 and 1912 is given by fig. 2.12 (*a*), which shows monthly averages of the measured intensity of the solar beam at observatories in North and South America, Europe, Egypt and India as percentages of the overall mean of the years 1883–1938. The monthly average of total solar radiation receipt (over all wave lengths) fell as much as 20–22% below the overall mean in some months, and the effect was substantial over 2–3 years after these eruptions. But the effect on total radiation reaching the surface is not as great as these pyrheliometric observations of the direct solar beam would suggest, the depletion of the beam being largely compensated by an increase of the scattered light and heat (as diffuse radiation or glare) reaching the Earth's surface from the sky on account of the dust veil: this is still solar short-wave radiation, though coming by devious paths through the atmosphere. The scattering is selective as regards direction, more of the radiation being scattered forwards towards the Earth than back to space. The heated dust also acts as a source of long-wave radiation, which is given out in all directions. The net effect appears to be that a volcanic dust veil reduces the total radiation penetrating to the surface by an amount that is a quarter (or less) of the reduction of the direct solar beam. The effect of a dust veil that was evenly distributed over the globe should increase towards high latitudes, because slanting rays from the sun have to pass through a much greater thickness of the dust layer than vertical, or near-vertical, rays.

The strength of the direct solar radiation falling on the Earth month by month from 1932 to 1954 as it appears from observations made at two observatories in southern Japan away from urbanized areas is shown in fig. 2.12 (*b*). (Scatter of the monthly mean values in all such diagrams is an inevitable result of the difficulties of selection and reduction of the observations to equivalence with clear sky conditions.) This figure suggests that there was no real variation until after 1949; a slightly falling tendency is discerned thereafter and a sharp drop by 5% or rather more to significantly lower values in 1953 and 1954. These cannot be attributed to any great extent to the volcanic eruption in Alaska in July 1953, because they begin too soon; effects of nuclear bomb tests in the equatorial Pacific in November 1952 and spring 1954 and in the Soviet Union in summer 1953 (which being ground bursts, unlike the air bursts of later years, may have yielded dust on a scale equivalent to volcanic eruptions) have been suspected (ARAKAWA and TSUTSUMI 1956). Since no persistent dust cloud was observed in those years, one must also consider the

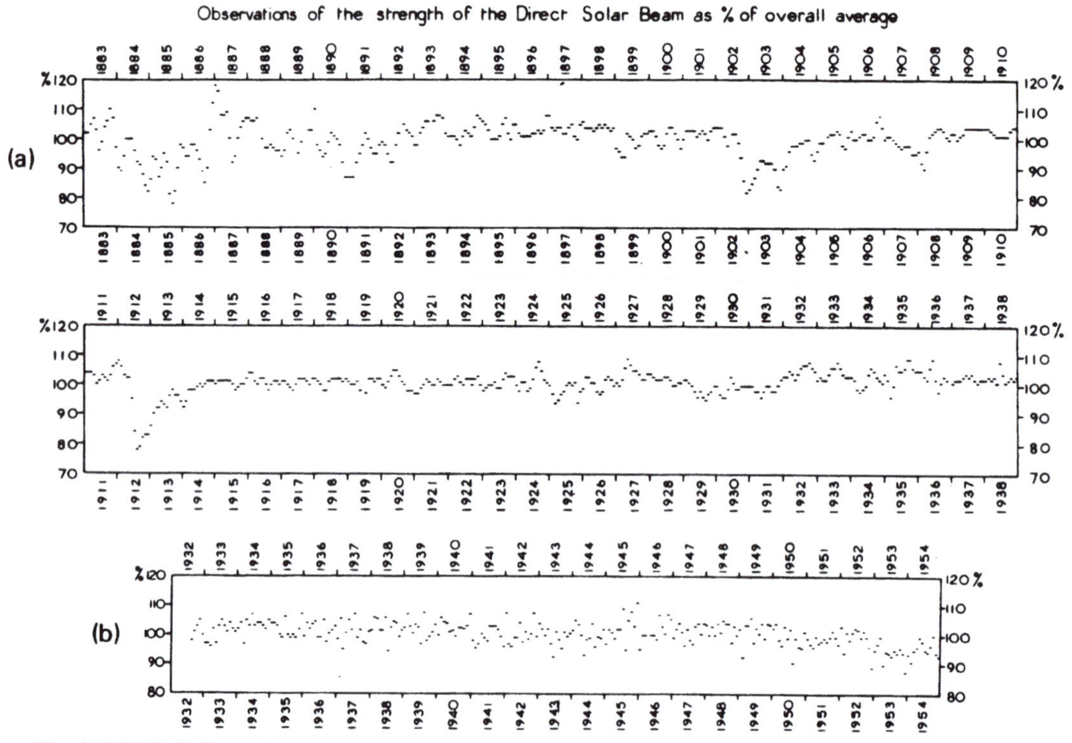

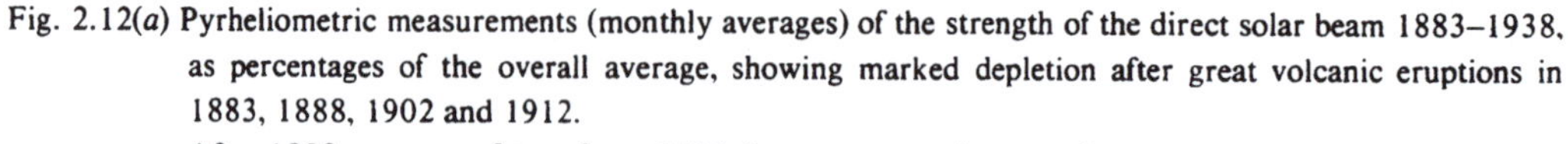

Fig. 2.12(*a*) Pyrheliometric measurements (monthly averages) of the strength of the direct solar beam 1883–1938, as percentages of the overall average, showing marked depletion after great volcanic eruptions in 1883, 1888, 1902 and 1912.
After 1892 averages of two, from 1896 three or more, observatories.
(*b*) Monthly means of the strength of the direct solar beam 1932–54, as percentages of the overall average of those years, as derived from observations at Matsumoto and Shimizu, Japan.
(*After* ARAKAWA *and* TSUTSUMI *1956*.)

possibility of other causes – e.g. assuming that instrument difficulties do not enter in, a real solar variation or pollution of the layers in the vicinity of the tropopause by exhaust gases and water vapour from high-flying aircraft.

Effects of absorption and re-emission within the atmosphere

The behaviour of the various substances of which the atmosphere is composed, and of those that are suspended within it, as emitters as well as absorbers of radiation, is governed by KIRCHOFF's law: the emission rate for radiation at a given wave length bears to the absorption rate a ratio that is constant for all substances, depending only on the temperature. In other words, whatever wave lengths a substance absorbs particularly strongly it also emits strongly for its temperature – i.e. for those wave lengths its approach to a 'black body' (perfect radiator) is good. Hence, the effect of water vapour, for example, is to absorb outgoing long-wave radiation within the lowest layers of the atmosphere and re-radiate it in all directions, so that some is returned to the Earth while a reduced amount is sent on to be similarly treated by the next higher layers, and so on. This amounts to blanketing the Earth, keeping in radiation that would otherwise escape – a radiation trap, often referred to as the 'greenhouse effect'. By this means, the balance between incoming solar and outgoing terrestrial radiation is attained at a higher surface temperature than would be the case in the absence of those substances that absorb terrestrial radiation in the atmosphere.[1] In the absence of water vapour the carbon dioxide would produce this effect on some of the same wave lengths, though less efficiently and allowing more of the terrestrial radiation to reach higher levels before being absorbed.

When there is cloud, terrestrial radiation is reflected as well as re-radiated back towards the Earth. Clouds therefore help to maintain higher surface temperatures at night, and in the polar winter darkness, than would occur without them – in the very same way as they reduce the gain of solar radiation by day. On balance, over the 24 hours and over the year as

1. See p. 18. The atmosphere is almost transparent to as much of the solar beam as is not reflected away by the clouds; it allows 80–90% of it to reach the Earth's surface. By contrast, only 10–15% of the long-wave radiation sent out by the Earth's surface escapes right through the atmosphere to space. The rest is largely trapped by the water vapour, though also by the other absorbers in the atmosphere mentioned in this chapter: so these substances together may be held responsible for the Earth being almost 40° warmer than it would be without them.

Temperatures near the ground fall by especially great amounts on clear winter nights in the cold air of the interiors of the continents where the air is very dry. Over the great continental territory of the U.S.S.R. the annual totals of outgoing terrestrial radiation are, because of the dryness of the air, latitude for latitude, one-third to a half greater in Siberia beyond 80–90°E than in European Russia; the lowest values are in the most maritime regions of the country near the Barents Sea, the Baltic and (for the latitude) the Black Sea, and over the Pripet Marshes. Correspondingly the annual net radiation gain of the surface (insolation less outgoing terrestrial radiation), which is more or less positive over the whole U.S.S.R., has values in the interior of eastern Siberia which are only about three-quarters of those for the same latitudes in the western part of European Russia (BORISOV 1959).

a whole, the effect of cloud between about latitude 35°N and 35°S (see later p. 59) must be to lower, and outside these latitudes to raise, the temperatures prevailing in the air below it. Any increase of cloudiness should, of course, operate in the same direction.

Because of the various ways in which radiation is absorbed or lost on its way through the atmosphere, places on high mountains have a different radiation climate from places at low levels in the same latitude. On the heights insolation by day and radiation cooling of the ground by night are fiercer than on the low ground. Borisov (1959) quotes figures from a study of places in the U.S.S.R. which show that between sea level and heights of 3000 m insolation received increases at a steady rate of about 10% for each 1000 m of ascent.

Ozone operates on the short-wave radiation in the same way as water vapour on the long-wave; selected wave lengths are absorbed from the solar beam, after passing through the ionosphere, upon reaching levels where ozone molecules can be formed – i.e. at, and near, the top of the ozone layer. The ozone then re-radiates energy in all directions, part being returned to space. Progressively depleted amounts of solar radiation penetrate to lower levels in the atmosphere, until all the energy in the wave lengths affected is soon absorbed.

Both ozone and volcanic dust may be said to have a 'reverse greenhouse effect' in that they exclude from the lower atmosphere solar radiation that would otherwise penetrate. This tends to maintain the balance between solar and terrestrial radiation at a lower temperature of the Earth's surface than would otherwise prevail. The proportion of the energy of the solar beam in the wave lengths affected by ozone is however very small: the more important effect of the ozone is that the radiation it absorbs warms the stratosphere. Variations of this absorption and of ozone concentration must affect the vertical distribution of temperature in the atmosphere above and below the ozone layer, and this in turn affects its stability as regards development of convective motions involving displacements of air in the vertical.

Absorption by the ozone in the stratosphere also removes from the solar beam ultra-violet wave lengths that would be lethal for living organisms.

Absorption of solar radiation at the Earth's surface and losses by reflection

The radiation which succeeds in penetrating right through the atmosphere to the Earth's surface is not all absorbed by the surface. Different surfaces have different characteristics as regards absorptive power and reflection. Moreover, these vary with wave length and angle of incidence. The general reflectivity (*albedo*) over the range of wave lengths composing the solar radiation for the various surfaces listed in Table 2.2 is taken from summaries of the work of many investigators by Alissow *et al.* (1956) and by D. M. Houghton (1958). The most general conclusions from Table 2.2 are:

(i) Snow cover brings about an enormous change on any type of surface, greatly reducing the intake of radiation.

(ii) Wetness and forest cover generally darken the surface and increase the intake of radiation.

The very wide ranges of albedo quoted for some types of surface probably depend mainly on wetness. The albedo of water surfaces ranges from 2% for the beam of the zenith sun at noon in low latitudes to 34% when the solar elevation is 10°, and 78% when the sun's rays impinge at an angle of only 2°.

Several of the surfaces listed as poor absorbers of solar short-wave radiation are good absorbers and radiators for long-wave radiation. Snow, sands, bare rock, grass and soil, wet or dry, forest and water all give out 95–98% of the radiation that would be emitted by a 'black body' at temperatures that occur at the Earth's surface. Most striking are the effects

TABLE 2.2

Percentages of solar radiation reflected by different surfaces (Albedos)

Surface	*Albedo* %	*Surface*	*Albedo* %
New-fallen snow	85 (90% has been observed in Antarctica)	Grain crops, depending on ripeness	10–25
		Clay (blue), dry	23
		Ploughed fields, dry	12–20
Old snow	70	Pine forest	6–19
Thawing snow	30–65	Oak tree crowns	18
Salt deposits from dried-up lakes	50	Granite	12–18
White chalk or lime	45	Other rocks, generally	12–15
Yellow deciduous forest in autumn	33–38	Stubble fields	15–17
Quartz sands, white or yellow	34–35	Clay (blue), wet	16
Clayey desert	29–31	Spruce tree crowns	14
Parched grassland	16–30	Wet fields, not ploughed	5–14
River sands (quartz), wet	29	Fir tree crowns	10
Green deciduous forest	16–27		
Green grass	8–27	Water, depending on angle of incidence	2–78

observed at a snow surface, which reflects away most of the sun's radiation and therefore heats up little by day, but gives off long-wave radiation almost like a 'black body'. Because of this, and because the air trapped in the snow, particularly if loose, impedes flow of heat from the usually warmer earth below, very low temperatures may be quickly produced at a snow surface in a single night.

H. G. HOUGHTON (1954) gives the reflectivity for incoming solar radiation of different types of cloud as listed here in Table 2.3. These figures mean that an increase or decrease of cloud cover causes a great change in the absorption of solar radiation, except over regions that are in any case snow-covered. The greater reflectivity of the lower cloud layers, and of heap clouds, is connected with their denser texture resulting from their greater water content. The average albedo of clouds appears to be about 55%.

The overall albedo of the Earth with its atmosphere, in the conditions of the present century (approx. 1920–50), appears from these studies to be in the region of 35–37%.

TABLE 2.3

Albedo of clouds

Type of cloud	*Albedo* %
High-level clouds and cloudsheets (cirrus)	21
Middle-level cloudsheets (between about 3 and 6 km)	48
Low-level cloudsheets	69
Heap clouds (cumulus types)	70

General survey of radiation exchanges and transmission through the atmosphere

A summary of the net effect, or balance, of all the radiative processes mentioned in this chapter is necessary to make the implications clear. The schematic vertical cross-section (fig. 2.13) and world maps for January and July (figs. 2.14 and 2.15) demonstrate the nature, locations and approximate magnitudes of the main effects as regards heat reaching the Earth's surface. It must be realized, however, that values attached to diagrams such as fig. 2.13 are overall averages for different places, times of the year and weather conditions, involving difficult and partly inconclusive estimation. They may serve nevertheless to indicate the possible effectiveness of changes in this or that item of the radiation budget in changing the surface climate.

Incoming solar radiation is partly (1) reflected, (2) scattered, (3) absorbed in the atmosphere, and (4) absorbed in the ground and sea. Terrestrial radiation given out from the ground, sea and atmosphere is partly (1) reflected back to the Earth, (2) scattered, (3) absorbed in the atmosphere, (4) lost to space. Also the energy absorbed in the atmosphere is re-radiated in all directions. In addition, the ground and sea pass heat to the atmosphere by conduction and convection, as well as energy taken up in the form of latent heat of evaporation which is released later as 'sensible' (i.e. feelable) heat elsewhere in the atmosphere, when the moisture is condensed to form clouds or fog. Fig. 2.13 shows the separate stages of this energy exchange that is constantly going on in the atmosphere. They may be made more specific and intelligible by quoting figures illustrative of the present epoch.

The first diagram of the type shown here in fig. 2.13 was presented by W. H. DINES in 1917, and since then many versions have appeared in the literature, most recently a usefully itemized diagram by MILLER (1966). Different authors give slightly different estimates for the various items, as is hardly surprising in view of the difficulties in estimation of such world-wide averages. It is certain, moreover, that there are real variations with time; although total incoming and outgoing radiation must nearly balance over any year, or short group of years, if drastic climatic variations are not to occur. The estimates to be quoted below are also summarized here in Table 2.4, which shows how the heat gain and loss at the surface are approximately balanced.

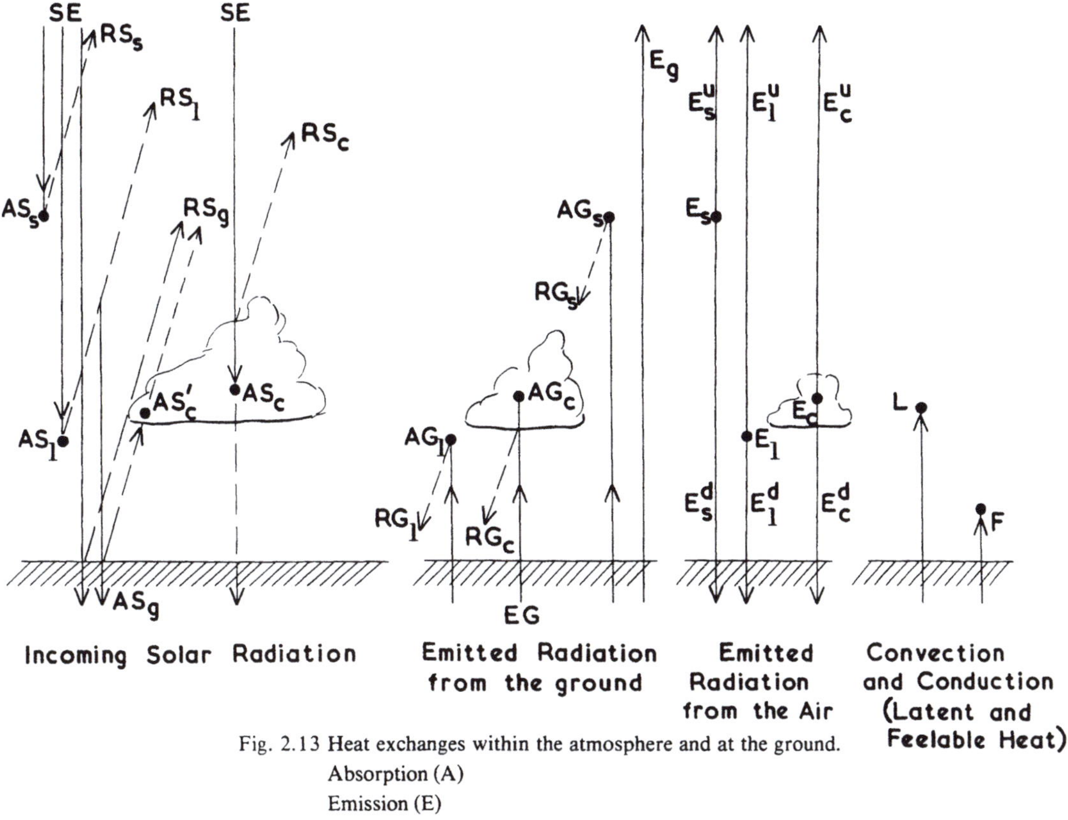

Fig. 2.13 Heat exchanges within the atmosphere and at the ground.
Absorption (A)
Emission (E)
Reflection or re-radiation (R)

In fig. 2.13 SE stands for the rate of supply of solar energy in the incoming beam, AS_s for the solar energy absorbed in the stratosphere, AS_l for the solar energy absorbed in the lower atmosphere and AS_g for the solar energy absorbed in the ground. AS_c is the amount of solar energy absorbed in clouds and AS'_c the solar energy absorbed by clouds after reflection from the ground. RS_s, RS_l, RS_g and RS_c are the amounts of solar energy returned to space after reflection and scattering from the stratosphere, lower atmosphere, ground and clouds respectively. Similarly, AG_l, AG_c and AG_s are the amounts of (long-wave) radiation emitted from the ground that are absorbed respectively in the lower atmosphere, the clouds and the stratosphere; and RG_l, RG_c and RG_s are the corresponding amounts returned to the Earth after reflection and scattering. In the next section of the diagram, E_l, E_c and E_s (with superscripts u for upward and d for downward components) stand for the amounts of (long-wave) radiation emitted by the lower air, the clouds and the stratosphere; E_g is the amount that escapes direct from the ground to space. Finally, L is the amount of energy picked up as latent heat by the atmosphere in evaporating moisture from the surface, and F is the feelable ('sensible') heat received by convection and conduction from the ground and sea.

TABLE 2.4

Summary of the Earth's overall average radiation budget

Present-day climates

(Units: per cent of the energy available in the solar beam)

Incoming			*Outgoing*		
Absorbed	Transmitted onwards by re-radiation, forward scattering etc. within the atmosphere	Returned Earth radiation	Earth radiation emitted (and lost) to space	Earth radiation from surface to atmosphere	Returned solar radiation
Ozone in stratosphere 2% Oxygen 2% Water vapour in lower atmosphere 7% Clouds 10% Earth's surface (from direct solar beam), 22%	Scattered from the blue sky and absorbed at Earth's surface 5–6% Absorbed at Earth's surface (from solar beam, i.e. short wave) after passing through clouds, 17%	Radiated (long wave) from the atmosphere and absorbed at Earth's surface 90–98% Returned to Earth's surface after reflection within atmosphere, 8–11%	Direct from Earth's surface 11% Radiated (long wave) from the atmosphere 54%	Absorbed largely by water vapour and carbon dioxide 100–107%	Reflected part of solar beam *c*. 35%
Total radiation received at the surface (a) short wave from the sun 44 to 45% (b) long wave from the atmosphere 98 to 109% (*a*) + (*b*) 142 to 154%			Total radiation lost from the surface 111 to 118% Heat lost from surface to atmosphere by conduction, convection, and in evaporation processes etc. (*a*) directly as feelable heat 6–12% (*b*) indirectly as latent heat, converted later into feelable heat 20% Total losses from the surface 137 to 150%		

Except for small residual accumulation (storage) or loss of heat in climatic changes, the three main energy fluxes must balance[1]

$$SE = R(S_s + S_1 + S_g + S_c) + (E_g + E_1{}^u + E_s{}^u + E_c{}^u)$$

Incoming solar energy — Reflected solar energy — Outgoing terrestrial radiation

These quantities can now be submitted to some attempt at measurement from artificial satellites, and it is hoped that elimination of calibration difficulties with the instruments will some day permit continual monitoring of the values with sufficient accuracy for any variations that occur to be observed. At present, neither the intensity of the solar beam (the 'solar constant') nor the proportion of it that is reflected, the overall albedo of the Earth and

1. The balance equations for radiation, heat and water vapour in the atmosphere have been set out and treated in rather more detail by FLOHN (1958, 1963).

its atmosphere (the ratio of the total R terms to SE in the above equation), can be regarded as firmly established to within several per cent. Yet, a change of either by even one per cent would certainly have a big effect on world climates. So far, most of the reliable measurements of the solar beam have been made by instruments at high altitude observatories on the ground, and errors must be introduced in correcting for the effects of the atmosphere. Russian calculations by BUDYKO (1955) and others (ALISSOW *et al.* 1956), based partly on measurements and partly on estimation of the various elements of the radiation balance over the world, gave the overall albedo of the Earth and its atmosphere in 1950 approximately as 37%. The Earth's albedo can also be observed more or less directly, since the dark part of the moon is illumined by sunlight reflected from the Earth: from DANJON's careful measurements of this faint-moonlight made in Europe over nine years about 1930, FRITZ (1949) calculated that the overall albedo of the Earth as a planet is about 35%. A margin of uncertainty is introduced by the fact that the measurements of the 'grey moonlight' were confined to the visible spectrum, and an adjustment had to be made because the albedos of the Earth and clouds are known to be lower for infra-red than visible wave lengths. Also, observation from Europe meant that the reflectivity of the extensive oceans on the Pacific side of the world did not enter into the light reaching the moon.

Taking 720 g cal/cm²/day as the solar constant averaged over the year and over the world, evidently about 250 cal are on average reflected and wasted under present climatic conditions. Most estimates (e.g. by FRITZ *loc. cit.*, and by H. G. HOUGHTON[1] and others) indicate that altogether about 320 cal (44%) of solar radiation are absorbed at the Earth's surface (AS_g in fig. 2.13), 120 cal of this after passing through clouds and about 40 cal normally in the scattered short-wave radiation from the blue sky. Absorption direct in the atmosphere is a good deal smaller: 15 cal (2%), mainly by the ozone in the stratosphere (AS_s), about another 15 cal by the oxygen mainly in the lower atmosphere, about 50 cal by water vapour (these items together making up AS_l), and 70 cal (10%) in the clouds (AS_c) – altogether about 150 cal (21%). The overall average long-wave emission from the Earth's surface (EG) is estimated as 800–850 cal/cm²/day; of this only about 80 cal escape through the atmosphere to space (E_g) and 60–80 cal are reflected back to the Earth ($RG_l + RG_s + RG_c$). Total energy emitted as long-wave radiation from within the atmosphere itself is generally taken to be 1000–1100 cal, of which 650–700 cal are transmitted to the Earth's surface ($E_s^d + E_l^d + E_c^d$) and only 400 cal or so are given out to space from the top of the atmosphere ($E_s^u + E_l^u + E_c^u$). This difference is partly due to the fact that the lower layers of the atmosphere are better emitters of radiation because of their higher temperature and water content, and partly because there is most reflection and scattering also within the lower air. The net amount of heat given off by the surface to the atmosphere by convection and conduction (F) is put at only 45–90 cal (equivalent to 6–12% of the solar radiation) on an overall world average by most authors. The average heat lost by the surface

1. H. G. HOUGHTON. On the annual heat balance of the northern hemisphere, *Journal of Meteorology*, **11**, 1–9. Lancaster, Pa., 1954.

in evaporation of moisture is liberated within the atmosphere wherever condensation to form clouds and fogs occurs, generally at a distance of some hundreds or thousands of kilometres from the moisture source; the quantity is estimated at about 140 cal (L in fig. 2.13).

Changes in the mean values of any of the items discussed should bring about changes in the mean temperature of the Earth, and in several cases it is clear that the temperature of some particular layer of the atmosphere should be chiefly affected. In those cases there should also be changes in the rate of change of temperature with height, which affect the stability of the atmosphere – i.e. its resistance to vertical motion and convection currents – and may thereby influence the forms and energy of the large-scale circulation of the atmosphere.

Most obvious, perhaps, are the changes that must occur through differences of total cloud cover or the extent of snow and ice. The average albedo of clouds (cf. Table 2.3) is considered by HOUGHTON to be about 55%. That of snow in winter may be taken as almost 80% (locally less over forested areas where winds may blow and shake the snow off the trees); in summer the albedo remains about 80% in the case of snow covering great elevated ice sheets such as at present exist in Antarctica and Greenland, but falls to 40–60% generally over sea ice – the summer figure is, however, likely to remain higher than this in cold years and cold epochs. Increase of cloudiness or snow cover therefore means increased loss and wastage of incoming radiation by reflection to space.

In the case of a change of either cloudiness or snow cover, it is simple to calculate the change of the Earth's total albedo that should occur as a direct result, and therefrom to derive a figure for the change of prevailing temperatures to be expected. However, the real problem is rather more complicated because of secondary effects for which allowance should be made: an increase of extent of snow surface, for example, must alter the geographical distribution, and probably the total amount, of cloud. An increase of cloud cover must have effects upon evaporation and the wetness of the ground, producing some (probably small) further change of total albedo. Moreover, a good deal of time, possibly running into decades or centuries, must be required for a new temperature level to be reached which is in equilibrium with the altered radiation conditions; this is because of the large amount of heat stored in the oceans.

Increased cloudiness at night, and especially in the long winter night in the polar regions, has an opposite effect at the surface. There the effect of increased reflection of the terrestrial radiation from the cloud (RG_c in fig. 2.13) is to raise the surface temperature. However, the radiation given out to space from the atmosphere is increased by the component from the cloud-top (E_c^u), and this means that cloudy situations in the polar regions are liable to produce the greatest cooling of the lower atmosphere in depth even though the temperatures at the surface are not as low as they would be if the sky cleared.

Changes of water vapour content in the atmosphere are also important, because (as already mentioned) water vapour is an efficient absorber, and radiator, of long-wave radiation. On cloudless nights more heat is lost from the Earth's surface if the air is dry. Since the total water vapour in a column of the atmosphere is always enough, except in the coldest regions, to absorb all the emitted terrestrial radiation over most wave lengths, the effect of less or more moisture in the lowest layers is to allow the outgoing radiation from the surface to be transmitted to a greater or less height in the atmosphere. The chief result to be expected is a change in the vertical distribution of temperature in the atmosphere as well as some slight change in the total radiation given off to space. Increased water vapour content should retain more terrestrial radiation in the cloud layers and below, raising the average temperatures near the surface and in the lower atmosphere. This may be expected also to increase the difference of temperature between the humid lower levels and the higher layers of the atmosphere, especially in high latitudes.

The water vapour content of the atmosphere may be expected to increase through increased evaporation from warmer seas when (for whatever reason) the average temperature over the Earth rises, e.g. because of some change either in the radiation supply or in the thermal condition of the surface brought about by albedo changes or by the vertical circulation of the oceans. But there are other ways in which the amount of water vapour in the atmosphere may be increased without any initial change of temperature level. A general increase of wind strength would produce greater evaporation from the rougher water surfaces. An increase of convective upward transport of moisture, allowing more to be taken up in the lower layers, should occur with any increase of the lapse rate of temperature with height (whether produced by heating at the surface or by cooling aloft). The results as regards the total heat exchange, and the radiation balance, are likely to be different according to which of these processes is responsible for the change of atmospheric moisture content. Cloudiness may or may not be affected: it will certainly be changed by different amounts according to how the change of atmospheric moisture content has come about and will acquire a somewhat different geographical distribution.

In relation to the very long warm eras that appear to have prevailed through most of geological time, BROOKS (1949) pointed out that the atmospheric moisture content was probably greater than now, not only because of the higher temperature but also because of the greater expanse of the oceans; the generally lower elevation of the land would mean less extensive cloud formation of the kind due to the winds being forced up over mountain ridges – so the increased water vapour content probably went with less cloudiness.

It is clear that any increase of ozone or volcanic dust in the stratosphere must result chiefly in the interception of more solar radiation in the layers concerned and a reduced lapse rate of temperature with height in the underlying atmosphere. Some indications suggest that the amount of radiation intercepted in the stratosphere may have varied by a factor of two or three over several periods lasting up to 2–3 years since 1880 (indicated in fig. 2.12 (*a*)) due to volcanic dust. Observations are so far insufficient to establish the possible variations of ozone amount, which presumably follow the course of solar disturbance. Professor C. W. ALLEN (1958) has, however, reported a secular variation of the amount of ultra-violet radiation received, presumably attributable to some variation in the amount of ozone present in the atmosphere between about 1920 and 1940.

It seems possible that carbon dioxide changes have a significant effect only in the stratosphere over cold regions where the moisture in the lower atmosphere is not enough to check all the outgoing Earth radiation on the wave lengths which CO_2 absorbs. The effect should then operate in the same sense as with changes of ozone or volcanic dust, increased CO_2 producing a rise of temperature in the stratosphere and greater stability (resistance to vertical motion) in the atmosphere below it. These effects have not, however, so far been demonstrated.

The actual world distribution of net radiation receipt, and subsequent redistribution of heat

The average distribution over the Earth of the balance of incoming less outgoing radiation each month of the year in the present epoch was first satisfactorily worked out by SIMPSON (1929), and the corresponding world maps for January and July are reproduced here in figs. 2.14 and 2.15 in the units of calories per day used elsewhere in this chapter.

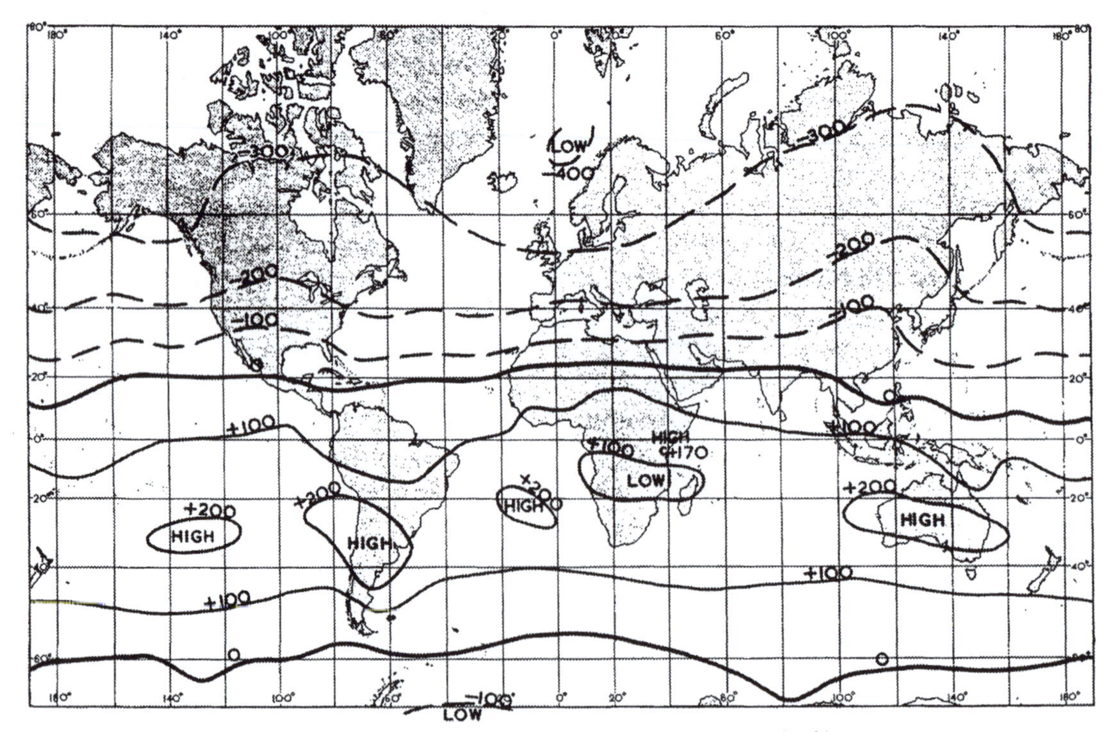

Fig. 2.14 Net radiation receipt in January, average in g cal/cm²/day.
Broken lines indicate net loss of radiation.
(*Adapted from* SIMPSON *1929*.)

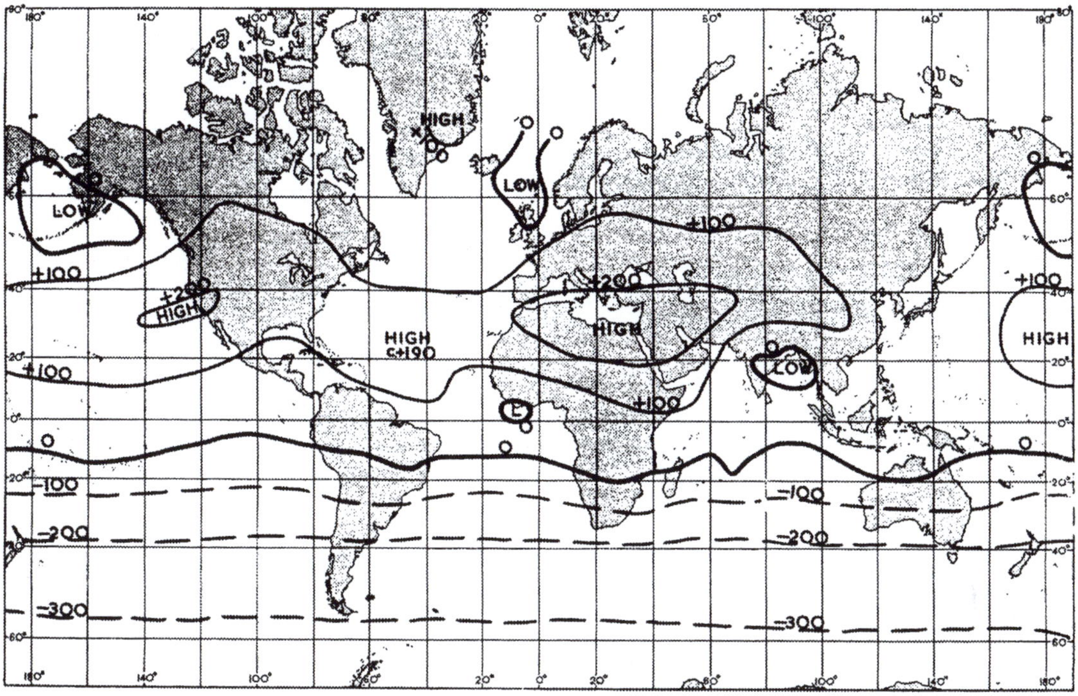

Fig. 2.15 Net radiation receipt in July, average in g cal/cm^2/day. Broken lines indicate net loss of radiation. (*Adapted from* SIMPSON *1929.*)

It is seen that:

(i) The distribution depends primarily on latitude and season. The isopleths nearly follow the latitude circles, despite all the irregularities of geography. This zonal distribution is most pronounced in the southern hemisphere.

(ii) The strongest gradients of net radiation receipt are in middle latitudes, especially in the hemisphere where it is winter.

(iii) In middle and higher latitudes there is a net loss of radiation, and over much of the Antarctic ice this is believed to be the case at all times of the year.

Over the year as a whole latitudes poleward of about 35° appear to show a net loss of energy. The highest figures of average energy gain are over the north African, Arabian and Australian deserts in their respective summers, the averages amounting to 250–60 cal/cm^2/day for the peak month. The net loss from parts of the Arctic and Antarctic may average as much as 350–400 cal in the mid-winter month. These figures represent the average net gains and net losses of heat for the Earth and its atmosphere combined in the regions mentioned. At the Earth's surface the net gains (radiation balance of the surface) are bigger and extend to higher latitudes, to about 75°N and 70°S. There is a net loss of radiation from the atmosphere in all latitudes, which is made good by energy picked up at the surface partly as latent heat.

Local or regional departures from the simple zonal pattern appear important in two main types of situation:

(i) Strong gradients mark the differential heating of land and sea at the coasts of the continents in summer – e.g. near northwest Africa and Europe, Chile and southern Australia. The strongest of all these local heating gradients – a drop of about 100 cal/day in net radiation receipt within five degrees of latitude – occurs in July just north of the isopleths shown on fig. 2.15 and is in fact zonally arranged, near 70–75 °N, at the edge of the Arctic ice and the heated northern continents;

(ii) the dense cloud cover of the equatorial and monsoon rain belt produces reversals of gradient in low latitudes, attributable to the large amount of solar radiation wasted by reflection from the cloud tops; there is even a slight net loss of radiation indicated in July over parts of the Bay of Bengal and the Gulf of Guinea, the latter area being almost on the equator.

Studies of the radiation balance using satellite measurements of the actual outgoing radiation (RASOOL 1964) confirm the importance of regions of cloudiness in low latitudes. Because of this India receives more radiation in March than in August, and the region of frequent cyclogenesis over the western Pacific shows a net radiation deficit at all times of the year.

In some connexions, e.g. when considering the effect of climate upon living things, the amount of solar radiation penetrating to the surface is more important than the balance of gain and loss by incoming, reflected and outgoing radiation, discussed above. Maps of average insolation reaching the surface for every month of the year have been given by BLACK (1956).[1] The pattern is extremely similar to that of figs. 2.14 and 2.15, save that there are no negative values. This similarity with the distribution of radiation balance implies that the amount of outgoing terrestrial radiation is remarkably nearly uniform over the whole world, summer and winter, as noted by SIMPSON. The outgoing radiation appears to average from 370 to 430 cal/cm^2/day in most parts of the world.

The energy effectively received from the sun, as mapped in figs. 2.14 and 2.15, is redistributed over the world by the winds and ocean currents. Figs. 2.16 and 2.17 show, from calculations by W. C. JACOBS (1951), the averages over the whole year of respectively the quantity of feelable heat given off by the oceans to the atmosphere between latitudes 0° and 60°N and the *total* heat given to the atmosphere – i.e. as feelable heat direct from the ocean surface plus latent heat to be released in the air later by the condensation of water

1. More recent calculations by BERNHARDT and PHILIPPS (1958) have been published as world maps, broadly resembling those of SIMPSON and of BLACK, but covering each separate item of the incoming radiation received at the Earth's surface for each month of the year. A similar treatment of the outgoing radiation by the same authors is promised.

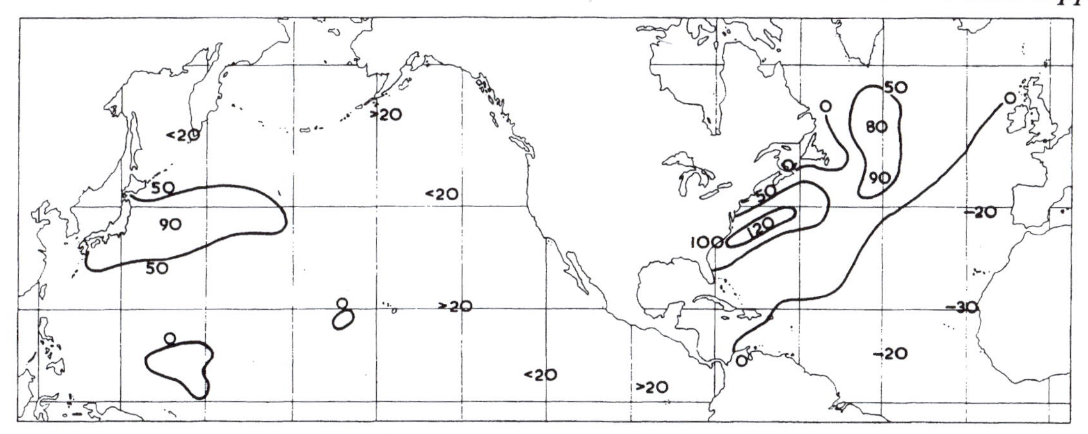

Fig. 2.16 Yearly average quantity of sensible (feelable) heat gained by the atmosphere from the sea, in g cal/cm^2/day.
(*After* W. C. Jacobs *1951*.)

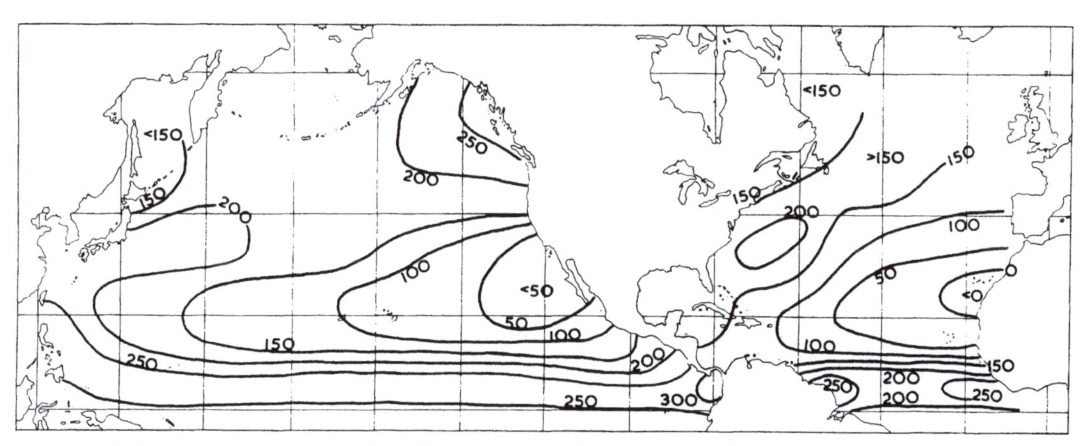

Fig. 2.17 Yearly average total quantity of heat gained by the atmosphere direct (as sensible heat) from the sea surface and by liberation of latent heat of condensation of water vapour, in g cal/cm^2/day.
(*After* W. C. Jacobs *1951*.)

vapour, often thousands of kilometres from where the evaporation took place. The greatest input of sensible heat into the atmosphere occurs (fig. 2.16) over the warm ocean currents in the Atlantic and Pacific, especially in winter when cold air from North America and Siberia comes directly over the warmest waters in the latitudes concerned. The average rate for the year amounts to about 100 cal/cm^2/day in parts of these regions, and probably rises to as much as 300 cal over open water near the edge of the ice in latitudes north of 60°N. A study by Craddock of the warming of polar airstreams over the ocean suggests that an uptake of 300 cal/cm^2/day is quite normal when northwesterly and northerly airstreams blow over the warm waters of the North Atlantic Drift between Iceland and Britain and in extreme cases heating rates as high as 1400 cal/day can go on while such a wind current lasts. Over the Gulf of Alaska, where airstreams from extensive Arctic ice surfaces may come directly over a warm ocean which is effectively out of communication with the cold Arctic waters, occasional heating rates up to 2210 cal/day have been reported by WINSTON (1955)[1] (e.g. in a northerly outbreak in February 1950). These figures indicate that the heating of cold polar airmasses over the warmest waters in high latitudes can be a powerful element in the heat budget of the atmosphere.

The amounts of latent heat of condensation liberated within the atmosphere produce averages of total heat supply (fig. 2.17) as high as 250–300 cal/cm^2/day over the year in the cloudiest regions, in the belts of frequent cyclonic disturbances – the main storm tracks – in high latitudes over the Pacific and Atlantic Oceans and in the equatorial rain belt. The balance is slightly negative – i.e. a net *loss* of heat – over upwelling waters in the Trade Wind zone off northwest Africa. The same must occur in more marked degree off the coast of Peru. The greatest of all releases of latent heat occur where mountain barriers cause depressions to slow up, and linger, while condensation is intensified in the moist airstreams forced to rise over the mountain wall – as at the coasts of British Columbia, northwest Europe, and doubtless similarly at the coast of southern Chile, and most of all in the monsoon rainfall over and near Assam and southeast Asia.

In the regions most affected by heat redistributed by winds and ocean currents and the wind transport of evaporated moisture, the heat arriving in these ways exceeds the effective solar heating there. Nevertheless, as we shall see in the next chapter, despite these local and regional complications, the overall average temperature distribution in the lower atmosphere is found to be graded from equator to pole very much like the distribution of net radiation receipt. This arrangement, broadly by latitude zones, of the radiation supply and of average air temperature, must be presumed to be a permanent aspect of world climate.

BUDYKO (1955) has calculated the mean yearly figures for the heat balance zone by zone in the northern hemisphere in the present epoch in Table 2.5 here, taken from ALISSOW *et al.* (1956). The units are kcal/cm^2/yr which can be reduced to equivalent average daily values in gramme calories by multiplying by 2·74. A +/– sign indicates gain/loss of heat by the surface (e.g. from/to the atmosphere), by the atmosphere or by the surface and atmosphere together; in the cases of ocean currents and atmospheric advection the +/– sign refers to heat transported into/out of the latitude zone. The column headed 'Latent heat balance' represents the difference between the effects of evaporation (entailing heat loss) and condensation (heat gain) within the zone. The most notable differences in the southern hemisphere from these figures are that in almost all latitudes more heat is lost (on average about 10% more) by the surface to the atmosphere in evaporation, and correspondingly more latent heat of condensation is liberated in the atmosphere; on the other hand, ocean currents transport much less heat across the latitude circles than in the northern hemisphere, except near 30°S – in other latitudes the ocean currents of the southern

1. J. S. WINSTON, Physical aspects of rapid cyclogenesis over the Gulf of Alaska, *Tellus*, **7**, 481–500. Stockholm, 1955.

TABLE 2.5

Heat balance of the northern hemisphere (kcal/cm²/yr)

Latitude	*Land surfaces*			*Ocean surfaces*				*Whole northern hemisphere surface*				*Atmosphere*		*Earth and atmosphere combined*		
°N	*Radiation balance*	*Feelable heat passed by stirring of air*	*Latent heat of condensation*	*Radiation balance*	*Feelable heat passed by stirring of air*	*Latent heat of condensation*	*Ocean currents*	*Radiation balance*	*Feelable heat passed by stirring of air*	*Latent heat of condensation*	*Ocan currents*	*Radiation balance*	*Latent heat of condensation released in the zone*	*Radiation balance*	*Latent heat of condensation balance of gain and loss in the zone*	*Feelable heat brought into the zone by the winds*
60–90°	+9	−3	−6	0	−36	−9	+45	+4	−19	−7	+22	−73	+16	−68	+9	+37
50–60°	+30	−16	−14	+44	−11	−36	+3	+36	−14	−23	+1	−76	+37	−40	+14	+25
40–50°	+44	−22	−22	+58	−9	−42	−7	+51	−16	−32	−3	−69	+45	−18	+13	+8
30–40°	+58	−34	−24	+72	−10	−69	+7	+66	−20	−50	+4	−59	+40	+7	−10	−1
20–30°	+61	−39	−22	+80	−4	−78	+2	+73	−17	−57	+1	−50	+35	+23	−22	−2
10–20°	+70	−37	−33	+88	−2	−81	−5	+83	−11	−68	−4	−54	+50	+29	−18	−7
0–10°	+80	−14	−66	+90	−2	−66	−22	+88	−5	−66	−17	−56	+87	+32	+21	−36
0–90°	+47	−23	−24	+69	−8	−61	0	+60	−14	−46	0	−60	+46	0	0	0

hemisphere flow much more nearly east or west and are believed to have much smaller southward and northward components than their northern hemisphere counterparts.

Long-term variations of radiation receipt and of the world distribution of heat

Extensive anomalies of ocean surface temperature, as well as of sea ice, may affect the world's heat budget significantly.

SAWYER (1965) quotes estimates that a positive (or negative) anomaly of sea surface temperature amounting to 1 °C over a wide area, if it persist for a whole season, with (for example) wind speeds of about 20 knots prevailing, means an average of about 100 cal/cm²/day extra (or deficient) energy passing from the sea to the atmosphere: about two-thirds of this anomaly is due to the change of evaporation and consequent increase or decrease of energy liberated in latent heat of condensation. Alternatively, an anomaly of 1 °C throughout a 50 m-deep warm surface layer of the ocean, acquired during the course of some particular summer, requires an anomalous heat transfer of 50 cal/cm²/day for 3 months to create or remove it.

An increase of ice affects the balance in at least three ways – by changing the albedo (reflectivity) of the surface, by reducing the rate of conduction of sensible heat from the waters below and by shifting the zone of extreme heating rates in air coming from the ice over open water. The loss of energy uptake from the surface brought about by an abnormal extension of sea ice probably varies rather widely according to the temperature of the open water normally present in the region, but may average 200–300 cal/day in some instances.

D. M. HOUGHTON (1958) has given estimates of the probable greatest anomalies in the rate of heating of the atmosphere over extensive land areas that occur over periods lasting from a few days to a few weeks at the present epoch. From his work it appears that the greatest changes must be those due to

(i) the high reflectivity (albedo) of snow covering a region which normally receives strong sunshine – as in middle latitudes in autumn or spring – reducing radiation receipt by up to 180 cal/cm^2/day;

(ii) the presence of green vegetation after rains in desert regions – the increased evaporation cooling, partly offset by lower reflectivity (darker surface) to incoming radiation, could mean a net reduction of heating of the atmosphere by up to 150 cal/cm^2/day;

(iii) parching of grasslands and steppe normally green can increase the heating of the atmosphere, by the converse processes to (ii), by up to 150 cal/cm^2/day;

(iv) thawing ground with a water content of, say, 4 g/cm^3 may for a few days absorb 150 cal of solar radiation/cm^2/day in sunshine without any rise of temperature or any heat passing to the air; under cloudy conditions 80 cal/cm^2/day may be taken from the air;

(v) the freezing of ground with the same water content may for some days put 80 cal/cm^2/day into the air, preventing the temperature falling any lower while the process continues;

(vi) wet ground has greater thermal conductivity than dry, and may give or take 30 cal/cm^2/day according as the air is colder or warmer than the ground.

When we come to consider the differences of climatic regime between one epoch and another, and even those between one decade or one year and another, we must distinguish between those aspects of the condition of the surface and of the response of the atmosphere to incoming radiation which are determined by the atmospheric circulation itself and those which it is powerless to alter quickly. The longest-lasting effects must be those associated with anomalies of the Earth's surface which require great quantities of heat for their removal – i.e. anomalies which may be said to possess great thermal inertia. The extreme cases of this are the 2000 to 4000-m thick ice sheets over Antarctica and Greenland and the other great ice masses which gradually accumulated over periods of the order of 20 000–30 000 years on land in the ice ages. A temperature anomaly extending through a great depth – e.g. hundreds of metres – of ocean water is also likely to require the heating or cooling of the atmosphere through more than one year to remove it. In some cases many decades or centuries may be required.

Of the figures quoted from D. M. HOUGHTON above only (i), (ii), (iii) and possibly (vi) are of a kind that could effect climatic differences on a time-scale of years or more.

Consideration of the world maps of normal net radiation receipt at the present time (figs. 2.14 and 2.15) suggests that the problem of long-term differences of climate may be simpler than that of the many short-lived fluctuations, at least in its bold features. Great differences in the effective heating of the Earth must clearly result from the albedo (reflectivity) changes brought about by:

(*a*) expansion or contraction of the zones of 'permanent' ice and snow;
(*b*) expansion or contraction of the cloud cover associated with the equatorial rain zone;
(*c*) migration to lower or higher latitudes of the zones of prevailing cloudiness associated with travelling depressions in temperate latitudes. This seems likely to be important because of the greater area of the latitude zones nearer the equator.

All these could account for a change of the average heating of the areas affected by from 100 to 200, (*a*) or (*b*) up to 300, $cal/cm^2/day$. Among them, (*b*) and (*c*) appear to be controlled by the circulation itself, though possibly initiated by (or also subject to) influences external to the atmosphere – from the sun or the sea surface or through effective screening of the sun by dust layers in the upper atmosphere. Anomalies produced by (*b*) or (*c*) can more readily disappear entirely within a few years than can occur with (*a*), though anomalous situations of these kinds have been observed to last for one or two decades and could during that time lead to changes of extent of the ice and snow in high latitudes that might persist longer. Some anomalies of the kinds mentioned under (*b*) and (*c*) may also be initiated by an expansion or contraction of the polar ice – owing to the prevailing sizes and spacing of the major systems of the atmospheric circulation.

In periods of so-called *meridional* circulation character – periods, that is, in which northerly and southerly windstreams are particularly prominent – the atmospheric circulation may affect the total heat budget of the Earth by albedo changes; 'tongues' of great cloudiness may stretch towards low latitudes over regions whose albedo would otherwise be slight. The greatest losses of available heating in this manner probably occur

(*a*) where the tongue of cloudiness – commonly in one of the 'upper cold troughs'[1] of great meridional extent – is maintained over a continent in summer and
(*b*) where deep 'pools' of cold air[1] arrive, and becoming cut off stagnate, over the subtropical regions of the oceans; this can happen in any season, but is commonest in the winters, springs and autumns. The storage of heat in the ocean to be released in later seasons must be affected.

Any changes which affect the total albedo of the Earth must alter the proportion of the available solar energy which is actually used to heat the Earth and drive the winds. Such changes can be looked on as variations in the efficiency of the Earth's surface and atmosphere in accepting the solar radiation and converting the incoming radiant energy to other forms. Hence changes of the overall average temperature of the Earth's surface and atmosphere, and of the total energy of the wind and ocean circulations, would be expected (however difficult they might be to measure) in such cases, even though the output of the sun

1. Explained in the next chapter.

remained constant. If, however, a slight solar variation should occur – and this might be responsible for initiating growth or shrinkage of the polar ice – changes of the internal heat balance economy of the Earth and atmosphere should produce additional effects in the ways we have described.

Summary: The roles of radiation, winds and ocean currents in climate

The climate of any place is produced by

(i) the radiation supply there;

(ii) the nature of the surface, whether it be water or ground with its vegetation cover (or lack of it), its slope and aspect to the sun – in effect, the ability of the surface to absorb and emit radiation;

(iii) the heat and moisture, etc., brought by the winds;

(iv) the heat and moisture supplied to the winds from the surfaces over which they pass, especially the surfaces of large bodies of water and most particularly through the agency of ocean currents, which themselves transport heat towards or away from the region in which the given place lies.

The radiation supply is graded mainly by latitude and season as shown in figs. 2.8, 2.14 and 2.15. This is a permanent arrangement that corresponds to the geometry of the Earth's orbit and the tilt of its rotation axis which, as we have seen, are subject to only minor variations. In the present epoch about two and a half times as much solar energy falls on the atmosphere over the equator in the course of a year as over the poles. And despite variable transmitting efficiency ('transmissivity') of the atmosphere in different places, seasons and epochs, owing to its varying moisture content, cloudiness, carbon dioxide, dust and pollution, etc., more heat always reaches low latitudes than the polar regions in the course of a year, in every epoch. This, it must be supposed, gives the stamp of permanency to an arrangement of climatic zones nearly parallel to the latitude circles and successively cooler towards the poles.

The radiation balance – that is the net gain or loss – and its seasonal changes are determined firstly by the latitude and secondly by the weather itself. Even in the central Sahara, the Australian desert and on the lofty ice plateau of Antarctica near the South Pole, the weather observations of one year differ from those of another because of what the wind circulation brings – variations of heat transport and of sky cover.

The winds are the principal mechanism of heat exchange. Indeed, the ocean, the only other large transporter of heat, is largely set in motion by the drag of the winds on the sea surface.

The winds transport heat from other places and other latitudes in quantities which may at times be comparable with that received direct from the sun. In high latitudes this advected heat plainly exceeds that coming direct from the sun for most of the year: at the poles there is no sun-heating at all for half the year at the surface and none for 5 months at a height of 21 km. (At the top of the ozone layer in the stratosphere there is no sunshine for about $4\frac{1}{2}$ months at the pole.) Much of the heat brought by the winds to places in middle and high latitudes is in the form of latent heat, only converted into heat that can be felt (sensible heat), and that can supply energy to the wind circulation, when the water vapour is condensed into clouds and rain or snow, usually thousands of kilometres from where the water vapour was picked up from the surface by evaporation.

Anomalous patterns of wind and ocean currents appear capable in some instances of producing anomalies of the heating pattern (distribution of heat sources and sinks) that render the arrangement self-maintaining over several weeks or a whole season. When sufficiently great anomalies of the temperature of a thick layer of ocean water, or of the quantity of the ice on the ocean surface in high latitudes, are set up the possibility is opened of a regime lasting for decades or centuries.

In the next chapter we must examine the actual, rather simple temperature patterns observed in the air, which are the end product of the heat supply and its redistribution. And we shall see how these account for the atmospheric circulation and its prevailing forms.

CHAPTER 3

The atmosphere in motion

(1) Convection

The winds that transport heat and moisture, and that drive the ocean currents, are at once the operating mechanism and the working substance of climate. Whenever the moving air is deflected upwards, or rises in a convection current, it undergoes 'expansion cooling' – i.e. its temperature falls 'adiabatically' as described in a later paragraph – so that clouds and mist are commonly formed from the moisture in it: and processes going on in the clouds are liable to produce drizzle, rain, snow, etc., and, when the vertical motions are violent enough, thunder and hail. To understand how the wind circulation is related to the initial heat supply from the sun, and to the pattern of derived heat sources on the Earth and within the atmosphere, is the foundation of all that follows – of the whole science of weather forecasting, of our interpretation of past climates and present climatic development, and of any prospect of rational prediction of future climates.

It is inequalities of heating that set the air in motion. Circulations of all sizes from the 'thermal' up-currents rising over a sun-baked pavement, or a ripening cornfield, to the great windstreams conveying warm tropical air towards the poles can be looked upon as convection currents. They have the common characteristic of being the means of transporting heat from places where it is hot (*source*) to where it is cold (*sink*).

The atmosphere receives most of its heat at the surface of the Earth; though there is another general heating surface high in the stratosphere at the top of the ozone layer about 50 km up, where a small part of the solar radiation is applied to heating the oxygen in the air in the course of converting some of it to ozone. There are other heating levels still farther up, in the outer atmosphere, but these need not concern us here. Away from the main heating surfaces the air's heat and radiation budget – the balance of gain and loss – is different; and, the local heat supply at such points being generally less, the temperature of the air is habitually lower the farther away the level considered is from one or other of the heating surfaces (cf. fig. 1.1). For this reason, prevailing temperatures normally decrease with increasing height above sea level, rates of the order of −0·6 to −0·7°C/100 m (or about −3 to −4°F/1000 ft) being the most commonly observed *vertical gradients* (lapse rates) of temperature in the lower atmosphere. The winds blowing past hills and mountain heights are therefore colder than the air near sea level on most occasions; though, when the ground is cooling directly by giving off radiation to a clear sky at night, and at times when the sun is very low in the sky, so that outgoing terrestrial radiation exceeds that coming in from the sun and sky, *inversions* of temperature are liable to occur. Places on the heights are then less cold than the valleys and extensive lowlands.

The case of extensive high-level plateaux bare of snow in moderate and low latitudes is exceptional. They represent a raised heating surface, making the air over them warm for the height above sea level. This occurs oftenest in summer and when the sun is high. On the other hand, the high ice plateaux of Greenland and Antarctica are cold, even for their latitude and height.

Thermal stability, instability, and vertical convection within an airmass

Small-scale convection and local air circulations may be started wherever the air over one bit of surface is heated more than the other air near by. This may be due to differences of albedo, as between cornfield and forest, or to such differences of thermal properties as between dry ground and swamp, lake or sea. When one comes to examine the heating of the Earth's surface, one has to take note of the specific heat and thermal conductivity of the different materials of which the surface is composed in different areas. Specific heats range from 1 for water (0·94 for sea water) to about 0·2 for most kinds of rock. The thermal conductivity of the solid materials of the Earth's crust is generally about 4×10^{-3} g cal/cm^2/cm/sec when the thermal gradient is 1°C per cm, but falls as low as $0{\cdot}13 \times 10^{-3}$ for sand where air is trapped in the interstices; hence incoming heat is effectively concentrated in raising the temperature of the actual surface of sand and the effect of outgoing radiation in lowering its temperature at night is also concentrated very close to the surface in the case of sand and snow. The thermal conductivity of water is about 15×10^{-3} g cal/cm^2/cm/sec (that of ice is 5×10^{-3}); but the *effective* thermal conductivity of water and of an air layer within the surface (within forest and high-standing crops, etc.) is enormously increased by eddy-conductivity (including convection), especially when the surface is cooling; the heat exchange is thus spread in depth and the temperature change at any given level correspondingly reduced. For all these reasons, the temperature of a water surface rises and falls less than that of any other for a given intake or loss of heat, and even wet ground experiences much smaller and slower temperature changes than a dry ground surface. Dry land surfaces, and the air over them, normally reach their highest temperatures of the day within about 3 hours after midday, and their highest of the year in the warm weather within a few weeks after the summer solstice; they usually reach their lowest temperatures about dawn and within a few weeks after the winter solstice. In oceanic climates the times are more variable, and especially where the ground is wet or there are many surface water bodies, the warmest and coldest times are liable to be delayed. Similarly also, temperature variations are greatly moderated within a forest or high-standing grass, though the actual surface (uppermost leaves and blades of grass) may experience a great range of temperature.

The air that is most heated (over the warmest surfaces) expands, becomes less dense than the surrounding air and tends to rise, especially when given an upward impulse by drifting against some obstacle or passing over uneven ground. As the air rises, it comes under less pressure from the atmosphere above and expands further: this is 'adiabatic' expansion – i.e. expansion without any further heat being put into the air; and in fact its temperature falls, because some of the internal energy of the molecules is used up in the expansion. This is the same process whereby low temperatures are produced in a refrigerator, by pumping gas through a nozzle and allowing it to expand. The rate of decrease of temperature of dry air rising in the atmosphere is nearly constant, the *dry adiabatic lapse rate*, equal to −0·986°C/100 m, to all intents and purposes −1°C/100 m (or −5·4°F/1000 ft).[1]

If the rising air in this way acquires a lower temperature than its environment at the same height, which will soon occur if the vertical gradient of temperature in the surrounding air is less than the adiabatic rate, it finds itself denser than the environment. In this case, the force of gravity brakes its rise and tends to make it fall back. This situation is said to be *stable* for dry air: in this case, the thermal condition of the airmass discourages vertical motion. It would be *unstable* only if the vertical gradient of temperature in the surrounding air exceeded −1°C/100 m, as occasionally happens when the whole airstream is being very strongly heated from below over the hot deserts and in outflows of Arctic air over the Gulf of Alaska or over the warm waters of the North Atlantic.

The situation is altered if condensation of water vapour starts. From then on, latent heat of condensation is liberated within the rising air. The resulting lapse rate of temperature with height, the *saturated adiabatic lapse rate*, is accordingly less than for dry air: its value depends on the pressure and temperature (which govern the quantity of water vapour that saturates the air) and on whether the condensation occurring produces water droplets or ice crystals. The rate ranges from about −0·4°C/100 m in warm tropical air near the Earth's surface to −0·7°C/100 m or more with sub-freezing temperatures and at greater heights. It approximates more and more nearly to −1°C/100 m at very low temperatures where the air's moisture content approaches zero.[2] Values of the saturated adiabatic lapse rate about −0·5°C/100 m are common in the conditions of the temperate zone. Thus, it often happens that the temperature of moist air rising, with condensation occurring, decreases at a rate rather less than the vertical gradient in the surrounding air. In this case, the density of the rising air becomes at successive heights increasingly less than that of the environment; its buoyancy increases and its rise is accelerated. This situation is said to be *unstable*, since vertical motion is encouraged.

1. The adiabatic lapse rate for dry air may be derived from simple physical principles, using only the gas laws and the mechanical equivalent of heat (first law of thermodynamics), as shown by BRUNT (1941, p. 39).

2. BRUNT (1941 pp. 65–6) derives the lapse rate of damp air rising by introducing the water vapour pressure, the quantity of liquid water condensed as droplets and the latent heat. The results are displayed graphically to cover a wide range of cases, which we have here briefly summarized.

These considerations explain many patterns of convection where abruptly rising currents of air transport heat upwards into and through the surrounding air. The columns of rising air, themselves moving nearly vertically upwards at speeds up to 5 to (in extreme cases) 50 m/sec in the centres of individual convection cells of the order of 5–50 km apart, are compensated by sinking ('subsiding') motions, generally at much slower rates but over wider areas, in the intervening air. Each convection cell is liable to produce a corresponding vertical growth of (cumulus) cloud, which may be small or large in keeping with the scale and spacing of the vertical motions.

We notice already from the various values of the saturated adiabatic lapse rate, and from the situations that produce the strongest vertical gradients of temperature, that instability and vertical motions should be most frequent and most strongly developed in the moist tropics and over surfaces that are warm for their latitude, whereas stability is likely over cold surfaces and in the polar regions (except such places as the Gulf of Alaska and the eastern Norwegian Sea–southern Barents Sea with their warm waters in the present epoch).

Tornadoes

The strongest vertical, or nearly vertical, up-currents developed in the most unstable air require a system of converging horizontal wind currents at the ground to supply them. This invariably takes the form of a rotating vortex, the rotation being usually in the cyclonic sense – i.e. the same sense as that of the largest cyclonic disturbances – despite the small horizontal dimensions of the tornado. Horizontal wind speeds as great as 50–100 m/sec (in round figures 100–200 knots), and more, may be briefly attained in the ring of fastest-moving air 5–50 m or so from the centre. Pressure may be reduced by 200–300 mb or more just at the centre of the vortex, where a column of water can be lifted 3 or 4 m out of an exposed water surface; heavy objects are lifted off the ground and carried along with the tornado, and condensation of water vapour brings the cloud down towards the ground in a twisting 'funnel'. Buildings over which the tornado centre passes are liable to burst outwards ('explode') owing to the sudden local reduction of pressure of the atmosphere. Trees and other objects are twisted to pieces (the trunk being commonly twisted off) by the rotating winds – by the great differences of horizontal wind speed over distances of a metre or two within the crown of the tree. Between the centre of the twisting column and the ring of maximum wind speeds 'solid' rotation is believed to occur – i.e. v the horizontal wind velocity is proportional to r the distance from the centre. Outside the ring of maximum speeds v falls away, and a law of the type $v = cr^{-k}$ holds (where c and k are constants and k has values between $\frac{1}{2}$ and 1). The radius of the ring of fastest-moving air varies as the tornado develops, decays and redevelops over periods of minutes. The actual wind speeds are not symmetrically distributed about the centre, because a general component in the direction in which the system is moving is added: this is the speed of translation of the system by the general wind over the area. This general wind is usually SW'ly in tornado situations in the northern hemisphere and NW'ly in the southern hemisphere, except in the tropics where tornadoes sometimes develop and are carried along by upper E'ly winds.

Tornadoes – also known as 'twisters' or, in milder cases, as 'whirlwinds' – are the most destructive wind phenomenon occurring on the Earth, although their size is small and objects more than 10–100 m from the path of the central funnel experience only moderate winds and escape damage. Tornadoes arise in the most violent convection – i.e. when there is abnormally great thermal instability for vertical currents and great moisture content, so that the saturated adiabatic lapse rate of temperature operates from near ground level in rising air – and are typically associated with severe thunderstorms. They also require the presence of something to start the air twisting: this may be a strong shear of the horizontal wind, as near a sharp front (strong thermal gradient zone) or the edge of a range of hills. Tornadoes are therefore commonest in continental interiors in middle latitudes, where fronts with great thermal contrasts occur (especially in spring and autumn, when strong heating of cold airstreams from high latitudes may introduce great instability). The valleys of great rivers where there is a plentiful moisture supply are particularly liable to tornadoes. The great instability in the convection clouds in the equatorial rains over West Africa and India also produces tornadoes, but they are commonest and most violent in the Middle West of the U.S.A. at the present epoch. Other, lesser tornado areas are found in Europe including England, China, Japan and in similar latitudes in the southern hemisphere.

Local convection and local winds

Two or three special types of local wind regime arise from differences of heating commonly produced by the local geography:

(a) *Land and sea breezes*

Dry land surfaces respond rapidly to changes of the radiation balance and become warmer by day and cooler at night than the surfaces of water bodies. The air over the warmer surface expands, becomes less dense, tends to rise and is replaced by a cool breeze cutting in from the cooler area nearby. Thus, at coastlines a cool breeze tends to blow in from the sea on warm days, and at night is replaced by a drift of air from the cooler land out to sea. Such breezes are best developed in clear weather favouring maximum exchanges of solar and terrestrial radiation at the surface. Their development is also facilitated by a vertical gradient of temperature making the heated air unstable (conducive to vertical motion), since the compensatory movements involve a circulation in the vertical plane with a return drift at some upper level. Wind speeds of 5–10 m/sec (10–20 knots) are attained at maximum development of surface sea breezes over coasts in temperate and subtropical latitudes. The night breezes from the land are usually lighter, except where other local effects are operative as under (*c*) and (*d*) below.

(b) *Lakes and estuaries, islands: local breezes*

A weaker development of the land and sea breeze type of circulation – *diurnal breezes* as they may be called for the sake of a more general name – occurs over and around smaller water bodies in land-locked areas and over small islands.

In all cases ((*a*) and (*b*)), convection (cumulus) cloud is liable to develop in the rising columns of air over the warmer surface and tends to become congested where the breezes from different coasts converge, producing a local concentration of vertical motion. The frequent occurrence of this pattern of organized convection over a heated island has been demonstrated, and mapped, in the case of Malta (36°N) by LAMB (1955*a*). Showers and even thunderstorms may be induced at this point, the moisture supply in the air from the water bodies contributing to their development. Close investigation of the diurnal breezes and local rainfall distribution over and around the great Lake Victoria (0–2½°S) in East Africa by FLOHN and FRAEDRICH (1966) revealed that the rainfall is greater over the lake than round about; this is associated with the peak frequency of thunderstorms in the area, occurring at night and, in particular, with the convergence over the lake of the nocturnal breezes from the surrounding shores. A marked rainfall maximum was indicated in a group of islands in the lake which receive 2300 mm/yr against 1000–1500 mm at places along the lake shores. Owing to the general easterly component of the large-scale (upper) wind circulation over that part of Africa, the area of heaviest rainfall and the convergence of the local night breezes were found to be displaced towards the western side of the lake.

The breeze from whichever surface is the cooler is maintained by a general subsidence from the upper levels over it: this tends to produce clear skies over the water in summer and by day and clear skies over the land in winter and by night – although the surface cooling sometimes produces fog.

(c) *Hill and valley* (*anabatic and katabatic*) *breezes*

When a hillside or mountain wall is heated by the sun, the air in contact with it becomes warmer than the free air at that height and so by expansion gains buoyancy and readily rises up the slope. Daytime *anabatic* (up-slope) breezes in quiet weather converging at mountain ridges from the slopes on either side are liable to form convection cloud, which 'sits' (remains stationary) on or over the ridge.

At night, especially with outgoing radiation to clear skies, the slopes chill the air in contact with them, so that its density increases and it slides down the mountainside. This is the *katabatic* (down-slope) breeze. Thus on clear nights the coldest air collects in the valleys and flows as a breeze that follows the natural drainage lines; where a valley narrows, this breeze may be funnelled into a mighty katabatic wind. Descending motion of the upper air over the mountains maintains the supply and tends to keep the sky clear there, though fog may form in the regions of lowest temperature in the valley bottoms especially where a valley is broad and the night wind neither fast nor turbulent.

(d) *Glacier winds*

Glaciers next to warm lowlands, sea or sun-baked mountain walls provide the temperature contrasts required to start air motions in the vertical plane and down the glacier surface, similar to the circulations discussed under (*a*), (*b*) and (*c*). The katabatic wind down a glacier is often channelled by a narrow valley or fjord; and such places at the foot of a steep glacier slope, as in Greenland, are liable to frequent katabatic gales of extreme violence. Wind speeds exceeding 50 m/sec (over 100 knots) have been measured.

So far we have only seen how various types of local wind are produced by local differences of heating. Next we must consider the effects of the main differences of heating between equator and poles, and the large-scale wind circulation.

(2) The general or large-scale wind circulation

Genesis of the main (planetary) circulation of winds over the world

Let us first consider what must happen if the atmosphere were at rest on the Earth's surface, initially under uniform pressure and at uniform temperature everywhere. Let the heating around each latitude zone be uniform, too – no geographical complications – but graded according to latitude. The air over the zone of greatest heating in low latitudes then becomes warm and expands. We must presume that the expansion takes place in the vertical plane, because near the surface the equal pressure of the air from either side (which we have postulated) prevents expansion sideways. Thus, the upper parts of the vertical columns of air over the warm zone are lifted. In this process, work is done against gravity, and so potential energy is put into the (upper) atmosphere.

The vertical expansion of the air columns and lifting of their upper parts means that, if we consider some particular height in the atmosphere above sea level, there should be more air (and therefore more mass) above that height over warm areas than over cold ones. This means that we should expect to find pressure differences aloft, such that there is a gradient of pressure from the tropics to the poles. And should there, after all, be warmer and colder areas within one latitude zone, the warmer parts are likely to be associated with ridges of high pressure aloft extending from the tropical zone and the colder parts to appear as tongue-like extensions of low pressure aloft from the cold polar regions.

Fig. 3.1 demonstrates that the expected vertical expansion of the air columns over the tropics and contraction over the polar regions correspond to observed differences of height interval between the levels in the atmosphere where pressure is 1000 and where it is 500 mb (millibars) – briefly called the thickness of the 1000 to 500 mb layer or *1000–500 mb thickness*. Since average pressure at sea level over the world is 1013 mb, this layer is roughly the lower half of the atmosphere. Fig. 3.1 also shows the latitude averages of net radiation receipt: there is a good deal of correspondence with the distribution of 1000–500 mb thickness in the atmosphere, particularly as regards the latitudes in which the strongest gradients are found. The chief difference is in the latitudes of maximum thickness, where the lower half of the atmosphere is warmest: these are closer to the equator than the position of maximum radiation receipt in summer – an effect of the atmospheric circulation itself, through the large amounts of latent heat of condensation liberated in very low latitudes (cf. fig. 2.17).

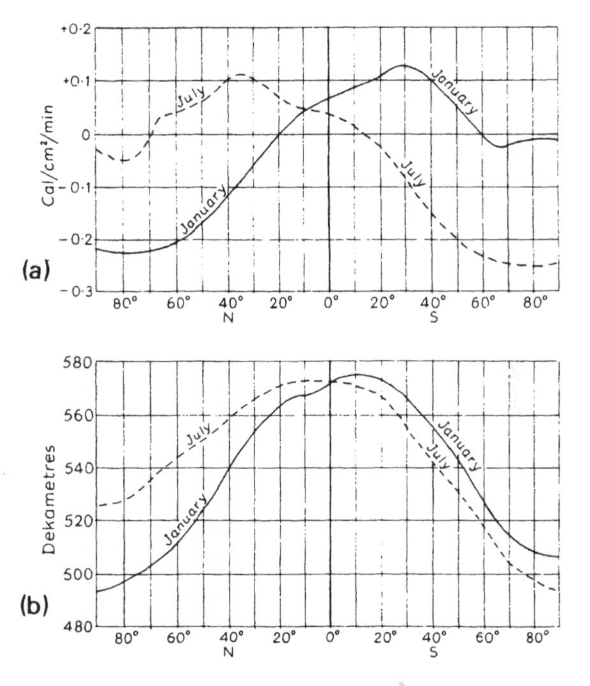

Fig. 3.1 Distribution by latitude at the present epoch of:
(*a*) Net radiation received. (*After* SIMPSON *1928*.)
(*b*) Vertical distance between the levels where the pressure of the atmosphere is 1000 and where it is 500 mb – usually known as 'thickness of the 1000–500 mb layer'.

Contour maps of average height of the 500-mb pressure level over the northern and southern hemispheres (fig. 3.2) reveal that the expected gradients of pressure are in fact observed in the atmosphere. These maps present the mean pressure distribution at a sample height of about 5 km, in the middle troposphere. The observed pattern is quite similar to this through a great range of heights, from about 2 km up to 15–20 km, including about 70% of the mass of the atmosphere. The pressure differences at each level between the warm zone and the cold zone should increase up to whatever height the temperature (or density) gradient continues in the same sense. This means that this pattern produces its strongest pressure gradients at the top of the troposphere, which (because of the intimate relation explained below between pressure gradient and wind strength) is therefore a level of maximum wind speeds.

The stratosphere above has its own heating pattern, different from that of the troposphere below and which therefore builds up a different pressure distribution. We shall consider this in a later chapter.

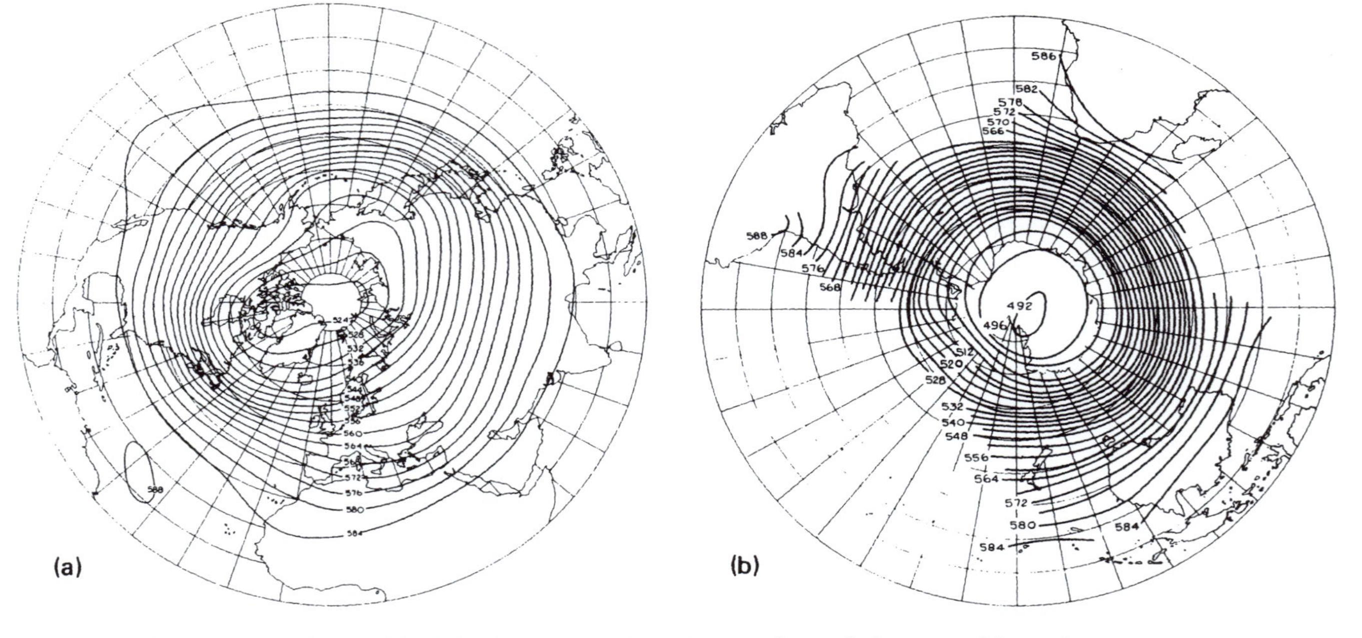

Fig. 3.2 Average height of the 500 mb pressure level in dekametres (mean of all seasons of the year):

(*a*) Northern hemisphere 1949–53.

(*b*) Southern hemisphere 1952–4.

The contours and gradients of the 500 mb pressure level have nearly the same significance as isobars and pressure gradients, at the 5–6 km level. The wind tends to blow along the contours. In accordance with Buy's Ballots law it goes counter-clockwise around the low-pressure regions over the northern hemisphere and clockwise around the low-pressure regions over the southern hemisphere. Note that both hemispheres show a single circumpolar whirl or 'vortex' of generally westerly winds at this height.

A pressure gradient represents a force tending to accelerate the air particles and direct them from high towards low pressure. Thus, the air is set in motion. And the stronger the pressure gradient – i.e. the closer the lines on a contour map such as fig. 3.2 (or the closer the isobars on a map of the pressure distribution at any fixed level) – the greater the accelerating force. The air's motion is, however, controlled not only by the pressure gradient force but by a balance between this and other forces which come into play on any moving particle owing to the rotation of the Earth beneath it, the curvature of the air's path and friction. So the air does not move directly from high to low pressure. As the Earth spins beneath it, moving air is deflected to the right in the northern and to the left in the southern hemisphere. It therefore passes counter-clockwise around centres of low pressure in the northern hemisphere, clockwise around low-pressure centres in the southern hemisphere (Buys Ballot's law). The deflecting force due to the Earth's rotation we will call D and the pressure gradient force G. Under conditions of equilibrium when the air is neither accelerated nor retarded but moves at a constant speed, in the free air in the absence of friction, these two forces D and G must clearly be equal and opposite to one another: D operates at right angles to the air's path, so it follows that both must be at right angles to the path of the air. Hence, under equilibrium conditions, the wind blows along the lines of constant pressure (isobars) or along the contours of a constant pressure surface like that in fig. 3.2. Where the wind blows on a curved path around a region of low, or high, pressure, a centrifugal force C also arises: this, too, operates at right angles to the air's path, but is always directed outwards from the centre of curvature. So we may write a simple equation for the balance of forces acting on the air at right angles to its path:

$$G = -D \pm C$$

The minus sign before D indicates that the forces D and G act in opposite directions; in the case of cyclonic curvature of the isobars – i.e. when the air's path curves around a centre of low pressure – the centrifugal force C is also in the opposite direction to the pressure gradient force G, and the minus sign is taken before C in the equation. The plus sign before C applies when the curvature is anticyclonic: for then the centrifugal force and the pressure gradient force act in the same direction.

The balance of forces acting upon the moving air is more fully expressed in the usual form of the gradient wind equation:

$$\frac{\partial p}{\partial s} = -2\rho\omega V \sin\phi \pm \rho\frac{V^2}{r}$$

in which

$\partial p/\partial s$ is the pressure gradient,
ρ is the air's density,

ω is the angular velocity of rotation of the Earth about its axis,
V is the wind velocity,
ϕ is the latitude,
and r is the radius of curvature of the air's path.

All these forces operate at right angles to the air's path, so long as equilibrium exists. The pressure gradient force $\partial p/\partial s$ is G in our simplifed statement of the equation.

D, the deflecting force of the Earth's rotation (the *Coriolis* force), is represented by the first term on the right-hand side: in this, the multiplier $2\omega \sin \phi$, called the *Coriolis parameter*, measures the acceleration which would be imparted by the Earth's rotation to a particle of unit mass moving at unit velocity at latitude ϕ, if there were no other forces acting upon it. C, the centrifugal force, is expressed by $\rho V^2/r$. The negative sign before this term is taken when the curvature is cyclonic; and since both the terms of the equation containing V then operate against $\partial p/\partial s$, a somewhat smaller value of V balances any particular pressure gradient when the path is cyclonically curved than in the anticyclonic case, when the positive sign before $\rho V^2/r$ is taken.

Frictional forces, whether at the ground or represented by internal friction within the depth of an airmass (due to turbulence, including vertical convection currents), operate in line with and directly against the air's motion. These and other circumstances that may occasion values of V that do not balance $\partial p/\partial s$ will concern us further in a later section of this chapter: they may permit the pressure gradient or other forces to deflect the wind from the line of the isobars and produce convergence and divergence of air in various places that change the barometric pressure distribution and pressure gradients.

In the common case of almost balanced forces when the wind is blowing over a nearly straight path along the isobars, r being big and the V^2/r term consequently small, the equilibrium wind speed is in direct proportion to the pressure gradient.[1] We can therefore construct a simple scale to indicate values of the wind that should balance the pressure gradient – i.e. a *gradient* or, strictly, *geostrophic wind scale* – to use by measuring the spacing of the isobars or contours. The constant of proportion ($2\rho\omega \sin \phi$) depends, however, on the density of the air and on the latitude. The equilibrium wind speed for a particular value of the pressure gradient is greater where the air is warm and in low latitudes than where the air is cold (dense) and in high latitudes. Gradient wind scales must therefore have separate markings for each latitude. In very low latitudes, the coefficient of V in the gradient wind equation becomes small (because $\sin \phi$ is small) and the V^2/r (centrifugal force) term is all-important if there is any curvature of the air's path; along paths of varying curvature equilibrium is then unlikely to be attained.[1] For these reasons, pressure gradients

1. The equilibrium wind for balance between pressure gradient and Earth-rotation forces only is often called the *geostrophic wind*. The term *gradient wind* is applied to the wind that corresponds to balance between the effects of pressure gradient, Earth-rotation and curvature (if there is any). The centrifugal force represented by the last-named is often called the *cyclostrophic effect*, because it arises from the cyclonic or anticyclonic curvature.

are usually small in low latitudes and within 5–10° latitude of the equator the pressure gradient is but a poor guide to the actual wind. Elsewhere the gradient wind usually gives a good approximation to the real wind in the free air beyond the reach of most of the effect of ground friction.

Corrections to the scale value of wind speed can be made for curvature of the path; the corrections vary for positive or negative curvature and the radius. Corrections or adjusted scales can also be used for cases of differing air density either in the upper air or at different temperatures: a drop of 30°C means about 10% less wind speed for equilibrium with a given pressure gradient at one and the same pressure level.

At the surface, friction always makes the actual wind less than the gradient wind – especially over land where the friction is greater than over the sea. Over land, reduction to about half the gradient wind speed is fairly typical; but in the shelter of mountains or even of

1. Gradient wind speeds for straight isobars at different latitudes with an isobar spacing of 1 mb per 1° latitude or 111 km (= 60 nautical miles), for normal pressures at the Earth's surface and temperatures about 15°C are as follows:

Latitude	90°	80°	70°	60°	50°	40°	30°	20°	10°	
Wind speeds										
for equilibrium	5·1	5·2	5·4	5·9	6·6	7·8	10·0	14·8	28·9	m/sec
with pressure	18·3	18·7	19·4	21·3	23·6	28·1	36·0	53·2	104·0	km/hr
gradient	9·9	10·1	10·5	11·5	12·8	15·2	19·5	28·8	56·4	knots

For isobars only half as far apart double the wind, and so on.

The gradient represented by the contours of a constant pressure surface in the upper air at a height interval of 4 dekametres being 600 km apart gives the same equilibrium wind speeds as in the above table.

Gradient wind speeds for cyclonically curved, straight and anticyclonically curved isobars at various latitudes are compared in the following table:

	Cyclonic			*Straight*	*Anticyclonic*		
Radius of Curvature	500	1000	2000	∞	2000	1000	500
Latitude 70°	10	10	10	10	10	10	10
	23	24	24	25	26	26	28
	43	46	48	50	53	56	69
45°	10	10	10	10	10	10	11
	22	24	24	25	26	27	30
	41	44	47	50	54	60	—
20°	9	10	10	10	10	11	12
	21	22	24	25	27	30	—
	35	41	44	50	60	—	—

The units in the above table are km/hr for radius expressed in km, or knots for radius expressed in sea miles.

Where a dash appears in the columns under anticyclonic curvature, no balance with the pressure gradient force is possible and the wind must blow outwards towards low pressure.

forests, and under inversions which suppress convection and turbulence between the surface and upper winds, the air near the surface may be brought to a standstill. The reduction below equilibrium speed means that the pressure gradient force becomes too strong for balance with the terms in the equation that include V (the wind speed): hence the surface wind is impelled to blow somewhat across the isobars from the higher towards the lower pressure. An angle of 30–35° to the isobar is fairly typical over land, but local shelter may produce light breezes quite unrelated to the pressure gradient; over the sea the surface winds are generally about two-thirds of gradient speed and blow at a smaller angle to the isobars. Thus, the winds at the surface, owing to friction, are continually tending to even out pressure differences.

In the free air it can happen, especially where the wind blows from a region of strong pressure gradients into a more uniform area, that the actual wind is too strong for balance with the pressure gradient. In the upper troposphere, at and near the maximum wind level, this situation can lead to winds exceeding the speed appropriate to balance the local pressure gradient by a substantial amount. At such places in the wind field, the moving air is deflected somewhat across the isobars towards the higher pressure side – i.e. against the pressure gradient force. This transfers an appreciable mass of air in such a way as to build up pressure differences. We shall study this later in this chapter.

Let us return now to consideration of fig. 3.2. With the winds generally nearly in equilibrium with the forces acting, and therefore blowing nearly along the isobars (or contours of a constant pressure surface), maps of average pressure, such as these, portray the pattern of prevailing winds.

We see that over each hemisphere there is a single circumpolar flow of (more or less meandering) upper westerly winds. This whirl of upper winds around the hemisphere is called the *circumpolar vortex*. It is the mainstream of the atmospheric circulation through a great range of heights, and carries most of the momentum of the wind flow over either hemisphere. Because it is brought into existence by the grossest differences of heating of the different latitude zones of the planet Earth without much regard to geographical complications, it is often referred to as the planetary circulation. Something similar must be supposed to have existed at all stages of the Earth's history, however much the geography may have changed; a similar arrangement should also occur in the atmospheres of other planets.

The winds are strongest where the pressure gradients are strongest – or where the contour lines, which show the topography of some constant pressure level, as in fig. 3.2, are closest together – generally over middle latitudes. There they are associated with the zone of sharpest temperature gradient between the equatorial and polar regions: this creates a marked slope of the constant pressure surfaces (such as that for 500 mb in fig. 3.2) in the upper air, owing to the different expansion of the air columns over the warmer and less warm regions.

The prevailing winds and pressure gradients are seen to be stronger over the southern hemisphere, particularly over the zone of unbroken ocean, than they are over the northern hemisphere. (Sections through the atmosphere from pole to pole, showing the locations and strengths of the main west and east wind currents up to heights of 20–30 km are shown in a later chapter, in fig. 4.1.) That stronger pressure differences are built up over the Southern Ocean can be largely attributed to there being less surface friction than over the corresponding zone of the northern hemisphere, where there is much more land. However, in the southern temperate zone south of 40°S no great warm ocean current like the Gulf Stream–North Atlantic Drift or the Kuro Shiwo in the North Pacific (both guided by bordering continents) transports heat poleward: this means that greater thermal contrasts can be sustained between the Antarctic and lower latitudes than in the northern hemisphere, where even the Arctic Ocean basin near the pole is accessible to sub-surface warm water of Atlantic origin. By contrast, the South Pole lies in the midst of a land-based ice sheet that is on average 2000 m thick and of continental extent. In consequence, the mean temperature of the air in the troposphere above is on average 11·5°C colder than over the North Pole: in January (high summer) it is only 3°C warmer than the corresponding layers over the North Pole (near their winter minimum) and in July the south polar tropospheric air is 27°C colder than that over the North Pole (FLOHN 1967). The southern westerlies are therefore on balance stronger than the northern hemisphere system, with the present geography of the Earth. Over either hemisphere, however, in the so-called polar front jet streams – the shifting zones of strongest flow, associated with the strongest thermal gradients – at heights of 10–12 km speeds of 40–60 m/sec (80–120 knots) are common. Extreme values reach 100–120 m/sec (over 200 knots) in many areas; and extreme speeds about 150 m/sec (300 knots) have been reported over Amsterdam Island (38°S) and Japan.

The angular momentum carried about the Earth's axis by the southern hemisphere westerlies, integrated through the bulk of the atmosphere, from the surface (approximately 1000 mb pressure) up to the 100-mb level, appears on a yearly average at the present epoch to be about one and a half times as great as that of the northern hemisphere westerlies (LAMB 1959). This must give rise to stronger centrifugal forces in the southern system, causing the southern hemisphere zones of wind circulation and climate to be slightly displaced towards the equator (compare, for example, the latitudes of maximum pressure in fig. 3.3); this may also be presumed to account for the deficit of atmospheric mass everywhere south of 35°S compared with corresponding northern latitudes: over the sub-Antarctic low-pressure zone this deficit is about 2%, mean surface pressures 60–70°S being rather more than 20 mb below those at 60–70°N – as illustrated in fig. 3.3. Between latitude 70° and the poles the discrepancy is similar. Much lower pressure values prevail in Antarctic anticyclones and ridges than in northern polar anticyclones. FLOHN (1967) estimates that the total effects make the mean pressure over the entire southern hemisphere 4 mb lower than over the northern and that the surface air averages 2°C colder over the southern hemisphere.

It may be the combined effect of this difference of centrifugal force and of greater heating over the northern hemisphere continents (especially Asia and North Africa) in summer that

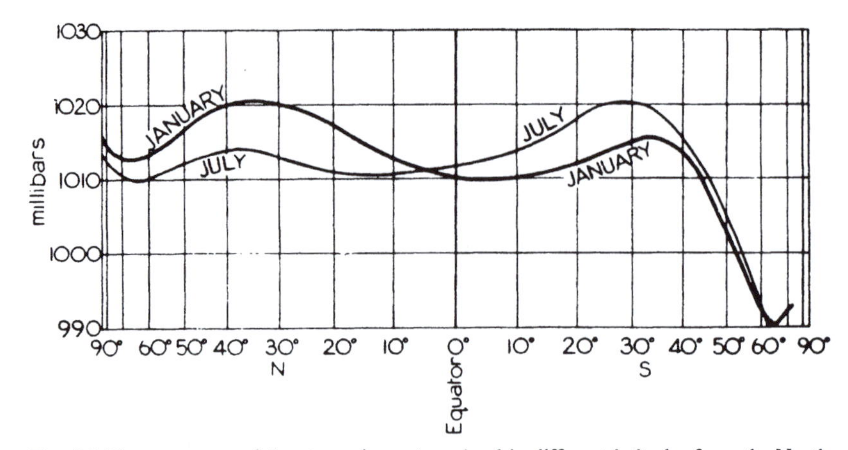

Fig. 3.3 Mean pressure of the atmosphere at sea level in different latitudes from the North Pole to the South Pole (average January and July values for an epoch about 1900–50).
The latitude scale is not linear but has been made proportionate to the areas of the Earth's surface in each zone.

makes the *meteorological equator* – the position of the convergence at the Earth's surface between air from the northern and southern hemispheres – lie generally north of the geographical equator. As observed between about 1900 and 1950, its average position was near 4 to 6°N. There is some evidence that in periods of weaker zonal wind circulation – as seen over extratropical latitudes in both hemispheres in the years 1960–5 and apparently about 1875–95 and in the early nineteenth century – the meteorological equator lay rather closer to the geographical equator.

During the maximum development of ice ages the zonal wind circulation over both hemispheres can be presumed to have been stronger than now, but displaced towards low latitudes because of the strong temperature gradients near the limits of the ice sheets. Evidence that in the last ice age the equatorial rain-forest extended some two degrees of latitude farther north than now over Africa (BÜDEL 1954, fig. 1) seems at first sight to imply that the meteorological equator – usually known in meteorology as the *Intertropical Convergence Zone* or *I.T.Z.* – lay at least as far north as in this century, but it can be accounted for just by the less extreme development of aridity at that time in the desert zone. Seasonal migration of the I.T.Z. north and south, in the northern summer and southern summer, must have been restricted in ice ages by deficient heating of the higher latitudes, with less seasonal migration of the zones of strong temperature gradient in each hemisphere. The great northward excursion of the equatorial rains over India, and over Africa into Abyssinia and the Sudan, in the summer monsoon was probably much reduced in ice ages compared with the present epoch and may, on the other hand, have been somewhat greater than now in the warmest post-glacial epoch about 5000 years ago.

The southern hemisphere offers a fair approximation to the conditions of a uniform globe without geographical features. About 81% of the total area of that hemisphere is ocean.[1] The observed pattern of the upper westerly winds prevailing over the southern hemisphere (fig. 3.2) is correspondingly closer to an ideal circumpolar ring than we find over the northern hemisphere, where the wind flow and pressure field are deformed by the geography and great troughs of low pressure extending from the pole towards low latitudes are a very noticeable feature. Nevertheless the existence in the Antarctic of a polar continent covered by lasting snow and ice heightens the thermal contrast between high and low latitudes, especially in summer. And the centre of the southern hemisphere's circumpolar circulation is seen to be displaced somewhat away from the geographical pole, to lie broadly about 82°S 60°E in the middle of the Antarctic ice surface.

A similar eccentricity of the northern hemisphere circulation (and its climatic zones) must have occurred during the Quaternary ice ages: the great Arctic ice surface of that time seems to have been centred well towards the Atlantic side of the pole – perhaps in the early stages towards the American side, about 85°N around 80°W, but ultimately in the region of 77–80°N 40–70°W at the time of greatest spread of the ice.

Course of the upper westerly winds around the globe: (*a*) The role of geography

If we examine the circulation over the northern hemisphere in fig. 3.2 (*a*) some effects of geography are readily seen. Great troughs in the pattern lie over Quebec–Labrador and over a broad area of northeast Asia. These may be understood as a product of the cooling of air towards the downstream end of its passage across the great continents. In the latitudes concerned, over most of the year, radiation loss of heat from the atmosphere (though not from the ground in latitudes less than 70°) exceeds the gain; but in winter, when there is strong radiation cooling of the ground as well as of the atmosphere, these two cold troughs appear as very great extensions of the polar regime over or about the regions indicated in fig. 3.2.

The two upper cold troughs appear early in the autumn over the northeastern parts of Asia and North America. By late September or early October they are associated at the surface with a spreading snow cover over the northernmost land areas. Once this snow surface and any adjacent sea ice and cold seas cover a broad enough sector in middle latitudes (usually in late October–early November over east Siberia and some weeks later over Canada, at the present epoch) the situation becomes stabilized to the extent that an upper cold trough will always be found within the snow-covered region from then on, and most of the time this trough in the upper flow constitutes a large feature dominating the region, until the spring thaw. Surface depressions tend to be steered by the upper winds around the southern limit of the cold region, which they help to maintain by adding to the depth of snow near its western, southern and eastern limits. Investigations (LAMB 1955*b*)

1. The distribution of land and water by 5° and 10° latitude zones at the present epoch is given in Table App. 1.7 (*a*). The overall proportion of ocean in the northern hemisphere is about 61% (or 57% if the ice-covered Arctic Ocean be disregarded).

have shown that snow and ice surfaces of more than about 2400 km (1500 miles) west to east extent in latitudes 45–60° in winter tend to stabilize the circulation pattern in this way. In summer wide oceans and the larger land-girt seas (Hudson's Bay–Baffin Bay, Okhotsk Sea–Bering Sea and, to a much less extent, the North Sea–Baltic) are liable to have a like effect by presenting much colder surfaces than the neighbouring continental areas.

Passage of the upper west winds over and around great mountain barriers – the Rocky Mountains and the mountains of Asia, also Greenland whenever it lies in the path of the flow – must also produce major troughs somewhere in the lee, at a distance downstream that depends on the strength of the general wind current. So the two main troughs in the regions observed may in both cases be attributed to a combination of thermal and dynamical effects that are commonly superposed in the same regions. One trough lies over or near northeast Canada, or over the Canadian Arctic Archipelago, almost all the time, summer and winter. The other trough commonly shifts out over the cold waters of the northwest Pacific in summer, or else becomes almost confined to high latitudes over the Arctic Ocean itself at that season, only to re-form over northeast Siberia early in the autumn. These seasonal movements emphasize the importance of the location of the coldest surface. But there is no mistaking the occurrence of displacements from week to week, and from one year to another, that are provoked by stronger or weaker upper wind flow and its dynamics: again and again these are seen to determine which areas of the land surface shall be cold and which shall be warm, because surface northerly winds tend to be developed in the upper troughs.

We have by now discovered two great roles of the geography in influencing the layout of the atmospheric circulation and, through that, the distribution of climates and weather.

(1) The disposition of land, sea and ice determines zones of specially strong thermal contrast across short distances – as at a continental coast or the edge of the pack-ice emerging from the Arctic Ocean. This thermal contrast is transmitted to the overlying atmosphere by the differences of heating of the air, and produces correspondingly strong differences of pressure – and hence strong flow – aloft. The strong upper windstreams generated by surface thermal contrasts tend to blow parallel with the isotherms in the zone of strong thermal gradient and to propagate themselves in line downstream from this point of origin. (We shall discuss the downstream developments a little more closely later.) Some aspects of this geography – those zones of strong thermal contrast that are affected by ocean currents and ice, and by snow cover on land – may differ in strength, position and alignment from one year to another, and from epoch to epoch; and this accounts for some differences of prevailing atmospheric circulation pattern that determine the weather characterizing the given season or the given epoch in most parts of the same hemisphere.

(2) Great mountain ranges – at least those where the effective height of the crest is 2–3 km or more above sea level – act as barriers, deforming and deflecting the flow of the atmosphere over and past them at all heights up to 15–20 km. This commonly piles the

air up into ridges, which appear as poleward extensions of the high-pressure regime from low latitudes, over and before the mountains. Great troughs of relatively low pressure at all heights aloft are usually found in the lee. And the more or less horizontal flow of the upper wind meanders (*with* the constant-pressure lines) around the ridge and trough.

Great mountain ranges also act as barriers holding back the very cold, dense surface air developed by radiation cooling of the land in continental interiors in the winter time. They prevent this air from draining away out to sea and obstruct the progress of invading warm airstreams from the ocean: the latter, because of their lesser density, tend to be forced up over the mountains and on over the cold air beyond, which is often little disturbed thereby. Dissociation of the surface cold air in the interior of Canada and of Siberia from the systems of airflow over the Pacific and Indian Oceans is observed for much of the winter season; and MINTZ (1965) has shown by computer calculations of the behaviour of an idealized model atmosphere expressed by the fundamental equations of motion that the Siberian winter anticyclone – the surface high-pressure regime which is the climatic expression of the density and intense cold of the surface air – would not be present if the Himalayan mountain chain were not there.

The southern hemisphere also shows examples of geographical effects of these various types, but the resulting departures from the wind flow that would be expected over a uniform globe are less marked than over the northern hemisphere. This is due to the simpler geography and the narrowness of the land areas in the lee (east) of the Andes and the Southern Alps in New Zealand. Nevertheless, the thermal gradients south of the heated Australian desert and at the fringe of the Antarctic are strong; there is also a sharp difference of water surface temperatures at the Antarctic Convergence, a constantly maintained water current boundary that lies in the ocean along an east–west line near 55–60°S. Troughs in the upper air flow are quite conspicuous east of the Andes and the New Zealand Alps, though their amplitude is generally less than the northern hemisphere counterparts; however, the broad flat trough occupying the Indian Ocean sector (cf. fig. 3.2 (*b*)), owing to the bulge of Antarctica and of the main area of ice-covered surface towards that side of the pole, is the biggest and most permanent feature in the course of the southern hemisphere westerlies.

Thus the effects of geography modify the circumpolar circulation more over the northern than over the southern hemisphere at the present epoch. Moreover, any long-term changes in the external radiation supply should take effect more quickly over the northern hemisphere, because it has more land – whereas oceans redistribute heat and smooth out changes by spreading the effects gradually even to the depths of the sea. Thus, ice ages may be expected to be more quickly established over the northern hemisphere than over the southern, though the latter must in the end become involved through the interchanges between the hemispheres of both air and ocean water. Shorter-period variations too are likely to be greater over the northern hemisphere where the albedo of the great land areas can be much altered by snow and even by vegetation changes.

The geography of the northern hemisphere may also have important effects through guiding the ocean currents along meridional (poleward and equatorward) channels, so that the ocean sectors tend to be at a different temperature from the land sectors at the same latitude and even the waters near the pole receive heat transported from other latitudes. The effects of this are, however, partly nullified at the present epoch by the presence of thick floating pack-ice with surface radiative properties (and thermal conduction) somewhat akin to those of lands with a similar winter snow cover. In the warmer geological eras, however, the presence of an open polar ocean, the water of which is exchanged with other oceans in lower latitudes by means of ocean currents, should reduce the overall thermal contrast between tropics and polar regions (and between summer-heated continents and the polar ocean) in that hemisphere. This should, in its turn, produce some general weakening of the atmospheric pressure differences and so of the wind circulation over that hemisphere.

Changes of strength of the large-scale wind circulation accompanying changes of equator–pole thermal range, such as we have been discussing, must occur with the grossest differences of climatic regime – as, for instance, between ice ages and warm interglacial eras. (Nevertheless, in the ice ages the zone of strong thermal gradient and presumably the location of the strong upper winds was transferred far from the polar regions, and would lie somewhere near the edge of the expanded ice surface.) We have seen that a similar comparison applies (particularly in the summer season) to the difference between the southern hemisphere, with its glaciated polar continent surrounded by a cold ocean, and the northern hemisphere today. But quite different changes occurred between the warmer and colder decades of the present century, the key to which seems to have lain in changes of upper air temperature greater over the tropics than over the pole (cf. pp. 95–6): it was the warmer decades that were marked by strong wind circulation.

Finally, the thermal effects of mountainous lands must be noted:

(*a*) Mountain regions in high latitudes form the gathering grounds of snow from which glaciers and greater ice sheets grow, if the snowfall is sufficient and if summer melting is inadequate to get rid of all the winter's snow – as in Scandinavia, Scotland, the Alps and southern Andes in the last ice age – ultimately changing the albedo of large areas and lowering world temperatures.

(*b*) High table-lands, and even precipitous mountain ranges, in low and moderately low latitudes – as in Tibet, Mongolia and Mexico and the southwest United States, Bolivia and southern Africa – are especially effective heat sources, at least in summer. In the case of the highest ranges and plateaux, the solar radiation reaching wide areas has had to pass through 30–40% less of the Earth's atmosphere than on a path to sea level. Over such regions the highest positions of the freezing level, and of the 500 mb pressure level, in their respective hemispheres are attained. Freezing level averages 6000 m or more over the Himalayas in summer, 6200–6300 m over Tibet over the area between 31–33°N 82–95°E in July and August (FLOHN 1959), compared with 4500–5100 m over the region of the equator and the Sahara. Tibet is at that season the central area of a high-level anticyclone – a part of the subtropical zone of high pressure that appears as a distinct cell of maximum pressure because high-level pressures are rather lower over the oceans on either side within that latitude zone (*Academia Sinica* 1957, 1958). Over the Bolivian Altiplano between 16–20°S 65–68°W the freezing level also appears to be abnormally high – the highest in the southern hemisphere – averaging 5500–5600 m in summer and 5000–5200 m in winter (FLOHN 1955).

Course of the upper westerly winds around the globe: (*b*) Rossby waves

In considering the meanderings of the flow of the upper winds around the Earth – deviations that, in the case of the largest-scale 'waves' in the westerlies, are forced by mountain barriers or by the presence of a region of cold surface of vast extent – we encounter the tendency of the wind to retain a constant vorticity (or spin) about the Earth's axis. In consequence of this retention of spin, any broad windstream moving to a lower latitude (where the distance of the Earth's surface from its rotation axis is greater) acquires a cyclonic curvature of its path relative to the Earth's surface at the new latitude. This curvature of the path ultimately turns the wind back towards higher latitudes. The wind is then directed past its original latitude and, going on towards higher latitudes, becomes subject to an opposite (anticyclonic) curvature of its path relative to the Earth, because at this stage it is getting too near the (polar) axis of the Earth's rotation. This ultimately turns the air back again towards lower latitudes. In this way an endless succession of waves might be performed until other forces acting upon the windstream so altered the path as to make the wave-train no longer recognizable. ROSSBY (1939, 1941) was the first to draw attention to this straightforward principle as bound to induce long waves in the upper westerlies, which are now known in consequence as *Rossby waves*.

Some examples of air trajectories determined by retention of constant vorticity (constancy of spin about the Earth's axis) are illustrated in fig. 3.4: two waves of different amplitudes and one case of a large-amplitude north and south swing that results in a figure of eight path around a high-latitude anticyclone and a low-latitude cyclone. The characteristic dimensions of these features are large, commonly several thousand kilometres across.

The long waves move, in general eastwards, at a speed c that is less than that, u, of the west winds themselves. (u is measured at the point of inflexion of the path, where the curvature changes from cyclonic to anticyclonic or vice versa.) These speeds are related to each other and to the wave length λ by the expression

$$c = u - \frac{\beta\lambda^2}{4\pi^2}$$

where β stands for the rate of change with latitude of the Coriolis parameter – i.e.

$$\beta = \frac{d}{d\phi}(2\omega \sin \phi)$$

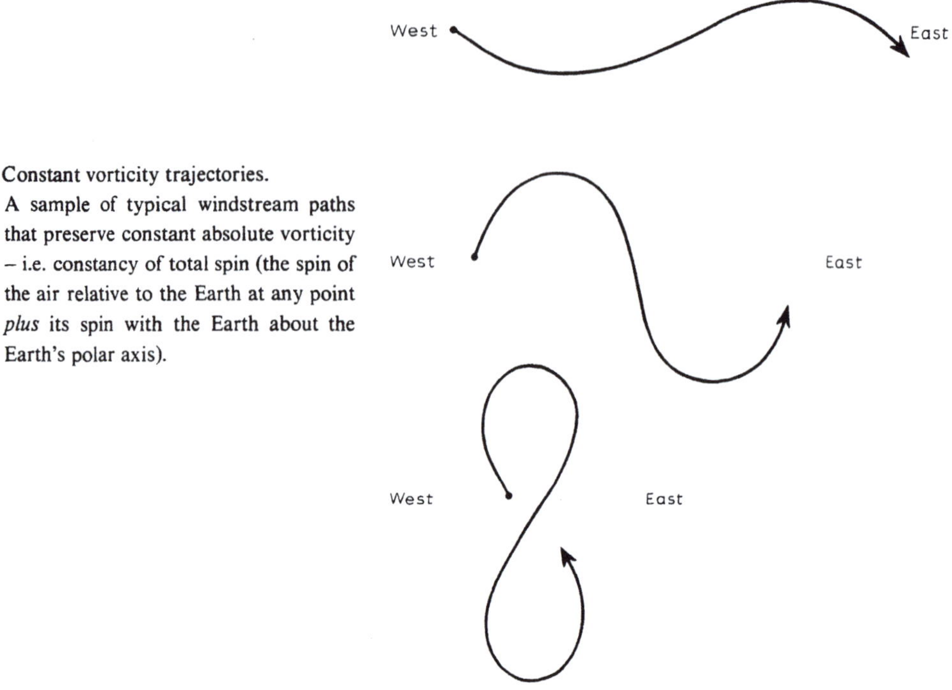

Fig. 3.4 Constant vorticity trajectories.
A sample of typical windstream paths that preserve constant absolute vorticity – i.e. constancy of total spin (the spin of the air relative to the Earth at any point *plus* its spin with the Earth about the Earth's polar axis).

The wave velocity c becomes zero when $u = \beta\lambda^2/4\pi^2$. This condition therefore defines a wave length that goes with stationary waves for any value of the zonal wind speed u at each latitude. This is important climatically, because stationary, and nearly stationary, waves linger and occupy one region with a ridge or a trough for a longer time than mobile ones. ROSSBY's principle led to the discovery that certain regions are repeatedly, and often persistently within one season, occupied by a cold trough or warm ridge in the upper westerlies – and are therefore habitually cold or warm for their latitude – just because the wave length proper to the wind flow downstream from some fixed (*anchored*) geographical disturbance as at the Rockies or the mountains and plateaux of Asia induces a trough or a ridge in that position. This is the reason for the secondary troughs in the mean flow pattern seen on fig. 3.2 (*a*) over eastern Europe–northwest Siberia and over the eastern Pacific: these are the second troughs in the wave trains downwind from the Rocky Mountains and the east Asian mountain ranges respectively – both are most pronounced in the regions mentioned in winter-time.

Fig. 3.5 displays the relationships between wind speed and stationary wave length and different latitudes of the main windstream. The wave length for 'standing' waves (expressed here as the number of degrees of longitude between successive ridges or troughs) is longer the higher the latitude. It also increases as the square root of the velocity of the upper westerly wind. As fig. 3.5 makes clear, these waves are large-scale systems. The wave

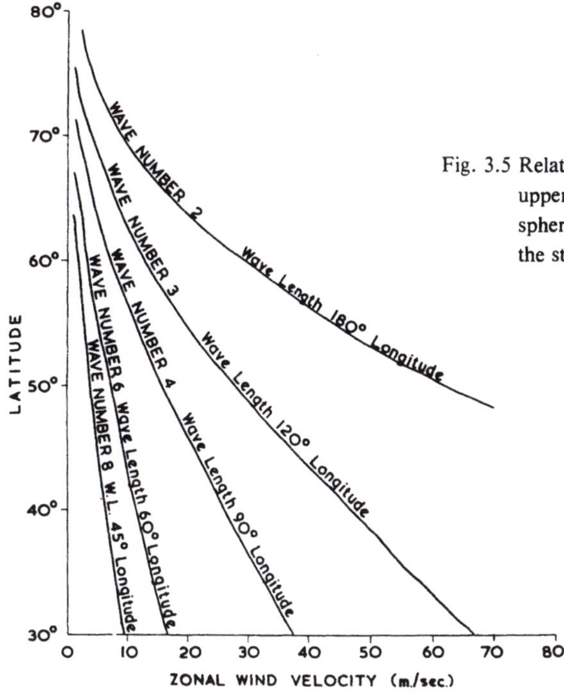

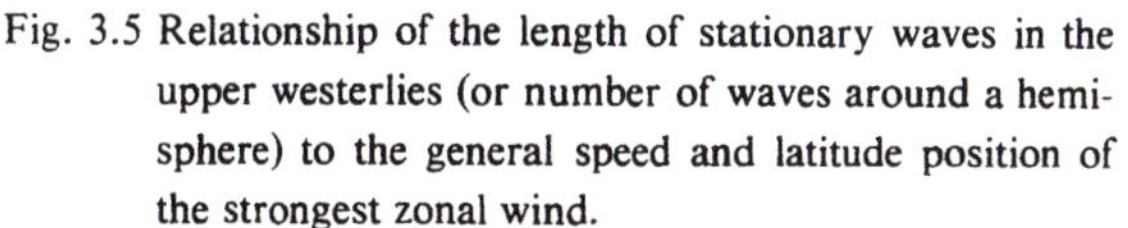

Fig. 3.5 Relationship of the length of stationary waves in the upper westerlies (or number of waves around a hemisphere) to the general speed and latitude position of the strongest zonal wind.

lengths are such that there are usually only 2, 3 or (at most) 4 troughs or ridges in a high-latitude flow – i.e. poleward of latitude 60° – and occasionally only one trough, representing an eccentric circulation (i.e. centred away from the geographical pole). Between latitudes 20° and 40° the number of troughs (or ridges) is more usually 4, 5, 6 or 7, occasionally even more. Sometimes when the number of waves in the windstream around the lower latitudes is just double that in the high latitudes a harmonious regime seems to be established with several of the troughs in two main wind currents in alignment with each other; and this may persist for many days, giving a long spell of persistent weather pattern.

The reality of the Rossby principle as the explanation of the long waves in the westerlies is verified by the observed wave spacings, wind speeds and wave speeds on daily synoptic maps of the upper wind flow. It also accounts for changes of wave spacing accompanying changes of strength and latitude position of the flow on time-mean charts of different months, seasons and climatic epochs. Similar patterns and wave numbers, and similar types of wave length change with changing flow speeds and heating gradients, can also be produced in laboratory models, in which the atmosphere is represented by water in a spinning cylindrical dish heated at the edge and cooled at the centre (e.g. FULTZ 1961).

A pretty example of standing waves in the upper westerlies is shown in fig. 3.6 by the flight of a constant level balloon at about 12 km altitude tracked over 33 days during which it passed three times round the southern hemisphere in the zone of westerlies. The phenomena are further illustrated here by the monthly mean pressure patterns at the 500 mb level over the northern hemisphere in two very different Januarys, each marked by a long spell of set weather type, in figs. 3.7 (*a*) and (*b*). January 1957 (fig. 3.7 (*a*)) was a month of vigorous westerlies over middle latitudes (close-packed lines indicating strong pressure gradients) and a long-wave length between the cold trough region over northeast Canada, in the lee of the Rockies, and the next cold trough induced downstream, over northwest Siberia and the Urals to near the Caspian Sea. In the highest latitudes the twin (cold) poles of the circulation mark an approach to a 2-wave pattern (*wave number* = 2). January 1963 (fig. 3.7 (*b*)) had a much weaker flow (weaker pressure gradients) and the strongest part of the flow was farther south than in 1957 in several sectors – e.g. North America and Asia Minor. The Rocky Mountains disturbance of the flow is there again, but the wave length is shorter downwind from the cold trough over northern Canada, itself nearer the Rockies than in 1957, and the next cold trough induced this time over the eastern part of central Europe. For most of the flow this is a 3-wave pattern, though in low latitudes additional troughs in the central Pacific and near the Persian Gulf suggest a 5-wave pattern. Another difference from 1957 is that the main upper wind flow wanders back and forth across a wider range of latitudes: it is said

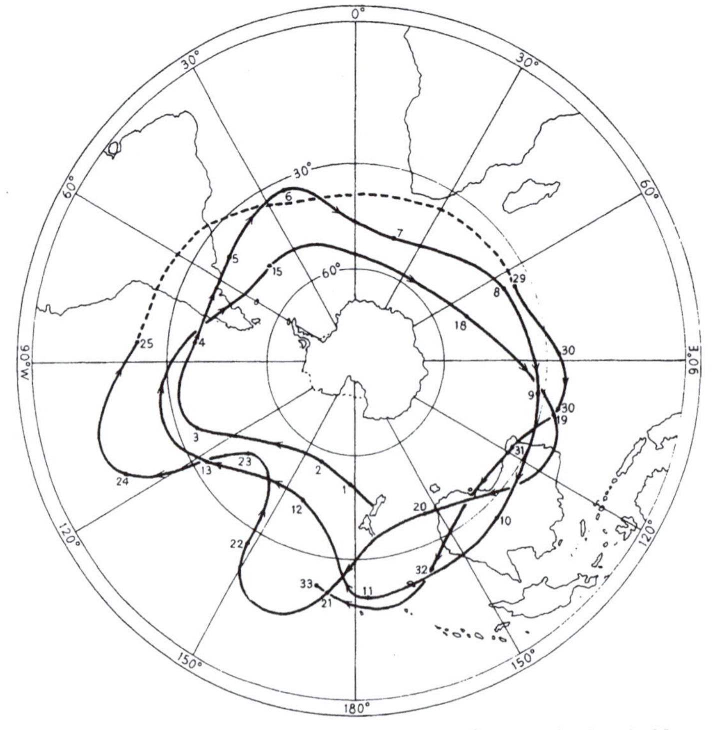

Fig. 3.6 Path over 33 days of a balloon launched from Christchurch, New Zealand, on 30 March 1966 and carried by the upper westerly winds at a constant height of 12 km around the southern hemisphere.

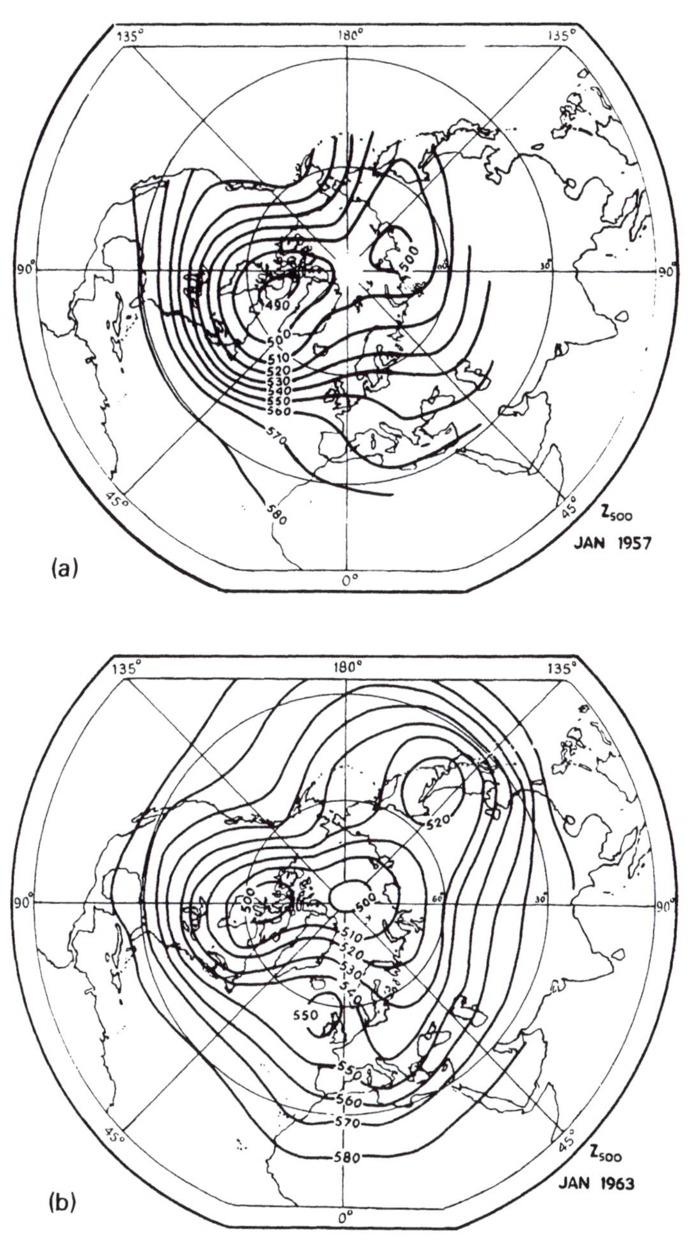

Fig. 3.7 Average height of the 500 mb pressure level in dekametres in January:

(*a*) 1957: Example of a strong zonal wind circulation.

(*b*) 1963: Example of a weak zonal wind circulation with large amplitude waves and prominent meridional currents.

to perform large-amplitude, meridional swings. And there is a complete anticyclone cell between the British Isles and Iceland cut off from the main tropical high-pressure region: such anticyclonic eddies cut off over moderate to high latitudes are called *blocking anticyclones*, because in the region concerned they block (or divert) the usual westerly winds. In this case the winds at the 500-mb level over much of Britain were northeasterly.

It seems likely that in the northern hemisphere during the ice ages, across most of that sector between about 100°W and 20–30°E where an ice surface protruded over the middle latitudes, waves in the upper westerlies were mostly of smaller amplitude than now and great mobility of weather systems travelling from west to east prevailed. All this resembles the situation between 40° and 60°S over the Southern Ocean today, where blocking is rare. The barrier effects of the great northern ice domes themselves, and the contrasting surface friction and thermal properties of the ocean and land sectors in lower latitudes, may still however have provided some locations favoured for the development of more stationary ridges or troughs than are seen over the Southern Ocean. Also the polar anticyclone was doubtless displaced towards the Atlantic–east Canadian sector at most times during the ice age.

Principal variations of the large-scale circulation

We may distinguish several types of variation that the circumpolar vortex of upper westerly winds can undergo:

(1) Changes of strength.

(2) Changes of latitude where the main wind flow lies.

(3) Changes of wave length (and therefore also of wave number).

(4) Changes of amplitude of the waves – i.e. of latitude range and of meridionality (the amount of north and south flow). In some of the more meridional (large-amplitude wave) situations, a complete anticyclone cell may separate out as a warm island in or from the poleward part (typically in latitude 50–70°) of a warm ridge, or a cyclonic vortex (a *cold low* or *cold pool*) may become cut off from the low-latitude tip (characteristically in latitudes 25–45°) of a cold trough. These cases are both described as *blocking* of the westerlies because of the wide diversion of the main upper westerly windstream from its usual latitudes.

(5) Eccentricity – the main centre of the circumpolar vortex may move away from the geographical pole to an eccentric position, sometimes as far away as latitude 60–70° for short periods, or it may split into two (or exceptionally three) poles.

Circulation strength and temperature distribution

ÅNGSTRÖM (1935, 1949), following the earlier view of F. M. EXNER (1917) and A. DEFANT (1921), treated the atmospheric circulation as effecting the equator–pole heat transfer by (large-scale) eddy conductivity. He concluded that the temperature difference between low and high latitudes must be reduced the stronger the circulation and that this could be expressed by the following equation:

$$Q_A = \frac{A\,.\,z}{R^2}\left(\tan\phi\,.\,\frac{dT}{d\phi} - \frac{d^2T}{d\phi^2}\right)$$

where Q_A is the quantity of heat introduced into an air column of height z at latitude ϕ by advection from lower latitudes; R is the radius of the Earth and A is the *Austausch* (or *exchange*) *coefficient* which measures the strength of the circulation.

Alternatively this can be written

$$\frac{d^2T}{d\phi^2} - \tan\phi\frac{dT}{d\phi} = -\frac{R^2}{A\,.\,z}Q_A$$

And if we presume that Q_A just balances the difference between incoming and outgoing radiation, Q_A can be substituted by $F(\phi)$ a function of the distribution by latitude of the radiation balance.

In these equations clearly $dT/d\phi$ is the gradient of temperature with latitude and $d^2T/d\phi^2$ is the local rate of change of that gradient.

ÅNGSTRÖM (1949) gave the following values as expressing the distribution of northern hemisphere climates then existing:

$$F(\phi) = 120\,.\,10^3\cos^2\phi - 80\,.\,10^3$$

This leads to the following solution of the above equation for T:

$$T = -\frac{a}{4}\sin^2\phi + k$$

where k is a constant and

$$a = 80\,.\,10^3\frac{R^2}{A\,.\,z}$$

This is very similar to the empirical formulas for observed overall mean temperatures:

$$T = 27 - 45\sin^2(\phi - 6{\cdot}5°)$$

for the northern hemisphere and

$$T = 23 - 48\sin^2(\phi - 11{\cdot}7°)$$

for the southern hemisphere.

So $k = 27$ and $\frac{a}{4} = 45$ approximately.

This gives $A = 1{\cdot}4 \times 10^7$, which is stated to agree well with eddy conductivity values derived from more local conditions within the range of latitude zones.

The existing average temperatures at the equator and (presumably north) pole are taken to be about 27° and −17° respectively, the difference 44°C. Assuming that the overall average temperature of the Earth did not change, and that total incoming and outgoing radiation remained the same, while only the strength of the atmospheric circulation varied, ÅNGSTRÖM deduced:

(i) with A reduced by 20%, the temperature at the equator should be 31° and at the pole −26°, the difference having become 57°C;

(ii) with A increased to 50% above its existing value, the temperature at the equator should be 22°, that at the pole −7°, the difference thus being 29°C;

(iii) the variations of temperature with time should be least, and should approximate to zero, near latitude 35°.

Probably, however, the calculations (and the values of the coefficient and constants derived) should not have been based on *surface* temperatures, which are strongly affected by inversions over the polar regions, but should be adjusted to upper air values representing the mean temperatures of the troposphere in depth. These were not sufficiently known when ÅNGSTRÖM wrote. FLOHN (1967) gives the following figures for the present average equator–pole temperature difference prevailing in the upper troposphere (700 to 300 mb layer) above the height of the south pole (2680 m a.s.l.): northern hemisphere January 32·9°, July 17·3°; southern hemisphere January 29·7°, July 44·1°. Thus, in July the southern hemisphere value is two and a half times that for the northern hemisphere, and the much stronger southern circulation does not succeed in obliterating the greater temperature range which is maintained by radiation.

Conclusion (iii), at least, does not entirely agree with the observed nature of climatic changes either at the surface or in the upper troposphere. The truth seems to be that variations are greatest where the formation or disappearance of an ice surface is involved and least near the equator, perhaps because of the extent of ocean and cloudiness there. Fig. 3.21 (p. 121) illustrates this for the extreme case of the ice ages compared with today; but the same verdict applies to the climatic variation that took place between the years 1900 and 1950. Probably some change of the radiation balance is involved in all climatic changes and the overall world average temperature does vary with it.

In reality, the eddy conductivity is only part of the story and only approaches adequacy when applied to the circulation and temperature gradients over rather short periods, such as a few days or a few weeks. Over such time lapses it can readily be observed that after a period of strong circulation the overall temperature differences are reduced. But, since the energy of the circulation is derived from the inequalities of temperature, the circulation then dies down. During the calmer period that ensues, temperature differences are once more built up between low and high latitudes until another bout of strong circulation occurs, characteristically 15–30 days after the previous one.

When we consider longer periods of strong circulation over many years or over a long climatic epoch, it is clear that the wind circulation must be transporting great quantities of heat yet failing to produce any lasting change in the situation because, for some reason or other, the temperature differences that supply the energy are continually renewed. This could, for instance, be due to (*a*) stronger radiation heating (more energy from the sun) which should increase the difference of energy supply between low and high latitudes,[1] or (*b*) warmer oceans in low latitudes, or (*c*) more extensive ice over the higher latitudes too thick to be melted quickly by warm winds. It is conceivable also that a self-maintaining strong circulation regime could be set up, depending simply on extra cloudiness over the higher latitudes and clearer skies in the tropics; though it seems likely that any such circumstances contrived by the atmosphere alone would be subject to disturbance and breakdown.

A long cycle seems possible in the case where a strong wind circulation, maintained for some years, transports enough heat polewards to melt much of the floating ice on the Arctic seas and to reduce the area of lasting snow and ice on land, thus reducing the temperature gradients between different latitudes, at any rate at the surface. This undoubtedly happened in the period of Arctic warming between 1900 and 1950, especially between 1920 and 1940. But some perplexing points remain: the circulation was at its strongest in the 1920s and 1930s when the ice was already much diminished, and there was even a secondary maximum of strength as late as about 1950, whereas the circulation was weaker in the 1880s and 1960s when there was much more ice. It is manifestly impossible to explain these circulation changes by the ice alone, and much more likely that the ice changes were merely responses to the circulation changes for which we have to find some other cause.

ÅNGSTRÖM calculated that the range of mean surface temperatures between equator and pole, which in the present epoch is about 45°C, would be reduced to 38°C if the circulation intensity – the *Austausch* – increased by 20%, and that it would wax to 60°C if the circulation became 20% weaker. Once such changes had been accomplished, however, it is hard to see how the strong circulation could be maintained with a weak temperature range between equator and pole – or vice versa.

During the long period of strong circulation between 1900 and the 1940s the overall range of *surface* temperatures did decrease owing to the warming of the Arctic and recession of the ice there – in agreement with ÅNGSTRÖM's hypothesis. The Arctic warming seems, however, to have been strongest just in the surface layer of the atmosphere – largely a matter of the strong winds of that time commonly blowing away the cold 'skin' of air beneath the low-level temperature inversion, or stirring up this air with the overlying layers and so destroying the inversion that is normally present over the ice. There is evidence that the overall range (poleward gradient) of *upper air temperature* between tropics and pole was

1. This is because for any given intensity I of the solar beam, the difference between the energy supply at any time falling upon a unit area of horizontal surface at two latitudes where the elevation angle of the sun is ϵ_1 and ϵ_2 is given by I ($\sin \epsilon_1 - \sin \epsilon_2$). So the difference of heating between any two fixed latitudes should vary in proportion to any variations of I.

greater in those years than in the period of weaker general circulation that had followed by about 1960. This was because the upper air temperatures changed most over a zone near the tropic (about 20–40°N) and changed comparatively little over the polar region. The effects upon upper air temperature of surface cooling of the polar region in the 1960s were doubtless partly offset by increased subsidence (compression-warming) of the air in stronger polar anticyclones, while less anticyclonic development in subtropical latitudes meant less subsidence (and hence a decline of the upper air temperatures) there. The circulation patterns in the 1960s resembled what is known of the regime of other periods of weak circulation before 1900, so that it appears that the decades of greatest strength of the planetary wind circulation between 1900 and about 1940–50 were marked by *increased* gradients of upper air temperature between low and high latitudes – quite unlike the changes observed at the surface. Indeed, this was to be expected. Because the difference of heating of the atmosphere in different latitudes and regions is the immediate source of the energy that drives the circulation, we must suppose that those places and epochs in which strong differences of heating are maintained continually generate strong flow of the winds around the circumpolar vortex.

ÅNGSTRÖM's treatment of the poleward transfer of heat as a function of the amount of turbulence that accompanies a mean zonal circulation of a given strength has since been carried further by PRIESTLEY (1949) who showed how this transfer might be analysed into three components:

(1) the eddy flux achieved by the travelling cyclones and anticyclones;

(2) a standing-eddy flux associated with the northward and southward components that actually exist in the mean circulation, i.e. in the semi-permanent (or 'standing') long waves in the westerlies;

(3) a third component due to the observed transverse or *toroidal* (mean vertical and meridional) circulation illustrated schematically here in fig. 3.19 (see later, p. 116).

Understanding can be improved further by evaluating the transports of sensible (feelable) and latent heat separately that are achieved by each of these components of the circulation. Complete hemispheric studies of these quantities over the duration of each of a variety of large-scale circulation patterns could do much to show how each is related to different situations of inequality of heating and heat redistribution between various latitudes and regions. It may be supposed, though it still remains to be adequately demonstrated, that the eddy fluxes transport heat at rates that have some more or less characteristic values for each type of large-scale mean circulation pattern and strength.

Circulation strength and the effect of moisture input and latent heat

A change of sea temperature changes the amount of water vapour passing into the atmosphere. When the water temperature rises, the saturation vapour pressure rises rapidly, and increasingly so the higher the temperature. The amount of water vapour present in saturated air almost doubles for each 10°C rise of temperature, and is about six times as great over the warmest oceans (surface 28–29°C) as over the coldest parts of the open sea (surface 0° to −2°C). For this reason, amounts of latent heat released (i.e. converted into feelable heat) in the air by condensation of clouds and rain are greatest over the equatorial zone (see fig. 2.17), despite much weaker wind circulation there than in latitudes outside the tropics.

A rise of temperature of the ocean surface, from whatever cause, must increase the input of water vapour into the air – and, for each degree rise, must do so most in low latitudes where the seas are warmest. This must put more latent heat into the air, but it may be converted into sensible heat – raising the air temperature and causing horizontal temperature gradients that generate wind circulation – *at some other place*, thousands of kilometres away, and often in another latitude, depending on where the moisture has been carried by the winds.

An increase of solar radiation falling upon the spherical Earth should increase the amount of heating available most in low latitudes. This might be expected to raise temperatures most in low latitudes, so increasing equator–pole temperature gradients and strengthening the general wind circulation. SAWYER (1963) has pointed out that this argument is only valid for a dry atmosphere, though it may always give the right answer in high latitudes where the humidity of the air is low. He shows theoretically, on certain assumptions, that the poleward transfer of latent heat by the water vapour in the real atmosphere should falsify the argument in latitudes below about 45°; so the temperature difference between latitudes 30° and 45° might be expected to decrease, not increase, if there were some general increase in the supply of solar radiation. At the same time, the poleward temperature gradient in latitudes higher than 45° – and the wind circulation there – should presumably increase all the more because of the extra latent heat received and converted into feelable heat in the condensation of clouds and rain, especially in the warmer airmasses in the travelling cyclones. The amounts of latent heat similarly liberated in the atmosphere in the equatorial rainbelt should also increase, unless the inter-hemispheric wind convergence that produces that rainbelt were to become less vigorous because of the removal of the most active zones of the wind circulation to higher latitudes. Observation of the warm decades of the early twentieth century suggests that this increased vigour of circulation and displacement of the main circulation zones towards higher latitudes did happen (and was most marked about the 1920s); at the same time the equatorial rainbelt became rather inactive and less latent heat was liberated there.

Perturbations causing shifts of mass in the atmosphere: development of surface wind and weather systems

The general wind circulation, which we have been considering in the preceding sections, differs from local wind circulations, and from most systems of breezes set up over mountains and coasts, because the scale is such that the effect of rotation of the Earth (ω) comes in. We shall meet it in a number of connexions in what follows.

There is some approach to a balance of forces – expressed by the gradient wind equation – acting on the air in motion, with the prevailing winds blowing nearly along the lines of constant pressure in a more or less zonal (west–east) flow around the Earth. For a variety of reasons, however, an unvarying zonal arrangement of the winds is not likely to persist, and indeed *could not* persist, for very long without a break. It must suffer perturbation:

(i) at mountain barriers, rather as illustrated in the schematic map (fig. 3.8);

(ii) where the friction changes, as on passing a coastline from sea to land, or vice versa, or on entering over a hot surface (which sets up vigorous local convection in vertical columns – internal friction in the airstream). The effects of these may be similar to the airstream encountering a solid barrier;

(iii) through differences of heating of large regions within the zone, causing vertical expansion or contraction of the air columns ('thickness' changes), which introduce warm ridges extending poleward and cold troughs extending equatorward, so that the airstream wends its way around them.

In the absence of all these, some local convection cell with a vertical current might suffice to set up a growing perturbation of the winds blowing into it, if the horizontal differences of heating were such that there was potential energy awaiting release – e.g. through rearrangement of the airmasses on either side so that the colder, denser air spread underneath the warmer, less dense air, lowering the centre of gravity of the system.

A purely zonal wind flow would have to break down at intervals, since the undisturbed system would transport no heat from the equator towards the poles, and very great temperature differences between different latitudes would be built up, representing enormous potential energy awaiting release. This store of potential energy should lead to the rapid growth and development of kinetic energy by any perturbation of the simple zonal wind flow.

There is a limit of dynamical stability which must prevent the so-called 'zonal' upper westerlies that arise through the latitude-to-latitude differences of heating from exceeding a certain strength. When this limit is reached turbulent eddies must develop. The shear at the equatorward side of the strong upper wind flow – i.e. the south-to-north rate of change of u, the west wind velocity – (du/dy, where y measures distances away from the equator along a meridian) cannot exceed the local value of the Coriolis parameter ($2\omega \sin \phi$) without breakdown of dynamical stability.[1] This shear is in the sense that tends to impart an

1. Values of the Coriolis parameter at different latitudes are, in units of 'per hour' (hr^{-1}):

Latitude	10°	20°	30°	40°	50°	60°	70°
$2\omega \sin \phi$	0·09	0·18	0·26	0·33	0·40	0·45	0·49

This means that near latitude 50°, at upper levels where friction can be considered non-existent, or near the surface of a calm sea, a west wind of 40 km/hr (approximately 25 mph) cannot blow, or cannot continue blowing, within 100 km (62 miles) on the poleward side of air that is at rest. At latitude 20° the greatest anticyclonic shear that can be sustained is less than half of this.

anticylonic rotation to air at the flank of the windstream. If the stage is reached at which breakdown of the zonal westerly current occurs for this reason, it should take place more or less simultaneously at random points around the zone – though especially in those sectors where the distribution of land and sea makes the poleward gradient of heating strongest. Perturbations produced by barriers and changes of friction seem likely, however, to disturb the zonal flow before the limit of dynamical stability is reached; these disturbances are plainly related to the large-scale geography and must be more frequent in some regions than others.

Northern hemisphere maps drawn by PETTERSSEN (1950) of the frequency of *cyclogenesis* – the first appearance of cyclonically rotating systems in the surface wind field – in the winters and summers of the period 1899–1939 showed:

(i) generally highest frequencies in a circumpolar ring about latitude 30–40° in winter and 40–50° in summer – along the warm flank of the zone of strongest thermal gradient which contains the mainstream of the upper westerlies. These cases are mostly new waves forming on the cold fronts trailing behind the fully developed cyclones in higher latitudes, the wave commonly first appearing where the front impinges upon a subtropical anticyclone (see later pp. 100 and 121, figs. 3.9 and 3.22);

(ii) more localized regions where all the very greatest frequencies of cyclogenesis occur –

(*a*) near, and especially just in the lee of, all the great mountain barriers – Rockies, Greenland, Scandinavia, Spain, the Alps, Himalayas and plateaux of east Asia;

(*b*) just off the east coasts of North America and Asia.

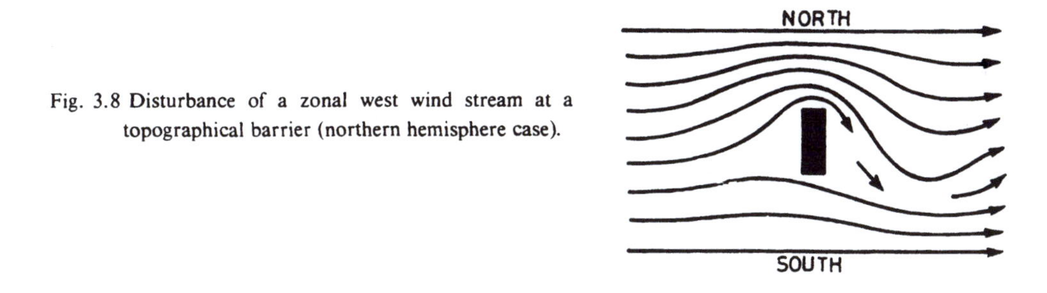

Fig. 3.8 Disturbance of a zonal west wind stream at a topographical barrier (northern hemisphere case).

Fig. 3.8 illustrates the nature of the disturbance of the mainstream of the upper westerly winds that is seen again and again at the great mountain blocks and at other barriers such as the 3000-metres-high Greenland ice cap, whenever the windstream is directed at them.

Where the streamlines are bunched together the flow is speeded up, where they open out it is slowed down. Such accelerations and decelerations destroy the balance between the forces acting on the moving air. Where the moving air enters a region of stronger pressure differences (stronger gradient), and has to be accelerated, it is for the time being moving too slowly for the V and V^2 terms in the gradient wind equation to balance the pressure gradient force: at such points therefore the air must be deflected somewhat towards the low-pressure side of the isobars (or – what amounts to the same thing – out of line with the contours and down the slope of a constant pressure surface). This is the same result as from friction slowing the air at the ground; and, in either case, transfer of mass of air towards the low-pressure region occurs, tending to even out the pressure difference. Where the air aloft emerges from a zone of strong flow (a *jet stream*) into a region of weaker pressure gradients, it is for the time being moving too fast for balance with the pressure gradient force and becomes deflected towards the high-pressure side of the isobars (or up the slope of a constant-pressure surface). This is a movement of mass which tends to increase the pressure differences. Such accelerations and decelerations may operate, and produce transfers of mass, throughout the deep layers of the atmosphere involved in the circumpolar vortex; but they are, of course, strongest near the level of maximum wind, at a height of about 10 km (12–15 km in subtropical latitudes) in the upper troposphere.

We shall next examine the pattern of departures from balance in the upper wind flow and how they determine the development and decay of surface high- and low-pressure systems – the *anticyclones* and *cyclones* (*depressions*) respectively that produce weather.

It must be understood that the moving air travels much faster than the big features of its flow pattern. The wind in a westerly jet stream may cross the Atlantic from America to Europe in 24–48 hours; whereas the features of the stream, the travelling ridges and troughs, and the jet-stream concentrations of gradient, normally take 4 or 5 days to cross the ocean and occasionally a good deal longer. A larger-scale thermal trough in the upper westerlies may be present all winter over the cold regions of North America and another one over Siberia, and a ridge may be present over the ocean, moving only temporarily forward or back a little. Thus, the air blows through the great features of the prevailing pattern of flow.

Fig. 3.9 portrays the effects in the most general case of acceleration and deceleration of a simple, almost straight, zonal westerly flow. This is the typical case of a concentration of the main thermal gradient, and associated strong pressure gradient and winds aloft, into a *frontal zone* and jet stream. Air at E in the confluence region at the entrance to the jet stream is moving too slowly for the increasing pressure gradient, and is deflected somewhat towards the low-pressure side – the deflection is exaggerated in the diagram. Air at O in the diffluence, or 'delta', where the jet stream fans out, is moving too fast for the decreasing

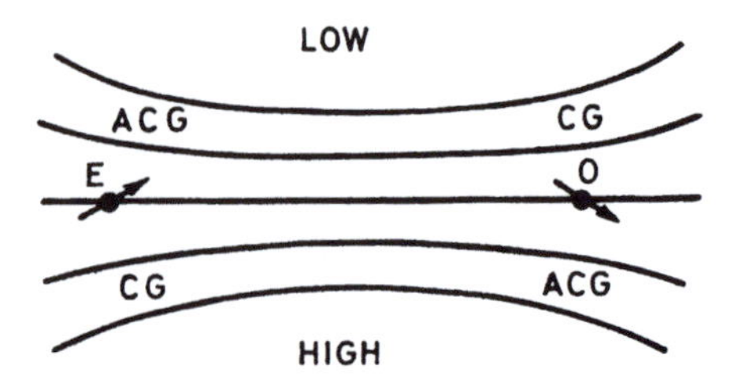

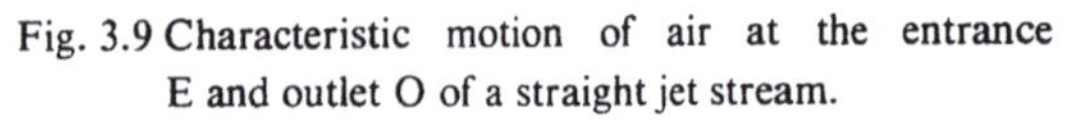

Fig. 3.9 Characteristic motion of air at the entrance E and outlet O of a straight jet stream. Anticyclogenetic effects in the areas marked ACG; cyclogenetic effects in the areas marked CG.

pressure gradient, and is deflected slightly towards the high-pressure side. The slower-moving air on either side, farther out from the axis of the jet, undergoes less acceleration and retardation and keeps more nearly in balance with the pressure gradient. Hence there is convergence of air somewhere at the left of the entry and also somewhere to the right of the jet exit. This piles up mass in these regions, leading to a build-up of pressure at all levels in the atmosphere beneath and hence to anticyclone formation or intensification, or to decay of any cyclonic features in these positions at the surface: the *anticyclogenesis* is indicated by ACG on fig. 3.9. At the other sides of the entry and exit from the jet, air is drawn away (divergence), pressure falls in the atmosphere beneath, and hence new centres of low pressure tend to form or intensify at the surface and any anticyclones in these positions decay or weaken: these areas of *cyclogenesis* are marked CG on fig. 3.9. That these characteristic departures from gradient wind balance actually occur at jet-stream exits and entrances has been demonstrated by MURRAY and DANIELS (1953).

Vertical motions, cloud development and weather

Near the surface, in the friction layer where the winds always blow somewhat across the isobars towards low pressure, convergence prevails in cyclonic regions (barometric depressions) and divergence of the surface winds in anticyclones. Hence, surface low- and high-pressure systems could not long exist without the contrary tendency at jet-stream level – divergence above the 'lows' and convergence above the 'highs'. The development or decay of these surface pressure systems is the resultant of high-level divergence and low-level convergence, or vice versa: indeed, these are quantities that can be computed to a useful approximation from the present-day observation network over the northern hemisphere, and this is done daily in numerical weather forecasting. From this convergence and divergence in the horizontal windfield comes the upward motion that must prevail in the troposphere between the surface and jet-stream level in cyclones (depressions) and the downward motion (subsidence) that must prevail between these levels in anticyclones. The maximum wind level itself, like the Earth's surface, must be a level where these broad-scale vertical motions decline to zero – such levels are sometimes called *Nullschichten* or 'zero'-layers (FAUST 1953, 1955).

Above the maximum wind level – at which the strongest accelerations and decelerations and shifts of mass relative to the existing pressure field occur – vertical motions in the opposite sense prevail: ascent above anticyclones and descent above depressions. As the tropopause is only a little above the maximum wind level, these latter motions are mainly in the lower stratosphere, though the effect is to lift the tropopause on average 2 km above the maximum wind in anticyclones and bring it down to only $\frac{1}{2}$ km above the maximum wind in cyclonic regions. These figures are from observations near latitude 50°N: there is also a latitudinal variation, with somewhat higher tropopause everywhere on the warm (usually low latitude) side of the jet stream and lower tropopause on the cold side of the jet (see later, fig. 3.19).

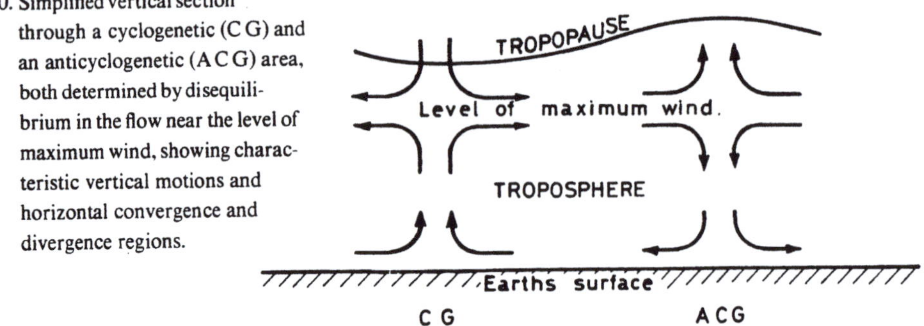

Fig. 3.10. Simplified vertical section through a cyclogenetic (C G) and an anticyclogenetic (A C G) area, both determined by disequilibrium in the flow near the level of maximum wind, showing characteristic vertical motions and horizontal convergence and divergence regions.

The prevailing vertical motions, illustrated in ideally simple form in fig. 3.10, account for characteristic differences of weather between cyclonic and anticyclonic systems, which may be summarized as follows in the following paragraphs.

Depressions and regions of cyclonic curvature of the surface isobars

Much cloud prevails throughout the middle and upper levels of the troposphere in the rising air currents below the maximum wind level: this is because the temperature of the rising air is brought down by the adiabatic expansion until the condensation point is reached. Further development within the clouds produces rain, snow and hail, etc. In the middle and upper troposphere the air is generally cooler in cyclonic than in anticyclonic regions over similar terrain at the same latitude.

The tropopause is lowered, however, and in the lower stratosphere descending motion, relative warmth and clear skies prevail.

The converging winds near the surface in middle latitudes cyclonic systems bring air-streams together that originated far apart from one another, in unlike regions – often from quite different latitudes. This sharpens the zone of thermal contrast between them into what is known as a *front*.

At the fronts the warmer, less dense air tends to overrun the colder in a slantwise upgliding motion: this often produces extensive cloudsheets of quite smooth appearance over the sloping upper surface of the colder airmass.

Anticyclones and regions of anticyclonic curvature of the surface isobars

Subsidence (sinking) of the air through most of the depth of the troposphere below jet-stream level down to the top of the ground convection and turbulence layer at about $\frac{1}{2}$–$1\frac{1}{2}$ km, produces rising temperatures. A temperature inversion develops in consequence just above the limit of the ground convection layer – a *subsidence inversion*. There is generally no cloud in the subsiding air – indeed, great transparency and long views develop. This is because of the adiabatic rise of temperature under compression, as the air sinks to lower levels, without addition of moisture: indeed, it can produce very low relative humidities (e.g.

5–15% saturation has been observed on the upper levels of the Scottish Highlands), which can parch the vegetation on mountain heights that come within the realm of the sinking air.

Below the subsidence inversion the thermal influence of the ground is dominant, particularly where the air stagnates in the central part of an anticyclone. And this may produce quite different weather. Over a warm surface thermal convection, often with small cumulus clouds (fair-weather cumulus) and much blue sky, is usual in the bottom kilometre. (This weather is characteristic of the Trade Wind zones over the oceans.) But over cold ground and seas extensive fogs and persistent low cloudsheets commonly prevail, and there may even be prolonged drizzle or fine snow. Industrial pollution also tends to collect into its highest concentrations. At such times the contrast on climbing through the inversion into the warmer, clear, subsided air above is extreme.

Above the maximum wind level, there is ascent of the accumulating air in developing anticyclones, and the tropopause is raised; cirrus cloud develops and may become widespread, its 'tangled' forms revealing vertical motion rather than sheet-like development. The lower stratosphere becomes cooler than it is over cyclonic systems.

That the unbalance between frictionally produced convergence at the surface and divergence at the level of the strongest upper winds must provide the dynamical reason for cyclonic development – and the contrary motions for anticyclonic development – was realized as early as 1914 by DINES. Subsequent work has explained the distribution of the effects in more detail (e.g. SCHERHAG 1934, BJERKNES 1937, ROSSBY 1941, SUTCLIFFE and FORSDYKE 1950). This is the foundation of most of the methods now developed for numerical weather forecasting with computers (e.g. CHARNEY and ELIASSEN 1949, BUSHBY and WHITELAM 1961).

Patterns of development, travel and blocking

It is observed that anticyclogenetic effects prevail along most of the warm side of the jet stream (i.e. of the 'frontal zone' associated with the main thermal gradient and generally strong upper winds) and cyclogenetic effects along most of the cold side. Contrary developments are more limited to the expected regions (fig. 3.9) near the jet entrances. Along the warm side of the jet is a belt of generally high pressure, the *subtropical anticyclones*. This belt is interrupted in certain sectors by low-pressure areas, that can often be regarded merely as cell divisions between the anticyclones, but which are associated with the right-hand side (northern hemisphere case) of the jet entrances, i.e. of the confluences to the strong upper winds. The cold side of the jet stream is where the *subpolar low-pressure belt* is formed; and this is commonly interrupted at certain longitudes by anticyclones or ridges formed at the left of the confluences in the upper westerly windstream.

Within the jet streams over Europe and Asia, predominance of motions that produce generally high pressure along the warm side, and low pressure along the cold flank, of the wind current has been demonstrated (SCHERHAG 1948, p. 29, REINEKE 1950, SONECKIN 1963). Winds in the core of the jet stream have been found to be generally stronger than required for balance with the pressure gradient: this is especially the case in jet streams that appear as sharply defined maxima in the vertical profiles of wind velocity. Commonly the gradient wind speed is exceeded by 10%, and in extreme cases by over 50%. The wind deviates towards the high-pressure side on average by 5–7° from the direction of the isobars, i.e. blows slightly up the slope of the constant pressure surfaces. The source of the energy represented by these over-gradient winds has been a matter of speculation: the neatest suggestion is that of HOLLMANN (1954) that the vertical motion prevailing in depressions, where the low-level winds are strong, may transport more kinetic energy up into the maximum-wind layer near the tropopause than is conveyed downwards in anticyclones, where the low-level pressure gradients are generally much weaker and there is little vertical motion other than widespread, gentle subsidence. In other words, the processes, to be examined later (pp. 120–2), which convert potential energy into kinetic energy within a depression, ultimately supply energy to the jet stream.

Next, we must examine the effects upon a moving airstream of changing curvature and of change of latitude. Both come into play as the air makes its way around the great ridges and troughs in the upper westerlies, the wave train already discussed in relation to mountain barriers and the broadest regions of warm and cold surface. Cyclogenetic and anticyclogenetic effects assert themselves as the air passes through these large-scale features in the flow, as well as when it rounds many more transitory, mobile ridges and troughs in the upper westerlies, the somewhat smaller-scale features that accompany individual cyclones and anticyclones, where the surface winds thrust warm air poleward and cold air equatorward before or behind the centre. In these cases, too, the wind travels somewhat faster than the features of the pattern described, and so a similar distribution of (cyclogenetic and anticyclogenetic) development effects occurs: it is these that are responsible for the forward development – i.e. the travel – of the cyclones and anticyclones concerned, steering them in the general direction of the upper windstream. And when the circulation itself or some accident of geography twists the thermal pattern, and consequently the upper wind flow, into a suitable shape, the system may be turned on to some new path or slowed up, and intensified or caused to decay.

Let us consider first the case of changing curvature of the air's path. We have seen how the gradient wind equation requires a larger value of V, the wind velocity, for balance with the pressure gradient when the curvature is anticyclonic than when it is cyclonic (cf. p. 78 and footnote p. 79). The equilibrium value increases the sharper the anticyclonic curvature (i.e. the smaller the radius) and decreases the sharper the cyclonic curvature. The resulting effects of this one influence upon the winds in and near the mainstream of the upper westerlies are displayed in fig. 3.11 through two ridges and a trough in the wave train. The wavy line marks the axis of the mainstream. Where anticyclonic curvature increases, the air which is for the time being moving too slowly for the pressure gradient deviates towards the low-pressure side: there it is likely to converge with air pursuing a less wavy path or moving more slowly outside the mainstream, and some accumulation of air and rising pressure

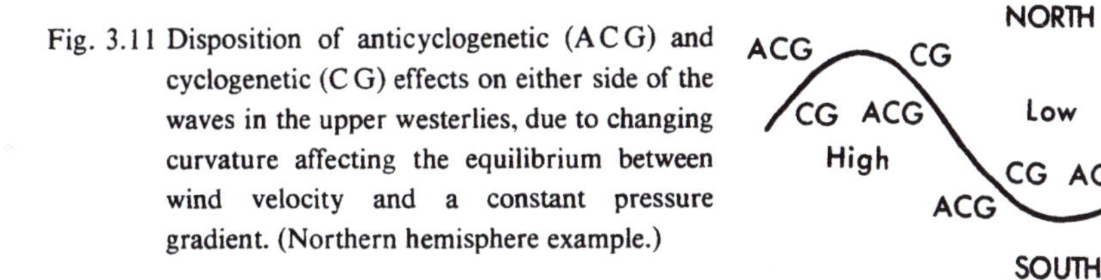

Fig. 3.11 Disposition of anticyclogenetic (ACG) and cyclogenetic (CG) effects on either side of the waves in the upper westerlies, due to changing curvature affecting the equilibrium between wind velocity and a constant pressure gradient. (Northern hemisphere example.)

(anticyclogenesis) results (ACG in the figure). At the other side of the stream at this point, the effect is cyclogenetic (CG in the figure). Where the anticyclonic curvature of the stream decreases, after rounding the crest of the ridge, the reverse happens; because, for the time being, the air is moving too fast for equilibrium with the pressure gradient. The dispositions are the same as this where cyclonic curvature begins, and increases, as the moving air enters the trough: here too the air is moving too fast for the pressure gradient. After the axis of the trough has been passed, cyclonic curvature decreases; and, as the equilibrium value of wind velocity increases, the air is for the time being moving too slowly, with effects as before the ridge – and so on, through the pattern.

The combined effects of changing curvature and changing pressure gradient along the wind's path through the wavy course of the upper winds gives rise to recurrent patterns of cyclogenetic and anticyclogenetic effects in the atmosphere below. The commonest are shown in fig. 3.12, with small arrows indicating the way in which the air entering and

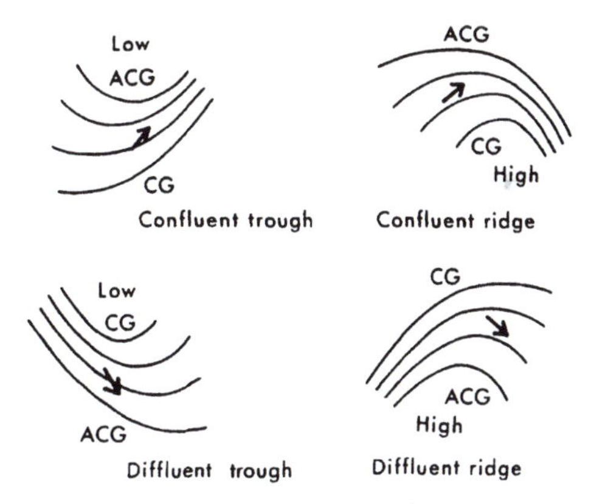

Fig. 3.12 Combined effects of curvature and pressure gradient changes at confluent and diffluent (fanning out) troughs and ridges in the upper westerlies. (Northern hemisphere examples.) The letters ACG (anticyclogenesis) and CG (cyclogenesis) have been entered at positions where the effects of changing curvature and changing pressure gradient both operate in the same direction.

leaving the fastest-moving parts of the windstream deviates from the line of the pressure field. As is seen, the pressure-rise (ACG) effect occurs where this deflected air converges with the less disturbed air just outside the strongest flow; and pressure-fall occurs beneath the area from which the deviating wind is drawn away. Thus, we find cyclogenesis just ahead (usually east) of a confluent trough, on the warm side, and also near the axis of a diffluent (fanning-out) trough, on the cold side of the flow. Both of these cyclogenetic effects, and the anticyclogenesis at the rear of the troughs, repeatedly produce new surface-pressure systems or cause old ones to decay. The same is true of the developments at confluent and fanning-out ridges. Fig. 3.13 is a hypothetical example of a wave sequence, showing the locations of the most marked pressure-change developments along the path of the upper westerly wind-stream. These are so disposed that the surface-pressure systems which they generate tend to increase the amplitude of the upper waves – both the cold trough and the warm ridge – as may be seen on fig. 3.13 by the way in which the schematic surface-wind arrows must tend to move warm and cold air; they also tend to propagate the whole situation eastwards.

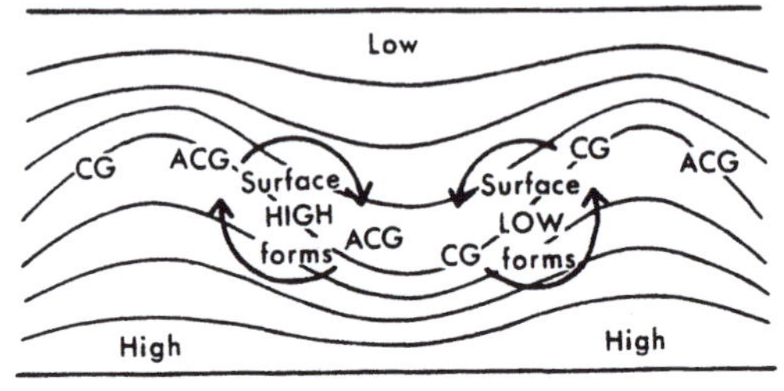

Fig. 3.13 Locations of principal anticyclonic and cyclonic development due to curvature and pressure gradient changes along the main-stream of the upper westerlies through a hypothetical wave pattern. (Northern hemisphere example.)
Surface wind directions in the pressure systems whose development is encouraged are illustrated schematically by arrows.
In situations where an already existing surface LOW happens to be passing through the area where high-pressure development is favoured, or vice versa, these previous systems weaken; but their presence and the time taken for them to decay may not allow time for the development tendency to produce a new surface weather system of opposite character before the upper wind flow over the region alters.

Ultimately all these tendencies are checked somewhere or other by geography. Warming limits the equatorward growth of cold troughs; cooling checks the poleward growth of warm ridges; and the eastward transport is liable to come up against some barrier, a mountain range or the thermal nature of an ocean or continent, that distorts the flow. In other cases, the advancing features catch up on an older, stationary circulation (cyclone or anticyclone) and are distorted by the flow around that.

The effects of changing latitude upon the main upper wind flow are best seen (fig. 3.14) by considering straight northerly or southerly windstreams (or streams with pronounced northerly or southerly components). For these cases the V^2/r term in the gradient wind equation is zero, and the equation reduces to

$$\frac{\partial p}{\partial s} = -2\rho\omega V \sin \phi$$

Since $\sin \phi$ is smaller the lower the latitude (ϕ), the coefficient of V is smaller in low latitudes and a bigger value of V is required to balance the same pressure gradient. Hence, towards the downstream end of a long northerly windflow (in the northern hemisphere case) the wind tends to blow too slowly for the pressure gradient and is deflected towards the low-pressure (eastern) side: there it is likely to converge with the slower-moving air outside the stream, producing an anticyclogenetic tendency, and there is likely to be a corresponding

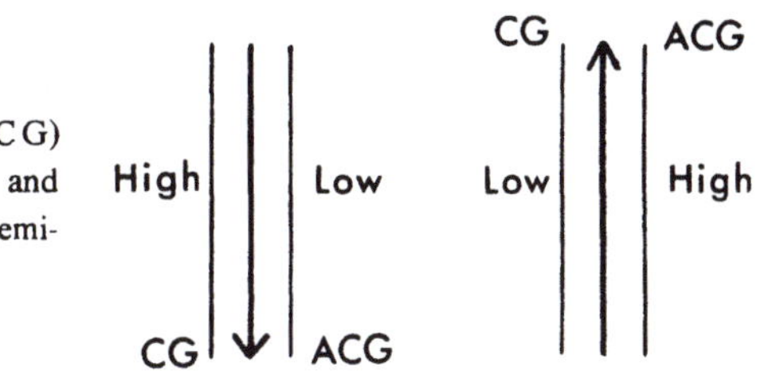

Fig. 3.14 Disposition of development (ACG and CG) effects in relation to long northerly and southerly windstreams aloft. (Northern hemisphere example.)

cyclogenetic effect (pressure-fall) at the western side of the stream. As with the other developmental mechanisms, the effects should increase with the strength of the windstream. The effects of changing latitude are greater the lower the latitude, because sin ϕ changes most rapidly in low latitudes (cf. footnote, p. 79). A long southerly wind flow arrives in high latitudes with a tendency to blow too strongly for the pressure gradient. The positions of the effects with these meridional airstreams are sketched in fig. 3.14. The effects with both northerly and southerly airstreams (or straight windstreams with much northerly or southerly component) are contrary to those induced by the changing curvature in a sinusoidal wave, as in fig. 3.11: the effects of changing latitude seem more likely to become important where the shape of the flow already differs from the sinusoidal form, and the wave is becoming distorted – e.g. in the formation of a *blocking pattern* and when large-scale eddies are becoming cut off north and south of the mainstream of the upper winds.

Blocking of the upper westerlies, and development of cut-off (warm) highs and (cold) lows. In upper ridges and troughs of greatly enhanced amplitude, these developments are what happens when the steady progression of the surface lows and highs and the (usually slower) eastward march of the upper waves is checked. The process usually develops in one or other of the three ways illustrated schematically in figs. 3.15 (*a*), (*b*) and (*c*). Again the examples are shown in the northern hemisphere orientation. The stages illustrated are such as might be reached at intervals of roughly one day. The first two sequences result in what are commonly called anticyclonic and cyclonic blocks respectively, according to which system is most prominent at the climax of the development. The anticyclonic block is sometimes called an omega block because of the resemblance of the pattern to the Greek capital omega Ω. The third sequence shows the development of a more extensive omega-type block, in which both the northern high-pressure system and the southern lows develop into big systems, through a characteristic repetition of the blocking development operating on successive travelling waves in the upper westerlies as each one comes up against the stationary blocking situation. The patterns are reminiscent of the appearance of breaking waves in the sea running up on a shore; and it seems likely that frictional or thermal factors inherent in the geography of land and sea play a part – especially since such situations are rare, and somewhat mobile or non-persistent when they do occur, over the southern hemisphere zone of unbroken ocean.

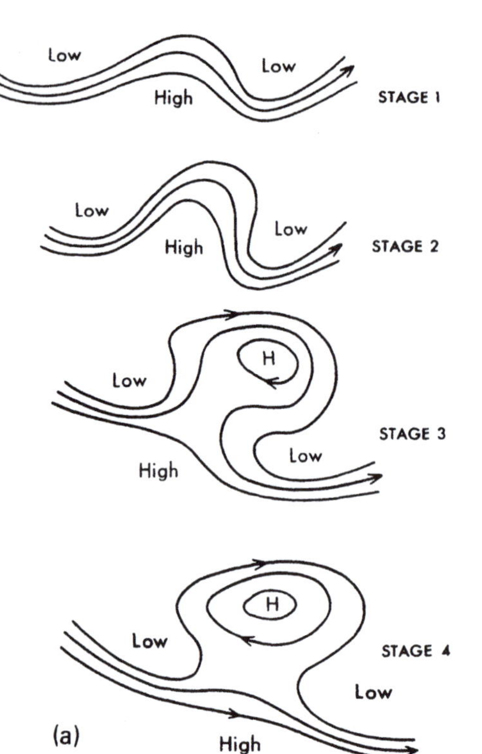

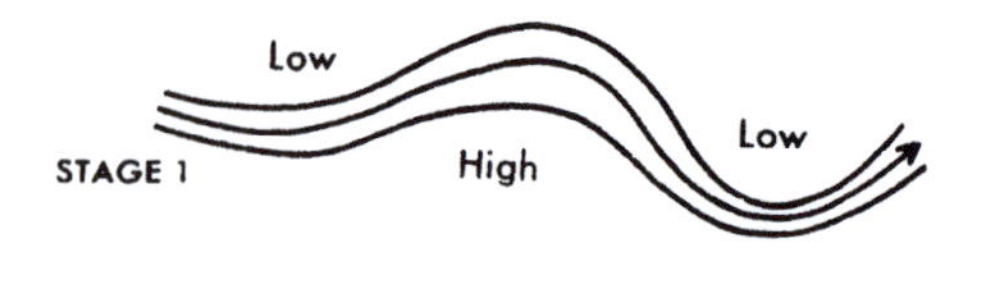

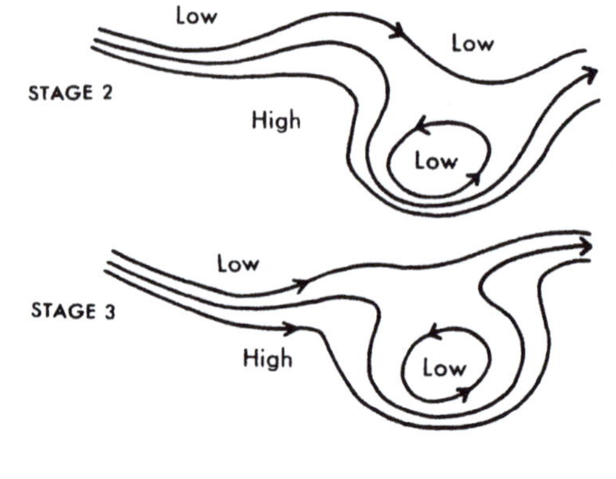

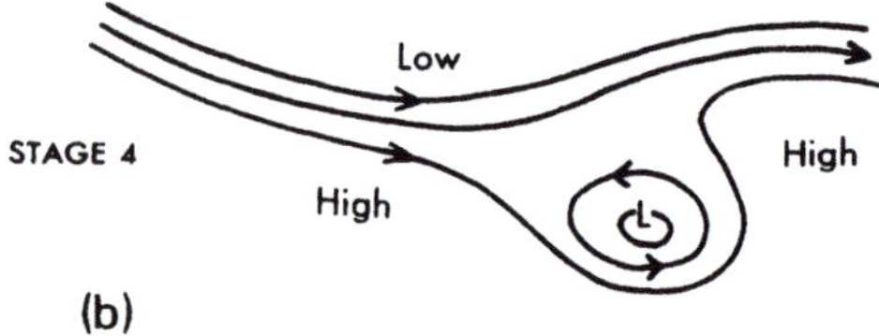

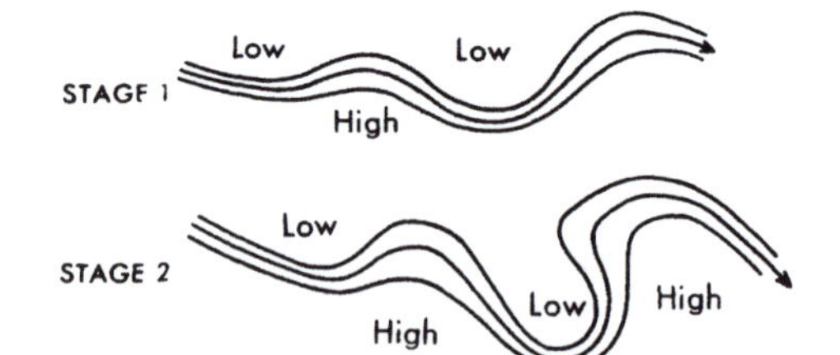

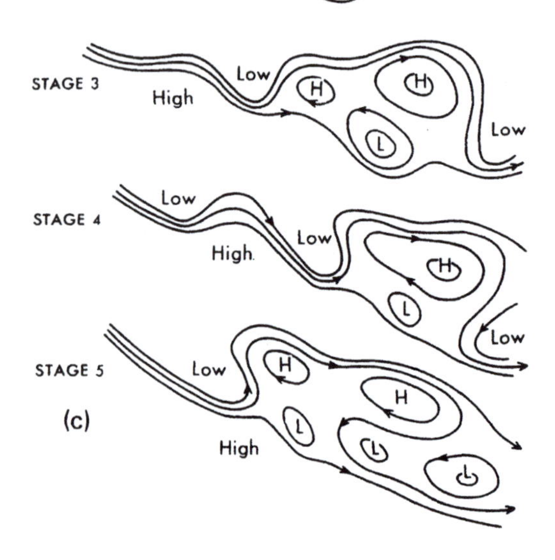

Fig. 3.15 Blocking.

Successive stages in the breakdown of an open wave pattern in the upper westerlies into 'blocked' situations, in which anticyclonic and cyclonic eddies are cut off and the mainstream of the upper westerlies becomes blocked and diverted into twin bands in high and low latitudes:

(*a*) Development of an anticyclonic or 'omega' (Ω) block.

(*b*) Development of a cyclonic (inverted 'omega') block.

(*c*) Repetitions of the blocking development producing an extended block ('omega' type) over a long sector of the hemisphere, with two bands of upper westerly winds and a band of zonal easterlies in middle latitudes.

Boyden (1963) established statistically a number of rules regarding the conditions which lead to the formation of *cut off warm highs* (*blocking anticyclones*) and *cut off cold lows*. The main requirement appeared to be a jet stream backing to a point from near south or veering to a point near north (320–360°) respectively, as a ridge and trough of increasing amplitude develop.

Fig. 3.16 presents the frequencies of occurrence of anticyclone centres in different parts of the northern and southern hemispheres in winter and summer. The almost unbroken ring of high frequencies in the subtropical zone (warm side of the jet stream) is prominent on all the maps. High frequencies are also produced by the anticyclones in very high latitudes – remote from the jet stream and its associated thermal gradient – over the northern polar ice in summer and over the central regions of the Antarctic ice cap at all seasons. In addition, there are frequency maxima produced by blocking anticyclones over Alaska, Greenland, Scandinavia and northeast Siberia as well as near the southern tip of South America and south or southeast of New Zealand: these positions all bear some relation to mountain barriers that interfere with the free flow of the upper westerlies and at times sustain great thermal contrasts between land and sea in their latitudes. However, the frequency pattern over the northern hemisphere actually shows another complete ring of maximum anticyclone frequency which includes the Azores (subtropical Atlantic) system and passes over high latitudes on the other side of the hemisphere, largely ignoring the geography of land and sea: this ring is centred near the isolated maximum of frequency of polar anticyclones over Greenland. It appears unlikely that the terrestrial geography can entirely explain the 'preference' of blocking anticyclones for positions on this ring, unless it be the integrated effect of what is more or less a continuous mountain wall that includes Spain, the Alps, the mountains of central and northeastern Asia and the Rockies (see also Chapter 7, p. 273 and Chapter 10, pp. 461–3).

A notable feature of all well-developed blocking situations is the splitting of the upper westerly windstream into two branches, one passing north and the other south of the block (cf. figs. 3.15 (*a*)–(*c*)). In between is a zone of easterly winds, light easterlies aloft but commonly strong easterlies at the surface. Fig. 3.15 (*c*) is a blocking situation of considerable zonal extent, spanning many longitudes with 'zonal' easterly as well as two bands of 'zonal' westerly winds. In the other blocking sequences illustrated (figs. 3.15 (*a*), (*b*)), and in the earlier stages of this same sequence (fig. 3.15 (*c*)), the zonal wind currents are weak and meridional flow (N and S components) prominent.

Development of the main wind zones and transverse (vertical meridional) circulations that link them

Long-term averages of the wind circulation and barometric pressure pattern always show the circumpolar vortex of more or less westerly winds aloft and, corresponding to it, at the surface:

(1) an almost continuous zone of high pressure – a ring or chain of anticyclones which girdles the Earth – in subtropical latitudes;

Fig. 3.16 Frequency of anticyclone centres (percentage of days with a centre within 100 000 km² centred on any point).

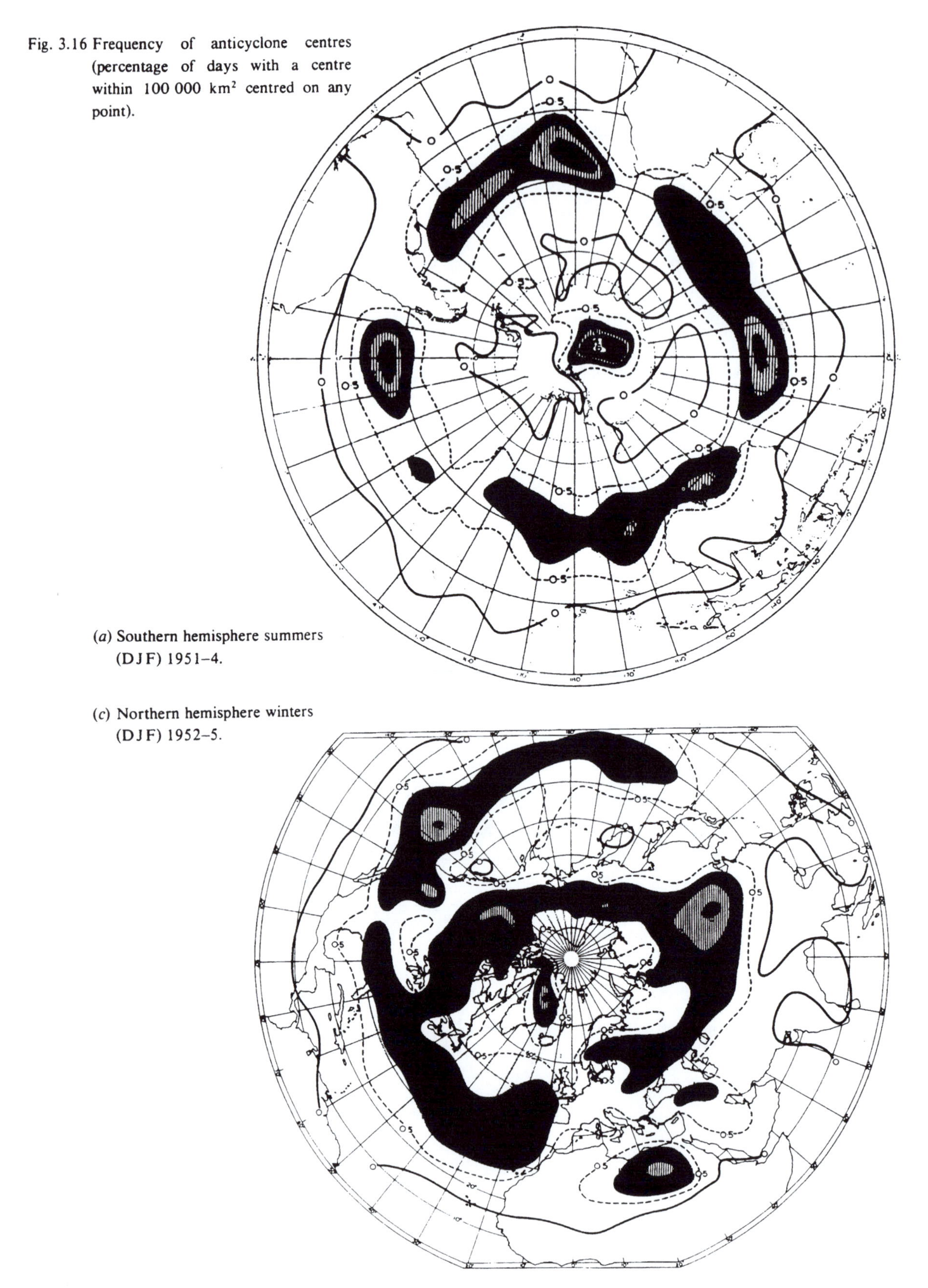

(*a*) Southern hemisphere summers (DJF) 1951–4.

(*c*) Northern hemisphere winters (DJF) 1952–5.

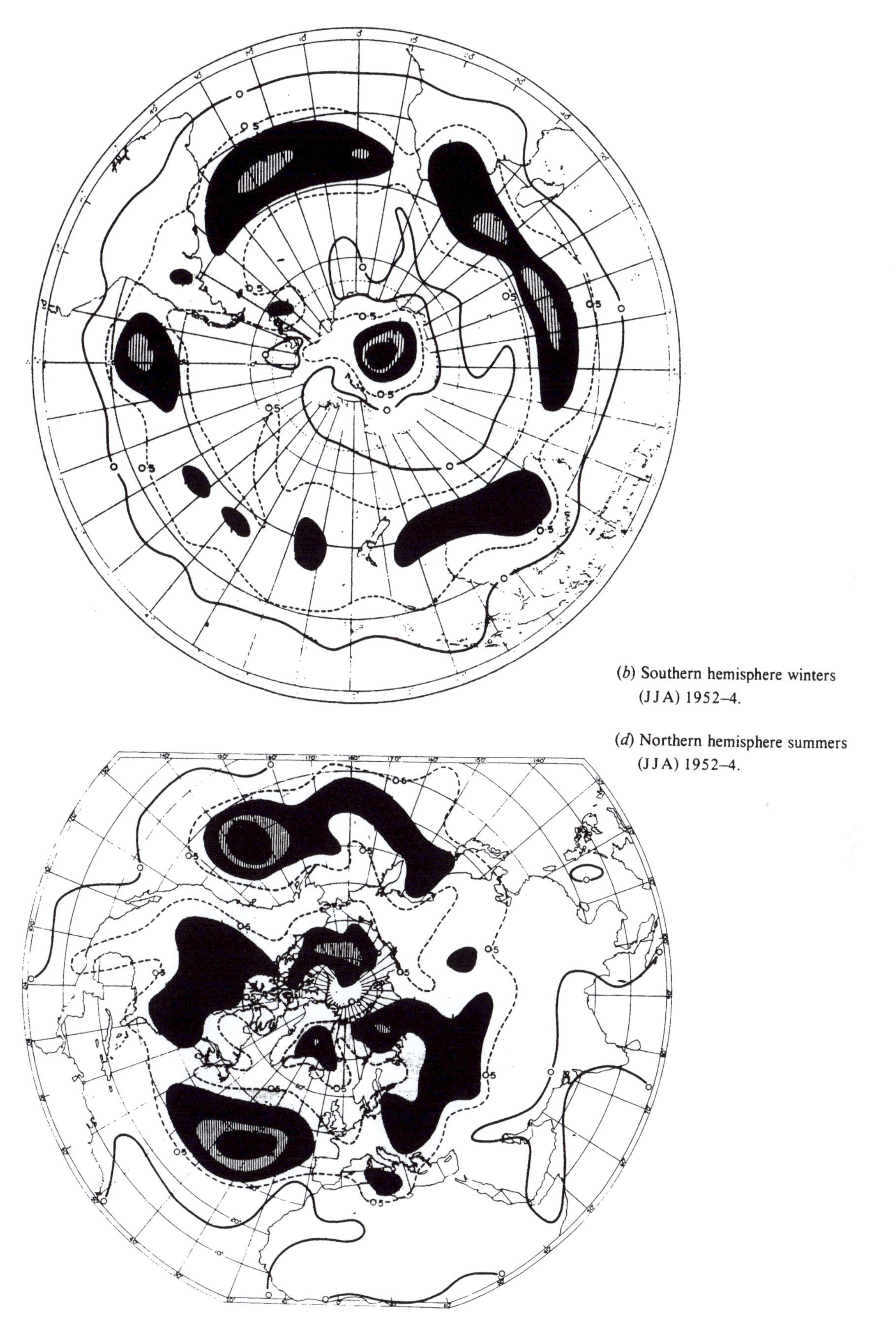

(*b*) Southern hemisphere winters (JJA) 1952–4.

(*d*) Northern hemisphere summers (JJA) 1952–4.

(2) an almost continuous zone of low pressure – with some distinct regions of maximum cyclonicity within it – in subpolar latitudes.

These features can be seen on the average pressure distribution maps for the first half of the present century, illustrated in fig. 3.17. In the northern hemisphere the subpolar low-pressure belt is interrupted because the main development of the cyclonic circulations is over the sea, especially in winter. The mean annual pressure gradients are also much weaker in the northern hemisphere, partly because of considerable seasonal shifts of the main features, owing to changes in the relative warmth of land and sea, and partly because of the greater friction over the extensive landmasses. The ring of high frequency of blocking anticyclones, seen in fig. 3.16 passing over northeast Siberia, Alaska and the Canadian prairies, has only a small effect on the mean pressure distribution because of the fairly high frequency of deep depressions crossing the high-latitude parts of the ring: nevertheless some influence of the ring can be traced in fig. 3.17 in the average pressure pattern over the regions mentioned.

In some epochs, characterized by weaker westerlies or more frequent blocking, the eccentric ring may become more prominent in the average pressure distribution. Nevertheless it is probable that, through all the variations of amplitude and position of the big waves in the upper westerlies, sometimes allowing greater prominence of meridional (north and south wind) than of zonal components, and despite the occurrence in blocking situations of complete cut-off cyclonic and anticyclonic eddies which may last up to several weeks, the broad zonal character of the long-period mean circulation is as permanent as the circumpolar arrangement of the heating and cooling zones and the circumpolar vortex aloft.

The prominence of northerly and southerly windstreams is reduced on long-period average maps, because even when such winds dominate a whole winter or summer in a particular area they rarely prevail in just the same places in the following year, except where some exceptionally strong geographical control is at work.

Fig. 3.18 is an idealized scheme of the zones of high and low pressure and prevailing winds that should arise at the surface of a uniform Earth, as a result of the development of a zone of mainly low pressure along the cold flank and high pressure along the warm flank of the main thermal gradient in middle latitudes. Away from the latitudes where these pressure systems are generated, one meets higher average pressure near the poles and lower average pressure near the equator. This determines the following sequence of zones of prevailing surface winds over either hemisphere:

(i) *Polar easterlies* – on the high-latitude side of the zone of cyclone centres, towards which cold Arctic or Antarctic air blows with easterly components. The actual direction of the wind veers and backs as each cyclone centre passes, except in some topographical channels and near the edge of the great ice caps where the wind is partly katabatic and may flow with great persistence from just one direction.

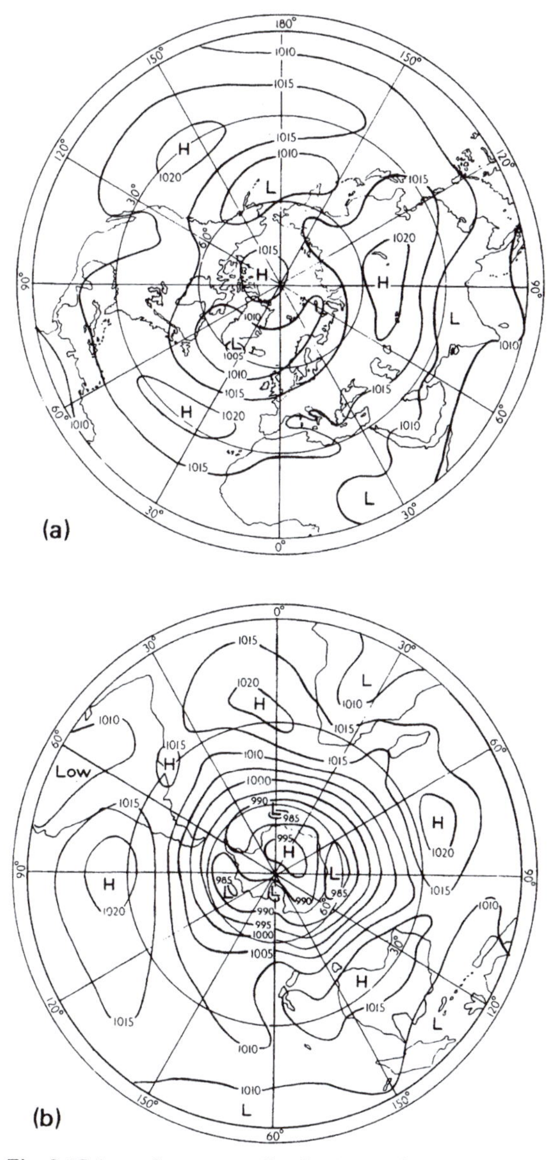

Fig. 3.17 Annual mean distribution of atmospheric pressure in millibars at sea level:
(*a*) Northern hemisphere 1900–40 approx.
(*b*) Southern hemisphere 1900–50s approx.

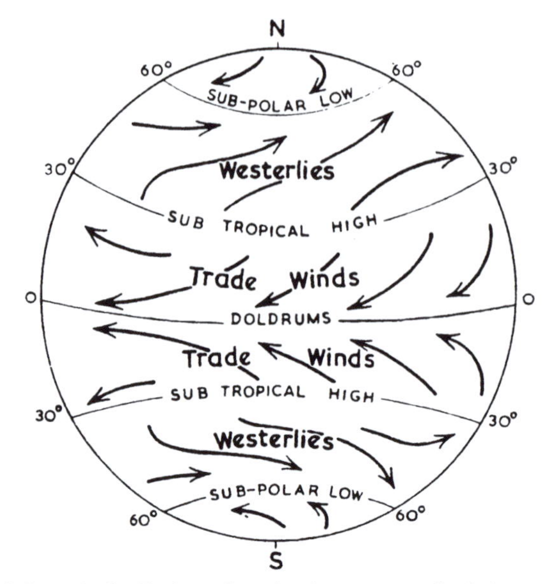

Fig. 3.18 General distribution of sea level pressure and wind zones over the world. (Schematic: as related to the mass displacements and barometric pressure developments determined by the flow in the upper westerlies over a uniform planet.)

(ii) *Middle latitudes westerlies* – the 'Brave West Winds' of the 'Roaring Forties' and Fifties over the Southern Ocean (where winds from westerly points are so prevalent that they used to be relied upon by the old sailing ships) and corresponding westerly winds of only rather less frequency in the same latitudes over the northern hemisphere, where they commonly reach gale force over the North Atlantic and North Pacific Oceans. Over the temperate zone of the northern hemisphere the surface winds are mostly from SW or WSW, and over the southern hemisphere temperate zone from NW or WNW.

(iii) *The Trade Winds* (so named for their reliability by the old merchant seafarers) – prevailing easterlies between the subtropical anticyclones and the meteorological equator, where the windstreams from the two hemispheres meet. They blow mainly from between NE and E in the northern hemisphere and between SE and E in the southern hemisphere and are not very strong except with the day-time sea-breeze effect added near the coasts of the hot continents or where a tropical storm happens to be developing.

In addition to the above three zones of well-marked prevailing winds, there are zones of variable winds about the latitudes of maximum and minimum pressure:

(iv) *Variable cyclonic winds*, often strong, in the zone of subpolar low pressure – i.e. along the paths of the cyclone centres.

(v) *Light variable airs and coastal sea breezes* prevail in the subtropical anticyclone belt, named the 'Horse Latitudes' in the days of sailing ships because it was necessary to throw the horses overboard if a ship was becalmed too long.

(vi) *The Doldrums* – mostly light variable winds, except in rain squalls, in the zone of the meteorological equator. This is often called the '*Intertropical convergence zone*' because it is where the airstreams from the tropical zones of the two hemispheres meet. On the one hand is the Trade Wind, and on the other the wind which has crossed the equator and begun to acquire a westerly component as it moves towards higher latitudes again.

A principle of some importance restricts the range of variations that can occur in the positions and extents of the main surface easterly and westerly wind zones. The total angular momentum of the Earth and its atmosphere about the rotation axis must remain constant. Friction of the Earth's surface on the wind in contact with it in the zones of surface easterly winds (where the wind is lagging behind the Earth in terms of the rate of spin about the Earth's axis) transfers some of the Earth's momentum to the atmosphere. This momentum must be restored to the Earth after no very great delay; otherwise the atmosphere as a whole would acquire increasing west-to-east momentum and the Earth itself (which spins in this direction) would slow down a little. The momentum must be restored to the Earth by friction in the zones where westerly winds prevail at the surface. Hence conservation of the balance of angular momentum about the Earth's rotation axis demands that there be as much total friction on west winds as on east winds. This, in turn, suggests that, unless the average speeds of the surface westerlies and easterlies differ greatly, the total areas of west winds and east winds prevailing at the surface should always be about equal. This condition is satisfied when the subtropical high-pressure maximum is on average just far enough on the equatorward side of latitude 30° that the areas of easterly winds between this maximum and the equator (the Trade Winds) and in the much smaller area of the polar regions (fig. 3.18), added together, about equal the area covered by the westerly surface winds in middle latitudes. Alternatively, if the total areas of prevailing easterlies were larger, these winds would have to be on average weaker than the westerlies.

Our discussion of how the balance of angular momentum is conserved, with the atmosphere acquiring westerly momentum from the Earth's surface in one latitude zone and losing it in another, has again shown that there must be a regular interchange of air between different latitudes. The necessary poleward transport of angular momentum by the winds, like the poleward transports of heat and water vapour, is achieved partly by eddy motions in the north and south directions and partly by prevailing northerly and southerly wind components in some parts of the mean circulation. The mean meridional circulation, seen in the vertical plane and averaged over all longitudes, is much as presented by PALMÉN (1951) in the diagram here reproduced in fig. 3.19. The characteristic speeds of the mean circulation in this plane are much less than (of the order of 1 m/sec, or one-fifteenth to one-fiftieth of) the mean zonal components of the circulation.

We see that the heat transport, the largest-scale convection, performed by the atmospheric circulation is far from having the simple form (gravitational circulation) of a single cell with warm air rising in the tropics and travelling to the pole aloft, while its place at the surface is taken by a cool, denser airstream that has come all the way to the equator from the pole.

If the Earth were not rotating, a single great convection cell over either hemisphere with 'Trade Winds' at the surface all the way from high latitudes to the equator and 'Counter-Trades' aloft carrying warmer air from the equator to high latitudes would presumably suffice to redistribute the heat received; such a circulation may be observed in any room with a heat source in one wall and cooling concentrated at the opposite wall. Laboratory experiments by FULTZ (1961) with water in a rotating dish (radius about 15 cm, depth 4·5 cm) heated at the edge (to correspond to the equator) and cooled at the centre (pole), as a model of the atmosphere over one hemisphere (in which the water movements are rendered visible by powdered aluminium and photographed from above by a camera rotating with the dish), suggest (if one accepts the limitations of a flat dish to represent the spherical Earth) that, if the Earth were rotating only slowly this type of single-cell convection might indeed prevail: the poleward-drifting upper layers arrive near the polar axis having acquired a considerable west-wind velocity relative to the dish below, because they retain the speed which they had near the rim (equator), and similarly the low-level drift of cool fluid outwards at the bottom of the dish arrives near the rim not as an axially outward-directed 'north' wind but having acquired a considerable east-wind component. So a slowly rotating

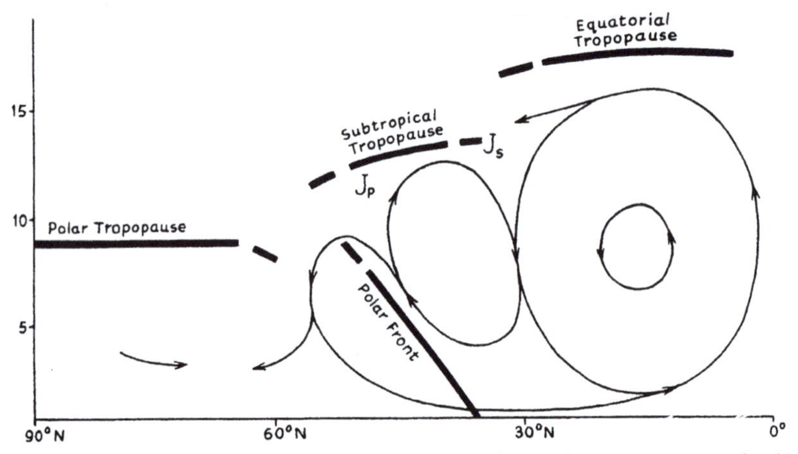

Fig. 3.19 Mean circulation in a vertical, meridional (N–S) plane between equator and pole. J_p marks the position of the polar front jet stream in the upper westerlies (blowing straight into the paper as the diagram is drawn) and J_s marks the very strong westerlies in the subtropical jet stream (also blowing into the paper).
The subtropical jet stream represents the extra W–E motion relative to the rotating Earth at latitude 30° possessed by air that has come (with the vertical and meridional circulation shown) from near the equator.
(Reproduced by courtesy of the Royal Meteorological Society, London.)

Earth would be expected to show Trade Winds moving equatorwards over both hemispheres and arriving with marked easterly components, as in fact observed in low latitudes, and Counter-Trades aloft becoming more and more a westerly windflow with increasing latitude. This is essentially the explanation advanced by G. HADLEY (1735) of the Trade Winds which had been ably mapped by his predecessor, E. HALLEY, as early as 1686. But it does not make clear why the surface Trades and upper Counter-Trades are not found over the entire range of latitudes. This does appear in FULTZ's experiments, because when the model is rotated faster the band of strong upper westerly flow becomes stronger and fails to reach as near to the rotation axis (i.e. fails to reach high latitudes). A system of waves appears ranging widely over the inner part of the dish and resembling the waves in the westerlies over the middle and higher latitudes observed in the atmosphere. Evidently the high speeds relative to the dish that were acquired by poleward transfer of fluid from the rim became dynamically unstable and could no longer transport enough heat to the cold region when the rotation of the dish was speeded up, so that another system of convection resembling the waves in the upper westerlies had to be developed. The simple Trade Wind and Counter-Trade set-up continued just over the 'lower latitudes', i.e. over the outer part of the vessel between somewhere near the strong band of upper westerlies and the rim of the dish. So this limited part of the circulation over the real Earth, between the subtropical anticyclones and the (meteorological) equator, is referred to as a 'HADLEY cell' of the circulation.

Another HADLEY-type circulation is found over the highest latitudes, between the polar anticyclone and the subpolar belt of low pressure. But in the realm of the strong middle latitudes westerlies – the main circumpolar vortex, which lies between these high- and low-latitude cells – the circulation is in the reverse sense: it is developed by the disequilibrium at various points in the course of the upper westerlies and by the gravitational overrunning and undercutting of the warmer and cold airstreams respectively at the polar front, as discussed on pp. 104, 120–2.

Air moving polewards from near the equator (as it does in the upper troposphere over low latitudes) must preserve the spin about the Earth's axis which it had near the equator; by the time it has reached latitudes around 30°, where the radial distance of the surface from the Earth's axis is reduced to $R \cos 30°$ (R being the equatorial radius of the Earth), this air is moving from west to east faster than the Earth's surface at that point by some 60 m/sec (about 115 knots). It therefore appears as a semipermanent west wind with speeds of this order, the so-called *subtropical jet stream*[1] (J_s in fig. 3.19), as in the experiment described

1. The subtropical jet stream typically presents a westerly wind moving at over 100 knots (over 50 m/sec) over a broad band, perhaps 3–5° of latitude wide, some thousands of kilometres in length, at a height of 12–15 km. There is a strong shear beneath it, and the air in the middle troposphere is not usually moving fast – though occasionally in winter the underside of the system is felt as a fierce storm on the high peaks of the Himalayas and central Andes. In extreme cases, probably always involving coalescence with a polar front jet stream that has moved to latitudes near 35° or less, speeds up to 150 m/sec (300 knots) have been reported – over southern Japan (30–35°N) and Amsterdam Island (38°S).

above with the fluid in the fast-rotating dish. This is a high-level jet stream concentrated in the uppermost part of the high-reaching troposphere of low latitudes near that level where the slow poleward transfer of air from above the equatorial rain zone is most marked, i.e. the top of the right-hand cell of the vertical–meridional circulation shown in fig. 3.19. The strongest west wind velocity relative to the rotating Earth beneath is attained at the limiting latitude reached by this poleward transfer, and the air involved in it is at that point about to begin its descent in the subtropical anticyclones. Hence, the subtropical jet stream, J_s, lies more or less directly above the subtropical high-pressure systems at the surface; the air concerned becomes involved in the general sinking motion ('subsidence') down through the troposphere in the anticyclones, and its energy (and momentum) are partly dissipated (and transferred to the Earth's surface) in friction and partly conveyed polewards and partly back towards the equator within the lower atmosphere (cf. fig. 3.19). The subtropical jet stream is thus linked in principle with the subtropical anticyclones rather than directly with the hemisphere's main thermal gradient zone, though it is an essential part of the hemispheric circulation generated by that thermal gradient. The subtropical anticyclones themselves are to be understood as a phenomenon of the warm side of the main thermal gradient and polar front jet-stream zone as already explained (e.g. p. 103).

Thus, the circulation in the vertical N–S plane between the equator and about latitude 30° is indeed a simple, direct gravitational overturning of warm air rising near the equator and sinking again when it has become cooler over subtropical latitudes, as supposed by HADLEY (1735). The circulation in the vertical meridional plane over the polar caps is also in this direct sense. But over middle latitudes the vertical circulation is indirect: air rises in the cool higher middle latitudes (though appearing in that locality as 'warm air' rising over the 'polar front' – as shown in fig. 3.19) and sinks in lower latitudes, joining in the subsidence over the subtropical anticyclones, where it is warmer. This circulation therefore uses up energy which it appears to derive from the gravitational processes (occlusion, lifting of warm air and sinking of cold air at the fronts) in the individual travelling cyclones (depressions) in the frontal zone. J_p in the figure marks the jet stream associated with the zone of main thermal gradient, the *polar front jet stream,*[1] whose origin lies in this frontal zone.

1. The polar front jet streams, though mainly westerly, may have any orientation according to where the concentration of gradient lies in the meandering course of the main upper wind flow. Occasionally, where this 'plunges' into low latitudes, it may merge with the subtropical jet stream and the two jet streams blow alongside in an unusually broad belt of very strong winds (cf. LAMB *et al.* 1957). Typically the maximum winds in the polar front jet stream exceed 25 m/sec (50 knots) over a belt 200–500 km broad and often attain 50 m/sec (over 100 knots) in the cooler seasons of the year. Extreme speeds over 100 m/sec (exceeding 200 knots) have been reported over various parts of Europe between 65°N and the Mediterranean. 300 knots is occasionally approached in the positions of strongest thermal contrast near the extremity of the great upper cold troughs where the cold air of the Siberian, Canadian or Antarctic winter approaches the warm regime over the subtropical oceans: this extreme value has been reported in one or two cases where the polar front and subtropical jet streams merged (see p. 117, footnote).

The mean latitude of the subtropical high-pressure belt, averaged around the northern or southern hemisphere, shows a seasonal movement north and south between summer and winter which is comparable with that of the limit of snow and ice surface but far less than that of the zenith sun at noon. This seasonal range averages about 10° of latitude at the present epoch in the northern hemisphere but less than this in the southern. SAWYER (1966) believes this indicates that the position of the subtropical jet stream and the size of the main Hadley cell of the circulation are fundamentally determined by the speed of the Earth's rotation and that they can move only very little north or south to accommodate changes in the heating and cooling budget. Nevertheless, the mean position of the axis of the subtropical high-pressure belts in the sectors for which data are adequate is seen to have varied by 1–4° of latitude as between different climatic periods within the last 100 years, and rather greater shifts in the poleward or equatorward limits of its influence in suppressing rainfall can be traced when warmer and colder epochs of longer duration are compared. The limits of Arctic and Antarctic sea ice seem to have varied since the year 1800 by 2–3° of latitude in the Atlantic sector, more or less in phase with the shifts N and S of the subtropical anticyclone and subpolar low-pressure belts: though it must be stressed that the increase of the Arctic ice after 1955–60 appears as a response to a southward displacement of the wind circulation zones that began some 20 years earlier.

The changes of the average heating pattern from one epoch to another are much smaller than those which occur between summer and winter every year. The observed latitude shifts of the zone of highest average surface pressure may not entail equally great displacements of the subtropical jet stream aloft, since periods in which the surface high-pressure zone appears displaced towards high latitudes may be times when the subtropical anticyclones were frequently replaced by 'blocking highs' covering latitudes between 45° and 70° (e.g. on the eccentric ring seen on fig. 3.16). Nevertheless most investigators of the evidence of past climates have concluded that the subtropical anticyclone belt was nearer the equator (and narrowed) in the ice ages and that it underwent a significant displacement polewards (in both hemispheres) in warm epochs.

Figs. 3.20 and 3.21 from FLOHN (1964), based on theoretical work by SMAGORINSKY, indicate that the main high-pressure belt may have been in latitudes 50–60° in the major warm, ice-free eras of geological time (e.g. Jurassic–early Tertiary). Such a position raises questions about the angular momentum balance, since the zone of prevailing surface westerlies would be limited to the small area between latitudes 60° and 80° or thereabouts. Presumably there was no room for more than a tiny zone of relatively high pressure and rather variable easterly winds near the pole. Under such conditions it is not possible that the whole wide zone between latitude 50° and the equator was occupied by regular easterly Trade Winds. Instead, one must suppose that mainly weak and chaotic, relatively small-scale, circulations with winds from all points of the compass were arrayed over these latitudes.

Before we can complete a survey of world climates in terms of the wind circulation that

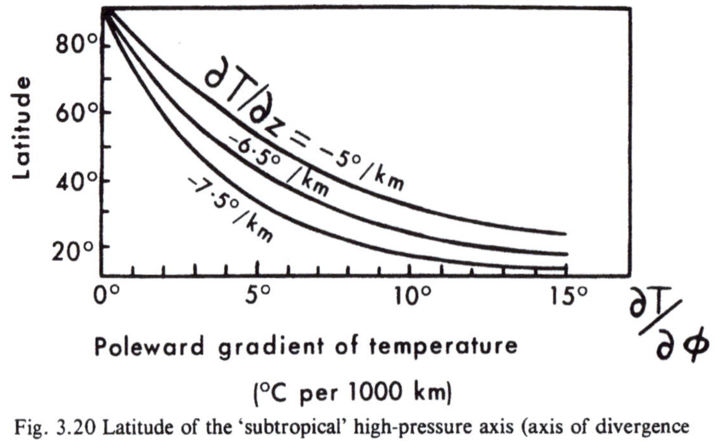

Fig. 3.20 Latitude of the 'subtropical' high-pressure axis (axis of divergence in the surface winds) to correspond to different values of the mean poleward gradient of temperature and the vertical temperature gradient.

(*According to* FLOHN *1964.*)

The vertical lapse of temperature $\partial T/\partial z$ (which determines thermal stability or instability, i.e. damps or encourages vertical motion) may not be an independent variable. It is probably determined by the same heating and radiation-loss situation that determines the poleward gradient of temperature. But the range of values provided by the three curves in the figure probably covers all likely values of the average lapse rate occurring in the widest range of climatic epochs. The middle curve corresponds to present conditions.

Global average (present epoch) $\partial T/\partial z = -6{\cdot}5°/\text{km}$.

(Note: 1000 km = 9·02° latitude.)

goes on in each zone or region, we must consider the characteristic life histories of individual cyclones and anticyclones. These are the systems that bring the variations of weather from day to day, and from week to week, the statistical effect of which produces the average pressure and wind distributions seen in figs. 3.17 and 3.18.

Depressions and anticyclones: life cycles

The normal life history of a frontal cyclone (or depression) in middle latitudes is seen in fig. 3.22 through its three main stages. First, a small wave, perhaps 200 km long and with amplitude 50 km, appears on the front which divides the warm and cold airmasses. The amplitude of the disturbance in the upper cloud and rain patterns is commonly rather greater than the deformation of the (surface) front itself at all stages of the development (as here shown), and this aids recognition. The first symptoms of wave formation are often the widening of the belt of cloud over the cold air side of the front. When the small frontal wave finds itself in a cyclogenetic area, as defined by divergence in the upper flow, and with

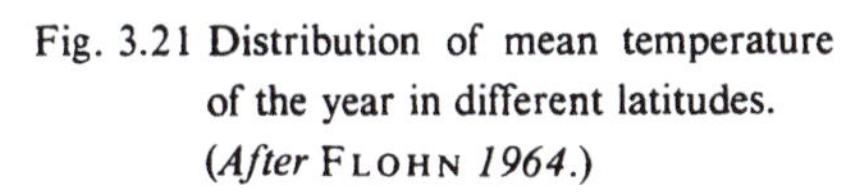

Fig. 3.21 Distribution of mean temperature of the year in different latitudes. (*After* FLOHN *1964*.)

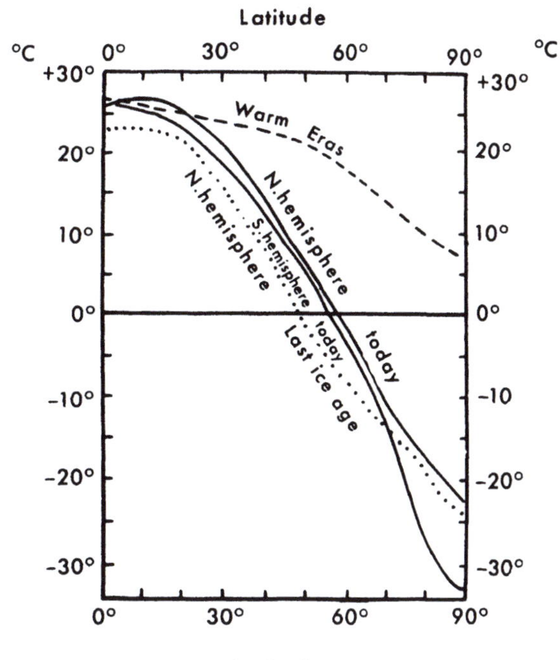

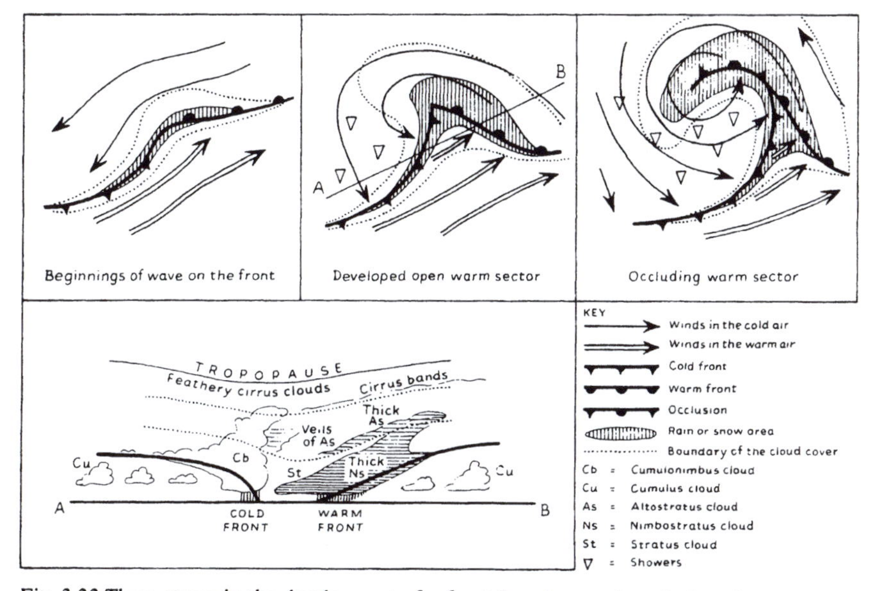

Fig. 3.22 Three stages in the development of a frontal cyclone and vertical section along the line AB.
(Invert the maps for southern hemisphere.)

converging surface winds (as shown in fig. 3.10), the amplitude of the disturbance grows. The next stage is an open warm sector depression, travelling in the direction of the warm air. The warm air overruns the (warm) front ahead of it and is also lifted off the surface by the colder, denser air cutting in at the cold front advancing behind it. In these ways, the warm sector air (or poleward bulge of warm air) is in the end lifted altogether off the ground, till it overlies the cold air that has spread underneath it. This process, known as *occlusion*, clearly lowers the centre of gravity of the system; and the potential energy of the former state, where cold (dense) and warm (lighter) air lay side by side, is thereby converted into kinetic energy: this is observed in the development of strong winds. The rising warm air sliding up over the sloping frontal surfaces – i.e. up over the cold air beneath – spreads clouds and rain some way beyond the ground-level frontal pattern and beyond the tip of the warm sector up to 100–500 km to the poleward side of the centre of the depression, as shown in the two later stages of the development in fig. 3.22. Finally, the circulation of the depression may reach a size of 2000–2500 km across.

Ascent of air over a gently sloping frontal surface may occur even in circumstances where both the warmer and the colder airmasses are stable for vertical motion, since poleward motion of the ascending warmer air is likely to bring it to positions where it is not only less dense than the cold air underneath but also in comparison with neighbouring samples of the same airmass at the height reached. A neat account of this has been given by SUTTON (1960), who calls it 'slantwise convection'.

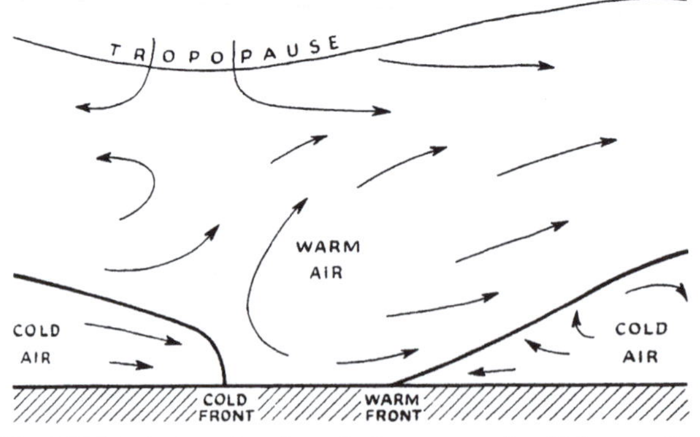

Fig. 3.23 Vertical section through the warm and cold fronts of a depression, advancing from left to right, and through the tropopause above: showing the characteristic air motions in this plane relative to the moving fronts.

Fig. 3.23 gives a somewhat more detailed view of the characteristic vertical motions in a developing cyclone than in the simplified scheme of fig. 3.10. In reading fig. 3.23, it should be borne in mind that the upward motions are commonly accompanied by cloud development and precipitation, the more so the stronger the air currents (as broadly indicated by the lengths of the arrows, though the vertical speed scale is exaggerated). These motions will be found to explain most of the cloud and weather development at the fronts and in the warm sector, shown in the vertical section in fig. 3.22, as well as the location of the lowering of the tropopause.

Fig. 3.24 shows the characteristic positions of a rather old occluding 'low' which still has some eastward movement, a warm ('steering') anticyclone, a small new frontal wave and a new anticyclone (just entering at the left of the diagram) with respect to the geography of the thermal pattern. The actual flow in the upper troposphere is usually similar to this thermal

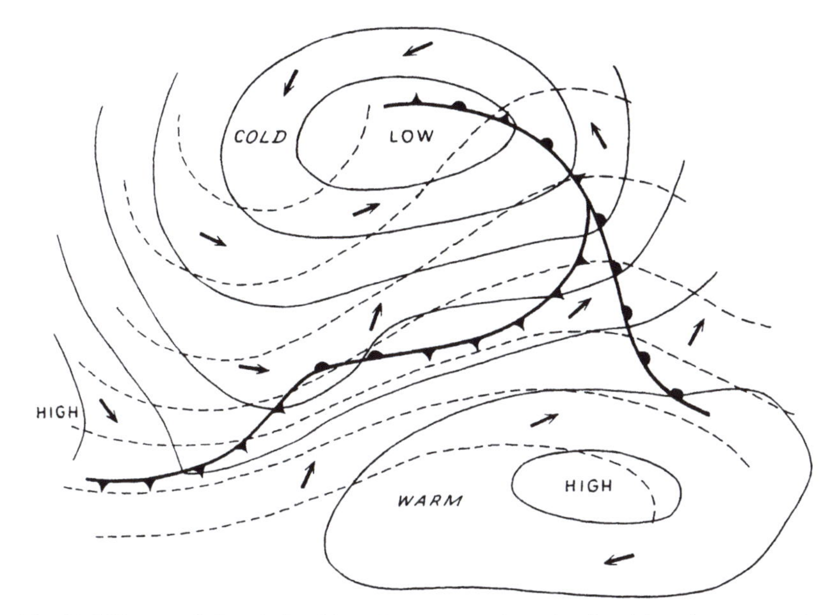

Fig. 3.24 Characteristic relationships of an eastward-moving frontal cyclone, a nearly stationary warm anticyclone, a small, new frontal wave and (at the left of the map) a newly developing anticyclone, to the thermal pattern. The arrows show surface winds. Northern hemisphere case.
The broken lines are isopleths of 1000–500 mb thickness.

pattern. The surface winds (arrows on fig. 3.24) move the warm and cold air: hence they carry the thermal pattern – e.g. the 1000 to 500-mb thickness lines in the illustration – forward with them, until the heating from the ground somewhere in low latitudes or cooling somewhere in high latitudes checks this progress. Thus the surface 'lows' and 'highs' move forward, generally in the direction of the strongest winds on their periphery, and their movement may be looked on as due to the forward movement of the pattern features to which they owe their development. They 'develop' forwards, while the air particles involved are constantly travelling into, up or down through, and out of the system. Sometimes the circulation of a 'low' or a 'high' develops enough to twist the thermal pattern and the flow of the upper winds so much that its own development is further intensified (e.g. by sharper curvature and more unbalance in the upper wind flow): in these cases, the movement of the system usually turns with the twisting thermal pattern over the centre and slows up or comes to a halt. This is therefore a common symptom of a 'stationary situation' or blocking developing.

The implications of figs. 3.22–3.24 for climate are that depressions develop from small waves on a front and are steered forward as they occlude, steered along the long flank of the warm anticyclone in the same general direction as the strong upper winds. The paths followed by the main cyclone centres are along, and towards the cold flank of the jet stream, while the (usually longer-lived) anticyclone centre lies beside the warm flank of the upper windstream, usually somewhere just outside the main current. The more mobile anticyclones and ridges of high pressure are steered along with the jet stream and tend to cross it gradually towards the warm side – unless a convergent configuration of the upper flow such as to maintain anticyclonic development on the cold side persists. Both the 'lows' and 'highs' are liable to become almost stationary when they get outside the realm of strong upper winds. For this reason, the cut-off cold-centred depressions (cold lows) and warm anticyclones associated with blocking, and many other systems at the end of their development, commonly remain nearly stationary. Stationary cold lows somewhere or other over high latitudes or in the subtropical zone may have their energy and existence maintained for some days, or even weeks, by sequences of 'secondary' frontal-wave depressions fed into the system. These secondary depressions originate as waves on the front in a zone of strong thermal contrast, wherever that for the time being lies; they are steered by and gradually across the upper windstream towards the central cyclone, to which they contribute the kinetic energy released by their occlusion process.

Travelling depressions can also be halted when the stronger thermal features of geography stop the progress of the warm and cold bulges of the thermal pattern or when this pattern is sufficiently twisted around by the circulation of an intense surface depression or anticyclone.

In general, places along the warmer side of the path of the surface depression centres experience a sequence of passing frontal rainbelts, alternating with periods of warm sector weather and periods of rather cooler, brighter conditions after the cold fronts, sometimes with showers. The warm sector weather is dull, muggy and may be drizzly near the tip (junction of the fronts), and at places on ground that slopes to meet the wind, but may be warm and bright farther away from the track of the depression centre. On the colder side of the track of the depression centre there is a long period of rain or snow and overcast skies, with the frontal surface overhead, as each 'low' centre passes near; but anywhere more than 100–300 km from the path of centre usually escapes with undisturbed, clear cold weather (or instability showers alternating with bright skies if the cold air be moving over a warmer surface).

Fig. 3.25 illustrates the normal life history of an anticyclone by means of the sequence at one place of clouds and weather, winds and vertical motion relative to the system, during its

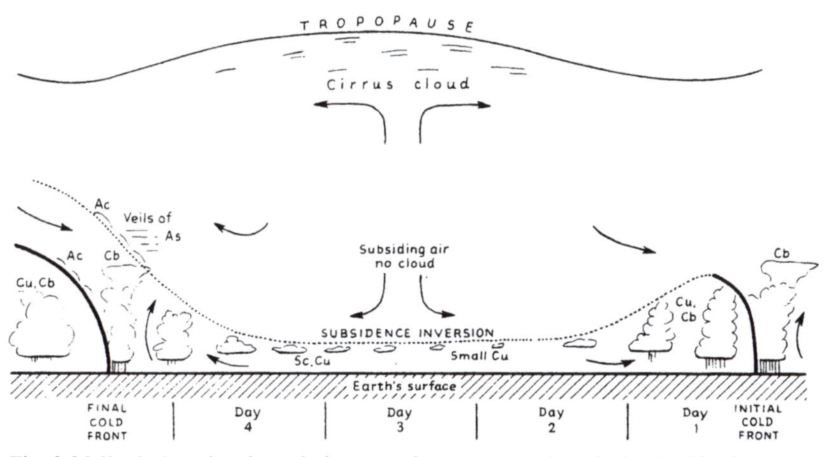

Fig. 3.25 Vertical section through the atmosphere over one place during the life of a typical warm anticyclone, which forms or arrives there after the passage of a cold front that introduces a cold airmass of no great depth.
The cold air subsequently becomes shallow and limited by one or more subsidence inversions at the base of the 'compression-warmed' (adiabatic rise of temperature under increasing pressure) air sinking in the core of the anticyclone.
Cloud development is checked at the inversion, and dust and haze put into the air at ground level collect there.
The development ends when a new cold front brings fresh, deeper cold air; and convergence of the surface winds ahead of the front lifts the subsidence inversion, allowing deeper cloud development once more.

arrival, sojourn and decay. After the passage of an initial cold front, the cold unstable air near the surface deepens temporarily (with generally northerly surface wind components in the northern hemisphere case) till there may be enough vertical cloud-growth to produce some showers. But this is soon eliminated: the growth of cumulus and shower clouds dwindles (and existing clouds may be seen to flatten) as the depth of the originally cool unstable airmass becomes shallower beneath the subsiding air coming slowly down all through the middle troposphere in the anticyclogenetic region. Throughout the middle days of the anticyclone's presence the surface air layer is shallow (typically about 1 km deep – rather more over regions of strong surface heating, rather less over cold surfaces in winter); the winds are light, and smoke, haze and moisture put into the air at the surface in the anticyclonic central region accumulate beneath the subsidence inversion. For this reason, although any clouds are thin and the weather may be fine and sunny in anticyclones over warm regions and in the warm season of the year, anticyclones over cool regions are liable to be foggy and haze always tends to thicken the longer the system lasts. Finally, as a fresh cold front approaches, the anticyclogenetic situation is destroyed and the sinking air motion is reversed; the subsidence inversion lifts and cloud growth increases once more. After the cold front the whole sequence may be repeated; or, if the cold air is deeper, quite different upper wind-flow patterns ensue and the weather may remain disturbed.

The characteristic life-span of a single anticyclone cell from the passage of the initial cold front to the destruction or rejuvenation of the system by the arrival of the next advance of cold air is about $4\frac{1}{2}$ days. The characteristic life of an individual depression is about $2\frac{1}{2}$ days from its first appearance as a frontal wave to its slowing up as an occluded system, decaying or being absorbed into some central cyclonic region which is maintained by a succession of lows steered one after the other into the same region.

A single anticyclone normally steers a succession of developing frontal waves along its flank, till it is finally dislodged or rejuvenated by incursion of the cold air behind the cold front of the last depression of the 'family'. The largest anticyclones are bigger than depressions and may measure 3000 km, or rather more, along their major axis.

In the subsidence region of an anticyclone the air generally sinks at rates of the order of 1 km/day in the middle troposphere (from heights of 6 to 2 km), slowing as it approaches the inversion at the base of the subsiding layer. Occasionally some of the dry, subsided air gets stirred down to the ground by turbulence within the bottom layers and suddenly clears any fog or cloud there; more frequently the subsided air is present on the upper levels of hills and mountains, where it may be recognized by its very low relative humidity, its exceptional transparency (strong sunshine by day and quick radiation cooling of the surface by night) and the deeper than normal blue of the sky (less scattered light).

The up-currents in cyclonic regions are of two kinds:

(i) upgliding motions at low angles of slope, from 1 in 50 to 1 in 500, over frontal surfaces;

(ii) almost vertical up-rushes in cumulus (heap-type) clouds, part of a congested pattern of turbulent upward and downward motion, in which there is a slight net upward component wherever cyclonic curvature of the isobars and convergence of the horizontal winds at the ground prevail.

The rates of ascent are quite different. The upgliding characteristically involves vertical components of the order of 5 cm/sec (about 4 km/day). The vertical currents in cumulonimbus (Cb) cloud-walls at cold fronts may be as strong as from 5 to 50 m/sec.

Climatic zones and their wind systems

Climatic zones are defined by the planetary wind circulation and the types of surface weather systems found within them. Later we shall study the variations introduced by the changing seasons and different epochs. These variations may be divided into shifts of the average position and changes of average intensity of the circulation systems. The main wind circulation features listed below should be found, subject to some shifts of position and intensity (and in some cases changes of number), in all climatic epochs. Indeed, parallel features might be expected on other planets where there is an atmosphere, subject to appropriate changes of scale brought about by different dimensions and rotation speeds of the planet, absence of water vapour and latent heat transport.

A. *Polar high-pressure regions*

The two polar caps are on average regions of relatively high barometric pressure (figs. 3.17, 3.18), partly by default in that the coldest regions are relatively infrequently visited by intense depressions (especially in seasons and epochs when the zone of strongest thermal gradient is far away in middle or lower latitudes), and partly because of the occasional development of intense anticyclones in which barometric pressure at the surface becomes very high because of the density of the very cold air in the bottom kilometre of the atmosphere.

In winter in the present epoch over Siberia, the Yukon and the central Arctic average temperatures of the lowest 2–3 km of the air columns are sometimes as low as −20° to −30°C: when this is so these lowest layers may contribute by their density as much as 30 mb more to the surface pressure than at temperatures of 0° to −5°C. If, at the same time, some wandering of the mainstream of the upper winds produces an anticyclogenetic pattern over the region – as can occur with a jet stream, warm ridge and cold trough pattern or a blocking eddy – a normal anticyclone is generated over the very dense surface air and pressure at sea level may rise to over 1060 mb, in extreme cases even to about 1080 mb.[1] The highest m.s.l. pressures occurring over the Antarctic are a good deal lower – apparently not above 1030–1040 mb – probably because of the general deficit of mass south of the zone of strong upper westerlies, a deficit amounting to about 2% of the atmosphere (rather over 20 mb) when the zones south of 60°S are compared with those north of 60°N (cf. fig. 3.3).

Long-term average barometric pressure over both polar regions, particularly over the sea, is reduced by occasional incursions of deep, often slow-moving, occluded (cold) cyclones and by some travelling, warm-sector or partly occluded, depressions (at a younger stage of their development), which having been generated in the zone of strong thermal gradient happen to be steered across some part of the polar region by the meandering course of the strong upper winds.

Most of the time quiet weather types prevail over the polar ice. Either type of cyclone is liable, however, to 'whip up' the surface winds to strong or gale force – to ferocious gales where the wind is 'funnelled' between the advancing front of the depression and a mountainous coast as well as in straight valleys and fjords in line with the wind. Snowfall is not great when the temperatures are low, but may be prolonged (if the situation is slow-moving) and accompanied by drift (blizzard).

1. The highest reliably reported value is believed to be 1083·8 mb at Agata (66·9°N 93·5°E 263 m a.s.l.) in Siberia on 31 December 1968 with a surface air temperature of −46°C.

B. *Subpolar low-pressure zones*

These are the zones (generally near the cold side of the strongest thermal gradient) where the average surface pressure is kept low (fig. 3.17) by the passage of frequent depressions. Over the fringe of the Antarctic the zone of low pressure is almost unbroken and the cyclone centres progress east in an almost unbroken sequence. In the northern hemisphere the lowest average pressure is over the oceans in the subpolar zone, and extends at the present epoch through the Norwegian Sea into the Arctic to the limit of open water near Novaya Zemlya. There are many (sometimes fast)[1] travelling depressions, accompanied by changing wind directions as they pass; but also stationary systems sometimes maintain very low pressure for days or weeks in one area or another – oftenest near Iceland, the Aleutians and Novaya Zemlya. There is some parallel to this in the frequency of slow-moving depressions over the Ross Sea and elsewhere in the great bights of the coast of Antarctica.

Passage of the fronts at the ground anywhere in the temperate zone or higher latitudes within the range of a cyclonic circulation and the associated airmass contrasts – commonly up to 1000 km and occasionally 1500 km (say, 10–15° latitude) on the warm side of the centre – often brings quick changes of wind and weather. Sometimes the fronts are strong, accompanied by great temperature changes and heavy rain or snow. In continental interior regions when the front is accompanied by the zone of main thermal contrast between tropical and polar regimes, especially between latitudes 35° and 70°N, its passage may in extreme cases change the surface air temperature by as much as 20–25°C within an hour or two. Similarly sharp changes are liable to occur when a front passes the coast of an ice-bound continent or ocean – e.g. the coast of Antarctica, when open water is near, or the Arctic coasts of Canada and Siberia in summer, when the land to the south is warm and the Arctic sea ice lies immediately to the north. Even in a maritime temperate region like the British Isles at the present epoch the air temperature occasionally changes by 10–15°C as a front passes and introduces an airstream of quite different origin. However, changes of only 1–5°C are commonest, and sometimes the change is spread over several hours by the passage of a zone of mixed or intermediate air. When the thermal contrast is sharp, this is properly called the *polar front*, as it then usually marks the meeting point of winds from regions of very cold surface with winds originating in the anticyclones in lower latitudes.

The deepest depressions in the subpolar low-pressure zone develop central pressures as much as 50–60 mb below the average for that zone; in consequence these and neighbouring latitudes are characterized by frequent strong pressure gradients, bringing strong winds and gales from various directions around the 'low' centres.

1. The speeds of translation of 'lows' in middle latitudes range from average rates of 500–600 km/day to extreme rates up to about 1500–2000 km/day, the fastest-moving ones being mostly of only small to moderate size, rather flat waves on the front (open warm sectors).

The wind directions are most variable in and near the paths of the centres. On the poleward side of the axis of lowest mean pressure, winds with easterly components prevail – the so-called *polar easterlies* (p. 112): these are somewhat variable and inconstant in the Arctic owing to the great variations of pressure that go on over the polar cap, but almost constant easterly winds are observed around the fringe of the great Antarctic ice sheet. In-blowing of the surface winds towards the region of lowest barometric pressure makes the zone of polar easterlies appear mostly as NE'ly winds in the Arctic and as SE'ly winds in the Antarctic.

Over the latitudes between the subpolar low-pressure axis and the subtropical 'highs' there is a great prevalence of westerly surface winds – known to the old sailing ships as the *Brave West Winds* – in middle latitudes. The strength of these winds in latitudes 40–60°S, near the deep depression centres over the Southern Ocean, gave rise to the names *Roaring Forties* and *Screaming Fifties.* The wind directions swing between SW and NW as each depression passes, and occasionally vary more widely; but the in-blowing component towards the low pressure region makes SW'ly and WSW'ly winds most prevalent at the surface over this zone in the northern hemisphere and NW'ly and WNW'ly winds in the southern hemisphere.

Blocking patterns, and anticyclones developed in the colder air at the entrances to jet streams or before confluent troughs and ridges in the upper westerlies, occasionally reverse the prevailing winds in middle latitudes. They are particularly common over Scandinavia, Greenland, Alaska and northeast Siberia: this is at least in some cases because of contortion of the upper wind flow near those regions due to the tongues of warm surface extending far north (the oceans in winter and heated continents in summer) just beside extensive regions of very cold surface in the same latitude.

Thus it happens (especially in the northern hemisphere) that, in the same latitude zone that is most frequented by the subpolar low-pressure systems, blocking anticyclones sometimes produce quite different weather, and they may be very persistent. Long spells of weather, of either cyclonic or anticyclonic type, or characterized by situations in which a given place finds itself in the sweep of the prevailing westerlies or alternatively of the polar easterlies, are therefore quite common in northern temperate latitudes – very much as the polar anticyclones and more occasional cold cyclones may give long spells of one character or another in high latitudes.

With blocking, E'ly winds may prevail for weeks at a time in parts of the zone between 45° and 60°N. This type of anomaly is rare, and in some sectors almost unknown, in the temperate zone of the southern hemisphere. This affects the average pressure distributions seen in figs. 3.17 (*a*) and (*b*) as well as the anticyclone frequencies shown in fig. 3.16.

The characteristically changeable weather of middle latitudes is related to the distribution of rain, snow and cloud-belts which accompany the travelling frontal cyclones (fig. 3.22). All these features are transferred to other latitudes north and south when the subpolar zone (or part of it) is occupied by a blocking anticyclone.

C. *Subtropical high-pressure belts*

The zones of highest average pressure near 30°N and 30°S (figs. 3.17, 3.18) are produced by the presence of the anticyclones continually generated and regenerated along the warm flank of the mainstream of the upper westerlies. The high-pressure belt is present at all heights in the troposphere, though some degrees nearer the equator aloft, and is recognizable up to heights well above 20 km. Its quasi-permanent existence accounts for the steadiness of the *Trade Winds* which blow at the surface from the anticyclone belt towards the equatorial zone of low pressure – generally NE'ly winds in the northern hemisphere and SE'ly winds in the southern hemisphere.

In warm latitudes high barometric pressure most readily occurs over the oceans, owing to the relatively lower temperatures and greater density of the surface air there. Also the lesser surface friction protects the accumulating mass of air in the region of anticyclone formation from dispersal by cross-isobaric flow towards neighbouring zones of low pressure (cf. p. 80). Regions of relatively lower pressure occur as cell divisions between the anticyclones in the subtropical zone, partly owing to land–sea differences of heating but mainly in association with the cyclogenetic tendency at the warm side of the confluences (jet entrances) in the flow of the upper westerlies – and therefore liable to progress, sooner or later, eastwards around the subtropical zone when each anticyclone cell is dislodged (with or without replacement by a new cell). The southern hemisphere subtropical anticyclones are generally kept somewhat mobile by the breadth and strength of the circumpolar westerlies aloft, so that *intercellular fronts* quite commonly cross the southern continents and may give intervals of rain in the subtropical zone.

Larger-scale interruptions of the anticyclone belt do occur at times of large-amplitude waves (meridionality) in the meandering upper westerlies and especially when blocking results from the increase of amplitude. At such times, the main surface anticyclones may be centred in 50–70°N (more rarely and briefly in 50–70°S) and slow-moving cut-off low-pressure systems, cold lows, typically about 1000–1500 km across, persistently occupy parts of the subtropical zone. The Trade Winds may then fail in the sector affected for days or weeks while the situation lasts. Cold lows are quite common in the Azores–Madeira region and near Hawaii: similar systems are sometimes observed in the South Atlantic near South America and St Helena and east of New Zealand, occasionally also in the Tasman Sea.

The weather in the zone of subtropical anticyclones is dominated by the subsidence of the air below the level of maximum wind; this means very little cloud and strong sunshine (except (*a*) where the warm air is advected over locally cool sea, e.g. in regions of upwelling cold water, to form fogs and low cloud, and (*b*) near the fronts and cyclonic developments in the cell divisions between anticyclones). Because subsidence is so usual, this is the arid zone, where the general lack of rainfall and strong evaporation produces the world's greatest deserts. It is also a zone of mostly light and variable surface winds, sometimes known as the

Horse Latitudes. Another consequence of the subsidence is the formation of a temperature inversion, the 'Trade Wind inversion', between the dynamically warmed, subsided air from above and the cooler air convected up from the Earth's surface – largely the ocean surface – below. Cloud growth, like the convection, is limited by this inversion, which is usually found at a height of about 1–1·5 km in those parts of the Trade Wind nearest to the centre of the subtropical anticyclone, though it may come down to below 500 m over the coldest upwelling water (e.g. near the Atlantic coasts of northern and southern Africa), and rises to 2–3 km towards the intertropical convergence or equatorial rain zone.

The Trade Winds (like the warm air in the zone of middle latitudes westerlies) are fed by air emerging from the subtropical anticyclones. Humidity increases with distance travelled by the air over the sea, and the vertical growth of the cloud increases with remoteness from the zone of strongest anticyclonic subsidence. The Trade Wind air arrives in lower latitudes over the sea as air that is cooler than the under-surface: it is therefore undergoing heating from below, is unstable for vertical convection and cumulus clouds are characteristic. These get bigger towards the equator, i.e. with increasing distance from the anticyclone, and give occasional showers there. The instability is liable to become concentrated in troughs in the upper easterlies (*easterly waves* (RIEHL 1954)), which travel westwards along the equatorial flank of the high-pressure belt, producing severe convection wind- and rainstorms.

D. *The Doldrums*

The Trade Winds from either hemisphere meet in a zone of relatively lower pressure, the *intertropical convergence* or meteorological equator. The convergence of these two airmasses, both unstable for vertical convection, gives rise to the heavy equatorial rains. The cloud systems, and the convergence which feeds the violent vertical currents and thunderstorms, are liable to be organized along one or two long bands, continuous over great distances, which sometimes can be traced through from an earlier history as fronts in the higher latitudes of either hemisphere (LAMB 1957). Some of these have passed, as intercellular fronts, through the subtropical zone (in the process of rejuvenation of the anticyclones there by development of new cells); and, although the contrasting thermal origins of the airmasses on either side of the polar front cease to be significant after a long history of travel into latitudes of nearly uniformly strong surface heating, the vertical motion at the old airmass boundary seems to remain in being as a locus of organized convection towards which the surface winds converge.

The strong convection over the warm ground and seas in low latitudes continues up to a tropopause that is higher than in other latitudes. Between the subtropical jet stream and the equator the average height of the tropopause is about 17 km, and the highest cumulonimbus clouds succeed in towering up to about 20 km.

Tropical cyclones, known in the Atlantic as *hurricanes* and in the western North Pacific as *typhoons*, are cyclonic circulations of medium size (commonly of the order of 1000 km across) but often of great violence,[1] which occasionally develop between latitudes 10° and 20°, when the zone of intertropical convergence has moved far enough to one side of the geographical equator for the Coriolis force ($2\rho\omega V \sin \phi$ in the gradient wind equation) to become important. This means that the horizontal windstreams at the surface, feeding the strong vertical convection currents in the convergence zone, become deflected by the rotation of the Earth and pass in a cyclonic whirl around a low-pressure centre instead of directly meeting there. The ocean surface must also be very warm, and the air's moisture content correspondingly great, for these violent storms to form: sea surface temperatures of 27°C or over seem to be required.

Tropical hurricanes may never have been appreciably more frequent than in epochs like the present. In the sixteenth and seventeenth centuries A.D., which we shall later identify as a period of rather colder climate than now over most of the world, West Indian hurricanes seem to have been fewer than in this century. In the ice ages the intertropical convergence hardly had freedom to wander far enough from the geographical equator for these storms to form, except in the Pacific (where conditions were much less different from now). And in the warmer climatic epochs, when there was little or no ice in the polar regions, horizontal temperature contrasts were probably insufficient to produce the required degree of vertical instability in the airmasses moving equatorwards over the sea – although increased water vapour input into the air may have made up for this.

All the 'breeding grounds', where tropical cyclones are born, or begin to grow from initial disturbances generated over Africa, Mexico, etc., are over the oceans: so the moisture input is obviously important. And it is also over the oceans that these storms, as they drift westwards with the moderate upper east winds over those latitudes, develop their greatest strength. Over land, despite the formidable destruction wrought by the winds, ground friction soon reduces the vigour of the circulation. A world map of present-day tropical cyclone activity, with characteristic tracks and annual numbers over each ocean, is given in fig. 3.26 (adapted from TANNEHILL 1956).

In several sectors – notably near the American North Atlantic seaboard, near Japan, near Madagascar and eastern Australia – tropical cyclones are sometimes steered into the surface pressure troughs which divide the anticyclone cells in the subtropical high-pressure zone. In these regions, the tropical cyclone is liable to induce a wave disturbance on the polar front; then the existing tropical storm acquires all the fresh energy of the developing frontal wave, its path 'recurves' polewards and eastwards, as it becomes engaged with the sweep of the upper westerlies of middle latitudes in the form of a frontal cyclone of bigger than usual size and energy.

Tropical storms most frequently develop in summer, when the equatorial trough makes its farthest advance into the hemisphere concerned. They are most liable to recurve, and form into a frontal cyclone of added energy, in the late summer and early autumn, when the

1. In a hurricane of extreme violence which struck Haiti in September–October 1963, a mean wind speed of 50 m/sec was sustained for some time near the surface and gusts of up to 70 m/sec were measured.

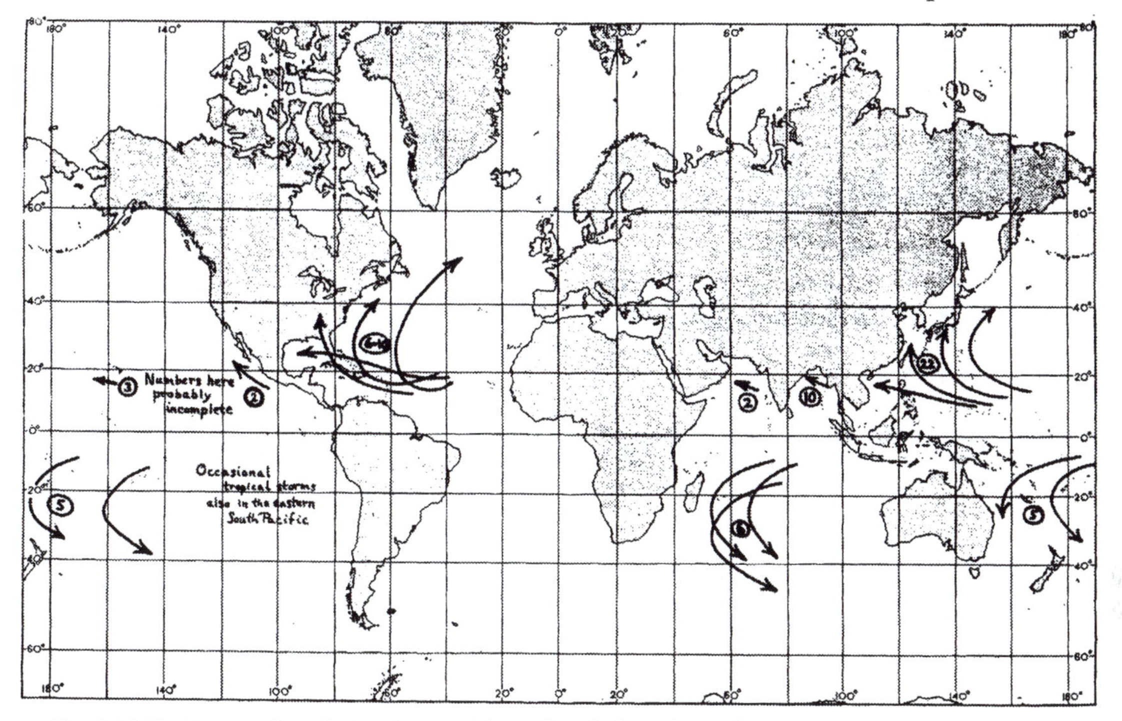

Fig. 3.26 World map of tropical cyclone activity and typical numbers of storms per year over each sea area in the first half of the twentieth century.
(*Adapted from* TANNEHILL *1956 in the light of discussion by* GABITES *1963.*)

polar cold air begins to thrust tongues forward into lower latitudes. In the present epoch there is one ocean over which tropical hurricanes never form: this is the South Atlantic, and the reason is probably that the shape of the continents on either side makes it improbable that the intertropical front will ever pass much south of the equator in the ocean sector – its farthest southward advances in the southern summer are over the neighbouring heated continents.

The Trade Wind zone most of the time has the most persistent weather type of any in the world, generally dry and sunny even over the oceans, though there are usually some cumulus clouds there; the weather may, however, be disturbed for a few days by an easterly wave or the development and passage of a violent tropical hurricane. Over land the Trade Winds spread the desert with its clear skies and scorching sunshine towards the equator – though how far they reach varies with the season as well as depending on the day-to-day variations in the subtropical high pressure belt as individual anticyclones intensify, pass by, decay or are rejuvenated.

List of types of climate

If we take account of the effects of terrain and season, we may recognize the following types of climate at the present day, most of which have a permanent validity though undergoing some shift from one climatic epoch to another:

(1) *Equatorial rain zone*

Always warm and moist, though wetter and drier seasons may be distinguished as the intertropical convergence moves north and south following the zenith sun.

(2) *Monsoon rain climates*

Regions reached by the equatorial rain belt only at one end of its seasonal range and having a dry season in the other part of the year. Temperatures range more widely, especially in the dry season.

(3) *Tropical and equatorial highlands*

Rainfall seasons as for the low ground and seas in the same latitudes, but rainfall amounts depend on slope and distance from the sea. Enormous rainfalls occur on the windward slopes of some mountain ranges. There are also drier regions sheltered by mountain ranges which stand between them and the windward coasts.

Prevailing temperatures decrease with height in the usual way and at some heights (e.g. 1500 m and above) are on average the same as in temperate or higher latitudes but without their characteristic seasonal variation.

(4) *The Trade Wind zone*

Pleasant, sunny, with moderate breezes and often rather cool for the latitude over the oceans, except when rudely disturbed by the occasional tropical cyclones. Windward slopes of the mountains on the islands may suffer prolonged cloudiness, hill fog and drizzle. Over more extensive land this becomes part of the arid zone.

(5) *Arid zone*

Realm of the subtropical anticyclones and Trade Winds over land. The dry climates extend almost to latitude 50° in central and eastern Asia and Argentina, where mountain ranges remove the moisture from the prevailing westerly winds.

In the great deserts rainfall occurs at any one place only very rarely, sometimes in torrential thunderstorm downpours. Evaporation is strong, and light or moderate rain falling from the clouds with the intercellular fronts often evaporates before or soon after reaching the ground.

Some low-lying islands in the Trade Wind zone are also largely desert.

(6) *Subtropical oceans and islands*

These regions are the home of the most frequent subtropical anticyclones; winds are mainly light and variable, temperatures and moisture in the surface air moderate. There is generally a seasonal rise and fall of average temperature of about 5 °C.

Various types of disturbance occasionally upset the regime:

(*a*) cyclonic, cloudy weather with invading fronts from higher latitudes, accompanied by rain or drizzle;
(*b*) more prolonged showery, rainy or thundery, disturbed weather, sometimes very windy and cool, with a slow-moving cold cyclone; or
(*c*) the passage of an occasional tropical hurricane, warm and wet, with violent winds, as it recurves to become involved in the zone of westerlies.

(7) *Subtropical desert fringe* (*Mediterranean climate*)

Warm rainless summers when the region comes within the desert regime. Cool, disturbed, windy, rainy winters with occasional calmer, sunny interludes. In winter the zone is affected by travelling frontal cyclones, developed from the airmass contrasts edging in from higher latitudes.

(8) *Steppes and prairies*

Between latitudes 35° and 50° in continental interiors remote or screened by mountains from the moisture sources of the prevailing westerly winds. The summers are hot, the winters cold or very cold. The summers bring rain in thunderstorms, the winters bring frost and varying amounts of snow.

(9) *Temperate and subpolar oceans and islands*

The zones of prevailing westerlies and of the cyclone centres over the oceans between 40° and 70°N and 35–65 °S are regions of prolonged and frequent storminess. The stress of the wind is the most prominent difficulty imposed by the climate. The weather is very changeable, cloudy and overcast skies are frequent (often accompanying travelling fronts) and rainfall is considerable, especially on windward slopes. The seasonal range of temperature is rather small; but the overall average level of temperature is almost subtropical in the lower latitudes and is low enough for frequent snow and sleet in the high latitudes. Over the warm waters of the northernmost Atlantic, in the Norwegian Sea and in the Gulf of Alaska great cumulonimbus clouds and instability showers and thunderstorms are produced in fast-moving outbreaks of very cold air from the Arctic.

(10) *Temperate* (*northern*) *continents and off-lying islands*

There is a transition from the frequent rains and small seasonal range of temperature near the oceans in the west to drier climates with extremely great seasonal range of temperature in the eastern and northern parts of the continents. The transition is sharp where mountain ranges stand athwart the invading oceanic winds from the west. Maximum summer temperatures differ little around the zone, but the winters are extremely cold in the central and eastern parts of the continents. Changeability of weather from day to day is a common characteristic of the zone. Occasionally, long spells of anticyclonic or easterly weather (blocking) extend the continental climate almost to the western seaboard and may maintain a settled unchanging regime for some days or weeks. Big differences from one year to another also result from the different long spells. Some of the summers are wet, and thunderstorms are frequent in the continental regions wherever moist unstable airstreams penetrate or where there is a local moisture supply. Winter snow depths also depend on an adequate moisture source and generally increase towards the coasts.

(11) *Northern lands*

Similar in climate to the central and eastern parts of the continents in the temperate zone, but the summer in the north is shorter and often more tranquil. High temperatures can be produced in summer by the 24-hour sunshine, if the wind be still and the sky clear. But near the coast of the Arctic Ocean and over the regions of swamp and frozen subsoil, overcast skies are common in summer, and then the temperature remains low.

Autumn, and to a less extent parts of spring, are disturbed periods and considerable falls of rain and snow then occur: autumn and early winter with rain and snow and frequent gales are the most trying time of year for men and animals. In summer and in the frozen winter, long periods of quiet weather prevail.

(12) *The Arctic Ocean*

A variable climate but with less frequent strong winds and gales than over the open ice-free ocean farther south. The peak of the summer from May to July brings much quiet, though often overcast, and sometimes foggy, weather with temperatures by July near or just above the freezing point and pools of water on the ice. The autumn is the most disturbed season with snow and gales. Quieter weather prevails in winter with clear skies and temperatures over the pack-ice only somewhat less severe than on the northern continents.

(13) *The high ice caps in Greenland and Antarctica*

The climate resembles that of the central Arctic Ocean, except that clear skies prevail at all seasons and there is prolonged sunshine in summer though temperatures remain below the freezing point. In Antarctica late winter brings the most disturbed regime. Snowfall is slight, but the light powdery surface snow commonly drifts on any wind exceeding 10–15 knots (5–7 m/sec). Fierce katabatic winds often rage at the edges of the ice cap, and in surrounding valleys and fjords, imposing the harshest conditions for life. In winter the lowest surface air temperatures in the world are produced when the air is still and the sky clear over the nearly level ice plateau, but the temperature may rise abruptly by 15–30°C when the wind increases or cloudy skies arrive.

(14) *Mountain climates in middle and higher latitudes*

Average temperatures are generally lower and amounts of rain and snow greater than on the neighbouring low ground. The seasonal range of temperature changes little with height. Nevertheless, there are variations. In winter temperatures are sometimes higher on the upper levels of the mountains, especially in continental regions, above the low-level temperature inversion produced by surface cooling of any extensive lowlands near by: at such times the higher hills and mountains rise above the low cloud, fog and haze trapped by the inversion. This may result in longer average sunshine in winter on the heights than on the low ground, as in central Europe. (Average sunshine figures for the Austrian Alpine summit stations Sonnblick (3106 m a.s.l.) and Obir (2044 m) from 1891 to 1950 were approximately double those for Vienna in December and January.) Much more rain and snow falls on the windward than on the leeward slopes of the mountains (except just near the summit); cloudiness is also less and sunshine more in the lee of a mountain range. The uplift of air caused by a mountain mass increases cloud development and precipitation from the rising air. The maximum rainfall does not always appear just at the ridge: in some situations the heaviest falls slightly overshoot the summit, but on great mountain ranges there seems to be commonly a height zone of maximum downput of rain and snow before the ridge – e.g. in some places in the Alps and more widely in Asia about the 1500 to 2000-m level (see Chapter 9, pp. 372–3). This may be partly an outcome of the interplay of local mountain-slope (anabatic and katabatic) wind systems with the general windstream approaching the mountains, but is perhaps mainly connected with the lesser water-vapour capacity of the colder air at greater heights.

CHAPTER 4

Seasonal changes

The large-scale wind circulation is subject to more or less regular seasonal shifts and changes of strength of its main features in response to the great changes of the heating pattern that accompany the migration of the zenith sun between about 23°N and 23°S. The circulation is also subject to apparently irregular shifts and changes on various time-scales: some of these account for the distinctive character of particular summers and winters or other seasons; others of much longer duration mark the character of a whole climatic epoch. In this chapter we shall study the tendencies that arise in the normal seasonal round of any year.

To see how the atmosphere responds to the great regular seasonal changes of the radiation budget is an essential first step before we can hope to understand the differences between one year or one epoch and another, which evidently correspond to much smaller changes of the radiation balance and heat input from the oceans into the atmosphere. Despite evidence, which must be examined in a later chapter, of disturbances in the atmosphere that appear to be responses to solar flares, faculae or sunspots, the more obvious changes in the heat input to the atmosphere when the same season in different years or epochs is compared seem to be associated with

(i) the thermal condition of the Earth's surface – e.g. due to cumulative effects of heat storage in the oceans over many years, or through redistribution of this heat by the vagaries of the (wind-driven) ocean currents and pattern of upwelling over some weeks or from year to year;
(ii) surface albedo changes, due to differences in the extent or distribution of snow and ice;
(iii) atmospheric albedo changes, due to different cloud amount or distribution, and possibly some attributable to dust, sea-salt and other impurities;
(iv) variations in the transparency of the atmosphere due to dust (notably volcanic dust), mist, sea-salt, smoke or man-made pollution, or to other variable constituents (CO_2, O_3, etc.).

First, let us see how the seasonal changes in the wind circulation appear to be related to the radiation supply, while these terrestrial variables are initially disregarded, and then when the more or less regular, or average, seasonal conditions of (i), (ii) and (iii) at the present day are taken into account.

Latitude and strength of the mainstreams of the circulation

The seasonal changes of radiation available in different latitudes at the present epoch are displayed in fig. 2.8. A very generalized view of the radiation actually received and its effects on the warmth (1000–500 mb thickness) of the overlying atmosphere is seen in fig. 3.1. Fig. 2.8 makes it clear that the greatest range between equator and pole in the available heat supply from the sun occurs about the March and September equinoxes. In mid winter in either hemisphere the overall radiation supply difference is rather less and is diplaced equatorwards, all of it between the equator and the polar circle. In mid summer the daily heat supply is nearly uniform from the tropic to the pole, and is actually greatest at the pole, so that the temperature gradient would be reversed but for

(i) the insolation wasted by reflection from the residual snow and ice, which are present in too great quantities to be melted, and
(ii) the impossibility of heating the ocean water in those latitudes sufficiently in the course of the season.

Dry land surfaces in high latitudes do, in fact, become very warm under conditions of 24-hour sunshine from cloudless skies provided there is no wind from neighbouring icy seas or glaciers. The maintenance of the poleward gradient of temperature in mid summer and of a ring of circumpolar westerlies, which in the southern hemisphere remain strong all the year round, is fundamentally due to the persistent cold water and ice surfaces.

From late July onwards in the Arctic, and from late January onwards in the Antarctic, however, the sun is too low in the sky to supply much heat per unit area in the highest latitudes, and cooling of the polar region once more contributes to building up an increasing poleward gradient of temperature.

Figs. 4.1 (*a*) and (*b*) show the main zonal features of the average wind circulation around the world in January and July in the present epoch: these are cross-sections from pole to pole through the prevailing westerly and easterly winds, from the surface up to 30 km height in the stratosphere. The left-hand side of the sections is along the meridian of 150°E near the east coasts of Asia and Australia; the right-hand side is along 30°W, in mid Atlantic. The shaded areas have prevailing E'ly winds, the unshaded areas W'ly. Unbroken solar warming of the ozone layer in the stratosphere over the hemisphere where it is summer does make the pole the warmest place at that level, and so the winds in the stratosphere reverse and constitute an easterly circumpolar vortex in summer. In the lower atmosphere the diagrams illustrate the seasonal displacements and changes of strength of the mainstream of the circumpolar westerlies. The seasonal changes of latitude of the mainstreams of the atmospheric circulation are far less than those of the zenith sun at noon.

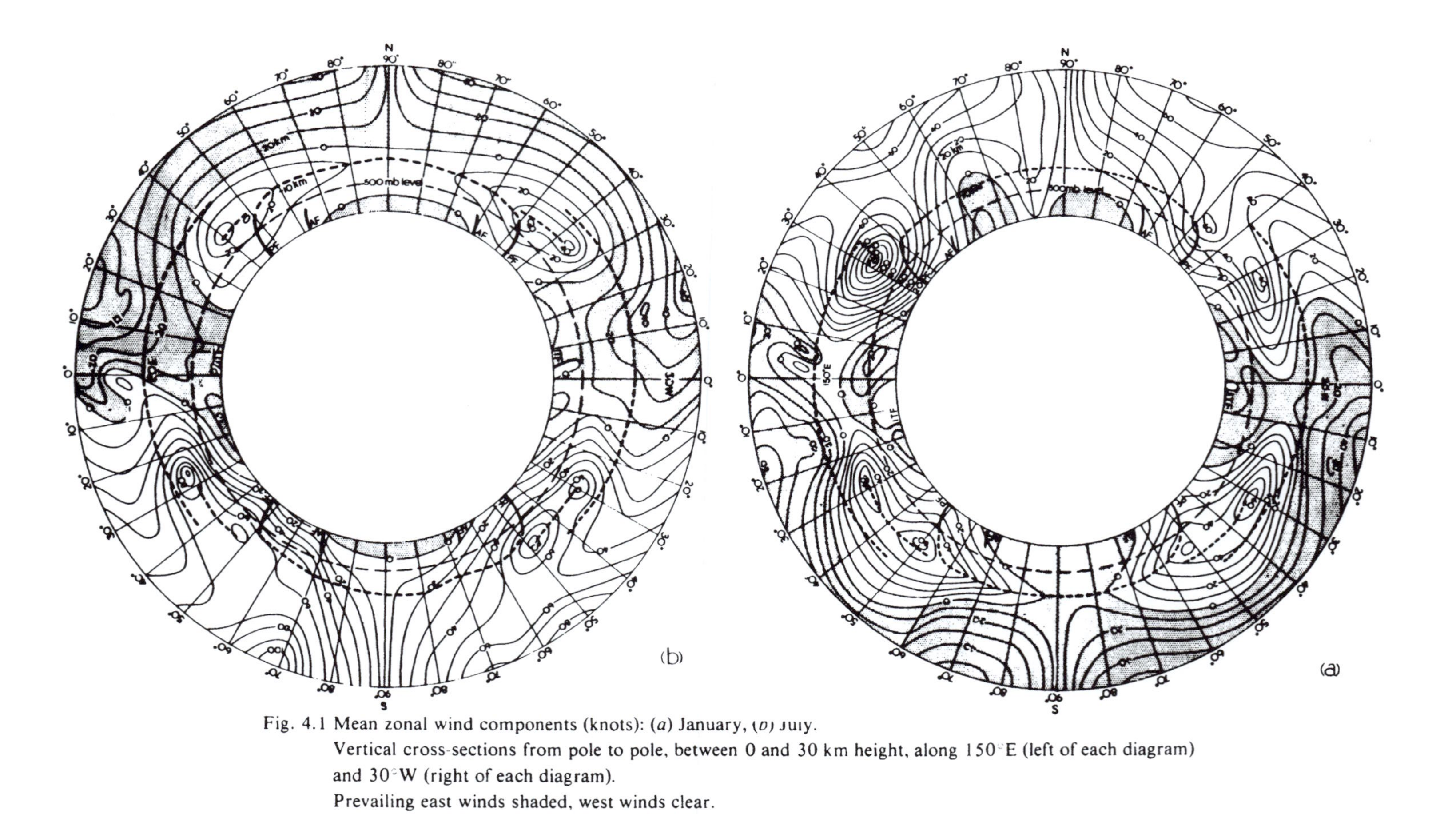

Fig. 4.1 Mean zonal wind components (knots): (*a*) January, (*b*) July.
Vertical cross-sections from pole to pole, between 0 and 30 km height, along 150°E (left of each diagram) and 30°W (right of each diagram).
Prevailing east winds shaded, west winds clear.
(*Reproduced by kind permission of the Comptroller of Her Majesty's Stationery Office.*)

These diagrams, like the best available maps of the upper winds around either hemisphere, indicate that, on balance, the core of the polar front westerlies moves 3–10° of latitude (the smaller figure applying to the southern hemisphere) nearer the pole in summer, as against the winter position, and lies about 8° of latitude outside the limit of snow and ice surface in winter and about 20° outside it in summer. There is in both seasons a second westerly wind maximum at a rather higher level in subtropical latitudes, except where the polar front wind maximum appears to have coalesced with this. These generalizations, however, serve for little more than to indicate the overall order of magnitude of the seasonal shifts. The actual changes are complicated from time to time in individual sectors of either hemisphere by less typical movements and notably by splitting of the vortex at times to give twin bands of westerlies in middle and higher latitudes. Although a poleward shift from January to July of the polar front wind maximum by about 10° of latitude characterizes the

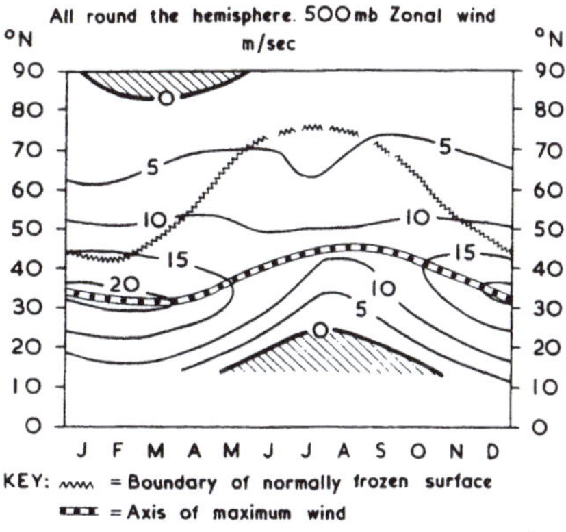

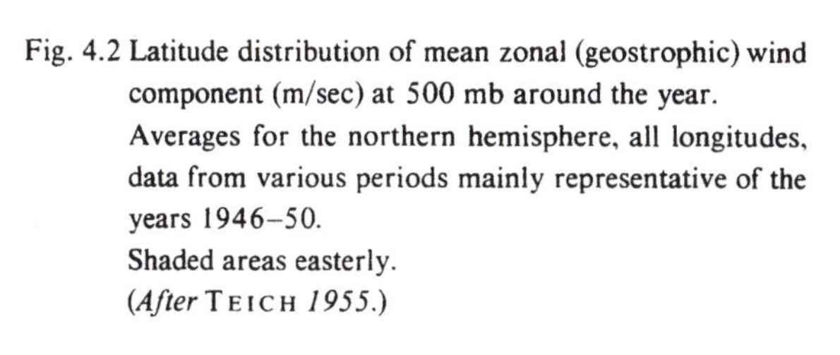
Fig. 4.2 Latitude distribution of mean zonal (geostrophic) wind component (m/sec) at 500 mb around the year. Averages for the northern hemisphere, all longitudes, data from various periods mainly representative of the years 1946–50. Shaded areas easterly. (*After* TEICH *1955*.)

northern hemisphere in general (fig. 4.2), the average shift amounts to 20° of latitude over Asia, where the jet stream 'jumps' from a position south of the Himalayas to north of the great mountain block some time during May and June and reappears as suddenly south of the Himalayas again about late September–early October (*Academia Sinica* 1957, pp. 443–6; REITER and HEUBERGER 1960, p. 29). By contrast, over the North Atlantic between 40°W and the British Isles a contrary shift takes place, so that the core of the polar front westerlies lies on average 1–4° farther south in July than in January: this counter-movement, however, more or less only affects high summer and results from a 'jump' south of the westerlies about mid or late June that may be an adjustment to the Himalayan–Mongolian highlands jump on the other side of the hemisphere. A sharp return to a northern position of the upper westerlies over the eastern North Atlantic and Europe usually takes place between late August and mid September. Thereafter the advance of winter and seasonal intensification of the circulation is accompanied by a more gradual southward displacement of the jet-stream zone in this, as in other, sectors – a general expansion of the circumpolar vortex. The southernmost position of the jet stream, and the main cyclone tracks associated with it, is normally reached at the latter end of the winter, between late January and the end of March.

Fig. 4.2 shows the average strength of the zonal westerly wind and the latitude of the strongest flow month by month at about the 5-km (500-mb) level in the present epoch. These are averages taken around the whole northern hemisphere. The corresponding picture for one particular sector, the middle of the great plains of North America, is presented in fig. 4.3 (*a*) and illustrates the extent to which the strong thermal gradient near the limit of frozen surface, which is sharply defined near 40°N in winter and near 70°N in summer in that longitude, is reflected in the upper wind field. 100°W has been chosen because it is normally somewhere in the middle stretch between a ridge and a trough in the upper westerlies at all seasons. Fig. 4.3 (*b*) shows the corresponding changes in the poleward gradient of surface

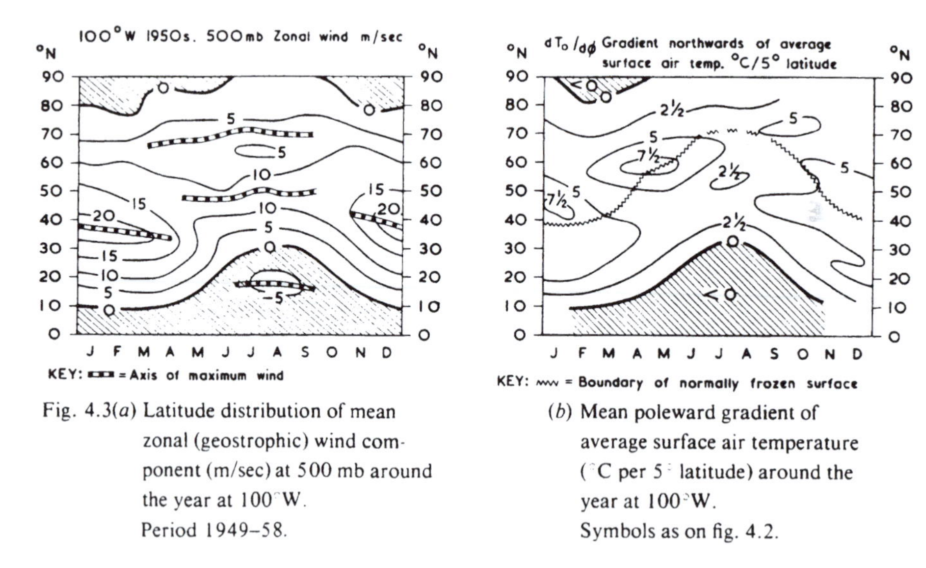

Fig. 4.3(*a*) Latitude distribution of mean zonal (geostrophic) wind component (m/sec) at 500 mb around the year at 100°W. Period 1949–58.

(*b*) Mean poleward gradient of average surface air temperature (°C per 5° latitude) around the year at 100°W. Symbols as on fig. 4.2.

air temperature and normal position of the limit of snow and ice surface in the middle of the American sector. We see that there is an upper wind maximum near, or just south of, the ice limit at each season; but in summer there is another wind maximum farther south, this being the main polar front jet stream and associated with the normal limit of warm, moist air of tropical or subtropical origin. The latter position must be determined by the dynamics of the wind circulation itself and corresponds to the summer position of the maximum wind averaged around the hemisphere, as seen in fig. 4.2.

The greatest strength of the northern hemisphere upper westerlies is attained between December and March when the snow surface extends nearest to the persistent warmth of the tropics. Over the Atlantic–European side of the hemisphere the greatest strength is attained in December or January: thereafter the circulation weakens, doubtless partly because of the meridionality (distortion of the flow into northerly and southerly directions) introduced by the great differences of latitude of the limit of frozen surface over continents and oceans in neighbouring sectors, and partly because the cooling of the seas to their seasonal minimum in February–March means less evaporation and less latent heat injected into the air. Between March and May the hemispheric temperature pattern is again distorted by the rapid heating of the land sectors while sea temperatures lag and the areas of deepest snow and ice (e.g. Labrador and parts of Siberia) remain cold. Weak and rather complex patterns of 1000–500 mb thickness and of upper wind flow result at that season. The month of May sees the weakest zonal wind flow in most years over the sector between eastern North America and Europe inclusive, but July–August is the minimum for the hemisphere in general.

Maps of the southern hemisphere westerlies are hardly yet adequate to discuss the seasonal shifts of the core of the upper westerlies, which apparently amount to only a few degrees of latitude. A study (LAMB 1959) of the main features of the southern upper westerlies, based on 500-mb maps of the period 1951–4 drawn in the South African Weather Bureau, Pretoria, indicated that the angular momentum carried by the system between the 1000- and 100-mb levels – i.e. between 0 and 16 km – through 90% of the mass of the atmosphere was one and a half times as great as that of the northern hemisphere westerlies, largely because the southern westerlies weaken much less in summer owing to the continued presence of a large ice surface in the Antarctic (but also partly because of less friction on the lowest air layers in the hemisphere that is predominantly ocean). The winter strength of the two wind systems was about equal. Winter–summer ratios of the angular momentum of the westerlies were found to be 4·1 to 1 for the northern hemisphere and 1·5 to 1 for the southern hemisphere wind systems (3-month averages, DJF and JJA compared). The seasonal behaviour of the northern hemisphere westerlies in the ice age was presumably more like that of the southern hemisphere today.

The belt of strongest winds becomes broader in winter and extends equatorward in the southern, as in the northern, hemisphere. But many studies (e.g. MEINARDUS 1929, VAN LOON 1965, 1966) indicate that between 40° and 50°S the surface and upper winds tend to be stronger in January (summer) than July (winter). The southern hemisphere upper westerlies have a marked half-yearly component in their seasonal round, the times of maximum development coming about March and September, close to the times when the overall difference of radiation supply between equator and pole is greatest (SCHWERDTFEGER 1960). The main maximum of the southern westerlies is, however, the late winter one when the ice surface reaches its greatest extent towards warmer latitudes, between August and early October.

Over a uniform Earth, whether its surface were all land or all water, the summer–winter range of prevailing temperature should be least at the equator and greatest at the poles. So the poleward gradient of temperature should be greater during winter and the circulation stronger then. Fig. 2.8 shows that the strong difference of radiation supply between latitudes

30° and 60° applies with little change throughout the winter half-year, from equinox to equinox. The overall equator–pole difference of radiation available is, as already explained, actually greatest at the equinoxes, 923 cal/cm^2/day, 6% greater than at the December solstice, 13% greater than at the June solstice, in the winter hemisphere. The influence of residual warmth in the seas at the end of the summer, and of spreading snow cover over the northern hemisphere continents during the course of the winter, suggest that the strongest poleward gradients of temperature should occur in autumn in the higher latitudes and towards the end of the winter in much lower latitudes.

Complications in the seasonal round of circulation intensity changes somewhat similar to those we see today in the southern hemisphere, though by no means identical, may well have affected the northern hemisphere both (*a*) in warm ice-free eras, when there was no ice on the polar seas, and (*b*) in the ice ages, when the thermal contrast between the open lands in low latitudes and the ice-covered sectors in middle latitudes was great in all seasons and may have been greater in spring and autumn than in mid winter. In mid winter, as we have seen (fig. 2.8), with the zenith sun at the southern tropic, the incoming radiation between latitudes 0° and 30–40°N is much less than in spring and autumn.

A study by VAN LOON (1966) has shown how the present geography of the southern hemisphere, and in particular the latitude disposition of ocean and landmasses, produces the tendency to rather stronger winds near 50°S in mid summer than in mid winter. Similar effects may operate in some years and epochs over and near the eastern Atlantic sector of the northern hemisphere, where the latitudinal layout of land, ocean and ice between North Africa and the east Greenland ice is not altogether unlike that affecting the Southern Ocean.

The bold lines in fig. 4.4 show the summer–winter (January and July averages compared) temperature ranges (ΔT) (*a*) for the most continental sector of the northern hemisphere, and (*b*) for the most oceanic sectors (combined into a single average) of the southern hemisphere. The thin line shows **ΔT** for 130°E in the southern hemisphere, the meridian which passes through the middle of Australia: at the latitude of Australia the seasonal range of temperature is seen to behave as it does in mid Asia. Over the ocean zone farther south the range is much less. The two broken lines show the January–July differences of the radiation supply which should reach the surface under a cloudless sky for either hemisphere at the present epoch.

The outcome is indicated in fig. 4.5. In the left-hand diagram the January–July temperature differences are given again for the mid-ocean longitudes and, by the bold line, averaged all round the hemisphere: strong features are the zone of minimal seasonal temperature range over the ocean in middle latitudes and the much greater temperature ranges (*a*) over the subtropical continents, and (*b*) (most of all) over the Antarctic ice. The other curves in the left-hand diagram indicate the poleward gradient of mean temperature per 5° latitude in January and July: between the zone of subtropical landmasses and the ocean in middle latitudes – i.e. between 30° and 50°S – the temperature gradient is slightly less in mid winter than in mid summer. Near the ice limit the reverse is true, and it is this strong

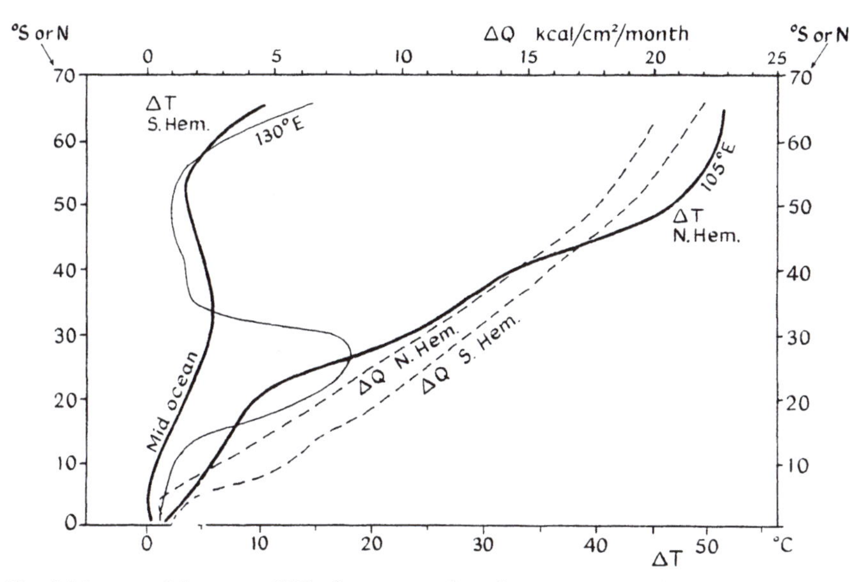

Fig. 4.4 January–July ranges (ΔT) of average surface air temperature, latitude by latitude.
Left-hand bold line, averaged for the middle of the ocean sectors of the southern hemisphere

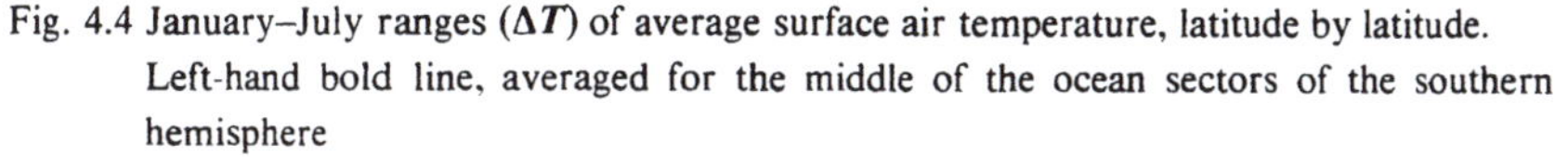

$$\frac{\Delta T_{20°W} + \Delta T_{70°E} + 2(\Delta T_{150°W})}{4}$$

Fine line, at 130°E in the southern hemisphere.
Right-hand bold line, at 105°E (mid Asia) northern hemisphere.
(The two bold lines respectively represent some approximation to the conditions of an all-water and an all-land hemisphere.)
Broken lines, ranges ΔQ January–July of short-wave radiation which would reach the surface under clear skies.
(*Adapted from* H. VAN LOON *1966 and reproduced by kind permission.*)

gradient near the ice which produces the strongest winds of all in late winter. The distribution of prevailing (geostrophic) zonal westerly wind speed in the upper troposphere, illustrated by the 500 mb level in the right-hand part of the diagram, corresponds well to these differences of thermal gradient. The zone of strong winds is much broader in mid winter than in mid summer, and there is a tendency for twin bands of strongest wind at that season, near the ice edge and near the tropic, but over the ocean in middle latitudes the winds are stronger in January (i.e. in high summer). The latitude of the strong winds in summer is similar to that of the polar front wind maximum in the northern hemisphere (figs, 4.2 and 4.3).

To see what factors produce the small seasonal range of temperature over the ocean near 50°S, VAN LOON invokes an empirical relationship found by BUDYKO (1963):

$$Q_t = Q_d + q = Q(1 - an - bn^2)$$

Q_t is the total amount of short-wave radiation actually reaching a unit horizontal area at the surface under average monthly mean cloudiness, n. (n is expressed as a fraction.)

Q_d is the direct radiation } reaching the surface
q is the diffuse radiation }

Q is the amount of radiation which would reach the surface under a clear sky. a and b are empirically established constants. The units of Q_t, Q_d, Q and q are kcal/cm²/month. The mean cloudiness, n, was obtained from CLAPP's (1964) analysis of satellite cloud photographs, giving some approach to complete coverage between 60°N and 60°S for one year (March 1962–February 1963). The computed results are set out in Table 4.1. Clear sky conditions would, as expected, give a stronger gradient of radiation receipt (Q) in winter. Actual radiation receipt (Q_t) has a rather stronger gradient over these latitudes in summer, owing to the difference between the cloudiness of the ocean zone and of the subtropical continents in the southern hemisphere.

The foregoing analysis of prevailing temperature gradients and winds in the present epoch indicates that the seasonal changes of latitude and strength of these over either

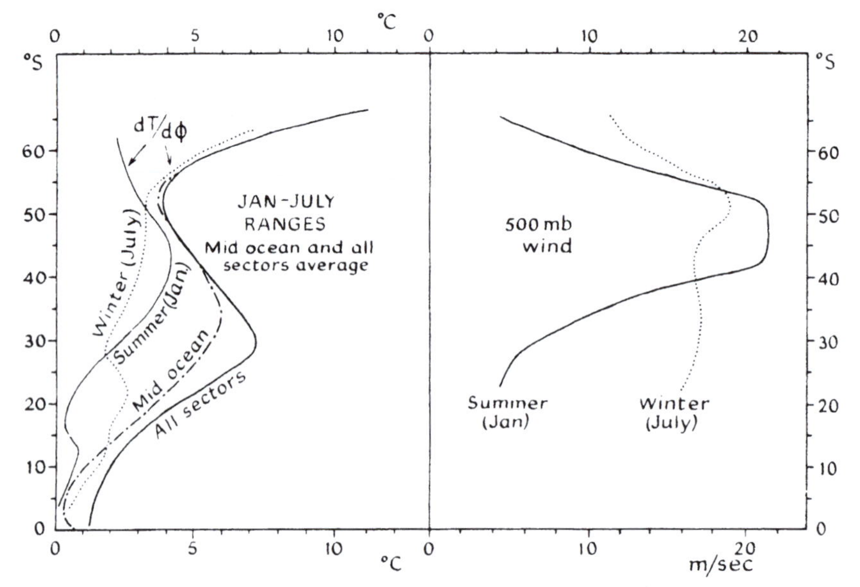

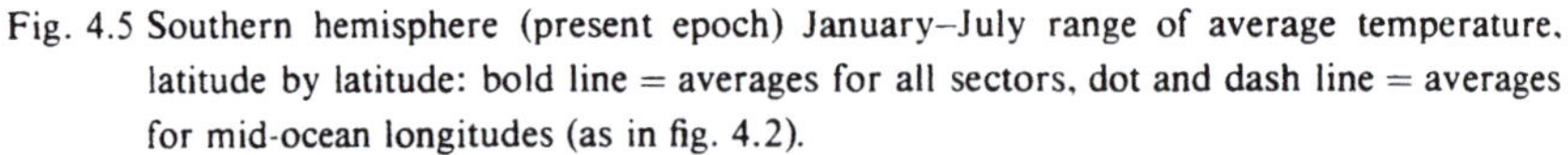

Fig. 4.5 Southern hemisphere (present epoch) January–July range of average temperature, latitude by latitude: bold line = averages for all sectors, dot and dash line = averages for mid-ocean longitudes (as in fig. 4.2).

Poleward gradients of (all longitudes) mean temperature ($dT/d\phi$): °C per 5° latitude: thin continuous line in left-hand diagram = in January, dots = in July.

Average 500-mb level westerly wind components (measured by geostrophic wind scale), continuous line = January, dots = July, are shown, latitude by latitude in the right-hand diagram.

(*Adapted from* H. VAN LOON *1966 and reproduced by kind permission.*)

TABLE 4.1

Radiation reaching a horizontal surface under average cloud cover at 35° and 50°S
(Units kcal/cm²/month)

	January		*July*	
	Q	Q_t	Q	Q_t
35°S	24·0	**14·2**	9·6	**5·4**
50°S	23·6	**9·8**	4·6	**2·1**
Difference 35°–50°S	0·6	**4·4**	5·0	**3·3**

hemisphere are partly, though not very strongly, correlated with the changes in the gradient of radiation supply. Besides this poleward gradient of the incident radiation, the temperature gradients created by albedo differences, due to the state of the Earth's surface and the cloudiness of the atmosphere, and located to some extent by the dynamics of the circulation itself, appear as strong controls. We may list the principal determining factors as:

(i) the distribution of incoming radiation available per day;
(ii) the geography of (*a*) ice and snow, (*b*) dry land and water;
(iii) the geography of cloudiness;
(iv) the limit of tropical air and distribution of release of latent heat of condensation.

Items (iii) and (iv) are largely controlled by the circulation pattern for the time being prevailing – an example of a 'feed-back' process, which may (by its own distribution of gain and loss of radiation and latent heat) either contribute to the maintenance of the wind-flow pattern prevailing or tend to change it. Naturally, the more stable patterns are the ones that characterize the longer spells of weather and, because of their longer duration, these account for most of the long-term averages.

Experiments by FULTZ (1956) with fluids in rotating dishes heated at the edge and cooled at the centre (to simulate the atmosphere over one hemisphere) support the theoreticians' suggestions that the mainstream of the upper westerlies meandering over a rotating Earth is likely to have a mean position somewhere over middle latitudes, the wave length (or the number of meanders) depending upon the magnitude of the overall temperature range and associated circumpolar wind speed. The 10° change of mean latitude of the polar front jet stream between winter and summer, seen in figs. 4.2 and 4.3 (*a*), which is accompanied by a displacement of the zone of highest surface pressure (the subtropical anticyclones) from about latitude 25–30 to 35–40°N, appears to be in reasonable agreement with the magnitude of the shift to be expected (fig. 3.20) by theory from the weakening in summer (roughly halving) of the poleward temperature gradient across the zone of strong winds.

Nevertheless, this very limited seasonal range of latitude observed today in the mainstream of the upper westerlies and features associated with it, when the average limit of snow and ice surface undergoes a seasonal shift approaching 30° of latitude and the thermal gradient over middle latitudes declines greatly in summer, suggests that the positions of the jet stream and subtropical anticyclones prevailing in the warmest ice-free eras may not have been more than 10° nearer the pole than they are now. Over the southern hemisphere temperate zone the seasonal changes of temperature gradient and of latitude of the main features of the wind circulation are both much smaller than over the northern hemisphere.

Seasonal shifts of longitude, wave length and geographical pattern of the circulation

The great differences in the heating and cooling properties of land and sea introduce great differences of surface temperature at the same latitude between the continent and ocean sectors of the northern hemisphere, as may be seen in figs. 4.6 (*a*) and (*b*). In the southern hemisphere, this effect can only arise in the latitudes where the lands are. In the higher southern latitudes the coldest sector is the Indian Ocean sector, where the Antarctic continent and the ice surface bulge out to the north. The influence of the differential heating and cooling of the oceanic and continental sectors of the northern hemisphere upon the thickness of the lower half of the atmosphere – i.e. 1000–500 mb thickness, which varies with the mean temperature of the air column – may be seen in figs. 4.7 (*a*) and (*b*).

The great cold troughs and warm ridges in the thickness pattern largely determine the flow of the winds in the upper troposphere. The average arrangement of the flow of the circumpolar westerlies in winter and summer presents patterns very similar to those seen in fig. 4.7.

The seasonal strengthening of the upper westerlies, and the spread or displacement of the mainstream towards lower latitudes as the winter cooling proceeds, introduce opposing tendencies in the wave length (spacing of the cold troughs and warm ridges). This may be understood by referring to fig. 3.5. Greater strength makes for longer wave length, a shift to lower latitudes tends to shorten it. How this works out in the upper westerlies at the 5-km level in the present epoch is shown in figs. 4.8 and 4.9, which give the average 500-mb heights in January, April, July and October in the 1950s around the northern and southern hemispheres. (Despite the sparsity of observing stations, there is no reasonable doubt about the average lie of the major features at 50°S, here derived from the Pretoria daily maps (VAN LOON 1961).)

Fig. 4.8 indicates near-constancy throughout the year of a ridge-deformation of the upper westerlies at the Rocky Mountains – the main mountain peaks near 50°N are between 121° and 114°W – and of the cold trough over eastern North America in their lee. This has often been taken to prove the primacy of dynamical influences upon the wind flow. But the thermal nature of the surface of the North American sector also plays a part: in summer, despite the weakening of the westerlies, the cold trough lies somewhat farther east than in winter over the region that is kept cold by the chilly waters surrounding northeast Canada, and the greatest warmth brings the crest of the warm ridge just east of the Rockies over the Great Plains and prairies around 110°W. The southern hemisphere westerlies, as depicted

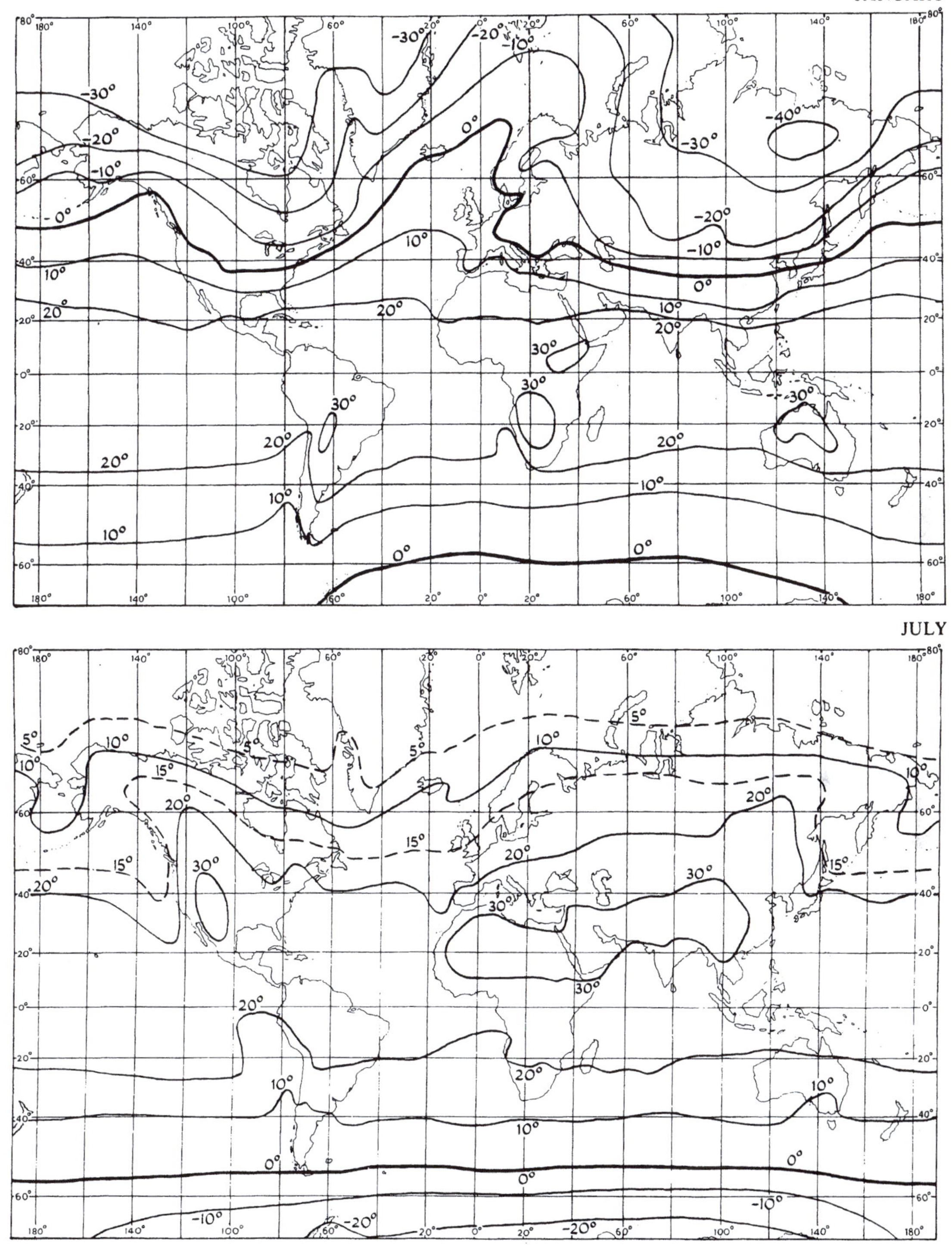

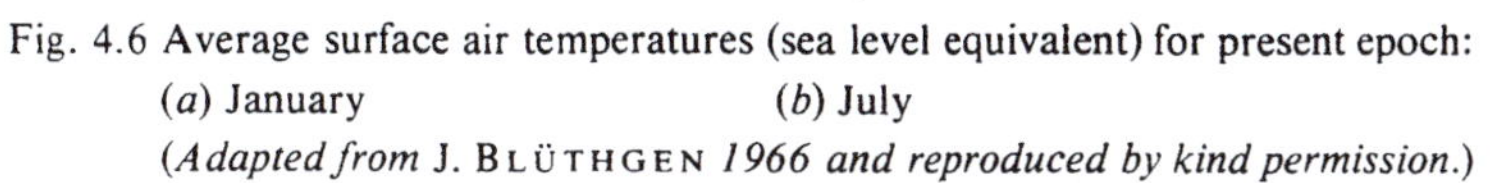

Fig. 4.6 Average surface air temperatures (sea level equivalent) for present epoch:
(*a*) January (*b*) July
(*Adapted from* J. BLÜTHGEN *1966 and reproduced by kind permission.*)

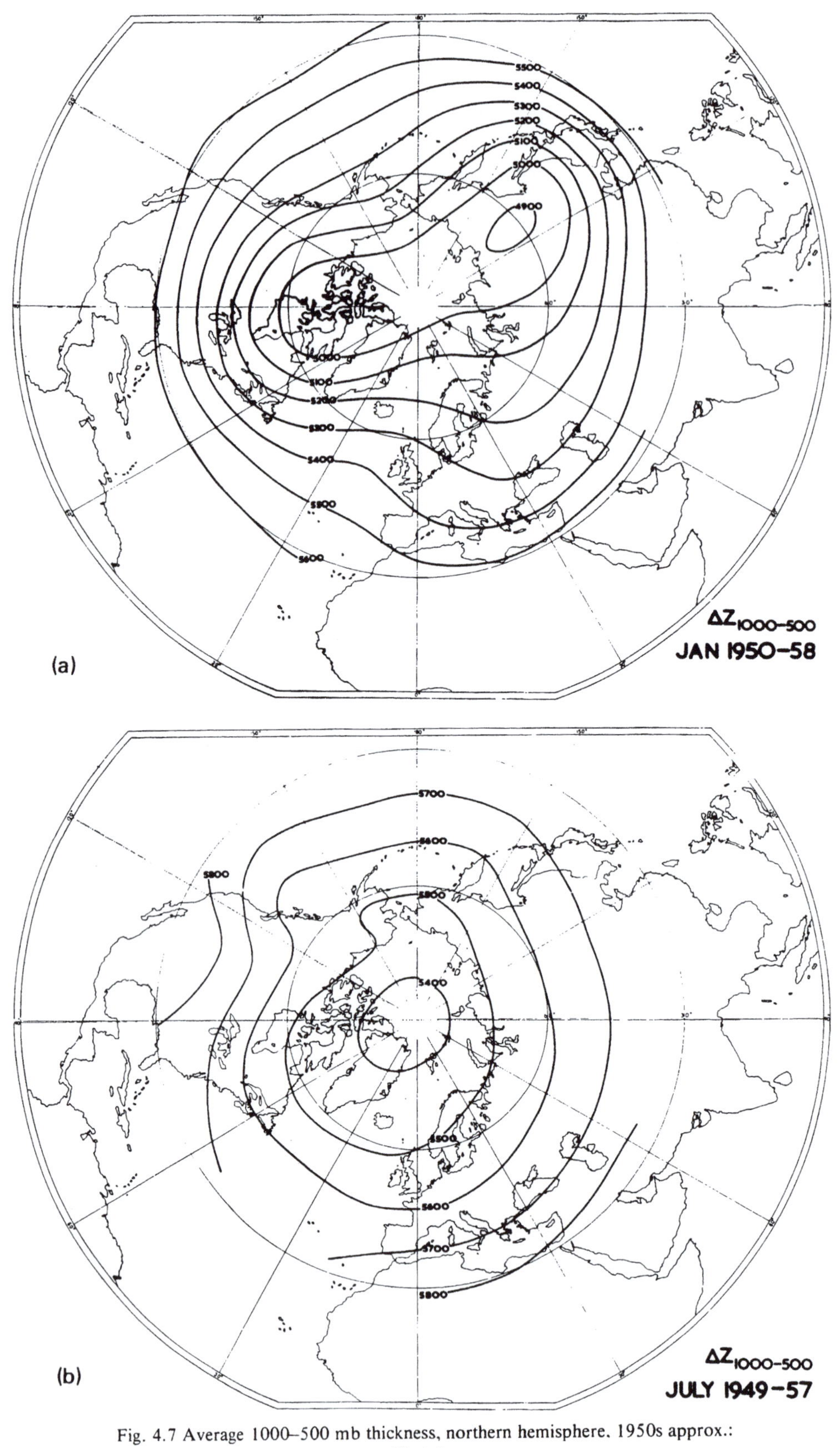

Fig. 4.7 Average 1000–500 mb thickness, northern hemisphere, 1950s approx.:
(*a*) January (*b*) July

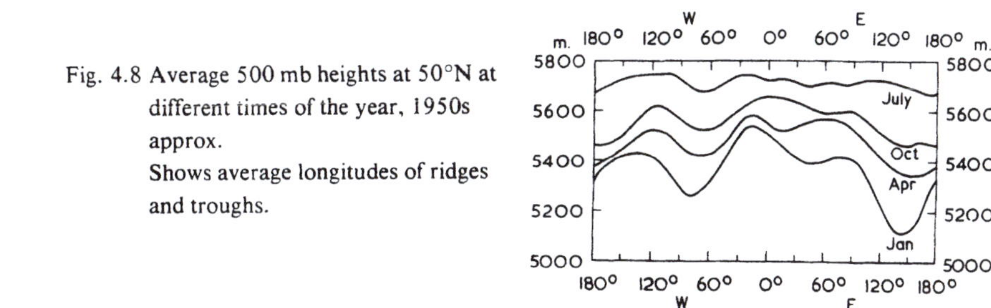

Fig. 4.8 Average 500 mb heights at 50°N at different times of the year, 1950s approx.
Shows average longitudes of ridges and troughs.

in fig. 4.9, show that the ridge and trough in the sector between about 90° and 30°W attributable to the Andes are of less importance than their northern hemisphere counterparts, doubtless because of the surface being mainly ocean both before and behind the barrier but also because the South Atlantic trough is somewhat overshadowed by larger features farther east. The broadest trough in the southern hemisphere westerlies is in the Indian Ocean sector and corresponds to the position of the greatest extent of the Antarctic ice surface; in the long run it amounts to an eccentricity of the entire circumpolar vortex towards that side of the hemisphere. This great trough in the Indian Ocean sector and the broad ridge which marks a more southern position of the mainstream of the upper westerlies across the South Pacific (where the ocean extends farther south) are the most nearly anchored and persistent features of the southern hemisphere vortex.

Figs. 4.10 (*a*) and (*b*) enable us to follow the preferred positions of ridges and troughs in the upper westerlies at 50°N around the year in more detail. These diagrams analyse the percentage frequencies of occurrence of (*a*) ridges, (*b*) troughs near each 10° meridian in each pentad (successive 5-day period) of the year for the years 1949–64. We see the great preponderance of ridges near the Rocky Mountains and at 90–100°E, the latter dictated by the mountains of Asia (though it is generally of small amplitude in winter). There is an almost equal preponderance of cold troughs over eastern Canada (80–50°W) and east Asia (120–150°E), though the latter becomes a minor feature in the heat of high summer and partly displaced out over the cool northwest Pacific. There is also, despite seasonal wavelength changes, more constancy than might have been expected in preferred positions of the secondary troughs. Throughout the year troughs are more frequent at 10°W–0°, at

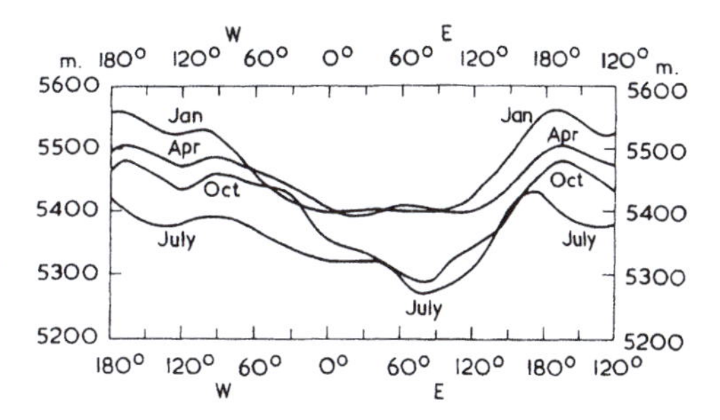

Fig. 4.9 Average 500 mb heights at 50°S at different times of the year, 1950s approx.
Vertical scale larger than in fig. 4.8 because the amplitude of the southern hemisphere troughs and ridges is less than their northern counterparts.

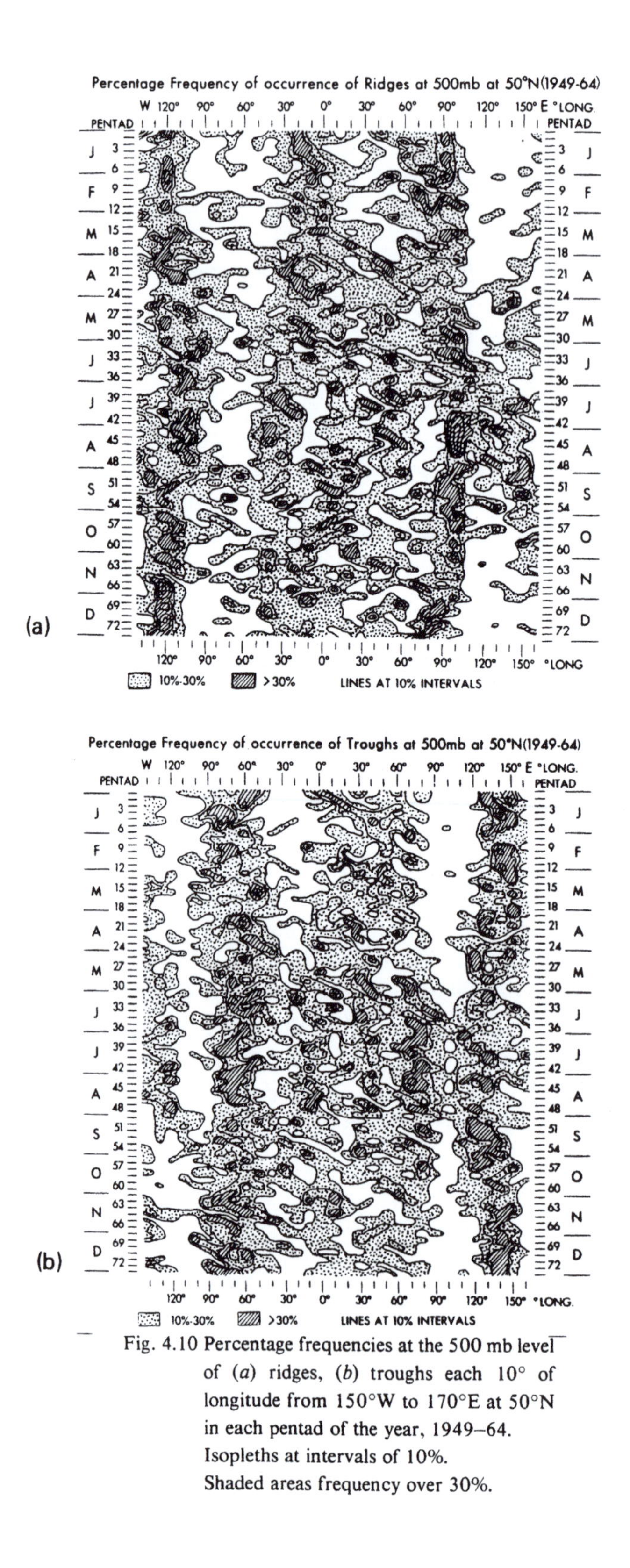

Fig. 4.10 Percentage frequencies at the 500 mb level of (*a*) ridges, (*b*) troughs each 10° of longitude from 150°W to 170°E at 50°N in each pentad of the year, 1949–64. Isopleths at intervals of 10%. Shaded areas frequency over 30%.

30–40°E and at 70–80°E than in neighbouring longitudes: the preferred positions are those most readily reached by cold N'ly winds at the surface, Arctic outbreaks that have come by the broad topographical channels of the Norwegian Sea, the Russian plain and the west Siberian lowlands.

One of the most marked seasonal shifts is the quick eastward march of the Canadian cold trough about March–April, caused by the seasonal warming of the North American continent spreading from the southwest.

The pentad by pentad frequencies around the year of different wave lengths measured at 50°N from the east Canadian trough to the downstream trough over, or near, Europe are similarly studied in fig. 4.11. The seasonal changes in this sector are a fair sample of those affecting the hemisphere. We see that there are seasonal changes of the whole distribution of probable wave lengths corresponding to the changing latitude and strength of the mainstream of the upper westerlies. There are also some shorter-term swings. The longest wave lengths occur with the greatest strength of the circulation in winter, but particularly about the month-ends in November–early December, and December–early January, January–February and February–March. In the autumn the build-up to longer wave lengths becomes prominent from mid September to late October, when thermal gradients are becoming strong but are still concentrated fairly far north around the cooling Arctic. In November there comes a marked discontinuity in this development: for an interval much shorter wave lengths become dominant, associated with blocking patterns and plunges of cold Arctic air at the surface to much lower latitudes (particularly over the continents) leading in the upper air to a much lower latitude of the jet stream and main thermal gradient. Through the principal winter months the wider spread of wave lengths occurring indicates at that time of year greater differences between different years in the strength and latitude of the main flow: long wave lengths prevail in mild winters with strong zonal flow in middle latitudes, short wave lengths prevail in cold winters when the westerlies are displaced farther south than usual. In late winter (February and March) the wave length shortens in most winters, as the strongest thermal gradient reaches its farthest position south and the westerlies weaken with the cooling of the seas (and consequently lessened supply of energy to the atmosphere). In April, May and early June the shortest wave lengths occur with the weakest westerlies on this side of the hemisphere and the most chaotic patterns of thermal gradient (cf. p. 143). At this time, and in winter when the jet stream and thermal gradients are farthest south, meridional swings of the upper wind flow tend to be prominent. In mid or late June another discontinuity of the circulation development affects most years: a shift to longer wave lengths and somewhat stronger upper westerlies near 50°N on this side of the hemisphere are aspects of the appearance of a better organized (less chaotic and more zonal) circumpolar flow, which tends to persist with rather little change through July and most of August.

In the southern hemisphere (fig. 4.9) the only seasonal wave-length changes which stand out from studies so far available are those which affect the spacing of the peak of the warm

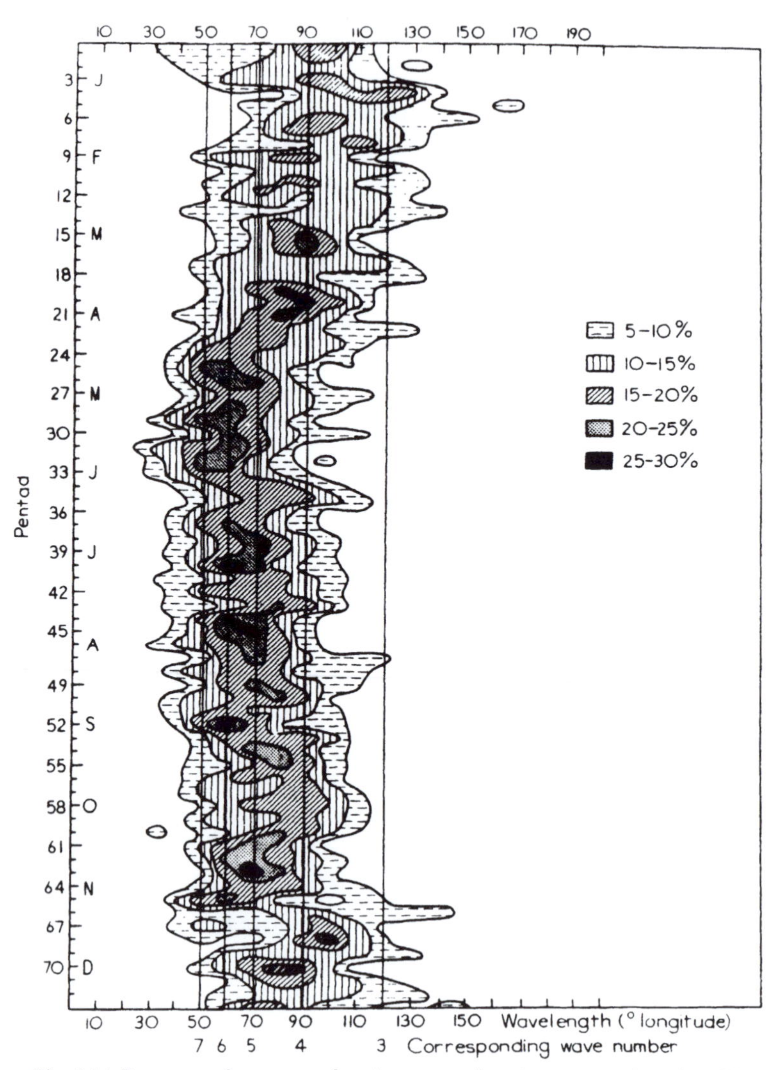

Fig. 4.11 Percentage frequency of various wave lengths, measured at the 500-mb level at 50°N from the Canadian trough to the next downstream trough in the upper westerlies, by pentads (successive 5-day spans) around the year, 1949–64.

Isopleths at intervals of 5%.

Wave lengths in degrees of longitude.

The vertical lines indicate the wave lengths corresponding to 7, 6, 5, 4 or 3 waves around the hemisphere.

ridge in the New Zealand sector downstream from the permanent cold trough in the southern Indian Ocean. Here, we see that the wave length is shortest in mid winter (cf. July positions), when the westerlies near 50°S are relatively weak and the amplitude of the waves (at this sample level) around the hemisphere is rather small. The strength of the westerlies and the wave length both increase from then on and reach their maximum in late winter, about September–October.

The surface regime

The average barometric pressure distribution at the surface in January, April, July and October over the northern and southern hemispheres in, or about, the 1950s is illustrated in figs. 4.12 (*a*)–(*d*), and 4.13 (*a*)–(*d*). These maps cover the entire Earth for about the same period of years as figs. 4.7–4.9. Their main features should therefore stand in an intelligible relationship to one another.

The seasonal shifts of latitude of the subtropical anticyclones and subpolar low-pressure areas are most clearly seen in the ocean sectors and correspond closely to the displacement of the polar-front upper wind maximum, as illustrated in the case of the northern hemisphere by figs. 4.2 and 4.3 (*a*) (pp. 141–2). They amount to only 5–10° of latitude in the northern hemisphere and less than this in the southern hemisphere (though in the Australian sector the summer–winter range is 5–10°). There is moreover a contrary movement in the East Atlantic–European sector of the northern hemisphere which brings the depression tracks farther south again for a season, the so-called high summer, in July and August than they were in May–early June or are in September–October. This, too, corresponds to the behaviour of the jet stream – already noticed on pp. 141–2. Such changes in the circulation pattern inevitably affect surface weather. Thus, the frequency of cyclone centres passing over the British Isles is markedly higher in late July and early August than in the preceding period from about 20 May to early July or in September, when the depression tracks again tend to be farther away. In most districts, but especially in the east and southeast of Britain, the average rainfall and frequency of rainy days in the middle two to three weeks of September are little more than half what they were between late July and late August. At the same time rainfall over the area from Iceland to north Norway shows a correspondingly sharp increase. A rather longer period around late May to early June in Britain has an even drier tendency, partly because the sea temperatures are still rather low and partly because the frequency of westerly winds is less than in either high summer or September. From the point of view of sunshine in Britain, May–June is normally the peak of the summer, especially in the districts nearest the Atlantic. The sudden onset, and more or less regular occurrence about the same date, of the southward displacement of the upper westerlies, accompanying a shift of the centre of the circumpolar vortex nearer the Atlantic sector, doubles the frequency of surface westerlies over the British Isles from 20 June onwards in the present epoch as compared with May and early June; it also brings an increase of cyclonic activity and raininess over most of northern and central Europe from that date on,

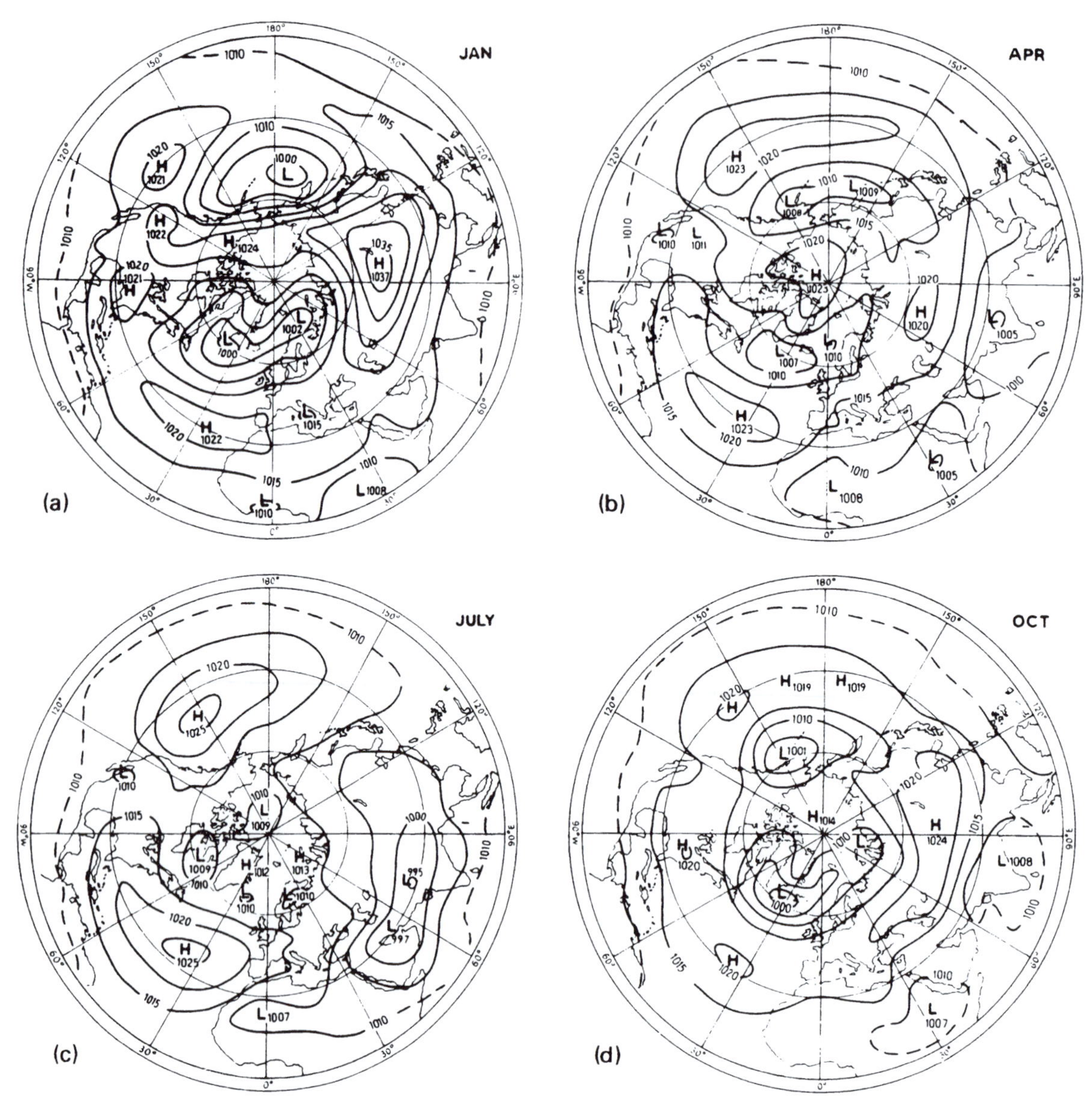

Fig. 4.12 Average m.s.l. pressure over the northern hemisphere, 1950s approx.:

(*a*) January (*b*) April
(*c*) July (*d*) October

Prevailing winds blow anticlockwise around the regions of low pressure, clockwise around high pressure.

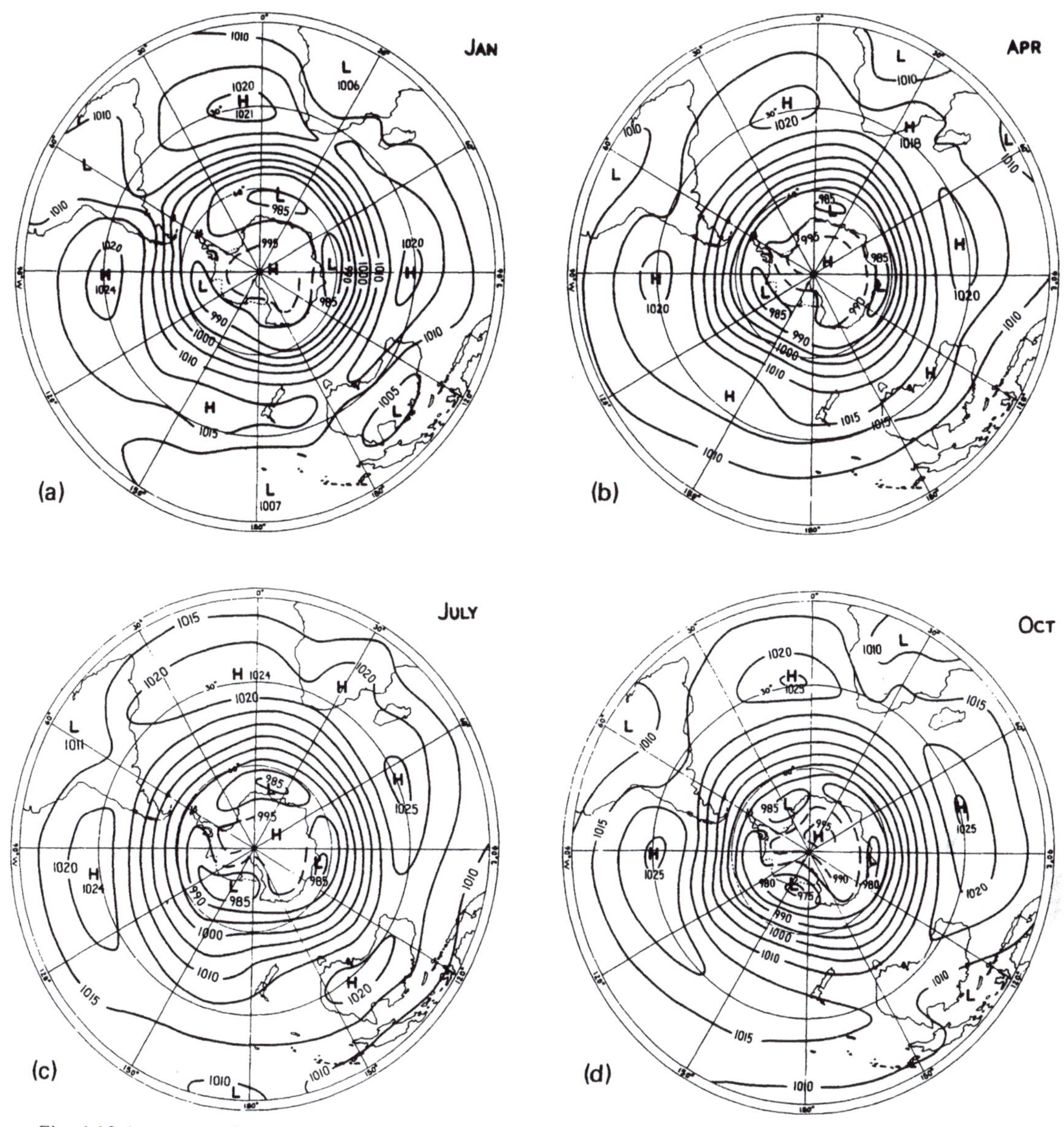

Fig. 4.13 Average m.s.l. pressure over the southern hemisphere, 1950s approx.:

(*a*) January (*b*) April
(*c*) July (*d*) October

Prevailing winds blow clockwise around the regions of low pressure, anticlockwise around high pressure.

so marked as to have acquired the name 'the European monsoon'. Equally obvious links between shifts of the prevailing depression tracks and weather may be looked for in other parts of the world.

The intertropical convergence zone, the belt of relatively low surface pressure which marks the meteorological equator, undergoes a greater seasonal movement – between mean positions, averaged around the globe, about 5°S in January and 15°N in July. This range is enhanced in the continental sectors, so that air flow from the other hemisphere plunges to about 15°S in South America, central Africa[1] and Australia in January and reaches almost 30°N over India in July. In this the Himalayas and the heated tableland of Tibet play a part in almost eliminating any incursions of cold air from the north. The drive of the stronger Trade Winds in the winter hemisphere seems also to be important.

The seasonal changes of strength of the prevailing surface winds and pressure gradients, indicated by figs. 4.12 and 4.13, parallel those described for the upper winds. Particularly obvious are:

(i) the summer weakening of the northern hemisphere circulation. It is weaker still in late April and May over much of the Atlantic and Europe as well as eastern North America, as indicated by the chart of average pressures for April (fig. 4.12 (*b*));

(ii) the strong pressure gradients and winds near 50°S in January. It can be seen that they are slightly weaker in July at that latitude, followed by the greatest development of the southern hemisphere surface wind circulation in late winter, as shown by the pressure differences on the October map;

(iii) the autumn strengthening in the higher latitudes of both hemispheres, seen at 50–70°N in October and around 60°S in April as compared with the preceding summer maps.

We can also trace in figs. 4.12 and 4.13 seasonal shifts of longitude of major features, answering to the wave-length changes and upper ridge and trough displacements illustrated in figs. 4.8 and 4.9 (cf. the discussion on pp. 148–55). Most noticeable are:

(i) easternmost extension of the North Atlantic depression activity into the Arctic in autumn (October map), corresponding to the maximum wave length at that season in the upper westerlies. (This regime however continues until January or after in many mild winters, when the westerlies maintain abnormal vigour and are not displaced as far south as in other years.)

(ii) indications of western positions of the North Atlantic and North Pacific pressure minima in April (and in those Januarys when the wave length in the upper westerlies is shorter than in October–December);

1. In the West African sector, where there is only sea south of about 5°N, the intertropical convergence seldom, or only briefly, passes south of this latitude over the cooler surface of the Gulf of Guinea even in January.

(iii) surface low pressure in high summer close to the north Canadian and northwest Pacific upper cold troughs, when the upper flow is so weak that the wave length is short despite its high latitude;

(iv) development of low surface pressure just east of New Zealand in June–July during the weak circulation and short wave-length phase about midwinter in the Southern Ocean westerlies. (In some winters this development occurs near enough to give notably low pressures in New Zealand and cold southerly winds there and in eastern Australia.)

The other features of these maps which show strong seasonal changes are the pressure systems near the North Pole and those that mark the Asiatic monsoons. To these we must next devote some attention.

The monsoons

'Monsoon' comes from an Arabic word meaning 'season'; but in meteorology and in general English usage it has come to be associated with prevailing winds, and wet or dry weather, which reverse with the seasons. The classic examples are the SW monsoon of southern Asia that brings the summer rains, on which much of India depends for its crops, and alternates with the NE monsoon of winter, which is generally the dry season in the sub-continent (but gives the wettest months in places from Madras to northeastern Ceylon exposed to winds that have crossed the Bay of Bengal).

In winter, radiation cooling increases the density of the lower air over the continents and ice surfaces so that, with the normal dynamical processes and exchanges of air associated with the jet streams going on in the middle and upper levels above the dense surface air layers, the total mass of the atmosphere over these regions increases, especially over Asia and the Arctic. This is seen in the average atmospheric pressures in winter (fig. 4.12) and in the magnitude of the seasonal changes of pressure over the continents and the polar ice (figs. 4.16 and 4.17, and pp. 167–71). In summer the heating of the extensive land surfaces reduces the density of the lower air and the total atmospheric mass over them, especially over southern and central Asia but also over the other continents. However, the pattern of low-pressure development is not explained either over Asia or the Arctic without taking account of the dynamical effects of the main upper wind flow. This is clearly needed to explain the cyclonic activity which normally produces low pressure over parts of the Arctic Ocean, including parts of the pack-ice, from about July to October or later. During most of this time the ice is colder than the surrounding regions. There is generally a polar anticyclone somewhere over the Arctic ice, or over the cold seas, in high summer: its average position in July in the 1950s was between northern Greenland and Novaya Zemlya or the Kara Sea – i.e. most characteristically in the European sector (fig. 4.12 (*c*)) – the position at different times and in different years being related to the centre of the circumpolar vortex and therefore, indirectly, to the lie of the jet stream around the polar cap. The dynamics of the upper flow also control the Indian monsoon low-pressure system and, together with the low density of the warm surface air, produce monthly mean pressures in July as low as in the Iceland or Aleutian low-pressure areas in winter.

On the southern flank of the strongest upper westerlies in April and May the seasonal heating of the land surfaces in lower latitudes produces centres of maximum 1000–500 mb thickness and 500 mb height near, or over, central America, the southern Sahara and near south India (cf. fig. 4.14 (*b*), p. 163): these positions determine confluent patterns of flow in the upper westerly winds over southwestern North America, over the western Sahara and over northwest India–Pakistan respectively. In the southern hemisphere spring, about October, confluent patterns develop for similar reasons near the low-latitude flank of the upper westerlies over the western sectors of South America, Africa and Australia. These confluences induce surface pressure to fall at their warm side (cf. fig. 3.9). Thus, low surface pressures tend to be produced, and maintained through the summer, towards the western sides of Australia, central South America and the southwestern U.S.A. There is, however, also a tendency for cyclonic developments to travel eastwards with the upper westerlies from these regions across the respective continents, at the same time crossing the mainstream of the upper westerly winds towards the higher latitude side, where they intensify at the jet exits: hence, the extension of the areas of prevailing low pressure towards the southeast over South America, South Africa and Australia in January and towards the northeast over North America in July (figs. 4.13 (*a*), 4.12 (*c*)). Over Asia the heating of the much bigger landmass and the elevated plateaux of Tibet and Mongolia introduces a further stage into the development of the summer monsoon, drawing the axis of the equatorial rains almost to 30°N and establishing an easterly jet stream more or less above it.

The SE Trade Winds of the southern Indian Ocean cross the equator and turn, as they go towards higher latitudes again in the other hemisphere, to appear as a SW'ly windstream in the lower air layers, which releases abundant rainfall in the intertropical convergence zone. This whole system moves north as the northern summer develops, but not in a continuous progress: it tends to halt over southern India and later leap forward to northern India. The onset of the SW monsoon, as defined by the rains, is on average about 30 May in southwest India, south of 15°N, but the sweep north to Delhi (29°N) and the Ganges plain does not take place on average until about 2 July (average dates for 1900–50 approx.). Both dates are subject to a standard deviation of 7–8 days.

When the axis of the upper westerlies has jumped, usually during May or early June, to the north of the great mountains of Asia, the heating of the massif can proceed to the point where (by about 10 June) the thermal gradient between 35°N and the Indian Ocean to the south is reversed, the oceanic air being the cooler. The temperature difference between the airmasses at low levels is slight, but in the upper troposphere the oceanic equatorial air is about 10°C cooler than the air heated over Tibet. The northward jump of the jet axis may be delayed in individual years by an unfavourable position of a slow-moving trough in the upper westerlies farther north and by lingering snow cover over the extensive mountain lands of middle Asia (Banerji 1950, Ramage 1952, Ramaswamy 1956); or the jump may be induced to take place early by the contrary conditions. The northward progress of the whole monsoon system over India may also be advanced or retarded by other factors much less understood, among them possibly things affecting the drive of the Trade Winds in the southern hemisphere. Interhemispheric reactions of one sort or another on various time-scales have been noticed in the east Asia–Australian sector. Examples are the 'Southern Oscillation' (WALKER 1924, BERLAGE 1957, TROUP 1965) and the changes of mean latitude of the subtropical anticyclones over Australia in the twentieth century (LAMB and JOHNSON 1959, 1961), though the latter seemed rather to be a response to changes in the vigour of the Asian summer and winter monsoons than vice versa.

It seems clear that the date of onset of the Indian SW monsoon in any given year (BHULLAR 1952, RAMDAS *et al.* 1954), or (better) the date when the jet stream passes north of the mountains of Asia, is symptomatic of that season's development of the wind circulation over a very wide area of the globe (WRIGHT 1967). A further (unpublished) study of the monsoon in the 5 years 1956–60 by P. B. WRIGHT revealed that the northern summers in these years were arranged in almost the same order as regards

(i) the date (earliness) of the northward jump of the jet stream;[1]
(ii) the longitudinal extent attained by the monsoon westerly current over southern Asia;
(iii) the extent and strength of the Madagascar Trade Wind stream in May and June;
(iv) the zonality (westerliness) of the flow over southern Australia about the same time;
(v) the latitude of the subtropical anticyclone in the North Pacific in May–June;
(vi) the meridionality (northerliness) of the flow over the Mediterranean in May and June;
(vii) the wetness and coolness (POULTER index[2]) of the summer in London.

700-mb level flow was used in the definitions of items (ii) and (iv)–(vi). It is believed that not only (vii) but also (iv)–(vi) define anomalies that tend to persist through the following July and August.

1. The jet stream jump can be observed in longitudes from the Mediterranean to China. The date of most rapid northward shift of the 200-mb level winds at 95°E appears most relevant in the connexion here considered.

2. R. M. POULTER (*Weather*, **17**, 253–5, 1962) has proposed a composite index (I_s) to measure the quality of a summer in London:

$$I_s = 10T + S/6 - R/5$$

where the June, July and August observations at Kew are used and

T = the 3-month average temperature in °F.
S = the 3-month total sunshine in hours.
R = the 3-month total rainfall in mm.

POULTER applied his summer index to the observations at London (Kew Observatory) from 1880 to the 1960s, and quasi-periodicities of about 2–2½ and 11 years were apparent in it. Nevertheless, with any such index which includes rainfall, somewhat unrepresentative values may be obtained in some years if sole reliance is placed upon

Once the south–north thermal gradient is reversed over India and southeast Asia, partly by the great amounts of latent heat liberated in the monsoon clouds as well as by the radiation heating of Tibet, the cooler air from the southern hemisphere, cutting in in the lower levels, readily carries the equatorial low-pressure axis as far north as the Ganges plain. The reversed thermal gradient means that the warmest air columns and the axis of a high-level anticyclone lie over, or near, the Tibetan highlands: south of this an upper easterly windstream blows at levels from about 6 km up to over 20 km, involving over 40% of the atmospheric mass over India, throughout the height of the monsoon season. The easterly current, at its fullest development, may be traced all the way from the China Sea to West Africa, and at times exceeds 50 m/sec (100 knots) over India in a jet core. Individual cyclonic disturbances at the surface are carried westwards by the upper easterly current, but they lose most of their moisture before they cross the watershed that divides the Ganges from the Indus valley. They develop and decay according to the same principles as apply in the case of the upper westerlies (KOTESWARAM 1958). Energy derived from the latent heat of condensation of the quantities of water vapour picked up from the tropical seas must play a big part. The frequency of cyclogenesis over the Bay of Bengal at this season is quite similar to that in the other regions of maximum frequency, in extratropical latitudes, over the Gulfs of Alaska and Genoa in winter and over the warm ocean currents in the western North Atlantic and western Pacific at all seasons (PETTERSSEN 1950). Tropical cyclones over the Bay of Bengal and the Arabian Sea, and troughs or 'waves in the easterlies' (RIEHL 1954, KOTESWARAM 1963), are also steered westwards by the upper easterly current.

The characteristic developments of the upper winds sampled at about the 10-km level over south and southeast Asia in winter, spring and summer are illustrated schematically in figs. 4.14 (*a*)–(*c*), together with typical associated positions of the surface fronts which explain the weather. Rather similar maps have been published by FLOHN (1960, 1963), and a fuller study by *Academia Sinica* (1957, 1958) has reviewed the stages of both the spring advance and the autumn retreat of the monsoon. The convergence and frontal activity in the intertropical zone is commonly concentrated along two lines, and two are indicated on these maps.

In winter (fig. 4.14 (*a*)) there is on average a trough in the upper westerlies about 90°E south of the Himalayas and some small depressions, known as 'western disturbances', travel

---

one observing station. It is better to construct averages for several stations within any fairly homogeneous climate area, such as southeast England. DAVIS (1967) obtained interesting results by comparing the behaviour of POULTER's index in different areas of Europe and places around the Norwegian Sea, and in a review of several such indices (1968) has proposed as best a new, slightly different, index:

$$I_D = 18T_x + 0{\cdot}217S - 0{\cdot}276R + 320$$

where T_x is the average daily maximum temperature in °C
S is the total sunshine in hours
and R is the total rainfall in mm
for the months June, July and August.

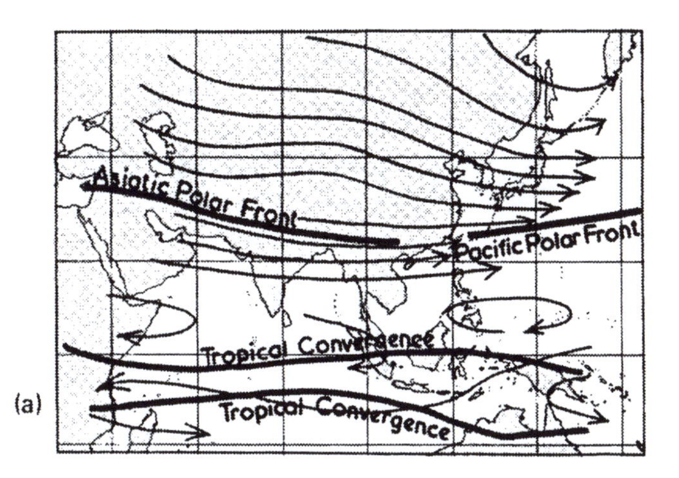

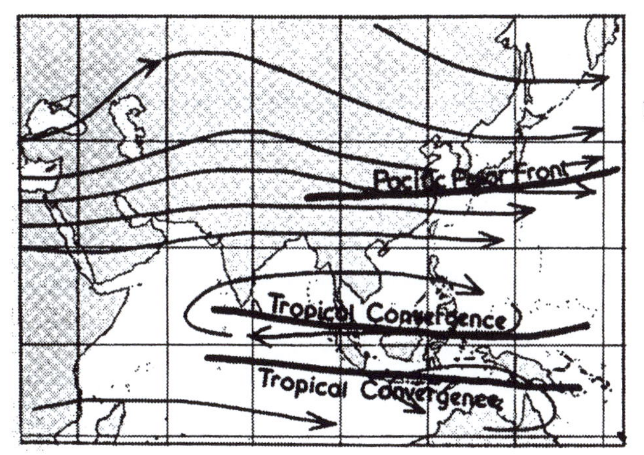

Fig. 4.14 Prevailing stream-lines of wind flow at about the 10 km level and general positions of the surface fronts over south and east Asia:
(*a*) Winter (approx. November–February)
(*b*) April
(*c*) July
(Schematic – 1950s approx.)

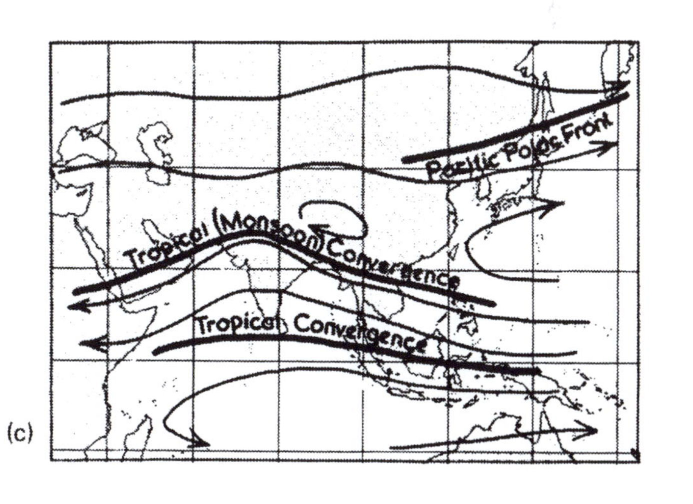

east along the polar front from the Mediterranean. These disturbances generally cause no appreciable precipitation anywhere on the warm side of the jet stream, but they produce a rainy season, 'the Christmas rains' (FLOHN 1963), in the northwestern part of the subcontinent of Pakistan and India, especially in Baluchistan, the northern Punjab and Kashmir. In Afghanistan and Baluchistan these are the principal rains and snows of the year. These disturbances are also the main source of the snow cover on the high levels of the Hindu Kush, the Karakorum and the northwestern Himalayas. The valley bottoms get much less rainfall because of the subsidence which is necessary to supply the daytime upslope breezes of the mountainsides. The rains vary from year to year, on rare occasions reaching as far as 15°N in India, and the snowfall occasionally reaches the valley bottoms in Kashmir and the Punjab. Over China mostly dry weather prevails under the cold side of the confluence in the upper westerlies, which brings forward the influence of the Siberian winter anticyclone.

In the pre-monsoon period, in the northern spring–early summer, India experiences sporadic showery rainfall from convection cloud but also much sunshine and the highest temperatures of the year. Sometimes divergent flow east of a trough in the upper westerlies, while these still blow south of the Himalayas, coming on top of very moist unstable air from the Bay of Bengal, produces severe storms and heavy rains over and near the northern part of the Bay (FLOHN 1963). Tropical storms also sometimes arrive over the Bay of Bengal from the east at this time of the year and are liable to intensify and be turned northwards by the trough in the upper westerlies, giving disastrous floods in Bengal and lower Burma.

When the jet stream first passes north of the Himalayas it becomes sharply defined over central China owing to the thermal contrast between the warm air from the south and the northern airmass supplied by an anticyclone over the cold waters of the Okhotsk Sea. The jet stream continues east over southern, and later central, Japan. This situation, which roughly occupies the month of June, develops to the east of the trough in the upper westerlies which then lies over Szechwan near 105°E: the confluent flow of the upper westerly windstream produces a cyclogenetic tendency over China near the warm side of the entrance of the strong stream. This accounts for the Mai-yü (in Japan Bai-Ü) rains, which are associated with frontal disturbances travelling east over central China and Japan. Later, around 10 July, this frontal activity passes farther north to Korea and north Japan, or beyond, and the high summer season sets in in eastern Asia: the Pacific anticyclone tends to spread in, bringing a period of great heat and dry weather to Japan and coastal China between mid July and mid August (as illustrated in fig. 4.15). At other times in summer, China experiences monsoon rains from convection clouds building in the moist, thermally unstable, equatorial airstream which arrives as a SW wind at the surface in south China and which flows from the S and SE over the rest of China and Japan. At the same season of the year also typhoons are liable to be steered in by the upper easterlies from the Pacific and recurve northwards over China and Japan, as they are carried around the western limit of the Pacific Ocean cell of the high-level anticyclone (cf. fig. 4.14 (*c*)). In those years when the

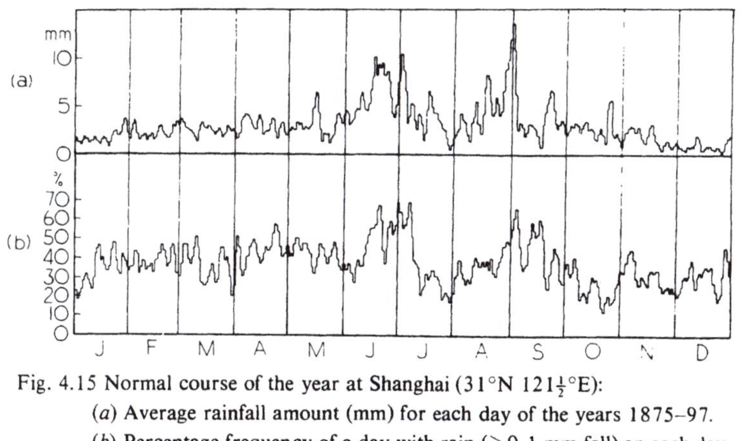

Fig. 4.15 Normal course of the year at Shanghai (31°N 121½°E):
(*a*) Average rainfall amount (mm) for each day of the years 1875–97.
(*b*) Percentage frequency of a day with rain (≥0·1 mm fall) on each day of the year, 1920–39.
(*After* FLOHN *1950, reproduced by kind permission.*)

upper easterlies are established early, the typhoon season starts early and these storms are more numerous than in other years over the Far East.

Over most of India the SW monsoon brings great cloudiness and rainfall, increased by convergence of the surface winds due to the cyclonic curvature of the wind flow. The northwestern area of the sub-continent, including West Pakistan and the region between Afghanistan and the Thar desert (Rajasthan), is an exception: there the flow of the low-level monsoon westerlies coming from the region of Aden and the Horn of Africa is anti-cyclonically curved and divergent, and so subsidence within it continually thins the cloud to broken stratocumulus, which gives no rain. Also, there are districts screened by mountains, e.g. the Ghats, from the prevailing wind: in those districts too the monsoon rainfall is slight. At the other extreme, the windward slopes of the eastern Himalayas and the ranges of Assam continually force the moist monsoon airmass upwards to yield its maximum down-put: thus Cherrapunji, 1300 m up in Assam (25·3°N 91·7°E) receives on average 11 000 mm rainfall in the year, the most regular heavy rainfall anywhere in the world.

Breaks sometimes occur in the Indian monsoon. One way in which this can happen in the present epoch is that occasionally, even in high summer, a trough in the upper westerlies of middle latitudes may penetrate far enough south to breach the Tibetan anticyclone for a few days. This then affects the steering of the travelling surface low-pressure centres coming from the east over northern India and turns them towards the north. Ahead of the tip of the upper cold trough, about 25–30°N, prolonged rain may then deluge the slopes of the Himalayas, while farther south over India the right-hand (anticylonic) curve in the monsoon windstream produces divergence of the surface winds, accompanied necessarily by subsidence and clearing skies (FLOHN 1963, see also RAMASWAMY 1962). Soon, renewed heating over the Tibetan highlands restores the monsoon situation, much as it existed before the cold intrusion.

In the autumn the same frontal belts, which passed north over southern and eastern Asia in spring, return south again. This brings another sequence of frontal depression rains travelling from the west or southwest over China and Japan, but before very long the increasing influence of the Siberian anticyclone is felt, first over inland China and later at times over Japan also. In the northern autumn tropical hurricanes reappear over the Bay of Bengal from the east, and may be steered westwards by the last of the upper easterly winds (now withdrawing southwards) to give rain in eastern India, or they may recurve north and shed their rain against the slopes of the great mountains.

Presumably owing to the constancy of operation of the control exercised by the Himalayan mountain chain over the atmospheric circulation patterns over India and south Asia generally, and indeed the tightness of that control (restricted freedom of flow directions) in the lower half of the atmosphere, there is no other part of the world which shows such small variation of the monthly mean atmospheric pressure and wind patterns at the surface from one year to another; indeed, this applies over the whole period since about 1800 for which barometer observations are available there. Standard deviation of monthly mean pressure for July is less than 1 mb over the central low-pressure region of the monsoon (despite the variable incidence of breaks in the monsoon) and is hardly greater in mid winter over the whole region that embraces Afghanistan and the Indian sub-continent.

Studies by RAO and JAGANNATHAN (1960) suggest that there has been no statistically significant secular change in the monsoon or average yearly rainfall in India, at least from 1870 to 1949, though further examination may be necessary (particularly in Orissa and Andhra Pradesh, between 15° and 24°N on the eastern side of India). Nevertheless, differences from one year to another do occur, and 'failures of the monsoon' mean famine to the densely populated lands of south and east Asia. Wherever there are such important year-to-year differences, cumulative effects of runs of bad years may be still more serious. Moreover, there appears to be no physical reason why the circumstances which produce either a 'dry' or a 'wet' year should not be more frequent in some epochs, giving rise to significant climatic variations. 1953 was a year of active monsoon over India and 1954 a year of weak monsoon. Comparison of these two cases by RAMASWAMY (1958) brings out what appear to be the main controlling factors in the atmospheric circulation. The weak monsoon season shows the same characteristic as the shorter monsoon breaks already mentioned, in that the upper westerlies of middle latitudes remain farther south than normal in that sector and the northern cold air manages to thrust an upper cold trough fairly far south. The characteristics are summarized in the table on p. 167.

The year 1953 which established the upper easterlies early and showed all the characteristics of a strong monsoon over south Asia affected China similarly, and the number of typhoons invading east Asia was higher than usual. 1954 showed the opposite characteristics over China too, and there was much less typhoon activity (*Academia Sinica*, 1958).

It seems clear that the characteristics of a weak monsoon year are those that should be more prominent in epochs of colder climate with an expanded circumpolar vortex, whereas

TABLE 4.2

Upper flow over Asia in strong and weak monsoon summers

Level	*Active monsoon*	*Weak monsoon*
100–400 mb	E'ly jet at 100–150 mb over India, undergoing some rather slight northward and southward fluctuations.	W'lies up to 200–150 mb, sometimes with a jet core, over West Pakistan and up to 350 mb over northeast India.
500–700 mb	Tibetan upper anticyclone well established. W'ly jet stream north of the Tibetan anticyclone, with something of a trough about 65°E and another towards Szechwan, 105–110°E.	Weak W'lies north of Tibet, but extensive large-amplitude trough deflecting the middle latitude westerlies about 90°E towards northeast India and the Bay of Bengal (rather like a weak version of the normal winter pattern).

the strong monsoon goes with early and strong establishment of the northern hemisphere summer. It may not be surprising therefore that there is evidence of greater moisture and more abundant trees in the now arid regions about the lower Indus valley from Afghanistan to the Thar desert in Rajasthan (SETH 1963) all through the warmest post-glacial climatic epoch, possibly from as early as 7000 or 6000 to 1000 B.C. The same is true elsewhere in the southern part of the desert zone, including the Sahara. (Some residual effect on the level of the sub-soil water table and extent of oases probably made possible Alexander's march to the Indus in 327 B.C.)

Seasonal shifts of mass: the continental monsoons and the ocean anticyclones

The development of the monsoons of Asia, represented by the summer low pressure centred over southern Asia and the Siberian winter high (figs. 4.12 (*a*) and (*c*)), is the biggest item in a great seasonal exchange of mass of air between continents and oceans and between the northern and southern hemispheres. Fig. 4.16 (from GORDON 1953) shows where the greatest seasonal changes between 80°N and 60°S take place – first and foremost over Asia. The secondary maximum over the North Pacific is accounted for by the rather big poleward displacement of the subtropical anticyclone in that sector, from 24°N in February to 37°N in August over the western Pacific and from 31° to 40°N respectively over the eastern part of the ocean. The high value of the seasonal range of pressure over Iceland and Greenland is associated with the displacement of the main polar high pressure towards that region in May. Fig. 4.17 (also from GORDON) shows that the greatest changes of mass of the atmosphere are concentrated about latitudes 20–40°N and 10–30°S and take place most rapidly around April–May and October–November.

The total seasonal transfer of air from the northern hemisphere to the southern hemisphere and back again has been variously calculated by BELINSKY (1957) and STEHNOVSKY (1966), respectively, as about 39×10^{11} or 104×10^{11} metric tons (t) in a normal year. The latter figure is about two-thousandths of the total mass of the atmosphere over the globe. The margin of uncertainty between the two calculations arises through the estimation of pressure values over the wide ocean spaces and the allowance that has to be made for mountain regions and the cold ice domes of Antarctica and Greenland, over which the surface air is very dense, occupying part of the space elsewhere filled by the lower atmosphere. It is unfortunate that these studies have used atmospheric pressure at sea level, a fictitious quantity wherever ground rises above sea-level, instead of pressure at the surface, a calculation which, when integrated over the various parts of the globe, involves allowing for all the complexities of topography but is the only one that clearly represents the actual mass of air. BELINSKY used the pressure charts in a standard Soviet world atlas for his calculations. The measurements made by STEHNOVSKY from his own carefully compiled (STEHNOVSKY 1962) world charts of monthly average sea level pressure for 1881–1940, for which the most comprehensive list of sources of barometric pressure data and earlier analytical studies were used, yielded the following figures for the mean mass of air over the northern hemisphere, 0–90°N, for each month in units of 10^{11} t:

January	**26 737·7**	July	*26 633·8*
February	26 725·2	August	26 637·3
March	26 710·3	September	26 664·4
April	26 687·2	October	26 682·4
May	26 672·6	November	26 712·4
June	26 640·6	December	26 722·0

The most rapid changes appear to occur in stages: between about the beginning of March and the beginning of May, and again from May to June, then from August to September and from October to November.

STEHNOVSKY's (1962) estimates of the average values of m.s.l. pressure month by month over the northern and southern hemispheres poleward of latitude $2\frac{1}{2}°$ and over the equatorial zone are reproduced here in fig. 4.18. The seasonal ranges are about 4 mb for the northern hemisphere, 3 mb for the southern and $1\frac{1}{2}$ mb for the narrow zone near the geographical equator. The maximum values of pressure over either hemisphere are in winter. The 5°-wide equatorial zone, as here defined, lies south of the general position of the meteorological equator (intertropical convergence) and accordingly shares the southern hemisphere's seasonal pattern of changing mass of the overlying atmosphere.

The seasonal flow of air from one hemisphere to the other is probably not a steady, progressive movement, but may be thought of as switching to and fro over periods of several days and with three or four stronger pulses in the course of the northern hemisphere's winter half-year – though these pulses also represent over-swings, followed by some return movement. Such is the impression given by STEHNOVSKY's monitoring of the mass of air over the northern hemisphere each day of the five years 1955–9, though the calculations were limited by the data available to the regions north of $27\frac{1}{2}°$N. Suggestions on the basis of these studies that there are characteristic differences in the mass of air over the northern hemisphere as between periods of strong zonal circulation and periods of blocking are probably premature because of this limitation of the study to extra-tropical latitudes.

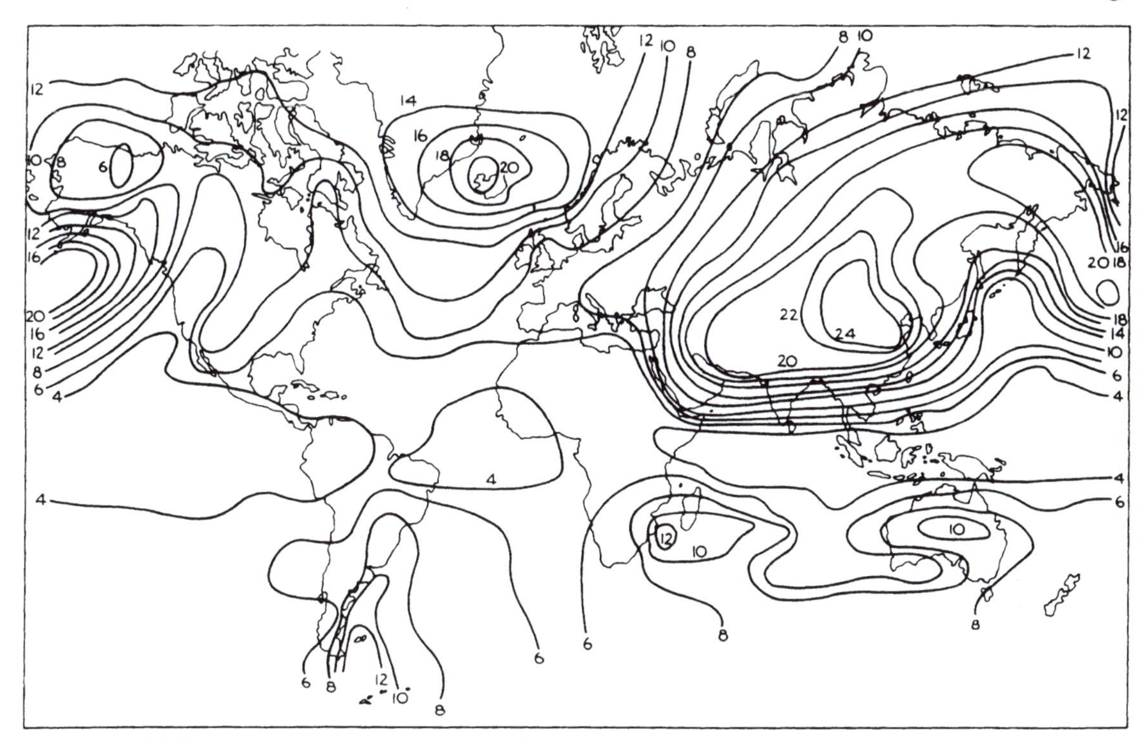

Fig. 4.16 Difference of prevailing m.s.l. pressure of the atmosphere (mb) between the months of highest and lowest average pressure in the course of the year. Data mainly 1921–38.
(*After* GORDON *1953, reproduced by kind permission.*)

The shifts of mass between middle and high latitudes within the northern hemisphere, when a strong zonal circulation gives way to blocking or meridional flow patterns, and vice versa, may be of the order of half the total seasonal interchange of air between the hemispheres. BELINSKY's (1957) figures of the change of mass of air over the broad belt between latitudes 10° and 70°N during one example of such a change from zonality to blocking give only an imperfect glimpse of this, because the real fulcrum of the oscillation was about latitude 50–55°. He found that on 18 February 1937, when the northern hemisphere westerlies were strong and a 1050-mb anticyclone centre lay over Siberia in middle latitudes while an intense low with central pressure below 980 mb was over the Arctic Ocean off northern Asia, the total mass of air between latitudes 10° and 70°N was 18×10^{11} t (metric tons) greater than it was 18 days later on 8 March 1937. On the latter date the middle latitudes westerlies were weak, an intense anticyclone occupied the polar regions with its centre about 1050 mb NE of Greenland, and there were slow-moving, meridionally (S–N or SW–NE) extended cyclones in five sectors between 45° and 60°N. Between these two situations less than 3 weeks apart, the average atmospheric pressure over the area between about 20° and 53°N (46% of the hemisphere) is estimated to have fallen

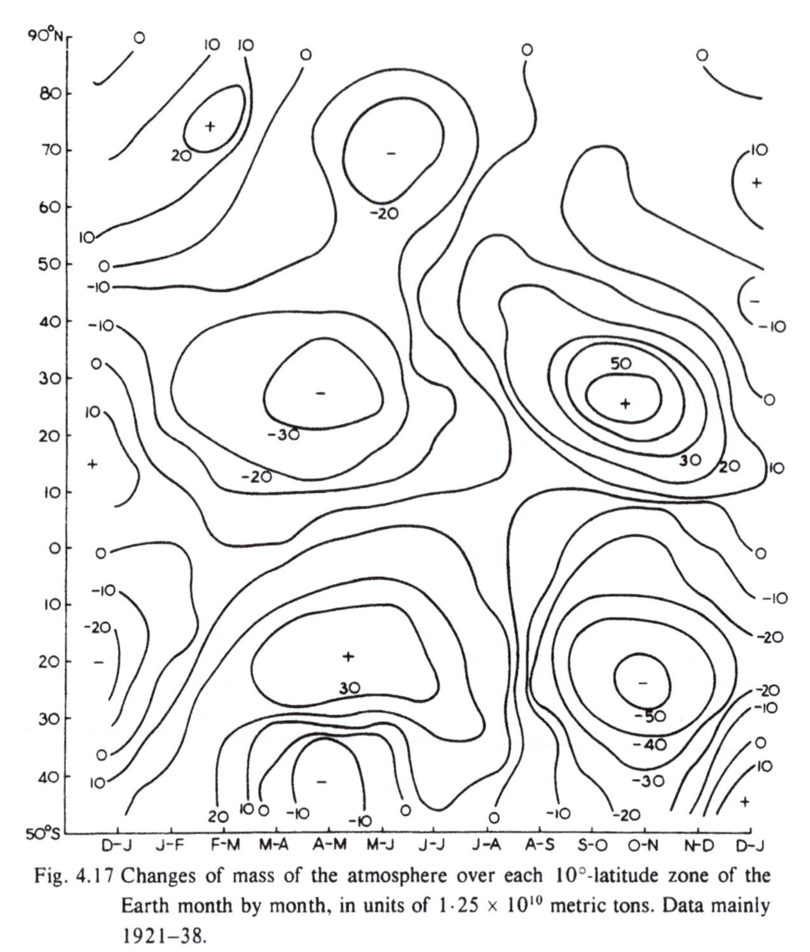

Fig. 4.17 Changes of mass of the atmosphere over each 10°-latitude zone of the Earth month by month, in units of $1{\cdot}25 \times 10^{10}$ metric tons. Data mainly 1921–38.
(*After* GORDON *1953, reproduced by kind permission.*)

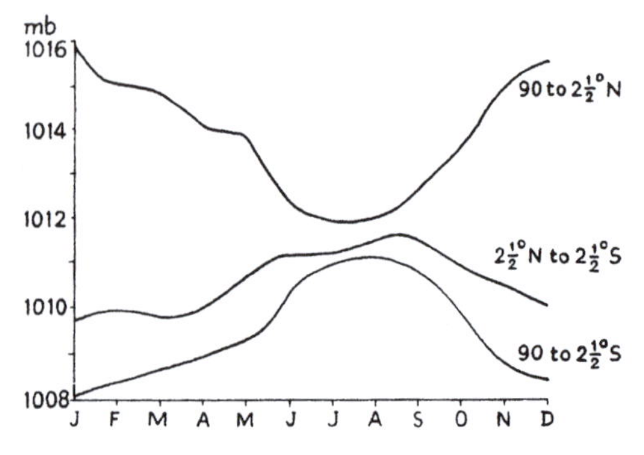

Fig. 4.18 Average pressure of the atmosphere at m.s.l. month by month over the northern and southern hemispheres and near the equator. Data mostly 1881–1940.
(*After* STEHNOVSKY *1962.*)

by about 3 mb, while the average pressure over the regions north of 53°N (20% of the hemisphere) rose about 15 mb.

In consequence of the world-wide seasonal displacement of air, the highest mean pressures in the regions of the subtropical anticyclones over the oceans of the northern hemisphere are attained in the summer half-year, when pressure is low over all the northern continents (including north Africa). By contrast, the subtropical anticyclones over the oceans of the southern hemisphere attain their maximum pressure values in the southern winter – i.e. more or less simultaneously with their northern counterparts. Thus, the southern hemisphere subtropical zone is also involved in compensating the displacement of atmospheric mass set up by the great monsoon over Asia. This presumably tends to suppress the monsoonal exchanges of air between Australia (or the other southern continents) and the ocean (cf. the smaller amplitudes in the southern hemisphere on fig. 4.16).

Over the Antarctic continent pressure is highest in the southern hemisphere summer and lowest in late winter. This regime, which runs counter to the character of the monsoonal changes over the other continents and the Arctic, may be explained by the size and vigour of the cyclonic developments arising from the potential energy in the thermal gradient around the limit of the Antarctic ice; the southern ice does not grow farther out from the pole in winter than an average latitude of 60°S in the present epoch, compared with the northern hemisphere average winter limit of snow and ice surface at about 49°N (LAMB 1958). In other words, the vigorous large-scale cyclonic activity of the winter half-year over the Southern Ocean, and the zone of the ice limit, remains too near to the central Antarctic to permit an extensive anticyclonic region there except in summer, when the depressions are smaller.

Examples of the main regimes of normal seasonal change of atmospheric pressure in different parts of the world that result from the processes we have been discussing are illustrated by the monthly mean pressure values in the present epoch in fig. 4.19. The regimes at other places and in intermediate regions may be worked out either from our discussion or by interpolation, using figs. 4.12, 4.13 and 4.18.

SCHWERDTFEGER and PROHASKA (1956) subjected the normal monthly changes of atmospheric pressure in all parts of the world to harmonic analysis. The first component – i.e. the single annual pressure wave – reaches high values (amplitude r_1 up to 15 mb) over Asia and (r_1 up to 7 mb) over the other regions of great overall range of pressure shown in fig. 4.16, probably also over the Ross Sea sector of Antarctica. The second harmonic component, corresponding to a half-yearly pressure wave, presumably called into being by the twice-yearly maxima of the gradient of solar heating (cf. p. 139), has maximum values (amplitude r_2 up to almost 3 mb) over Greenland, the Gulf of Alaska and the western Pacific near 40°N, as well as near the Caspian Sea and over most of Antarctica, and (exceeding 1·5 mb) over the regions of all the other oceans' subtropical anticyclones.

Over most of the world the first and second harmonics are of comparable amplitude. There are a few regions indicated by SCHWERDTFEGER and PROHASKA's maps where

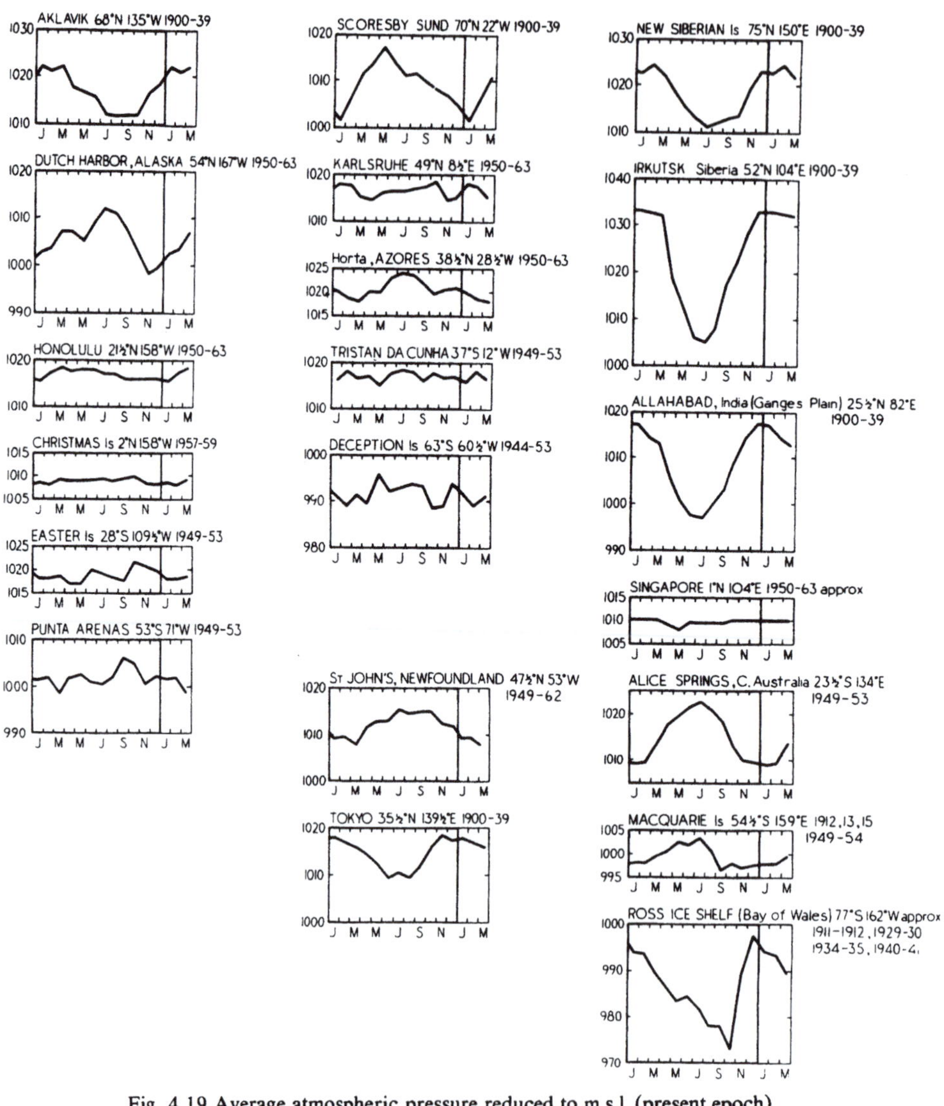

Fig. 4.19 Average atmospheric pressure reduced to m.s.l. (present epoch) month by month around the year.
Places selected to illustrate the principal regimes.
Scale uniform (to bring out amplitude differences).

both the first two harmonics are so small (about 1 mb or less) that higher frequency pressure waves, if there were any significant tendencies of that kind in the normal seasonal round, should be discernible. These regions include:

(i) middle Europe (between southeastern Britain, the Baltic and the Balkans);
(ii) eastern North America (between the Great Lakes and Newfoundland);
(iii) the region of the Arctic about Novaya Zemlya (between the White Sea and north of Taimyr);
(iv) the region of the Okhotsk Sea–Bering Sea–northern Alaska;
(v) about Tristan da Cunha in the South Atlantic;
(vi) South Island, New Zealand;

also, perhaps, the equatorial zone.

Pressure waves and singularities

The best-known higher-frequency pressure waves are:

(i) the approximately 30-day wave which seems to underlie the recurring seasonal episodes known as singularities in Europe, and
(ii) the 45-day interval between the Antarctic pressure surges noted by SIMPSON (1919).

Recognizably similar wind and weather sequences may be identified in Europe within ±3 days of the same date in from 30% to over 90% of the years in the case of different singularities.

Calendars of these singularities, or fairly regular seasonal episodes, as they affect central Europe and England (BAUR 1947, FLOHN and HESS 1949, LAMB 1950, 1964, BAYER 1959), indicate that much of the same seasonal sequence appears recognizable in probably any period of 20–50 years since A.D. 1550. The main features are periods of anticyclonic tendency, spreading over much of Europe over and north of the Alps, at more or less monthly intervals. The anticyclonic tendency culminates about the twentieth of each month from November to March; it appears to have forerunners in late August, mid to late September and mid October, and may be recognizable also in the smaller amplitude pressure changes which give rise to phases of higher mean pressure across south Germany or over a wider region of central and northern Europe about 20 April, 20 May and 5–15 June. About the month-ends, by contrast, pressure tends to be low and cyclonic influence with storms and rain and oceanic air spreads in from the Atlantic, notably about 23–30 September (over the northwestern Atlantic fringe, especially Scotland), 25 October–12 November, 26 November–10 December, 26–31 December, 26 January–4 February, 26 February–9 March, 24 March–10 April, 26 April–9 May. The FLOHN and HESS list of *Regelfälle* affecting the central European region, which seems to show this phenomenon most clearly, is given here in Table 4.3. These are, by the authors' definition, episodes which occurred in over 67% of the years 1881–1947, affecting central Europe with the given type of weather on 3 or more consecutive days within a range of at most 12 days of the calendar.

TABLE 4.3

The most frequent singularities (*Regelfälle*) in the weather of central Europe 1881–1947 (After FLOHN and HESS 1949)

Dates	*Type of weather*	*% of years*	*Mean duration*
9–18 June	Cyclonic, European summer monsoon	89%	7·3 days
21–31 July	Cyclonic, European summer monsoon	89%	7·2 days
1–10 Aug.	Cyclonic, European summer monsoon	84%	7·2 days
3–12 Sep.	Anticyclonic, late summer fine weather	79%	6·5 days
21 Sep.–2 Oct.	Anticyclonic, early autumn fine weather ('Old Wives' Summer')	76%	6·3 days
28 Oct.–6 Nov.	Anticyclonic, mid-autumn fine weather	69%	6·3 days
11–22 Nov.	Anticyclonic, late autumn weather	72%	6·4 days
1–10 Dec.	Cyclonic, oceanic, mild W'ly types	81%	5·9 days
14–25 Dec.	Anticyclonic, European winter monsoon	67%	7·4 days
23 Dec.–1 Jan.	Cyclonic, oceanic, mild W'ly types (the 'Christmas Thaw' and mid-winter mildening)	72%	6·2 days
15–26 Jan.	Anticyclonic, European winter monsoon	78%	7·4 days
3–12 Feb.	Anticyclonic, European winter monsoon	67%	6·1 days
14–25 March	Anticyclonic, early spring fine weather	69%	6·2 days
22 May–2 June	Anticyclonic, late spring fine weather	80%	6·4 days

The effect of these tendencies upon the average atmospheric pressure at m.s.l. for each day of the year over the 20 years 1919–38 at Lerwick (Shetland), Helsingfors (Finland) and Dutch Harbor (Alaska) is seen in fig. 4.20. In the case of the latter two places, the limits within which the pressure lay in 80% of the years are also shown. It seems that the whole run of years had some tendency to produce higher pressures at the European stations about the middle to latter part of the months of November, December, January, February and March than at or just after the month-ends; the same is true between mid November and January at the Alaskan station.

The first of the monthly cyclonic episodes in the eastern Atlantic–European region in the autumn do not commonly extend their influence beyond the Atlantic fringe of Britain and Scandinavia, though in some cases the cold fronts of the Atlantic cyclones penetrate far enough south to set off the first storms of autumn in the western Mediterranean, and in late October–early November, about 26 October–12 November, this happens often enough to give the highest weekly average rainfall figures of the year in Rome and Florence (FLOHN 1948). At the other end of the winter, the cyclonic periods in spring are more erratic in date than the rest of the sequence, perhaps because of year to year differences in the distribution of deep snow and 'stubbornly' cold surfaces left over from the winter. These must affect the progress of the circulation development. In late April and May the depressions reach their culminating depth and become slow-moving in two regions, one over the western Atlantic

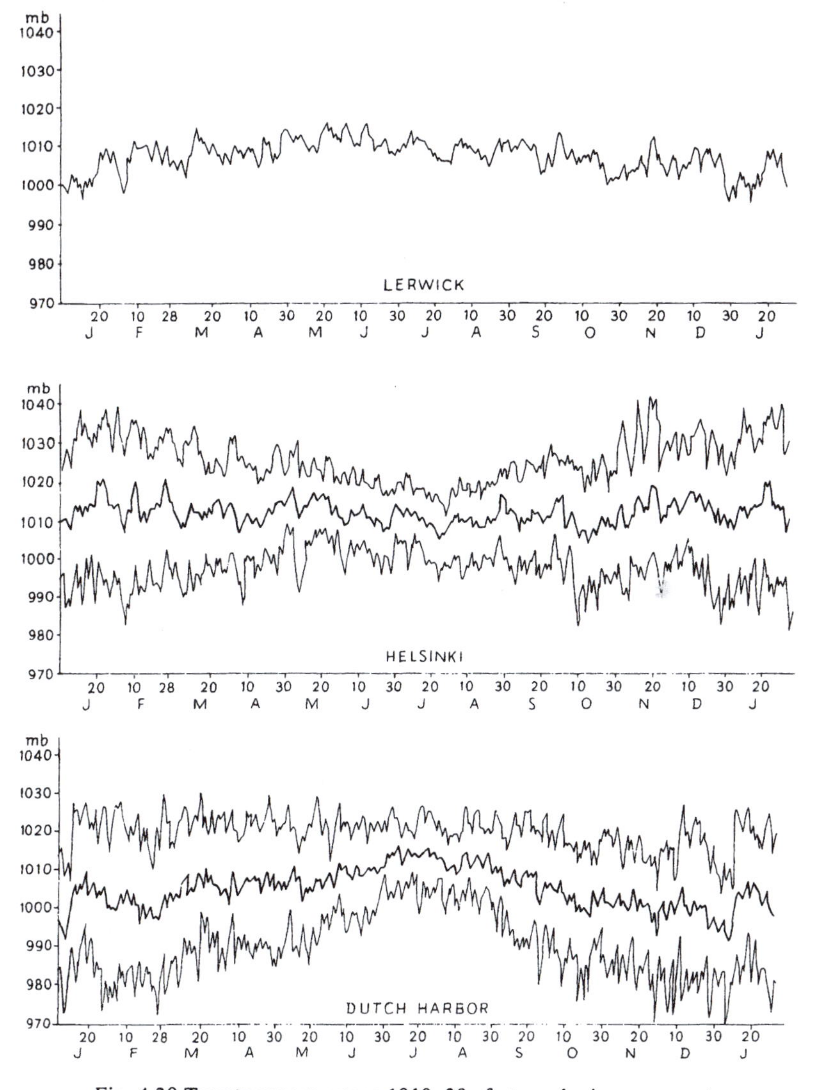

Fig. 4.20 Twenty-year averages 1919–38 of atmospheric pressure at m.s.l. on each day of the year at:
Lerwick, Shetland (60·1°N 1·2°W)
Helsinki, Finland (60·2°N 25·0°E)
Dutch Harbor, Alaska (53·9°N 166·5°W).
The outer curves indicate the values between which the pressure lay in 16 years out of the 20.

close to Newfoundland and Labrador, the other as often over the North Sea or Baltic as over the eastern Atlantic: hence, northerly and northeasterly winds are a common accompaniment in Britain and Scandinavia, a situation that is liable to give late spring snow and frosts. The roughly mid-month anticyclonic singularities through the winter are prone to give spells of frost on the European lowlands, but show up on the Alpine summits as liable to sunshine, warmth and low relative humidity [1] within the subsiding air of the anticyclone core. The breakdowns from the anticyclonic periods to the cyclonic ones are in general (though not always) quicker than the reverse changes: the accompanying storms and inrush of Atlantic air have struck the consciousness of generations of observers alert to the weather, giving rise to such well-known names as the 'Christmas thaw' and the 'equinoctial gales' of late September and March. (These last occur as noticeable changes from the weather tendencies of the weeks just before, though gale frequencies over most of the North Atlantic region are 20–30% higher in mid winter than in September or March.) The sudden return of the westerlies in Europe about 20–24 June has much of the same character and may be in some way related to the winter 30-day wave, but is undoubtedly also connected with the high-summer displacement of the circumpolar vortex towards this side of the hemisphere.

The effects of this calendar of semi-regular seasonal episodes on fairly long records of the frequencies of various phenomena in the heart of the European singularity region are illustrated in fig. 4.21. The periods of years represented differ from fig. 4.20, yet most of the same tendencies are again seen. The barometric pressure curves for Aachen and Karlsruhe show clearly the tendency for more anticyclonic conditions about, or just after, the middle of each month from October to March to be followed each time by an abrupt fall of pressure, ushering in cyclonic conditions near the turn of the month. The frequency of low relative humidity (under 60% saturation), indicative of subsidence, at the summits of the Zugspitze and of the Brocken, corresponds with the anticyclonic periods. The frequency of rainy (or snowy) days at Zürich has its main peaks of the winter half-year in the cyclonic periods around 25 October, 5 December, 30 December, 2 February, 1 March and 23 March, and again with the very usual further pulse of cyclonic activity over central Europe about 29 April–4 May. The last of these apparently also marks the climax of the spring tendency for the eastern Atlantic cyclones to reach their mature stage and slow up over the European sector: this produces a generally low level of average pressure over central Europe in March, April and early May (cf. the curves for Aachen and Karlsruhe). Most of the same sequence of times of peak frequency of rain or snow appears in a Zürich weather diary for the years 1546–76 (FLOHN 1949, LAMB 1964, App. I), peaks around 25 October, 5 December, 2 February and 27 April (all dates adjusted to the modern calendar) being particularly prominent though the high peak after Christmas was displaced into January (5–15). Finally, some effects upon the seasonal course of the prevailing temperature

1. FLOHN's (1943) analysis of the Zugspitze and other mountain observations.

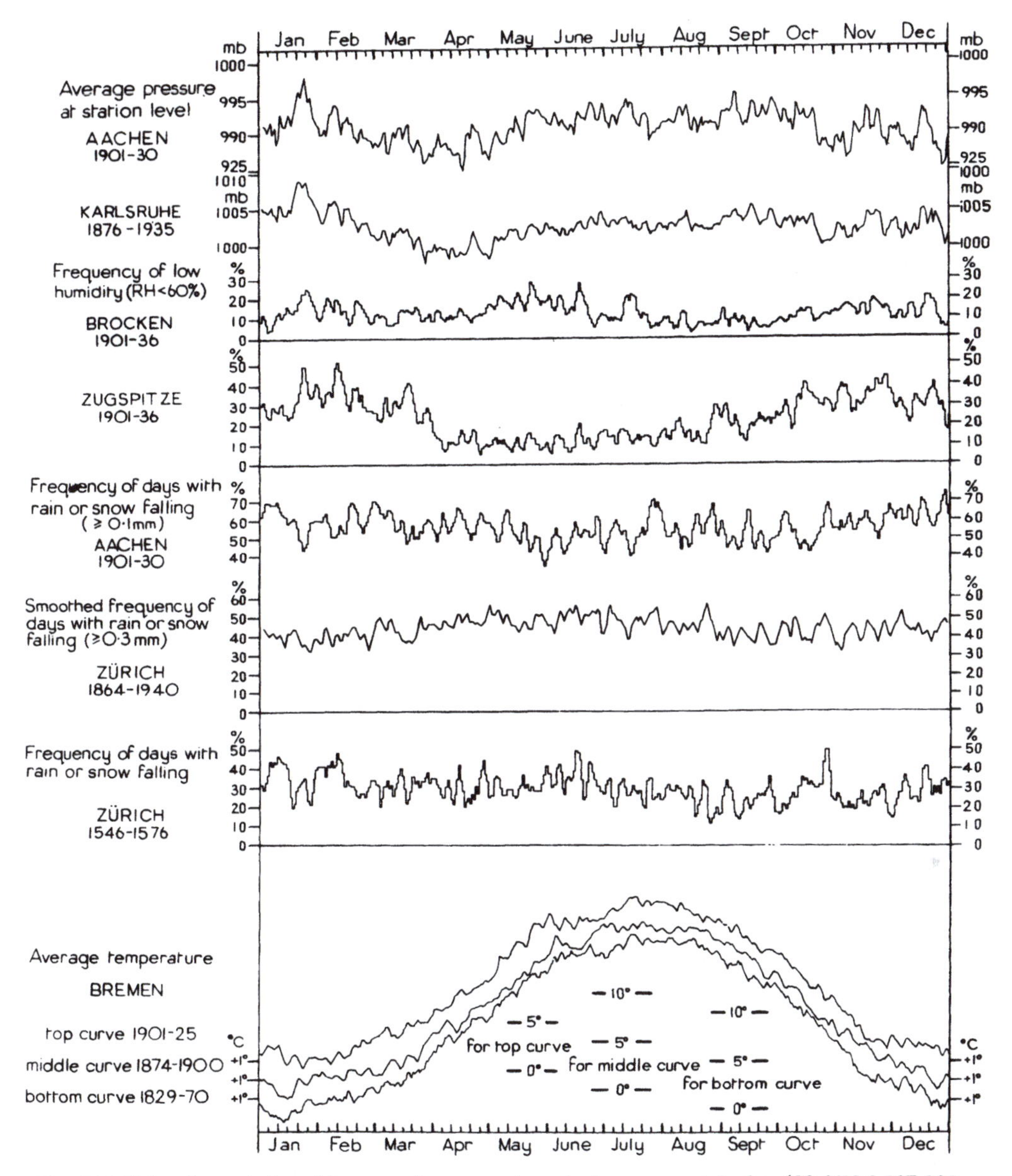

Fig. 4.21 Values for each day of the year of: average atmospheric pressure at Aachen (50·8°N 6·1°E 205 m a.s.l.) and Karlsruhe (49·0°N 8·4°E 116 m a.s.l.); frequency of days with low relative humidity (under 60% saturation) at the top of the Brocken (51·8°N 10·6°E 1153 m a.s.l.) and the Zugspitze summit (47·4°N 11·0°E 2962 m a.s.l.); frequency of days with ≥0·1 mm downput of rain and snow (water equivalent) at Aachen and ≥0·3 mm downput of rain or snow at Zürich (47·4°N 8·6°E); average air temperature at Bremen (53·1°N 8·8°E).

(Curves reproduced from FLOHN *1954 by kind permission.)*

level can also be distinguished. In spite of the dominance of the main annual temperature wave, with its maximum in July and its minimum in January, the curves of average temperature for each day of the year in Bremen (fig. 4.21) for three different groups of years since 1829 show a tendency for levelling off, or temporary reversals, of the seasonal cooling in the cyclonic periods about the end of November–early December and between Christmas and early January. Between about the 15th and 25th of November, December, January and perhaps February the average temperatures tend to lie below the general seasonal trend that would be indicated by any smoothed curve.

Similar phenomena presumably underlie the few significant calendar-bound recurrence tendencies which have been recognized in other parts of the world – e.g. the 'January thaws' (about 21–2 January) in New England (WAHL 1952, 1953) and the mid-winter mildenings (WEXLER 1959) and 45-day pressure waves (SIMPSON 1919) around the fringe of Antarctica, especially near the Ross Sea. FLOHN (1947) has noticed that the amplitude of the 30-day pressure wave in the European singularities (if measured in terms of the height above sea level of a given atmospheric pressure level) increases with height up to the stratosphere where the seat of the phenomenon may lie. SCHERHAG (1943) found evidence that the 30-day wave operative through the 1941–2 winter in Europe had its maximum amplitude in the upper stratosphere, presumably in or about the ozone layer.

Though 30 days seems to be the most commonly noticeable oscillation period, especially in the cooler months of the year, other rhythms (or none) are apparent in some years or seasons. There is some evidence (LAMB 1964) that different periodicities may represent the normal life-cycles of characteristic developments within the various principal regions of high or low atmospheric pressure, and these life-cycles may differ somewhat with the time of year.

The 30-day wave may be most of all typical for the northern continental anticyclones – hence its prominence in the colder seasons of the year – while at one time or another, 15-, 30- and 45-day oscillations are each traceable in the regions of the northern polar and Greenland high-pressure region. For the North Atlantic cyclonic activity phases of culminating vigour seem to be reached at 20- to 25-day intervals in summer and 15- to 20-day intervals in the winter half-year, the period length shortening as and when the general vigour of the circulation increases. This suggests that the cycle length is related to the build-up of potential energy during a period of markedly zonal flow (only small amplitude waves) in the circumpolar westerlies, which allows the polar regions to become colder owing to the absence of in- and outflow of air (i.e. little exchange with other latitudes): this is usually followed by breakdown to a period with large amplitude waves, meridional flow and blocking eddies, during which the thermal contrast between high and low latitudes is diminished till the circulation weakens. This is the 'index cycle', more or less as described by ROSSBY and WILLETT (1948) and NAMIAS (1950) – so-named because of the accompanying fluctuation of the zonal index which measures the general strength of the upper westerlies around the hemisphere between 35° and 55°N, although these authors did not discover any clearly preferred lifetime for the phenomenon in terms of zonal index averaged around the hemisphere.

The suggested different time interval between successive bouts of activity of the Atlantic cyclones and that of the successive pulses of the continental anticyclones may explain why different years show different rhythms in western–central Europe and why some show no rhythm at all. It is also widely suggested that the singularities best known in Europe appear more strongly in years when the planetary (circumpolar) circulation is weak and monsoonal tendencies over the continents are strong, or that different singularity sequences appear in years of prevailing high-index (strong circumpolar westerlies) and low-index (weak westerlies). The former would expose Europe more to the Atlantic cyclonic activity, the latter more to continental or polar anticyclones.

BAUR (1964) has drawn attention to the exceptional character of the 1963–4 winter in Europe, in that it showed the entire sequence of alternating continental anticyclones with frosty weather around the twentieth of each month, punctuated by incursion of Atlantic storms and mild air about the turn of the month. The dates of the sequence were just those defined by the average frequencies of various weather types and other phenomena over many years, but the amplitude of the variations of pressure, temperature and rainfall was extreme. For example, the 1919–38 average m.s.l. pressure over part of central Europe (Bohemia) for 21–5 January slightly exceeds 1025 mb, higher than at any other time in January. A peak value (1023 mb) occurs just then in the longer-period average for 1883–1944. In 1964 the pressure over that area was 1040 mb. The following minimum is represented by average pressure over the same region falling to 1017 mb about 31 January–4 February: in 1964 the average for those dates was 1011 mb. Mean temperature in Berlin was about 5°C below normal around 20 January and 4°C above normal from 28 January to 4 February. The January thaw in New England was also well marked. The completeness of the winter singularity sequence in Europe in 1963–4 was only matched in three other winters in the last 200 years (1786–7, 1878–9 and 1891–2), and the amplitude of the swings was probably quite unmatched in all that time.

Prevailing winds and natural seasons

The winds that prevail at the surface in different parts of the world in January and July in the present epoch, as a result of the atmospheric pressure regimes we have discussed, are shown in figs. 4.22 and 4.23. The winds at other times of the year are equally controlled by the patterns of barometric pressure for the time being prevailing, including those that go with the shorter-lived episodes or singularities.

The arrows in bold print in figs. 4.22 and 4.23 indicate where the frequency of winds from one rather narrowly defined direction exceeds 50% – i.e. predominates over all other wind directions. The greatest persistence, or 'steadiness', of wind direction is shown by the winds

(i) in the Trade Wind streams: though these may be disrupted during periods of blocking in higher latitudes (if the subtropical anticyclone is displaced) and they are completely

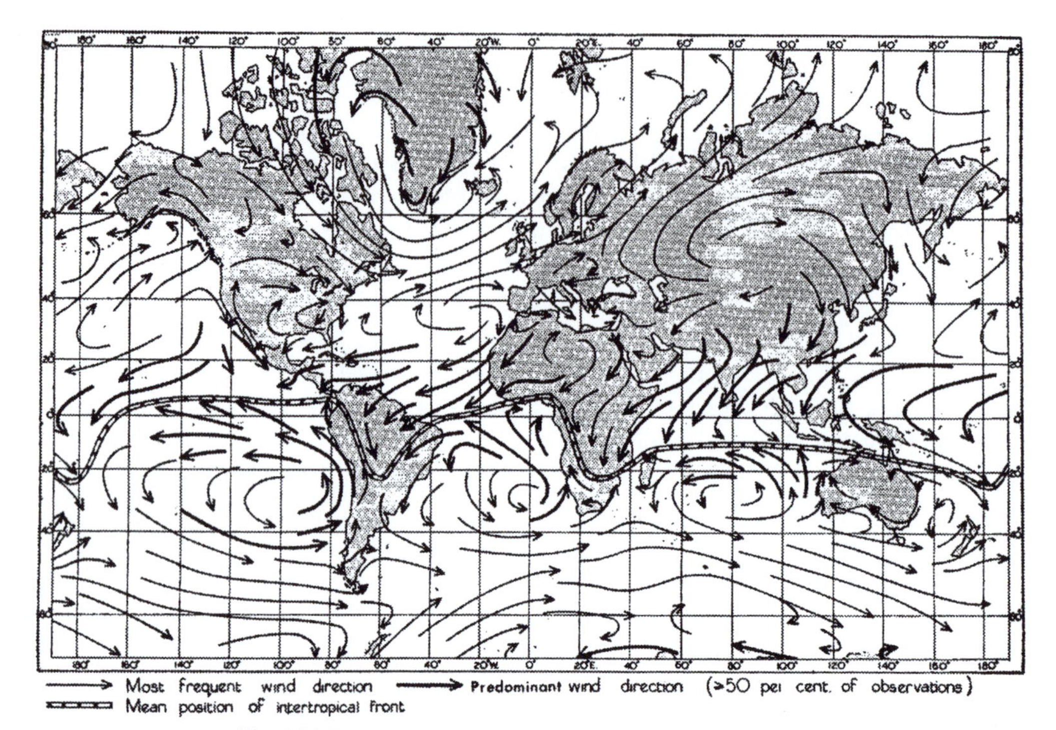

Fig. 4.22 Prevailing surface winds in January (approx. 1900–50).

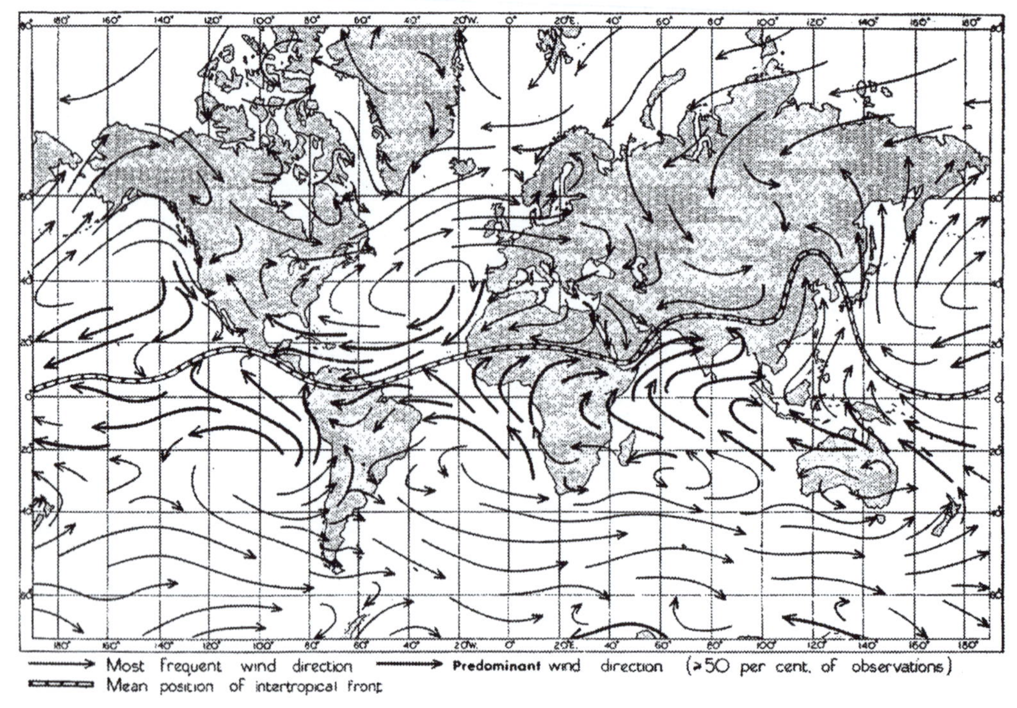

Fig. 4.23 Prevailing surface winds in July (approx. 1900–50).

upset for a few days at a time during the development, and near the path, of any tropical cyclone (typhoon or hurricane);

(ii) in the monsoons of south and east Asia;

(iii) over and near Greenland and Antarctica, where the katabatic flow of very cold surface air down the slopes of the ice cap, through the natural drainage channels of the peripheral valleys and fjords, and thence along the coastal mountain walls, exercises strong control over the wind direction.

In the regions of more variable winds in middle and high latitudes, study of the frequencies of various weather types – defined by the for-the-time-being prevalent surface wind direction and upper wind pattern steering the surface depressions and anticyclones – indicates that long spells of over 25 days' duration of persistent, or repetitive, weather character are frequent at certain times of the year. Displacements, maybe only by a few hundred kilometres, of the position of the main thermal gradients and the course of the upper westerlies – as well as occasionally big differences of prevailing speed, wave length and amplitude of the waves in the upper westerlies – when the same season in different years is examined, account for the different character of the long spells in different years. This is illustrated (figs. 4.24 (*a*) and (*b*)) by the monthly mean pressure maps for July in 1954 and 1955: in the British Isles, the former was persistently cool and rainy, the latter as markedly warm and sunny. Similar differences affected much of northern Europe and also eastern North America. Unlike Januarys are illustrated in figs. 4.25 (*a*)–(*d*): the two charts on the right, 1928 and 1957, with long sweeps of vigorous westerlies, were very mild Januarys in Europe, the two on the left, 1917 and 1940, were severe. In other cases of Januarys with weak westerlies over the Atlantic the realm of northerly winds corresponding to that seen in 1917 over Scandinavia may be so far west as to affect the Norwegian Sea and British Isles: this happened in January 1945 and 1959, when the western Atlantic low was farther south than in the cases here illustrated. It is characteristic that the blocking situations were more diverse, though both showing low pressure over the western Atlantic, whereas the two cases of vigorous long-fetch westerlies were very much alike.

Fig. 4.26 shows the frequency with which long spells, of whatever character, lasting more than 25 days in the British Isles, were in progress on each day of the year in the period 1898–1947. On the basis of this prevalence of long spells, and the times when they most commonly begin and end, LAMB (1950) has defined five natural seasons in the course of the year in Britain in the present epoch. From a study of the frequency of cyclonic and anticyclonic development in all the middle and higher latitudes of the northern hemisphere, BRADKA (1957) arrived at a closely similar division of the year. These observations mark out natural seasons which derive their character directly from the wind-pattern types and paths of frequent cyclonic or anticyclonic activity during the long spells and only indirectly from the solar radiation available at the given time of year.

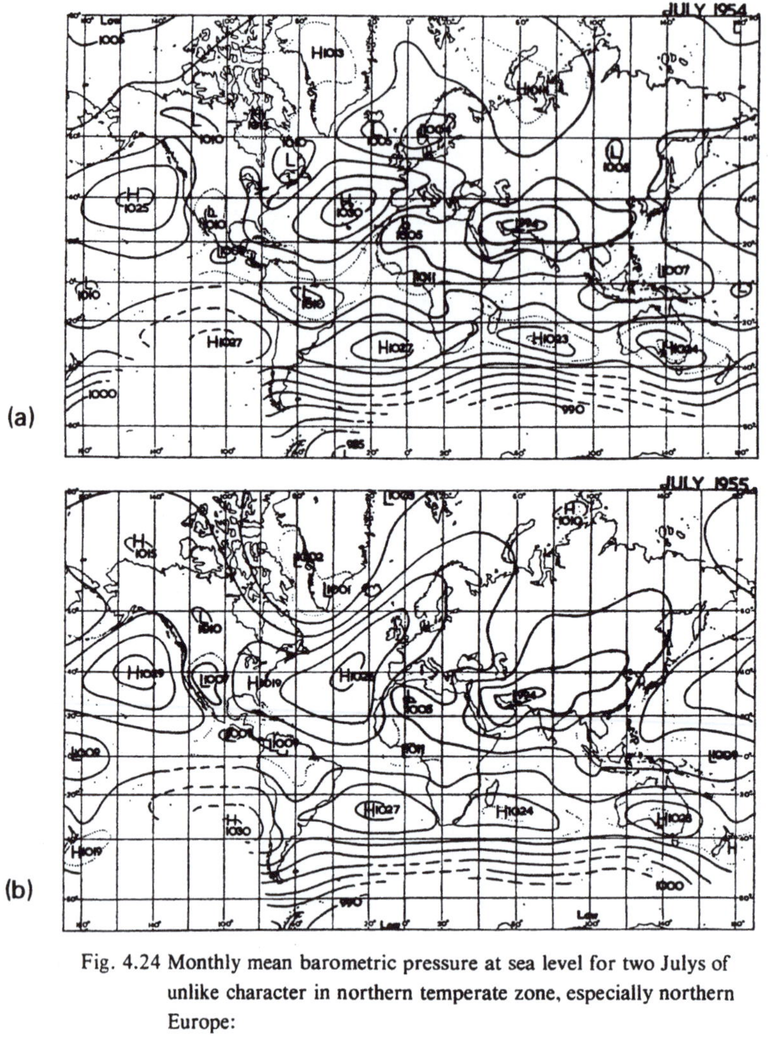

Fig. 4.24 Monthly mean barometric pressure at sea level for two Julys of unlike character in northern temperate zone, especially northern Europe:
(*a*) July 1954 (*b*) July 1955

In some cases the relationship of the long-spell tendencies of the season to the large-scale circulation development which we have traced in this chapter is very clear. The High Summer period, which produces a secondary maximum of the westerlies and the year's maximum frequency of cyclones centred over the British Isles about the end of July–early August, is the time when the centre of the circumpolar vortex is normally displaced towards this side of the northern hemisphere. The season which follows is associated with the usual early September swing-back of the jet stream farther north over the East Atlantic–European sector, which steers the Atlantic depressions northeast to enter the Arctic in that sector, with anticyclones becoming frequent, first over France and Biscay and then progressively farther north-

TABLE 4.4
Natural seasons in the British Isles

AUTUMN	Early September to mid November (INDIAN SUMMER and EARLY AUTUMN until mid October, thereafter LATE AUTUMN)
EARLY WINTER (or FOREWINTER)	Late November to mid January
LATE WINTER (or EARLY SPRING)	Late January to late March
SPRING (or EARLY SUMMER)	April to early June
HIGH SUMMER	Late June to about the end of August

east (as blocking 'highs') over central, eastern and northeastern Europe, as the wave length in the upper westerlies increases. The Autumn season could be subdivided for Britain and western Europe, the period till nearly mid October having a high frequency of anticyclonic character ('Indian Summer' and Early Autumn) and the period after mid October being commonly cyclonic and rainy owing to the gradual expansion of the circumpolar vortex and long wave length carrying depressions across the Atlantic on long ENE-ward tracks towards Scotland and Norway. The Early Winter, or Forewinter, period has a much less easily defined character, being dominated by alternations of progressive (W'ly or cyclonic) and blocked (anticyclonic) periods in Europe, though in some years the W'ly dominates in long spells at this time; it sees the circumpolar westerlies increasing to their year's maximum but also the great, nearly fixed cold troughs over Siberia and North America attaining a big amplitude. The Late Winter period is clearly associated with the southernmost position of the circumpolar westerlies over the northern hemisphere and therewith increased frequency of blocking anticyclones in sub-Arctic latitudes. This season could be subdivided for western Europe by the rising surface temperatures in March (or, given snow-free ground, from late February onwards), though the wind-pattern tendencies do not revert to any increase of westerlies until the last days of March. After a rather brief period around 24 March–10 April, in which westerlies are common over Britain and western Europe (effected by the seasonal northward progress of the zone of strongest thermal gradient), the remainder of Spring–Early Summer is mostly given to weak circulation patterns and blocking, with frequent N'ly and E'ly winds in that sector. Over the northern hemisphere generally the circumpolar westerlies tend to be weak at that time and thrown into complex (short wave-length) patterns by the temperature differences between land and sea and residual snow, or ground waterlogged by snow-melt, in the same latitude. Thus, Spring–Early Summer is a time of year when long spells of settled character are rare, except when associated with blocking (most commonly N'ly or E'ly surface winds or anticyclonic type in the British Isles); the weather more commonly changes from week to week. In the month of November also there is a period when long spells are rare and week by week changes often remarkable; after the progressive intensification of the upper

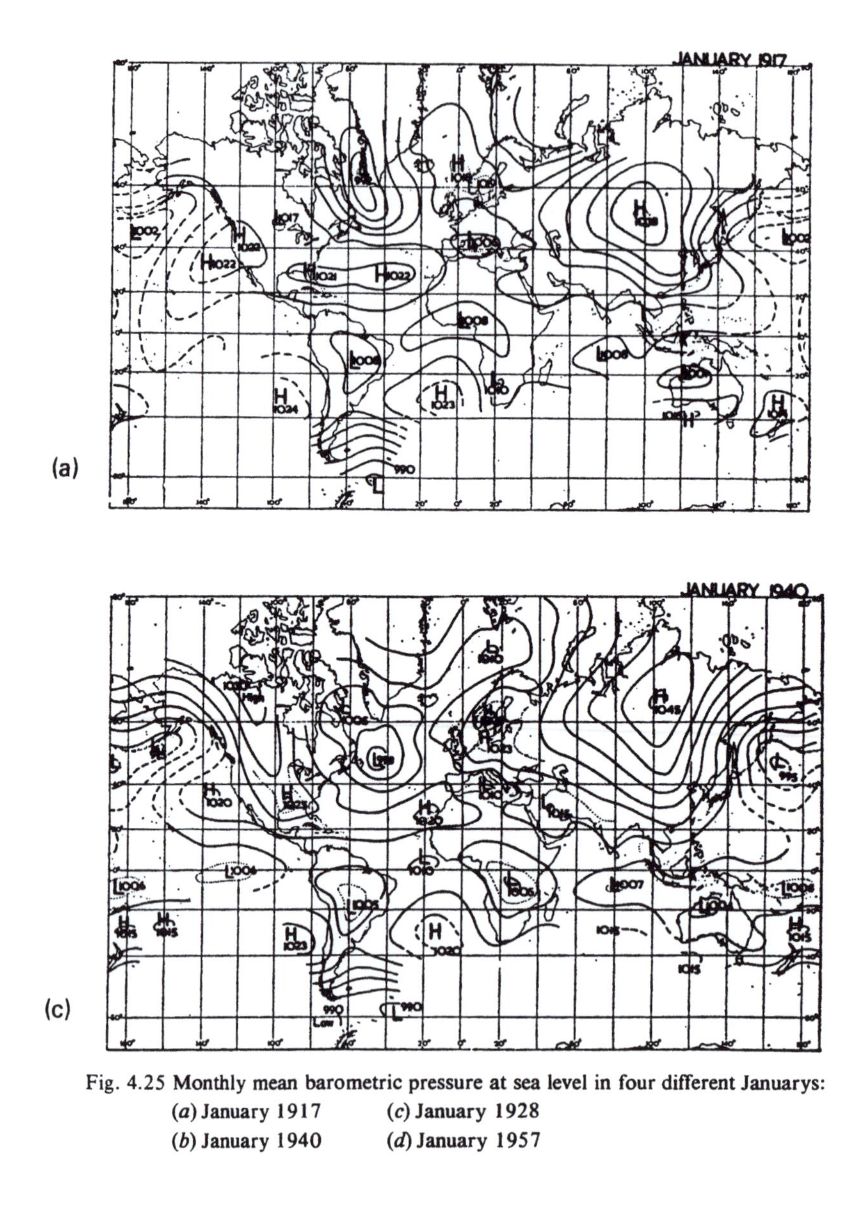

Fig. 4.25 Monthly mean barometric pressure at sea level in four different Januarys:
(*a*) January 1917 (*c*) January 1928
(*b*) January 1940 (*d*) January 1957

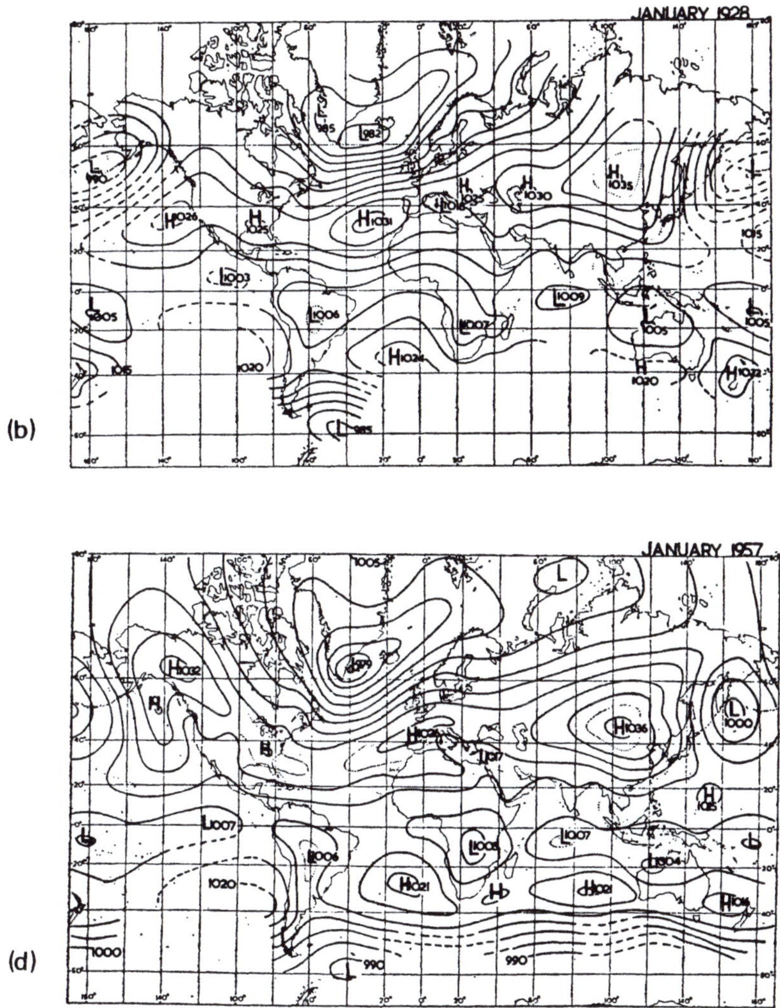

JANUARY 1928
(b)
JANUARY 1957
(d)

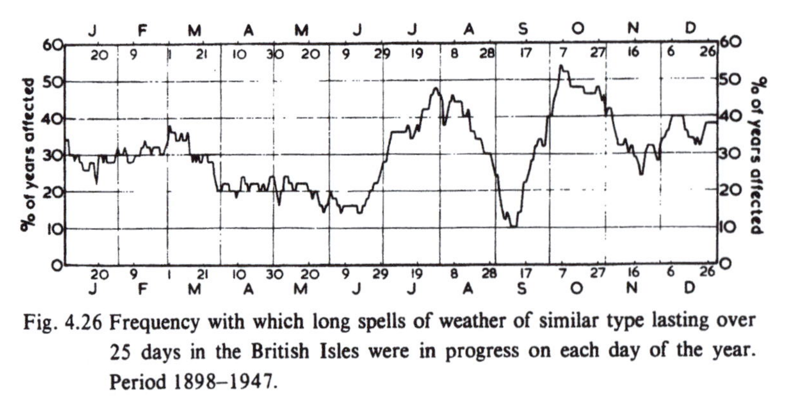

Fig. 4.26 Frequency with which long spells of weather of similar type lasting over 25 days in the British Isles were in progress on each day of the year. Period 1898–1947.

westerlies through September and October, seen in the wave-length diagram (fig. 4.11), there comes a sharp discontinuity accompanied by a throw-back to shorter wave lengths and high frequency of a blocking episode in mid to late November, before the more intense zonal circulation and longer wave lengths of winter are established.

As the wave-length diagram indicates, these seasonal divisions of the year, here described for Britain, correspond to circulation changes affecting the northern hemisphere temperate zone in general.

CHAPTER 5

The stratosphere

Although this book is concerned with the weather and climates which are a vital part of the human environment at the ground, it is necessary to pay some attention to the stratosphere. This is partly because of the likelihood of (*a*) some interaction between the independently generated circulations of the stratosphere and troposphere, and (*b*) variable damping effects upon the circulation in the lower atmosphere (troposphere) due to some (probably slight) variation of long-term mean tropopause height associated with variations in the direct heating of the stratosphere. But also there is reason to suppose that some grand-scale effects of the circulation and patterns of vertical motion in the troposphere may reveal themselves in simpler shape in the lowest layers of the stratosphere that are affected by that vertical motion.

The technical difficulty of developing adequate observational coverage, particularly the difficulty of developing balloons capable of carrying instruments regularly to heights of 30 km and above in the coldest climates on Earth, has meant that daily surveys of the stratospheric circulation over any large part of the northern hemisphere began only in the early 1950s and have – all too sparsely – covered the whole world still more recently. Observations in the stratosphere above 30–35 km mainly depend on rocket firings, made regularly only since 1959 and more or less limited to North America though now being made – at least from time to time – in many other areas: e.g. the Hebrides, the equatorial Indian Ocean and Antarctica. So research on possible 'coupling' between the tropospheric and stratospheric circulations is as yet in its infancy. On many aspects only tentative interpretations can be put forward so far; and, particularly in connexion with the longer-term anomalies and changes of climate, we can speak only of probabilities and what may reasonably be supposed from physical principles and the relevant phenomena so far observed.

The large-scale wind circulation in the stratosphere and how it is generated

The stratosphere possesses its own pattern of wind flow, generated by the distribution of heating and cooling – chiefly of the ozone layer – within it and largely independent of the troposphere below. The main features are a circumpolar vortex of westerly winds about a cold polar core in winter and of easterly winds about a warm polar core in summer (figs. 5.1, 5.2). The form of the summer vortex (fig. 5.2 (*b*)) ignores the geography of land, sea and mountain, and is plainly determined by the astronomical arrangement of maximum radiation

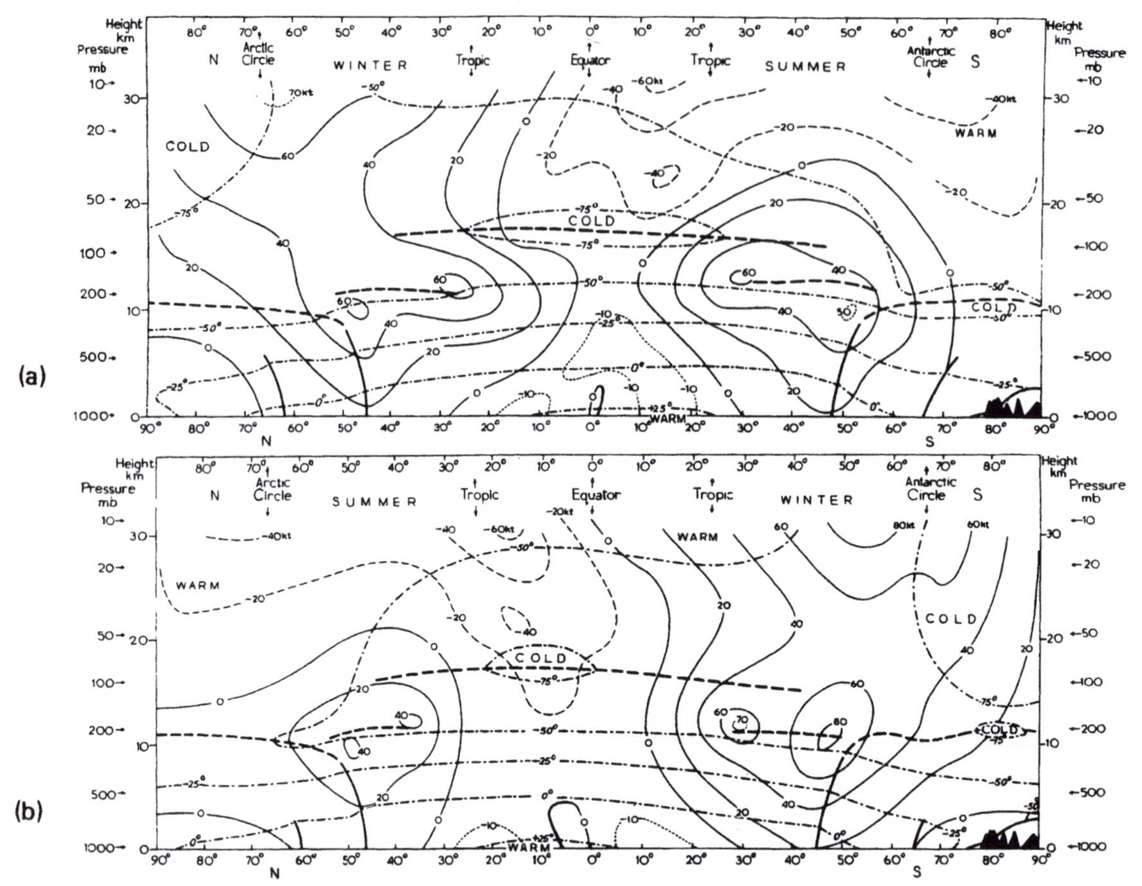

Fig. 5.1 Cross-sections from pole to pole near 30°W showing average zonal wind speeds (knots) and temperatures (°C) from the Earth's surface up to about 35 km:

(*a*) in January (*b*) in July

Period 1950s.

Westerly winds are indicated by isopleths shown as solid lines, easterly winds by broken lines and negative numbers.

The dot and dash lines represent mean temperatures.

(*Reproduced by courtesy of the Comptroller of H.M. Stationery Office.*)

supply over the highest latitudes where there is perpetual day (cf. fig. 2.8). The northern hemisphere winter pattern (fig. 5.2 (*a*)) shows just one geographical intrusion upon the symmetry of the westerly wind circulation surrounding the region of coldest stratosphere developed by the polar winter night: the stratospheric centre at this season tends to lie somewhat away from the North Pole in the sector between Greenland and mid Siberia, while there is a persistent high-pressure ridge or anticyclone over the Alaskan–Aleutian Islands sector of the northernmost Pacific. That the influence producing this situation passes up from the lower atmosphere, and is in some way associated with the great winter cooling

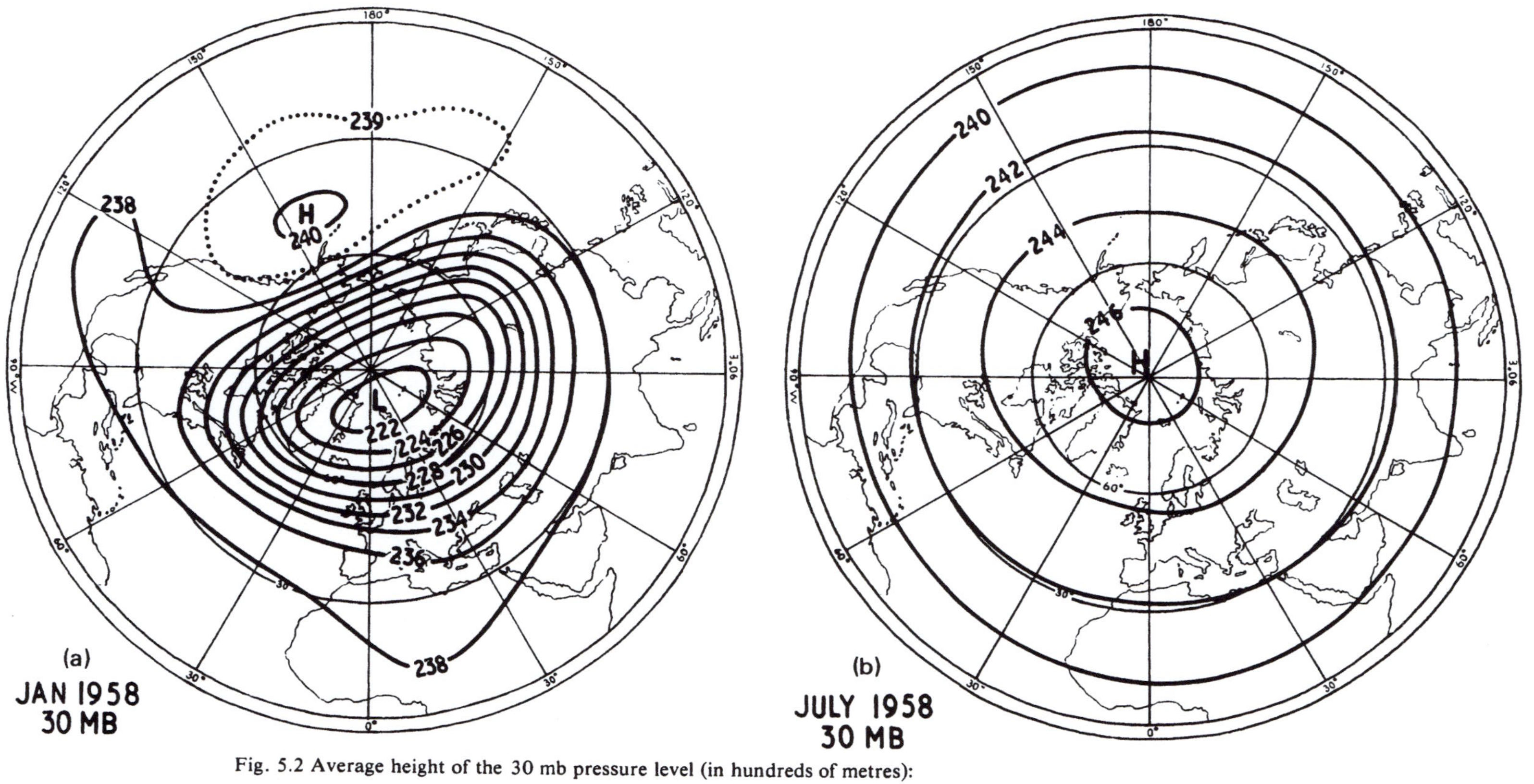

Fig. 5.2 Average height of the 30 mb pressure level (in hundreds of metres):

(*a*) January 1958 – a characteristic, fully developed winter circulation. The whole-winter average picture for the 5 years 1957–61 shows virtually identical positions of all the main features.

(*b*) July 1958 – a characteristic, fully developed summer circulation. The whole-summer average picture for the 5 years 1957–61 is extremely similar.

of Siberia, is shown by its developing from the moment when the tropospheric jet stream passes south of the Himalayas in autumn and the warm surface airmasses from the south are excluded from inner Asia. No corresponding feature is found over the southern hemisphere, where the stratospheric vortex is nearly circular and centred close to the South Pole (or, perhaps, with a displacement even slighter than in the troposphere towards the Indian Ocean side, where the broadest Antarctic ice surface lies).

In each season the maximum winds in the stratosphere or mesosphere (wind speeds in winter approaching double those in the tropospheric jet streams) are attained at heights between 50 km and about 70 km above the Earth (see, for example, MURGATROYD 1957, 1965). This high-level circumpolar vortex is subject to occasional major disturbances ('sudden warmings') in winter, accompanied by spectacular rises of temperature at the levels concerned and leading in extreme cases to a period of easterly flow, rather as in summer. The stratospheric circulation, with its disturbances and the characteristic seasonal reversal, extends up to about 80 km and down to 15–20 km; but, because of the low density of the atmosphere at those heights, this deep layer represents only 5–10% of the atmosphere by mass. The momentum carried by the entire stratospheric wind system when near its full seasonal development is also probably at most times only a tenth to a twentieth of that developed in the troposphere, though the ratio clearly varies and may sometimes be rather greater. Any dynamical effects of the stratospheric circulation upon the troposphere are unlikely to be great unless they be cumulative during some persistently anomalous stratospheric regime (see SUN CHU CHING *et al.* 1964) or tip the balance between two possible courses of tropospheric development when this is delicately poised.

Above the level of maximum winds in the troposphere at heights around 10–12 km, which is the locus of greatest accelerations and decelerations and hence of the departures from gradient wind balance which produce divergence or convergence in the horizontal wind field, the prevailing vertical motion of the air over cyclones and anticyclones is opposite to that in the atmosphere below (cf. fig. 3.10). In the lower stratosphere and about the tropopause the prevailing motion is upward over anticyclones and downward over depressions, and correspondingly over the anticyclogenetic and cyclogenetic regions along the warm and cold sides of the tropospheric jet streams respectively. The tropopause itself is liable to be carried up and down by a kilometre or two in these situations. Through the accompanying temperature changes in the ascending and descending air, at rates of the order of 1°C/100 m, the air in the lower stratosphere tends to be relatively warm over tropospheric low-pressure regions and cold over tropospheric high-pressure regions. This reversal of tropospheric temperature associations and thermal gradients generally decreases the (horizontal) wind speeds with increasing height above 12 km. The fading influence of the tropospheric winds continues up to a level of minimum wind at about 20 km in the lower stratosphere (and occasionally higher in summer and autumn). Below this level the air is at least partly carried along with the tropospheric circulation.

Above 15–20 km different vertical motions prevail once more. These are associated with the development of the stratospheric circulation proper (see diagrammatic representations derived by FAUST and ATTMANNSPACHER 1961, FAUST 1967, TEWELES 1964, KULKARNI 1966).

Ozone and the heating of the stratosphere

The stratosphere is the seat of a heating region quite separate from that at the Earth's surface and located mainly in the upper part of the ozone layer, with its top about 50 km above the Earth. This heating causes expansion of the air columns affected, thereby lifts the overlying atmosphere and, because of the heating differences between one region, or latitude zone, and another, sets the circulation of the stratosphere going in the same way as we have seen in the troposphere (Chapter 3).

Solar radiation is absorbed in the production of ozone (cf. Chapter 2, pp. 43–4) and in heating (as well as in destroying by molecular dissociation) the ozone already formed. Atomic oxygen (O) and thence ozone (O_3) are formed from some of the oxygen molecules (O_2) by absorption of the energy in ultra-short wave lengths less than 0·24 μm in the so-named Schumann, Herzberg and Hopfield bands (GOODY 1954). Ozone is destroyed by radiation in the Hartley and Huggins bands around 0·25 and 0·32 μm and in some longer wave-length bands (e.g. the Chapuis bands around 0·6 μm). So the equilibrium concentration at any time depends on the ratio of the intensities of the incoming radiation in these shorter and longer wave-length ranges. Since the energy in the solar beam at wave lengths less than 0·3 μm varies, increasing considerably in the shortest wave lengths with the state of disturbance of the sun, it seems that the amount of ozone produced in the atmosphere and the heating of the ozone layer should vary somewhat over the years.

There is probably much less variation in the rate at which ozone is destroyed. Ozone, once formed, is carried about the Earth, and ascends or descends with the winds in the layers concerned. Any ozone which rises too high will be destroyed by the solar ultra-violet radiation reaching levels in the atmosphere where the ozone is already at the equilibrium concentration; but any conveyed downwards will be protected by the ozone in the layers above it from much of the short-wave radiation, which is also depleted by scattering. Below about 30 km ozone already takes longer to form and decay, and it becomes almost stable below 20 km. Destruction of O_3 molecules liberates free oxygen O atoms, which at heights where O_2 molecules are abundant soon recombine on impact, so reconstituting the ozone O_3. Usually the greatest concentration of ozone is at heights between about 25 and 30 km (RAMANATHAN and KULKARNI 1960) where (in terms of mixing ratio by volume of air) it is of the order of 10 parts per million: between 12 and 15 km it is about a tenth of that, though occasionally as much as 2 ppm. Below 10 km, in the troposphere, amounts are very small, ozone being destroyed in oxidation of substances at, or derived from, the surface. There are evidently great differences between the northern and southern hemispheres in the rate at which ozone is destroyed at and near the surface. The amount of ozone present in the surface air in Antarctica in summer is found to be more than twice that in the Arctic air in summer; the difference is presumably to be attributed to the lack of oxidizable minerals and the absence of vegetation in the Antarctic.

The ozone in the Earth's atmosphere is important to life. It is itself highly poisonous, and dangerous to human beings in concentrations exceeding 0·1 ppm, although very small amounts of it – i.e. in still greater dilution – are stimulating and beneficial. Moreover, its presence in the stratosphere protects life at the surface by intercepting and absorbing the lethal ultra-violet radiation from the sun.

Considerable variations are observed in the total amount of ozone present in the air columns from time to time, especially seasonally, and from one part of the world to another. These variations mark the results of transport of ozone by the winds, particularly in the lower stratosphere below 30 km, and the same winds transport the heat acquired in the production of the ozone. Through this varying transport of ozone the level of maximum ozone concentration occasionally comes as low as 22 km over low latitudes (e.g. Delhi, 28°N); and seasonally high concentrations appear between 15 and 25 km over the Arctic in spring (RAMANATHAN and KULKARNI 1960, KRAUS 1960). The maximum similarly descends to about 22 km in spring at Melbourne, latitude 38°S. KULKARNI (1962), noting this, points out that the facts seem to indicate a mean drift of subsiding air in the lower stratosphere in the winter from the equator towards higher latitudes, a conception of meridional circulation in the stratosphere originally suggested by DOBSON (1929) and BREWER (e.g. 1949) – in this the ozone serves as a tracer of the air movement from the latitudes where the ozone is produced. Our emerging understanding of meridional drifts in the stratosphere has been worked out partly from theoretical considerations – e.g. the latitude distribution of heat gain and loss, and conservation of angular momentum – and partly from the study of tracers such as ozone, nuclear fission products from bomb tests and volcanic dust injected at known latitudes on known occasions (LAMB 1970). There is a (slow) mean circulation in a north–south–vertical plane, which seems commonly to consist of three cells over each hemisphere as seen in fig. 5.3 (MURGATROYD 1969, 1970), the direction of the general drift in the lower stratosphere below 30 km being poleward over low and high latitudes but equatorward over middle latitudes, 30–60° according to MURGATROYD but 50–75° according to VINCENT (1968). There are eddies, however, which are probably more important than the mean circulation outside the tropics and transfer injected matter in both directions. There seems also to be a rather complex succession of month by month changes in the mean circulation but, on the whole, the latter is directed poleward over most latitudes most of the year, being poleward even over middle latitudes in some months in late summer–early autumn.

The average seasonal variation in the ozone amount over all the higher latitudes, differing somewhat between the northern and southern hemispheres, is seen in fig. 5.4. It is not accounted for by the seasonal progression of the radiation supply, and must indicate poleward transport, varying seasonally, of ozone in the lower stratosphere from the latitudes of fairly steady ozone production in the tropics. The ozone which reaches higher latitudes in

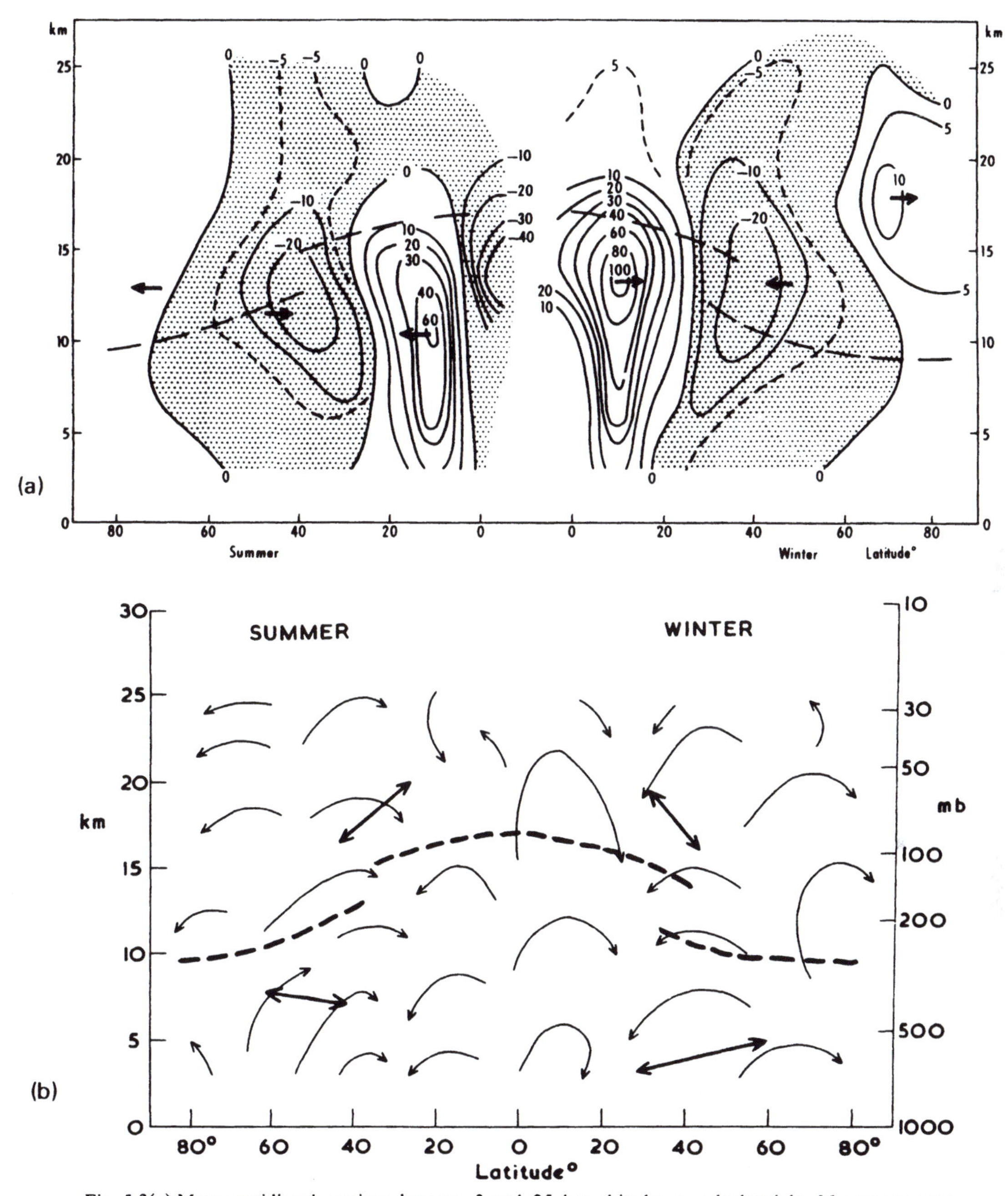

Fig. 5.3(*a*) Mean meridional motions between 3 and 25 km altitude, as calculated by MURGATROYD (1969).

Velocities in cm/sec.

Shaded areas – equatorward motion.

(*b*) Mean meridional motions (thin arrows) and preferred paths of eddy exchanges (bold arrows) in troposphere and stratosphere. The broken lines mark the tropopause.

(*Adapted from a diagram by* MURGATROYD (*1970*) *by kind permission.*)

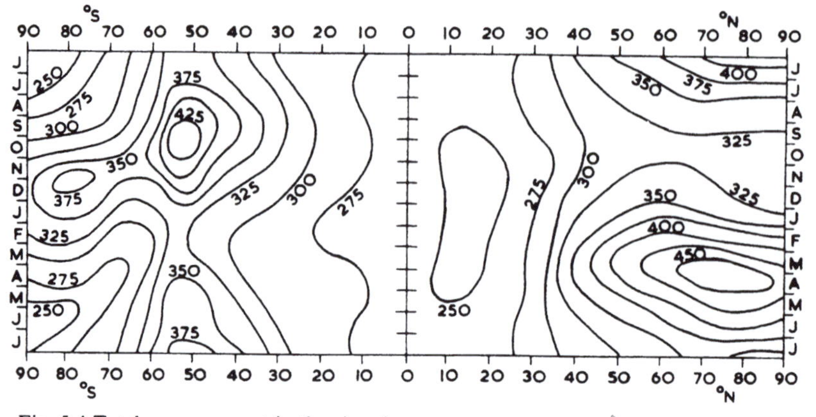

Fig. 5.4 Total ozone present in the air columns.
Averages for different latitudes month by month (various periods, mainly around 1956–60).
Units: equivalent depth of ozone in 10^{-3} cm at standard temperature and pressure.
(*Adapted from* HARE *and* BOVILLE *1965.*)

this way leaks down into the troposphere through interchange of stratospheric and tropospheric air through the tropopause disruptions (shown in figs. 5.1 (*a*), (*b*)) near the jet stream; after entering the troposphere the ozone is gradually destroyed. In the summer most ozone is produced over the polar regions; in winter none is produced there. Some variations of ozone from year to year are also indicated (e.g. by the observations from 1955 to 1964 over Australia shown by KULKARNI 1962, fig. 2; 1966, fig. 6), and may affect the temperatures prevailing in the stratosphere, but have received little study so far.

Such variations of ozone amount as occur over low latitudes, largely a-seasonally and in the longer term (whether or not through solar variation), seem likely to affect the general equilibrium height of the boundary layer between the vertical components of the circulations generated in the stratosphere and troposphere. Hence, they should affect mean tropopause height and the temperatures attainable by radiation cooling in the uppermost reaches of the troposphere; thereby some effect upon the energy developed by the large-scale wind circulation in the lower atmosphere seems likely.

SCHERHAG has remarked (1963) that the temperatures observed in the lower stratosphere at 20- to 27-km heights over the Antilles, California, the Canadian Arctic and Berlin from 1951 to 1962, all showed a long-period fluctuation which more or less paralleled the course of a geomagnetic index which responds to solar disturbance (though apparently with a double wave in the 11-year cycle). The temperatures were lowest in 1954–5 (about sunspot minimum) and highest around 1958 (about sunspot maximum), the amplitude amounting to 2° or 3°C. Such a fluctuation may be attributable to variation in the production of ozone and/or in the frequency of the great winter disturbances of the stratospheric circulation and temperature regime. Temperatures at 30 km over Ascension Island, just south of the equator, since 1955, suggest the same trend; but they stepped up sharply in 1963, after the arrival of volcanic dust in the stratosphere (see below), to values exceeding the 1958 maximum by several degrees.

Heating of volcanic dust and substances concentrated in a layer near 20 km

Radiation is intercepted in the stratosphere by other substances besides ozone, notably by volcanic dust when present, but perhaps most usually in the layer between about 16 and 23 km which is always rich in aerosols and suspended particles of tropospheric origin as well as meteoric dust, all of the order of one micron (10^{-6} m) diameter or less (JUNGE *et al.* 1961, HARE 1962). The aerosol layer may be partly a result of convergence in the vertical circulation in the stratosphere, but some of the suspended substances – e.g. minute H_2SO_4 globules – are formed there through oxidation of SO_2, etc., carried up from the surface, on encounter with increasingly abundant free oxygen (O) atoms. In connexion with the climatic *differences* from one year to another, or over longer spans of time, the most important of these items is likely to be volcanic dust occasionally present as long-lasting veils after great, explosive eruptions. These veils, actually formed from fine volcanic dust and vapours, have also been most commonly reported at heights around 20 km; though some of the strongest initial injections go much higher, and some particles and vapours of volcanic origin may even be carried up to 80 km by convection above the heated ozone layer. The volcanic particles in persistent veils also appear to be characterized by diameters about one micron. Dust veils in the stratosphere after great eruptions have been observed to last up to 3 years or so over low latitudes, and durations up to 3–10 years seem to occur over middle and higher latitudes.

Abnormal interception of radiation in the stratosphere after the great eruption of Mt Agung in Bali in March 1963 seems to be demonstrated by a +3°C anomaly of the stratospheric temperatures at 20–25 km over Ascension Island (8°S 14½°W) later in 1963 and 1964, including the highest temperatures ever observed there at the heights concerned (EBDON 1967); there was an apparent deficit of temperatures in the troposphere below, amounting to 0·2–0·4°C over the whole Earth, in the first years after the eruption.

The volcanic dust constituting veils in the high atmosphere is liable to be unevenly distributed by latitude at various stages after any eruption. First it is carried by the winds, and spread fairly uniformly within a month or two, over the latitude zone into which it was injected. The spread to other latitudes may take half a year or more and appears to be accomplished by a net poleward drift of the air in the lower stratosphere, though in a series of seasonal surges. Studies of the radiation effects of the dust in the stratosphere after the Bali eruption in 8°S in March 1963 (DYER and HICKS *personal communication*) show that it drifted poleward in a series of annual pulses bringing maxima of the dust over middle latitudes in each hemisphere in the winters and proceeding to appear as maxima over the polar regions in the following spring–early summer. Successive years' maxima showed a fall-off in the southern hemisphere and in low latitudes generally south of 20°N, but no decline was discernible in the annual maxima of the first three years after the eruption in the middle and higher latitudes of the northern hemisphere. A similar progress has been suspected after other great eruptions in earlier years. Thus, after a great volcanic explosion in low latitudes there may be a stage in which there is a dust veil in the stratosphere over the tropics but not over higher latitudes, followed by a year or two with a fairly uniform veil over all latitudes and ultimately by a situation in which the dust remains over the polar regions but has cleared from the tropics. The intercepted solar radiation in the initial stage should boost the winter thermal gradients and circulation strength in the stratosphere, while the loss of insolation at lower levels should at the same time weaken the equator–pole thermal gradient and circulation strength in the troposphere. By contrast, when the stratospheric dust is concentrated over high latitudes, the opposite tendencies should be induced.

When the input of dust into the stratosphere is in high latitudes, as with eruptions in Alaska, Iceland, Kamchatka or the Antarctic, the dust apparently fails to spread in significant quantity to low latitudes beyond the reach of the meandering upper westerlies of the hemisphere concerned. This is in agreement with expectation from the meridional drifts in the stratosphere already derived (fig. 5.3), which should spread dust in the stratosphere from eruptions near latitude 60–70° both poleward and equatorward, though there should be little or no transport to latitudes beyond about 30° except by eddies.

A dust layer of any given thickness and concentration will, of course, be most effective in reducing the solar radiation penetrating to the surface over high latitudes because of the oblique path of the rays of the sun through the layer.

Any epochs of abnormally frequent great volcanic explosions should be marked by some overall reduction in the amount of solar energy fed into the tropospheric circulation, at any rate when an average is taken over many years of such an epoch, and extra strength of the stratospheric circulation owing to the radiation intercepted by the dust there. (Dust in the troposphere is continually washed out by rain and can hardly remain suspended in the atmosphere for more than 10–20 days.) The upper and lower circulations should also undergo opposite trends during the years immediately following individual eruptions, depending upon the concentration of the stratospheric dust now over the tropics and now over high latitudes.

Seasonal variations in the stratosphere

The geographical distribution of the direct heating in the stratosphere, normally operating mainly on the upper part of the ozone layer, depends almost solely upon the astronomically determined distribution of incoming solar radiation, uninterrupted by any significant clouds or haze higher up in the atmosphere.

The essentials of seasonal variation of temperature in the stratosphere are that (i) in summer the large daily totals of incoming solar radiation in the 24-hour day over the polar regions produce the highest temperatures there and not over lower latitudes, and (ii) in winter uninterrupted radiation loss produces a very cold stratosphere over the polar regions.

Temperatures in the equatorial stratosphere have little seasonal variation and are lower throughout the year than those observed in a 'warm' belt produced over middle latitudes by the prevailing downward motion over some degrees of latitude on the poleward side of the jet streams in the troposphere.

The prevailing temperatures are crudely illustrated by the few isotherms in figs. 5.1 (*a*) and (*b*). The normal situation in the stratosphere shows little change of prevailing temperature with height, i.e. approximate 'isothermalcy', up to about 30 km, though in the sunlit season temperature rises above that to a maximum at about 50 km (cf. fig. 1.1) and the cooling in the winter night over the Antarctic at its extreme (rarer over the Arctic) produces a slight fall of temperature right up to the ozone layer or beyond, causing the tropopause virtually to disappear (as first noted by COURT 1942). Characteristic temperatures at the 20- to 25-km level in winter are around −80°C in the cold polar core over the Arctic (occasionally −90°C over the Antarctic), −45° to −60°C in the warmer belt over middle latitudes (about −40°C in the Aleutian anticyclone region), and −55° to −65°C near the equator. In summer, temperatures at the same level are usually about −40°C in the 'warm' polar core, −45° to −50°C over middle latitudes and still −55° to −65°C near the equator.

The corresponding wind speeds developed in the stratospheric winter vortex are westerlies up to 175 m/sec (over 350 knots) at the 60-km level and sometimes exceeding 100 m/sec (200 knots) at the 25-km level, where speeds of over 60 m/sec (120 knots) are usually to be found in the middle of the strong wind belt near the limit of the polar darkness. The stratospheric easterlies in summer are notably steady and persistent but characteristically only attain about 25–30 m/sec (50–60 knots) at heights around 25 km, and up to 40–60 m/sec (80–120 knots) at the 40- to 60-km levels, over subtropical latitudes (about 30°) where they are strongest and no more than half these values over middle and higher latitudes (cf. HARE 1962, MURGATROYD 1965).

Perturbations and breakdowns of the stratospheric circulation and the sudden warmings associated

Wave-like perturbations and eddies, which constitute one of the processes whereby heat and momentum are conveyed from latitude to latitude, are also observed in the circulation in the stratosphere, though it is seldom that any of much significance disturb the relatively sluggish flow of the summer easterlies over middle and high latitudes: one rather notable summer disturbance which brought temperature rises of up to 6°C in five days at the 25-km level over middle latitudes immediately after a solar eruption in July 1958 has, however, been described by SCHERHAG (1959). The main locus of travelling perturbations analogous to the mobile waves and cyclones and anticyclones in the troposphere is believed to be in the layer above the ozone, where temperatures decline with increasing height from 50 km up to 80 km – i.e. in the mesosphere (cf. fig. 1.1). Those disturbances of the winter westerlies which filter down into the stratosphere are limited, because of the 'static stability' – i.e. resistance to vertical motion in an environment where there is little change, or an increase, of temperature with height – and because of the high wind speeds prevailing in mid and late winter, to very long wave lengths. The significant waves around the hemisphere are rarely more than three in number, and more usually only one or two. In the early part of the winter these waves are of small (lateral) amplitude and move only sluggishly. They are therefore rather ineffective as a means of transporting heat polewards, and the polar region becomes extremely cold. This causes a gradual build-up of the south–north thermal gradient in the stratosphere surrounding the region of winter night, with corresponding increase of strength of the westerly winds, leading to breakdowns towards the end of the winter in the form of very large-scale eddies which carry a great injection of heat (and ozone) into the higher latitudes and are liable to reverse the stratospheric circulation, at least for some time. In the course of these breakdowns the temperatures between 20 and 30 km in the region up till then occupied by the cold pole may rise in a few days from almost –80°C to –30°C or thereabouts at 20 km and even to 0°C at 30 km, i.e. to well above summer values. Such extreme events do not occur every winter: since January and February 1952 when they were first observed by SCHERHAG, only the winters of 1957, 1958, 1963 and 1968 have provided cases of comparable magnitude. Lesser warmings to between –50° and –40°C are more common and may occur several times before the final warming which ushers in the summer circulation.

It is now understood that these 'explosive' warmings, which not uncommonly raise the temperature by some 30°C in 24 hours in the lower stratosphere, are directly due to vertical motion of the air. Nevertheless SAZONOV (1967) believes it possible to link their initiation with the arrival in the region affected of a burst of 'soft' cosmic rays of solar origin. Descent of the windstream by about 3 km in a day (3–4 cm/sec), in an originally isothermal environment, would achieve a temperature rise of some 30°C owing to the increasing atmospheric pressure upon it ('adiabatic compression'). The wind in the stratosphere, as in the upper troposphere, generally moves faster than the pattern of the flow even where there are developing or travelling eddies. Thus, the downward motion is to be found on the upwind (i.e. usually western) side of a stratospheric warm area, where the wind flows down into it. And, depending on the vigour of the flow and the rate of descent, the warm patch is capable of retrogressing towards the direction from which the wind is coming, or remaining stationary, or moving slowly forward in about the same direction as the wind.

Fig. 5.5, from HARE and BOVILLE (1965), shows the gradual build-up of strength and the various breakdowns of the Arctic winter stratospheric westerlies through three winters, and the undisturbed course of the easterlies in summer. The final warming of the polar stratosphere, which brings the establishment of the easterly circulation that lasts through the summer and accompanies the rapid increase of solar heating over the Arctic in April–May, but may also be partly due to vertical motion ('adiabatic compression'), usually exceeds in magnitude all but the more extreme winter warmings; but the spring warming over the Antarctic seems to be achieved in a number of smaller steps.

The final breakdown of the winter circulation over either pole does not occur at the same

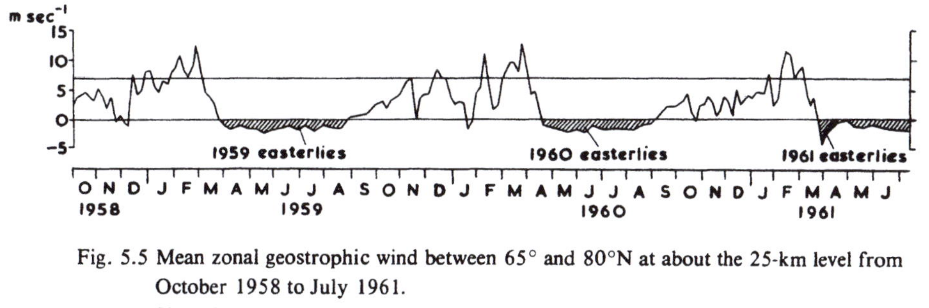

Fig. 5.5 Mean zonal geostrophic wind between 65° and 80°N at about the 25-km level from October 1958 to July 1961.
Shaded area – easterly winds. Clear area – westerly winds.
(*Adapted from* HARE *and* BOVILLE *1965*.)

date each year nor when the same wind speed has been reached. Thus, a very strong stratospheric vortex around the Arctic in February 1964 led up to one of the earliest final change-overs to summer circulation, but a still stronger vortex in February 1967 continued in being until just past the average date of the previous ten years. Different wind speeds prove critical in different winters presumably because of variations in the temperature gradients with height and latitude leading to differences of shear.

Studies of the evolution of the stratospheric circumpolar vortex at the 30-mb level in two cold seasons (September–April), 1965–6 and 1966–7, by PČELKO (1967) led him to define four stratospheric pattern types: (1) a circular cyclone centred near the pole and without any marked troughs, (2) a cyclonic circulation with 1, 2 or more troughs, (3) a split vortex with two independent cyclonic centres, (4) a circulation consisting of three or four separate cyclonic centres. The percentage frequencies and average durations of spells of these types were as follows: (1) 12·5%, 6·5 days; (2) 39·5%, 12 days; (3) 43·5%, 12 days; (4) 4·5%, 9 days. Types (3) and (4) tended to appear increasingly as the winter season went on. Each stratospheric pattern type appeared to go with a characteristic steering pattern in the troposphere below: viz. (i) predominantly zonal flow in all sectors around the pole; (ii) more meridional components of the circulation in and about the sectors affected by the stratospheric troughs and sometimes a displaced centre of the hemispheric circulation; (iii) split vortex in the troposphere also, and more azonal, meridional and, over high latitudes, recurving flows; correspondingly also with pattern (iv) which, however, never occurred during the winter of persistently strongest zonal circulation, 1966–7. The breakdowns and changes of type took place essentially simultaneously in troposphere and stratosphere, 60% within ±1 day, 95% within 2 days.

Culmination of the winter circulation, interactions with the troposphere and dates of seasonal change-over

One way in which interaction between the circulations in the stratosphere and in the atmosphere below could be important is if, as seems likely to be the case, the tropospheric wave patterns were more settled (persistent) when the wave length prevailing there was a sub-multiple (harmonic) of the wave length prevailing in the stratosphere and/or when one or more of the major trough positions in the upper and lower circulations coincided, especially if the vertical motion in the trough were in the same sense at all heights between the ozone layer and the tropopause. Observation suggests that the greatest storm cyclones developed in the lower atmosphere may attain a few per cent more energy if the associated jet stream and vertical motions happen to be in line with those in the stratosphere. The North Sea flood catastrophe in the lowlands around Hamburg produced by the great NW'ly storm of 16–17 February 1962 may have been an example of this (SCHERHAG 1963). Thus, winters when the Arctic stratospheric vortex is so strong that no wave number greater than two can produce a stationary wave length may, through the operation of the Aleutian–Alaskan ridge, require a trough in or near the western Atlantic; and in the extreme case of a one-wave system the trough may have to lie broadly over Europe and Asia; a 3-wave system would favour other positions with troughs of more limited extent and presumably less capability of affecting the troposphere.

The dates of the breakdowns of the stratospheric winter vortex, and of the summer easterly regime setting in, in the years 1956–65 are shown in fig. 5.6. In the Arctic, cases of

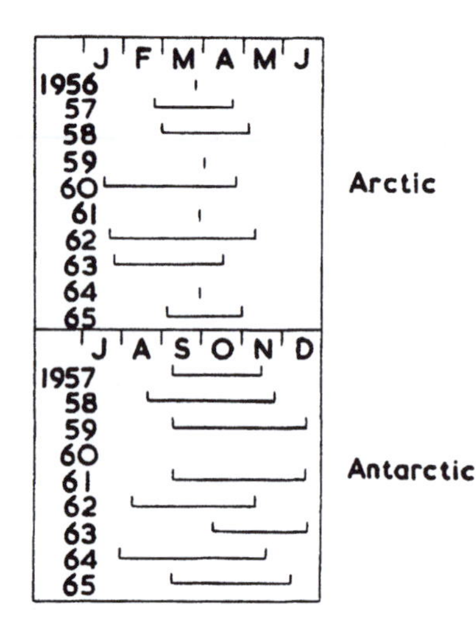

Fig. 5.6 Periods over which the spring transformation of the stratospheric circumpolar vortex took place at the 50 mb (approx. 20-km) level 1956–65. Arctic and Antarctic cases, months arranged so that the (radiation) seasons correspond.
(*Adapted from* GAIGEROV *1967*.)

early breakdown and polar warming were always followed by renewal of the cold polar core through radiation cooling, reconstitution of the winter westerly vortex and a later than average completion of the change-over to persistent easterly wind. In those years (1956, 1959, 1961, 1964) when there was no breakdown until late March, the final change-over was accomplished by a single warming event so that the summer easterlies arrived early. In the Antarctic, by contrast, the warming normally did not start at such an early stage of the season as over the Arctic and proceeded more smoothly or in a succession of small to moderate steps/jumps without full recovery of intensity of the winter regime between them (cf. LAMB 1960). This means that in the Antarctic a later than usual start led to a later than usual finish of the process. A tendency for late and early years to alternate has been suggested, though the Arctic cases here displayed show more examples of two late years followed by one early year. It has also been suggested that an early date of the spring change-over in March–early April in the Arctic stratosphere may go with early and prolonged establishment of summer-type circulation in the northern hemisphere troposphere, particularly of a northern position of the subtropical anticyclones and finer than average summers in the temperate zone. (According to British data, fig. 5.6 needs amending in that the Arctic spring change-over in the stratosphere was not early in 1956.)

The highest temperatures attained in the stratosphere over high latitudes in the sudden warmings in late winter and spring, exceeding the mid-summer values by up to 20°C or even 30°C, require vertical motion to produce them and are not held in being once the subsidence ceases. The actual summer regime in the stratosphere is maintained by the production and direct heating of the ozone by the constant insolation in summer over high latitudes. The season of unbroken sunshine in the stratosphere over the pole lasts about 7 months (from 6 March to 9 October over the North Pole) at 20 km and half a month longer (from 28 February to 17 October) at the top of the ozone layer at 50 km, the remainder of the year being the polar winter night. At latitude 80°N the period of darkness lasts about 14 weeks at 20 km and 10 weeks at 50 km. In agreement with the tendency of the radiation supply, the hemispheric thermal pattern in the stratosphere tends to produce easterly winds in summer: the summer easterly wind vortex normally appears first in the upper stratosphere and spreads downwards. Correspondingly, in the cooling season from autumn onwards the thermal pattern should produce westerly winds, appearing first and becoming strongest at the top of the stratosphere or in the mesosphere; but, because the tropospheric westerlies are strengthening at that season, westerly winds spread up through the lower stratosphere also. Hence the autumn change-over spreads both downwards and upwards, the latter development being characteristically fitful. During the autumn change-over, quite minor wind fluctuations of tropospheric origin may affect the date of its completion by several weeks; hence this date seems unlikely to be of any predictive value regarding either stratospheric or tropospheric development in the following winter.

The average (1957–66) span of the season of summer easterlies at the 20-km level near 60°N 0°E was from about 26–30 April to 4–8 August, the onset varying from about 20 March in 1959 and 1964 (perhaps also 1956) to after mid May in 1958, 1960, 1962, 1965 and 1966. The autumn change-over was spread by gentle fluctuations over the period between 20 July and 30 August, being earliest in 1956 and 1962. (Notice that the winters that followed these two early change-overs were near the extremes of unlike character.)

Stratospheric–tropospheric interplay and some cases of blocking

A strong appearance of linkage between events in stratosphere and troposphere has been noticed by LABITZKE (1965) in the case of premature (i.e. winter) stratospheric warmings in the Decembers, Januarys and Februarys of the eleven years 1954–64. In the years for which daily stratospheric charts of the northern hemisphere exist, from 1956 onwards, the events concerned could all be seen to go with great stratospheric warmings in which the temperatures at the 25- to 30-km level in the region affected rose 30–40°C, to above −50°C and in two cases over Europe to about −20°C. In all cases the warmings occurred in regions occupied immediately before by an extremely strong jet stream in the troposphere and great cyclonic activity at the surface, which presumably initiated, or amplified, the vertical motion in the stratosphere that produced the warming there. The situations concerned were, however, sharply divided into two groups:

(1) In one set of cases, called by LABITZKE the 'American type', the events described took place over the western Atlantic with a W'ly-NW'ly tropospheric jet stream emerging from America over, or near, Nova Scotia, associated with an intense upper cyclone centred (e.g. at the 500-mb level) over Quebec–Labrador. The positions of the surface lows, which were also intense, were more various: the extreme cases were 1957 when the surface low pressure was concentrated southeast of Greenland and 1963 when it was over Baffin Island and the Canadian Arctic – two radically unlike situations. In all these cases of American-type stratospheric warming – which were observed in alternate winters, in 1956–7, 1958–9, 1960–1 and 1962–3, but subsequently in 1965–6 and 1967–8, sometimes with repetitions during the same winter and starting in one case as early as the beginning of December (1965) – the stratospheric anticyclone formed by the warming proceeded to move (as shown in fig. 5.7 (*a*)) across the Atlantic to Europe or, in the case of the two strongest warmings, in 1957 and 1963, to turn northeast about mid Atlantic towards the northeast Greenland coast.

(2) In the winters in between, as well as in 1951–2, 1953–4 and 1963–4 (again with a tendency for repetitions within the same winter), the stratospheric warmings took place over Europe with a strong NW'ly jet stream over the region of the North Sea and 500-mb level cyclone centre over northern Europe. In these cases the sea level situations were more uniform, with intense depressions centred over Scandinavia and generally high pressure over north Greenland and the Canadian Arctic. The stratospheric anticyclones formed, or intensified, between 30° and 70°E by these European-type warmings all moved west or northwest (fig. 5.7 (*b*)), the warming itself sometimes being traceable over a month-long course circuiting the Arctic completely in the westbound direction.

Several of the eastbound stratospheric anticyclones formed after American-type warmings could likewise be followed (cf. fig. 5.7 (*a*)) most of the way around the northern polar region. With these, as with the westbound cases already noticed, and as with the sample movement of tropospheric cold centres followed by SCHERHAG (1958), the rate of movement is such that about a month – actually from 25 to 33 days – seems required for a complete circuit of the Arctic.

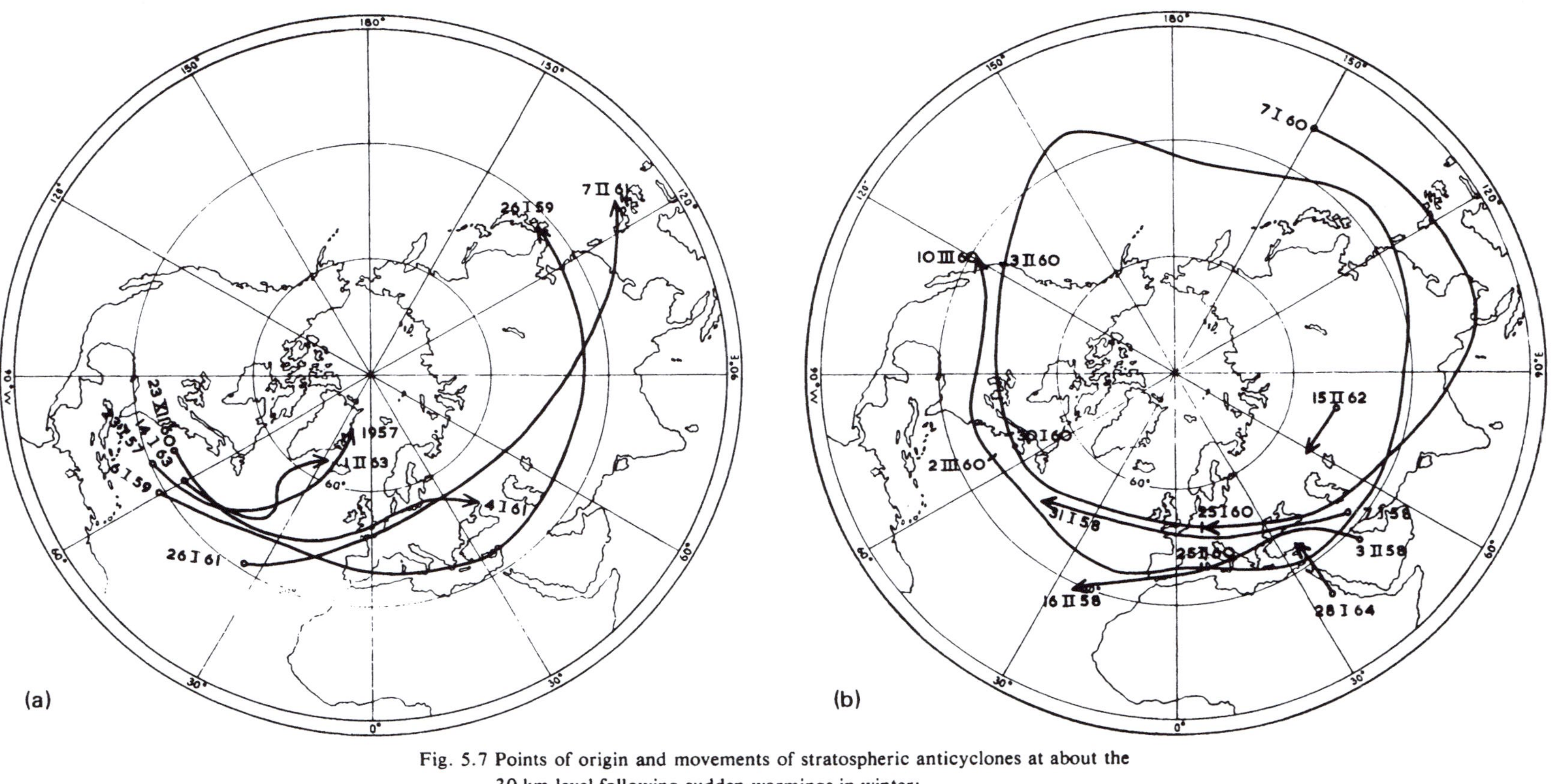

Fig. 5.7 Points of origin and movements of stratospheric anticyclones at about the 30-km level following sudden warmings in winter:

(*a*) 1957, 1959, 1961, 1963
(*b*) 1958, 1960, 1962, 1964

(*After* Labitzke *1962*, *1965*.)

In the case of all the European warmings there was an important sequel in the troposphere in that within 10 days after the beginning of the stratospheric warming sea level pressure rose 30–70 mb in the region of northern Europe previously occupied by low pressure, producing a blocking anticyclone there which reversed the surface winds over much of Europe and dominated the weather situation for the next one to three weeks. Usually the surface anticyclonic regime was at first nearly stationary but subsequently showed a net retrogression to Greenland: this was brought about through several successive renewals of the main high-pressure cell, the old cell generally moving east or south as it weakened and being replaced by a new one from the west or northwest. Events in the troposphere after the American-type warmings in the other years followed no such regular pattern; though in about half the cases there was some appearance within about ten days of an anticyclonic regime at the surface over Canada or New England where the main cyclonic area had been, it did not persist as in the European cases. This difference may be an outcome of the habitually great mobility of tropospheric systems in the American sector near, and south of, the great winter cold trough.

There are several respects in which these interesting cases of coupling between the stratospheric and tropospheric circulations leave the picture unsatisfactory or incomplete. The radical difference between the surface weather patterns in January in 1957 and 1963, respectively the mildest and coldest Januarys in Europe for some long time previously, yet both showing strong American-type stratospheric warmings which pursued a notably similar course, has already been noticed. Moreover, the circulation link in these American-type cases appears to go from troposphere to succeeding events in the stratosphere, not vice versa. It is a case of the stratosphere reacting strongly to events in the troposphere and producing what may be the clearest available indicator (a so-called 'strong signal') of the vertical motion developed. The blocking sequel to the European-type stratospheric warmings seems to be a genuine and dependable case of linkage from stratosphere to subsequent happenings in the troposphere, unless it is really a much delayed sequel to the earlier events in the troposphere which triggered off the stratospheric warming: either way the stratosphere provides a useful forecast indicator. But although the resulting blocking anticyclones usually produced a week or two of east winds and bitter winter weather in Europe, none of the winters concerned was a great winter of prolonged severity (as 1963 was). In other words, the blocking anticyclone did not remain stationary for very long.

There are, moreover, remaining difficulties about identification of the level of origin of the rapid stratospheric winter warmings. Despite the indications that they regularly follow the same situation in the troposphere over America or Europe in the days before, the warming in the stratosphere is nearly always first apparent at high levels, 30 km or above, and descends towards the tropopause. I am indebted to Mr C. L. Hawson of the Meteorological Office for the suggestion that this may only mean that the layer of warmest air produced by the phenomenon in the stratosphere leans forward with height (i.e. in the direction of motion of the system) like the slope of a warm front surface in the troposphere.

However, at least one case of the contrary slope has been observed. Also, there are several times as many warmings in the course of a winter above the 27- to 30-km (10- to 20-mb) level than ever get below that. These high-level warmings are probably related to circulations developed in the mesosphere above the heated ozone layer. It is the less frequent warmings of the whole depth of the stratosphere – usually occurring only one to three times in a winter – that raise the question of coupling with tropospheric events. In these cases it seems possible that the vertical motion induced by tropospheric circulations may happen to get in phase with that of a high-level stratospheric warming, with the result that the vertical motions are accentuated and give rise to systems occupying the whole stratosphere and undergoing some life-cycle that lasts many days.

The '26-month' alternating wind regime in the equatorial stratosphere

The opposite motions around the hemisphere of the stratospheric anticyclones formed by the winter warmings in (roughly) alternate years (figs. 5.7 (*a*), (*b*)) have kept in phase, over the years for which observations are available, with the winds at the same height in the lower stratosphere over the equator where westerly and easterly winds alternate in an oscillation usually of 25–28 months' duration, which 'swamps' completely all seasonal and lesser variations. GEB (1966) has also demonstrated from data for the years 1950–65 that surface blocking anticyclones over Europe and the eastern Atlantic were significantly more frequently renewed from the west, were oftener centred north of 60°N and that the situation had significant tendencies to last longer and move retrogressively – i.e. westward – when the winds at the 25-km level in the equatorial stratosphere were easterly. The oscillation is illustrated in fig. 5.8 by the winds at about 20 km over Canton Island, near 3°S in the central Pacific. The movement of the stratospheric anticyclones around the Arctic was westwards in those winters when east winds prevailed at 20–30 km in the equatorial stratosphere or when the winds there were just changing over and becoming increasingly easterly (cf. fig. 5.9). Because the period of the oscillation is somewhat over 2 years it should get out of phase with the northern winter season at intervals: there are, however, indications

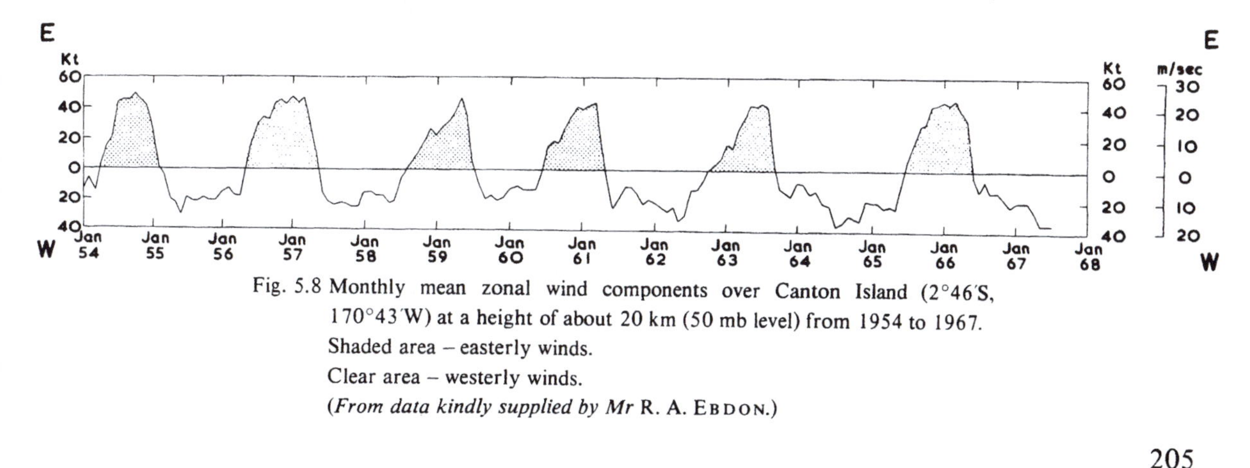

Fig. 5.8 Monthly mean zonal wind components over Canton Island (2°46′S, 170°43′W) at a height of about 20 km (50 mb level) from 1954 to 1967.
Shaded area – easterly winds.
Clear area – westerly winds.
(*From data kindly supplied by Mr* R. A. EBDON.)

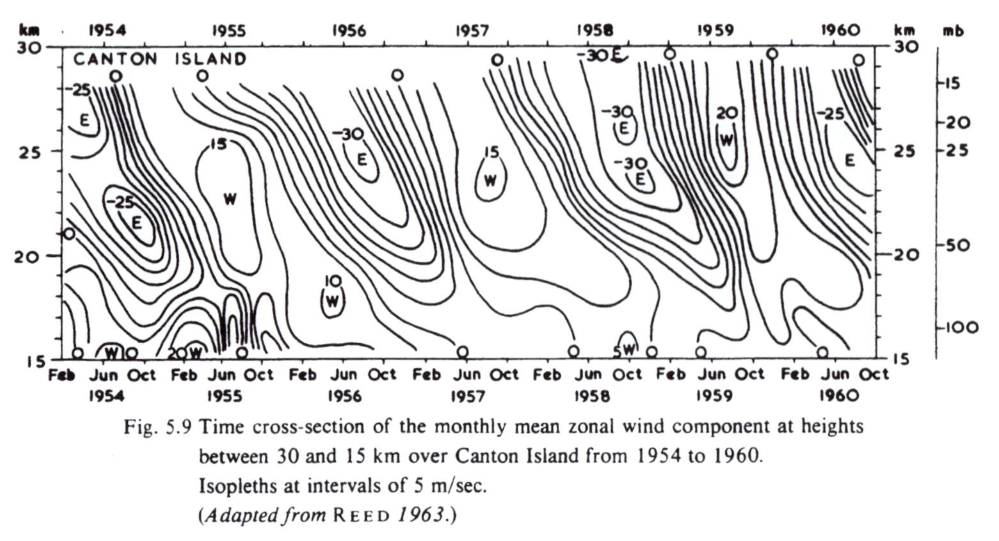

Fig. 5.9 Time cross-section of the monthly mean zonal wind component at heights between 30 and 15 km over Canton Island from 1954 to 1960. Isopleths at intervals of 5 m/sec. (*Adapted from* REED *1963*.)

from statistical studies by LANDSBERG *et al.* (1963) of long series of surface data which show some association with the stratospheric oscillation, that the latter tends to become locked with the season of the year, lingering for some years in phase with the northern winter, especially January, February and March. If the period of the oscillation were strictly 26 months, its phase should only coincide with the middle of the northern winter at intervals of 13 years (six cycles being completed in 13·0 years). For a 25-month cycle the coincidences would be only once in 25 years, whereas, for a 27-month cycle coincidences would occur every 9 years.[1] The problem of the spacing of groups of years with the equatorial stratospheric easterlies in phase with the European winters is one familiar from the theory of sound – that of 'beats' between waves of different frequencies or period lengths – in this case beats between the 12-month and the approximately 26-month oscillations. In reality, however, the period of the latter is not constant. EBDON (1963) reports that it varied between 22 and 29 months as shown by the dates indicated for the change-overs in the equatorial stratospheric winds during the years 1908–18 and 1954–62, and the next two cycles after 1962 were 32 and 35 months long.

The roughly 26-month alternation of the winds in the equatorial stratosphere seen in figs. 5.8 and 5.9, was discovered in 1961 by VERYARD and EBDON. It is, despite all variations, the most regular rhythm firmly established in the atmosphere by observation apart from the daily and yearly ones. It is best seen between latitudes about 5°N and 5°S, where its

1. The problem may be stated generally that for any periodicity of length p (measured in months) the number of years n that elapse between coincidences of phase with exactly the same time of year is given by

$$12n = mp$$

where n and m are both integers and are respectively the lowest numbers that will give a solution to the equation.

amplitude is greatest and seasonal oscillations are small. The winds concerned are nearly due east or due west, encircling the globe, though there is evidence of some cellular structure in differences of wind speed at different longitudes. The change-overs from E to W and W to E appear to take place almost simultaneously in all longitudes at the equator.

Phase lags arise as the associated phenomena, including changes of ozone amount and temperature, are propagated to other latitudes, first at 30 km or above in the ozone layer. At latitude 30° lags of 3–6 months have been found at various heights; at higher latitudes the amplitude of the wind oscillation is small and overlaid by seasonal and other variations, but the lag at latitude 50–60° may be nearly zero – i.e. in phase again with the equator.

As fig. 5.8 shows, periods of rather steady westerly wind alternate with easterlies that increase to a usually sharp maximum and then are fairly quickly replaced by the westerlies. The overall duration of west winds and east winds in each cycle is normally about equal – on average, roughly 13 months of each. In fig. 5.9 it can be seen that the amplitude of the wind-changes decreases from the 25- to 30-km level down to the tropopause, where it is small. Also, with considerable regularity the new E or W wind regime appears first at the highest levels surveyed and proceeds slowly downwards, onset of the westerly regime taking 8–10 months to descend from 30 to 20 km and onset of the easterly regime taking 8–15 months in the cases shown. Indirect statistical techniques of analysis (PROBERT-JONES 1964) have indicated the likeliest origins of this sequence to be at 43, 61 or almost 80 km – i.e. in the upper part of the ozone layer or in the mesosphere above it.

The causation and mechanism of the 26-month oscillation are still matters of debate (see, for instance, the summary given by MURGATROYD 1965). The mechanism can be discussed in terms of the thermal wind relationship. The accompanying temperature variations are in opposite phase at the equator and at latitude 30°, the amplitude being about 2°C in the 20- to 30-km levels over the equator and half that at latitude 30° but passing through a minimum in between (at about 17°N). Over the equator in the lowest part of the stratosphere the westerly winds more or less accompany the maximum temperature, though at higher levels the temperature oscillation precedes the wind oscillation by several months. There is, however, approximate agreement with the geostrophic winds derived from the density, and corresponding pressure, gradients between latitudes 0° and 30°. The temperature variations concerned are such as could be produced by vertical motion reaching a maximum sustained rate of about 10 m/day and meridional drifts at up to 1–2 km/day between the equator and latitude 30°. There is thus a case for believing that the phenomenon represents slow overturning of the stratosphere between these latitudes in rather shallow meridional cells, operating in alternate directions and descending bodily one after the other from the height at which they are formed. In the lowest part of the stratosphere, however, below about 25 km, this process must merge in some way into the general net poleward transport indicated by the movement of ozone, volcanic dust and other trace substances (see, for example, GOLDSMITH and BROWN 1961, KULKARNI 1966).

The suggestion that the phenomenon is set in motion because for some reason the atmosphere is specially responsive to the fifth harmonic of the 11-year cycle of solar disturbance seems inappropriate in that from about 1900 to 1960 the solar cycle was more nearly 10–10½ years long. It seems clear that, for whatever reason, the atmosphere does not complete its budget of heat transport within each year but needs this longer period, which actually does not harmonize with the solar year, to do so. The idea of some, possibly 26-month, variation in the solar ultra-violet producing a corresponding variation in the heating of the ozone layer remains attractive, however, because it operates close to the apparent level of origin of the changes in the stratospheric circulation. Indeed, SHAPIRO and WARD (1962) have reported a minor, but significant, 25-month periodicity in 200 years of sunspot number data. The most notable disturbance of the 26-month regularity so far observed in the stratospheric winds followed the formation of a world-wide dust veil in the stratosphere from the volcanic explosion in Bali (8°S) in March 1963. Abnormal heating of the stratosphere was observed and the duration of the next cycle of west winds in the equatorial stratosphere, which set in at the 20-km level over Canton Island in September 1963, was prolonged to 22 months. It must be admitted, however, that an abrupt change (weakening and lengthening) in the '11-year' sunspot cycles also seems to have taken place from 1963 to 1964 (see p. 210).

We have referred to intermittent occurrences of blocking in alternate winters in Europe, the behaviour of which seems to show links with the phase of the equatorial stratospheric winds. There are other items of surface weather in most parts of the world in which traces have been found of what appears to be the same periodicity, though the manner of the linkages running through the world-wide atmospheric circulations in troposphere and stratosphere is not at present understood.

LANDSBERG (1962) has listed many aspects of the surface weather, as well as related wind and ocean surface conditions, in many widely separated parts of the world which have been reported to show statistically significant quasi-biennial – i.e. alternate-year – rhythms over long runs of years. The precise periods suggested ranged from 1·9 to 2·7 years.

PROBERT-JONES, from data for 1954–61, considered the apparent 25- to 26-month oscillation in the equatorial stratosphere here discussed to be really made up of two preferred periods of 22½ and 27 months, though several individual places' data examined yielded maximum amplitudes at 25–26 months and the values from such a short span of years must be subject to sampling error. LANDSBERG's list, summarized here in Table 5.1, includes several examples of approximately 2-year periodicities registered in geological and biological material from past ages.

N. E. DAVIS (1967) has examined the summers of the years 1880–1966 in Europe. This showed that over this 87-year span average day maximum temperatures and sunshine amounts in summer were significantly greater in the odd years than in the even years over an area covering Britain, southern Scandinavia, central Europe and France while the reverse was true for the region covering Iceland, Spitsbergen and north Norway. The temperature difference was 0·5°C at London (Kew). The frequency of blocking anti-

TABLE 5.1

Reported indications of approximately alternate year rhythms in surface weather

Place	*Element*	*Periodicity indicated (years)*	*Observations covering*
Iceland	Pressure	2·2–2·4	1884–1925
Bergen, Norway	Temp., precip., pressure	2·0–2·3	1861–1914
Stockholm	Winter temp.	2·0	1757–1906
Uppsala, Sweden	Precip.	1·9–2·2	1836–1912
Russia	Snow cover, ice	2·0	1874–1893
U.S.A.	Temp.	2·1	1839–1873
U.S. Pacific coast	Precip.	2·4	1850–1922
Europe	Temp.	2·0–2·5	Eighteenth century to 1922
Britain	Various elements	2·2–2·4	ditto
Edinburgh	Precip., pressure	2·1, 2·4	Eighteenth and nineteenth centuries
Central Europe	Temp., precip., pressure	2·0–2·4	Various periods: eighteenth century to 1927
Azores	Pressure	2·25–2·4	1884–1920
N. Italy	Precip.	2·2	1764–1863
Punjab	Precip.	2·1	1863–1918
Batavia/Jakarta	Pressure	2–2·5	1866–1960
Southern Oscillation	Pressure gradient	2–2·5	1866–1940
Whole Earth	Dec.–Feb. temp.	2	1901–1912
Norway coast	Sea surface temp. in winter	2·0	1874–1904
N. Atlantic	Sea surface temp.	2	1883–1904
Nile river	Floods	2·0	641–1451
Western U.S.A.	Tree rings	2·1, 2·7	Last 3000 years
N. & S. America Scandinavia, India	Varves (sedimentation layers in former lake beds)	2·2	Last 18 000 years

cyclones over the northeast Atlantic and northern and central Europe was also significantly greater in the summers (June, July and August) of the odd than the even years. In detail the data might be better fitted by a periodicity of between 2·0 and 2·1 years (about 25 months), leading to phase changes through which the better summers in Europe tended to come in the even years at intervals of about 30 years (around 1870, 1900, 1934 and 1964 though each time only over runs of a *few* years). Connexion of this alternating-summers oscillation in the troposphere with the stratospheric wind oscillation over low latitudes cannot be regarded as firmly established, but seems rather similar to the relationships with blocking behaviour in alternate winters over Europe noted by LABITZKE (1965) and GEB (1966). (See also next chapter.)

It should not be thought, on the basis of Europe's summers between 1880 and 1964, that the summers in the odd-numbered years *always* tend to be the better. At least between 1866 and 1880 the opposite was true, and again between 1902 and 1910; in the 1930s and early 1940s, as in the 1960s, differences between the odd and even years were slight or non-systematic. The warmest summers of the nineteenth century in England were 1826, 1846,

1868 and 1899, mostly even years. A study by P. B. WRIGHT (1968) of the long series of reports of the annual vintages in Luxemburg since 1626 shows that periods in which the odd years had generally the better summers alternated with those in which the even years were the better.[1] The phases with the better summers in the odd years returned at about 30-year intervals since 1850. This is what should happen if the true period of the basic oscillation were not exactly 2 years but 2·15 years, i.e. almost 26 months, or almost the same as for the stratospheric oscillation as observed between 1954 and 1964.[2] The Luxemburg vintages further show, however, that between the start of the record in 1626 and the mid-nineteenth century the phases in which the odd (or even) years' summers tended to be the better came round more frequently, about every 16 years. This is what might be expected if the basic oscillation averaged not 2·15 but 2·3 years or 27–28 months. It may be significant that the average length of the roughly 11-year sunspot cycles from 1914 to 1964 was 10·2 years, and over the whole time since the 1820s–1850s it averaged 10·8 years, as compared with 11·4 years between 1626 and 1823 – i.e. a change in the same ratio and about the same time as that implied in the mean length of the quasi-biennial oscillation.[3] In both cases the latter appears to have been just one-fifth of the length of the sunspot cycle.

An apparent reversion to a slower time-scale seems to be affecting both the quasi-biennial oscillation and the sunspot cycle since about 1964: indeed, the next two repeats of the equatorial stratospheric wind oscillation, and of some surface phenomena which may be associated with it, approximated to a 33 to 35 months, or more nearly a triennial, oscillation. These observations suggest that the atmospheric and oceanic processes involved in the generally 2- to $2\frac{1}{2}$-year oscillation depend on, and are adjusted to, some type of solar disturbance as trigger.

It has been suggested that the behaviour of the meridional circulation cell, or tier of descending cells, in the stratosphere between latitudes 0° and 30°N is in some way linked with a vacillation of similar period in the Hadley cell of the tropospheric circulation (REED 1963), thereby possibly affecting the troposphere in middle latitudes also. This is, however, hard to reconcile with the dwindling amplitude of the 26-month oscillation below 25 km in the lowest part of the stratosphere and its virtual disappearance near the tropopause. Quite another interpretation of how this oscillation is linked with events of similar period-length in

1. The earliest period for which the apparent biennial oscillation of the European summers has been reported is in the years 1529–41, for which LADURIE (1967) found it strongly marked in the vintages in France, all the even years being the better.

2. Refer to footnote, p. 206. When we further demand that coincidences between that phase of the supposed approximately 2-year periodicity which tends to give fine summers in, for instance, England should occur only in the odd years (or only in the even years), the lowest number of years *n* that can separate such occurrences must itself be an even number.

3. Since this paragraph was written D. J. SCHOVE has traced the variations of cycle length, sometimes approximating to biennial oscillations and sometimes more nearly triennial oscillations, in various types of data extending back over nearly 500 years. (See SCHOVE, D. J. (1971) Biennial oscillations and solar cycles, A.D. 1490–1970. *Weather*, **26**, 201–9. London 1971.)

the troposphere is possible. There is some evidence (ANGELL and KORSHOVER 1962) that the stratospheric oscillation may actually originate in middle to high latitudes[1] over North America (its phase is early there (cf. p. 207)); and this may be related to RIETSCHEL's (1929) discovery that the surface temperatures in winter, studied for the years 1901–12, hinted at a 2-year oscillation of the cold pole of the lower atmosphere between Canada and Siberia. This investigation should be extended over a much longer span of years. The result could mean that it is, at least partly, at the pole of cold air in the troposphere that some linkage between the stratospheric oscillation and the troposphere operates.

We shall examine further aspects of the world-wide biennial oscillation in surface weather in the next chapter together with evidence of other quasi-periodicities, including some on much longer time-scales. Of any long-term variations in the circulation, or of temperatures prevailing, in the stratosphere nothing is at present known. It has been suggested (LANDSBERG *et al.* 1963) from an examination of supposedly related surface data back to 1840 that in some runs of years, or longer epochs, the change-overs from E to W and vice versa in the equatorial stratospheric wind may not take place simultaneously all the way round the equator, and this could be due to a greater eccentricity developing in the circumpolar circulations of either hemisphere. Clearly also the prevailing dates of the seasonal change-overs of the stratospheric vortices in either hemisphere might vary, and this could have repercussions in the troposphere, as could any long-term variations of tropopause height. Solution of these problems must await the future accumulation of much longer runs of reliable stratospheric data, though the longer past history of tropospheric weather data should be explored in as much detail as possible for variations of circulation vigour and pattern which might have a bearing upon the subject. A survey of the past history of the lower atmospheric circulation and weather will be found in the later parts of this book.

1. The suggestion that the oscillation originates in higher latitudes seems likely because of the theoretical impossibility of generating anticyclones or a high-pressure zone near the equator such as would be needed to produce the westerly winds there. At very low latitudes when the cyclostrophic term (V^2/r) in the gradient wind equation exceeds the Earth rotation term ($2\omega V \sin \phi$), with anticyclonic curvature (r negative) or with straight isobars and $2\omega \sin \phi$ very small, there is no balance possible with the pressure gradient force directed outward from a high-pressure region. Any accumulation of pressure will therefore be dissipated and any westerly winds observed near the equator must represent westerly momentum acquired in other latitudes and transported thither.

CHAPTER 6

Cyclic and quasi-periodic phenomena

In earlier chapters diurnal and seasonal rhythms in the behaviour of winds and weather have been noticed. That these are responses to the regular 24-hourly and yearly astronomical variations of the radiation supply needs no elaboration, and the chains of thermal and dynamical causation that produce them can be fairly simply followed. The semi-diurnal pressure wave (see later, pp. 217–18) which is very prominent in low latitudes, where longer-term variations of pressure are slight, is of partly tidal – i.e. gravitational – and partly thermal origin. No other rhythms are known in the atmosphere which could have such straightforward origins and which can be observed to show a similar approach to regularity in operation. Even the semi-diurnal, diurnal and annual (i.e. seasonal) rhythms are liable to fade, and may be lost for a time, especially in those latitudes where their amplitudes are weak – the former two near the poles, the last-named in the tropics. Variations of wind and heat transport, cloudiness, sky haze, etc., due to other causes, may for a time so affect the radiation receipt at the surface as to suppress, or mask and overlay, the expected rhythm.

The three very long cyclic variations in the Earth's orbital arrangements first calculated by MILANKOVITCH (see Chapter 2, pp. 30–37 and Appendix II) must also be supposed to have effects upon the radiation regime, and through that upon the wind circulation and other climatic conditions, as inescapable as those that accompany night and day, winter and summer. These are:

(i) the varying ellipticity of the orbit – period *circa* 96 000 years,
(ii) the varying obliquity of the ecliptic (tilt of the Earth's rotation axis) – period *circa* 40 000 years,
(iii) the precession of the solstices and equinoxes – period *circa* 21 000 years.

The effects of these variations on the input of solar radiation at various latitudes in the summer and winter respectively correspond to superposed sine curves of differing wave length, amplitude and phase, producing an at first sight irregular sequence of fluctuations of varying amplitude (cf. figs. 2.5–2.7). Because the time-scale of these variations is so long, their effects upon the atmosphere cannot be demonstrated with certainty nor in detail to compare with those that accompany night and day or the seasonal round of the year.

Nevertheless, it is widely accepted that they have to do with the sequence of ice ages and interglacial periods and with the warmer and colder 'interstadials' (i.e. stages) within each.

BERNARD (1962*a*, *b*; 1964) has taken the MILANKOVITCH variations of the radiation budget, together with our present understanding of the derivation of the general atmospheric circulation and moisture cycle, as the basis of a formally complete theory of palaeo-climatology, by means of which the time sequence and characteristics of different regimes can be calculated. BERNARD's logical derivation leads him to a confident assertion that interglacial warm periods in high latitudes must correspond to maximum obliquity and to pluvial periods in Africa, whereas the ice ages should be periods of minimum obliquity and of aridity in Africa. This type of rigorously theoretical approach must be regarded as preliminary, if not premature, until it can be demonstrated as fitting and explaining all well-verified observations. Similarly KUTZBACH *et al.* (1968) have used the MILANKOVITCH summer and winter half-year radiation calculations as a basis for deriving thermal gradients (1000–500 mb thickness) between 20° and 70°N at 1000-year intervals over the last half million years. But, inasmuch as no account is taken of the secondary, but at times large-scale, effects of the albedo variations associated with waxing and waning of the snow and ice area of the polar caps, such calculations can only be a first stage in derivation of the probable variations of the atmospheric circulation. The latter may even be forced by the changing geography of snow and ice to show a radically different time-scale or to suppress almost altogether some of the theoretical oscillations of circulation strength.

ZEUNER (1958, 1959) actually adopted MILANKOVITCH's calculations as a proposed chronology into which the observational evidence of the climatic sequence of the last million years must fit. This gave a theoretical method of dating the evidence of past climates and their effects upon the landscape and its inhabitants, before modern physical dating methods depending on measurement of the decay of radioactive elements had been discovered. Radiocarbon datings of relevant material from the last 50 000 years or so – the approximate limit of practical applicability of the technique – have now indicated a chronology of the last glaciation and the post-glacial period which seems understandably related (FAIRBRIDGE 1961, LAMB *et al.* 1966) to MILANKOVITCH's calculated curve of the summer sun available in the Arctic and sub-Arctic, where weakened insolation may be most directly connected with the growth of ice sheets; though it is necessary to admit some 5000 years delay in accomplishing the melting and disappearance of the former great northern ice sheets. In fact, the last maximum of the summer radiation, about 10 000 years ago, seems to have been the time of most rapid melting of the ice caps, as indicated by the rate of rise of sea level (LAMB *et al.*, *loc. cit.*).

The relationship of the dates of the MILANKOVITCH radiation fluctuations to the full terrestrial sequence of ice ages and interglacial warm periods in the Quaternary era cannot yet be adequately tested, because independent methods of dating the ground evidence from more than 50 000 years ago still depend too much on extrapolating sedimentation rates in the sea-bed and lake-bed deposits examined or growth rates in bogs. Nevertheless, the assumption of a terrestrial response to these long astronomical cycles affecting the radiation supply, with whatever lag is imposed by the inertia of large accumulations of ice, rests on the firmest possible theoretical justification.

Between these long (10^4–10^5 years) and short (1 day and 1 year) periods no such regularity is apparent in observed atmospheric behaviour. Hints of recurrence tendencies at intervals not strictly constant but of the order of 5, 15, 30 and 45 days, which appear for a time and then are lost again, may correspond to the life-cycles (maturing and decay) of an individual anticyclone, of one mode of poleward transport of heat, or of the migration of some tropospheric or stratospheric feature around the Arctic or Antarctic respectively. Such evanescent rhythms have, at best, a very limited predictive value: even this can only be attained by monitoring the progress of some flow-system in the atmosphere, the 'mechanism' of which is either reasonably well understood or at least familiar in performance. The nearest approach to regular, and continuous, operation may be that of the 'Southern Oscillation', to be described later in this chapter, and the attendant (approximately) 2-year fluctuations in surface weather some of which have already been mentioned (Table 5.1, p. 209).

A search for dependable periodicities – inspired by the hope of finding some predictable element in the year-to-year and decade-to-decade, or longer, variations of climate – has long provoked effort out of all proportion to the results so far attained. Lists of periodicities reported present in long series of weather observations were published by SHAW (1936, p. 320) and by BERLAGE (1957, p. 13); these indicated that almost every conceivable period from 1 to 36 years had been suggested by investigation of one observation series or another. New statistical methods, and computers to tackle hitherto unmanageable calculations, now promise to put this field of research in perspective.

In the following paragraphs we survey the matter afresh from two opposite aspects:

(i) known oscillations (of whatever period length) inherent in any external circumstances that may be presumed to affect the atmosphere and oceans,
(ii) oscillatory tendencies of various period lengths, and apparently of significant amplitude, found in the behaviour of the atmosphere.

In reality all the phenomena we shall be discussing are no more than quasi-periodicities, since all are subject to variations of amplitude and some modulation of period length.

Periodicities inherent in influences outside the Earth or external to the atmosphere

First, we may list certain astronomical periodicities not so far mentioned which seem liable to influence the atmosphere directly or indirectly:

(a) *Tidal forces of the planets operating upon the sun* ('*sun tides*')

The varying tidal pull (or, more accurately, tide-raising force, in the manner of the moon's differential attraction on the Earth and its fluid envelope (see pp. 218–19), exerted by the planets on the sun is thought to affect – or, as some have suggested, control – the amount of disturbance on the sun's face in the form of sunspots, etc. Other, less likely, suggestions as to sunspots include one arrived at by H. H. TURNER in 1913 (described by BROOKS 1949, p. 367) attributing them to the impact of meteors in a so far unidentified swarm following an elliptical orbit around the sun with a period of about 11 years and subject to interference by the Leonid swarm which has a period of about 33 years. Alternatively, the various observed disturbances of the sun, and such variations of its energy output as are known, may be entirely due to powerful thermal and thermo-chemical processes inside the sun or to some interplay between these and the varying tidal force of the planets.

The gravitational attraction of a planet P with mass m at a distance s is proportional to m/s^3; i.e. it varies directly with the mass of the planet and inversely as the cube of the distance separating sun and planet during the planet's passage along an elliptical orbit. Of the planets which might be important in this connexion, only Venus has a nearly circular orbit and therefore exerts an almost constant force at the sun. (Data on planetary orbits are given in Appendix III.)

The combined tide-raising force F on the sun of planets P_1, P_2, P_3, etc., acting along the line towards one of them, P_1, may be written as

$$F = k\left\{\frac{m_1}{s_1{}^3} + \frac{m_2}{s_2{}^3}(\cos \widehat{P_1SP_2}) + \frac{m_3}{s_3{}^3}(\cos \widehat{P_1SP_3}) + \ldots\right\}$$

where k is a constant, $\widehat{P_1SP_2}$ is the angle subtended at the sun between the planets P_1 and P_2, and so on. The combined force must attain maximum values when two or more of the planets are in line. Jupiter and Venus, at their mean distances, contribute most to this tide-raising force on the sun, respectively about 34·6 and 33·0% of the total (BOLLINGER 1964, 1968). The contribution of the Earth at its mean distance is approximately 15·5%. Together these three planets account for 83·0% of the total. Conjunctions of Jupiter and Venus occur at 237-day (0·65 year) intervals and oppositions at half this time. Conjunctions of the Earth, Jupiter and Venus occur at intervals of 24·00 years; alignments, with either Jupiter or Venus in opposition, occur every 12·00 years. These alignments and conjunctions are reinforced by alignments with Mercury and Saturn at intervals of 60 and 120 years respectively, and by alignment with Uranus at intervals of 84 and 168 years respectively. It is evident that all these planets return to similar positions, with the major ones being in conjunction, at intervals of 1680 years (the least common multiple of the figures mentioned). Alignments, with one of the main contributing planets in opposition, occur at the half-period of 840 years.

It is necessary to examine whether any of these period lengths correspond to significant variations of either solar disturbance or weather. Clearly, the most prominent and best known solar disturbance cycle differs slightly from 12 years; though WOOD and WOOD (1965) derived an 11·08-year period (identical with the mean separation of sunspot minima between 1610 and 1960) as likely by considering the interacting effects on the sun of the

Earth, Venus and Mercury and those of the four large outer planets as two separate groups affecting the centre of gravity of the solar system.[1] BOLLINGER states that the proposed 84- and 168-year period lengths can be identified in the recorded heights of the annual floods of the river Nile, available from A.D. 620. (S. M. WOOD 1946, 1949, believed that cycles of about 737 years were detectable in the Nile record and that these could be associated with the return of certain groups of planets to the same relative positions at intervals of this length. BOLLINGER's argument is the more convincing, because it depends not on all planets alike, but on sorting out which planets are most effective in the combined tidal pull upon the sun.) We shall come across suggestions of identifiable periodicities of 80–90, about 170 and 1700–2000 years in other observed data. It happens that 1680 years is rather close to the roughly 1800-year rhythm found by ŠNITNIKOV (1949) in lake levels and wetness in Eurasia; though the quality of the data has allowed other investigators of post-glacial climates to suggest various periods in the range 1800–2200 years.

The wandering lake, Lop Nor, in Sinkiang, in the heart of Asia, which according to the evidence of SVEN HEDIN (1940) alternates between a northern and a southern position, shifting at intervals of 1700 years, presumably completes its cycle in about 3400 years or just two of the longest periods we have been considering. This resembles the roughly 3500-year intervals between the forward stages of the glaciers in the latter part of the last ice age and afterwards, as first noted by SUESS (1956) and still more or less apparent in the latest radiocarbon datings of the evidence in North America and Europe – see Table 6.1.

There is, thus, a prima facie case for accepting the existence of some long cycles in weather phenomena which may be of about the right order of length to agree with the planet-induced sun-tide cycles noted. The amplitudes of the climatic fluctuations mentioned on time-scales of about 1700–2000 and 3500 years appear to be great in terms of rainfall in the heart of Asia and of summer and winter temperatures in Europe and central North America. These fluctuations presumably operate through changes in the prevailing latitude, strength and wave length of the mainstream of the upper westerlies and the associated thermal gradient. Unfortunately, there is little prospect of adequate precision in dating the evidence to establish how nearly the long periods in weather and external phenomena agree. We shall examine later in this chapter the possible existence in weather phenomena of periodicities agreeing with the shorter planet conjunction/alignment cycles mentioned. Some indication of whether cycles of the greater lengths mentioned in these paragraphs occur in the behaviour of the sun might be obtained by periodogram analysis of the apparently somewhat varying quantity of radiocarbon (^{14}C) present in the Earth's atmosphere over the past 7000–8000 years as derived by SUESS and others (e.g. RALPH and MICHAEL 1967); though even these data constitute too short a run to probe for a solar oscillation to correspond to the suggested 3500-year weather cycle.

1. The centre of mass of the solar system can at times lie outside the sun: i.e. the sun may be centred more than one solar radius from the centre of mass of the system when Jupiter and Saturn lie in the same general direction.

TABLE 6.1

Radiocarbon dates of glacial stages in North America

(Adapted from SUESS (1956) in the light of subsequent datings)

Stage in North America	*^{14}C date*	*Equivalent cold stages in Europe*
'Tazewell' ice maximum	*c.* 17000–15000 B.C.	End of the main Würm III (or Weichselian) ice age maximum phase.
'Cary' readvance	*c.* 14500–13500 B.C.	
'Mankato' readvance	*c.* 11000 B.C.	End (?) of the first notable warmer phase, the Bølling interstadial.
'Valders' readvance	*c.* 9000–8200 B.C.	Just after the Allerød warm period (*c.* 10000–9000 B.C.) there came a temperature drop and glacier readvance unequalled in post-glacial times (Zone III in British pollen stratigraphy diagrams).
'Cochrane' readvance	*c.* 6500–4700 B.C.	No known equivalent, though FRENZEL (1966) refers (p. 103) to evidence of short-term cold spells in the Alps between 5500 and 4500 B.C.; the so-called Boreal period of marked dryness and warmth in Europe approximately coincided with this time span and was linked, according to the evidence advanced by LAMB *et al.* (1966), with a persistent anticyclonic influence covering much of northern and central Europe. This pattern was called into being by the continuing ice age in North America when Europe was already clear of ice, a situation determining a prevalent SW to NE jet stream and steering of Atlantic disturbances. The same situation may well, however, have maintained a cold regime just farther east over northwest Asia.
'Neoglacial': end of the post-glacial warmest epoch, reappearance of many glaciers in the Rockies	*c.* 1000 B.C.	End of the post-glacial warmest time. Glaciers advancing and many reappearing in the Alps. Wetness setting in abruptly in northwest Europe. Several centuries of very wet conditions are registered in the Irish, Welsh and North European peat bogs (1000–50 B.C.)

Note: Recently evidence has been found in Europe and in North and South America (see, e.g., FRENZEL 1966, p. 104) of an important cold oscillation in the midst of the post-glacial warm millennia at a date just intermediate in the above series, about 3000 B.C.

The possible relevance to weather prediction of many and diverse short periods in planetary tidal forces on the sun, between 12·6 and 17·5 days and 1·9–13·3 months long, has been discussed by TAKAHASHI (1957) and need not concern us here.

(b) *Tidal forces of the sun and moon operating upon the Earth*

Sun and moon both produce semi-diurnal tides in the atmosphere as in the oceans. The double wave is attributable to the gravitational attraction of either body, acting upon the atmosphere and ocean on the near side of the Earth and acting upon the Earth itself to pull it away from its fluid envelope on the far side. The lunar tide in the atmosphere – its amplitude was established by Chapman (1919) – is, however, so minute (amplitude everywhere less than 0·2 mb) as to be of no practical importance. The observed semi-diurnal variation of pressure over a range amounting on average to about 4 mb in the tropics (Chapman gives the amplitude, or half range, at the equator where it is greatest as 2·6 mb) and 1 mb in middle latitudes, with maxima around 10 h and 22 h local time, is due to the sum of components produced by the sun's gravitational (i.e. semi-diurnal tidal) effect and its heating effect. The average tide-raising force of the sun on the Earth is about half that of the moon, and in the ocean the sun raises a tide about half the height of the lunar tide; when the tidal pulls of sun and moon reinforce each other, about full moon and new moon, the total range of the ('spring') tide is therefore about three times that of the neap tides at the lunar quarters.

The combined tidal force undergoes quasi-periodic variations as the Earth and moon go round in their orbits and as their positions combine with different declinations of the sun and moon. The moon goes about the Earth, from full moon to full moon, in 29·531 days. This is known as the lunar 'synodic' month, marked off by conjunctions of moon, Earth and sun. But this period involves the Earth's own progress in its orbit round the sun. If we regard the moon's progress around the Earth from perigee to perigee in its slightly elliptical orbit, these times of perigee (proximity) being when its tide-raising force is greatest, the period is 27·553 days. This is the tidal month, also known as the lunar 'anomalistic' month. Thirteen tidal months make 358·2 days. The interval between successive crossings of the equator, the moments when the moon's declination is zero, is also slightly different, 13·661 days, the cycle length between equator crossings in the same direction being 27·332 days: 355 days, or thirteen of these intervals, may be regarded as the most real approach to a tidal year, when sun, moon and Earth return to similar, though not identical, positions.[1] The eccentricity of the moon's orbit (Allen 1955) averages 0·055 and is greatest when the major axis of the ellipse, known as the 'apsidal line', between the perigee and apogee positions, is in line with the sun: the moon comes nearest the Earth when this is so at full moon. Similarly, the sun's tide-raising force on the Earth is greatest when its distance from the Earth is least – i.e. at 'perihelion' – which at present falls in the middle of the northern winter about 3 January (cf. pp. 30–32). As recently as about A.D. 1250 according to the Brouwer and van Woerkom (1950) calculations of the Earth's orbital cycles (see

1. These, and all other, astronomical periods *as expressed in days* have changed in the course of the Earth's history, because the day changes its length as the spin of the Earth upon its axis slows down through the dissipation of the energy by tidal forces. The year has hardly changed its absolute length, the Earth's time to revolve round the sun being almost constant: but the year, which now has 365·242 days, appears to have had about 370 days 100 million years ago and 424 days 600 million years ago. Palaeontological evidence exists which seems to confirm the astronomical calculations (Berry and Barker 1968). Patterns of the growth rings formed each day, each 'fortnightly' tidal cycle and each year in corals agree with the astronomical results at least as far back as the Cretaceous and suggest that the number of days in the lunar month has changed in the same proportion, i.e. 29·65 days in the Cretaceous 100 million years ago compared with 29·53 today. This implies that the time taken by the moon to revolve about the Earth remains constant. Hence, periodicities of tidal origin, as measured in years, should also stay constant.

Chapter 2 and quotation by BERNARD 1962*a*, p. 16), perihelion coincided with the northern winter solstice. Hence, there should have been great maxima of the combined tidal force upon the higher latitudes of the Earth at extreme phases of the moon about that time.

The moon's orbit is tilted at 5°9′ to the ecliptic, and for this reason the extreme declination which it attains on its way round the Earth varies. The maximum declination in the present epoch is about 28°40′N and S: the moon ranges this widely at intervals of 18·6 years. Midway through each 18·6-year period the moon's declination ranges only between about 18°20′N and 18°20′S. Recent maxima of lunar declination were in or about 1876, 1894, 1913, 1931, 1950 and (October–November) 1968. The tidal force of the moon acting upon high latitudes is greatest when its declination is greatest, and particularly when its perigee position in its orbit coincides with maximum declination. MAKSIMOV and SMIRNOV (1965) have calculated the changes in the mean slope of the surface of the North Atlantic Ocean from 45° to 75°N for the Januarys and Julys of the period 1870–1970 caused by lunar declination: the slope varies from about 6·5 cm upward in the years of maximum lunar declination named to 6·5 cm downward in the intervening minimum declination years. The authors believe that the corresponding differences of gradient force in the ocean surface must entail differences in the poleward component of ocean currents, and they found oscillations of about 19-year period and amplitude generally 0·2–0·4°C superposed on the longer-term secular changes in the water surface temperatures of the North Atlantic between 55° and 67°N.

RAWSON (1907, 1908, 1909) drew attention to the apparent existence of approximately 19- and 9½-year cycles in the average latitudes of the subtropical high-pressure belts of the northern and southern hemispheres and in the occurrence of droughts recorded in South Africa from A.D. 1622 to 1900 and in Argentina from 1827 to 1900, such droughts being presumably connected with a persistent north or south displacement of the subtropical anticyclone belt. An 18- to 20-year oscillation also seems detectable in the level and salinity of the Caspian Sea (ZAYTSEV 1965), though superposed on important longer-term trends. RAWSON associated these phenomena tentatively with the effect of lunar declination, though CHAPMAN's result referred to above (pp. 217–18) suggests that this is unlikely to be a simple tidal effect in the atmosphere. BRYSON (1948), proceeding from the fact that other forces acting on the air in the central regions of the subtropical anticyclones are in a state of near balance, argued that the horizontal component of the lunar tidal force was unbalanced and might therefore, after acting similarly for some days, cause measurable displacements of the anticyclones. He was able to demonstrate a remarkable parallelism between the latitude of the subtropical anticyclone centres (*a*) over the eastern North Pacific, and (*b*) averaged around half the northern hemisphere and the equatorward component of the lunar tidal force 6 days earlier. (The data in case (*a*) were for the months of May to August in the years 1919–39 and in case (*b*) for all months between January 1944 and April 1946.) The association appeared to be statistically significant. The amplitude (half range) of the anticyclone shift was about 1° latitude within the average lunar month. G. W. BRIER (*personal communication*) believes this offers a practical method of forecasting week by week variations of rainfall at places in the temperate zone.

BRIER attaches importance to yet other lunar cycles of 13·6, 27·2 and 44 years. These periods arise in the following ways. After 162–163 months, i.e. 13·6 years, comes the first near approach (actually within $1\frac{1}{2}$ days) to simultaneous completion of whole numbers of (*a*) 29·53-day lunar 'synodic' months, (*b*) 27·55-day lunar 'anomalistic' months and (*c*) half-years of 182·62 days in which the Earth moves exactly to the opposite side of its orbit. 27·2 years brings the corresponding first near-repetition of the phase of the lunar month with the Earth in its original position in its orbit. 44 years brings the first near coincidence between the completion of a whole number (45) of 358-day cycles of 13 tidal months and of calendar years. The horizontal and vertical components of the tidal force, with their semi-diurnal and fortnightly fluctuations, might be expected to show recognizable similarities for some weeks or months at these intervals; and effects upon the latitudes of the subtropical anticyclones and of the jet stream might be looked for. The effects seem not to be strong, however, or often masked by other phenomena. These period lengths seem not to be prominent in meteorological time series, of which illustrations are given later in this chapter.

In the present state of our understanding of tidal phenomena in the atmosphere, two general points are worth noting. Analogous states of the tidal force at intervals such as 9·3, 13·6, 18·6, 27·2 years and so on must comprise the week by week variations in the course of the lunar month. If, as most investigators suppose, the effects operate through a tidal pull exerted northwards or southwards upon the mass of atmosphere in the subtropical high-pressure belt, this should show a poleward or equatorward displacement chiefly in alternate weeks. There should be a distinctive week by week sequence of a tendency to drought or raininess in latitudes intermittently affected by the high-pressure system. Secondly, these period lengths could account for some analogous sequences at odd times of the year apparently unrelated to natural seasons.

The combined tide-raising force of sun and moon must produce the greatest tidal range in the oceans when Earth, moon and sun are in closest alignment and when the moon's perigee and Earth's perihelion coincide; for the greatest tides in high latitudes, sun and moon should at the same time be at maximum declination. This can only happen near the solstices. This gives rise to a long interval of approximately 1800 years (PETTERSSON 1914, 1930) between great maxima of the tides in northern waters.[1] The last time there was a near approach to the exact alignment required to produce maximum tidal force in high latitudes was, according to the calculations used by PETTERSSON, in A.D. 1433. The Earth, sun and moon were then in line, and the moon like the sun at maximum declination, at the northern winter solstice (see fig. App. III.2). The Earth was at perihelion only 2–3 days later (compared with the present occurrence of perihelion 12–13 days after the solstice). Rather

1. According to a recent recalculation by C. M. STACEY (quoted by FAIRBRIDGE 1961) the average interval should be nearer 1670 years.

similar conjunctions of the moon, Earth and sun should have occurred at about 9-year intervals (see Appendix III) before and after the year 1433. In some previous 21 000-year precession cycles the Earth's perihelion position may have fallen even nearer the solstice when the moon's tidal effect was at its maximum, though the combined tidal force would not be greatly increased thereby. The tidal maximum of the late Middle Ages was clearly the greatest for many thousands of years. Greater maxima must have occurred, however, at times when the ellipticity of the Earth's orbit was greater (p. 31) and the perihelion position correspondingly nearer the sun.

The intervals between great maxima of tidal force are about 1600–1800 years but are not quite equal: the last maxima have been calculated as falling around 3500 B.C., 1900 B.C., 250 B.C. and A.D. 1433, the next one as about A.D. 3300. The minima of tidal force, and hence of range of the tides, are calculated to have fallen about 2800 and 1200 B.C. and A.D. 550, the next minimum expected around A.D. 2400. Secondary maxima occur at intervals of 84–93 years and tertiary maxima every 9 years.

If variations on the long period lengths suggested by tidal theory are in fact detected in climate, and therefore in the atmosphere, the effects may be brought about through the abnormally great ocean tides around the maxima of the tidal force allowing more interchange of water across the 'sills' (submarine shelves and ridges) at the entrances to the Arctic Ocean, and to the Bering and Baltic Seas; this may also break loose large quantities of the Arctic sea ice. The latter suggestions have been strongly put forward by PETTERSSON (*loc. cit.*), I. V. MAKSIMOV and others. It seems to be a fact that abnormally great quantities of ice were emerging from the Arctic in the late Middle Ages; but this tendency could be observed from about A.D. 1200 onwards, probably attained its maximum in the 1690s and 1780s, and is more reasonably attributed to climatic conditions favouring an excessive production of ice on the polar seas. The tidal theory would suggest an abnormal influx of warm water from the Atlantic into the Arctic compensating any abnormal outflow from the Arctic in 1433 and about that time.

The importance of the maxima of tide-generating force and tidal range is, however, not limited to greater height of the surface tides passing over the Iceland–Faeroes–Scotland ('Wyville–Thomson') ridge in the ocean bed or through the Danish Sound (Øresund) into the Baltic nor even to flooding of coastal lowlands such as those around the North Sea. PETTERSSON studied not only the sea-surface tidal wave but also that affecting the internal surface between the Atlantic salt water below and the Baltic fresh water above: the internal tidal waves on this inner surface, or density discontinuity, in Gullmarsfjord at the entrance to the Baltic, were found to attain in extreme cases 25–30 m – a much greater range than that of the surface semi-diurnal tides which amount to no more than 30 cm at that point. 'Submarine' tides measured in the Straits of Gibraltar and Messina appear to have an even greater range, up to 60 m. NANSEN observed similar internal tides on the upper boundary of warm saline, Atlantic water entering the Arctic Ocean. Owing to the much smaller difference of density between the different watermasses than between either and the overlying atmosphere, the force required to raise a tidal wave of similar height on the internal surface bounding two watermasses is of the order of a thousandth of that required to produce the sea-surface tide.

Thus, a periodicity of about 1700 to 1800 years, as well as the 18·6-year periodicity mentioned, and the harmonics represented by just half these periods, might be expected in the amount of warm Atlantic water penetrating the Arctic and the Baltic. PETTERSSON (1930) appears to have succeeded in demonstrating the occurrence of abnormal amplitude of the submarine tide and inflow of saline Atlantic water at submarine levels into the Baltic at appropriate dates about 9 years apart in 1909–10, 1917–18 and 1928, but it is not clear whether these occasional great tides have any important lasting effect on sea temperatures or weather. Any independent long-term variation in the power of the wind circulation which drives the Atlantic ocean currents would be liable to have enormously greater sustained effects both in the Baltic and in the Arctic polar basin.

BROOKS (1949, p. 369), like PETTERSSON (1914), believed that maxima of the tidal force should lead to more breaking of the Arctic pack-ice, of which increased quantities could therefore reach the Atlantic. This reasoning may have been chiefly an attempt to account for the undoubtedly rapid spread south of the Arctic pack-ice between A.D. 1200 and 1700 and much circumstantial evidence that another abrupt increase of the Arctic sea ice had taken place about 2000 years earlier, between 1000 B.C. and about 330–320 B.C., when Pytheas is believed to have visited Iceland. Alternative theories are, however, available to explain both these cold climatic episodes. Because of the well-established changes in ^{14}C-content of the atmosphere around those times some corresponding, or at least contemporaneous, changes in the amount of solar disturbance are suspected. Moreover, suggestions of a 1700- to 2000-year periodicity in the planet-induced sun-tides have already been noticed together with certain climatic variations on about the same time-scale.

(c) *Nutations of the Earth's axis of rotation*

The Earth's rotation axis is subject to an oscillation forced by the regular variation of the pull of sun and moon upon the equatorial bulge, a pull which tends to tilt the Earth's polar axis more the greater the declination of sun or moon (JEFFREYS 1954). The motion of the Earth's celestial pole which results can be resolved into two components:

(i) the progressive, nearly uniform motion of a fictitious average pole around the sun – called *precession*;
(ii) the revolution of the Earth's actual pole around the fictitious mean pole. This motion is called *nutation*.

The nutation is nearly elliptical, though in detail many minor wobbles are superposed by the variations of lunar and solar distance due to the ellipticity of the orbits of moon and Earth. One nutation is completed in 18·6 years. The observed amplitude, which is only 9 seconds of arc (about 300 m shift of the pole), is reduced by the inertia of the Earth's molten core.

In addition, the Earth's polar axis performs a free oscillation or nutation as predicted by EULER, in the course of which it describes from west to east a narrow-angle nearly circular cone. This free nutation, with a period of about 14 months – it varies from about 1·1 to 1·2 years – and an amplitude of about one second of arc (i.e. latitude) was first detected by S. C. CHANDLER in 1892 and is sometimes known as 'CHANDLER's wobble'. In detail, the path of the pole seems to be elliptical with an ellipticity that varies: the ratio of major to minor axis averaged 1·06 between 1903 and 1916 and 1·22 between 1922 and 1935 (MAKSIMOV 1958). The major axis retrogressed from 75° to 45°E when these two periods were compared.

The nutations of the Earth's pole produce minute and unimportant variations of sea level, known as the polar tide. A 14-month periodicity has been demonstrated in them, though whether their phase in the different oceans can be intelligibly related to the instantaneous position of the pole is not clear. MAKSIMOV claims that a polar tide with 14-month period of west to east rotation, of significant amplitude and with intelligible phase relationship to the movement of the pole, can be detected in the atmosphere. If so, interference ('beats') between the approximately 14-month period and the 12-month seasonal variation of atmospheric pressure should produce a 7-year periodicity in the behaviour of various climatic quantities. MAKSIMOV mentions in this connexion the continentality (July–January temperature difference) of the climate of Europe, which bears an inverse relationship to the strength and persistence of the prevailing west winds from the ocean. According to MAKSIMOV the 14-month pressure wave attains its maximum amplitudes near latitudes 70°N and 70°S, i.e. just in the zones important in connexion with blocking anticyclones: the mean values of the amplitude in terms of monthly mean pressure near latitude 70° were about 1–1·3 mb. Over the 7-year cycle the amplitude was liable to reach 2·7 mb. It should be noted, however, that these amplitudes are rather small in comparison with the standard deviation of monthly mean pressure (comparing months of the same name in different years), which averages 4·6 mb at 70°N. A higher, but not altogether dissimilar, proportion of the variability of pressure at 70°S might be accounted for by this regular fluctuation which MAKSIMOV associates with the free nutation. The pressure wave concerned has nodes at the poles and near the equator (actually at 0–30°S according to the investigation reported). More recently MAKSIMOV *et al.* (1966, 1967) have suggested the existence of an elliptical migration of the Iceland low-pressure area with a 14-month period associated with the free nutation of the Earth's pole. The amplitudes of the two movements are, however, several orders of magnitude different, that of the Iceland low between 1892 and 1962 averaging plus or minus 100–150 km and up to 450 km over some periods 20 years long. The nutational migration of the Iceland low according to MAKSIMOV follows (i.e. the anomaly is towards a meridian about 90° to the west of) that of the Earth's rotation pole. The actual migration of the Iceland low is the sum of this movement and its normal seasonal migration, which is about twice as big. When the pole moves towards the 0° meridian, the Iceland low is displaced SW and when the pole moves towards the 180° meridian, the Iceland low is displaced NE. The amplitude (half range) of this movement, shown by harmonic analysis of the observed month by month positions of the Iceland low for the 14-month periods between 1945 and 1951 (MAKSIMOV 1970), was about 2° of latitude and 5–6° of longitude. The Azores high underwent a contrary movement with about two-thirds of this amplitude, as the theory suggested.

(d) *The rotation of the sun*

The approximately 27-day rotation of the sun about its axis, is liable to bring round the same disturbed regions of the sun's face to the central meridian facing the Earth at about this interval – 27 days for features near the sun's equator, 28 days near solar latitude 30° (see Chapter 1) – provided the solar disturbance lasts long enough. A periodicity of this order should therefore operate from time to time in any terrestrial responses to solar disturbance.

So far in this chapter, we have considered what cyclic processes there are going on in elements of the external circumstances that may affect or control weather and climate. These have covered the observable, or readily calculable, orbital cycles and such tidal variations as they must produce on the sun or in the oceans and atmosphere on the Earth. These suggest a variety of time scales for periodicities which might be looked for in weather and climate. The responses in the case of diurnal and annual variations of weather are familiar. Suggestions of responses seen in some much longer-term weather variations have been mentioned. The processes inside the sun which produce the observed disturbances of its outer layers, and such variations of solar output as occur with them, as well as the processes in the Earth's atmosphere and oceans involved in long-term variations of climate, are difficult or, in some cases, impossible to observe directly. If there are any periodic, or quasi-periodic, elements in these – such as the normal time taken to complete some circulation – we must probe for them by some form of statistical analysis of periodicities.

Methods for detecting oscillatory tendencies or characteristic time-scales of fluctuation in an observation series

The techniques most used are:

(1) Filtering
(2) Harmonic analysis
(3) Various forms of correlogram or periodogram
(4) Power spectrum analysis.

All these methods involve amounts of arithmetic that are best handled on an electronic computer. The methods have been conveniently described in a number of places (e.g. BRIER

1961, CRADDOCK 1965, 1968*b*, MITCHELL *et al.* 1966, PANOFSKY and BRIER 1958). The observations in the series to be investigated should be at uniform time intervals.

(1) *Filtering* is a simple procedure, though usually rather coarse in its resolution of quasi-periodic tendencies that have different period lengths. A smoothing function, or (weighted) moving average, called the 'filter' – of the type $\frac{1}{4}x_1 + \frac{1}{2}x_2 + \frac{1}{4}x_3$ or, more generally, $W_1x_1 + W_2x_2 + \ldots + W_nx_n$, where $W_1 \ldots W_n$ are arbitrarily chosen weighting coefficients – is applied to the successive terms, x_1 to x_n, and then x_2 to x_{n+1}, x_3 to x_{n+2}, and so on, of the observation series. The filter can be designed (by choosing the coefficients and the number of terms treated by them at one time) to suppress the effects of variations on either short or long time-scales, while preserving those at the other end of the spectrum, or to preserve with little change of amplitude just those variations which occur within some band of time-scales thought to be of interest. In this case it is called a 'band-pass filter'. For a given interval between the observations constituting the series, the number of terms (n) of the weighting function must increase with the period lengths which it is desired to study. The effect of any smoothing function is, of course, to reduce variance, but this may be done selectively as regards different wave lengths (or the corresponding frequencies). With any filter the nature of the response curve – i.e. of the modification of amplitude for different frequencies in the original data – needs to be studied.

The simplest of all filters is the unweighted moving average or 'running mean' of n terms $\bar{x}$ (from b to $b + n$), i.e. $\frac{1}{n}(x_b + x_{b+1} + \ldots + x_{b+n})$. The properties of this type of filter, as neatly expounded by LILJEQUIST (1949), will serve as a simple model of what happens in such operations. Consider an observed element x which undergoes a natural wave-like variation of some period length P. Its value at any time t may be expressed as

$$x_t = \bar{X} + a_o \sin\left(2\pi\frac{t}{P} + \varphi\right)$$

where $\bar{x}$ is the overall mean value of the element, a_0 is the amplitude of the periodic fluctuations and ϕ is the phase angle.

The sine can take positive or negative values according to the value of ϕ. This is the simple harmonic wave formula. If p is the time interval covered by the n observations used for the moving average and $\bar{x}$ is the mean value of x over the time interval t_0 to $(t_0 + p)$, we can write $n\delta t = p$, where δt is the unit time interval between successive observations (e.g. one month or one year). Then

$$\bar{x} = \bar{X} + \frac{1}{n}\sum_{t=t_o}^{t=t_o+n\delta t} a_o \sin\left(2\pi\frac{t}{P} + \varphi\right) u$$

For δt small enough in relation to p and P this is nearly the same as the integral expression:

$$\bar{x} = \bar{X} + \frac{1}{p}\int_{t_o}^{t_o+p} a_o \sin\left(2\pi\frac{t}{P} + \varphi\right) dt$$

If we rewrite this expression in terms of τ, the middle instant of each time interval p, such that $\tau = t + \frac{p}{2}$, it will be found to reduce to

$$\bar{x} = \bar{X} + a_o \frac{\sin\left(\frac{\pi p}{P}\right)}{\frac{\pi p}{P}} \sin\left(2\pi\frac{\tau}{P} + \varphi\right)$$

Then $a_o \frac{\sin\left(\frac{\pi p}{P}\right)}{\frac{\pi p}{P}}$ is the amplitude of the oscillations of the running mean $\bar{x}$, which we may call a_m and compare with the amplitude a_o of the oscillations of x. The period length of the oscillations of $\bar{x}$ is given by 2π times the reciprocal of the coefficient of τ and is seen to be equal to P – i.e. the same as for x in the original observations. But the phase will be reversed if the coefficient $\frac{\sin\left(\frac{\pi p}{P}\right)}{\frac{\pi p}{P}}$ has a negative value. This means that for some values which we may choose for p, the running mean may be exactly out of phase with the oscillations in the original observation data.

Fig. 6.1 shows how the amplitude a_m of the oscillations of the running mean compares with that present in the original observations for different ratios of p to P, P being the periodicity which we wish to examine. We see that so long as p is much smaller than P, the amplitude of the fluctuations of the running mean is not much smaller than a_o – i.e. a_m/a_o is nearly 1. The amplitude a_m is zero when $p = P$ or any whole multiple of P, so that with these values of p the periodicity P would disappear in the running mean. And for values of p between P and $2P$, between $3P$ and $4P$, and so on, the running mean oscillates out of phase with the unfiltered original observation series.

In meteorological and climatological observation series we are usually confronted with a certain range of more or less random variations superposed on quasi-periodic fluctuations of different period lengths and not necessarily of sinusoidal character. Our model discussed above is therefore oversimplified. The general problem is to disentangle, or isolate, any more or less periodic oscillations or time-scales characteristic of identifiable processes of climatic variation and investigate how big a part these play in the whole variability observed.

When some particular filter is chosen for its property of suppressing fluctuations on time-scales other than those it is intended to probe for, the filter should first be applied to a series of random numbers: the ratio of the variance of the filtered version to that of the original numbers is then calculated. This indicates the general reduction of amplitude which results

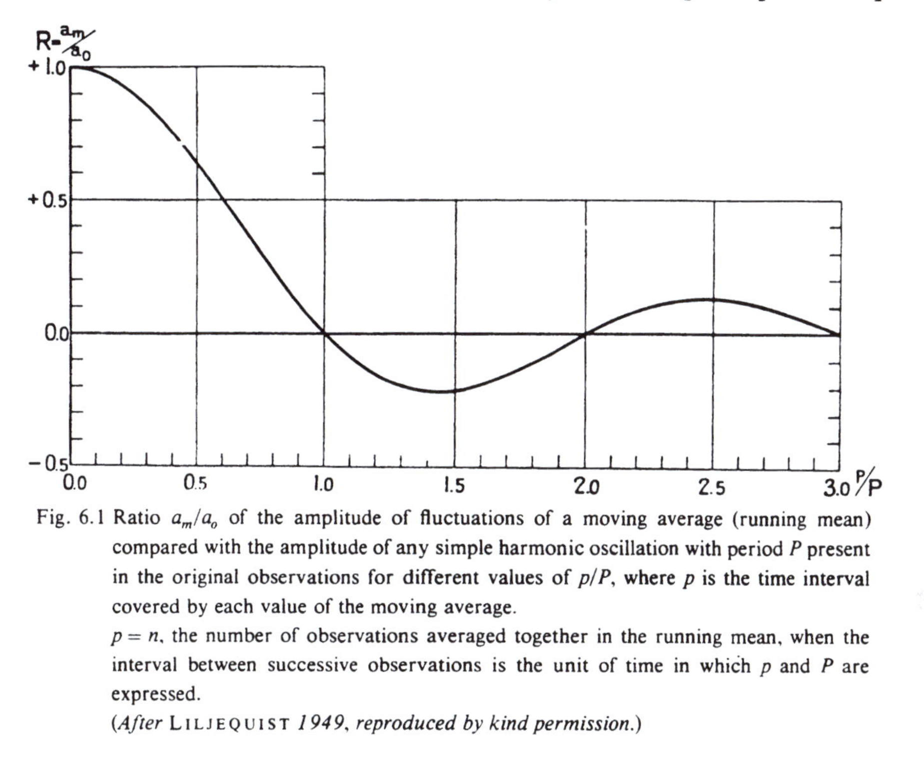

Fig. 6.1 Ratio a_m/a_o of the amplitude of fluctuations of a moving average (running mean) compared with the amplitude of any simple harmonic oscillation with period P present in the original observations for different values of p/P, where p is the time interval covered by each value of the moving average.
$p = n$, the number of observations averaged together in the running mean, when the interval between successive observations is the unit of time in which p and P are expressed.
(*After* LILJEQUIST *1949, reproduced by kind permission.*)

from the filtering operation (corresponding to a_m/a_o in our simple discussion of the running mean). Next the chosen filter is applied to the observation series, and the ratio of the variance of the filtered series to that of the actual observation series is calculated. If this ratio differs significantly from that obtained by operating the filter upon random numbers, it indicates the probable existence of a real repeating or oscillating tendency of about the period length under study in the observation series.

(2) In *harmonic analysis* of a series of observations of a variate $x_1, x_2, x_3 \ldots x_N$ with mean value $\bar{x}$, the complete series may be represented by

$$x = \bar{x} + \sum_{h=1}^{h=N/2} \left\{ A_h \sin\left(\frac{360^\circ}{P} ht\right) + B_h \cos\left(\frac{360^\circ}{P} ht\right) \right\}$$

where P is the total or 'fundamental' period of the function, t is the time at which x is observed and h is the number of the 'harmonic'. Borrowing the language of sound, the total period length covered by the observations analysed [1] is what is called the 'fundamental period' regardless of whether it gives an amplitude important in the weather data under study. The first harmonic is the total period, the second harmonic half of that, the third harmonic one-

1. Corresponding to the length of a piano string.

third of the total period and so on. The amplitudes of the sine and cosine terms, A and B, are given by

$$A_h = \frac{2}{N}\Sigma\left\{x \sin\left(\frac{360^\circ}{P} ht\right)\right\}$$

$$B_h = \frac{2}{N}\Sigma\left\{x \cos\left(\frac{360^\circ}{P} ht\right)\right\}$$

except that A for the last harmonic is 0 and B for the last harmonic is half the value given by the above expression. Finally, the sine and cosine terms are combined and the series is represented by

$$x = \bar{x} + \sum_{h=1}^{h=N/2} C_h \cos \frac{360^\circ h}{P} (t + t_h)$$

where $C_h = \sqrt{A_h^2 + B_h^2}$ is the amplitude of the hth harmonic and t_h the time of maximum value of that harmonic is $\frac{P}{360^\circ h} \tan^{-1} \frac{A_h}{B_h}$. The contribution of each harmonic to the total variance of the series of observations is $C_h^2/2\sigma_x^2$, except in the case of the last harmonic analysed where it is double this.

In its nature, harmonic analysis indicates periodicities which are exact submultiples of the total period of observation: it then assigns amplitudes and phases to each. The nature of the observed variations usually differs from this analysis in a variety of ways – e.g. by containing both periodic and aperiodic elements, by any periodic element not conforming to a sinusoidal oscillation, or in that the true period lengths represented do not coincide with those of any of the harmonics of P. Moreover any, or all, of the harmonics may be a fiction devoid of physical meaning, though sometimes particular harmonics do correspond to particular physical processes.

(3) *Correlograms* explore the tendency for recurrence of high or low values of x at all possible time intervals, or lags L, within the available series $x_1, x_2, x_3 \ldots x_N$ of observations, by calculating the (auto- or) lag-correlation coefficients r_L, where

$$r_L = \frac{\Sigma(x_i - \bar{x})(x_{i+L} - \bar{x})}{N.\, \sigma_x^2}$$

Here x_i stands for the individual observation of x, x_{i+L} for the Lth observation later (the observations being made at equal intervals), N is the total number of observations of x and σ_x is the standard deviation of x over the whole series.[1] The correlogram is a graph of

1. Properly, the denominator of the expression should be replaced by N times the product of the standard deviations of x over the two stretches of the series used, i.e. from x_1 to x_{N-L} and from x_{1+L} to x_N. The expression as printed is, however, the formula generally used as a reasonable approximation.

the r_L values plotted against lag, i.e. for all values of L (starting with $L = 0$, where the correlation coefficient is 1). Negative values may occur at some lags. Testing for statistical significance is difficult, because in general in meteorological series $r_1 \neq 0$ (it may be positive or negative): this means that there is some persistence or anti-persistence, and successive observation values cannot be treated as independent of each other. r_L^2 may be studied if it is desired to consider only the magnitude and ignore the sign of the responses at different lags. The distribution of high and low values of lag correlation within a single time series may be more rapidly revealed by computing only the mean value of the products in the numerator $\frac{1}{N}\Sigma(x_i - \bar{x})(x_{i+L} - \bar{x})$: this is called the 'autovariance'. It has the disadvantage that the units cannot be compared with those of other time series or even those applying to subdivisions of the same series.

A simplified periodogram, known as 'ALTER's correlogram' (ALTER 1937), consists of a graph of the mean differences between the observations of x at x_i and x_{i+L}. The sign of the difference D is ignored

$$D = \frac{\sum_{1}^{N-L} |x_{i+L} - x_i|}{N - L}$$

This has the virtue of rapid calculation, making it a useful aid in initial exploration for periodic or quasi-periodic elements in a series of observations.

(4) *Power spectrum analysis* of a series of observations x proceeds by applying harmonic analysis to the correlogram. PANOFSKY and BRIER (1958, p. 140) define the spectrum in this way:

> Suppose we have a record of N observations of a meteorological (scalar) variable taken at uniform time intervals Δt. Let the overall length of record, $N\Delta t$, be denoted by P. We subject this record to harmonic analysis, choosing P as the 'fundamental period' or 'period of the first harmonic'. $N/2$ harmonics could be calculated, the last of them having a period $2\Delta t$. The fraction of the total variance σ_x^2 contributed by each harmonic except the last [see p. 228] is given by $C_h^2/2\sigma_x^2$, where h is the number of the harmonic.

A common practice is to plot the actual variance $C_h^2/2$ contributed by each harmonic against h. The area under the spectrum then gives the sum of the variances contributed by each harmonic, adding up in all to the total variance. When the ratio $C_h^2/2\sigma_x^2$, or *relative variance*, is plotted against h, this amounts to a 'normalized' spectrum and the total area under it is unity.

As in the case of simple harmonic analysis of an actual observation series, $C_h^2 = A_h^2 + B_h^2$, where A_h and B_h are the amplitudes of the sine and cosine terms respectively. However, PANOFSKY and BRIER (*loc. cit.*, p. 143) point out that, since the lag correlation coefficients computed for positive lags from $L = 0$ to $L = +m$ are the same as those that would be obtained for negative lags from $L = 0$ to $L = -m$, the correlogram would be symmetrical about $L = 0$ and in such cases the sine terms vanish. All the A_h's are zero, and $C_h = B_h$. The B_h terms are given by

$$B_h = \frac{r_o}{m} + \frac{2}{m} \sum_{L=1}^{L=m-1} \left\{ r_L \cos\left(\frac{360°}{2m} hL\right)\right\} + \frac{r_m}{m}(-1)^h$$

except that the B_o and B_m coefficients are just half those given by the formula. The total, or 'fundamental', period P is taken as $2m\Delta t$, Δt being the time interval between observations. The spectrum may be plotted with $B_h{}^2$ as ordinate and $h/2m\Delta t$ as the abscissa. The smaller m is, the less information about any long periods will be visible in the spectrum. In practice, smoothed estimates of B_h such as $S_h = 0{\cdot}5B_h + 0{\cdot}25(B_{h-1} + B_{h+1})$ have to be used instead of B_h to get rid of the wildest scatter produced by most samples of any time series studied. The resulting S_h is a smoothed version of the normalized amplitudes that would be given by harmonic analysis of the original series, i.e. of $C_h{}^2/2\sigma_x{}^2$. PANOFSKY and BRIER have explained how its values can be tested for statistical significance by calculating values of chi squared with appropriate numbers of degrees of freedom. S_h measures the 'power' or 'relative variance' of fluctuations with characteristic time-scale around the period of harmonic h. The areas under the curve indicate the proportion of the total variance contributed by fluctuations with frequency about each value of h (i.e. period length about $2m/h$), and the area under the whole curve represents the total variance.

Periodicities observed in the disturbances of the sun

Fig. 6.2 illustrates the results of power spectrum analysis applied to solar disturbance, as measured by the Zürich sunspot numbers (annual values) from the year 1700 to 1960. Period lengths of about 90 years and 11 years are prominent. The small hump at 5·7 years may be regarded as a harmonic of the 11-year cycle or as expressing the fact that the latter is not a truly sinusoidal oscillation. Another minor peak found by SHAPIRO and WARD (1962) between 1·9 and 2·4 years may also be statistically significant. Strong fluctuations occur on shorter time-scales of the order of a few months, particularly in the weaker sunspot cycles, though apparently without regularities such as facilitate prediction (VITINSKII 1965). The well-known 22- to 23-year cycle, known as the HALE cycle after its investigator who established its undoubted physical reality in that the magnetic polarity of the spots reverses from one 11-year period to the next, is notably missing from the spectrum. This may be because it also shows itself in a tendency for the 11-year cycles to occur in pairs of somewhat similar amplitude not conspicuously related to the amplitudes of the preceding and succeeding pairs.

Let us now turn to examples of spectra obtained from weather data and long series of observations of other things which register responses to weather differences from one year to another.

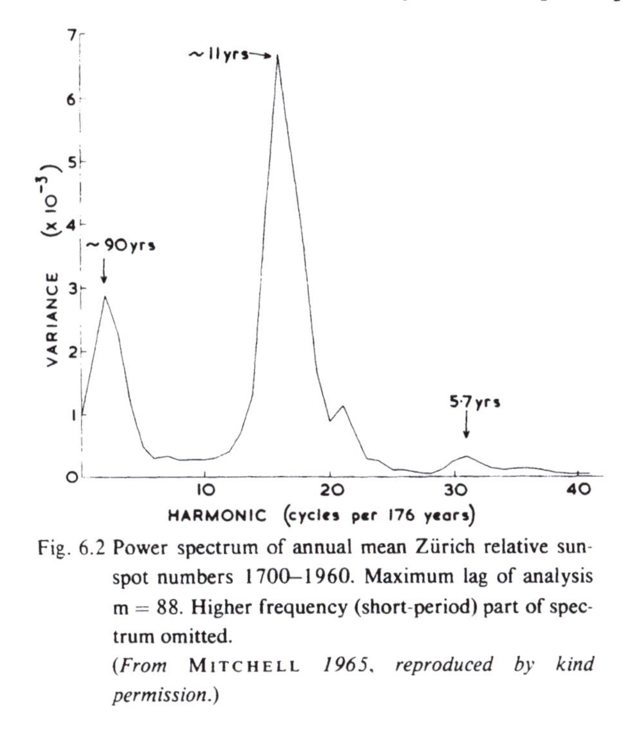

Fig. 6.2 Power spectrum of annual mean Zürich relative sunspot numbers 1700–1960. Maximum lag of analysis m = 88. Higher frequency (short-period) part of spectrum omitted.
(From MITCHELL *1965, reproduced by kind permission.)*

Periodicities registered by the behaviour of the atmosphere or oceans

Fig. 6.3 shows periodograms for lags up to 25 years obtained by VOLTZINGER *et al.* (1966) from the mean values for each January from around 1850 to 1950 of the barometric pressure difference between Madeira and Iceland and between central Siberia and Iceland, as well as for the maximum area attained each winter by the ice on the Baltic Sea. (The horizontal lines indicate the average amplitude for all the different lags studied.) Period lengths around 21–23 years and 5–6 years are consistently shown.[1] There are only hints in some of these curves of other features: around 18 years, 14–15 years, 11–12 years and 2–3 years.

Spectra, illustrated here in fig. 6.4, obtained (*a*) by BRIER (1961) from 100- to 160-year runs of annual rainfall figures for nine places[2] between 35° and 60°N since about the year 1800, and (*b*) by LANDSBERG (1962) from the temperatures in central Europe[3] from 1761 to 1953, are consistent only in according some importance to a cyclic tendency at

1. See also WAGNER, A. J. Long-period variations in seasonal sea-level pressure over the northern hemisphere. *Monthly Weather Rev.*, **99**, 49–66. Washington 1971.

2. The observations were made at Albany (N.Y.), Budapest, Copenhagen, Frankfurt, Greenwich, Helsinki, Malta, Milan and Rome.

3. BAUR's series of values for Berlin, De Bilt (Holland) and Vienna averaged together.

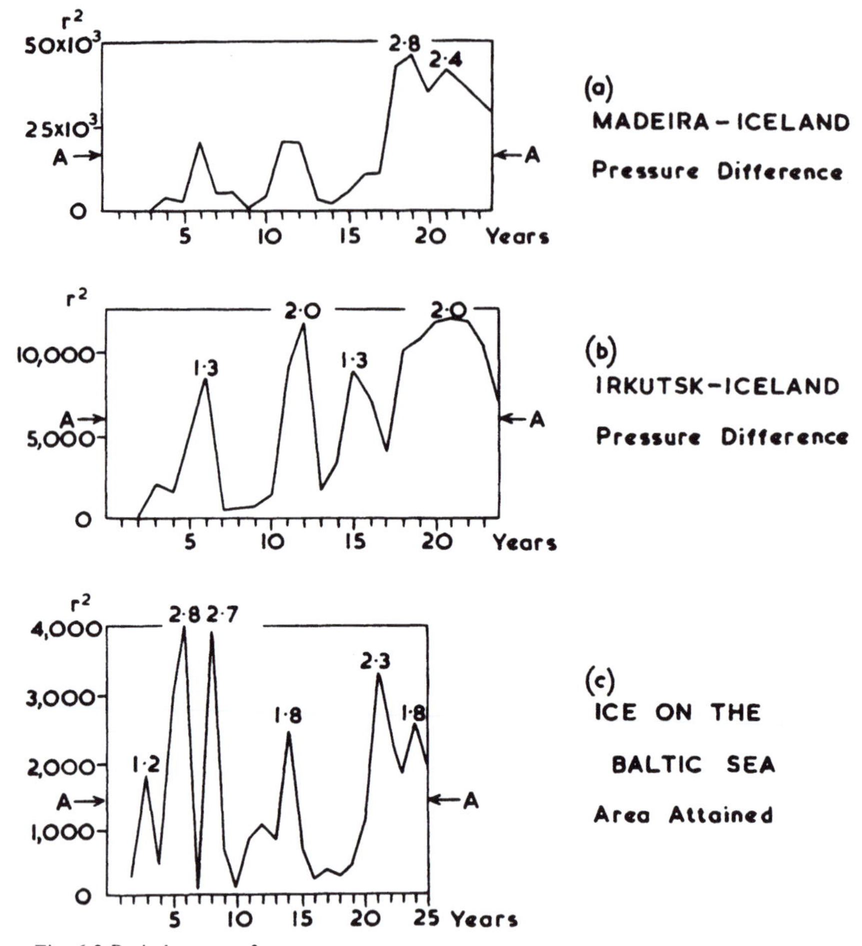

Fig. 6.3 Periodograms of:

(*a*) Difference of mean January pressure (in mm Hg) Madeira *minus* Stykkisholm, Iceland: approx. 1850–1950.

(*b*) Difference of mean January pressure (in mm Hg) Irkutsk *minus* Stykkisholm, Iceland: approx. 1850–1950.

(*c*) Maximum area of ice formed on the Baltic Sea (in km²): winters 1900–50.

Note: A → marks the average value of r^2 in each diagram.
(*After* VOLTZINGER *et al. 1966*.)

around 2·2–2·5 years; even there the amplitude is small. There are fairly frequent hints besides of power in the spectra at around 3·5–4 years and some other periods already mentioned above.

The well-known early probe by BRUNT (1925) for periodicities with cycle lengths up to 35 or 40 years in a dozen long series of European weather data spanning 100–150 years

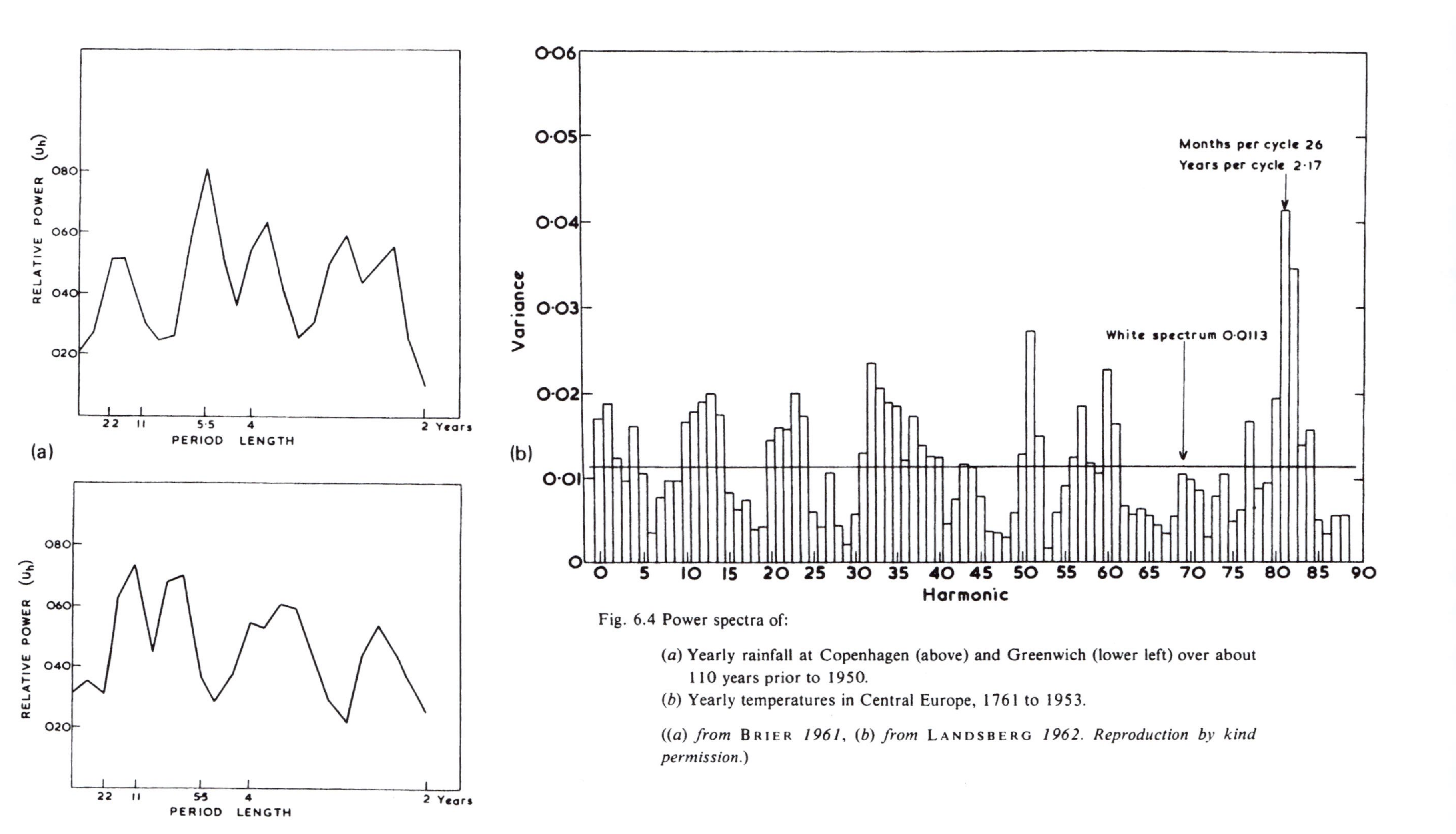

Fig. 6.4 Power spectra of:

(*a*) Yearly rainfall at Copenhagen (above) and Greenwich (lower left) over about 110 years prior to 1950.
(*b*) Yearly temperatures in Central Europe, 1761 to 1953.

((*a*) *from* Brier *1961*, (*b*) *from* Landsberg *1962. Reproduction by kind permission.*)

between about 1760 and 1920, also showed little in the way of consistent periodic tendencies. The series examined were: rainfall at Edinburgh, London, Milan and Padua; temperatures at Berlin, Edinburgh, London, Paris, Stockholm and Vienna; atmospheric pressure at Edinburgh and Paris. A periodicity of about 13–14½ months was among the 3 or 4 with biggest amplitude in 10 of the 12 series examined; one of 2·0–2·5 years was about equally general and prominent. A 13- to 14½-month periodicity (cf. MAKSIMOV's free nutation period mentioned earlier in this chapter) was also identified in the long records of rainfall at places in England since 1727 and of rainfall in other, widely scattered parts of the world (including the U.S.A., Siberia, India, Chile, South Australia) by ALTER (1922, 1933) – who connected it with the ninth harmonic of the '11-year' sunspot period. The next most consistent indications of BRUNT's analysis were of almost equally big amplitudes at period lengths around 23 and 35 years,[1] each of these in 5 or 6 of the 12 series and one or other of them in 10 cases. The amplitudes of all these periodic elements found by BRUNT – the greatest ones up to about 7 mm in the monthly rainfall figures, up to 0·4°C in monthly temperatures and 0·7 mb in monthly mean pressures – were small in comparison with the total year-to-year variability (in terms of the standard deviation) when the same months in different years are compared. Standard deviations of the monthly values at the places concerned are generally 1–3°C and 2–7 mb and equivalent to 10–20 mm of rainfall (though this element has a skew distribution). It is clear, however, that the amplitudes attached to the periodic (i.e. cyclic) elements in the case of some longer periods may account for a large fraction of the variability of the long-term average values. This is further, and more plainly, suggested by some of the evidence of period lengths of around 90–100, 170–200, 800 and 1700–2000 years mentioned elsewhere in this chapter.

The power spectrum (fig. 6.5) obtained from the thicknesses of the annual growth rings of trees near the northern forest limit in Finland from 1463 to 1960 is readily compared with that from solar disturbance data (fig. 6.2). The tree-growth spectrum is based upon SIRÉN's (1961) analysis of the yearly growth since the year 1180 of numerous trees representing the forest population in an area where the chief control is the prevailing summer temperatures. Here periodic recurrences around 90, 30 and 23 years are most prominent. Peaks at 90 and 30 years and the minor peak at 11–12 years, repeated themselves in a separate analysis of the tree-ring series from 1700 to 1960, though none of these attained statistically significant levels. The entire tree-ring series for Finnish Lapland 1181–1960 and SCHOVE's sunspot series (see Appendix I) over the same time-span were also treated by MITCHELL (1965, and unpublished personal communication 25 July 1966) with a filter designed to suppress shorter-period fluctuations: in both series variations of period length around 200 years and 90–100 years were the most prominent features.

Numerical series of actual weather data from Europe from the early Middle Ages to the present have been derived by LAMB (1963, 1965, 1967) from diaries and from the

1. This is one of the rather few independent indications found of the 34·8-year cycle derived by BRÜCKNER (1890) from many types of European weather data including some records centuries long.

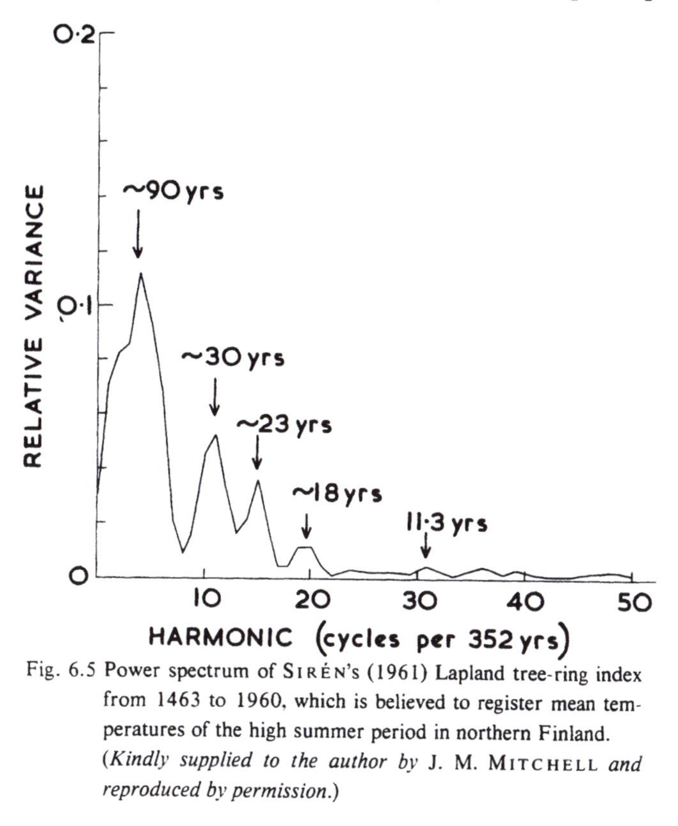

Fig. 6.5 Power spectrum of SIRÉN's (1961) Lapland tree-ring index from 1463 to 1960, which is believed to register mean temperatures of the high summer period in northern Finland. (*Kindly supplied to the author by* J. M. MITCHELL *and reproduced by permission.*)

frequencies of extreme months of various kinds indicated in available manuscript annals, audited accounts of manors, state papers, etc. The derived sequences of 50-year average temperatures and rainfall figures for England and the apparent course of the (10-year average) frequency of SW'ly surface winds in eastern and southern England are shown by the graphs in figs. 6.6, 6.7, 6.8. The SW'ly wind has probably at all times in this epoch been the prevailing wind in the British Isles; but there are large-amplitude changes to be observed in its frequency level (fig. 6.8), several of the changes being remarkably abrupt. These changes must correspond to similar changes in the amount of air transported from the ocean across Britain by the general westerly winds aloft that accompany SW winds at the surface. Hence, many other aspects of the climate must be affected. There are step-like variations seen in the temperature and rainfall diagrams (figs. 6.6, 6.7) at intervals of about 200 years, and a few suggestive of a roughly 100-year oscillation; the variations at the 200-year intervals are, however, superposed on an apparently larger amplitude wave of the order of 800 years long, of which the series are only adequate to cover one cycle.

J. M. CRADDOCK (1968*a*, *b*) has recommended the use of SHERMAN's statistic (SHERMAN 1950) to test the statistical significance of an apparently fairly regular spacing

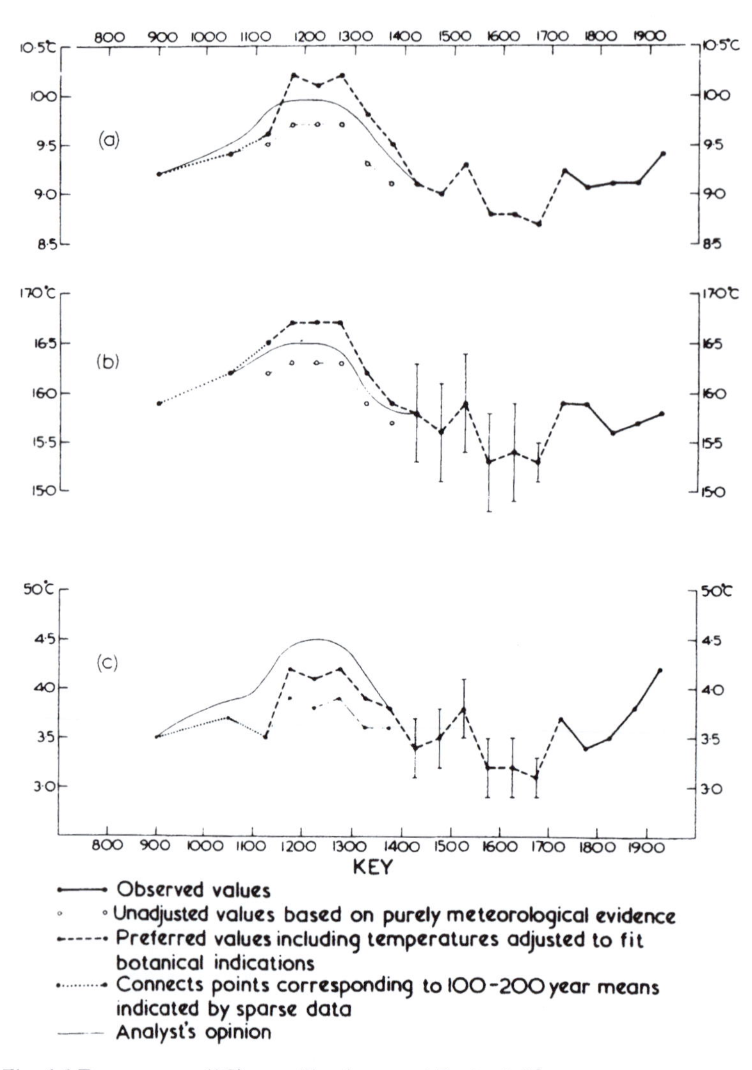

Fig. 6.6 Temperatures (°C) prevailing in central England, 50-year averages:

(*a*) Year
(*b*) High summer (July and August)
(*c*) Winter (December, January and February).

Observed values from 1680, as standardized by MANLEY. Values for earlier periods as derived by LAMB (1965). The ranges indicated by the vertical bars are three times the standard error of the estimates.

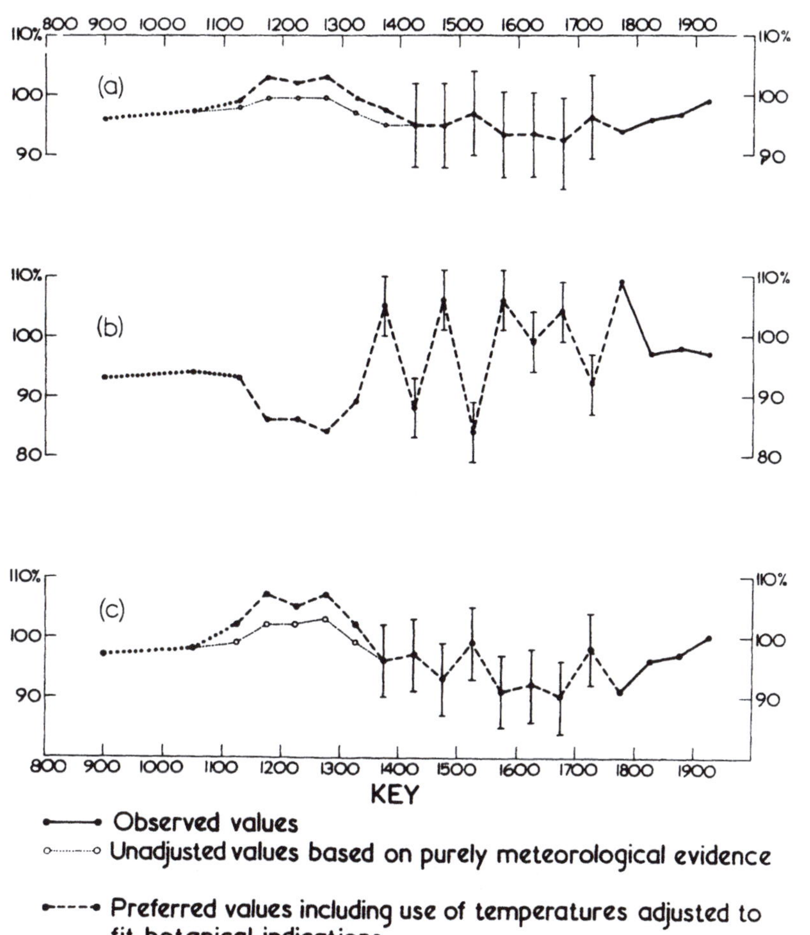

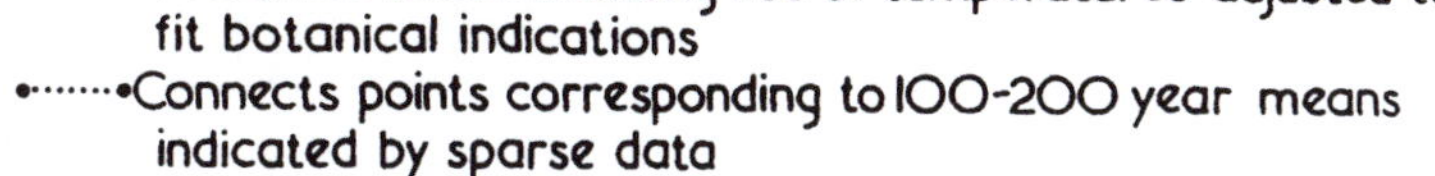

Fig. 6.7 Rainfall amounts (as percentages of the 1916–50 averages) over England and Wales, 50-year averages.

(*a*) Year
(*b*) High summer (July and August)
(*c*) Cooler 10 months of the year (September–June).

Observed values from 1740, as presented by NICHOLAS and GLASSPOOLE. Values for earlier periods as derived by LAMB (1965). The ranges indicated by the vertical bars are three times the standard error of the estimates.

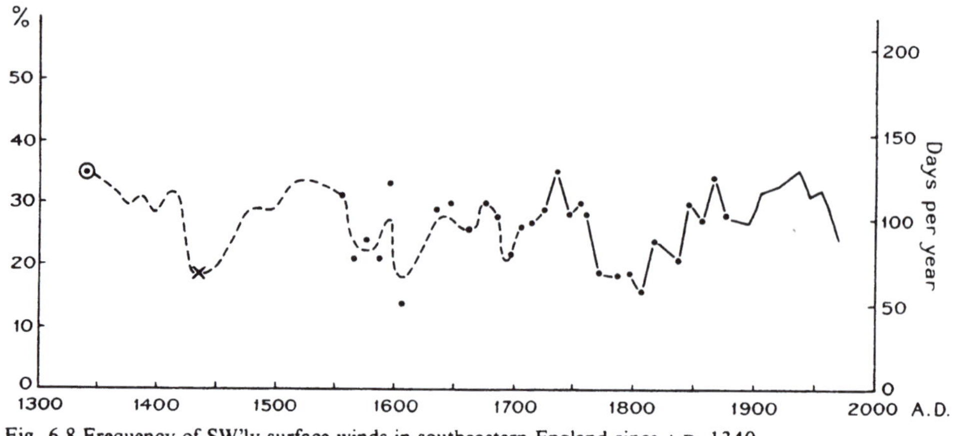

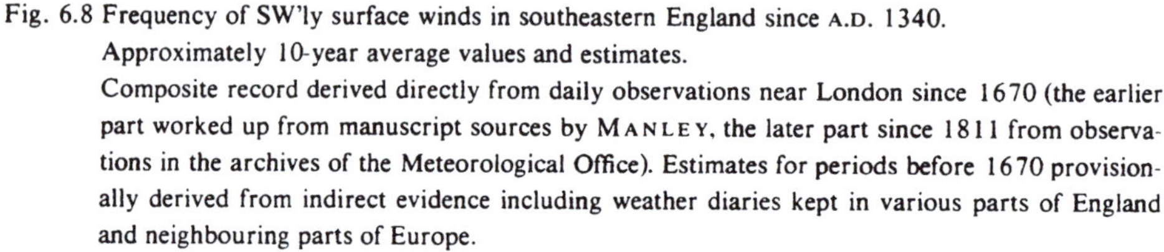

Fig. 6.8 Frequency of SW'ly surface winds in southeastern England since A.D. 1340.
Approximately 10-year average values and estimates.
Composite record derived directly from daily observations near London since 1670 (the earlier part worked up from manuscript sources by MANLEY, the later part since 1811 from observations in the archives of the Meteorological Office). Estimates for periods before 1670 provisionally derived from indirect evidence including weather diaries kept in various parts of England and neighbouring parts of Europe.

in time of abrupt phenomena in crude series of this sort such as the step-like changes occurring just three or four times in figs. 6.6 and 6.7. SHERMAN's statistic is

$$\omega = \frac{1}{2T} \sum_{j=0}^{n} \left| t_{j+1} - t_j - \frac{T}{n+1} \right|$$

where n events $k_1 \ldots k_n$ occur at times $t_1 \ldots t_n$ and T is the total time spanned by the data from the beginning, i.e. from the earliest possible moment for an occurrence t_o to the last possible moment called t_{n+1}, at the end. The n observed events divide T into $n + 1$ segments. Thus $\frac{T}{n+1}$ is the average time between occurrences, as shown by the available sample of observations. CRADDOCK (1968*a*) has worked out the probabilities of various values of ω occurring by chance in a series of up to 31 events. Its value always lies between 0 and 1. ω measures the regularity of the interval between successive events: high values of ω are produced by wide variations of the interval length, low values by little variation.

Application of SHERMAN's statistic to the sequences of 50-year mean values since A.D. 1100 read off figs. 6.6 and 6.7 yielded the following results:

(i) The turns towards milder winters (more frequent westerlies?) in 1150–1200, 1500–50, 1700–50, 1900–50 (fig. 6.6), when the mean winter temperatures in England appear $\geqslant 0{\cdot}3°C$ above those of the previous half century, gave $\omega = 0{\cdot}21$ (10% probability of occurring in a random sequence). But if we treat the temporary halting in 1350–1400 of the steep long-term decline of winter temperatures as another occurrence[1] of the same kind of event, ω becomes 0·157 (a value with only a 1 in 100 chance of occurring in a random sequence). The timing of these events seems to be closely tied – as might be expected – to the phases of maximum frequency (fig. 6.8) of the SW and W winds in Britain.

(ii) The turns towards wetter summers in England indicated by high-summer rainfall $\geqslant$5% greater than in the previous half century, occurring (fig. 6.7) in the second halves of most centuries, gave $\omega = 0{\cdot}167$ (probability of arising in a random series of numbers only 1%).

These two tests therefore afford some statistical verification of quasi-periodicities around 100 and 200 years, possibly the same as the approximately 90- and 170- to 200-year tendencies noted in other data cited earlier in this chapter. Unlike most of the shorter-period variations discussed, these long-term variations appear to have large amplitudes in the occurrence of some weather phenomena.

An approximately 200-year period is also prominent in the calculated amount of radioactive carbon (^{14}C) in the Earth's atmosphere (STUIVER 1961), production of which appears to vary inversely with the amount of solar disturbance. SUESS reports that radiocarbon measurements on tree-ring chronologies established in bristlecone pine make it possible to trace this approximately 200-year oscillation back to about 300 B.C., and before that as a 400-year oscillation back to around 5000 B.C.

LINK (1958, 1964) has derived, from a painstaking survey of the frequencies of observation of aurora borealis and of the making of astronomical discoveries, rather clear evidence of an approximately 400-year oscillation in the cloudiness of the nights in China since 2300 B.C. (This investigation was a fascinating exercise, in which a method was devised of making due allowance for secular changes of cultural interest and technical competence in making the observations.) Neither this nor the one 800-year wave-like variation of temperature and rainfall in England seen in the derived values (figs. 6.6, 6.7) from the early Middle Ages to date, nor the evidence of 1700- to 1800-year cycles in the rainfall and lake levels in central Asia and of 3500-year variations of the glaciers cited earlier in this chapter, can as yet be submitted to statistical tests because of the inadequate length of the records.

A number of workers have applied power spectrum or other forms of periodicity analysis to the sometimes very long sequences of varves in lake deposits and shales of recent and earlier geological eras. The individual varves are believed to be produced by changes of temperature, rainfall, evaporation and storminess in the course of one year. Even though the effects of these factors singly have not been, and probably cannot be, isolated, the varying thicknesses of the varves can reasonably be supposed to reflect weather differences from

1. This may seem like an inadmissible use of a statistic, or special pleading, but may be justified in this case by the quality of the medieval data; a 0·1° error or change in the derived mean temperature for 1350–1400 could mean that that was a period of rather higher temperature than either before or after. Such a temporary recovery of the level of winter temperatures seems in fact to have occurred between about 1340 and 1390.

year to year. Fig. 6.9 shows the spectra, analysed for periodicities of under 100 years, obtained by WARD (in ANDERSON 1961) from the two halves of a more than 900-year sequence of varves in shale found in Alberta, Canada and attributed to the Upper Devonian – i.e. deposits laid down some 300 million years ago. The prominence of period lengths of about the same numbers of years as obtained in the spectra of recent weather data, notably the peaks at about 2·2 (extreme right of the diagram) and 23 years, also the indication of some feature at about 90 years, is striking. In the spectra of Upper Jurassic limestone varves from New Mexico, analysed for periods under about 40 years, a 2·2-year peak was again shown, but there was no consistency in the longer-period parts of the spectra.

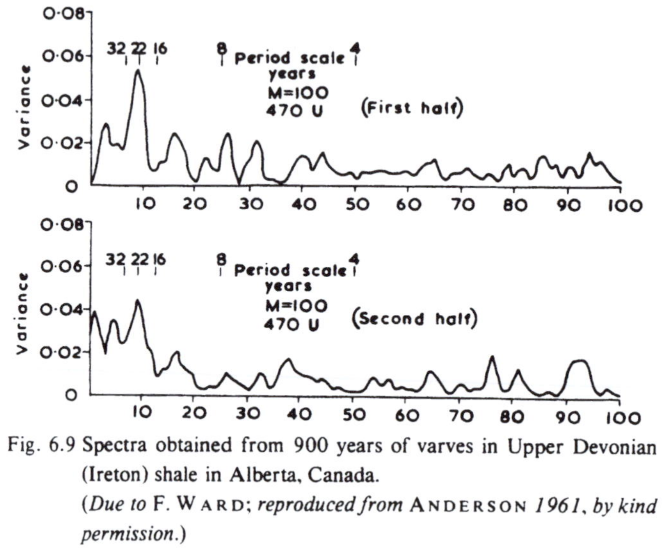

Fig. 6.9 Spectra obtained from 900 years of varves in Upper Devonian (Ireton) shale in Alberta, Canada.
(*Due to* F. WARD; *reproduced from* ANDERSON *1961, by kind permission.*)

Varves laid down in post-glacial times in sixteen lakes in European Russia, from Karelia in the north to Crimea in the south and in Siberia near 90°E, as well as in four lakes in or near the Alps and three in Indonesia, as analysed by SHOSTAKOVICH (1931, 1944), all showed their most prominent rhythmic elements at period lengths 2–3, 5–6 and 10–12 years. He found the same periods prominent in strata of Pleistocene (glacial) and earlier geological ages back to the Pre-Cambrian, over 500 million years ago. Other workers have found less concentration of the evidence on periodicities of just these lengths.

BERLAGE's (1957) list of periodicities found by research on various modern meteorological observation series suggested a continuous spectrum from 1 to at least 36 years cycle length. He suspected, however, that the amplitude of quasi-periodic variations corresponding to the following two cases, or groups of cases, was greater than for other period lengths.

(i) The somewhat irregular but approximately 2·3-year period of the Southern Oscillation. This was described as a 'fundamental terrestrial period', apparently the most fundamental under this heading, and tentatively identified with the large-scale circulation of the waters of the South Pacific Ocean between about the equator and 40°S and between the South American coast and Java.

(ii) Beats arising between this 2- to $2\frac{1}{2}$-year cycle and the also irregular but approximately 11- and 22-year cycles of solar activity and their higher harmonics around 3·7, 5·5 and 7·4 years.

In the apparent absence of any comprehensive survey or systematic card index of power spectrum analyses of series of weather data from the present day, and of series produced by parameteorological fossil evidence from past ages, we may do best to take note of the figures summarized in Table 6.2 on p. 242 and the various much longer cycle-lengths mentioned elsewhere in this chapter.

A biennial or, rather, a somewhat variable 2- to 3- year oscillation clearly emerges from this table as the most widely identified feature. It does not necessarily have the greatest amplitude. One must, however, be careful about judgments of amplitude based on study of observation series for any one place. Until the geography, or manner of operation of an oscillation in space and time, is known, it may happen that studies of it at this or that point fall in nodal regions where the amplitude is smaller than elsewhere. The amplitudes of some of the long-term fluctuations in the various elements of surface weather, whether strictly periodic or otherwise, on time-scales from 90 years and upwards to those of the ice ages and interglacial warm periods, appear to exceed those of any of the shorter-term pulsations with periods of more than one year. And in other cases besides, the statistical significance of the long-term fluctuations is greater. In the lower stratosphere, as we have seen in the previous chapter, the amplitude of the quasi-biennial oscillation is big and reverses the winds from west to east and vice versa.

Quasi-biennial oscillations – the Southern Oscillation and others

WALKER in his studies of world weather relationships (1923, 1924, 1927, 1928, 1929, 1933) and the Indian monsoon named three 'oscillations', which would nowadays be regarded as indices monitoring respectively the prevailing strength of (1) the North Atlantic (temperate zone) westerlies, (2) the North Pacific westerlies and (3) the easterly Trade Winds blowing across the Pacific Ocean, and into the Indian Ocean, on either side of the equator.

Owing to the limited observation network available in their day, WALKER and BLISS (1933) defined their indices numerically, for the purpose of studying their correlations with contemporaneous, and subsequent, temperature and rainfall in different parts of the world, as follows:

(In these expressions, P stands for the departure of pressure from the overall mean, T for the departure of temperature and R for the rainfall departure.)

TABLE 6.2

Periodicities most frequently and prominently indicated in meteorological and related data

Item analysed	*Period lengths (years)*								
Sunspot numbers (direct obs. since *c*. 1700, indirect data from antiquity)	*c*. 5½			10–12		*c*. 22–23 'Hale cycle' (see text) p. 230)	80–90	*c*. 170–200	(400?)
Radioactive carbon in the Earth's atmosphere (clear indication since 200 B.C.)								*c*. 200	
Temperatures in Europe (1760 ff.) (Brunt, 1925; Landsberg 1962)	*c*. 2·2								
Temperatures in England (*c*. 1100 ff.) (Lamb 1965, 1966)								170–200	
Rainfall in Europe (*c*. 1800 ff.) (Brier 1961; Brunt 1925)	2–2·5								
Rainfall in England (*c*. 1100 ff.) (Lamb 1965, 1966)							*c*. 100	170–200	
SW winds in England (1340–1965) (Lamb 1967)								*c*. 200	
Baltic ice (1900–1950) (Voltzinger *et al*., 1966)	3	5–6	8	11–14		21–24			
Severity of European winters (1215–1905) (Easton 1928)							89·5		
W. and E. Greenland and Barents Sea ice (1820 ff.) (Maksimov 1952, 1954)							*c*. 71–77		
Greenland ice cap, thickness of annual layers (A.D. 1200 ff.) (Johnsen *et al*., 1970)							78	181	
Night cloudiness in China (2300 B.C. ff.) (Link 1958, 1964)									*c*. 400
Pressure differences in January (1850 ff.) Madeira–Iceland and Siberia–Iceland (Voltzinger *et al*., 1966)	3	6		11–12	15	18–24			
Pressure differences in January and July (1750 ff.)									
for W'ly winds over Britain						18–23		(*c*. 190?)	
for S'ly winds over the North Sea (Lamb and Johnson 1966)						18–23	55–80		
Latitude of the anticyclone belt:									
(*a*) over Siberia in January (1836 ff.)							80–85		
(*b*) over the N. Atlantic in July (1865 ff.) (Lamb and Johnson 1966)							85–110		
Zonal index* at 500 mb measured right round the northern hemisphere between 35° and 55°N in January (1943 ff.) (Brier *unpublished*)	2·2								
Meridional circulation pattern frequency over Europe (1881–1964) (Böhme 1967)	2·1–2·2								
Pleistocene ice-age varves (cited by Anderson 1961)	2–3		7–8	10–12	15–20				
Tertiary strata (e.g. Miocene, Oligocene and Eocene, various strata in Russia and U.S.A.) (cited by Anderson 1961)	2–3	5–6		11–12		23			
Miocene (Sarmatian anhydrite in Sicily) (Richter-Bernburg 1964)				9–10	15	21 and 32–33			
Mesozoic strata (e.g. Cretaceous, Jurassic) (cited by Anderson 1961)	*c*. 2½	6	8	10–13			170–180		
Palaeozoic strata (including Permian, Carboniferous, Devonian) (cited by Anderson 1961)	*c*. 2½	*c*. 5½	7–8	10–14					
Upper Permian (Zechstein) salt deposits in Germany (Richter-Bernburg 1964)		5–6		10–12		23	85–105	170–210	400
Pre-Cambrian (cited by Anderson 1961)	*c*. 2½	*c*. 5½		11–12					

* See footnote p. 247.

(1) *North Atlantic Oscillation*

$P_{Vienna} + 0{\cdot}7\ P_{Bermuda} - P_{Ivigtut} - P_{Stykkisholm} + T_{Bod\o} + T_{Stornoway} + 0{\cdot}7$ (arithmetic mean of $T_{Hatteras} + T_{Washington}) - 0{\cdot}7\ T_{Godthaab}$

The units were so chosen that the standard deviation of each of the constituent terms was $\sqrt{20}$.

(2) *North Pacific Oscillation*

$P_{Honolulu} + \frac{1}{3}(T_{Qu'Appelle} + T_{Calgary} + T_{Prince\ Albert}) - \frac{2}{3}(P_{Sitka\ or\ Fort\ Simpson\ or\ Juneau} + P_{Dawson} + P_{Nome}) - T_{Dutch\ Harbor}$

The fractions apply to the values for places which are not far enough apart to be independent of each other.

(3) *Southern Oscillation*

$P_{Santiago} + P_{Honolulu} + 0{\cdot}7\ P_{Manila} - P_{Batavia/Jakarta} - 0{\cdot}7\ P_{Darwin} - P_{Cairo} - T_{Madras} + R_{India}$ + Nile flood $- 0{\cdot}7\ R_{Chile}$

These were called oscillations because

(*a*) with big values of the indices, corresponding to strong zonal winds, pressures tended to be above normal in the subtropical anticyclone belts and below normal in the subpolar or equatorial low-pressure belts,

(*b*) with low values of the indices, and weak zonal wind flow, an inverse distribution of pressure anomalies occurred.

This suggested a see-saw of atmospheric mass between the high-pressure belts near latitudes 30°N and S and the low-pressure belts near latitude 60° and the equator. This is one way of describing in simple terms the observed alternation of periods of intensified zonal wind flow and periods of blocking or meridionality.

The expressions WALKER used are only of historical interest today when the pressure gradients can be measured almost anywhere in the world as a direct index of the strength of this or that windstream.

WALKER's definition of the Southern Oscillation, however, seems also to take account of an asymmetric tendency for the northern subtropical high-pressure belt to be weakened over Africa and Asia when the Pacific sector anticyclones, north and south of the equator, are expanded and the southern one spreads over Chile. This is generally taken as an oscillation of atmospheric mass between the subtropical anticyclones over the eastern Pacific, particularly the Easter Island anticyclone, and the equatorial-cum-Indian monsoon low-pressure belt in the East Indies and right across the Indian Ocean sector. BERLAGE (1957, 1961) regards the Indonesian low-pressure area about Batavia/Jakarta (near 6°S 107°E) and the South Pacific high-pressure area around Easter Island (near 27°S 109½°W) as if they were the scales of an air pressure balance and has given world maps of correlation coefficients connecting the local seasonal and annual air pressure anomalies with those at Jakarta. The annual maps for two separate 8-year spans, in the 1930s and 1950s respectively, are shown here in figs. 6.10(*a*) and (*b*) for comparison. The negative correlation coefficients between the two key areas are so strong and the pattern of significant relationships is so extensive over the world, that it is reasonable to use the pressure at Batavia/Jakarta alone as an index of the Southern Oscillation.

The oscillation is expressed in both these maps by a negative correlation coefficient as high as −0·9 between the pressures for the time being prevailing in areas about Easter Island and at Jakarta together with an equatorial strip of the Indian Ocean. This exceeds the magnitude of the correlation coefficient between the annual, or monthly, pressures at the Azores and Iceland, for which values ranging from −0·40 to −0·75 obtain in different decades (see, for example, the maps given by BERLAGE and DE BOER 1960). So the interchange of atmospheric mass between the low-pressure belt in the equatorial Indian Ocean and the southeast Pacific anticyclone appears to be an even more regular pattern of variation than the well-known alternation of intensified zonal westerlies and blocking in the North Atlantic. Its most prominent time-scale (2–2½ years) is also less obvious in blocking.

The maps (figs. 6.10 (*a*), (*b*)) appear to express a link between the so-called Southern Oscillation and the occurrence of blocking in high latitudes. This shows itself by correlation coefficients exceeding +0·7 linking pressure at Jakarta with the pressures in certain sectors of the Arctic/sub-Arctic and of the sub-Antarctic. But there is a surprising aspect of this: the sectors of the Arctic where the pressures, and presumably the frequency of blocking, varied in phase with pressure at Batavia/Jakarta differed between the two periods 18 years apart, seen in fig. 6.10: in the 1930s it was primarily Greenland, in the 1950s Novaya Zemlya, Kamchatka and a region west of Hudson's Bay. Moreover, apart from the Canadian region mentioned, the 10-year mean pressures in these regions seem to have been below normal in the periods when the value varied in phase with Jakarta. In the 1930s there was a tendency for vigorous circulation over the Atlantic and an extension of the Azores anticyclone over Europe, alternating with rather frequent blocking anticyclones over northern Europe and the Novaya Zemlya region; pressure was below normal over Greenland, Iceland and Spitsbergen and in these areas apparently fluctuated with the Jakarta pressure.

In the 1950s, by contrast, average pressure was higher over northern Greenland with blocking anticyclones there commoner than over Scandinavia and northeast Europe; cyclonic activity seems to have been increasing in the Novaya Zemlya region (though this became more pronounced in the 1960s) and near Kamchatka on the northwest Pacific, and it was principally near Novaya Zemlya that the pressure level fluctuated with that at Jakarta. The longitude shift between the 1930s and 1950s from Novaya Zemlya to Greenland of the region most affected by high pressure in the sub-Arctic, seems to be connected with much longer-term variations in the strength of the westerlies, and corresponding changes of prevailing wave length in the upper westerlies, which will come to notice in later chapters of

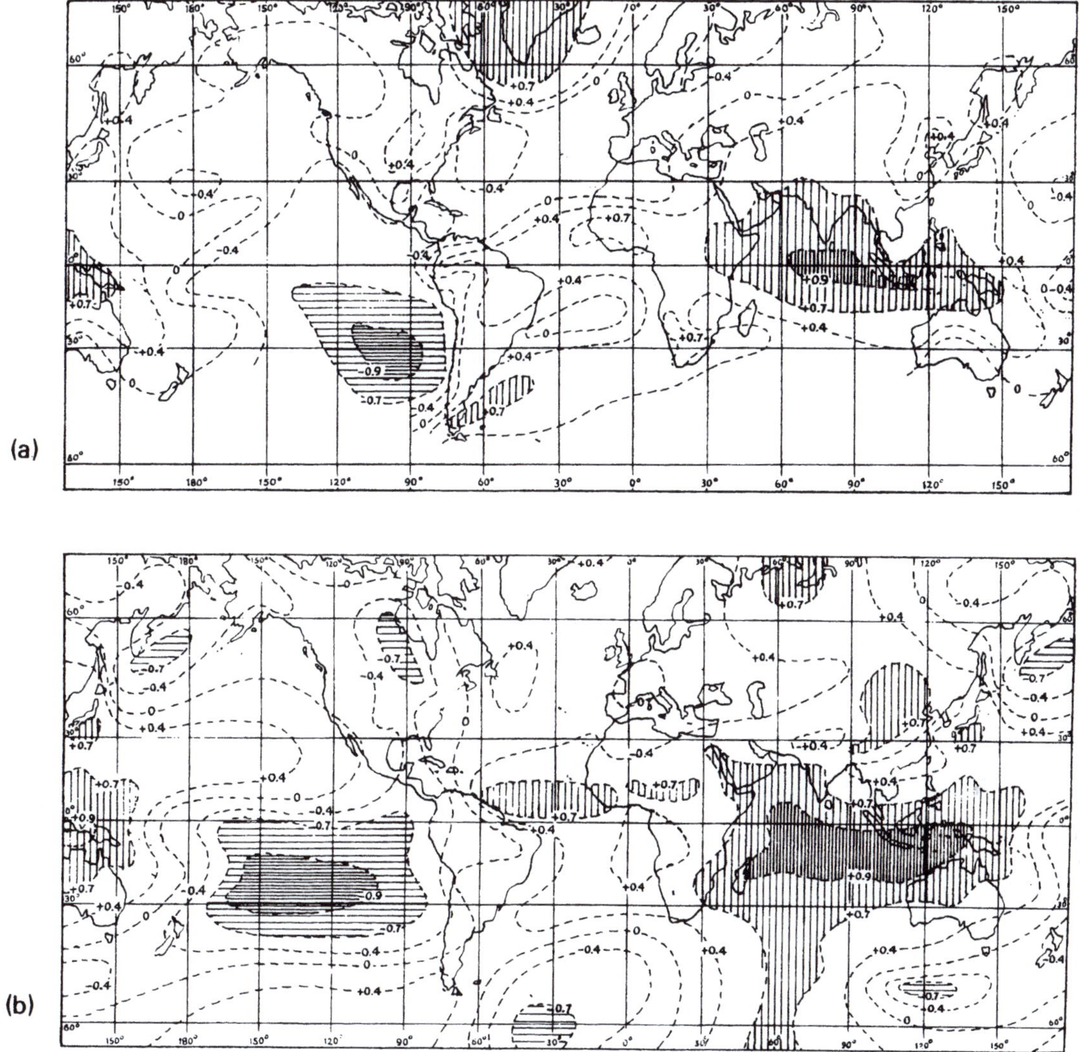

Fig. 6.10 The Southern Oscillation.

World distribution of correlation coefficients connecting local 12-monthly air-pressure anomaly with that at Batavia/Jakarta (approx. 6°S 107°E):

(*a*) 1931–8

(*b*) July 1949–June 1957.

(*After* BERLAGE *1961, reproduced by kind permission.*)

this book. A reverse shift from Greenland to north and northeast Europe appears to have taken place between the 1880–95 epoch and the 1930s.

The history of the pressure level prevailing at the Batavia/Jakarta end of the Southern Oscillation since 1866 is reproduced in fig. 6.11 in terms of 6-monthly departures from the overall mean. This displays the oscillation clearly: its period is subject to many irregularities, the time elapsed between successive peaks ranging from under 2 to about 5 years, but most commonly 2–2½ years. The amplitude also varies greatly, but is generally a large fraction of the total variability of the pressures prevailing over that area. Comparison of the Batavia air-pressure curve (fig. 6.11) with the equatorial stratospheric winds at the 50-mb (approx. 20-km) level over Canton Island (fig. 5.8) from 1954 to 1965 shows the variations of the latter as much more regular than those in the surface pressure regime. But the maxima of westerly wind at that height in the lower stratosphere fell generally 6–12 months after maxima of the 6-months' average pressure level at Batavia/Jakarta.

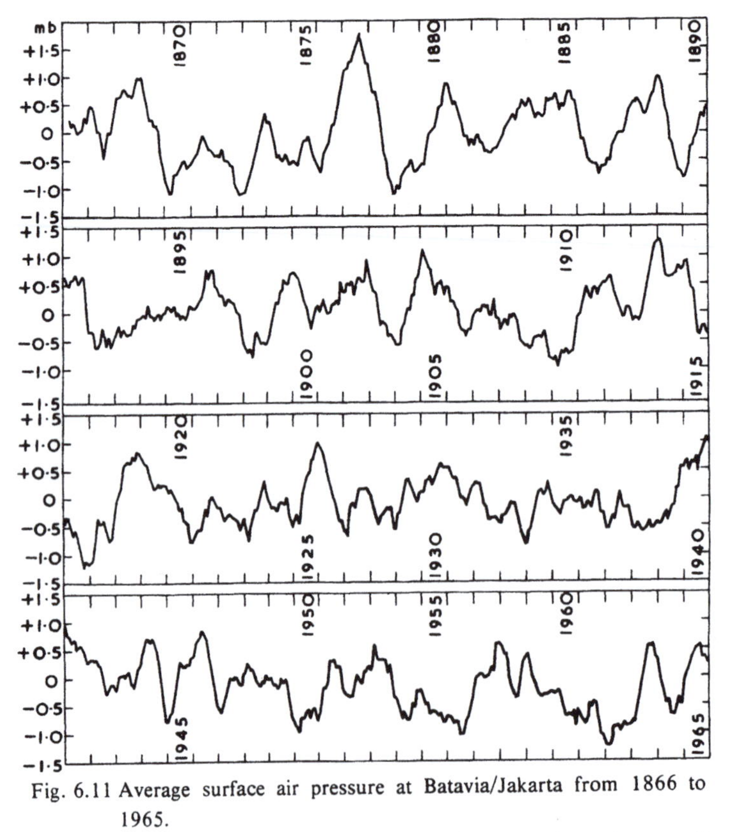

Fig. 6.11 Average surface air pressure at Batavia/Jakarta from 1866 to 1965.
Six-monthly running means of the departure from the overall average.
(*After* BERLAGE *1966, reproduced by kind permission.*)

Thus, the Southern Oscillation, the stratospheric wind oscillation and their links with the intermittent occurrences of blocking of the northern hemisphere westerlies by anticyclones in the higher latitudes, referred to above and in the last chapter (p. 209), may perhaps be regarded as all part of a single process with an underlying time-scale of the order of 2–2½ years. BÖHME (1967) has found an oscillation with an average period of 25·2 months in the frequency of meridional circulation patterns over Europe between 1883 and 1962; from 1948 to 1962 the period averaged 25·8 months. Moreover, comparison with data on the alternation of W'ly and E'ly equatorial stratospheric winds at about the 30-mb level (*c.* 24 km) for 1911–15 and 1950–61, showed that the two variations were coherent with the hypothesis that the maximum frequency of meridional circulation patterns at the surface over Europe tends to occur when the equatorial stratospheric wind at this height changes from W to E.

BERLAGE (1966, p. 32) quotes 21 months as the average period length of the North Atlantic Oscillation which expresses the fluctuations of blocking in this part of the world. However, BÖHME's findings regarding the varying frequency of meridional circulation forms over Europe, and BRIER's observation of a 2·2-year oscillation in the 500-mb zonal index[1] in January around the entire northern hemisphere at 35–55°N, suggest that a more representative value would be 2–2½ years. These fluctuations in the zone of prevailing westerlies may therefore be further manifestations of the same process as the other oscillations of similar mean length here discussed. BRIER's (*personal communication*) finding of an oscillation of this period, averaging 2·2 years, in the strength of the northern hemisphere upper westerlies in winter is supported by an independent investigation by VOROBEVA (1967), who remarks that as long ago as the 1880s and 1890s A. I. VOEIKOV established that long and short duration of frozen rivers and of snow cover in European Russia tended to occur in alternate winters, with the northwestern and southwestern parts of the country in antiphase. HELLAND-HANSEN and NANSEN (1917, p. 129) found traces of 2- to 3-year alternations in the prevailing level of water temperatures in the North Atlantic Ocean, which we may now presumably connect with the fluctuation of wind circulation strength. These authors regarded this oscillation, and others related to it, as not of really constant period but now about 2 years, now about 3 years (a description equally applicable in the 1960s); like VOROBEVA, they also tended to connect these oscillations with a corresponding minor periodicity in solar disturbance (cf. SHAPIRO and WARD 1962). SCHERHAG (1967) has found that a biennial oscillation also appears to affect the overall mean temperature of the entire (northern hemisphere) atmosphere from the surface up to 16 km.

Two somewhat different types of explanation have been suggested for a large-scale quasi-biennial oscillation in the atmosphere. Both regard it as probably originating in air–sea interaction but subject to disturbance of its amplitude and period-length by external

1. The *zonal index* is the strength of the zonal component of the mean circulation, measured in this case by the average pressure difference between latitudes 35° and 55°N or by the mean geostrophic west wind corresponding to that.

influences, for instance by any variation of the intake of solar energy affecting either the temperature of the tropical seas or the vigour of the large-scale wind circulation. BERLAGE (1957) suggests that the phase of the Southern Oscillation during which the easterly Trade Winds in the tropical Pacific are strong – i.e. when pressure is higher than average at Easter Island and below average in Indonesia – accelerates the cold Humboldt Current from the south along the coast of South America, also its extension along the coast of Peru and out into the equatorial Pacific, thereby increasing the upwelling off the coast of Peru and the strength of the South Equatorial Current across the Pacific Ocean. Hence these waters become colder than normal, and this negative temperature anomaly is conveyed westwards across the equatorial Pacific to arrive in the Indonesian low-pressure area on average 6 or 7 months later. This induces a gradual increase of atmospheric pressure in the Indonesian low over the following half year or so, slackening the Trade Winds and thereby allowing water temperatures gradually to rise in the eastern Pacific where the high-pressure system weakens. The resulting area of positive water temperature anomaly is, in its turn, conveyed westwards across the equatorial Pacific and arrives in the Indonesian area 7 or 8 months later, inducing a gradual lowering of atmospheric pressure there, which completes the cycle. The whole round of developments would take on average about 24–30 months to complete.

This view of the associated variations of atmospheric pressure and ocean temperatures over the South Pacific and Peru seems largely confirmed by a subsequent investigation of the ocean data by SCHELL (1965), which showed additionally the importance of the strength of the winds in the long fetch of westerlies in latitudes 40–45°S along the southern flank of the Easter Island anticyclone driving the water of the temperate zone towards the South American coast. (The westerlies and Trade Winds in any one sector are positively correlated with one another and vary with the intensity of the subtropical anticyclone.) A similar association between a subtropical anticyclone and the drift of a large area of warmer or cooler water across the ocean, with noteworthy effects upon the rainfall over it, has been observed in the South Atlantic where the warm water and positive rainfall anomalies take 3–4 months to pass from Angola to Brazil (EICKERMANN and FLOHN 1962).

About every third or fourth cycle of the Southern Oscillation has markedly greater amplitude (cf. fig. 6.11). The interval actually varied from 4 to 14 years in the time since 1866. The years of weakest drive of the winds in the South Pacific and warmest water off the coast of Peru are unusually favourable for convective (cumulus type) cloud development there and are liable to be accompanied by disastrous rainfall in that normally dry country. For this, however, it seems necessary that the South Pacific situation described should come around New Year or in the first months of the year – i.e. in the depth of the northern hemisphere winter. At this season meridional airstreams from the north cross the equator and penetrate the southern hemisphere over a limited sector, carrying the intertropical front before them with its great cumulus and cumulonimbus cloud lines able to maintain their development over any unusually warm water off Peru, whereas they decay over the cold water in other years. Perhaps because this coincidence of season is required, and because such plunges of the intertropical convergence lines into the southern hemisphere bring intense rainfall whenever and wherever they occur, the association of these so-called *El Niño* years of outstanding rainfall with the phase and amplitude of the Southern Oscillation is by no means as clear as BERLAGE and DE BOER (1960) seem to have expected.

At the 'desert islands' of normally very low rainfall strewn across the equatorial Pacific, gross rainfall anomalies occur apparently in a similar manner (PALMER and PYLE 1966). Thus, Chatham Island in the Galapagos (near 1 °S 89½°W) received a total of 35 mm rain in 1950 but 1420 mm in 1953. Malden Island (4°S 155°W) averaged 75 mm yearly in 1906–8, but over 2500 mm in 1914–15. In these extreme cases, the heavy rainfall coincided with abnormally high pressure at Batavia/Jakarta (cf. fig. 6.11), i.e. with abnormally weak phases of the Southern Oscillation. This might be held to apply also to the 1891, 1896, 1899–1900, 1902 and 1925 rain years in Peru, but hardly to the wet years in 1893, 1907–8, 1917, 1920–1. The so-called 7-year rhythm of the *El Niño* is by no means readily identified: BERLAGE (1966) gives the dates of more recent notable *El Niño* seasons as 1925–6, 1930, 1932, 1939–41, 1943, 1951, 1953 and 1957–8. The variations referred to are gross, and the phenomenon needs further investigation as to its rhythmic (statistical) and meteorological-oceanographic nature. DOBERITZ (1967*a*, *b*; 1968) has established by statistical analysis of data stretching in some places from 1890 to 1965 that the rainfall anomalies in the arid zone of the equatorial Pacific tend to be coherent all the way from the South American seabord to about 165°E. He confirms that the anomalies appear in the northern winter season and adds that they are seen first at the northern limit of the region near 4°N: from there they proceed slowly southwards, reaching the centre of the arid zone in about 2 months and occasionally penetrating as far as 9–10°S by the time 5 months have elapsed.

From the diverse approaches by SCHELL and DOBERITZ described, it appears that the *El Niño* phenomenon depends upon a chain of circumstances in the course of which an anomaly of the Easter Island anticyclone and the winds over the South Pacific in the preceding year prepares a warmer than usual water situation in the Pacific Ocean off Peru and along the equator; the northern hemisphere winter wind circulation must then be such as to penetrate unusually far southwards over this water.

The other, more generalized, interpretation of a fundamental 2-year rhythm in world weather is suggested by BRIER (*personal communication*). Air temperatures prevailing in winter influence the ocean surface temperatures (positive correlation). This ocean temperature anomaly is spread by convection and wave-stirring through a layer of some depth and tends to persist to the following summer (positive lag correlation), when it affects the temperature of the overlying air in the same sense (positive correlation). It seems that there must be a negative correlation between both air and ocean temperatures of the summer and those of the following winter; otherwise the whole anomaly system would persist over successive years, whereas in fact they seem more nearly independent of each other. TROUP's (1965) summary of the lag correlation coefficients connecting WALKER's values of the Southern Oscillation from one season to the next and on to the next-but-one shows in fact all positive correlation coefficients; but whereas the coefficient is + 0·8 to + 0·9 from the southern hemisphere winter to the following summer, it falls to + 0·2 from the southern summer to the following southern winter. BRIER's argument, in fact, suggests the existence of a (weak?) negative correlation between corresponding seasons in successive years and hence a tendency for some positive correlation between alternate years. External influences would be liable from time to time to accelerate or delay the processes involved in the alternation, so that the average period would vary and would be unlikely to be exactly 2 years, though the variations would presumably be random and it seems unlikely that they would tend systematically to delay the cycle. The regions, wherever they were, in which the links in this chain of correlation coefficients were strongest, would have to produce coefficients great enough to guarantee at least the sign (direction) of the relationships and the overall nature of the sequence. Further, the region of origin would be bound to lie in latitudes high enough for summer and winter to have their usual meanings: within the tropics an essentially two-season structure of the year could not apply. TROUP regards the SE Trade Winds of the Pacific as the seat of the oscillation; though these in turn may reasonably be regarded as an expression of the strength of the subtropical anticyclone generated by the westerly circumpolar vortex of the southern hemisphere.

Despite these uncertainties, the Southern Oscillation and its roughly 2- to 3-year periodicity can clearly be taken to involve an alternation in the pattern and rate of heat distribution affecting much of the world and its wind circulation. The oscillation appears at least as firmly established as that between periods of blocked or meridional flow and strong westerlies in the northern and southern hemisphere temperate zones. And in some ways these two phenomena are linked. Certainly, the Southern Oscillation should be seen as related to the liability of the equatorial Pacific to some of the biggest and most extensive anomalies of sea surface temperature in the world, about which more is said in Chapter 10.

A study by BAYER and BAYEROVA (1969) seems to throw some light on part of the mechanism by which the quasi-biennial oscillation is spread around the middle latitudes of the northern hemisphere: this oscillation is found to be very marked in the zonal index (sampled at the 500-mb level over the years 1949–66) and to have by far its greatest amplitudes over the western Pacific and over the western Atlantic, in both cases in the longitudes of the strongest part of the jet stream. The time-scale of these variations slowed up in the 1960s as did that of other aspects of the 2- to 3-year oscillation.

Oscillations of period length about 11 years

Various workers interested in the possible importance of the '11-year' sunspot cycle have investigated differences of average pressure level prevailing in different parts of the world within the sunspot cycles of the present century. WEXLER (1956) produced northern hemisphere maps which showed prevailing January pressures all over the Arctic about sunspot maximum to be higher than about sunspot minimum in the cycles between 1900 and 1939 (i.e. just 4 cycles investigated), the peak values of the difference being + 9·5 mb near Alaska and + 5·1 mb over northeast Greenland. These figures suggest more blocking anticyclones over the sectors mentioned in the Januarys around sunspot maximum, but from so few cases their statistical significance is questionable. More recent studies, however, suggest a general tendency for positive pressure anomalies over high latitudes around sunspot maximum. MAKSIMOV and SLEPTSOV (1963) studied annual mean atmospheric pressures all over the northern hemisphere over 3 sunspot cycles between 1913 and 1942, and over the southern hemisphere for one cycle 1932–42. The result revealed simple patterns of maximum amplitude over high latitudes; however, the highest values of the '11-year' cycle amplitude were only about 1·5 mb over the Arctic and over the Antarctic. Over all the lower latitudes between about 0° and 35° the amplitude was less than half a millibar. Another recent study of yearly mean pressures over the northern hemisphere from 1899 to 1939 by GASJUKOV and SMIRNOV (1967) yielded the pattern of amplitudes of variations with period about 11 years shown here in fig. 6.12 with a maximum of about 2·0 mb over Alaska. The distribution of the higher values is interesting because of its relationship to the zone of maximum frequency of aurorae (solid line on the map), which are known to be

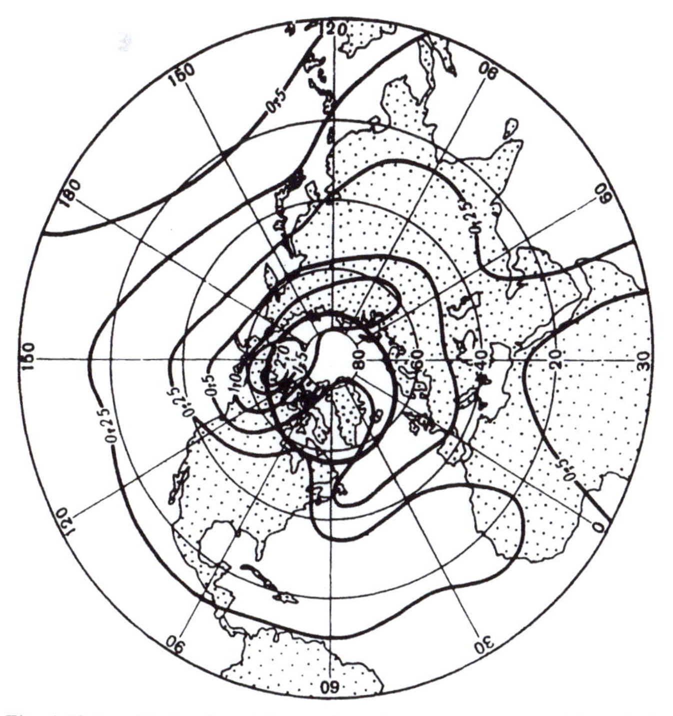

Fig. 6.12 Amplitude of variations of yearly mean pressure with period about 11 years, northern hemisphere 1899 to 1939: millibars. (*After* GASJUKOV *and* SMIRNOV *1967*.)

related to solar activity.[1] The standard deviation of annual mean pressure (1951–66) in the area of the GASJUKOV and SMIRNOV 11-year oscillation's maximum is about 1·0 mb, and the amplitude is about twice the standard deviation from northeast Siberia near 150°E to Alaska. Over the European part of the crescent of relatively high amplitudes, the amplitude of this oscillation is only about half the standard deviation of annual mean pressure.

On the basis of the statistical investigations reported earlier in this chapter, no very strong pressure-pattern effects with an 11-year periodicity should perhaps be expected. WEXLER's and the Russian authors' results reported above, and much work by BAUR (e.g. 1959), suggest that significant anomalous tendencies of the atmospheric circulation and world weather patterns in summer and winter may be associated rather with particular phases of the roughly 11-year cycles of solar disturbance. The associations of summer rainfall in Europe with different phases of the 11-year sunspot cycle found by BAUR (*loc. cit.*) would be liable to emerge from correlogram analysis as quasi-periodicities of about 3–4 and 5–5½ years rather than 11 years. We shall have more to say on other aspects of these and similar reported relationships in the next chapter and in Chapter 10.

Present knowledge of tendencies in the northern hemisphere circulation and surface pressure patterns apparently associated with different phases of the 11-year solar disturbance cycle, based on a variety of mostly preliminary investigations and indirect approaches, but in many cases extending back over about 10–15 cycles, suggests a rather complex sequence. Some features at least and some of the effects in terms of temperature, rainfall, windiness, etc., seem likely to be statistically significant. These tendencies may, perhaps, apart from the effects of biennial oscillation, be summarized as follows:

(*a*) high pressure over the polar regions around and somewhat after sunspot maxima;
(*b*) strong development of the middle latitudes westerlies, subpolar lows and subtropical anticyclones around the middle of the declining phase of solar activity;
(*c*) weakened circulation around sunspot minima;
(*d*) strong meridional and cellular circulation systems at some stage during the more rapid rises of solar disturbance and greatest frequency of very strong anomalies of pressure and temperature (intense systems);
(*e*) also (perhaps later) during the ascent, a phase of strong middle latitudes westerlies, strong subpolar lows and subtropical anticyclones, seems particularly liable to occur.

Besides these general tendencies, mostly suspected by BAUR (1956) and now largely substantiated by a recent investigation (results in Appendix IV), it seems clear that there are

1. One aspect of an unusually long run of yearly data reported by S. I. KOSTIN strongly suggests that appearances of an approximately 11-year cycle in certain weather series may be related to the sunspot cycle of similar length. The yearly layers (varves) in the silts laid down at the bottoms of two lakes in the Crimea (Lake Pert and Lake Saksij or Saki) over the last 4000 years show cyclic variations of thickness, the period length averaging 11·2 years and varying in individual cycles between extremes of 7 and 17 years corresponding to the variations of the solar cycle itself.

differences of behaviour between the stronger and weaker cycles of solar disturbance and also that shorter-term variations, including individual solar eruptions and sudden declines of activity, may have significant effects on weather. For example, BAUR (1956, p. 131) found over the period 1761–1952 (17 cycles) that when after rising solar disturbance in the previous year up to about July–September a sharp decline ensued in October and November, the strength of the zonal wind circulation over the North Atlantic and Europe tended to undergo a parallel sequence, but with sufficient lag to give weak circulation in the following winter, which therefore tended to be severe. As a logical counterpart to this, an unpublished examination of the (southern hemisphere) westerlies over New Zealand over the 108 months 1960–8 showed a significant tendency for positive anomalies of strength about 10 months after a short-term maximum of sunspot activity (measured by departures of solar disturbance from the running 9-month mean). Such associations may be useful in forecasting, and some have been used to frame statistical 'rules' about subsequent weather. We shall examine these aspects further in the next chapter.

Among the few weather elements in which an approximately simple 11-year period and probable linkage with the solar cycle have been reported is thunderstorm activity. First noted by VON BEZOLD (1884) and MYRBACH (1935) from analysis of data on frequency of lightning strikes in central Europe over periods of up to 134 years, and extended to over 2000 places throughout the world for the very active sunspot cycle 1954–64 by KLEJMENOVA (1967), it appears that in most regions the amount of thunderstorm activity varies in antiphase to the sunspot cycle. The existence of complications, however, has to be recognized: some regions and some runs of years show either no relationship or even an opposite relationship, as in some of the regions of the world surveyed by BROOKS (1934). Moreover, the curve of thunderstorm activity may be double-peaked with its main maximum during the ascending phase of the sunspot cycle.

The operation of longer-term climatic fluctuations

We need to know more about the origin and mechanism of the Southern Oscillation and other similar repeating fluctuations, if we are to be able to watch intelligently, and – over short stretches – predict, their progress from year to year. For the same reason, no purely statistical identification of the still longer-term variations, with time-scales ranging upwards from 23 years to the ice age–interglacial cycles, which last tens of thousands of years and can be established through perhaps four to eight repetitions, affords an acceptable basis for forecasting even a fragment of a cycle ahead without physical understanding. We must seek to know why they occur and establish how they operate, for example by observing evidence of changes in the distribution of heating and cooling, or in the vigour and pattern of heat transport by the atmosphere and oceans, shifts of the main windstreams and cyclone tracks, and so on. Much of the later parts of this book will be devoted to this quest, through an examination of past climatic patterns and the slower evolutions observable in the present epoch.

CHAPTER 7

Anomalous patterns of atmospheric circulation, weather and climate

This chapter deals with observed variations of the large-scale wind circulation, variant patterns that differ in greater or less degree from the simple zonal circulation model (with its well-developed circumpolar vortex, long waves of no great amplitude in the upper westerlies and at the surface general westerly winds in middle latitudes balanced by polar and tropical easterlies) described in Chapter 3. We are not in the first place concerned here with the duration of an anomaly, nor with any tendency for regular periodicity or otherwise, but with identification of modes of behaviour and, where possible, characteristic successions and their time-scales. Full understanding and confidence in prediction demand knowledge of the physical processes at work.

The character of one season, or sometimes the greater part of one year, of a group of years, or even of a whole climatic epoch, may be defined by some long-lasting anomaly in the position and alignment of the mainstream of the upper westerlies. This steers the sequences of surface depressions and anticyclones and even controls the places where they develop. The anomaly may be fundamentally a matter of the strength and prevailing latitude of the upper flow, as well as the corresponding wave amplitude, wave lengths and wave positions which for the time being characterize the circumpolar vortex.

Some spells of persistent weather pattern can be attributed to an evident harmonious adjustment of the circulation to the thermal condition of the undersurface. The flow of the surface winds, together with the distribution of cloudiness, etc., that accompanies them, may be seen to transport heat, drive the warm and cold ocean currents, and determine areas of radiation heating and cooling, in such a way as to maintain the particular hemispheric thermal pattern which continually regenerates the same circulation regime. This sort of closed circuit of cause and effect underlies those summers and winters that are dominated by long spells of this or that weather character in many areas in temperate and high latitudes. Such self-maintaining regimes are liable to occur whenever the waves in the circumpolar vortex place the mainstream of the upper westerlies in line with the belts of strongest thermal gradient for the time being established in the Earth's surface. But the basic conditions of the stationary pattern tend to fade away with the changing radiation input (which changes the energy available and the wave-length probabilities) as the given season ends; and the long spell normally ends with it.

There are plainly other self-maintaining arrangements which give to some successive years, and even long runs of years, a partly similar character. This can happen through the accumulation, or loss, of heat stored in the sea or of snow and ice on land. The ice ages are the extreme example.

Proceeding from this observation, some have argued that no change in the external energy supply arriving from the sun need be postulated as the cause of ice ages, far less of any of the numerous smaller, and shorter-lived, climatic fluctuations (e.g. CURRY 1962; SHEPPARD 1964, pp. 330–1). All that is needed to start development of an ice age or of a warm ice-free epoch, according to this view, is some more or less fortuitous variation in the heat exchange between ocean and atmosphere, perhaps arising from an inhomogeneity in the heat stored in the ocean depths and brought to the surface in the course of the very slow vertical turn-over of the oceans; the anomaly might also be created at the ocean surface through some change of prevailing wind speed in the vast Trade Wind zone, where the present average (about 6·5 m/sec) is critical for breaking waves on the sea surface and any change could perhaps have a great effect on the amount of spray and evaporation, and hence on the general moisture content of the atmosphere, as well as on the temperature, salinity and density of the ocean surface waters (KRAUS 1955, WEYL 1968). The time-scale of the oceans' vertical circulation is dealt with in a later chapter, but is at least of the order of 500–1000 years.

There is no reason to doubt that climatic shifts, and possibly some gradual evolutions stretching over years, decades or longer, may be triggered off in ways like these. The atmospheric circulation, and its heat and moisture transport, must respond to the postulated anomalies of ocean surface temperatures and of its own moisture content. We shall notice some examples of linked ocean and atmosphere anomaly patterns in a later chapter (pp. 394–410).

Random variations of the combined heat economy of ocean and atmosphere need not, however, be the only cause of either short- or long-term climatic differences. They probably have more to do with climatic variations of more modest scale than ice ages. Similar variations could also be triggered by variations in the external energy supply arriving from the sun, and the ocean–atmosphere exchanges would then play a contributory part in the course and development of any observed climatic change. Indeed, the operation of some external trigger, whether acting on the ozone heating in the stratosphere and affecting overall mean tropopause height or seen more directly in some initial change of the prevailing strength of the mainstreams of the wind circulation in the stratosphere or lower down, would conform well with the suggestions of KRAUS and WEYL referred to above. The suggestion that external influences are irrelevant carries less weight insofar as it can be demonstrated that:

(i) the external conditions do vary (e.g. the Earth's orbital variations and the indications of solar variability discussed in Chapters 2 and 10);

(ii) the timing of climatic changes answers to the calculated (or observed) timing of changes in these external conditions;
(iii) physically reasonable cause and effect linkages can be traced between some relevant external events observed and apparent responses in the atmosphere and oceans.

Example of a self-maintaining, self-amplifying process: BROOKS' ice age theory

BROOKS (1925, 1949) worked out a hypothesis regarding ice ages, whereby a very small initial change in the temperature of the polar ocean, whether imposed by the incoming radiation supply or brought about within the circulations of the atmosphere and oceans, should be greatly amplified. The proposed effect should arise from the change of albedo when an area of previously open water freezes over and becomes covered with snow, as well as ice, for much of the year. This is largely equivalent to the substitution of a new land surface for open sea in high latitudes. From comparisons of the surface temperatures prevailing in oceanic regions and over continents at various distances from the sea, and in different latitudes, BROOKS arrived at estimates of:

(*a*) the poleward gradient of temperature that would prevail over an open ocean of unlimited extent;
(*b*) the 'cooling power' (i.e. temperature depression) induced by land areas of various sizes;
(*c*) the 'cooling power' of a floating ice cap, of various sizes, centred near the pole.

Whereas (*a*) is nearly constant, estimated at 0·5°C per degree latitude, (*b*) and (*c*) increase with the square of the radius of the snow- or ice-covered area (since they are governed by its area) until this reaches 8–10° of latitude: beyond this point the cooling power increases much more slowly in proportion to further growth. In the highest latitudes, the temperature depression produced at the edge of the pack-ice seemed likely to be about 0·25°C for every 1% of the area within a circle of 10°-latitude radius that was covered with ice. Assuming that, with the present geography of land and sea, the mean annual temperature near the North Pole in any ice-free epoch would be very close to the freezing point of sea water (−2°C), BROOKS found that a very small initial cooling from that point, by perhaps only a fraction of a degree, should have a very big effect once a floating ice cap was established and should lead to the growth of the ice-covered area. Equilibrium would be reached at some point well beyond the 10°-latitude radius where the effectiveness of further ice growth in depressing the surface temperatures begins to decrease. Taking 7·5°C as the normal January to July range of temperature at the edge of the Arctic sea ice, it did not appear that summer melting would suffice to halt the waxing of the ice sheet wherever the prevailing winter temperature was below −4°C. The calculations indicated that the ice sheet might attain nearly 25°-latitude radius before equilibrium was reached, though this might not be until many winters had passed. The prevailing temperatures should be lowered by 25°C near the pole and by 12°C at the edge of the ice. The present mean annual surface air temperature in the neighbourhood of the North Pole is −20°C: some (including BROOKS) have therefore inclined to the belief that if the Arctic sea ice could be artificially removed it would not now re-form.

The equations formulated by BROOKS and the results obtained should not, however, be taken as more than a rough indication of the order of magnitude of the end-products of unstable growth (and decline) of the Arctic sea ice under favourable external conditions. BROOKS remarked that the same principles, but not the same figures, should apply to the establishment of an ice sheet on land in high latitudes. For this, in reality, however, increased *thickness* of the snow and ice cover, affecting its ability to survive summer melting, must be vital. Some reduction of the summer insolation might also be necessary, since most land in the latitudes concerned is normally snow-covered in winter even in interglacial climates.

Under conditions favouring decline of the Arctic sea ice area by progressive melting from year to year, as with some increase in the radiation supply or of the heat transported by the ocean currents from other latitudes, the ice sheet should for a long time retreat only slowly, since its cooling power would be maintained by the high albedo over the large area which it still occupied. In the last stages, however, when the area had shrunk to less than 10° radius, when its thickness had also declined and its surface become darkened by melt-water pools, the final decrease should be rapid.

As BROOKS' reasoning implies, the radiation and thermal conductivity properties of snow and ice work the same way in individual winters and summers. LAMB (1955) found that a snow surface of about 3000 km west-to-east extent in middle and high latitudes was sufficient so long as the snow lasted to induce a large cold trough over it in the upper westerlies, even in the face of the most vigorous warm air transport, and hence tended to divert the depression tracks around its perimeter, where further snow would be deposited.

Twentieth-century changes of prevailing circulation and the climates generated

Cases are also known, however, when a circulation anomaly sets in and persists from summer to winter, or vice versa, and even over a run of years, without there being at the outset any apparent thermal anomaly in the Earth's surface to produce it. In these cases it seems necessary to invoke some change in the external energy supply as the first cause. This seems all the more necessary since the northern hemisphere thermal patterns in summer and winter differ from each other too radically to induce cyclonic (or anticyclonic) developments in the same regions in both seasons. Nevertheless, after some years the new wind circulation characteristics are liable to induce changes in the thermal condition of the surface such as tend to maintain, or intensify, this circulation regime. An example is the weakening of the general wind circulation and transference of the mainstream to lower latitudes, the first symptoms of which may have been seen in the northern winters around 1940, but which set in persistently over the North Atlantic from the mid 1950s onwards, and which by the 1960s seems to have affected the whole world. The onset of this change

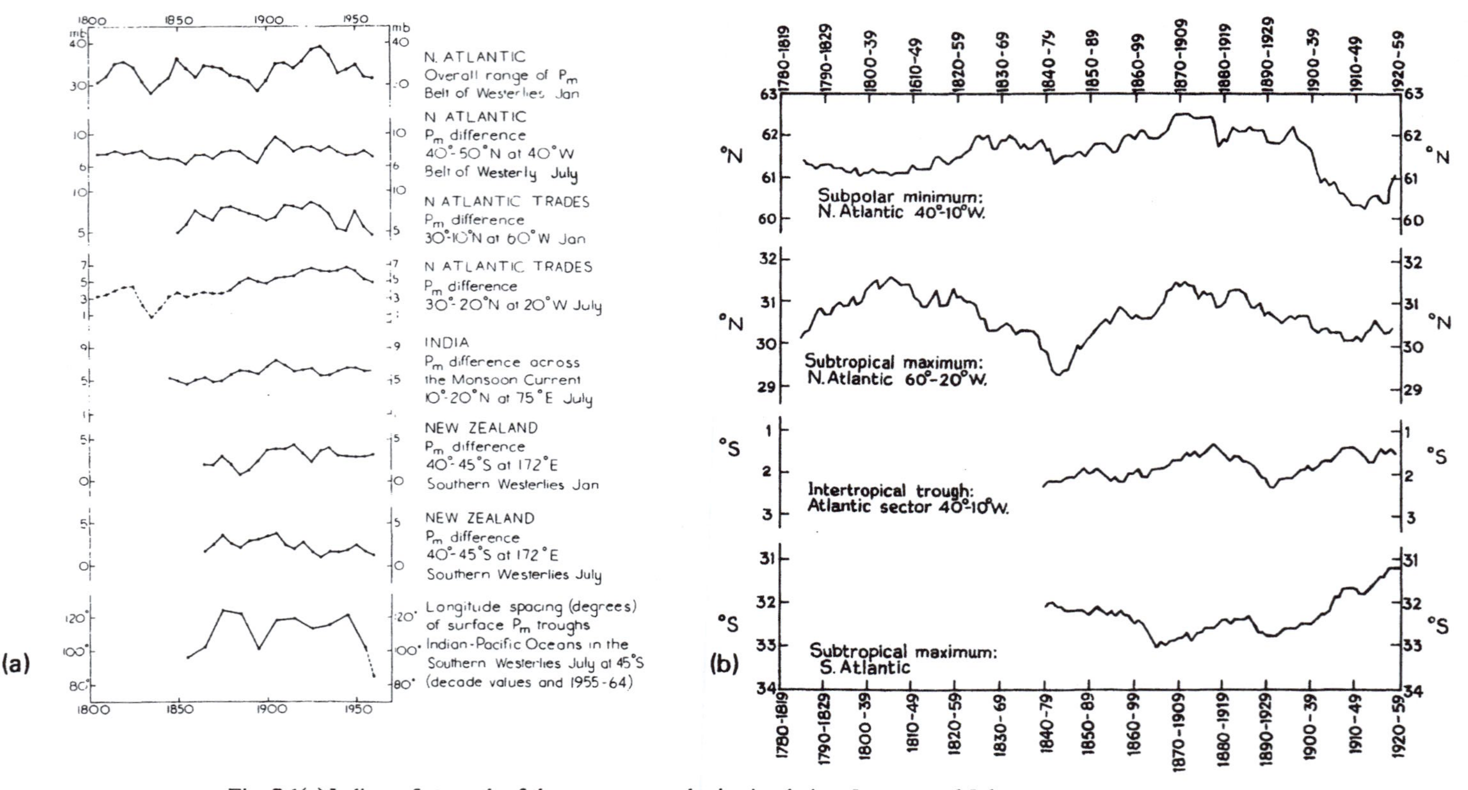

Fig. 7.1(*a*) Indices of strength of the mean atmospheric circulation, January and July.
(Ten-year mean values, mostly plotted at 5-year intervals, against the middle of the period to which the values apply.)
(*b*) Indices of prevailing latitude of main features of the world distribution of barometric pressure at sea level: North and South Atlantic, mainly between 40° and 10°W, January.
(Forty-year running means.)

occurred just when the oceans had been warmed to high latitudes by the long epoch of strong circulation preceding it, the limit of the Arctic sea ice had withdrawn far to the north, and climates in middle latitudes on land and sea had become warmer generally than at any time since the invention of meteorological instruments. Not until the late 1950s or early 1960s did it become clear that the Arctic sea ice was increasing again and that sea temperatures were falling in wide areas of the North Atlantic. But by 1964–8 the extent of the northern ice and prevailing water temperatures in most of the Atlantic were once more similar to what they had been in the nineteenth century before the decades of strong, and northward-displaced, wind circulation.

Figs. 7.1(*a*) and (*b*) illustrate by various indices the world-wide regime of strengthened circulation that culminated between 1900 and 1950 and the marked shift of the main features back towards lower latitudes that has been in progress since about 1940 (in some parts of the world since the 1920s). Fig. 7.2 shows the geography of this change of mean barometric pressure distribution over the northern hemisphere in the 1950s and 1960s as compared with 1900–39, i.e. the departure of the surface pressures prevailing in the later years from the 40-year average shown in fig. 3.17(*a*), which is now known to have represented a time of strong circulation and exceptional predominance of the westerly winds in middle latitudes. (Broken lines have been used for the isopleths of pressure change over the north polar basin in fig. 7.2 because of uncertainty about the true mean pressures there between 1900 and 1939, when the region was little visited: the main reference series of daily weather map analyses certainly implies too high average pressure during those years, since the quasi-permanent presence of a polar anticyclone, with pressure every day 1020 mb or more, was assumed. A careful attempt was made to eliminate the effects of this in the average pressure maps adopted for 1900–39 in the light of observations made in some later years around 1950 (*Berlin, Free Univ.* 1953; *U.S. Weather Bureau* 1952).) The principal change observed in the 1950s and 1960s seems to be an expansion and intensification of the Arctic high-pressure regime. Average pressure has probably changed relatively little near the pole, and this is clearly the case at Spitsbergen, whereas Greenland shows a strong pressure rise, as to a less extent do other sectors (particularly of the land ring) between 60° and 80°N. The correspondingly increased N'ly surface winds over the sea east of Greenland have promoted the increase of ice there. Over the Atlantic sector generally north of about 50°N the northern pressure rise has meant decreased westerlies and a southward displacement of the so-called Iceland low.

Fig. 7.3 records the changes in extent of Arctic pack-ice, as surveyed about the end of the melting season, at the end of August. The 1938 extent of open water (point stipple) was the greatest ever known, except locally off east Greenland where the exceptionally strong circulation was debouching the ice from the central Arctic. Unlike the situations in the 1960s and around the end of the last century, this was not due to greater ice production but to rapid removal of ice from the central polar region. The total area of the semi-permanent Arctic pack-ice is estimated to have shrunk by more than 10% between 1920 and 1938, the

Fig. 7.2 Departure of overall average pressure at sea level for the years 1951–66

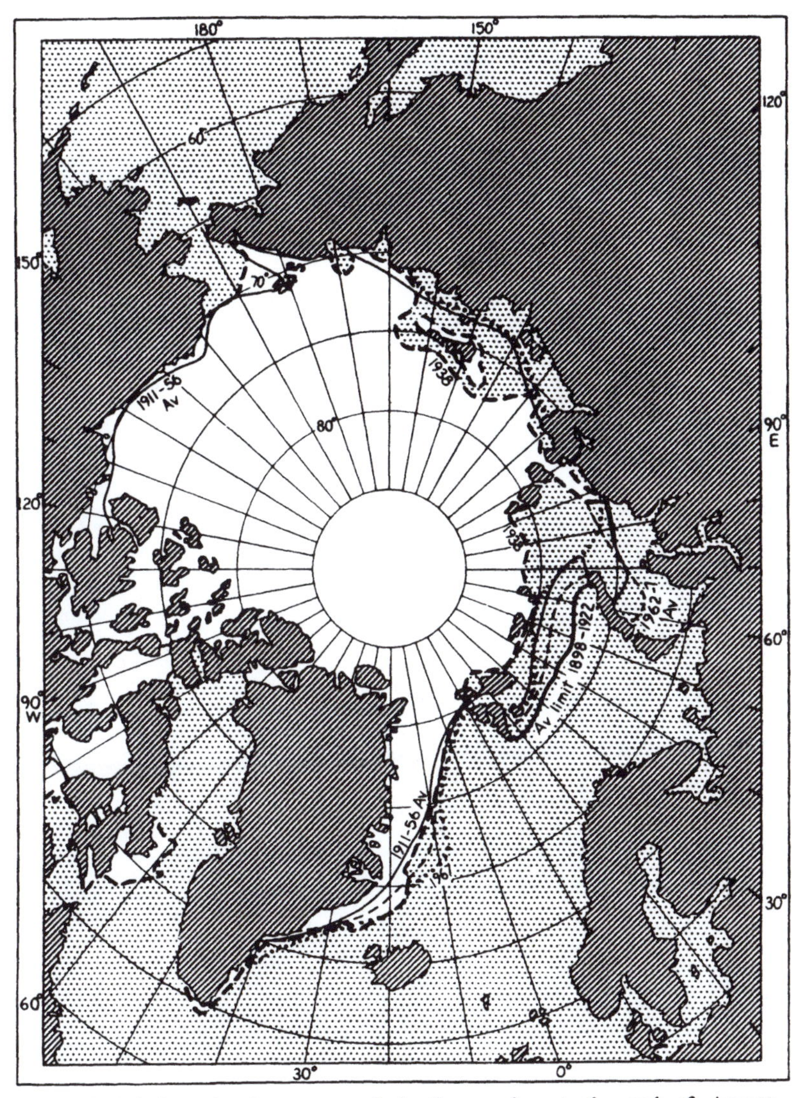

Fig. 7.3 Variations in the extent of Arctic sea ice at the end of August (approximately the end of the melting season). Open water at the end of August 1938 (greatest known extent except off east Greenland) shown by stipple, bounded by bold broken line.
Bold continuous line 1898–1922 average.
Thin continuous line 1911–56 average.
Thin broken line 1962–7 average.
Dotted line 1967.

years in which melting and removal, both due to the vigorous wind and ocean current circulation, were most rapid. The effects upon the movement of the floating ice may be seen in fig. 7.4: note the quick movement of the first Russian camp NP I established on the drifting ice near the North Pole in 1937 to where the party had to be rescued off east Greenland only 9 months later. By contrast, in the weaker circulation and more anticyclonic regimes near the pole in the 1950s and 1890s the drifts (e.g. of the ice islands T-1, etc., the Russian drift ice stations NP II to NP VI and Nansen's ship *Fram*, all seen in fig. 7.4) are much slower and tend to go into repeated circuits of the polar basin north of Greenland and Canada. The greatest lowering of surface temperatures prevailing before 1900–20 and since

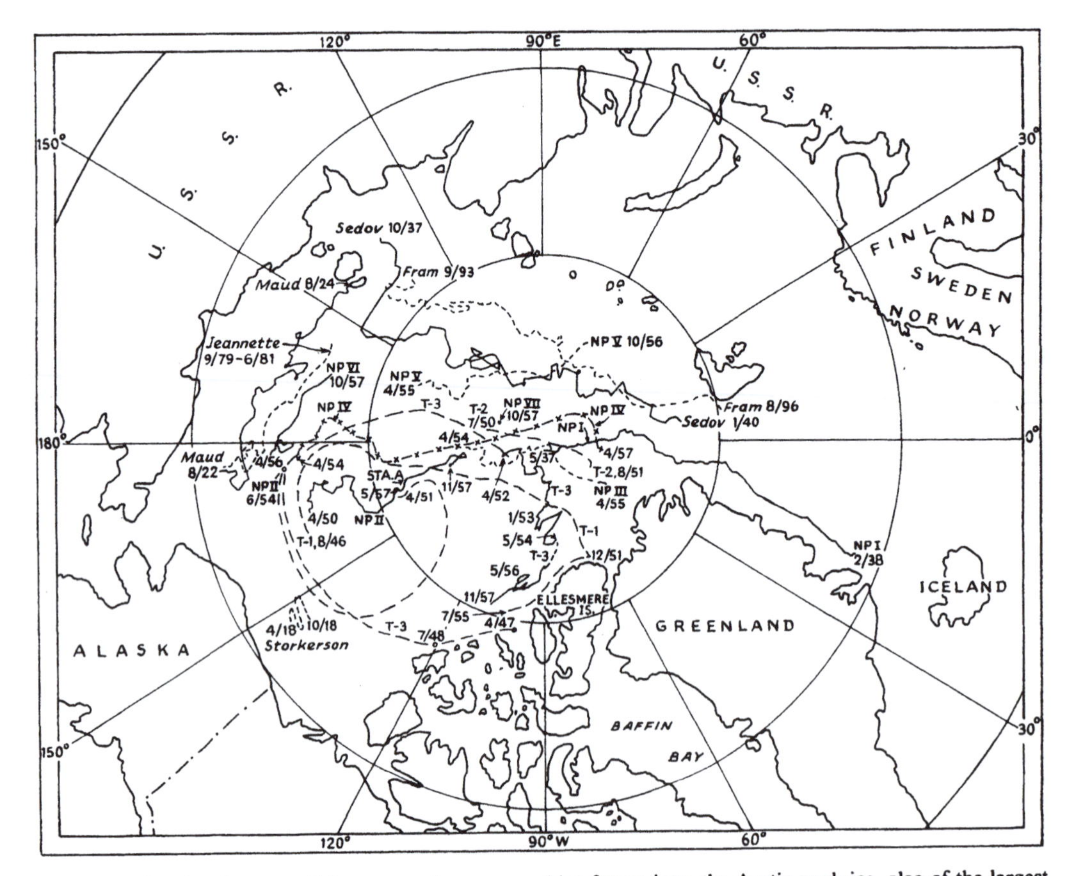

Fig. 7.4 Drifts of various expeditions camped on, or on ships frozen into, the Arctic pack-ice, also of the largest known ice islands in the Arctic Ocean.
Notation for dates against positions shown:

9/93 = September 1893.
5/37 = May 1937, etc.

(*After* BROWNE *and* CRARY *1959*.)

1950, as compared with the years of vigorous circulation in between, has been observed over the sectors (*a*) around Novaya Zemlya and Franz Josef Land to west Spitsbergen, which were between 1900 and 1950 in the path of the main warm air and warm water supply, and (*b*) between the pole and northern Canada.

We must examine further what is to be learnt from the present-day variability of weather pattern and from the changes of climatic regime in recent times for which we have adequate observation coverage. We shall notice cases in which one or more of the following seem to provide the key to the behaviour of the wind circulation and the climatic patterns it generates:

(i) internal relationships within the atmospheric circulation, including inter-hemispheric relationships, but perhaps especially breakdowns of regime following build-up of excessive heat storage, momentum or shear in some part of the system;
(ii) characteristic persistences and successions within the atmospheric circulation, which may involve effects upon the sea surface temperatures or state of ground prevailing over wide areas and a subsequent 'feed-back' from these;
(iii) ocean temperature anomalies of wide extent and affecting sufficient depth to have great inertia (persistence) because of the large heat exchange required to obliterate them;
(iv) extent of ice on the polar seas;
(v) state of ground: general dryness, waterlogging or (deep) snow cover over areas comparable in extent with a great ridge or trough in the upper westerlies;
(vi) volcanic dust veils in the stratosphere;
(vii) solar disturbance (or the lack of it).

In general, the time-scale is important and may affect such questions as whether the atmosphere determines the changes in the ocean or vice versa.

The present chapter is mainly concerned with (i) and (ii), the internal relationships within the atmosphere. The other items will be dealt with in Chapter 10.

Weather-type classifications

Daily circulation map sequences, even when reduced to the codified form of pattern classifications, may reveal climatic processes affecting the heat and moisture exchanges and transports which a monthly or seasonal mean map fails to show. Some areas in the latter which appear featureless or inactive, because of a flat or nearly uniform field of mean pressure, may be shown by the daily situations to have experienced an alternation of N'ly and S'ly, or E'ly and W'ly winds, meaning that evaporation of surface water and input of moisture and latent heat into the atmosphere were much more than the period-average map implied.

For study of the mechanics of climatic variation, including the amount of alternating transport through eddy activity, daily circulation pattern classifications, though a rough tool, are a help in submitting the enormous mass of data to analysis. Analysis by matrix algebra (using computers to extend the method to very large matrices) indicates that some 50 eigenvectors are needed to account for 99% of the variance of the large-scale circulation at the 500-mb level over the northern hemisphere (CRADDOCK and FLOOD 1969), each actual chart being represented by the sum of various multiples of the 50 functions concerned, i.e. an almost infinite variety; though one eigenvector alone, representing the main seasonal variation, accounted for 59% of the variance and 15 eigenvectors sufficed to account for 85% during the period 1965–7 studied. Several much simpler schemes of classifying the circulation according to a limited number of types are available, chiefly relating to the middle latitudes of either hemisphere: see particularly, for the northern hemisphere (or parts of it), BAUR (1947), DZERDZEEVSKI (1968), HESS and BREZOWSKY (1952), KATZ (1960), LAMB (1971), WANGENHEIM (1964) and for the southern hemisphere, ASTAPENKO (1960), DAVIDOVA (1967). The simplest are the daily weather-type classification for the British Isles by LAMB (*loc. cit.*), which covers more than 100 years from 1861, and the classification of each day's 500-mb level flow pattern over the northern hemisphere by WANGENHEIM (*loc. cit.*) from 1891, which is believed to be kept going to date by GIRS and the staff of the Leningrad Arctic and Antarctic Institute.

The definitions of the British Isles types, which incorporate both circulation pattern and characteristic weather resulting in and around the 10-degree 'square' 50–60°N 0–10°W, are as follows. The 'square' is centrally placed in the zone of prevailing westerlies though in one of the two sectors (Greenland–Scandinavia and Alaska) most affected by blocking, the variations of which should therefore be well registered.

Definitions of the British Isles weather types

The following types are recognized in the classification here presented:

AC. *Anticyclonic type*

Anticyclones centred over, near, or extending over the British Isles; therewith also cols situated over the country, between two anticyclones. *Mainly dry with light winds (though thunder often occurring in the cols in summer). Usually warm in summer, cold or very cold in winter; mist and fog frequent in autumn.*

C. *Cyclonic type*

Depressions stagnating over, or frequently passing across, the British Isles. The further criterion is used that a depression should be centred, or its central isobar on a 4- to 5-mb spacing should extend, over the mainland of Britain or Ireland at some time during the day. Small cyclonic features may be overlooked if they constitute merely details within a col when most parts of the British Isles are under anticyclonic influence. Further, to reduce the fortuitous element in the positions on any one day of the cyclone centres of a single cyclone sequence, individual days between the passing of separate centres over Britain are also counted as cyclonic if the isobars retain cyclonic curvature or if the main (surface) low-pressure axis remains over the country. *Mainly wet or disturbed weather, with very variable wind directions and strengths. Usually mild in autumn and early winter, cool or cold in spring, summer and (sometimes) in late winter. Both gales and thunderstorms occur.*

W. *Westerly type*

High pressure to the S (also sometimes SW and SE) and low pressure to the N of the British Isles. Sequences of depressions and ridges travelling E across the Atlantic. (This general type has been described by some authors as Southwesterly because the surface winds blow from about SW for more of the time than from other directions. The steering of pressure systems is generally from about W or WSW. This can be taken as the most mobile, or 'progressive', of all the types.) *Generally unsettled or changeable weather, usually with most rain in northern and western districts of the British Isles. Winds shifting rapidly between S and NW, occasionally SE or even E for a short time. Cool in summer, mild in winter with frequent gales.*

NW. *Northwesterly type*

Azores anticyclone displaced NE towards the British Isles or N over the Atlantic west of our coasts, or with extensions in these directions. Depressions (often forming near Iceland) travel SE or ESE into the North Sea and reach their greatest intensity over Scandinavia or the Baltic. *Unsettled or changeable weather, particularly in northern and eastern districts of the British Isles, sometimes with fresh or gale-force winds from between W and N. The warm sectors may contain unstable air especially in late winter and spring. Cooler than the westerly type and milder than the northerly type.*

N. *Northerly type*

High pressure to the W and NW of the British Isles, particularly over Greenland, and sometimes extending as a continuous belt S over the Atlantic Ocean towards the Azores. Low pressure usual over the Baltic, Scandinavia and the North Sea. Depressions move S or SE from the Norwegian Sea (sometimes having formed in the Iceland–Jan Mayen region, sometimes having come through from farther north, sometimes having entered the Iceland–Jan Mayen region by way of a col near south Greenland). *Cold, disturbed weather at all seasons, especially in eastern and northern districts. Snow and sleet common in winter; also associated with late spring and early autumn snow on high ground in the north and with late spring frosts in all districts. The onset of northerly type weather is often accompanied by high winds.*

E. *Easterly type*

Anticyclones over, or extending over, Scandinavia and towards Iceland. Depressions circulating over the western North Atlantic and in the Azores–Spain–Biscay region. *Cold in autumn, winter and spring; sometimes intensely cold in southern districts and suitably exposed localities elsewhere, with occasional snow in the south and snow or sleet showers in eastern and northeastern districts; fine in the west and northwest. Warm in summer, sometimes thundery. Very dry weather in western districts, relatively dry in many other districts except in the east and south.*

S. *Southerly type*

High pressure covering central and north Europe. Atlantic depressions blocked west of the British Isles or travelling N and NE off our western coasts. (Seems less persistent than the other types, occurring mainly as occasional variations within spells that are predominantly either westerly or easterly; very rare in summer.) *Warm and thundery in spring and summer, mild in autumn. In winter mild or cold according as the airmass carried over the British Isles is oceanic or continental in origin.*

Hybrid types (e.g. cyclonic westerly or anticyclonic northwesterly) are recognized on days when two or more of the above definitions are satisfied. This increases the number of variants to twenty-six, but reduces the unclassifiable days (weak, chaotic or quickly changing patterns) to about 4% overall, the proportion ranging from 2·4% in some of the most westerly years to 6·5% in some of those with most blocking. There is in reality a much greater variety of blocking than of low-amplitude zonal patterns, the latter with their mobile weather systems continually advancing from west to east.

The general westerly type (**W**), which corresponds mostly with SW'ly winds at the surface, especially in southern Britain, is the most frequent single type, as would be expected in this part of the world, and has probably been so at all times since the end of the last ice

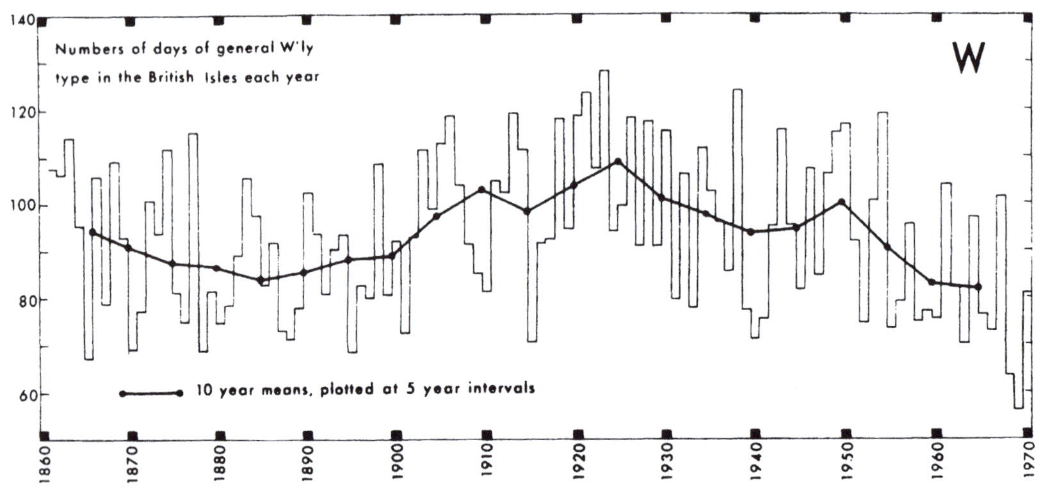

Fig. 7.5 Frequency (number of days per year) of Westerly type shown by the British Isles daily weather type classification from 1861 to 1970.
Individual years and 10-year mean values plotted at 5-year intervals against the middle of the decade to which they apply.

age. Nevertheless, it undergoes substantial variations of frequency (fig. 7.5) from year to year, from decade to decade, and probably over much longer periods also (cf. fig. 6.8). The build-up of frequency of the W'ly type during the nineteenth century to the high level sustained between about 1900 and 1950, especially around the 1920s, and its subsequent decline, usefully confirms that the changes of overall vigour of the global wind circulation shown by the sample indices for January and July in fig. 7.1 may be taken as representative of the year as a whole.

Other classifications

The WANGENHEIM/GIRS classification consists of just three types of hemispheric flow pattern, illustrated in fig. 7.6 by a characteristic 500-mb isopleth in the middle of the mainstream of the meandering upper westerlies. These are essentially: one 4-wave pattern, consisting of rather long waves of small amplitude, i.e. the zonal type **W**, and two 5-wave

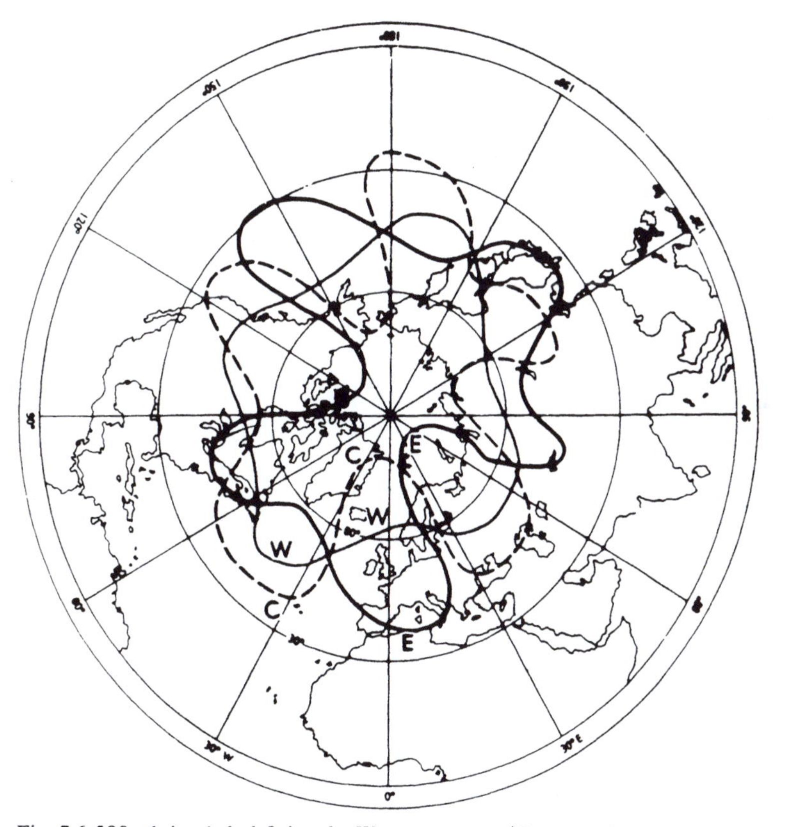

Fig. 7.6 500-mb isopleth defining the WANGENHEIM/GIRS northern hemisphere circulation pattern types W, C and E.
W = thin continuous line. E = bold continuous line.
C = broken line.

situations with shorter wave length and greater amplitude, **C** and **E**, distinguished from each other by different positions of the cold troughs and warm ridges. Linkage with the surface weather distribution is determined by the prevalent positions of development and steering of depressions and anticyclones, as described in Chapter 3. Some distortion may be introduced by classifying each day's hemisphere map on this scheme, since we have seen (fig. 4.11) that a greater variety of wave lengths and wave numbers in fact occurs; in practice this classification has been taken as applying chiefly to the Atlantic and Eurasian sectors, while the Pacific and American sectors are independently allocated to a zonal or one or other of two meridional types. Nevertheless, the varying frequency of the zonal type **W**, which has kept broadly in phase with the variation of the British Isles W'ly type, seems an indisputable index of an important aspect of circulation behaviour.

KATZ (1960) has devised a type classification for the North Atlantic–Eurasian sector 30°W–110°E which defines types, or groups of types, very like the WANGENHEIM/GIRS three and adds a group of more meridional types, **M**, with shorter wave lengths than **C** and **E**, i.e. wave number greater than 5.

DZERDZEEVSKI's (1968) classification defines forty-one types or sub-types to cover the northern hemisphere. These are chosen to represent patterns ranging from ideal zonality to violation of zonality in this or that sector or in up to four sectors simultaneously. Two types are also distinguished respectively by depressions travelling more or less straight across the polar basin from ocean to ocean and by anticyclones passing, or extending, from the pole over the two northern continents. The scheme is illustrated here by schematic surface maps of some sample types in fig. 7.7 and by sample correspondences with ASTAPENKO's similar classification of the southern hemisphere in fig. 7.8.

No hemispheric classification scheme can cover adequately all the situations which occur, and the combinations of unlike situations in different sectors, by a limited number of types. In particular, these schemes described hardly cover those situations where there are eastward-travelling depression sequences both in high and low latitudes simultaneously, and twin jet streams, north and south of a blocking anticyclone of considerable west–east extent. Nor could they be applied without modification to the ice ages (or warm ice-free epochs) with their greatly different latitudes of the main zone of cyclonic activity and (in the ice ages) probably marked eccentricity of this zone with respect to the geographical pole.

KATZ (*loc. cit*) has devised simple indices for measuring the intensity of the zonal (I_Z) and meridional (I_M) components of the general circulation. Each is taken as the average, for the region considered (in practice regions 35–70°N, or 52–70°N from 30°W to 110°E were chiefly studied), of the number of 500-mb contours intersecting the meridians (in the case of I_Z) or the latitude circles (in the case of I_M) per equatorial degree (i.e. per 111 km). The ratio $I' = I_M/I_Z$ defines the general character of the circulation; cases where $I' > 1$ may be called meridional, $I' < 1$ zonal, in general type. A closer relationship to actual mass transport of air may be obtained by distinguishing I_{ZW} and I_{ZE}, representing respectively the extent of zonal westerly and zonal easterly components, and defining $I_Z = I_{ZW} - I_{ZE}$: this I_Z is then a measure of the net flux of air from west to east between the overall latitude limits considered. Similarly the segments, or areas, of this zone showing net northerly and net southerly components may be counted separately to give respectively the indices I_{MN} and I_{MS}. But since, totalled around one latitude zone of the globe, $I_{MN} - I_{MS} = 0$, KATZ takes his overall meridional index I_M as the sum of the northerly and southerly transports: $I_M = I_{MN} + I_{MS}$, and this I_M is a measure of the interlatitudinal exchanges of air due to eddy activity. The appellation zonal or meridional is then given to cases where the ratio I_M/I_Z is less than or greater than its overall average value. Statistical averages of the index values may be found for each of the hemispheric pattern types and yield information about circulation strength and eddy activity. DZERDZEEVSKI and his co-workers have applied the same principles.

PČELKO (see Chapter 5, p. 199 herein) has calculated average values of KATZ's 500-mb indices for three sectors all round the northern hemisphere and found that distinctively low values of the ratio I_M/I_Z (i.e. markedly zonal flow patterns) accompany the single-centred circular form of the winter circumpolar vortex in the stratosphere. There was a much greater tendency to meridionality in the 500-mb level flow in the troposphere, particularly in the eastern North Atlantic–Europe–West Siberian sector, when the stratospheric vortex had two or three centres or strong troughs.

TRENKLE (1956) has used the average zonal index (gradient wind in m/sec) at the 500-mb level between 50° and 60°N from 60°W to 60°E (i.e. across the entire North Atlantic and Europe), month by month from October to March, for each of the *Grosswetterlagen* defined by BAUR, and also used by HESS and BREZOWSKY (1952), to derive the probable variations of strength of the upper westerlies since 1881 from the HESS-BREZOWSKY daily classification. The *Grosswetterlagen* define patterns covering the North Atlantic and Europe.

The results of all these approaches as regards the overall variation of strength of the zonal circulation from around 1880 to the 1960s agree with the general trend of the selected indices shown in figs. 7.1 and 7.5. As examples, TRENKLE's values of the zonal index for the winter six months each year are graphed in fig. 7.9; the frequencies of DZERDZEEVSKI's zonal (full line) and meridional groups of types are graphed in fig. 7.10. KATZ's indices alone give a different result, perhaps just a peculiarity of the broad zone 35–70°N over the sector between mid Atlantic and mid Asia; even these indices, however, agree regarding the increasing meridionality from the earliest years (1938–42) to the latest years (1953–7) KATZ studied.

Distributions of variability

We now turn to the characteristic geographical distributions of variability of individual elements of the weather – pressure, temperature, rainfall and windiness – on different time-scales.

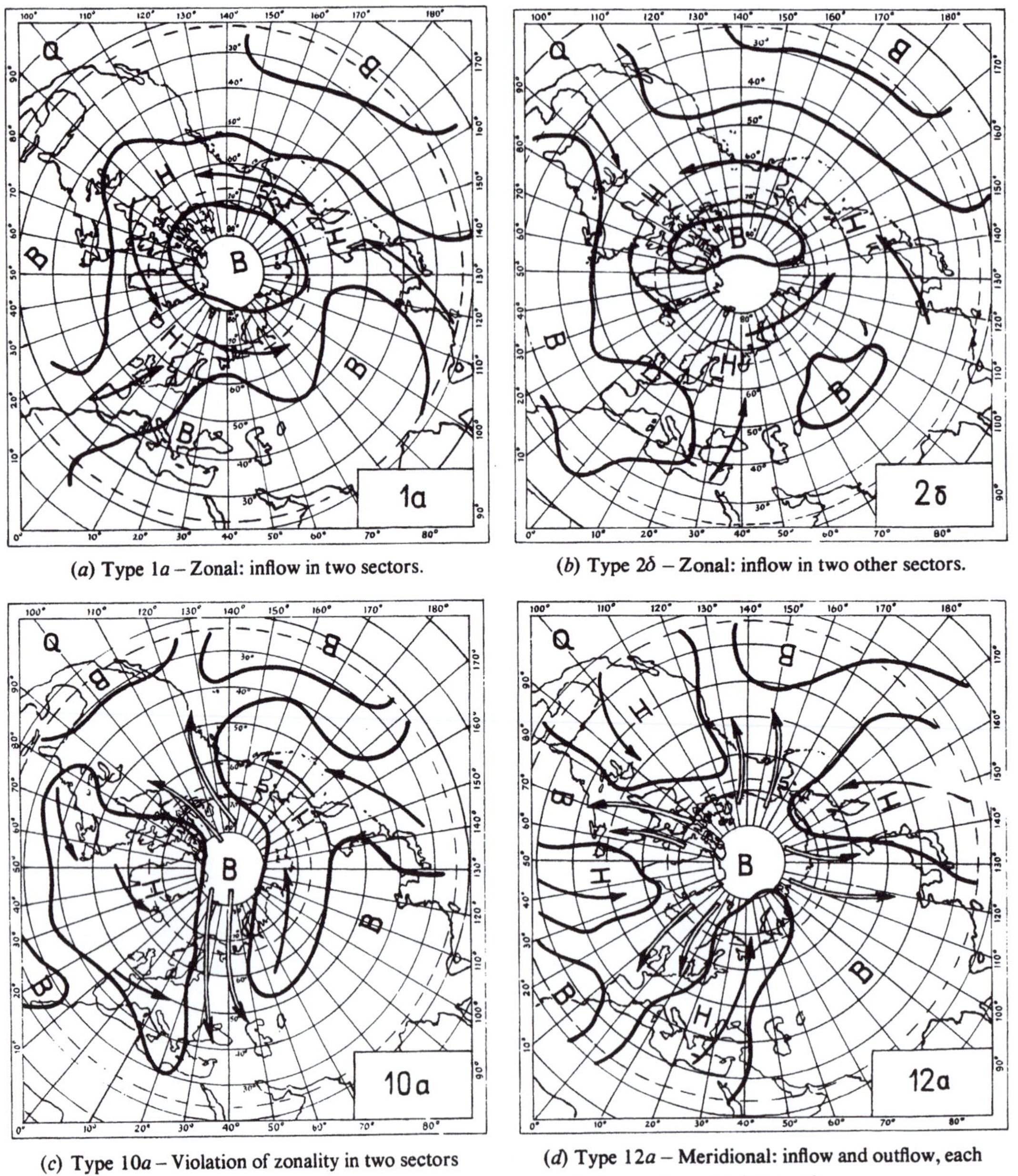

(*a*) Type 1*a* – Zonal: inflow in two sectors.

(*b*) Type 2*δ* – Zonal: inflow in two other sectors.

(*c*) Type 10*a* – Violation of zonality in two sectors simultaneously.

(*d*) Type 12*a* – Meridional: inflow and outflow, each in four sectors simultaneously.

Fig. 7.7 Schematic surface maps illustrating some DZERDZEEVSKI northern hemisphere circulation types:

B indicates High pressure/Anticyclonic region.

H indicates Low pressure/Cyclonic region.

Solid arrows = common tracks of travelling depressions.

Double arrows = common paths of high pressure centres and ridges advancing.

(*After* DZERDZEEVSKI *1968*.)

Fig. 7.8 Corresponding types of circulation over the northern and southern hemispheres, DZERDZEEVSKI and ASTAPENKO classifications respectively.
Schematic surface maps.
(*After* DZERDZEEVSKI *1966*.)

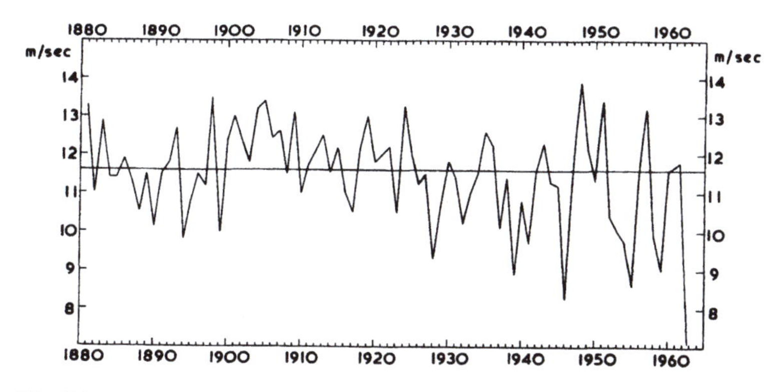

Fig. 7.9 Average zonal westerly component of the geostrophic wind at the 500 mb level 50–60°N over the sector 60°W to 60°E in the months December–February each winter 1881–1963.
(*Data kindly supplied by* H. TRENKLE.)

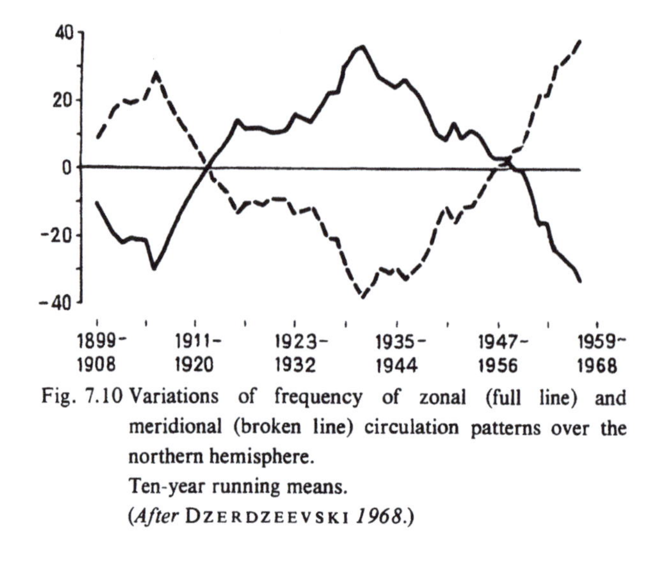

Fig. 7.10 Variations of frequency of zonal (full line) and meridional (broken line) circulation patterns over the northern hemisphere.
Ten-year running means.
(*After* DZERDZEEVSKI *1968*.)

(1) *Pressure at sea level*

The average changes of pressure from one day to the next (*interdiurnal variability*) in winter and summer over the northern hemisphere have been studied and mapped by BAHR (1911) for a 10-year period about the beginning of the century and by BERGER (1961) for the period 1951–7. The zone of maximum variability closely follows the main paths of travelling subpolar cyclones as mapped by KLEIN (1957) and indicated here in figs. 4.12(*a*) and (*c*) by the axis of lowest average surface pressure. The greatest values of all, over 11 mb, appear on the winter maps just off the Atlantic seabord of Canada (Nova Scotia–Newfoundland) and near east Greenland–Iceland and, about 9 mb, off the coast of northeast Siberia–Alaska: in these areas the particularly sharp temperature contrasts between airmasses from the warm ocean and from the snow and ice surfaces sharpen the pressure changes accompanying each passing cyclone with the addition of great changes in the density of the air in the bottom 1–2 km of the atmosphere. The zone of maximum variability in winter extends northeastwards with the depression tracks past Iceland to the Barents Sea and on to the northwest coast of Asia, but from there east and southeastwards the variability rapidly diminishes into regions where the cyclonic storms rarely penetrate and always weaken. Comparing the July map we find that the maximum day-to-day variability has shifted south over the eastern Atlantic, with the peculiar seasonal shift of the depression tracks and circumpolar vortex in that sector (see Chapter 4), so that a zone with average changes exceeding 5 mb/day passes from west to east across the Atlantic about 55–63°N: the greatest average value of all exceeds 6 mb over the southern part of Hudson's Bay. Separate zones of high variability, with averages over 4 mb/day, mark the frequent paths of travelling cyclones in summer along the Arctic fringes of Asia and Canada and in the North Pacific, roughly along the line of the Aleutian Islands.

The fact that maximum interdiurnal variability of pressure is linked with the jet stream through the sequences of surface frontal cyclones travelling principally along its cold flank, makes it possible to use the average day-by-day changes of surface pressure in individual months or years to reveal variations in the proximity of the jet stream to this or that observing station (BAYER 1965). This seems to offer a method of reconstructing the position and course of the depression tracks, and thereby of the mainstream of the upper westerlies, in past years long before the establishment of an upper air observation network and possibly for the European sector back to the eighteenth century.

The year-to-year variations of surface pressure (*interannual variability*), when the same months or seasons in different years are compared, follow a quite different geographical pattern from the day-to-day variations. The distribution of these longer-term variations is illustrated here by the maps of standard deviation of monthly mean pressure for May, July and December 1951–66 in figs. 7.11 (*a*)–(*c*). Most obvious in these (and in the other months of the year not illustrated) are the maxima in the regions where blocking anticyclones sometimes occupy the usual zone of subpolar cyclonic activity and the middle latitudes westerlies, particularly between Greenland and the British Isles, Scandinavia–northwest Siberia and over or near Alaska and the Aleutians. The large variability extending to the lower middle latitudes near the Azores and Hawaii may be attributed to the dislocation of the subtropical anticyclones during spells of blocking. In every month except July, however, there is some approach to a ring-shaped, eccentric zone or crescent of maximum values, reproducing the eccentric ring of high anticyclone frequency already noticed in figs. 3.16 (*c*), (*d*). This zone of greatest standard deviation (S.D.) of monthly mean pressure is seen as a wide area across, or near, northern Asia in the winter months, especially December (fig. 7.11 (*c*)), when its radius is greatest; in spring it contracts to a narrower ring, still centred about the Canadian Arctic, passing in ever higher latitudes across the Asian sector of the polar ice until June, and then expands again from August onwards. The maximum S.D.s are over 10 mb at some point in the high latitude part of this ring in each month from December to March, being near Iceland in January and over the pack-ice near 84°N 120°E in March. The lowest S.D.s are seen in July (fig. 7.11 (*b*)). At all times of the year S.D.s are low over Canada; and the ring of high values, as it dwindles and passes to its lowest latitudes, tends to merge with the subtropical high-pressure belt across the American sector.

Our knowledge of longer-term variations of pressure regime going back over as much as 200 years is confined to the regions about the North Atlantic ocean (LAMB and JOHNSON 1959, 1961). The incidence of large pressure anomalies (up to 6 mb in January), affecting periods from a decade to half a century long, during that time, indicates that these too have much to do with the variations of blocking and changes of preference as between Greenland and Scandinavia for the main region for the blocking anticyclones. There are indications of corresponding longitude shifts of the region of lowest pressure south of the anticyclone. Changes of prevailing wave length, north–south amplitude and latitude of the features most closely associated with the mainstream of the upper westerlies are also clearly involved.

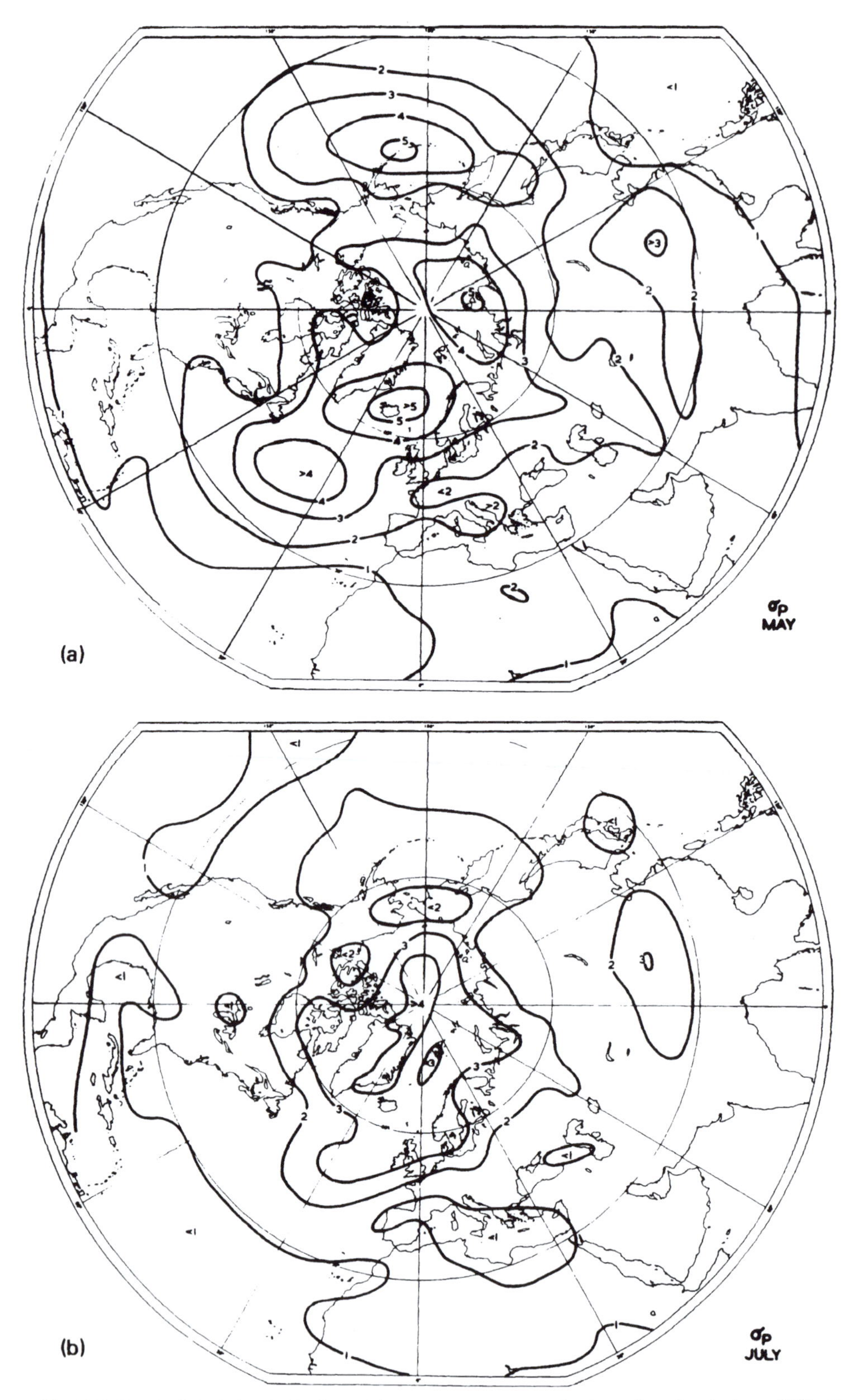

Fig. 7.11 Standard deviation (mb) of monthly mean pressure at sea level, northern hemisphere, 1951–66.
(*a*) *above* May (*b*) *below* July (*c*) *opposite, top* December

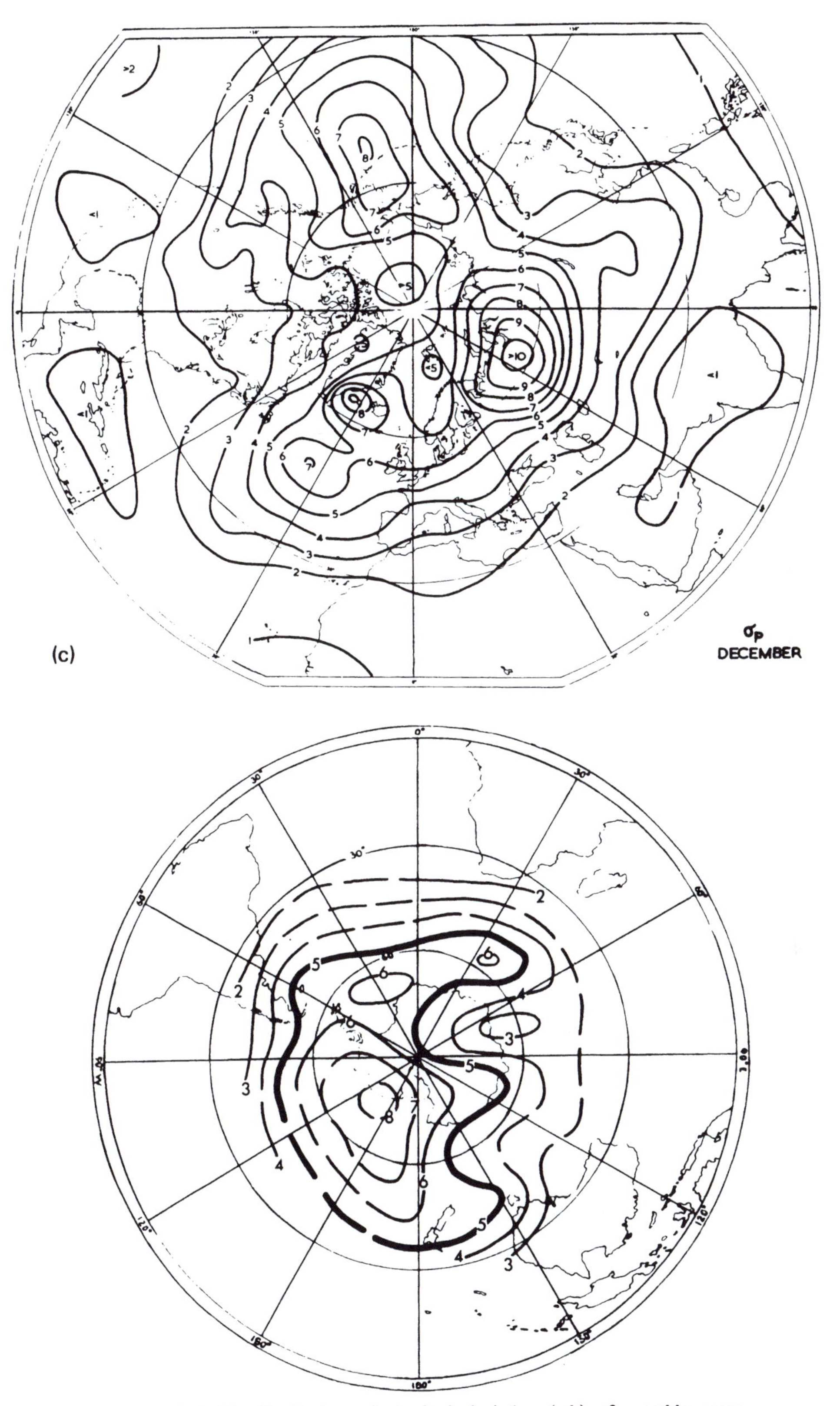

Fig. 7.12 Probable distribution of standard deviation (mb) of monthly mean pressure at sea level, southern hemisphere, July.

The frequency of blocking in the European sector, indicated by the HESS–BREZOWSKY daily *Grosswetterlage* classification from 1881, shows fluctuations which approximate to a 23-year oscillation with peaks around 1892, 1915 and 1937 (BREZOWSKY, FLOHN and HESS 1951). The frequency of Scandinavian anticyclones in each month of the year is presumably reflected in the average pressures for Trondheim available from about 1770 to date (BIRKELAND 1949): the curves for most months (see, for example, January and July in LAMB and JOHNSON *loc. cit.*) show an irregular series of peaks at about 18- to 25-year intervals. The amplitude varies widely from one swing to another, and the phase is not the same in all months of the year. A feature of a different nature is that the curve for May, which has latterly (especially since 1930) been the month of highest pressure, shows a fairly steady rise of about 5 mb in the 10-year means from low values around 1780. The May averages then were not as high as for several other months between April and October, though for 1768–77 (the first decade mean) pressures in May averaged as high as in recent decades. September shows a partly similar history, and it seems that these two months, which in the twentieth century have been associated with a tendency to fine anticyclonic weather in northwest Europe, including Britain, have been more given to Scandinavian blocking in or about the times of highest frequency of the zonal westerly circulation type over the year as a whole; May and September gave more westerly, disturbed weather in periods when meridionality was most prominent over the rest of the year. The relationship may have something to do with an extended ice surface and a more nearly east–west zone of strong thermal gradient near 70°N in the springs and autumns of the 'meridional' years – a situation which returned, and apparently produced the same effect, in the 1960s.

Very much less is known about variability of pressure, and hence of the large-scale wind and weather patterns, over the southern hemisphere. Average interdiurnal variability of surface pressure appears to reach its highest values, generally 9–10 mb, at the coast of Antarctica or between the coast and the axis of the main path of depression centres near the ice limit; slightly lower figures prevail over the open Southern Ocean within the same distance, but on the warmer side, of the depression centres (cf. figures given by HOFMEYR 1957, PEPPER 1954 and LAMB 1956). The variability of monthly mean pressure will be discussed in terms of standard deviation, as in the northern hemisphere paragraphs above, though few figures are available. Fig. 7.12 presents a tentative sketch of the distribution over the middle and higher latitudes for the mid-winter month of July, believed to be similar to June and August also, and which may be as representative as December in the northern hemisphere. The data were put together for this map by the author and were such as were available, i.e. for various years, mainly 1957–68, but in some cases for much longer periods. As in the northern hemisphere the largest variability is found in a roughly crescent-shaped area, or eccentric arc, which merges with the subtropical high-pressure belt in one sector: in this case it is in the Indian Ocean–Australian sector, where the most permanent upper cold trough lies and where the subtropical anticyclone belt is on average in its lowest latitude, somewhat as in the American sector of the northern hemisphere. It is more difficult, but as in the northern hemisphere not altogether lacking in plausibility, to suggest (see Chapter 3, p. 109, and Chapter 10) that the arc of high variability in some degree approximates to constant geomagnetic latitude, its centre related to the position of the geomagnetic pole. (This may be more nearly true of the upper tropospheric high-pressure ridges associated.)

As in the northern hemisphere, the belt of high variability of monthly mean pressure is sometimes occupied by the main depression tracks, even where it crosses near or over West Antarctica, and by slow-moving cyclonic centres where it enters the lower latitudes; at other times it is occupied by anticyclones. This map clearly marks out the area between the Andes and the Antarctic Peninsula, as well as the seas on either side, as one favoured region for blocking anticyclones. But it appears that the South Pacific Ocean from southeast of New Zealand to Antarctica, and the region between South Africa and Antarctica, may also experience more interruptions of the usual progression of cyclonic centres than other sectors of the Southern Ocean. These interruptions apparently come about mainly through ridges of high pressure protruding south or southeast and occasionally linking the subtropical anticyclone belt with Antarctica, where a separate centre of high pressure more readily forms than over the ocean. The tendency to blocking in the sector between New Zealand and West Antarctica (east of the Ross Sea) shows itself also in a skew distribution of the mean pressure values for the winter months at Chatham Island (44°S 177°W) and in South Island, N.Z. (e.g. Dunedin 46°S 171°E): mean pressures for July are normally in the range 1005–1020 mb and some influence of the fringe of the subtropical anticyclone is felt, but in a small minority of Julys the mean value is as low as 990–995 mb. This must represent a shortening of wave length and increase of amplitude in the upper westerlies just as in the more meridional and blocked circulation patterns in the northern hemisphere.

More space has been devoted to these pressure variations than will be given to the corresponding variations of the other elements, because the barometric pressure and wind pattern represents the mechanics of the distribution of heat, moisture and weather, and its variations explain their variations. A particularly gross example must be the differences of snowfall and windiness over West Antarctica, in the region of great variability of regime shown in fig. 7.12, between those months when it is under the influence of travelling depressions and other times when the Antarctic anticyclone extends all over that part of the continent.

(2) *Surface air temperature*

This element is only partly controlled by the wind circulation and weather processes; it responds also directly to the local balance of incoming and outgoing radiation and to local heat storage, especially in water bodies. The normal *diurnal range* of temperature between night and day increases with distance from the sea. Within the British Isles, its average value ranges from about 3°C in the most maritime districts (e.g. Hebrides, Scilly Isles) in the depth of winter, and 5°C at the height of summer, to 6°C in mid winter in the broadest inland area far from the western sea (e.g. Cambridge) and 10–11°C in the same area in high summer. The range is greatest when the wind is still, under cloudless skies and when the atmosphere is most transparent. For these reasons the diurnal range of temperature is especially big on high mountains in clear weather. Unpolluted air of great transparency commonly reaches Europe in Arctic N'ly and NW'ly windstreams in spring, the vigorous convection to great heights induced by heating over the ocean helping to keep the pollution and moisture content sparse: in such situations, when the wind falls light, extreme diurnal ranges of surface temperature may be observed, e.g. 24°C in southeast England one day in March 1929. In the Sahara, under favourable conditions of sunshine and with air of similarly cold northern origin in winter–early spring, diurnal ranges may be almost 40°C, from frost at dawn to afternoon heat. Under skies overcast with low cloud in high latitudes in winter, or in oceanic regions at any season, surface air temperature varies little from day to night (characteristically under 1°C); changes of airstream are then likely to make more difference than time of day.

The response to changes in the direct radiation budget was well shown by the sudden rise of 8·3°C within 3 hours, from −73·6° to −65·3°C, observed at the South Pole in fine weather with little air movement on 11 May 1957: there was no change of wind and the temperature change could only be attributed to an increase of cloud from nil to sky half covered. The effect of wind changes is often demonstrated over the Arctic and Antarctic ice when the turbulence of strong winds breaks down the surface inversion: thus, in late May 1957, also at the South pole, surface air temperature rose from −69° to −46°C in 24 hours as the wind increased from 5 to 20 m/sec. In still weather in the polar regions in winter the surface air is normally 20–30°C colder than the air 1–2 km overhead (extreme inversions amounting to over 40°C within 700 m have been observed over the Antarctic plateau); even partial breakdown of the inversion by strong winds, including katabatic gales in the peripheral fjords of Antarctica and Greenland, always brings strong rises of surface temperature.

Day-to-day changes of temperature are generally great in the middle latitudes zone affected by the passing warm and cold fronts of travelling cyclones and are greatest in the subpolar zone of lowest pressure, where the depression centres pass now on one side and now on the other. The variations are especially great near the limit of polar ice and open sea. At such places the temperature may change by as much as 30°C in less than an hour as the wind shifts, coming first from the open water and then off the ice or vice versa. Similarly great changes of surface temperature may accompany the passage of fronts in continental interior regions, especially where a warm, turbulent *Föhn* wind, or *Chinook*, descends from the mountains, breaks the inversion layer over the lowlands and displaces the cold air, as well as when the latter returns. Calgary (51°N 114°W), at the foot of the eastern slopes of the Rockies in Alberta, is particularly liable to these violent temperature changes, presumably because the winter snow cover is often thin in such a dry place and the warm Chinook wind may remove it completely. Thus, the surface air temperatures at Calgary in the morning hours of the last days of December 1962 were about −20°C; from 2 January onwards with westerly winds they were above the freezing point and on the 7th +11°C was observed, but on the 9th the reading was back to −21°C and on the 10th, −32°C. Similar sequences were noticed twice in the 1964/5 winter (−37°C on 9 January, +10°C on 15 and 16 January) and in 1966/7.

There is, in general, an increase in the range of day-to-day variations of temperature as one goes away from the thermostatic influence of the oceans towards a continental interior as well as on going towards high latitudes. Even so, *average* values of interdiurnal variability of temperature are nowhere very large and bear witness to the frequent persistence of weather type from one day to the next. Bigger changes, accompanying larger-scale changes of wind flow at intervals of some days, or when a much longer spell breaks, contribute a good deal to the average variability. HANN (1932) quotes the *average interdiurnal variations of temperature* (between one day's mean temperature and that of the next) in England as about 2·1°C in winter, 1·5° in summer, and averaging 1·7° over the year; the averages for the year are given as 1·5°C in western Ireland, 2·3° in central Europe, 3·0–3·5° in western Siberia and central North America.

Year-to-year variations of temperature as, for instance, registered by the standard deviation of a monthly mean (figs. 7.13(*a*), (*b*)), also increase towards continental interiors but are amplified most of all in high latitudes where shifts of the boundary of persistent snow cover and sea ice from one year to another go with the variations. Within the region where the ice regularly persists throughout the summer on the polar sea, apart from open leads and melt-water pools on the ice surface, however, the temperature in the warmest month is constantly between about −1° and +3°C and can hardly vary from year to year. What the maps imply in terms of the differences between the warmest and coldest Julys and Decembers in the regions of greatest variability may be seen from the following examples in Table 7.1:

TABLE 7.1

Average temperatures of the warmest and coldest Julys and Decembers (1900–50) at Archangel and Verkhoyansk

	Average temperature (°C) of the	
	Warmest month	*Coldest month*
Archangel (64½°N 40½°E)		
Julys	21·3° (1938)	11·8° (1926)
Verkhoyansk (67½°N 133½°E)		
Decembers	−35·9° (1901)	−51·9° (1904)

A survey covering the northern hemisphere from 1881 to 1960 of great anomalies of monthly mean temperature exceeding 10°C (GEDEONOV 1967), which all occurred between November and April and were most frequent in January, indicated that great positive

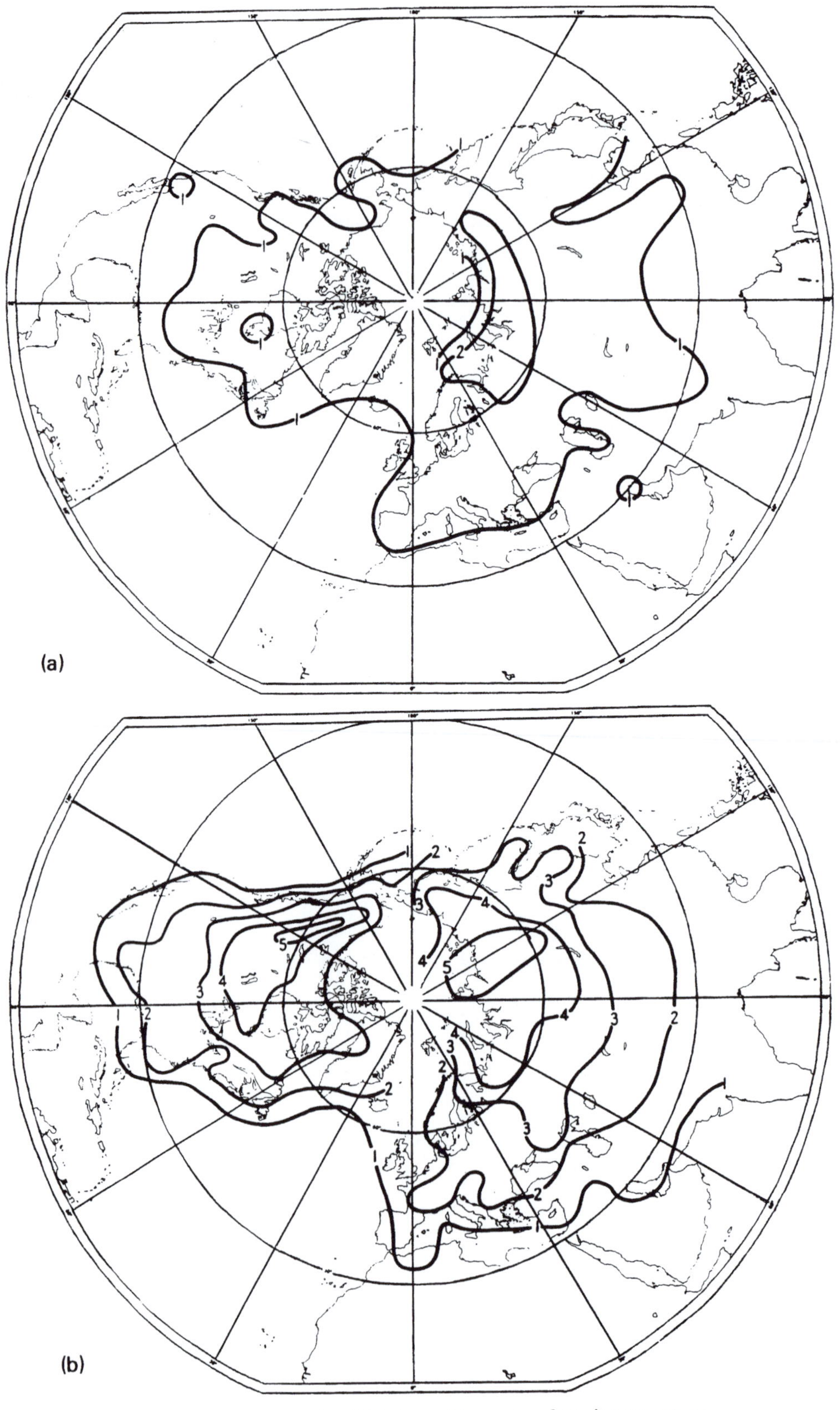

Fig. 7.13 Standard deviation (°C) of monthly mean surface air temperature:
(*a*) July (*b*) December

anomalies occurred oftenest[1] in northern regions near 70°N in central northern Siberia and west Greenland; great negative anomalies occurred oftenest[1] farther south, near 50°N in the Rocky Mountains, 45°N in Kazakhstan and 55°N between the Yenisei and Lake Baikal. However, a notable feature of the climatic cooling associated with weakened and southward-displaced wind circulation since 1960 has been the frequency of very large negative anomalies of surface temperature over the sector of the polar ice about 80°N near Franz Josef Land and Novaya Zemlya (covering Franz Josef Land eight times in the 9 years 1960–8, the extreme anomaly there being for January 1964, −18°C). During the earlier part of this century this region was in the path of frequent warm air advection and frontal cyclones, and presumably an intensified warm water current, advancing from the Atlantic: this was the most notable period of warming of the Arctic.

When the limit of 'permanent' pack-ice, or of lasting snow and ice on land, shifts in the course of a longer-term climatic change, a zone near this limit is where much the greatest changes of prevailing surface temperature are found.

Shifts of the boundaries of warm and cold ocean currents, or changes in their vigour attributable to changes of the winds that drive them, may also have big effects on prevailing temperatures. There seem to have been changes of 50- to 100-year mean temperature by as much as 2–3°C in some parts of the North Atlantic at times within the last few centuries.

(3) *Rainfall*

Rainfall distribution is characteristically intricate, with strong local differences, due to the windward and leeward effects of even minor hills and mountains. The same is true of rainfall anomalies. The variations are nevertheless responses to changes of prevailing wind, thermal stability or instability (temperature lapse with height), and moisture content of the air, over wide regions.

Day-to-day variations of rainfall are greatest in the zone of travelling depressions, owing to the changing wind directions and passing fronts (except occasionally in spells of a few days duration when the successive disturbances happen to arrive at nearly 24-hour intervals instead of the more usual 2- to 3-day spacing).

Rainfall anomaly patterns are greatly simplified if we deal in percentages of the normal fall at each place. This eliminates many of the big local differences. The wider regions of excess rainfall (or snow) thus revealed are in the path of (usually repeated) anomalous cyclonic activity, and may register a displacement of the depression tracks from their normal areas; the wider regions of deficiency mark the experience of an unusual amount of anticyclonic influence. Smaller areas of anomaly may mark a shift of the windward and leeward ('rain-shadow') effects of a mountain range owing to prevalence of an unusual wind direction. Thus, monthly and longer-term rainfall anomalies are closely linked with shifts and anomalies of strength in the general wind circulation.

1. Over ten times in the 80 years in the regions here named.

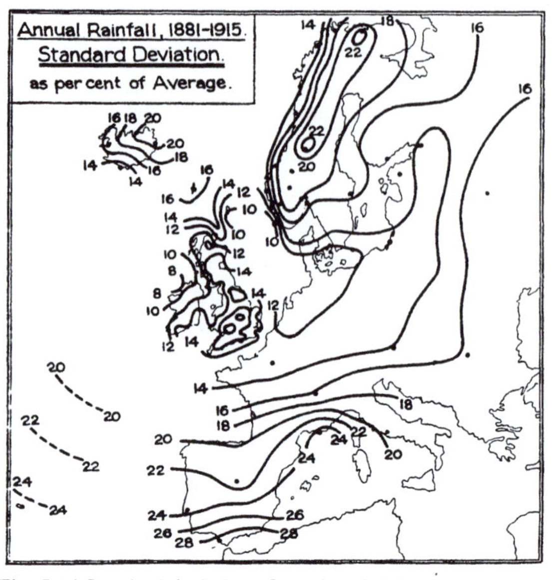

Fig. 7.14 Standard deviation of yearly rainfall, as a percentage of the overall average, 1881–1915.
(*After* GLASSPOOLE *1925.*)
Note that small areas enclosed by unnumbered lines over England approach 20%.

Fig. 7.14 illustrates the distribution of standard deviation of yearly rainfall over the East Atlantic–European sector of the northern hemisphere. The greatest year-to-year percentage variability is characteristically in two parts of this map: (*a*) at the fringe of the subtropical desert zone,[1] and (*b*) near mountain ranges athwart the prevailing winds in latitudes 50–70°N, the zone of blocking where the prevailing wind direction varies greatly in frequency and may even be reversed in some years. There is a third zone of very great – by some measures the greatest of all – year-to-year variability of rainfall, (*c*) along the equator in the central-eastern Pacific with extensions to the coasts of Peru and California (see references elsewhere to *El Niño*).

These zones of high variability, (*b*) and perhaps also (*a*) applying equally generally in southern and northern hemispheres, dominate the world maps (e.g. BIEL 1929, BERRY *et al.* 1945) of rainfall variability (despite disagreement of the figures, doubtless due to use of different years' data, among the few such maps that have been published).

1. In the desert zone itself average rainfalls are almost meaningless. Whatever figure is quoted results from just a few falls of rain, sometimes separated by many completely dry years, though the individual falls may be heavy and produce temporary rivers and even flooding.

(4) *Windiness*

Surface wind strength generally undergoes a diurnal variation, wind speeds being greatest in the afternoon when surface heating makes convection most active and so puts the surface air in communication with the faster moving air above the surface friction layer; at night, when the vertical lapse rate of temperature is less, and particularly when an inversion is formed, there is little convection and the surface air is liable to be brought to rest, particularly in sheltered places. The same principle operates in the long-term changes of average surface wind strength (and of gale frequencies) accompanying climatic variations. This is clearly seen in high latitudes and continental interiors, when the general circulation and radiation conditions cause the frequency of formation of temperature inversions to change: in such regions, especially in winter-time, a weakening of the upper wind flow may, by allowing more development of surface inversions, be accompanied by a much greater weakening of the surface winds.

Only katabatic winds are an obvious exception. The cold air drainage off ice caps, glaciers and mountain heights, is most active at night and in winter, when radiation cooling is most rapid. And on slopes and in valleys liable to katabatic gales these may be accelerated by a wind flow above the friction layer moving in about the same direction, especially the vigorous circulation of the winds around a deep frontal cyclone passing near. These winds are therefore strongest and most frequent at night, and in winter, and when the main depression track is near.

In general, windiness and gale frequencies are greatest in and near the zone of travelling depressions and their maximum is displaced with any shifts of this zone. They are also, of course, increased by intensification of the general circulation or in any circumstances (thermal instability of cold air over warm sea) giving rise to specially deep cyclone centres.

Local maxima, depending on the contribution of local topography, occur over mountain ridges and tops, in valleys and fjords, and along coasts that approximate to a mountain wall (whether of rock or ice) in high latitudes; in the latter position, especially when the strong wind flow on the cold (forward) side of a warm front or warm occlusion advancing from the ocean is constricted between the mountain or ice face and the frontal surface (the latter overhead as well as out at sea), the cold winds blowing along the coast and additionally fed by katabatic drainage from the interior are accelerated in a sort of funnel ('frontal funnel effect'). In these circumstances, mountainous sections of the open coasts of Greenland and Antarctica are exposed to more frequent and fiercer gales than out at sea (especially open waters beyond the front of the depression).

Displacement of the zone of most frequent gales when the cyclonic belt shifts in the course of a climatic change is illustrated by comparing figs. 7.15(*a*) and (*b*), which show the distribution of gale frequencies about Greenland, Iceland and the northernmost Atlantic in 1930–5 and 1945–9 respectively. The earlier period was distinguished by an abnormal frequency of depressions from the Atlantic passing north of Iceland and penetrating far into

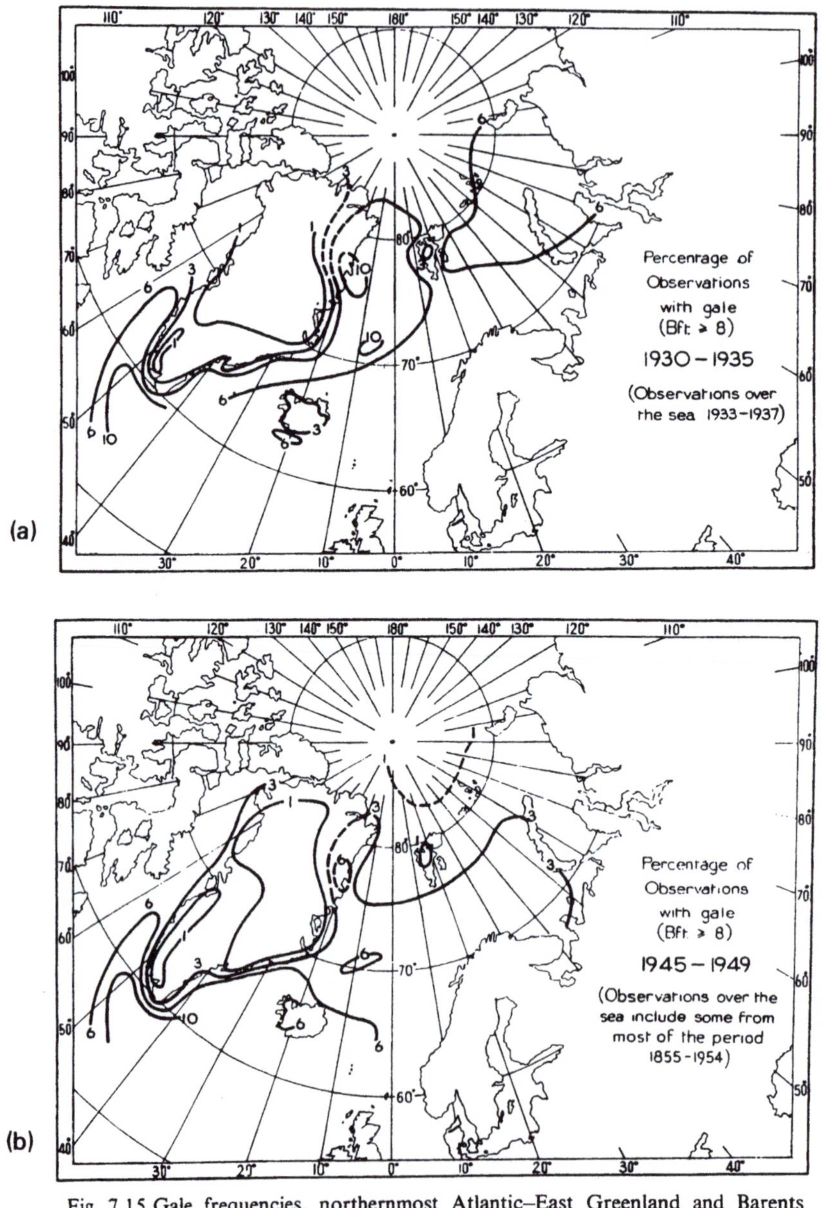

Fig. 7.15 Gale frequencies, northernmost Atlantic–East Greenland and Barents Seas: percentage of observations with Beaufort force ≥8.
(*a*) 1930–5 (*b*) 1945–9

the Arctic (PETTERSSEN 1949) at the height of the Arctic warming. In the later period the shift of the main Atlantic cyclonic activity farther south, to latitudes between 45° and 65°N, which has characterized the 1950s and 1960s, was already beginning; gale frequencies increased over Iceland, and between there and the British Isles, but were greatly reduced north of Iceland and over the polar ice. The total area of high gale frequencies may also have contracted even though they became more frequent in about 50–63°N.

Weather pattern shifts: some examples

The hang-together between anomalies of the hemispheric thermal pattern[1] and the distribution of sea-level pressure and surface winds, as well as the consequential weather and surface temperature regimes, can be seen in the charts of sample Julys and Januarys in figs. 7.16–7.20. Each of the months chosen was marked by a long spell of some kind in middle latitudes. The darkened areas on the composite maps were where abnormally high surface temperatures prevailed (over 2·5°C above normal in July, over 5°C above normal in January); point stipple is used to mark areas of temperatures correspondingly far below normal. On the rainfall charts the darkened areas had more than twice the normal total for the month.

July 1955 (fig. 7.16)

With an upper cold trough over Greenland and the western Atlantic which was unusually sharp for summer-time, associated with some increase of ice and cold water in that sector, the North Atlantic cyclonic activity was steered far to the northeast and the Azores anticyclone spread a strong ridge along the right-hand side of the jet stream. This surface anticyclonic regime covered the British Isles and northern Europe, where it was a very warm, sunny month and most of the area had well under half the normal rain. It was also unusually anticyclonic over much of the U.S.A. and southeastern Canada in the rear of the upper cold trough, and the month was very warm over a wide area of the Middle West where the surface winds were southerly.

July 1960 (fig. 7.17)

This July had a strong zonal westerly upper wind flow over the Atlantic, with the mainstream unusually far south; there were cold troughs over eastern Canada and Britain, but a large-amplitude warm ridge over European Russia and northern Scandinavia. The Atlantic depressions followed tracks mostly south of 60°N, and there were concentrations of cyclonic activity over eastern Canada and Britain. An anticyclone dominated the Barents Sea just east of the crest of the upper warm ridge. The month was unusually warm over Lapland and northern Russia as well as in the sharp upper warm ridge region over the U.S. Rockies; it was rather cool over Britain and eastern North America. Twice the normal rain

1. The thermal pattern is commonly, and conveniently, represented by the isopleths of 1000–500 mb thickness, a concept explained in Chapter 3 (p. 74). The run of the thickness isopleths resembles the flow of the winds in the upper troposphere.

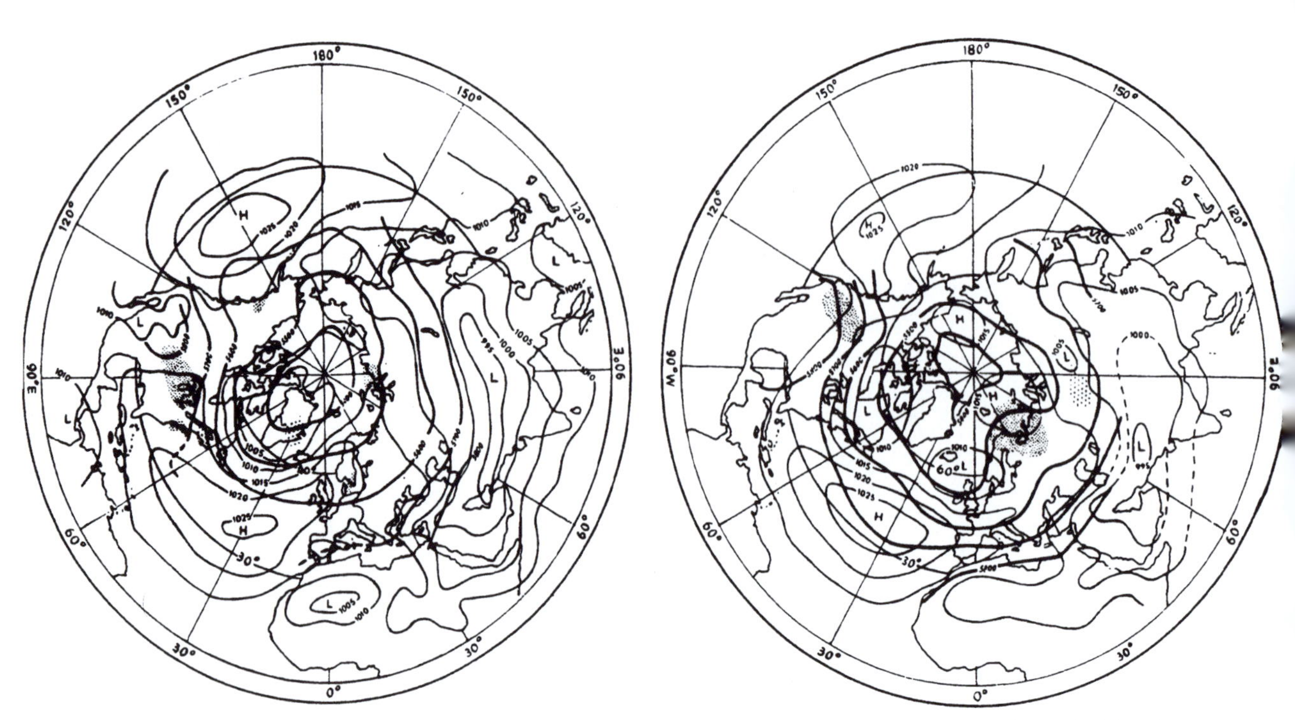

Fig. 7.16 July 1955:

Fig. 7.17 July 1960: Key as for Fig. 7.16.

Average m.s.l. pressure (thin lines numbered in mb). Average 1000–500 mb thickness (bold lines numbered in m). Regions of anomalous warmth (shaded where mean temperature over $2\frac{1}{2}$°C above normal) and anomalous cold (point stipple where over $2\frac{1}{2}$°C below normal).

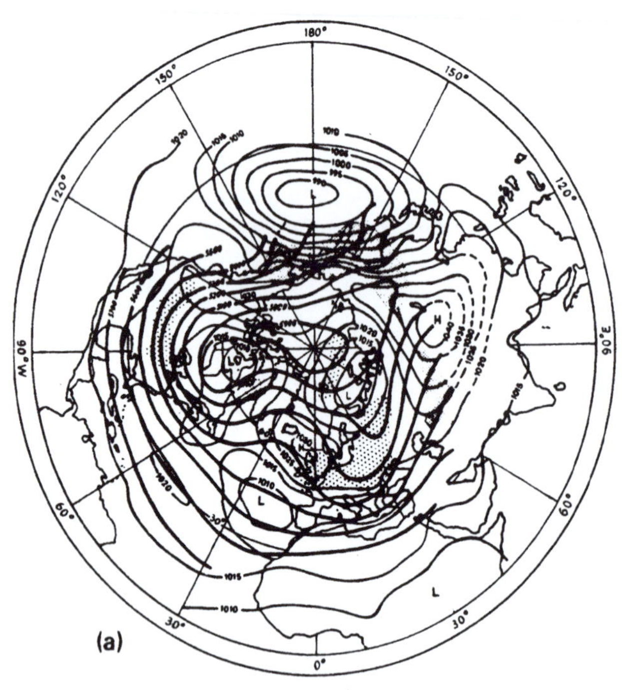

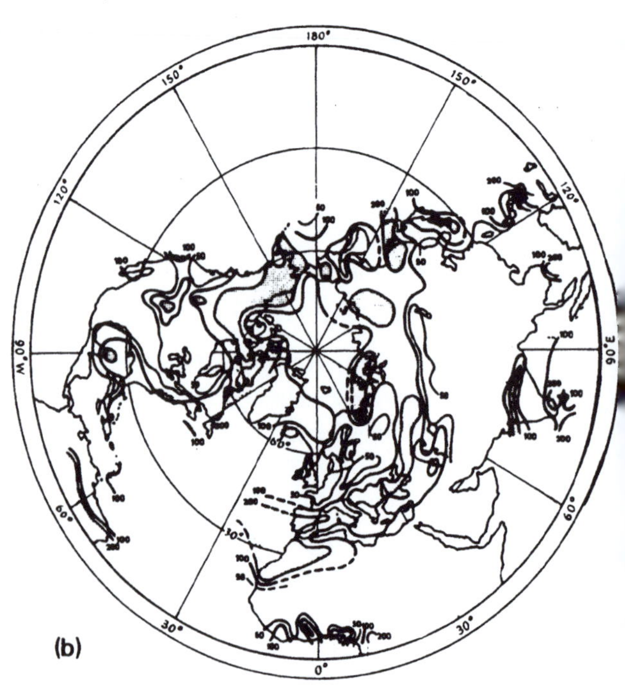

Fig. 7.18 January 1963:

(*a*) Average m.s.l. pressure (thin lines numbered in mb). Average 1000–500 mb thickness (bold lines numbered in m). Regions of anomalous warmth (shaded where mean temperature over 5°C above normal) and anomalous cold (point stipple where over 5°C below normal).

(*b*) Downput of rain and snow (converted to equivalent rainfall) as percentage of normal. (Shaded areas over 200%. 100% and 50% isopleths also shown.)

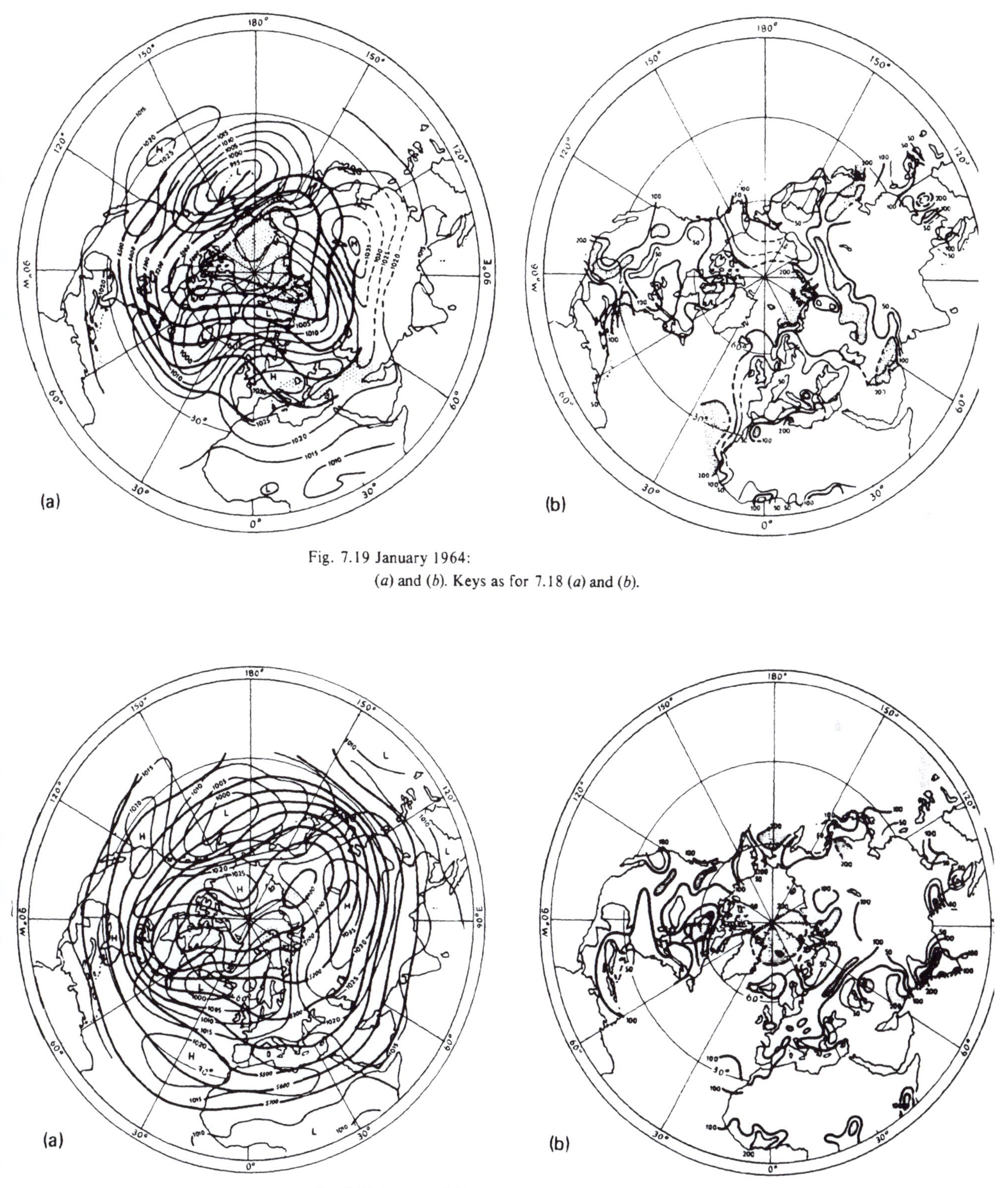

Fig. 7.19 January 1964:
(*a*) and (*b*). Keys as for 7.18 (*a*) and (*b*).

Fig. 7.20 January 1965:
(*a*) and (*b*). Keys as for 7.18 (*a*) and (*b*).

fell in the eastern United States and in south Norway, Denmark and parts of central Europe; above average rainfall also covered the British Isles, France and central Europe.

January 1963 (fig. 7.18)

With large-amplitude upper cold troughs over North America and Europe and an upper ridge in between, which divided the upper westerlies in the Atlantic sector into two streams, one near 80°N and the other in the latitude of the Azores, the mobile cyclonic activity was partly in the inner Arctic and otherwise crossed the Atlantic towards the Mediterranean. A blocking anticyclone was centred between Scotland and Iceland, and NE'ly surface winds prevailed over Britain and central Europe. In the latter regions the month was exceptionally cold, in England the coldest since 1740 and in part of Germany the anomaly reaching −10°C. Rainfall was over twice the normal from the Azores to Spain and North Africa; less than half the normal downput was recorded in England and neighbouring lands across the North Sea, but this was sufficient to give exceptional depths of snow. There was ice on parts of the North Sea, and many ports were frozen.

January 1964 (fig. 7.19)

The amplitude of the upper cold troughs was again abnormally great, also that of the Atlantic warm ridge, but the positions were somewhat east of those in the previous year. The month was dominated by a stationary anticyclone covering most of Europe east of the upper warm ridge, and England had its driest winter for over 200 years (since 1743). The cold N'ly and NW'ly winds over Russia, east of this anticyclone, brought frosts as far south as the Persian Gulf. There was exceptional snowfall in northwest Siberia with the mobile depressions plunging southeastwards from the Barents Sea. S'ly winds over the Atlantic between the European anticyclone and the stationary cyclonic region over the western Atlantic brought a very mild January in Ireland and Scotland, but the month was cold farther east and the Zuyder Zee was frozen.

January 1965 (fig. 7.20)

This was the most normal of the three Januarys depicted in the Atlantic sector. The upper westerlies were again somewhat south of normal around much of the hemisphere, over Japan, much of the Pacific and in the American sector, with a certain southern displacement of the cyclonic activity. This allowed the Arctic to become colder than normal, but the amplitude of the waves in the upper westerlies was not great, and there were few noteworthy weather anomalies in middle latitudes.

Other examples of the very unlike cyclone tracks and distributions of temperature and precipitation accompanying zonal and blocking patterns of the upper westerlies in the Atlantic and European sectors may be found in the literature (notably REX 1950). WAHL (1968) has demonstrated how the temperature and rainfall anomalies of the 1830s in the

United States corresponded to a more southern position of the commonest cyclone tracks and penetration farther south than in recent decades (1931–60) of the cold air outbreaks over the centre of the continent; symptoms of reversion have been noticed with the increased blocking and meridionality since about 1957, and 1966 reproduced the conditions of the 1830s for the first time for long since.

Observed successions and interrelationships within the atmospheric circulation

Forecasting rules applying to periods which range, according to circumstances, from 3 or 4 days to a few months ahead can be formulated on the basis of recognition of specified early symptoms of development of (*a*) large amplitude waves in the upper westerlies, (*b*) blocking, in the full sense of stationary cut-off large-scale eddies or (*c*) the mainstream of the upper westerlies and associated cyclone and anticyclone paths being directed in such a way as to have characteristic effects in the region of interest. The usual technique is to list many like cases, apply statistical criteria of significance to the observed frequencies of various sequels, and accept as 'rules' only those circumstances which are followed sufficiently often (of the order of 90% or more) by one type of outcome. Any such rules which are found to be valid for more than about 5 days ahead presumably involve regeneration of the circulation pattern concerned through the agency of a suitable layout of thermal anomalies in the undersurface, since the atmosphere's store of energy is used up in about this length of time. The surface thermal pattern concerned may, however, be itself brought about, or encouraged, by the given winds and weather.

A forecasting 'rule' of this type was found by GALLÉ (1916), using ideas developed much earlier still by SABINE (1846) on the probable effect of anomalous strength of the Gulf Stream upon the temperatures prevailing in the following winter in Europe. GALLÉ found that anomalies in the strength of the Trade Winds in the North Atlantic 15–25°N 25–45°W, taking data for the years 1899–1914, were correlated with air temperature anomalies over most of Europe some time later. The strongest association was between the Trade Wind departures from normal averaged over the months June–November and the temperatures prevailing in the winter months December–February immediately following, such that the correlation coefficients (which were positive all over Europe except in Iceland, northern Scandinavia and northern Russia, Ireland, northwest Scotland and Spain) exceeded +0·7 from Holland to the Black Sea and were about +0·8 in central Europe (Berlin–Prague). Forecasts that the winter would be warmer, or colder, than the overall mean would be right in about seven cases out of eight. Reasoning about the supposed chain of causation, somewhat elaborated since Gallé's time but in its essentials already grasped by him, goes as follows. Strengthened Trade Winds go with an intensified subtropical anticyclone and impel more tropical Atlantic water towards the Gulf of Mexico and the coasts of the southern United States; the Gulf Stream is intensified in consequence; the intensified anticyclone also tends to drive the warm water faster (i.e. with less modification of its temperature) towards the south of the Newfoundland Banks and on into the central Atlantic. This not only maintains higher temperatures in the SW'ly and WSW'ly winds reaching Europe but is liable to strengthen the thermal gradient between mid Atlantic and Iceland, intensifying the upper westerlies and supplying more energy to the travelling cyclonic storms.

We shall in this chapter consider what interrelationships, in time and space, are observed within the atmospheric circulation itself. Relationships clearly involving the undersurface will be dealt with in Chapter 10.

ELLIOTT and SMITH (1949) noted that blocking in the northern hemisphere, lateral mixing and northward heat transport were all greater in the winters of those years in which south–north thermal gradients were stronger than in other years. Accumulation of heat in low latitudes, i.e. failure of the circulation for a time to transport enough heat polewards, implying increasing south–north thermal gradient accompanied by increasing upper westerlies, apparently necessitates large-scale readjustment of the circulation to redistribute the heat. ESSENWANGER (1953) demonstrated from a statistical study the quasi-regular build-up of strength of the upper westerlies in latitudes 50–60°N to a sharp maximum about two days before blocking development. The blocking anticyclone later tends to collapse when another wave pattern is set up in the atmosphere upstream from it which is out of phase with the wave positions associated with the block: statistical examination by RÖDER (1966) indicated that this oftenest occurred because of new blocking development farther west, otherwise mostly through renewed jet stream propagation from the west towards the block or through a cold pool approaching it from the west.

The blocking phenomenon here described clearly differs from the case of an expanded polar anticyclone covering an extended region of fairly uniform cold surface. In that situation the hemispheric circulation and its active heat transfer processes may be largely confined to a lower range of latitudes. The size of the polar cap, as delimited either by the position of the zone of strongest thermal gradient associated with the circumpolar vortex or by some selected 1000–500 mb thickness isopleth, varies from year to year and between times of warmer and colder climate, as well as seasonally; it is habitually greatest in late winter.

Poleward heat transport in the southern hemisphere plainly proceeds with much less blocking than in the northern, though average temperatures over Antarctica are lower than in the corresponding seasons over the Arctic and a stronger thermal gradient and stronger upper westerlies are maintained over middle latitudes in consequence. Large-amplitude ridges do commonly develop over the Southern Ocean, usually with their axes aligned somewhat from northwest to southeast, and link the subtropical high-pressure zone temporarily with Antarctica. Over the Antarctic continent an (initially warm) anticyclone cell is liable to split off from the tip of the ridge and become stationary there, either renewing the polar anticyclone after a cyclonic interval or joining up with a pre-existing high-pressure regime over other parts of the continental ice sheet. But over the surrounding ocean anticyclones and ridges soon drift away east, as the westerlies are continually renewed in the zone of ever strong thermal gradient between 40° and 60°S. Evidently the geography of land, sea and mountain ranges that straddle most latitudes in the northern hemisphere plays a vital role in allowing stationary blocking patterns to develop and enabling the observed poleward heat flux to be achieved (LAMB 1959).

Forecast 'rules' connected with wave amplitude, blocking, zonality, etc

Increasing amplitude of a ridge or trough in the northern hemisphere upper westerlies is an early symptom of sufficient blocking, or poleward shift of a subtropical anticyclone cell, to give a dry spell of at least several days duration in the sector affected in middle latitudes, given certain further conditions about the situation in neighbouring regions. Thus, MILES (1961) found that nearly all winter anticyclones over Scandinavia, i.e. in one of the regions most frequently occupied by blocking highs, formed east of a large amplitude ridge in the 1000–500 mb thickness distribution over the Atlantic; in half the cases the surface anticyclone was just 30° ± 5° of longitude east of the axis of the thermal ridge. Further characteristic aspects of the situation were that a deep upper cold trough, the deepest in the American–Atlantic sector, should lie east of 70°W, there was usually a deep surface low about 60°N between 30° and 45°W, a well-marked upper southerly steering current at 500 mb north of 50°N between 20° and 35°W and a corridor of surface S'ly winds centred a little farther east. The upper winds over the developing anticyclone were usually from between 320° and 360°, i.e. about 40° veered from the normal winter WNW'ly flow over the region. The anticyclone was likely to remain stationary for some days, given the required meridional cold trough over the western-central Atlantic east of 70°W. On the other hand mobility, with the Scandinavian anticyclone moving away southeast, was likely when there was continued strong upper flow around the northern side of a warm anticyclone or a warm ridge of rather less meridional character about the longitude of the British Isles. The seasonal distribution of blocking anticyclones in this sector (see also EVJEN 1954, JOHANSEN 1958, SUMNER 1959) suggests that the thermal contrast between neighbouring ocean and land areas in the latitudes 55–70°N concerned plays a part in achieving the stationary upper ridge and trough pattern but that the winter snow cover over Scandinavia itself is of less importance. At some times of the year (especially March–May and sometimes in October–November) surface anticyclones are commoner over Greenland, Iceland and the Norwegian Sea than over Scandinavia; this also appears to be true of some years and groups of years as a whole and must depend on prevailing wave lengths and wave positions (as well as amplitude) in the upper westerlies.

Spells of at least 3 days without rain in southeast England, which also require some amplification of an upper warm ridge but not necessarily a blocking anticyclone, can be forecast with a high percentage of success under conditions specified by LOWNDES (1964, 1965), given (i) a fairly large amplitude upper trough (500-mb level used) over the western Atlantic between 50° and 60°W, (ii) that the flow round its base is centred south of 52°N in the summer half-year or south of 47°N in the winter half-year (49° in April), and (iii) provided that (*a*) the zonal index is low, i.e. not too much flow around the south of the trough, and (*b*) there is neither any easterly component in the southerly upper flow, nor much northward elongation of the surface anticyclone, nor (iv) must there be so much meridionality and so short wave lengths that the next downstream trough could develop over the British Isles. RATCLIFFE (1965) further finds that southwesterly upper flow over the area 55–65°N 20–40°W, provided that the flow is the strongest in the Atlantic area and also that there is no equally strong NW'ly flow immediately farther west, is sufficient for 3-day dry spells in southeast England and accounts for about half the cases. Fronts cross the area of interest but are normally weak and give no rain.

Rainy spells of at least 4 days duration in southeast England ($\not<$ 15 mm total rain and $\not<$ 1 mm on most days of the period) can be predicted (LOWNDES 1962) when there is a rather deep trough over the eastern Atlantic between 10° and 30°W and surface pressure is already low ($\leqslant$ 1009 mb at Valentia for the more western trough positions, $\leqslant$ 1007 mb at London for the more eastern trough positions); a common preliminary is a N'ly outbreak responsible for forming the trough. RATCLIFFE and PARKER (1968) have, however, noted that more than half the wet spells in southeast England at all times of the year are accounted for by zonal situations in which the area lies near to the core or slightly to the cold side of a straight or cyclonically curved Atlantic jet stream, which should be the strongest flow over the Atlantic area between 40° and 70°N and 60°W to 10°E at the time. There are two further conditions, namely that there should not be any ridging of the flow over the British Isles and that there should be no cut-off low just south of it over Spain.

The various wet spell criteria mentioned are clearly in line with the distribution of cyclogenetic tendency, accompanied by prevalence of some upward component of air motion and hence cloudiness and rainfall, in relation to the jet stream and the forward side of upper cold troughs; though they imply that the jet stream or cold trough concerned undergoes no important shift other than slight forward propagation during the 5 days.

It has been found possible to formulate a few somewhat similar rules, including both persistence relationships and sequences, that cover considerably longer periods. The meteorology of these is not yet completely understood, but probably involves persistence of some other factor such as ocean temperature anomalies, solar behaviour or a volcanic dust veil, or a combination of these. Nevertheless, some can apparently be stated in terms of purely meteorological variables. The following are examples.

For forecasting the next month's rainfall over England and Wales in broad categories 'wet', 'average' or 'dry', defined by terciles of the distribution of monthly rainfall totals observed in past years, RATCLIFFE (1968), using 22 years' data (1946–68), has formulated rules based solely on statistically significant relationships with the average 500-mb contour pattern of the preceding month: the forecasts indicated were right for over half the months and were completely wrong (2 categories out) for under 10% of the months. One set of rules depends on the previous month's positions and spacing of the 'Canadian' and 'European' 500-mb troughs (defined by the longitude of lowest 500-mb height at 50°N), when both were present. The overall average positions of the 'Canadian' trough were about

65°W from May to August, 70°W from September to November and 70–75°W from December to March; the 'European' trough was generally much less well defined except in winter and spring and was more irregular in its positions, but in more than half the months could be clearly distinguished somewhere between 25°W and 25°E. In general, a western position of the Canadian trough and a trough west of Britain indicated a wet month to follow, eastern positions indicated a dry month to be expected. For cases when no trough near the British Isles could be distinguished the position and orientation of the strongest flow over the Atlantic, as seen on the mean 500-mb map of the previous month, provided rules. The normal direction of this 'monthly mean jet stream' was found to be from 260–270° March–October and 250–260° November–February; over the zone 40–50°N it was from between 230° and 290° in nearly every month of the period. When the strongest flow lay over the area 40–50°N 40–70°W in the preceding month and was veered from the limit of the normal range specified above (i.e. >270° or >260° according to the month), a wet month generally followed in England and Wales; when the direction was backed from the limit of the normal range (i.e. <260° or <250° according to month), the next month was usually dry. More generally, if the strongest 500-mb level flow were southwesterly, and especially if this were over the area 40–60°N 20–40°W, a dry month usually followed if the core of the flow passed north of 55°N 20°W and a month with about average rainfall if the core were nearer Britain than this. Other rules were derived to cover more complex and rarer cases. All these forecasts were least successful for April and May, the time when average positions change most rapidly, and did best for August and September, when the seasonal changes are much slower.

Forecast 'rules' connected with stages of the seasonal development

At various times around the year forecasting 'rules' based on the observed behaviour of surface wind and weather systems can be similarly established. These are related to the incidence of long spells and natural seasons referred to in Chapter 4 (e.g. fig. 4.26), and through them to the upper westerlies, and some are related to the tendency for characteristic shorter seasonal episodes ('singularities') around particular dates. Thus, BAUR (1956, 1958) has noted that when the mild westerly weather commonly experienced in the first 10–15 days of December in western and central Europe is sufficiently marked to put the average temperature in Berlin over those days more than 2·5°C above the mean of the period surveyed (1775–1950), the winter (December–February) can be expected to be mild (twenty-five cases out of twenty-eight gave temperatures above the overall mean) and characteristically owing to more than usual W'ly winds. The contrary situation, the years in which this early December singularity failed, showed less regular sequels. However, in those winters in which the next normal bout of W'ly winds and cyclonic situations, frequent in western Europe from 26 December to 12 January, fails or is cut short so that there are no W'ly days between 4 and 12 January in Britain, the temperature of January and of January and February taken together is usually low. The temperature was below the average of the period (1873–1965) surveyed in twenty-eight cases out of twenty-nine; either January or February, or both, was severe (anomaly exceeding −3°C) in eleven of these years.

The blocking tendency in the European sector increases in late winter–spring, from January onwards: further statistically significant relationships connect the stage reached about the beginning of February with the weather of the weeks that follow. Glimmering recognition of these tendencies is enshrined in ancient weather lore: thus, from England

'As the days lengthen,
the cold strengthens.'

and from Scotland

'If Candlemas day (2 February) be dry and fair,
Half the winter's to come and mair.
If Candlemas day be wet and foul,
Half the winter was gone at Yule.'

(The former Candlemas day couplet signifies an anticyclonic situation, possibly E'ly wind in southern Britain; the latter points to W'ly cyclonic situations still continuing.) BAUR (*loc. cit.*) has framed 'rules' based on observations of the period 1891–1955 that show the meteorology of this: cases with a Scandinavian anticyclone established and snow cover over the German lowlands at the end of January can usefully be taken to presage a cold February, those with no snow cover in that area and a markedly cyclonic situation over (southern) Scandinavia indicate a mild February to follow. Late winter is the time of strongest tendency for month-to-month persistence of temperature anomalies over north-west Europe (WIESE 1925, CRADDOCK and WARD 1962, MURRAY 1967*a*). This is no more than a statistical persistence, exceeding what would be expected to occur by chance in a random series of values, but subject to rather frequent failures. Cold or very cold winters which continue into February in England also tend to be followed by a cold spring, mild or very mild winters and Februarys by a warm spring (MURRAY 1968*a*).

BAUR (*loc. cit.*) has shown that 'rules' of the same style can also be established for parts of North America, and studies (*U.S. Weather Bureau* 1965 and NAMIAS 1952, 1954) of surface and upper air temperature persistence tendencies at different times of the year over the United States may be compared with those cited above for Europe and the eastern Atlantic. Similar research could be expected to yield results of predictive value for other regions in both hemispheres.

The circulation patterns associated with the spring warming in April over Europe and North America may serve as significant indicators of the character of the following summer (HAY 1966, RATCLIFFE and COLLISON 1969). The data studied cover the years from about 1900 to the 1960s. Summer rainfall in England and Wales tends to be highest in years when the European 500-mb trough in April lies between about 10° and 25°E. From this position the normal seasonal shortening of wave length in the upper westerlies brings the summer position of the trough near, over, or just west of, the British Isles. The April position of the Canadian trough also appeared to have some effect: when it was east of 60°W, implying southwesterly upper winds and an upper ridge between there and the British Isles, the summer rainfall was less than otherwise expected with the given European trough position; when it was west of 65°W the summers were wetter than otherwise expected. Underlying these relationships, one may suppose, eastern positions of the Canadian and European troughs in April occur when there has been early and extensive disappearance of the winter snow from Canada and Europe, and from these positions the later retrogression westwards of the European trough is unlikely to bring it over or west of Britain. It evidently does not matter that the retrogression concerned is not a continuous process between April and summer: in fact, the shortest of all wave lengths in the Atlantic sector are observed in May–early June and there is a certain renewed increase and abrupt, though limited, eastward shift of the circulation features about mid to late June (cf. fig. 4.11), accompanying the well-known 'return of the (surface) westerlies' and onset of the 'European monsoon'. Over the southern part of the U.S. Great Plains, by contrast, a real persistence tendency from spring to summer is observed (NAMIAS 1963, p. 355).

The changes to which Europe is liable between late May and high summer (July and August), referred to above, themselves lead to dependable 'rules' (BAUR 1958) concerned with the more extreme years. A sharp change of character in many years accompanies the wave shift that underlies the June singularity. When the last days of May (27th–31st) are over 3°C warmer than normal in central Europe, a drastic fall of temperature by at least 4°C within 24–48 hours can be expected in early June (before the 10th). And when the first half of June is very warm (mean temperature at Berlin over 2°C above normal) the high summer months have been wet in fourteen cases out of fifteen (between 1848 and 1952). In England those summers (ten cases between 1873 and 1965) in which there is great warmth in early June, with the temperature over not less than one pentad between 31 May and 14 June at least 2°C above average, both July and August have been without exception cooler than average and have had more than the 2-month average rainfall, though one or other of these months might not be wet. July–August is another period of marked persistence tendency in northwest Europe affecting temperature (see references already cited) and rainfall (MURRAY 1967*b*, 1968*b*); this is associated with the peak incidence of long spells of persistent circulation type (Chapter 4, pp. 182, 186) which defines the natural high summer season. It is this persistence which presumably gave rise to the ancient legend that the weather of St Swithin's day (15 July) marks the character of the next 40 days. Nevertheless, there seems to have been a time, embracing most of the eighteenth century and perhaps longer, when there was much less persistence from July to August or possibly only persistence from July to the first half of August. This was a curious climatic phase when Europe, from about 1740 to 1810, experienced on the whole slightly warmer summers than now, but evidently rather short summers since the temperatures prevailing over the other 9 or 10 months of the year were lower than in recent times.

The most stable (persistent) type of warm summer in England, those summers associated with an extension of the Azores anticyclone right across the British Isles and northern Europe, tends to be followed by warmth in September also. But warm summers in which the anticyclone only just reaches to the British Isles tend to break down earlier and be followed by a cold September; this is particularly so if low pressure develops over Scandinavia as the summer goes on (HAY 1968). Cyclonic westerly or southwesterly cool summers in Britain tend to be followed by cool Septembers, perhaps because of the colder than normal surface waters spread over the Atlantic generally north of 50°N by this pattern.

Circulation patterns and the distribution of warm and cold surface developing in the early months of the cooling season in high latitudes doubtless have an increasing likelihood, as the autumn goes on and passes into forewinter, of setting the character of the ensuing winter. Nevertheless, it has proved easier to find indicators of predictive value for the European winter in October and in early December than in November. This must be because of the sequence of big changes of circulation character, from progressive zonal to blocked and vice versa, that habitually takes place during November. This is a hemispheric disruption of the progressive seasonal intensification and wave-length increase, which is probably due either to a fairly regular failure of the poleward heat flux during the prevalence of intensifying zonal flow about late October–early November or to interference between the increasingly vigorous circulations set in motion by the intensifying thermal gradients over or near east Asia and Canada.

HAY (1967) found that strong westerlies and a deep centre of low monthly mean pressure near Iceland in October usually precede mild westerly winters in Britain. Enhanced frequency of northerly weather, with the low pressure centred over the region between the Norwegian Sea and Kara Sea, in October tends to be followed by cold or rather cold winters in England with continued excess of N'ly wind components. The data used in these studies covered the years 1873–1963. During those parts of this period when blocking was most frequent, 1875–95 and 1940–63, above-normal pressure to the northwest of Britain (0–50°W at 55°N) in October was associated with very cold winters to follow. HAY (1970) has further noted that Octobers with 3 or more days with anticyclones centred near Iceland (60–70°N 5–35°W), also cases when the average pressure over Iceland (at 65°N 20°W) for the 5 days 11–15 October is high ($\geqslant$1015 mb), there is a statistically significant tendency for a very cold winter (December–February lowest quintile) in England. This was based on data for 1873–1968, but was verified also in earlier periods for which the requisite daily data were available (1779–84 and 1823–36). BAUR (1958), using the data of the period 1802–1954 to verify a rule enshrined in central European weather lore, found that dry, warm (over 2°C above normal), blocked (anticyclonic and/or S'ly) Octobers in central Europe were (in all the eleven cases in which the preconditions were satisfied) followed by cold Januarys in that area. These October preconditions for cold winters presumably point to an already developed meridional circulation tendency, with cold snow or ice surfaces protruding south in or near the east Greenland and west Siberian sectors.

November is essentially a changeable month in Britain and most of Europe. This conceals how it comes about that a cold winter in England is significantly more probable than usual after a November in which the overall 'westerliness' (measured by the zonal index: monthly mean pressure difference between the Azores and Iceland) is very close to normal (within 2 mb). In the ten winters concerned between 1865 and 1954 there were only 5 very mild months with temperatures above 5 °C (most of these being extremely mild, i.e. 6–7°) against 14 that were very cold with mean temperature below 3 °C and only 11 months between 3° and 5°: this tendency to the extreme indicates that the winter period December–February was marked by an abnormal amount of blocking. There seems to be a tendency for the November blocking to come earlier in the month, and to break down early before cold winters, a mild westerly period following the November block but itself liable to be finished by about the end of November. The extreme winters of 1794–5 and 1962–3 provided examples.

Besides the foregoing persistence and sequential rules, growing attention is paid in Russia to the recognition of advance symptoms, or 'forerunners', of the circulation developments characterizing a coming season (BORISOVA and RUDIČEVA 1968). One, or sometimes two, pentads are noted in which the northern hemisphere circulation development is most sharply out of character with the remainder of the season in which they occur. This development is considered likely to typify the circulation pattern prevailing in the following, or some later, season in the same year, particularly in those cases where the forerunner is manifested at or about certain characteristic dates. Table 7.2 below gives the average dates of the *natural synoptic seasons* recognized in Russia and the most frequently reported dates of advance symptoms according to BORISOVA and RUDIČEVA.

TABLE 7.2

Russian natural synoptic seasons and forerunners: mean dates

Natural season		*Mean dates of out-of-character episodes 4–9 days long in previous seasons: regarded as forerunners*
Spring	10 March–6 May	10 November and 24 January
Foresummer	7 May–30 June	17 January and 26 March
High summer	1 July–22 August	19 February and 30 May
Autumn	23 August–15 October	31 May and 18 July
Forewinter	16 October–24 December	24 July and 15 September
Winter	25 December–9 March	17 September and 7 November

Synchronous relationships in the circulation development over a hemisphere and over the globe

Besides the tendencies for interrelationship of circulation and weather developments around much of the hemisphere that are called forth by the dynamics of the upper westerlies, producing changes of wave length or of wave form, there are clearly times when like changes of pattern take place more or less simultaneously over both ocean sectors of the northern hemisphere. Various interrelationships between contemporary tendencies over the northern and southern hemispheres have also been reported. All such developments affect the positions and development of the main climatic zones, and it is important to assess these tendencies further. Rather little exploratory work in this field has been done so far, yet it is already clear that the nature of such relationships may differ according to the time-scale considered. Particularly, those which concern poleward or equatorward shifts of the main circulation and accompanying climatic zones, whether over opposite sectors of the same hemisphere or in the northern and southern hemispheres at the same time, in response to some variation of the heat supply or of heat storage in the lower latitudes, may only work the same way so long as there is no great change in the geography of 'permanent' snow and ice or of the extent of warm ocean surface. When there is such a change, e.g. a lasting increase in the extent of ice in one hemisphere, or in one sector of one hemisphere, the range of movement of the atmospheric circulation zones may be damped or distorted accordingly.

The degree of coupling between the Atlantic and Pacific sectors of the northern hemisphere, i.e. the frequency and the nature of nearly simultaneous developments over the two oceans, has been studied by SORKINA (1966) in terms of monthly anomalies of the number of days on which various broad classes of wind circulation pattern occurred during 545 months (in 47 years) of the present century. These classes were defined as (1) well-developed (zonal) subtropical anticyclones, (2) incipient weakening and breakdown of the subtropical anticyclones, (3) breakdown of the anticyclones and meridional movement of cyclones between the high-pressure cells, (4) quasi-stationary cyclonic cells or 'central depressions', (5) Arctic intrusions. Two peculiarities of the Atlantic sector affected the degree of correspondence with the Pacific that was possible: (*a*) greater anomalies occur in the Atlantic sector, (*b*) the Atlantic is not so much affected by monsoon currents as the Pacific is, so that winter-like circulation patterns can occur over the Atlantic in summer and summer-like patterns are occasionally seen in winter, but not over the Pacific.[1] Despite these limitations, anomalies of the frequency of some particular class of circulation over the one ocean were mostly accompanied within one month by a similar anomaly over the other ocean: this was true for over 90% of the cases where the anomaly amounted to over 9 days of the month, and in only about a quarter of these cases was the anomaly in the other ocean of low

1. In the overall mean, the Atlantic sector circulation does not weaken in summer as much as that over the rest of the northern hemisphere. Its (500-mb level) zonal index 30–70°N 1957–64 for July was 123% of the mean for the hemisphere; corresponding figures for the Asian and Pacific sectors were 86 and 103%.

intensity (amounting to under 5 days of the month). Among the great anomalies of this order about half occurred in the very same month over both oceans. Taking all cases of anomalies exceeding 4 days of the month and affecting both oceans within ± 1 month, the peak anomaly occurred first in the Atlantic 1·3–1·5 times as often as it did in the Pacific. It seems therefore that both oceans tend to respond alike to some influence that affects both around the same time, but that the Atlantic characteristically reacts more quickly and with greater amplitude.

Analysis of the southern hemisphere circulation during the International Geophysical Year expeditions 1956–8 indicated that nearly half of all blocking developments (defined by break-up of the sub-Antarctic low-pressure zone into a number of cells with meridional components) affected all sectors at once; it was comparatively rare (under 10% of the cases) for such developments to occur in one sector alone (TAUBER 1964).

There is some evidence of a quite different relationship between events in the two ocean sectors of the northern hemisphere when longer time-spans are considered. Thus SCHELL (1956) reports that while the extent of ice in the Atlantic sector (western and northern Atlantic plus Greenland and Barents Seas) decreased by 18% from 1901–20 to 1921–39, ice extent in the Pacific sector increased by 22% over the same years. This seems to imply that while the circulation zones, and in particular the subpolar cyclonic activity, penetrated increasingly far north in the Atlantic sector and along the north coast of Asia, the secular trend was southward in the Pacific sector.

The fact that interhemispheric reactions must and do take place has been demonstrated in general terms by KIDSON and NEWELL (1969) from a study of momentum and heat transports between 40°N and 30°S in each season of the year for 1957–64. Total angular (westerly) momentum in the atmosphere was also summed for each 10°-latitude band from 80°N to 80°S for January, April, July and October. It was shown that more momentum is acquired by the atmosphere (from the friction on the E'ly Trade Winds) in the winter hemisphere, where the circulation is strong, and that atmospheric angular momentum is exported across the equator, this export being particularly great in July. It was also found that this export of momentum from the southern hemisphere is greatest just when the number of tropical cyclones forming per month is greatest. Hence the southern hemisphere circulation is liable to have a strong influence on the northern hemisphere in summer.

The total (westerly) angular momentum content of the atmosphere relative to the Earth is greatest in the latter part of the northern hemisphere winter, when the circulation over both hemispheres is strong. For the same reason, the relative momentum of the atmosphere was presumably greater (but more nearly in balance between the hemispheres) over the year as a whole in the ice ages than in interglacial and non-glacial times.

Study of simultaneous developments in the subtropical anticyclone belts of the North and South Pacific over the years 1954–9 (BURMISTROVA 1965) showed that relationships can only be found within certain narrower longitude sectors. Over the extreme western Pacific, events in the subtropical belts north and south of the equator appeared quite independent of each other. Over the central parts of the ocean, 160°E to 160°W in the North Pacific and 160°E to 120°W in the South Pacific, *coupling* (i.e. simultaneous development of anticyclones in these longitudes north and south of the equator) predominated in spring (71%) and autumn (61%) but was only present in 45% of the occurrences in January and 58% in July. When the eastern North Pacific east of 160°W was compared with the same longitudes (160–120°W) in the South Pacific, unlike developments north and south of the equator predominated (over 60% in all seasons and 77% in July): i.e. when an anticyclone was present in these longitudes on one side of the equator it was usually absent in the other hemisphere. The eastern South Pacific east of 120°W is rather more to be considered in connexion with events in the American and Atlantic sectors of the northern hemisphere (see p. 301).

A more general study of interhemispheric relationships in January and July 1957–64 by TAUBER (1967) indicated that in most individual months when the zonal index, measured by pressure differences at various heights between latitudes 30° and 70°, is above average in the northern hemisphere it is below average in the southern hemisphere, and vice versa. The correlation coefficients between the zonal indices averaged around either hemisphere over the years studied were $-0{\cdot}75 \pm 0{\cdot}08$ at the 700 mb level, $-0{\cdot}56 \pm 0{\cdot}12$ at 500 mb and $-0{\cdot}44 \pm 0{\cdot}14$ at 300 mb. But there seemed to be longer-term changes going on, even within the 8-year period studied, these having a parallel (in-phase) trend over both hemispheres.

The course of the surface pressure-difference indices of circulation strength illustrated in fig. 7.1(*a*) indicates some general parallelism in the long-term trends over the northern and southern hemispheres, at least from 1850 to the 1960s. Most indices of this kind show a general rising trend up to a maximum some time in the first half of the twentieth century and a tendency to decline after that. Increasing vigour of the wind circulation over the 100–150 years up to 1925 ± 15 years seems to have been a world-wide phenomenon, paralleled by increasing moisture transport over the continents and increasing transport of heat and moisture to high latitudes. The world-wide nature of these longer-term changes is eloquently demonstrated by a correlation coefficient of $+0{\cdot}75 \pm 0{\cdot}10$ found between successive 10-year values of snow accumulation at the South Pole and the number of days a year with SW'ly surface wind at London from 1760 to 1957 (LAMB 1967).

Studies by LAMB and JOHNSON (1959, 1961) of variations of the mean latitudes of the main limbs of the atmospheric circulation at the surface over long periods of years revealed a number of interhemispheric relationships. Forty-year averages of the latitudes of the main features in the southern Indian Ocean and Australasian sector show displacements amounting to 2–4° latitude (fig. 7.21) in phase with the variations of strength of the east Asian monsoon currents, i.e. northward in July and southward in January when these monsoon currents (in common with the mainstreams of the world's wind circulation elsewhere, cf. fig. 7.1(*a*)) were becoming stronger. The tendency for reversion in the last 20–40 years, as the

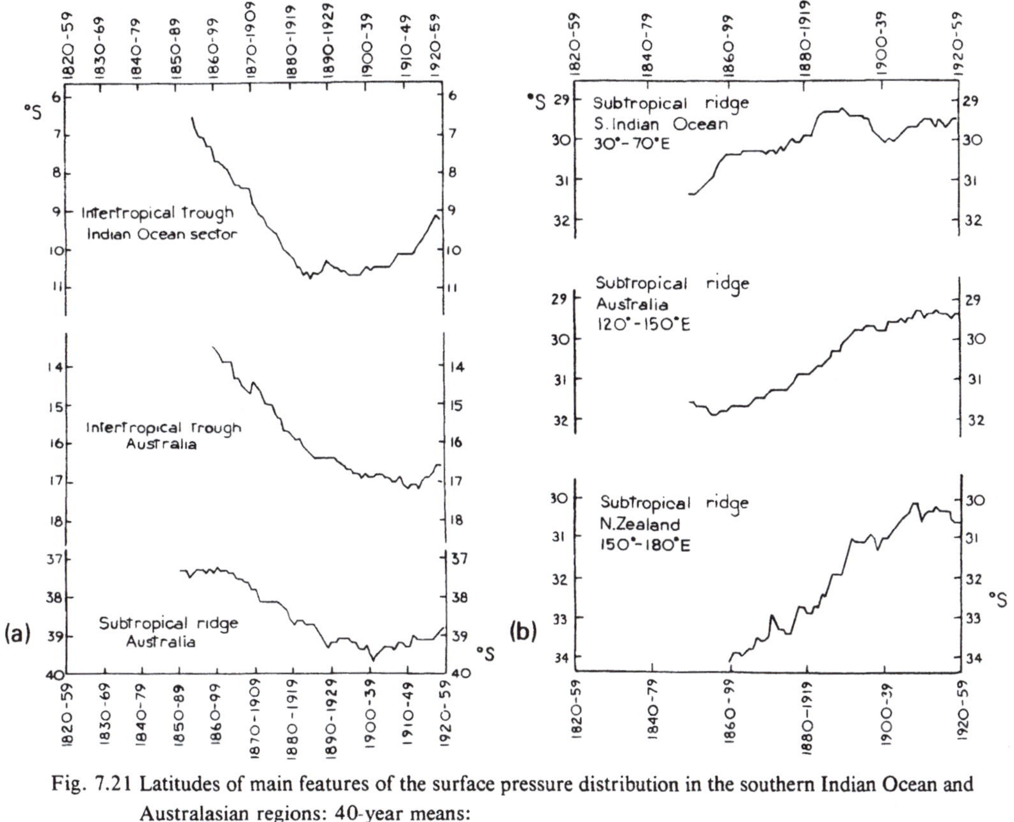

Fig. 7.21 Latitudes of main features of the surface pressure distribution in the southern Indian Ocean and Australasian regions: 40-year means:
(*a*) January (*b*) July

circulation has begun to weaken, is also apparent in the latitudes of these southern hemisphere features.

At the other side of the globe, a long history of broadly parallel movements north and south of the 10- to 20-year mean positions of the North Atlantic cyclonic activity and that in the eastern South Pacific and South American sector, between about 120° and 60°W, is suggested by the records of southward protrusions of the Arctic sea ice affecting the coast of Iceland and of rainfall at Santiago (33°S) in Chile from the 1540s to the 1950s (fig. 7.22). Dry periods at Santiago mean that the sub-Antarctic depression tracks are keeping too far south, even in the southern winter, to affect the place. This interpretation is verified by comparison (fig. 7.23) of the rainfall decade by decade at Santiago with the measurements of snow accumulation at 80°S 120°W in Antarctica: the inverse relationship of the downput of rain and snow at 33° and 80°S in this sector is clear from fig. 7.23 and gives a correlation coefficient of $-0{\cdot}67 \pm 0{\cdot}10$. This prevailing parallelism of the longer-term shifts of the main circulation zones all the way from the Arctic to the Antarctic in the American sector and neighbouring longitudes over the oceans has interesting further

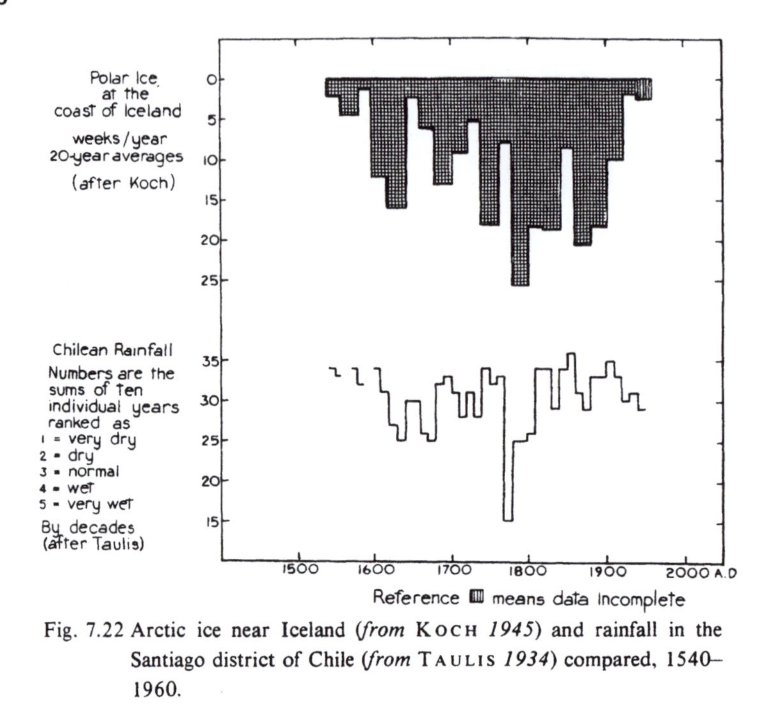

Fig. 7.22 Arctic ice near Iceland (*from* KOCH *1945*) and rainfall in the Santiago district of Chile (*from* TAULIS *1934*) compared, 1540–1960.

implications. It has been noticed (e.g. LAMB 1967) that the limit of Antarctic sea ice was apparently 1–2° latitude farther south at the time of the voyages of discovery between 1770 and 1830 than around 1930–50 despite evidence of somewhat lower temperatures (average anomaly of the order of 1°C) prevailing over most of the world in those earlier years. It seems possible that the storminess near the ice edge associated with a more southern position of most of the sub-Antarctic depression tracks may have been responsible for maintaining open water farther south than now. The geography of the Southern Ocean affords no protection for great expanses of pack-ice against the onslaught of storms, and growths of the Antarctic ice in calm periods are doubtless readily broken up whenever storms and great ocean swells are renewed. There is some ground in this for thinking that even in the ice ages the sea-ice limit and the sub-Antarctic cyclonic activity may have remained not too far from its present latitude, and perhaps at times, a little farther south, despite the generally lower temperatures. It seems likely that growth of the Antarctic sea ice and shifts of the main ocean current boundaries in the Southern Ocean must always have been slight by comparison with the northern hemisphere (especially the North Atlantic), apart from temporary growths of ice shelves soon broken up by storminess and by tidal waves (*tsunami*) of seismic or volcanic origin, leading to a few years of exceptional numbers of icebergs; some of the largest tabular bergs might wander, however, as erratics, to quite low latitudes before melting.

Fig. 7.1(*b*) showed the latitude changes of the main zones of the circulation in the North

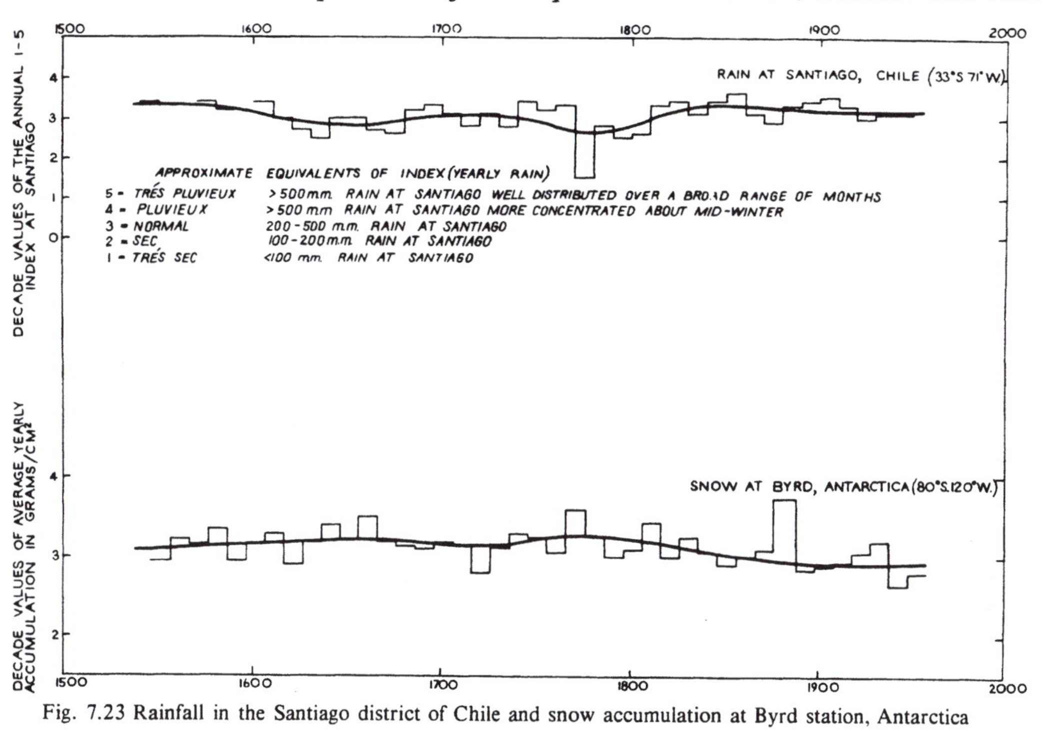

Fig. 7.23 Rainfall in the Santiago district of Chile and snow accumulation at Byrd station, Antarctica compared, 1547–1957.

and South Atlantic over the last 120 years. In this sector also there was probably a general northward shift of most, or all, of the features concerned by just 1–2° of latitude between the time around 1800 and 1900 or a little later, if we judge by the ice limit changes, though the curves in fig. 7.1 (*b*) indicate some equally marked shorter-term fluctuations. From some time about the 1920s onwards, however, the main features in both North and South Atlantic move progressively equatorwards; this new trend, which has become even more marked in the 1960s, is already the most pronounced feature of the curves and amounted to 2° of latitude by 1959 (when the graphs end). The prevailing positions in the North Atlantic have returned to roughly where they were about 1800, and the quieter regime in the Arctic has permitted the Arctic ice to extend south again, particularly in the Atlantic sector. The accompanying tendency for expanded polar anticyclones in both hemispheres is illustrated here (fig. 7.24) by the surface-pressure anomaly maps for two extreme winter months, January and August 1963, in the northern and southern hemispheres respectively: the similarities of the overall pattern in both hemispheres need no stressing, but it needs to be added that the equatorward shift of the low-pressure zones in the 1960s has affected all seasons of the year (LAMB 1966). The accompanying meridionality in the higher latitudes has brought more frequent snowfall in winter than for many decades past in Britain, as it has

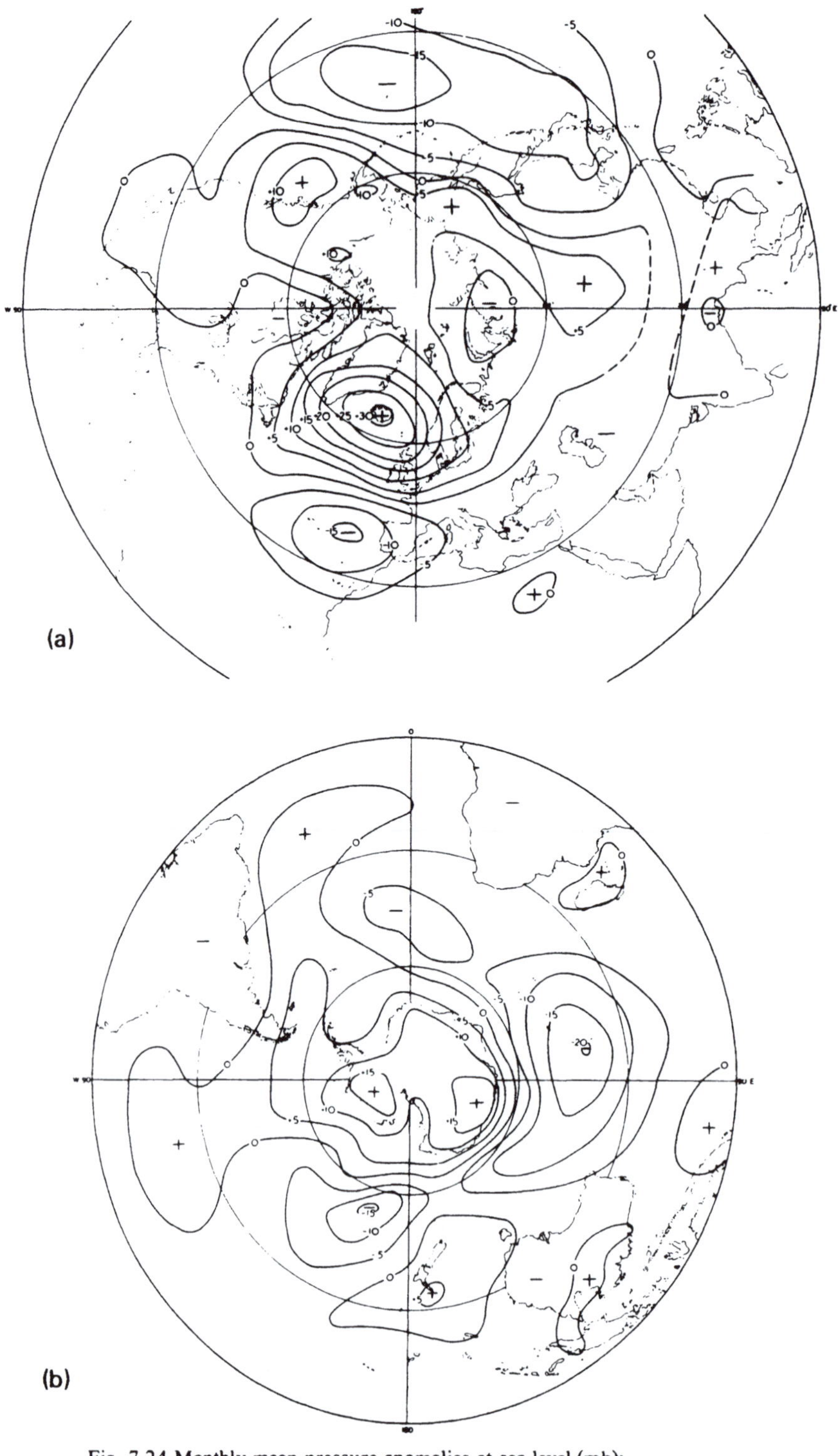

Fig. 7.24 Monthly mean pressure anomalies at sea level (mb):

(*a*) January 1963 northern hemisphere (departures from 1900–39 average).

(*b*) August 1963 southern hemisphere (departures from average values believed to apply to 1949–53 but any available data between 1900 and 1953 used for points south of 40°S).

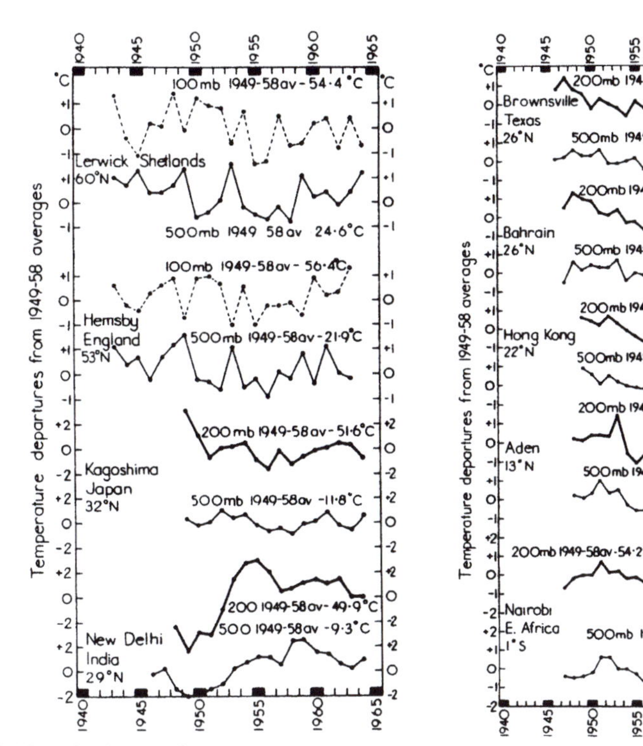

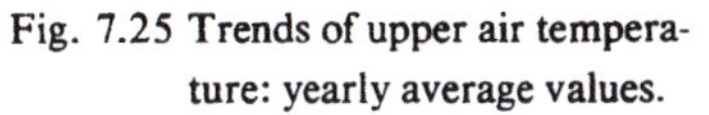
Fig. 7.25 Trends of upper air temperature: yearly average values.

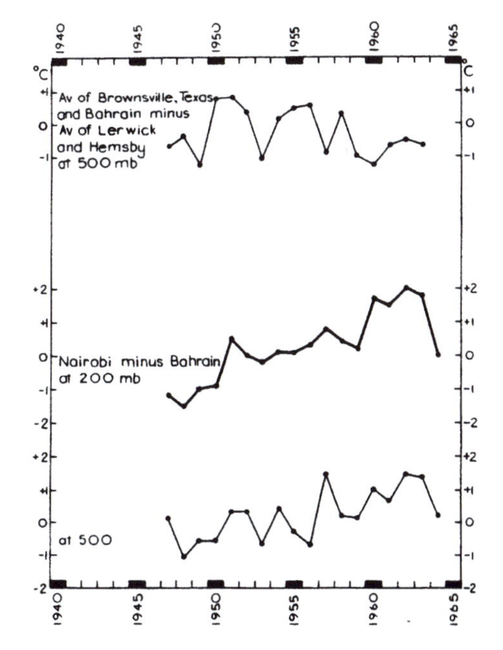

Fig. 7.26 Differences of upper air temperatures prevailing at different latitudes, variation 1947–1964: top curve – *c.* 26–56°N, lower curves – *c.* 0–26°N.

also on the higher ground in South Africa and Australia; the summer weather has been very warm or very cool in different sectors according to whether S'ly or N'ly winds prevailed; and the short wave lengths in the upper westerlies have changed the positions of greatest wetness and dryness in the temperate zones as well as increasing the year-to-year variability of rainfall.

The three-dimensional structure of this equatorward shift of the climatic zones can be seen in the changes of upper air temperature during the 1950s and early 1960s at widely different latitudes over the northern hemisphere (figs. 7.25, 7.26). Attention must be confined to places where no changes of practice regarding instrumentation or radiation corrections affect the results. It turns out that the great fall of surface temperatures in high latitudes must be attributed to more development of surface inversions in the calmer climate prevailing and is not matched by more than a very slight fall of upper air temperatures, e.g. at 71°N at Jan Mayen (not illustrated). Upper air temperatures have fallen more at places near 30°N, presumably owing to weakening of the subtropical anticyclones (less subsidence) and their shift towards lower latitudes. So the thermal gradient between 30° and 70°N has declined (fig. 7.26), as would be expected when the westerlies weaken. There was some increase of this gradient from 1946 to 1950 when the westerlies were rising to a secondary peak. By contrast, the thermal gradient between the equator and 30°N has tended to increase, partly owing to the cooling near 30°N and perhaps partly owing to greater release of latent heat in the equatorial rains, which have intensified and remained more concentrated near the equator than in the preceding years.

CHAPTER 8

The oceans

The workings of the ocean have been too little considered in most texts on climatology. Though the thermal state of the ocean surface is the result of winds and weather controlling the water flow and radiation budget over some time previously, the heat stored and transported in the surface layer of the ocean – and the moisture given off and later condensed in clouds – introduce great regional anomalies in the heating of the overlying atmosphere, as we have seen in Chapter 2 (figs. 2.16, 2.17). The strong thermal gradients at the cold boundaries of the great warm ocean currents create strong horizontal differences in the heating of the air over those regions, thereby tending to concentrate the flow of the upper winds (and most effectively when these are in the same alignment as the ocean current boundary). Hence, quite small displacements of these current boundaries, and the corresponding anomalies of sea surface temperature in the regions affected, may, if they persist, have big effects on the vigour and steering of the atmospheric circulation overhead and developments in it downstream (see Chapter 10).

Heating and cooling of the ocean

It is widely recognized in climatology that the difference between oceanic and continental climates depends upon the sluggish response of the temperature of the sea surface to heat gain and loss. This results from the large specific heat of water (compared to rock and all common substances) and the vertical transmission of heat in the sea by wind stirring (wave action) and convection. Thus the surface waters of the oceans in middle latitudes generally reach their yearly maximum and minimum temperatures about February and August, and the greatest and least extent of the Arctic sea ice are reached about March–April and September. By contrast, the warmest and coldest months in the heart of the great continents come immediately after the solstices and the reversal of the trend of mean temperature thereafter is soon sharp. The average diurnal range of temperature of the sea surface in the tropics under clear skies is only 0·7°C with a moderate to fresh breeze and 1·6°C with still air; under the most favourable circumstances it reaches about 2°C. In the North Sea the average diurnal range of the water surface is only 0·1–0·2°C in winter and 0·4–0·7°C in summer. The overall, all-weather average for the world's oceans may be only 0·3°C.

These figures contrast with average diurnal ranges of the ground surface in England –

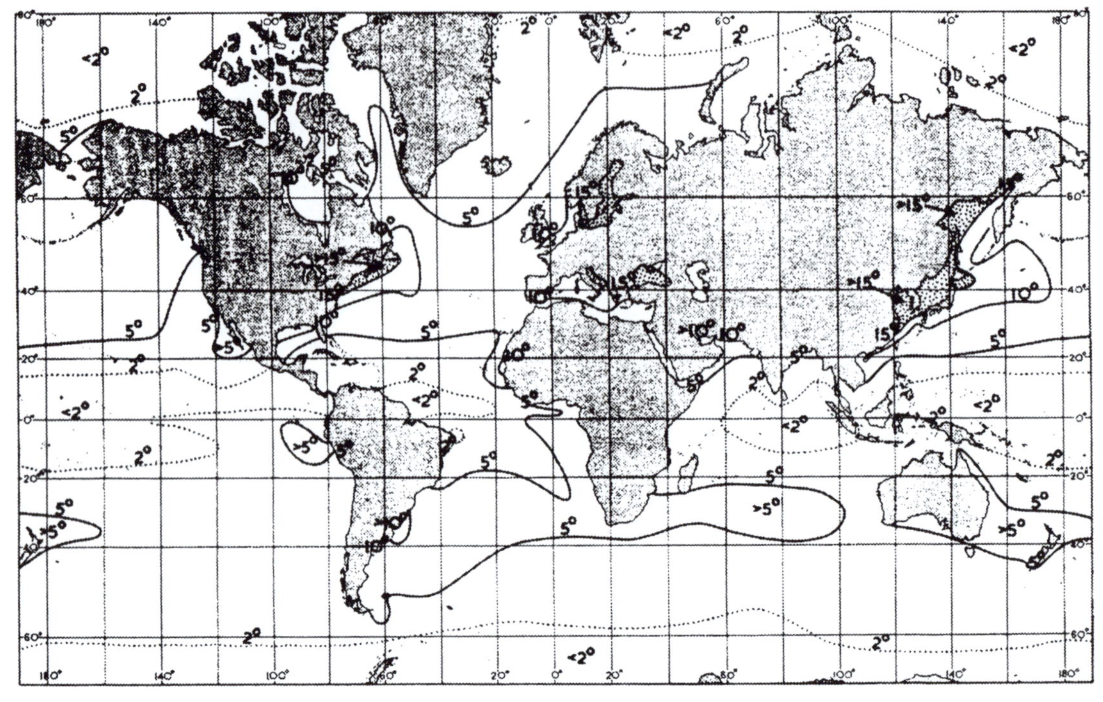

Fig. 8.1 Average yearly range of sea surface temperature (°C). Warmest – coldest month. (*Adapted from* Dietrich *1957*.)

depending somewhat on the type of soil (see Lamb 1964, pp. 90–2) – of about 5°C in winter and 20°C in summer, the extreme being about 45°C. In the tropical deserts daily ranges of the surface amounting to 40–50°C are probably common over nearly half the year and in extreme circumstances 60°C may be exceeded. (An extreme diurnal range of 84°C has been measured in the surface at Khartum.)

The annual range of sea surface temperature between the mean values of the warmest and coldest months is displayed in fig. 8.1; and fig. 8.2 shows the annual mean values. Annual ranges exceeding 15°C are confined to a few more or less land-locked seas and to waters over the continental shelf off eastern North America and east Asia[1] in middle latitudes, where they are mainly due to the effect of cold winds blowing off the continents in the winter. Over most of the oceans of the world the annual range is less than 5°C, greater values being almost only found in middle latitudes. The ranges are generally greater close to land and (no doubt, due to the land effect) are, on the whole, greater in the northern hemisphere. The pack-ice surface in the central Arctic undergoes a range of about 40°C.

Mean sea surface temperatures (fig. 8.2) are generally higher than the temperature of the air over the water, by about 0·8°C on an average over many latitudes in the case of careful measurements on the German research ship *Meteor* at a height of about 8 m above the

1. The annual range approaches 20° near Korea.

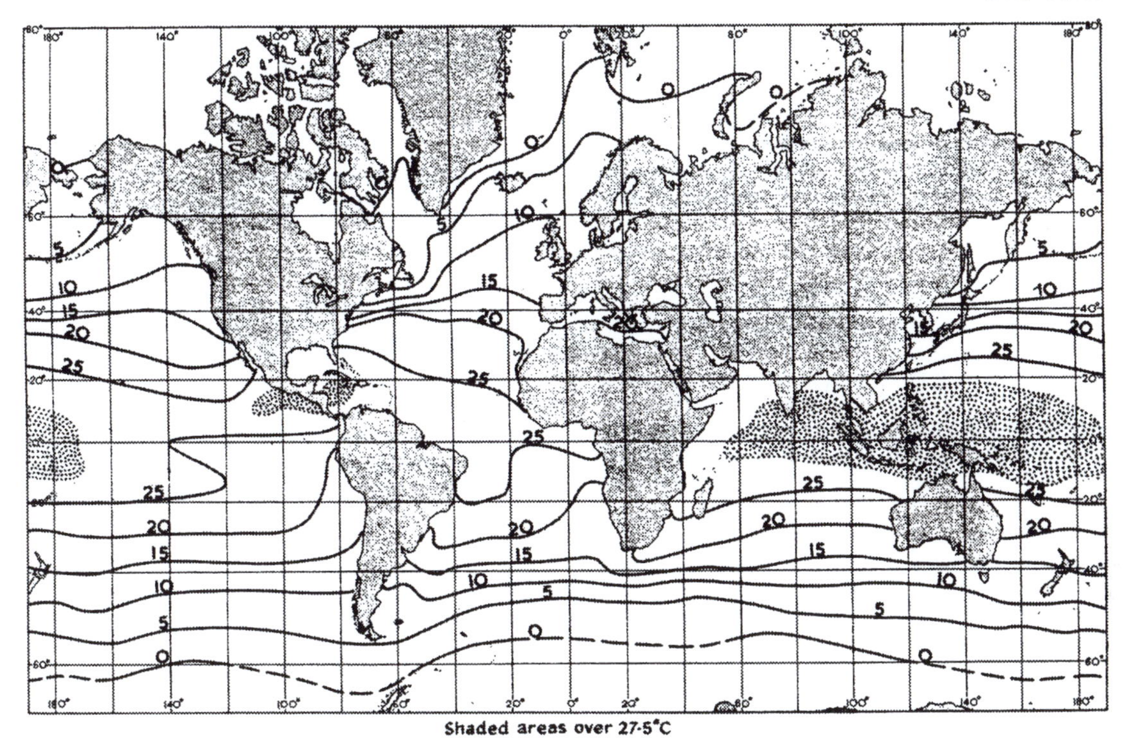

Fig. 8.2 Annual mean sea surface temperatures over the world (°C). (*Adapted from* DIETRICH.)

water. This arises because the sea in most latitudes absorbs more radiation than it gives out to the sky; therefore in most places the sea surface must give off heat to the atmosphere, from which the surplus is ultimately lost to space. Only over cold ocean currents, and over waters drawn to the surface by upwelling at coasts where the prevailing winds blow out to sea, is the air commonly warmer than the sea and a net flux of heat passes from the atmosphere to the water. Fogs and low cloud are very frequent over such waters, the prime examples being the region of the North Pole in summer, with its many seasonally open leads among the ice, as well as the waters off Labrador and Newfoundland (especially the Grand Banks) and off Peru. Great areas of open, ice-free water maintained in high latitudes, as in the Norwegian and Barents Seas, the Irminger Sea and Davis Strait, and in the Gulf of Alaska, are associated with heat transport by warm ocean currents; over these waters enormous rates of heating of cold airmasses from the neighbouring ice are sometimes observed in winter (see Chapter 2, pp. 60–1). The sea surface is cooled in the process.

As in the atmosphere, heat is spread down and up through the ocean mainly by convection. The extent to which this is possible depends upon the density layering (i.e. the water density variation with depth). Water sinks when it is denser than its surroundings, rises when it is less dense. At the temperatures prevailing in the upper 1000 m of the ocean in most areas of the world (i.e. above 5 °C) water expands when it is heated; for water with the salinity general in the sea this is true at all temperatures above its freezing point, which is −2°C for salinity 35 parts per thousand. (The relationship between freezing point and salinity is linear.) Pure water has its maximum density at +4°C,[1] but for sea water with the salinity values common in the ocean density increases all the way down to its freezing point (see Table 8.1 on p. 312). These figures mean that in the tropics and in middle latitudes surface heating causes expansion and the warmest water tends to stay on top. A layer of low density is created above the colder, denser water below; this is a stable stratification. If a strong vertical gradient of density is created towards the warmest layer at the top, absorption of the incoming heat is thereby effectively limited to the uppermost layers which cannot mix with the denser layers below (except so far as wave action penetrates). This density gradient is called a *thermocline*, and its presence means that the temperature of the top water rises much more rapidly than it could otherwise. Below the thermocline the temperature generally changes little with depth (an 'isothermal' distribution). When the surface water is cooled, its density generally increases; ultimately the thermocline disappears, and the surface water may become dense enough to sink to great depths. When this deep convection stage is reached, any further fall of temperature of the surface must become very slow.

The compressibility of water is slight. At a depth of 1 km, where the pressure amounts to about 100 atm (atmospheres), the density is about 10% greater than at the surface; at the greatest depths in the ocean, at pressures of 500–1000 atm, the further increase of water density is no more than a few per cent. Hence, also, the adiabatic changes of temperature of water which changes its level in the ocean are slight by comparison with equivalent changes in the atmosphere. Sea water sinking from the surface has its temperature increased generally by rather less than 0·2°C/km depth. Water coming up towards the surface is adiabatically 'cooled' at the same rate. Since most upwelling brings water to the surface from no more than 100–300 m depth, the adiabatic temperature change can be disregarded.

In general, therefore, vertical columns of the ocean through some range of depths in which conditions are approximately isothermal and isohaline (i.e. provided that there are no water layers of differing salinity present) are similar to those parts of the atmosphere in which an adiabatic lapse rate of temperature prevails, in that parcels of the fluid displaced up or down continue to have nearly the same density as their surroundings (or preserve a nearly constant density difference). Such layers of the ocean tend to become well mixed. But where the water density decreases towards the surface, as when the surface is heated or is occupied by a layer of fresh water, the stable stratification hinders mixing of the upper and lower water.

Heating of the surface layer of the ocean in calm weather in summer, or wherever and whenever the radiation balance is positive, produces a thermocline. This happens above all

1. Cooling of the surface of fresh water, or water of low salinity, below +4°C leads to a density minimum in the coldest water which therefore tends to remain at the surface. Ice is therefore more readily formed on the surface of such waters than of saline ocean water.

in the subtropical anticyclones and seasonally in middle and higher latitudes, especially in fine spells of weather. In the cooling season the thermocline is as easily destroyed. Rough weather at any time reduces it, if the preceding conditions had been calm enough to produce a sharp temperature maximum just at the surface. But so long as the stable stratification persists, cooling by the atmosphere can abstract heat only from the uppermost layer; so the surface water temperature falls rather fast until the thermocline is obliterated. Thereafter, the water rendered dense by cooling at the surface can sink to whatever depth water of similar density is encountered. This convection brings about vertical exchanges through great depths of the ocean, which must all be cooled with the surface; hence, from this stage on, the surface temperature falls much more slowly (and in some cases hardly at all during the remainder of the winter).

Salinity

Differences of salinity such as occur near river mouths, near the floating ice on the polar oceans, and in nearly isolated gulfs or seas in low and high latitudes create great differences of density. Salinities widely found in the ocean range from about 32 to 37 parts per thousand: with no change of temperature, the difference in density is as great as between water at 25° and at 8–10° respectively.

The prevailing salinity of the surface water in different parts of the world is surveyed in fig. 8.3. The regions of greatest salinity are all in, or near, the subtropical anticyclone belts, where evaporation from the heated surface of calm waters raises the salinity of the residual water above the thermocline. Mean salinity values reach 39‰ in the eastern Mediterranean and exceed 40‰ in parts of the Red Sea and Persian Gulf. Low salinities are found near the mouths of the great rivers (e.g. Amazon, Mississippi, St Lawrence, as well as the Siberian Arctic river mouths where the value approaches nought), in the Baltic and in the melt-water of the ice-bearing East Greenland Current and the Labrador Stream.

The variations of density for water at surface pressure over the full range of temperatures and salinities occurring in the ocean are set forth in Table 8.1. From this it can be seen that the salinity difference between the cold waters near the southern limit of the Labrador Stream, off eastern U.S.A. and Newfoundland, and the warm North Atlantic water of Gulf Stream origin is not sufficient to prevent the cold water being the denser of the two watermasses at the temperatures occurring: it therefore tends to penetrate forward underneath the warm water in this area, while the latter overlies it in a wedge-shaped layer that thins towards the northwest. Farther north, in the Labrador Sea and south and southeast of Greenland, the density difference between the warm and cold water currents is slight enough for a good deal of mixing to take place, forming a watermass which sinks and penetrates beneath the warmer parts of the North Atlantic surface drift. But in seas at temperatures below about 5 °C temperature differences affect density less than do the salinity differences occurring. Hence, the cold, ice-bearing water of the Arctic Ocean and the East Greenland Current north of Iceland, in which salinity is about 32‰ and the temperature about −1 °C,

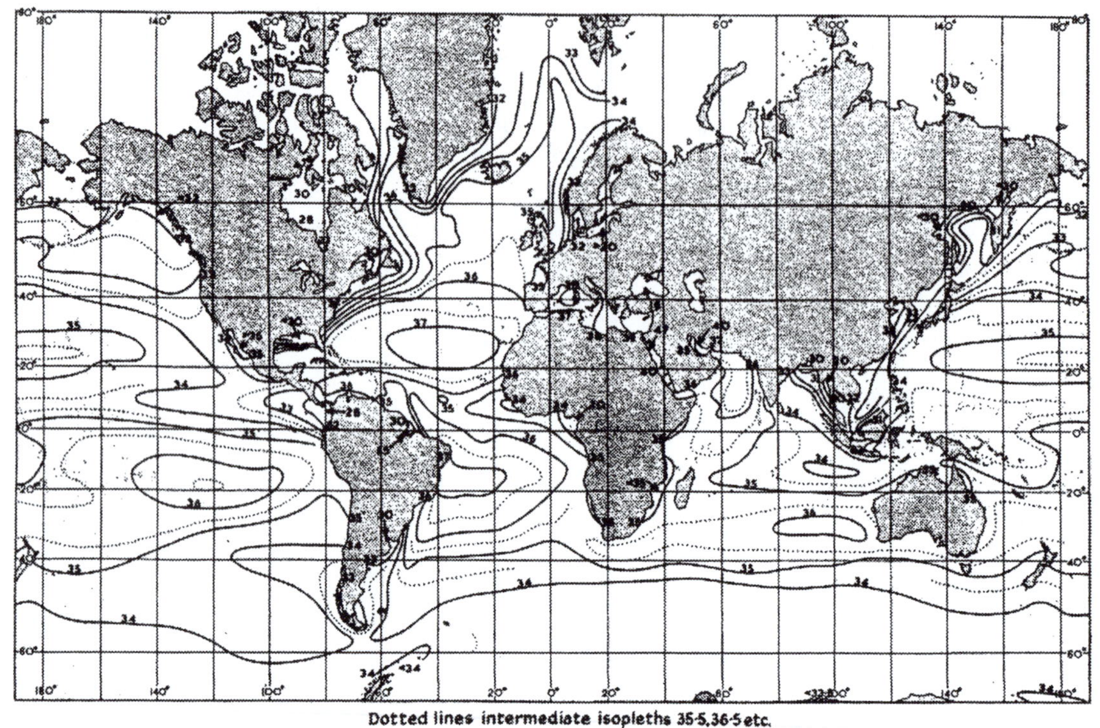

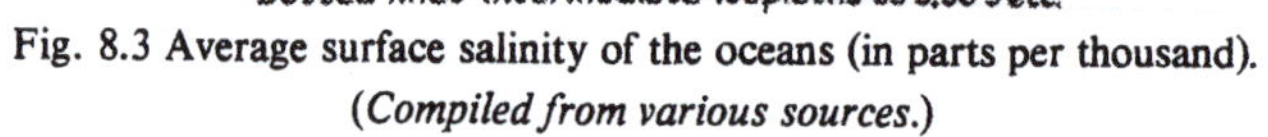

Fig. 8.3 Average surface salinity of the oceans (in parts per thousand).
(*Compiled from various sources.*)

TABLE 8.1

Density of water of different salinities at different temperatures

(Density units sigma values,* at atmospheric pressure)

°C \ ‰	0	5	10	20	30	32·5	35	37·5	40
30	−4·3	−0·6	3·1	10·6	18·0	19·9	21·7	23·6	25·5
25	−2·8	0·8	4·6	12·1	19·6	21·6	23·4	25·2	27·1
20	−1·6	2·0	5·8	13·3	21·0	23·0	24·7	26·6	28·5
15	−0·7	3·1	6·8	14·5	22·2	24·2	25·9	27·8	29·8
10	−0·3	3·7	7·6	15·2	23·1	25·2	26·9	28·8	30·8
5	0	4·1	8·0	15·8	23·7	25·7	27·6	29·6	31·6
2	0	4·1	8·0	16·0	24·0	26·0	28·0	30·0	32·0
0	−0·1	4·0	8·0	16·1	24·1	26·1	28·1	30·2	32·2
−2	−0·3	3·8	7·9	16·0	24·2	26·2	28·2	30·3	32·3

* The sigma (σ) values are a commonly used abbreviated way of expressing the density (ρ) (or specific gravity of sea water relative to that of distilled water at 4°C):

$$\sigma = 1000(\rho - 1)$$

Thus, the specific gravities in the above table range from 0·9957 to 1·0323.

remains on the surface; and the northernmost penetration of the warmer North Atlantic water with salinity about 35‰ passes underneath it, with temperature +3·5° near Spitsbergen and +1 °C at 200–400 m depth beneath the pack-ice near the North Pole.

The process of freezing of salt water sheds salt into the immediately underlying water layers, which thereby acquire a high density, being both saline and cold, and sink, to supply the deepest watermasses of the world's oceans. This process is most productive in the Antarctic zone, particularly in the Weddell Sea (in the South Atlantic sector) from where the bottom water in all the oceans except the Arctic and northern Atlantic is derived.

Density differences determined mainly by salinity cause Atlantic water to enter the Baltic below the surface, while the fresher Baltic water emerges in the top layers. They also enable ice and cold melt-water to float far out over the warmer waters of the world's oceans in exceptional circumstances.

At the mouth of the Mediterranean the less saline Atlantic water enters at the surface through the Strait of Gibraltar, while water rendered dense by evaporation in the Mediterranean flows out underneath. This Mediterranean water, though mixing with the Atlantic waters, remains recognizable by its high salinity, producing a salinity maximum at about 2000 m depth, where it spreads far and wide above the Antarctic bottom water and even round the Cape of Good Hope into the Indian Ocean. Ultimately its identity is lost by mixing.

Movement of the ocean waters: forces acting

Agencies which set the sea in motion may usefully be distinguished as follows:

(*a*) Pressure differences at the surface.
(*b*) Density differences which create horizontal pressure differences within the ocean.
(*c*) Wind drag on the sea surface.
(*d*) Tidal forces.

We shall disregard tides here, because tidal currents are reversible and indeed return within a few hours. The implications of long-term variations in the strength of the combined tidal force of sun and moon, presumably affecting the rate of exchange of water across the shallow entries to the Arctic Ocean and the Baltic Sea, have been considered in Chapter 6.

(a) *Atmospheric pressure: hydrostatic effect*

In oceanography, the sea surface is generally considered as practically a constant pressure surface. This is, of course, not strictly accurate. Atmospheric pressures over the ocean have been observed to range from about 1060 to 880 mb (the latter in the centre of a typhoon) in different regions and at different times, though most of the time the overall difference is nearer 100 mb. These pressure differences are slight by comparison with those that may be developed within the body of the ocean, and the accelerations they cause in the great mass of ocean water must be correspondingly small. They should, however, if acting long enough for equilibrium to be obtained, produce a rise of the sea surface by more than 1 m in the centre of a typhoon, and by 30–50 cm in many a cyclone in middle and higher latitudes, and depress the surface in anticyclones, were this effect not counteracted by dynamical effects of the drag of the wind on the sea surface and the centrifugal and Earth-rotation forces acting upon the moving water, which we shall discuss below. It seems possible, despite this, that the hydrostatic rise of sea level may be a significant element in some sea floods of low-lying coasts in intense cyclonic storms – e.g. around the North Sea – where lateral, or centrifugal, motion of the water is restricted.

(b) *Thermohaline (density) currents*

Under this heading we are considering how the water must be set in motion by the pressure distribution developed at any depth by the differing density of the overlying water columns. Because the ocean surface is some approximation to a constant-pressure surface, such density currents (associated with differences of temperature or salinity) can only develop at some depth within the ocean and not at the sea surface. The case is similar to that of the thermal differences within the atmosphere which result in prevalence of upper westerly winds through a great range of heights starting some way above the surface. The greatest horizontal differences of density are to be found in the upper part of the ocean (the *ocean troposphere*). But, though the near uniformity of the cold water in the deeper layers of the ocean has led to some use of the term *ocean stratosphere* to describe these layers, there is no apparent tendency (like that in the lower stratosphere in the atmosphere) to develop a contrary thermal or density distribution: hence, the horizontal pressure distribution developed in the upper part of the ocean presumably continues little changed to great depths.

The pressure p at any depth z in a water column, in which the density ρ is a variable, neglecting the pressure of the atmosphere above, is given by

$$p_z = -g\int_0^z \rho \, dz$$

or this can be written

$$p_z = -g\bar{\rho}z$$

where $\bar{\rho}$ stands for the mean density of the overlying water column and g is the gravitational acceleration. The minus sign indicates that the z values in the downward direction are taken as negative.

The density-gradient current v is then given by exactly the same equation as applies in the atmosphere:

$$\frac{\partial p_z}{\partial s} = -2\rho_z \omega v \sin \phi \pm \rho_z \frac{v^2}{r}$$

In this equation s measures distance down the pressure gradient, ω is the angular velocity of the Earth's rotation, ϕ is the latitude and r is the radius of curvature of the water's path. The negative sign before the last term applies when the curvature is cyclonic, the positive sign when it is anticyclonic. ρ_z is the water density at depth z. As in the atmosphere, the pressure gradient and other (i.e. Earth-rotation or 'Coriolis' and centrifugal) forces represented by the respective terms of this equation (cf. Chapter 3, pp. 77–8) are at most points approximately in balance with the water flowing nearly along the lines of equal pressure, the low pressure (under the warmer or less dense water) being to the left of the flow in the northern hemisphere. The values of v required for balance with any particular pressure gradient are greater in low latitudes and with anticyclonic curvature.

The above equation can, however, only be taken as representing the forces associated with the density gradient alone if the sea surface is strictly level. That this may not be so we shall see in the next section (see p. 318). Pressure differences due to the density gradients in the upper part of the ocean must increase towards the bottom of the layer in which the density gradients exist, and are presumably transmitted on down through the nearly homogeneous layers below, but it is clear that at and near the surface the actual current distribution is explained by the wind drag. In the upper part of the ocean the wind-drag effect completely overrides any component attributable to inequalities of density; it even produces variations of height of the sea surface which enter into the pressure distribution at all depths below. Hence, it has never been possible to determine by simple observation the magnitude of the density currents.

We conclude this section therefore by noting that from physical principles:

(i) where two watermasses of different density exist side by side at the same level, forces come into play which must produce relative motion – i.e. there must be a current;

(ii) the internal boundary surface between two watermasses of unlike densities ρ_1 and ρ_2, and which in the simplest case flow parallel to one another at velocities v_1 and v_2, has a slope given by

$$\frac{dz}{ds} = \frac{2\omega \sin \phi}{g}\left(\frac{\rho_1 v_1 - \rho_2 v_2}{\rho_2 - \rho_1}\right)$$

In this expression z, s, ω, ϕ and g have their usual meanings, and ρ_2 is greater than ρ_1. Notice that if $v_1 = v_2 = 0$, the surface becomes horizontal; but with $v_1 >> v_2$ and only a small difference in the densities the slope can become nearly vertical. For similar values of $(\rho_2 - \rho_1)$ and $(v_1 - v_2)$ the slopes are steeper the higher the latitude;

(iii) convection in the ocean, particularly the largest-scale convection represented by the relative movements of the different watermasses, must often be very slantwise, indeed at even lower angles to the horizontal than most frontal surfaces in the atmosphere. This is seen for example in the central regions of the North Atlantic where water of Labrador Sea origin underlies the warm North Atlantic Drift water;

(iv) the current system described by the density distribution is very unlike, and nearly opposite to, that observed at the ocean surface. In the case of the Gulf Stream it would describe a return current at some depth in the ocean, drifting southwestwards with the warmer water at the left.

(c) *Wind-driven currents*

The swiftest ocean currents are observed at, and near, the surface. Observation further shows that these are accounted for by the wind drag on the sea surface, and that they change with the changes of the wind. These are the currents which achieve the greatest heat and mass transports in the ocean and which therefore directly affect weather and climate.

The stress of the wind τ_a, measured in dynes/cm^2, upon the sea surface, has been found independently by EKMAN (1905) and ROSSBY (1936) to be given approximately by

$$\tau_a = 2{\cdot}5 \times 10^{-3} \rho_a W_{15}{}^2$$

where ρ_a is the density of the surface air and W_{15} is the wind velocity measured 15 m above the sea surface. For light winds, under 7 m/sec, when the sea surface is smooth, this expression gives too great stresses. The important points to notice are that there is a sudden increase of stress from the moment the wind becomes strong enough to make breaking waves on the sea surface, and that from this point on the stress increases with the square of the wind velocity. The following figures are given by SVERDRUP (1945):

Wind (W_{15}) m/sec	2	4	6	8	10	12	14	16	18
τ_a g/cm/sec^2	0·04	0·16	0·34	1·81	2·83	4·09	5·56	7·25	9·20

These figures suggest that when the surface water motion attains equilibrium – i.e. with the stress of the wind, the friction (represented chiefly by the eddy viscosity) of the underlying water and the Earth rotation (Coriolis), centrifugal and other forces in balance – the ratio of the water current velocity to the wind velocity is likely to increase with the wind. This must in any case be true as one goes from smooth to rough sea. However, EKMAN obtained from observation the simple relation (for rough surface conditions):

$$\frac{v_o}{W} = \frac{0{\cdot}0127}{\sqrt{\sin \phi}}$$

where v_o is the surface water current velocity and the other symbols have their usual meanings. v_o and W must, of course, be measured in the same units. For W, EKMAN used the observations māde with a hand anemometer on Nansen's ship *Fram*, probably only 3–4 m above the sea surface. The range of the ratio is probably mainly between about one-twentieth and one-hundredth; DIETRICH and KALLE (1957) take 1·5% as typical. Notice, however, that the ratio increases towards low latitudes.

The moving water is deflected to the right of the surface wind direction (northern hemisphere case; to the left in the southern hemisphere) by the Earth's rotation. The angle of deflection should be 45° regardless of wind strength, according to EKMAN (1905), though differing values have been suggested by other theoretical approaches. Observation suggests a deflection in this sense by 20–50°, and this applies also to the drift of the polar pack-ice.

Icebergs and the bigger broken-off segments of ice shelves which drift about the Arctic as 'ice-islands', may deviate from the drift of the surface water and ice, because their greater depth causes them to respond to the subsurface currents.

The subsurface layers are set in motion by the drag of the water above; this drag is due to the eddy viscosity and becomes very slight when the sea is calm and the density stratification strong. In general, however, each layer will be set in motion by some drag from the layer above it and always deviated by the Earth rotation effect from the direction of that stress. So the direction of motion should change with depth in the manner indicated by the 'EKMAN spiral' (fig. 8.4). The diagram also indicates the decrease of the drag-current velocities with increasing depth, as the stress transmitted decreases from layer to layer. Finally, according to EKMAN's theory, a depth should be reached at which the current caused by the wind drag is almost reversed and reduced to near zero. This depth D is called the 'depth of frictional resistance': its value in metres is given by

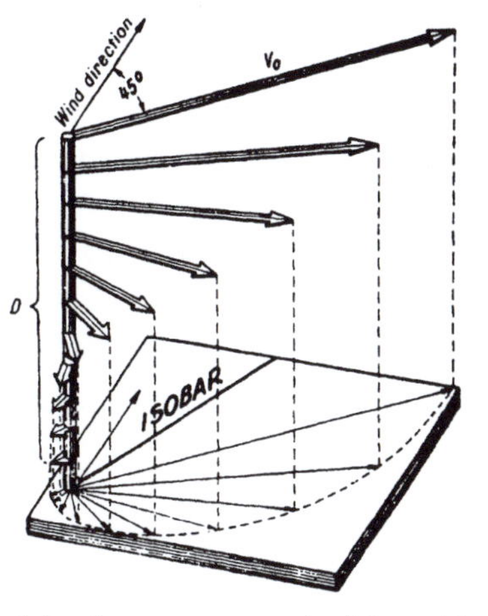

Fig. 8.4 EKMAN's spiral: schematic representation of how the surface wind-drag produces a current which is transmitted from depth to depth, the movement of each layer being less than that of the one above it and turned to the right (in the northern hemisphere) by the effect of the Earth's rotation.

$$D = \frac{7{\cdot}6W}{\sqrt{\sin\phi}}$$

where W is the wind velocity in m/sec, or

$$D = 600v_o$$

where v_o is the surface water current velocity in cm/sec. For a wind of 7 m/sec, D is 180 m at latitude 5° and 60 m at latitude 50°. For light winds and smooth sea, however, the depth of frictional resistance is given by

$$D = \frac{3{\cdot}67\ W^{\frac{3}{2}}}{\sqrt{\sin\phi}}$$

The vertical exchanges in the form of eddies within the friction layer, measured by the eddy viscosity μ_e in g cm/sec, are forced by the wind; these also therefore depend on the surface wind velocity W and are given by

$$\mu_e = 1{\cdot}02\ W^3$$

for light winds < 7 m/sec and smooth sea,

and
$$\mu_e = 4{\cdot}3\ W^2$$

for stronger winds and rough sea.

The motions which we have so far been considering in this section are the direct, or primary, effect of wind stress. There is, however, also a secondary effect of considerable importance. The deflection of the primary wind-drag current by the Earth rotation or Coriolis force produces a resultant drift of the water in the friction layer to the right (northern hemisphere sense) not only of the surface wind but somewhat to the right (i.e. high atmospheric pressure side) of the isobars of the sea-level pressure distribution. This must cause convergence of surface water in the central regions of stationary anticyclones, and tend to raise the surface or build up a head of water there (cf. pp. 314, 315). In cyclonic regions it produces horizontal divergence of the surface water, which is partly compensated by upwelling. Some downwelling probably occurs in anticyclonic regions, tending to deepen somewhat any warm surface layer there.

The absolute topography of the sea surface far from any coast cannot be surveyed. But by considering the relative topography (which should result from the observed water density distribution) above some level of (approximately) no flow in the deep ocean – i.e. above some constant pressure surface which must be very nearly horizontal – height variations of the sea surface may be deduced. By this method DEFANT (1941) has mapped a supposed mean topography of the surface of the Atlantic Ocean, finding it generally raised in the western half of the subtropical anticyclone zone, the greatest anomaly there being +40 to +50 cm all along the warm flank of the Gulf Stream from the Bahamas to near 38°N 55°W. Negative height anomalies appear all along the cold flank of the Gulf Stream–North Atlantic Drift, ranging from −50 cm near 35°N to −100 cm south of Newfoundland and in the sea between Labrador and South Greenland. Such height differences must create pressure differences which are transmitted down through the ocean and enter as a component of the horizontal pressure gradient $\partial p_z/\partial s$ (cf. p. 314) observed at any depth z within the ocean. This pressure gradient must, in fact, be the determining factor in the continuance of a pattern of actual currents which, though much weaker with depth, resemble the pattern of the surface drift down to levels far below the friction layer. Calculations of the total water transport in the main ocean currents usually proceed from the assumption, increasingly verified by observation, that it is at great depths between 1500 m and the ocean bottom that the motion induced by the surface wind stress is reduced to nothing (cancelled out) by the nearly opposite current-producing tendency of the water density distribution (SVERDRUP 1945, WELANDER 1959).

The fact that the greatest piling up of water in the subtropical anticyclone region is carried close to the western limit of the ocean is explained by STOMMEL (1958). It is due to the latitude variation of the Coriolis parameter, and is connected with the fact that the ratio

v_o/W is greater the lower the latitude: hence the mass deviation in the water associated with the easterly Trade Wind stream is more important than that associated with the prevailing westerlies farther from the equator. These features are seen in all the oceans, so that the centres of the water-current gyres are in each case displaced far to the west and the pressure gradients associated with the topography of the sea surface produce the strongest flow in a poleward-moving current close to the western boundary of each ocean.

A head of water is also built up wherever the effect of the surface wind drag impels the water against a coast or into an enclosed sea. Precise levelling by surveys along the coasts reveals this. The equatorial current driven by the easterly Trade Winds through the Caribbean into the Gulf of Mexico builds up the water level there. The average sea level in the Gulf of Mexico on the Florida coast at Cedar Keys is 19 cm higher than at St Augustine on the Atlantic coast. Sea level is 45 cm higher at the Cuban side of the Florida Strait (SVERDRUP 1945). This head of water determines the powerful 'Florida Current' through the strait, which marks the beginning of the Gulf Stream. The current transports on average 26 million m^3/sec of water through the strait, and has been known since the year 1519 when it carried a becalmed Spanish ship (the *Donna Isabella*) through into the Atlantic. Farther out in the Atlantic, where it is conventionally called the Gulf Stream, the current is maintained as a western boundary current (by the head of water built up at its warm flank in the anticyclonic region) in the manner described in the preceding paragraphs. More water is driven into the flow, until near 38°N 69°W calculations indicate that 74–93 million m^3/sec are being transported. At this point the current has already left the continental shelf. Flow in the same general direction is now known to extend almost to the bottom of the deep ocean. Once 60°W is passed, water leaves the current in eddies to the right until by 45°W (off the tail of the Newfoundland Banks) the total transport is estimated as rather under 40 million m^3/sec. East of this the surface current is a gentler drift which varies more widely with the varying winds and has several main branches.

From time to time NW'ly or N'ly storm winds blowing water into the North Sea (particularly when closely following W'ly gales over the Atlantic) raise the sea level to abnormal heights there, up to a probable maximum some 3 m above the normal tide level in the shallowest southern and eastern extremities of the sea. This is chiefly important for the special cases when the storm surge coincides with the spring tide, causing disastrous flooding of the coastal lowlands in eastern England or on the continental side.

A map computed by REID (1961) of the approximate topography of the surface of the entire Pacific Ocean indicates that the continual east-wind drift along the equatorial trough in the atmosphere, which produces the westward-moving Equatorial Current in the ocean surface, causes the surface to rise some 70 cm higher at the western end near Borneo and the Moluccas than it is at the coast of Ecuador. (The water is in fact driven uphill by this amount.) This head of water and the lateral height anomalies built up north and south of the equatorial trough drive a strong (up to 1·5 m/sec at its core) eastward-moving return current, which stretches in a narrow corridor (300–450 km broad) right across the ocean just 50–100 m below the surface at the equator, where it is known as the 'Cromwell Stream' or 'Equatorial Undercurrent', and occupies the ocean surface between about 5° and 7°N as the so-called 'Equatorial Countercurrent' (MAGAARD 1964). The total water transport of about 40 million m^3/sec makes this one of the mightiest currents in the Pacific. (The Kuro Shiwo, the North Pacific equivalent of the Gulf Stream, carries up to 70 million m^3/sec.)

Precise levelling along the American Atlantic seabord about 1927–32 (see SVERDRUP 1945, pp. 167–8) unexpectedly revealed a 35-cm rise of the sea surface northwards from Florida to Halifax, Nova Scotia that is not understood with certainty. The upward slope of the sea surface is similar to that along the Equatorial Current through the Caribbean Sea into the Gulf of Mexico and may mean that the Gulf Stream tends to pile the water up against and over the Newfoundland Banks. This head of water may help drive the southern extremity of the Labrador Current water southward beyond this point as well as to drive the eastward drift near 50°N beyond the Grand Banks.

Ocean currents and general circulation

Ocean currents, unlike winds, are conventionally known by the directions towards which they are moving.[1] This usage, so awkward in meteorological and climatological discussion, is modified as far as possible in these paragraphs.

Fig. 8.5 maps the world's main ocean currents of the present day and those convergence lines in the surface water flow which occupy quasi-permanent positions and along which strong thermal gradients are located. In a few areas there are important seasonal reversals (Indian Ocean, east Asian waters, north Australia): these are indicated by broken arrows for the water movements in the northern winter. The currents are listed in Table 8.2 below, giving characteristic figures for volume of water transported and for velocity of the surface water in the swiftest-moving part of the stream, where these are available. In the many cases where ~0·1 m/sec is entered, the available information only indicates the approximate order of magnitude. (The table and map have been compiled from a variety of sources quoted in the bibliography for this chapter.) It will be noticed that the water going into the Arctic by way of the Norwegian Sea plus the branch of the Irminger Current which usually rounds the northwest cape of Iceland, and that which moves north with the West Greenland Current, do not appear to be fully compensated by the outflow with the East Greenland Current and the Labrador Stream respectively. The balance is made up by the sinking and spreading out into the Atlantic at depths of 1000–3000 m of the denser water resulting from cooling and mixing of the warm saline inflowing water with the colder Arctic waters: this is the origin of

1. Confusion could be lessened by always preferring the adjectives 'eastward', 'northward', etc., for ocean currents to the forms 'easterly', 'northerly', etc.

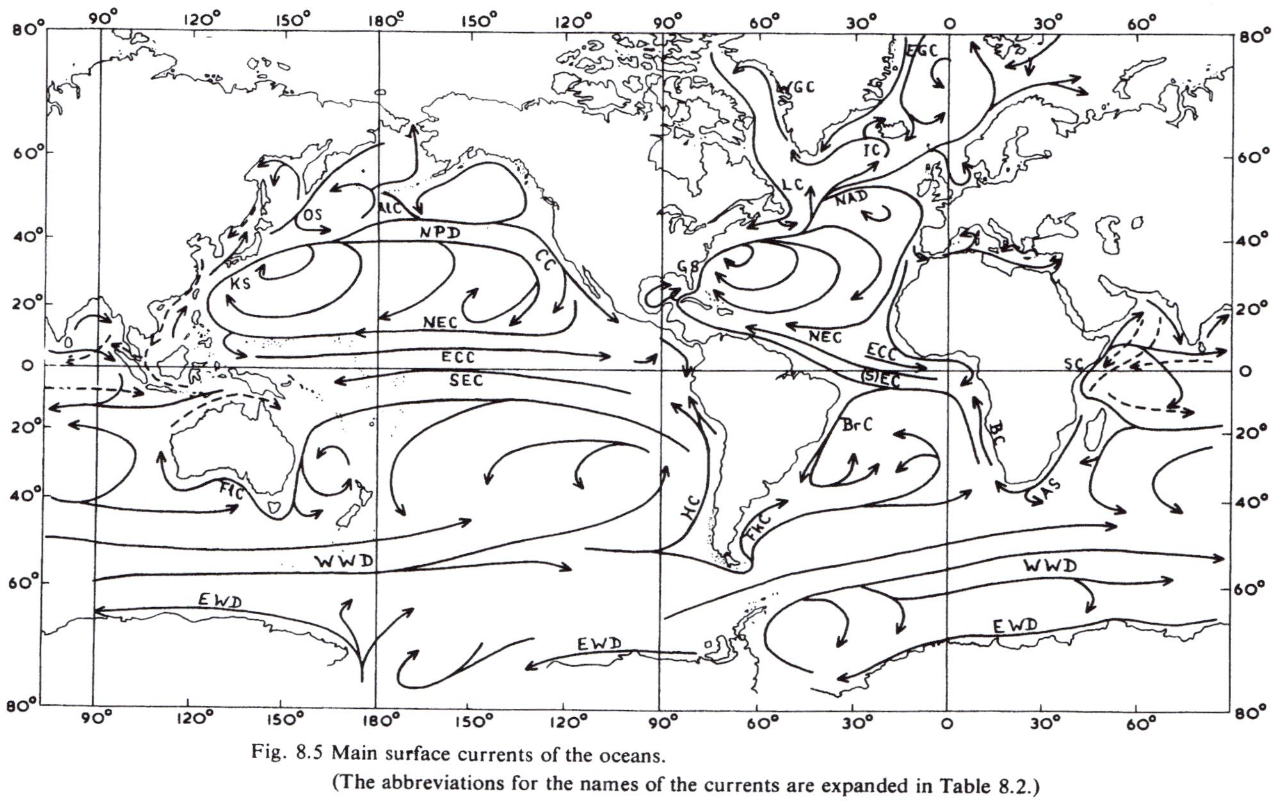

Fig. 8.5 Main surface currents of the oceans.
(The abbreviations for the names of the currents are expanded in Table 8.2.)

TABLE 8.2

List of the world's main ocean currents

Current	Designation on Map (*fig.* 8.5)	Estimated average volume transport of water 10^6 *m³/sec**	Estimated average surface water velocity in core of stream *m/sec*
Gulf Stream	GS		
Florida Strait 23–24°N		26	1·0
near 32°N		38	1·2
,, 36°N 73°W		55	1·2
,, 38°N 69°W		75–90	1·4 (max. *c.* 2·5)
,, 40°N 50°W		40	0·3
North Atlantic Drift	NAD		
main branches			
,, 45°N 30°W		14	~0·1
,, 52°N 30°W		10	~0·1
,, 60°N 0°		3	
Irminger Current near 62°N 20°W		3	
East Greenland Current	EGC		
Denmark Strait 67°N 30°W		3·5	0·1–0·4
West Greenland Current	WGC		
near 60°N 50°W		7·5	~0·1
Labrador Current	LC		
near 60°N 60°W		5·6	~0·1
Equatorial Current	EC		
Northern branch: South Atlantic water flowing along Guiana coast into the Caribbean		6	0·5–1·0
Brazil Current	BrC		
near 25°S 40°W		35	0·3–0·5
Benguela Current	BC		
near 30°S 10–15°E		35	
Agulhas Stream	AS		
near 30°S 35°E		20	0·5–1·5
Somali Current	SC		
seasonal, northern summer only, near 5°N 50°E (full development July–August)		50	~2·0
Antarctic Circumpolar Current or West Wind Drift	WWD		
broad zone 45–60°S near 60°E		160	0·1–0·2
,, 120°E		125	0·1–0·2
,, 170°E		140	0·1–0·3
,, 90°W		175	0·1–0·2
near Drake Passage ,, 70°W		125	0·1–0·5
,, ,, ,, ,, 60°W		75	0·1–0·5

TABLE 8.2—*continued*

Current	Designation on Map (*fig.* 8.5)	Estimated average volume transport of water 10^6 m^3/sec*	Estimated average surface water velocity in core of stream *m/sec*
Antarctic East Wind Drift	EWD		
mainly within 200 km of coast of Antarctica			0·3 (but up to 1·5 in 180–120°E)
Humboldt Current	HC		
near 37°S 75–90°W		10–15	
South Pacific Equatorial Current	SEC		
0–5°S		20–30	0·3–0·7
Equatorial Counter Current (Cromwell Stream)	ECC		
near 5°N		40	1·0
North Pacific Equatorial Current	NEC		
10–20°N		45	0·2
Kuro Shiwo	KS		
near 23°N		35	0·7
,, 33°N		70	1·1
North Pacific Drift	NPD		
main branches			
near 37°N 170°W		20	0·2
,, 47°N 170°W		15	0·1
Aleutian Current	AlC		
near 50°N 180°		15	0·1
Oya Shiwo	OS		
near 50°N 160°E			~0·1
California Current	CC		
near 30°N 120°W		10	0·1

* The unit one million cubic metres per second is sometimes called one *sverdrup*.

the North Atlantic Deep Water and Bottom Water (see fig. 8.6 later) characterized by salinity about 34·9‰ and temperature about 3·0°C, giving sigma values up to 27·9.

SVERDRUP (1945) gives the following figures for the water budget of the Arctic Ocean:

Inflow northwest of Scotland	3·0 × 10^6 m^3/sec
Net inflow through Bering Strait	0·3 × 10^6 ,,
Run-off from rivers	0·16 × 10^6 ,,
Excess precipitation	0·09 × 10^6 ,,
Total inflow	3·3 × 10^6 ,,
Total addition of fresh water	0·25 × 10^6 ,,
Outflow through Denmark Strait (East Greenland Current)	3·55 × 10^6 ,,

These figures are presumably estimates relating to the period 1900–40 approximately.

The layer of low salinity in which the ice forms on the surface of the Arctic Ocean is evidently constituted of a mixture of the Pacific water of salinity about 32‰ which flows in through the Bering Strait and the fresh water derived from the rivers and directly from precipitation over the Arctic. The salt rejected in the freezing process enters the water below the ice, affecting its density probably insufficiently to produce mixing with the much more saline Atlantic water beneath. Of the 160 000 m^3/sec run-off mentioned in the above table, the rivers of the Soviet Union contribute 100 000, the flow of the Yenisei alone varying seasonally between 17 000 and 20 000 m^3/sec and the Ob transporting about 12 000 m^3/sec. Should it ever become necessary to tap off any large proportion of the flow of the Siberian rivers to irrigate the arid lands in central Asia, where increasing population, cotton growing and industrialization demand more water than is currently available, such action might therefore lead to a significant increase in the salinity of the surface water in parts of the Arctic Ocean and reduce the ice cover there, since the water vapour evaporated in central Asia would presumably be carried away to the Pacific by the upper westerly winds.

It will be seen in Table 8.2 that the velocity of the fastest-moving water at the surface is not strongly correlated with the total volume of water transported by the current. The latter depends greatly on the breadth of the current and depth of the ocean. Estimates of the breadth and depth of the West Wind Drift in the Southern Ocean show this, when the figures in Table 8.3 (mainly after Treshnikov *et al.* 1966) are compared with those in Table 8.2.

The maximum velocity at the surface is commonly developed only in a rather narrow filament within any current. Particularly sharp velocity maxima commonly occur just near the limits of the western boundary currents. Farther from the boundary velocities are lower, and some water swirls away from the current in eddies of sizes ranging in diameter up to 100 or 200 km. The Gulf Stream–Labrador Current boundary itself becomes contorted into meanders up to 100–300 km across, and great 'globules' (or pools) of the warm or cold water of this size are liable to be cut off and surrounded by the other watermass.

The general speed of the Gulf Stream and its continuation in the North Atlantic Drift is such that the surface water typically crosses the ocean from near 40°N 60°W to the approaches to the British Isles in 3–6 months. If, therefore, an anomaly, cold or warm, of ocean surface temperature persists in the general region of the Newfoundland Banks, this must be due either to persistence of a lateral shift in the position of the current boundary, or to a change in the radiation budget and heat exchange at the water surface in the area, or else to a change in the heat flux from greater depths in the sea: in other words the anomaly must continually be created in the region concerned, for the water is continually flowing through the pattern. Some water temperature anomalies are, however, seen to spread across the ocean with the winds, as when cool water spreads eastwards across the Atlantic in the fifties (north latitude) with the summer westerlies after a severe ice season in Labrador–Newfoundland waters.

TABLE 8.3
Dimensions of the West Wind Drift in the Southern Ocean

Longitude	*Breadth, km*	*Max. depth of the eastward drift, m*	*Mean latitude of the core of the drift*
60°E	1500	2500	46·2°S
90°E	2650	1400	51·8°S
120°E	2250	2000	48·6°S
170°E	1450	2750	53·0°S
90°W	1250	3200	58·5°S
60°W	570	2400	57·5°S
0°	1900	2000	48·0°S

TRESHNIKOV *et al.* (1966), proceeding from the figures quoted in Table 8.2 for the great West Wind Drift which occupies the breadth of the Southern Ocean surface between about 45° and 60°S and induces an eastward drift in latitudes near 60°S as deep as the ocean bed, have suggested that, since more water appears to be carried eastward from the Atlantic to the Pacific than passes through Drake Passage, there should be a net meridional circulation through the Bering Strait and the Arctic into the Atlantic.

The net water movement through Bering Strait at the present time is believed to be in this sense, but it must be very small and of the order of magnitude already quoted (p. 323) from SVERDRUP. This can be no more than a small residual entering into the water budget of gain and loss from all causes in the various oceans. It could not be calculated reliably from study of the much more massive flows in distant oceans, and may well have been reversed in the climatic regimes of some other epochs. Hence, the effective compensation of the apparent loss from the South Atlantic to the West Wind Drift must take place within the Southern Ocean itself, and this is shown in the movement (see maps by WÜST 1957) of the deepest layers between 40° and 60°S from the southeast across the meridian of 20°E.

Fig. 8.6 has been constructed to show the general drift of the circulation in depth in a meridional plane about mid Atlantic, from Greenland to the Weddell Sea, from data given by WÜST and SVERDRUP (*loc. cit.*). The mean salinity isopleths show the concentration of the greatest salinities in a shallow layer near the surface in the realm of the subtropical anticyclones, where evaporation is high. Rather lower salinity appears in the zone of equatorial rains, also in the subpolar rainbelts. Much lower salinities are seen in the polar waters produced by melting ice, although greater salinity is transferred to the water layers immediately underneath where ice is forming on the sea.

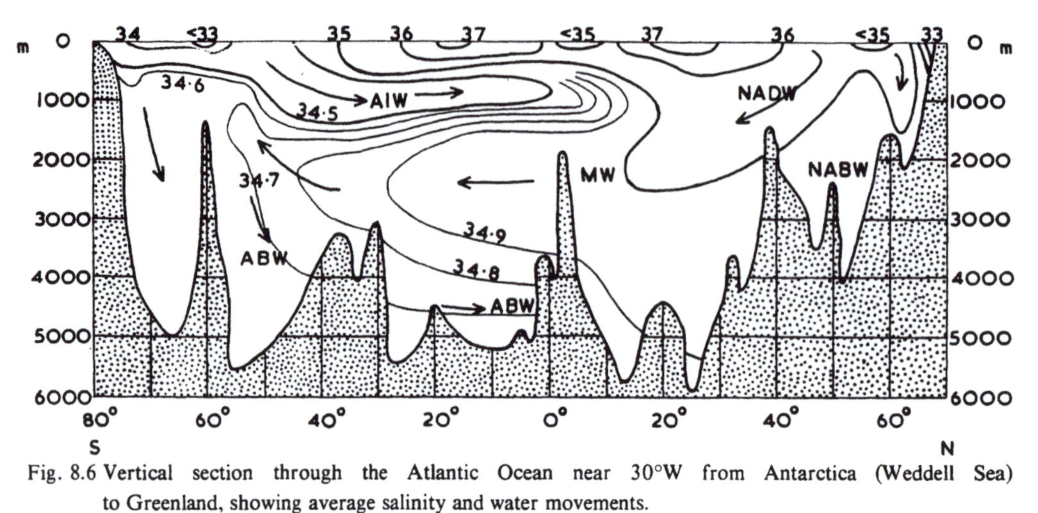

Fig. 8.6 Vertical section through the Atlantic Ocean near 30°W from Antarctica (Weddell Sea) to Greenland, showing average salinity and water movements.

The letters on the diagram identify the following watermasses:

(1) **NADW** – *North Atlantic Deep Water*, and
(2) **NABW** – *North Atlantic Bottom Water*

These are both produced by sinking in the sea areas between Labrador and Greenland and near Iceland (see pp. 320–2) and are distinguished from each other by the greater density of the **NABW** (sigma 27·8–27·9), which is only produced in the coldest seasons, in comparison with that of the **NADW** (sigma broadly about 27).

(3) **MW** – *Mediterranean Water*

The deeper water emerges from the Mediterranean, where it has temperature about 13° and salinity 38·4–38·7, to mix and spread among the Atlantic waters at intermediate depths, the temperature falling to about 4° in the process and the salinity to 34·8 (sigma values about 27·7).

(4) **AIW** – *Antarctic Intermediate Water*

This is the water that sinks at the Antarctic Convergence, the more southerly of the two convergence lines in the Southern Ocean shown in fig. 8.5. The initial surface temperature of the water at the Convergence is about 2°C and the salinity 33·8. This gives sigma values about 27·0–27·5 as the watermass spreads north, with some mixing, readily identifiable by a salinity minimum at about 1000 m depth. The temperature rises to about 4·5°C as the equator is reached.

(5) **ABW** – *Antarctic Bottom Water*

This is the densest of all the watermasses and ultimately supplies the water in the abyssal depths of all the oceans of the world except the Atlantic north of about 30°N and the Arctic. It is formed only in the Weddell Sea, through the freezing of the moderately salt water on the continental shelf adding salt to the water just below the ice. This water acquires salinity about 34·6, with temperature −1·5° or rather below (sigma value about 27·9).

The routes by which the Antarctic Bottom Water spreads from the Weddell Sea into the Atlantic and eastward around the Southern Ocean into the Indian Ocean and the Pacific can be traced by the isopleths of temperature of the bottom water, which rises slowly with increasing distance from source to reach about 0° near 60°S in the Indian Ocean sector and +0·9° at the same latitude in the Pacific. It is about +2·3° by the equator in the Atlantic and + 1·7° near the equator in the other oceans.

The rate at which these circulations in the vertical and meridional planes proceed, and the progress of the Antarctic Bottom Water around the Earth, is indicated by ^{14}C age determinations (i.e. radioactivity measurements) on the carbon in the bicarbonates dissolved in the water (BIEN *et al.* 1963). The enrichment of the radioactive carbon in the atmosphere by the nuclear bomb tests around 1960 made it possible to test the rate of mixing downward in the surface layers: it was found that even the uppermost 200 m of the oceans above the thermocline, despite wave action, require several years to become thoroughly mixed. At all depths below the thermocline, ^{14}C measurements have indicated great lengths of time elapsed since the water was in communication with the atmosphere. This must mean that owing to the existence of a warm layer above a thermocline over most of the oceans of the world, and a stable stratification near the surface wherever there is fresh water from ice-melt or from rivers, the rate of exchange of the deeper waters with the surface layers is slow except in a few favourable areas and situations. Typical ages for the Atlantic Deep Water appear to be 500–800 years and at the corresponding levels (700–2000 m) in the Pacific 1000–2000 years. In the Antarctic, south of about 50°S, even the surface water gives apparent ages of over 1000, and sometimes in the Far South over 2000, years. This must be chiefly due to the water making its way to the surface there from greater depths, e.g. as shown in fig. 8.6, but also due to some admixture of water that has been locked up in the Antarctic ice cap. The Antarctic Bottom Water below 3000 m depth near 60°S gives its youngest apparent ages – in the order of 1700 years – near the Weddell Sea, and the ages become consistently greater with increasing distance eastward and northward from there. It appears to take this water 500 years to reach 14°N in the Eastern Pacific and about half this length to reach the equator in the Indian Ocean. The bottom current speeds suggested are in the region of 0·05 cm/sec in the Pacific and 0·03 cm/sec in the Indian Ocean.

Heat transport by the ocean circulation

The quantity of heat transported by ocean currents varies with the total volume of water moving, particularly in the top 100–200 m, rather than with the speed of the swiftest part of the current at the surface. SVERDRUP *et al.* (1942) estimate that the total heat transported by the North Atlantic Ocean northwards across the 55°N parallel averages $0{\cdot}3 \times 10^{16}$ g cal/min, and MODEL (1950) gives about $0{\cdot}1 \times 10^{16}$ g cal/min for the amount entering the Norwegian Sea. The figure given for 55°N is about 10% of the total heat transported polewards

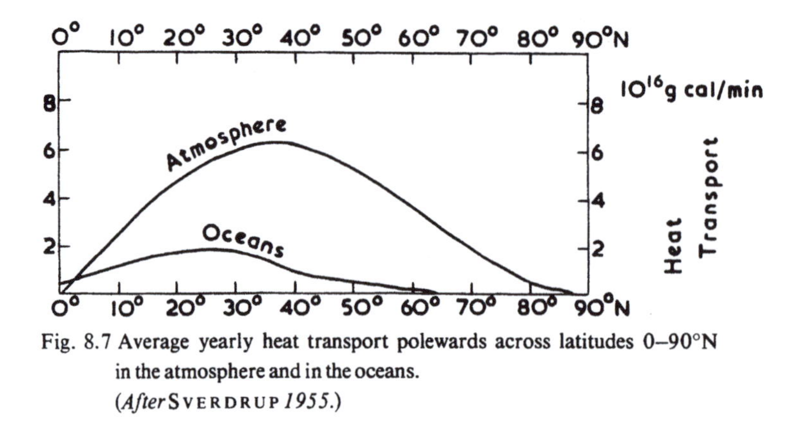

Fig. 8.7 Average yearly heat transport polewards across latitudes 0–90°N in the atmosphere and in the oceans. (*After* SVERDRUP *1955*.)

across that parallel by the oceans and atmosphere together. The comparison between the oceans' and atmospheric heat transport for other latitudes in the northern hemisphere can be seen in fig. 8.7. The combined heat transport is greatest at about latitude 35°, the latitude which divides the zones that have an overall radiation surplus from those with a radiation deficit. The greatest contribution from the ocean currents is, however, between about 20° and 32°N, where both the Gulf Stream and the Kuro Shiwo are directed more nearly northward. At these latitudes the oceans account for about a quarter of the total heat transport. This may be true at latitudes about 20–35°S also; but farther south the poleward component of the ocean currents is slight, and the ocean's heat transport is likely to be a minute proportion of the whole.

Heat exchanges between ocean and atmosphere

In most parts of the world heat passes from the sea to the air most of the time, the temperature of the sea surface being found to be, on an overall average, about 0·8°C warmer than that of the air at a normal ship's deck-height of about 8 m above sea level. Energy is carried away from the sea surface into the atmosphere partly as feelable ('sensible') heat, but even more in the latent heat of the moisture evaporated. In these processes the temperature of the air registers much bigger changes than the sea, since the same amount of heat would raise the temperature of over three thousand times the volume of air as of sea water of salinity 35‰ by one degree at S.T.P. (the specific heats near 0°C being 0·97 and 0·00031 respectively).

The rate of supply of sensible and latent heat to the overlying air depends on the temperature of the water surface and the temperature difference sea–air, also on how nearly saturated with water vapour the air is already.

(1) The amount of feelable heat given off per unit area of the sea surface in unit time is

$$Q_f = c_p \mu_e \left(\frac{dT_a}{dz} - \alpha \right)$$

where c_p is the specific heat of the air at constant pressure, μ_e is the coefficient of eddy conductivity applicable to the turbulent motion of the lowest air and $(dT_a/dz - \alpha)$ expresses the amount by which the actual temperature gradient with height dT_a/dz exceeds the adiabatic lapse rate α. In practice, α is negligibly small by comparison with dT_a/dz in the lowest few metres of air and is therefore omitted from any calculations. Also the finite difference between the temperature of the sea surface T_s and the air temperature T_a measured at a height z on (or from) a ship has to be used: so the expression becomes

$$Q_f = c_p \mu_e \frac{1}{z}(T_s - T_a)$$

(The dimensions are $ML^{-2}T^{-1}$ and are usually calculated in c.g.s. units.)

(2) The heat abstracted in evaporation per unit area of the sea surface in unit time is

$$Q_e = L \cdot \mu_e \cdot \frac{0{\cdot}621}{p} \cdot \frac{de}{dz}$$

where L is the latent heat of evaporation, p the atmospheric pressure and de/dz the vertical gradient of water vapour pressure in the lowest air. In practice, this has to be similarly simplified to

$$Q_e = L \cdot \mu_e \cdot \frac{1}{z}(e_s - e_a)$$

where e_s is the saturation vapour pressure at the temperature of the sea surface and e_a is the actual vapour pressure measured at height z on (or from) a ship.

The saturation vapour pressure over sea water of 35‰ salinity is about 2% less than over a surface of pure water. The figures in Table 8.4 show how greatly this source of energy increases with the water temperature.

TABLE 8.4
Saturation vapour pressure in air over water of 35‰ salinity

Temp., °C	−2	0	5	10	15	20	25	30
Vapour pressure, mb	5·19	5·99	8·56	12·05	16·74	22·96	31·12	41·68

The air immediately in contact with the water approximates closely to the temperature of the water surface. Hence, the drier the air at ship's deck height (and above) – low values of e_a – the greater the rate of moisture supply to the air and the greater the energy used in evaporation.

The total energy passing from sea to air is

$$Q = Q_f + Q_e$$

The ratio Q_f/Q_e, which is often known as BOWEN's ratio, varies from about +0·5 for cold air over warm sea in winter to −0·2 in some extreme cases of warm air over cold sea in summer. Its overall average value is about +0·1, i.e. only about 10% of the energy supplied by the sea directly heats the air, whereas 90% is used up in evaporation and realized as feelable heat elsewhere, when the moisture condenses.

BOWEN's ratio is found to have mean values that are characteristic for various climates and seasons. It may therefore be used, with caution, to calculate Q_e from Q_f, i.e. from the temperatures.

Notice that where a cold sea cools the air, and where warm air passes heat to the sea, a stable stratification of density is induced in both air and sea, inhibiting convection and tending to limit the temperature changes and heat exchange to layers close to the sea surface. (The same is largely true in the sea in weather where the warming of the sea surface is primarily due to insolation, particularly in shallow turbid waters where the extinction coefficients for solar radiation are highest.)

In the contrary case, where the sea is warmer than the air, the heat passing into the surface air makes for an unstable density stratification and convection to great heights, while the heat loss from the sea surface (once any thermocline is eliminated) induces an unstable stratification and deep convection in the sea. The heat exchange thus proceeds rapidly and through deep layers of both sea and atmosphere. The temperature changes near the surface may, however, be much slower than with stable layering.

The most general climatic consequences seem to be:

(1) Temperature changes just close to the ocean surface are likely to be most rapid when a thermocline is present, with clear skies and strong radiation flux, with large values of air–sea temperature difference and strong winds in the warm air.

(2) At greater heights and depths temperature changes depend upon convection and are likely to be most rapid when the sea is heating the air and the air is cooling the sea, especially when the air–sea temperature difference is great and with strong winds.

(3) The water will be best mixed with depth in areas, seasons and climatic epochs when the sea surface is being cooled.

(4) Upwelling of cold water should be important chiefly in restricting the area over which the ocean heats the atmosphere (rather than for the quantity of heat abstracted from the atmosphere, which is likely to be small and confined to a shallow surface layer).

Observed variations of the ocean currents

It was FRIDTJOF NANSEN, drifting in the *Fram*, beset in the ice, across the north polar sea from 1893 to 1896, who first reported that the current varied with the wind, though deviating through the effect of the Earth's rotation up to 50° to the right of the wind. Though the general drift was from the East Siberian Sea towards the outlet between Spitsbergen and Greenland, the cyclonic situations and westerly surface winds common in high summer and early autumn reversed the drift while they blew. More recent observations of the drift of the Russian camps on the floating pack-ice, mainly in the 1950s and since, and of the ice island T-3 between 1952 and 1954, have shown that the drift is generally in line with the gradient wind (i.e. with the isobar direction) at any given time or deviates on average by no more than a few degrees clockwise from this direction. The considerable scatter observed in this relationship can be put down to the convergence and divergence of the wind systems and currents, and the changes of the wind with time, over the almost land-locked Arctic Ocean. The ratio of the speed of the current to that of the gradient wind is about $\frac{1}{100}$ (see BROWNE and CRARY 1959).

The composite map of drifts across the Arctic Ocean observed at various times between 1879 and 1957 (fig. 7.4, p. 262) reveals the following.

(i) The anticyclonic (clockwise) gyre centred in the remotest part of the ice from the Atlantic, north of Canada and Alaska.

(ii) The now numerous drifts recorded across the Arctic on the Siberian side of this gyre.

(iii) There is great variation, however, in the speed of these drifts, much the fastest being that of the first Russian drifting ice-floe camp NP1 from near the pole to 71°N off east Greenland in 9 months in 1937–8. The drifts of the *Fram* in 1893–6 and the *Sedov* in 1938–40, across the same area near the pole, made only about a quarter of the progress of NP1.

(iv) Apart from the prominent week by week, and month by month, changes of direction and speed, there is a longer-term change of behaviour by which the drifts across the pole in the 1950s tend to turn clockwise into a further circuit of the polar basin rather than come out between Spitsbergen and northeast Greenland. This change can, doubtless, be associated with the increased frequency of anticyclones over the polar region in the 1950s compared with earlier decades.

On the fringe of the Arctic, particularly in the Greenland–Iceland–Spitsbergen sector, there have been great changes in the extent of the polar sea ice in the twentieth century. These have already been discussed in Chapter 7, and illustrated in fig. 7.3. In the long recorded history of the ice affecting the coasts of Iceland, as seen in fig. 8.8, there have been other rapid increases of the ice equalling or exceeding that in the 1960s, probably in all cases associated with periods of high frequency of N'ly winds over the Norwegian Sea. The case in the 1960s has been shown by MALMBERG and STEFÁNSSON (*private communication* 1969) to be associated with increased volume of flow, and further southward penetration than for many years previously, of the polar water of low temperature and low salinity in the east Iceland branch of the East Greenland Current. This current had been strongly marked at times in the nineteenth century, but had apparently weakened – the east Iceland branch failing – in the warm decades between 1920 and 1950. It returned in the 1960s. In 1965 the average flow of the East Greenland Current at 75–72°N, where it runs strongest (especially when a simple cyclonic gyre forms in the Norwegian Sea), was

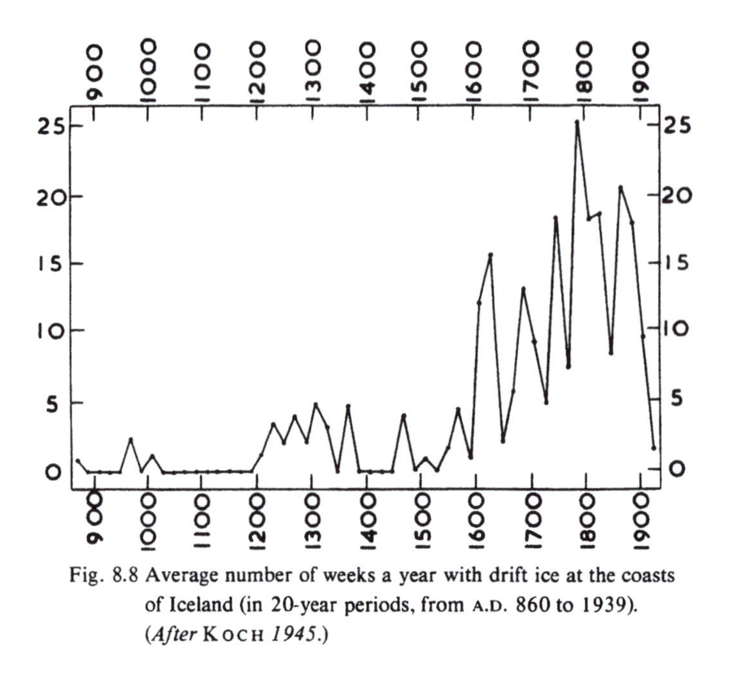

Fig. 8.8 Average number of weeks a year with drift ice at the coasts of Iceland (in 20-year periods, from A.D. 860 to 1939). (*After* KOCH *1945.*)

35 × 10^6 m^3/sec and in some months equalled that of the Gulf Stream at its strongest (AAGAARD 1969). Though most of the intrusion of this water in the area off northeast Iceland in the 1960s was in the top 30–70 m, all depths down to 200 m were affected. At the same time, there has been evidence of a somewhat enhanced, compensating transport of warm water northwards along the Norwegian coast right into the Barents Sea. The same pattern may perhaps be traceable in the very first stages of the climatic change in Europe that ended the early medieval warm epoch: the Norwegian historian, Professor A. HOLMSEN (*personal communication*, 2 October 1964) attributes the permanence of the cessation of farming after the Black Death in Norway at the heights occupied between A.D. 800 and 1300 to a climatic cooling that had been under way from 1250 onwards, continued more or less unbroken to 1500 and was resumed around 1550–1600; an increase of ice was noticeable at the Iceland coast (cf. fig. 8.8) from as early as 1200, yet the cod fishery off north Norway was particularly abundant between 1300 and 1500, indicating a tongue of warm water reaching there.

These current variations have been established by observations of temperature and salinity, or in some cases calculated from the wind stress, not by direct measurements of water movement. There is, however, one testimony to the latter in the arrival in storms on the shores of Shetland and the northern Hebrides in 1966, and after, of quantities of driftwood never experienced in the previous 70 years or more: the tree species represented have been found to be of Siberian, not North American, types. That the wood had not spent long in the warm water of Gulf Stream origin is shown by the attacks of the cold-water marine borer organism, *Psiloteredo megotara* (Hanley), and the absence of any by warm-water organisms.

Variations of the East Greenland Current may be brought about in different ways. Thus, the swift flow of the surface layers in 1938 (which brought a great 'outburst' of ice to Iceland in the midst of the almost ice-free years) was clearing large areas of the polar basin of ice; it seems to have been accompanied by a quickened turn-round of the warm, saline Atlantic water right round the Arctic basin. By contrast, the extra supply of cold, polar water of low salinity reaching Iceland in the 1960s (also bringing ice, though some of this may have formed not far from Iceland) seems to have been the product of the run of cold years in the Arctic which set in in 1960–1. Perhaps partly through an increased flow of fresh water from the Siberian rivers, on which ice forms easily, but also through atmospheric circulation patterns favouring heat-loss from the Arctic and less advection from other latitudes, a situation seems to have been set up which greatly increased the output of polar cold water by comparison with the preceding decades and doubtless contributed to the cooling observed in the Atlantic itself. The 1960s also saw a great increase in the frequency of surface N'ly winds over the East Greenland Sea.

The surface currents in the ocean respond even to week-by-week and day-by-day changes of wind. This was well seen in the quick change in the North Sea once the east and northeast winds set in in the severe winter of 1963. The transport of warm North Atlantic Drift water through the straits of Dover failed and within 2 months the surface water temperature fell 6°C below normal there. The mean water temperature for the month of February was −1·2°C in the open North Sea 20 km NW of Borkum; the 0°C isotherm lay 60 km off the Dutch coast and up to 20 km outside the Wash and the estuaries of eastern England (ELLETT 1963).

Over the years 1945–65 there were five oscillations of sea water temperature measured in the top 200-m depth off Kola in the Barents Sea and also of salinity in the German Bight area of the North Sea, the ranges being about 1°C and 1‰ respectively. Both these variables had their maxima and minima in phase with the 9-month totals of S'ly component of the gradient wind over the British Isles (R. R. DICKSON, Fisheries Laboratory, Lowestoft, *personal communication*).

The Gulf Stream–North Atlantic Drift system is no less subject to variations. It is clear that these variations depend on the winds and weather prevailing, though they also react upon the weather by changing the pattern of heating of the atmosphere and, hence, the steering of depressions and anticyclones (see Chapter 10).

The sea surface temperature anomalies of the order of ±2°C, and more, which commonly persist for months at a time in the region of the Gulf Stream–Labrador Current boundary south of Newfoundland probably mark lateral displacements of that boundary over a range of 200–300 km. Deviations from its usual course of the much weaker drift in the central and eastern Atlantic may be much greater than this.

Figs. 8.9 (*a*) and (*b*) show that such anomalies can persist for at least 40 years. These

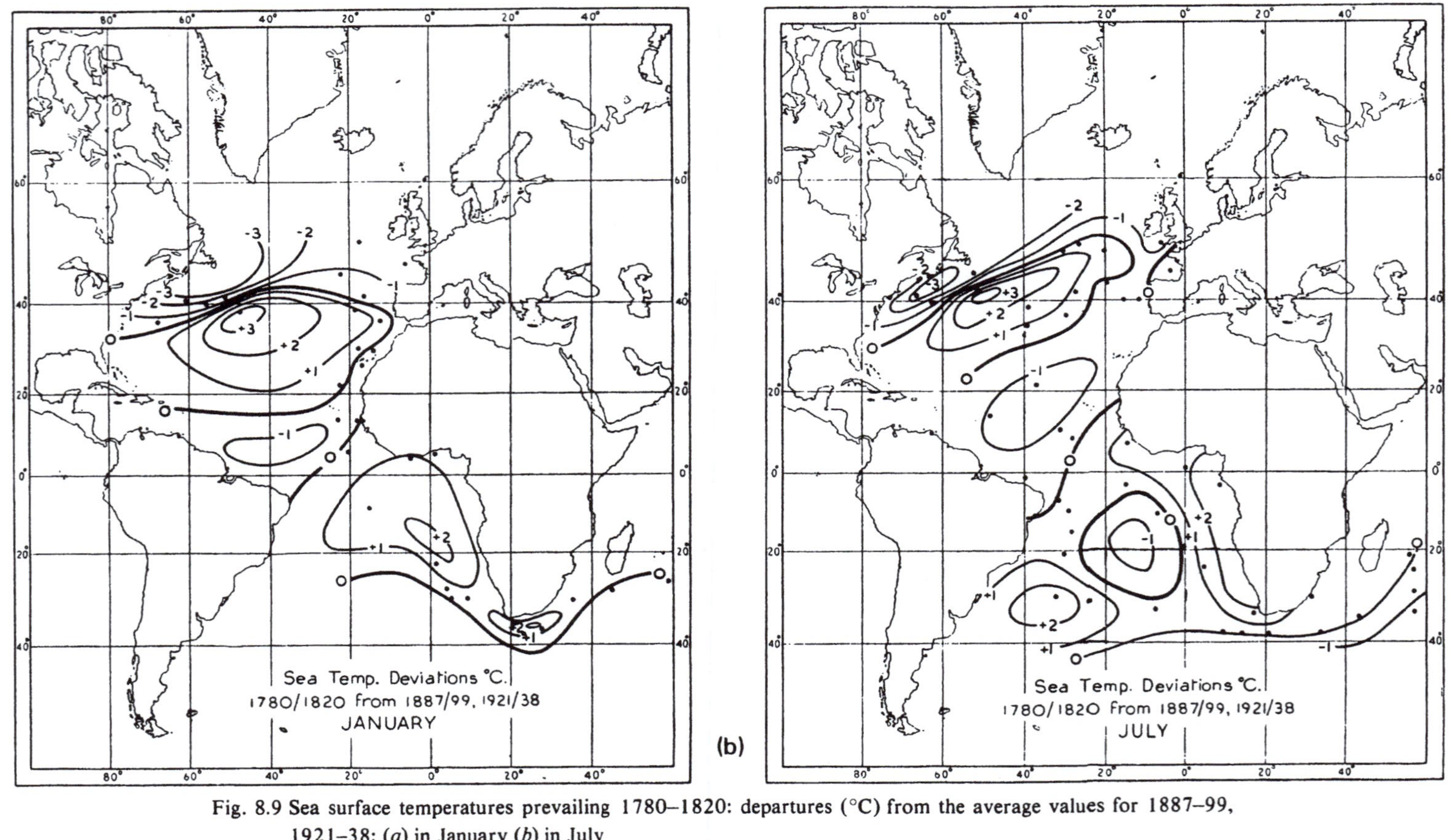

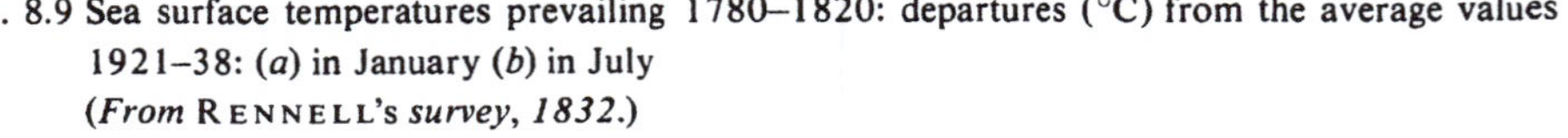

Fig. 8.9 Sea surface temperatures prevailing 1780–1820: departures (°C) from the average values for 1887–99, 1921–38: (*a*) in January (*b*) in July
(*From* Rennell's *survey, 1832.*)

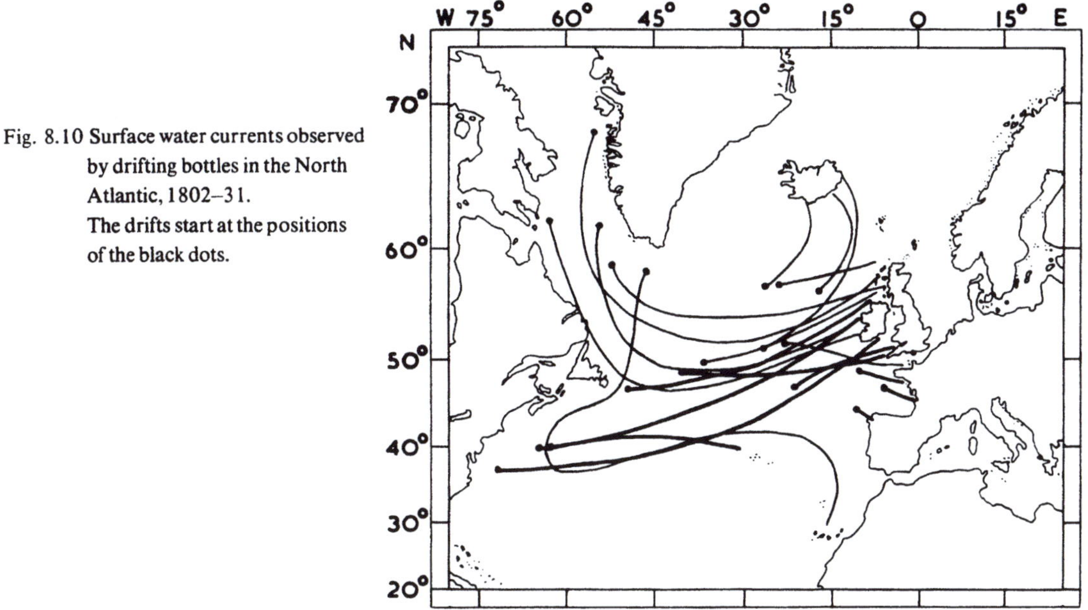

Fig. 8.10 Surface water currents observed by drifting bottles in the North Atlantic, 1802–31.
The drifts start at the positions of the black dots.

maps of how the sea surface temperatures prevailing in the Atlantic in the Januarys and Julys of the period 1780–1820 departed from the averages about the early part of the twentieth century are derived from the first survey carried out for the British Admiralty by Rennell (1832). Comparison of these maps with fig. 10.12, which gives the Atlantic sea temperature anomalies over the years 1780–1848 indicated by American ships' observations, supports belief in the reliability of these early data. The maps appear to indicate a marked southern displacement, and possibly strengthening on an eastward course more towards the Azores, of the Gulf Stream–North Atlantic Drift by comparison with the early twentieth-century years represented in fig. 8.5. The same impression is gained when Rennell's drift-bottle drifts of the years 1802–31 (fig. 8.10) are compared with fig. 8.5. Sabine in 1846 called for investigation of the 'cause of the remarkably mild winters which occasionally occur in England', such as the one experienced in 1845–6, and alluded to observations in November 1776 and in the winter of 1821–2 that 'the warm water of the Gulf Stream spread itself beyond its usual bounds . . . to the coast of Europe, instead of terminating as it usually does about the meridian of the Azores'.

Overall the North Atlantic showed a cooling trend after about 1950 till at least 1968, though the subtropical latitudes (especially between 20° and 30°N) became slightly warmer than before and there were positive anomalies along the Norwegian coast, in the Irminger Current and up the west coast of Greenland. The continental shelf waters off Nova Scotia and Long Island were cooling right to the bottom. These changes were undoubtedly associated with the south, and sometimes southwest, displacement of the 'Iceland depression' into the fifties north latitude (LEE, RODEWALD and others 1967). As a result, after 1964 the prevailing sea temperatures came to resemble those of 1780–1850.

The importance of such variations to the heating of the atmosphere over the ocean off northwest Europe is shown by figures found by BOYUM (1966) within quite a short run of years. The 1948–58 average total heat ($Q_f + Q_e$) given off by the ocean to the atmosphere at 66°N 2°E was 432 g cal/cm²/day in January and 105 in July, but the average was as high as 643 g cal/cm²/day for January 1955 and under 30 in some summer months.

It has been suggested that the boundary with which the strongest gradient of sea surface temperature in the Southern Ocean is associated, the Antarctic Convergence, maintains a constant position from year to year and from era to era. Such behaviour would not be surprising in the Drake Passage, where the topographical control must be strong; though the proposition has been applied much more widely. It seems, however, not to be in accord with observed week-to-week changes of position, strength (and sometimes development of double or triple structure) of this thermal boundary, nor with the fact that in a sequence of remarkably cold years between 1895 and 1907 the Antarctic ice was sometimes no more than 100 miles (160 km) south of Cape Horn. DEACON (1945) stated that the position of the Antarctic Convergence seems to be determined chiefly by the latitude in which the warmer Atlantic Deep Water current climbs steeply above the Antarctic Bottom Water (cf. fig. 8.6). He also noted that the biological species represented in the ocean bed sediments mark the boundary, a diatom ooze typical of the Antarctic zone extending to just north of the present Antarctic Convergence and beyond this limit globerigina ooze typical of warmer waters; but sediment cores taken from the sediment just north of the limit, particularly near the Crozet Islands (46°S 50–53°E) show diatom ooze underlying the globerigina ooze. This is thought to be fossil evidence of a somewhat more northern position of the Antarctic Convergence at some stage during the ice age and a greater spread of the Antarctic Bottom Water at that time.

Climatically sensitive points

There seem to be a number of points at which small changes in the ocean, or in the ocean–atmosphere relationship, could set up changes in the world climatic regime that would have a marked persistence or even a self-amplifying tendency.

(1) *The branching of the Atlantic Equatorial Current at the 'nose' of Brazil*

As seen in fig. 8.5, the Equatorial Current in the South Atlantic divides into two streams at the nose of Brazil between 5° and 8°S. Any shift, north or south, of the axis of this current must affect the amount of southern hemisphere water, and the heat transport from the warmest latitudes of the South Atlantic, that flows with the northern branch into the Caribbean to feed the Gulf Stream (BROOKS 1926).

This is probably an essential element in the guiding effect of the present geography, with the North Atlantic providing the one wide open channel for currents in and out of the Arctic

Ocean, whereby climatic shifts north and south attain a larger amplitude in and about the Atlantic sector than in any other sector. This seems to apply alike to the shifts of the climatic zones and prevailing wind belts between ice ages and warm interglacial periods and to the smaller shifts that distinguish the climatic differences between different millennia, centuries or decades.

Though it cannot yet be substantiated by ocean current measurements, altogether new, indirect evidence of its working has been found since BROOKS' time in climatic history studies. The most striking aspect is the extent to which, despite broadly parallel temperature trends in both hemispheres, the wind belts, water current boundaries and pack-ice limits seem to have moved north or south in parallel north and south of the equator. Fig. 8.11 illustrates the point, with the ice on the Weddell Sea increasing from 1910 to a maximum in the early 1930s, as world temperature rose (about equally in the averages for both hemispheres) and as the northern polar ice retreated (cf. fig. 8.8). All these trends have since reversed. These changes are fairly closely paralleled by the 10-year mean latitude of the intertropical trough in the atmospheric pressure field over the Atlantic sector in January and July (LAMB and JOHNSON 1966). This trough lay on average 2–4° latitude farther north in the 1930s–early 1940s than it did in the middle of the last century or in the 1960s. The gradual displacement north or south of this feature of the atmospheric circulation, which presumably exercises the most direct control upon the course of the Equatorial Current,

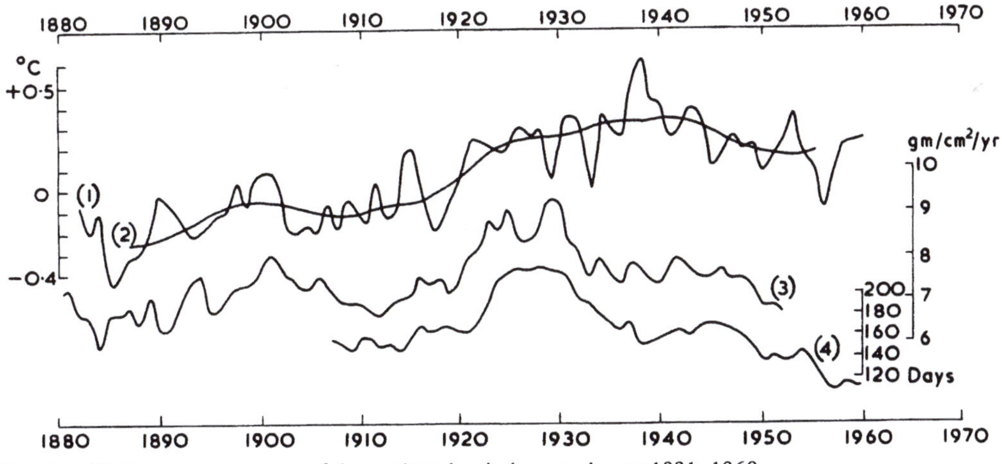

Fig. 8.11 (1) Average temperature of the northern hemisphere, each year 1881–1960.
(2) Ten-year running means of (1).
(3) Yearly accumulation of snow at the South Pole, 10-year averages.
(4) Number of days per year with the Weddell Sea ice closing Scotia Bay, South Orkneys; 10-year averages.
(*After* FLETCHER *1969*.)

seems to have presented long-continued trends despite variations of the subtropical anticyclones which are sometimes clearly independent of it, e.g. the equatorward tendency of both the North and South Atlantic anticyclones in the 1960s (in some months of the year traceable from about 1940 onwards) and their opposite tendency between about 1850 and 1900.

Fig. 7.22 has already indicated how the long history of the Arctic sea ice at Iceland and the rainfall from the South Pacific westerlies at Santiago, Chile from A.D. 1535 points to both features moving at first generally south, and then from 1780 to the early 1900s generally north, in parallel. Moreover, the southernmost positions reached by the early Antarctic exploration ships (in most sectors) from Captain COOK around 1770 until the 1830s were 1–2° south of the normal latitude of the Antarctic ice limit at the end of the melting season in the period 1900–50, though the Arctic ice in the Greenland–Iceland sector spread farther south between the 1770s and 1820s than at any other time except around 1695 (see map, fig. 8.12).

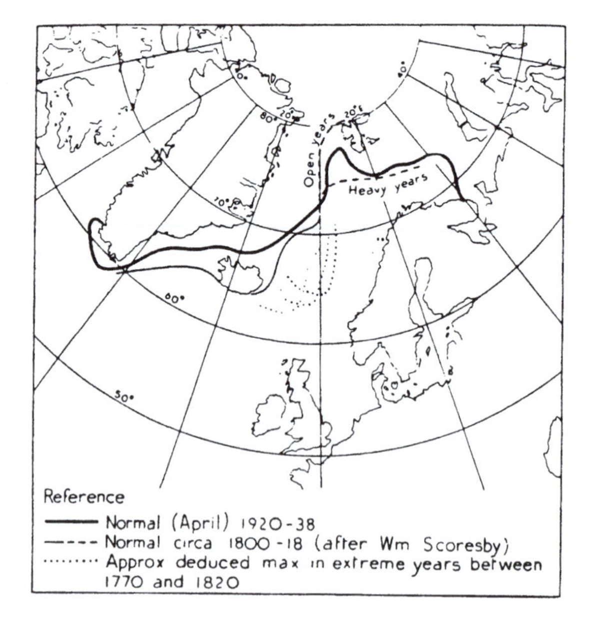

Fig. 8.12 Extent of the Arctic pack-ice at the spring maximum in various years.

Over the period from 1750–1800 to the 1960s, covered by LAMB and JOHNSON's (*loc. cit.*) chart reconstructions, the latitude (and vigour) of the Iceland low (i.e. of the North Atlantic subpolar cyclonic activity), which must be more closely affected by variations of radiation input and the flow of the upper westerlies over neighbouring continents than in the case of any other ocean, looks like the prime predictor. The retreat of the Arctic pack-ice from 1800 to the 1950s, and its later readvance, and even the trend of latitude of the intertropical trough in the equatorial Atlantic, seem to follow the shifts of the Iceland low, with lags of 15–30 years.

(2) *Ice cover, or lack of it, on the polar seas*

We have already explained in Chapter 7 how Brooks worked out the extra cooling of the atmosphere introduced by the presence of a floating ice cover on the polar ocean. The pack-ice thereby tends to maintain itself. It also increases the thermal gradient in the atmosphere across a zone near the ice limit, and this must intensify the large-scale atmospheric circulation over latitudes near that zone. At the same time the thermal conductivity of sea ice is low, owing to the air trapped in it, so that the warmer ocean beneath is to some extent shielded from the heat loss which the low temperature at the surface would otherwise induce. Despite wide fluctuations in the spread of the ice towards lower latitudes, particularly as between its ice-age maximum and the warmest millennia since, analysis of the ocean bed deposits in the central Arctic (Hunkins and Kutschale 1965, Ku and Broecker 1965) indicates that the ice cover of the northern polar sea has had a continuous existence over at least the last 70 000 years, and it appears stable today.

If the Arctic pack-ice were to be broken up (e.g. by storms or by human action) without any change in the present solar radiation regime, or in the composition of the watermasses present, the saline Atlantic water which has nearly constant density with depth would readily develop wind-stirring and convection to considerable depth. Evaporation, associated more with the effect of windiness on an open sea than with heating, would increase the salinity of this water near the surface. At temperatures near 0°C a change of salinity by only 0·035 parts per thousand affects the density as much as a temperature change of one degree. Due to the convective mixing any temperature changes in this Atlantic water within the Arctic would be extremely slow. Moreover, the fresh water supplied by the Siberian rivers would still tend to spread over the surface and to freeze in winter. From a careful consideration of all the variables, and of the radiation and heat fluxes available within the ocean and atmosphere, Doronin (1968) concluded that the pack-ice cover of the Arctic Ocean would form again within a few years.

In so far as the changes in the currents and the increase in the extent of ice since 1955, and especially in the 1960s, are a response of the Arctic to a reduction in the radiation budget – a decline by some 4% in the average intensity of the solar beam as measured at the surface of the Earth between 1945 and 1962 is reported (Pivarova 1968) – this indicates, as did the decrease of the ice between about 1920 and 1945, that the polar sea ice is in a state of fairly sensitive and immediate balance with the radiation available.

The history (Lamb 1967) of several occasions in the nineteenth century when enormous ice islands and numbers of bergs appeared on the Southern Ocean in this or that sector, presumably due to the breaking up by waves or volcanic action of considerable floating ice-shelf extensions of the Antarctic continental ice cap, the bergs sometimes drifting to remarkably low latitudes but all being dispersed and melted each time within 5–10 years, suggests that mechanically induced changes of the extent of the Antarctic ice would similarly quickly revert to the position and balance appropriate to the present radiation climate.

(3) *The atmospheric moisture transport across Panama from the Atlantic to the Pacific*

WEYL (1968) also emphasizes the stabilizing effect of the oceans, with their enormous heat capacity and possibilities of development of convection, upon world climate. He stresses, however, that at low temperatures the occurrence or absence of convection, and the liability of the surface water to freeze, are more affected by salinity differences than by small changes of prevailing temperature. From this point of view the much higher salinity of the Atlantic than of the other oceans at the present epoch (fig. 8.3) is clearly important. The density of these saline waters increases all the way down to their freezing point, so that radiation or wind cooling of the surface produces deep convection and the decline of surface temperature becomes very slow. Moreover, ice does not form until the temperature reaches about −2 °C. It is at least partly for these reasons, as well as the freer interchange of water currents between the wide Atlantic Ocean and high latitudes, that the Atlantic water does not readily freeze over in winter south of about 75 °N, compared with the seasonal advance of ice on the Pacific water in and about the Bering Sea to near 60°N. It is suggested, doubtless rightly, that a change in the salinity of the Atlantic water would affect the limits of the ice in the far north and at the American fringe.

The present great salinity of the Atlantic also means that the fresh water from the Siberian rivers, and from summer melting of some of the existing pack-ice, remains in a layer of low density (sigma values approaching 24, compared with about 27·8 in the underlying Atlantic water) only about 100 m thick, at the surface of the 3000-m-deep Arctic Ocean. Any convection started at the surface cannot extend below this layer: so the heat loss (and seasonal gain) at the surface are concentrated within it, as if the ocean were only 100 m deep. The surface readily freezes, at temperatures which range from 0° near the river mouths to about −1·5 °C farther out.

WEYL also points out that the salinity of the Atlantic controls the limit of growth of the Antarctic ice on the Southern Ocean. This operates through the salinity and density characteristics of the Atlantic Deep Water circulating southward and coming up in latitudes south of the Antarctic Convergence, i.e. south of 50–55 °S in most sectors (cf. figs. 8.5, 8.6). The density versus depth stratification in this zone where Antarctic sea ice is formed is very different from that in the Arctic Ocean. Because of the ring of ocean circulating round the Antarctic continent, which itself has no rivers but thrusts ice only slowly out into the ocean, the surface salinity in the zone we are considering ranges from 33 to over 34‰ (by no means as low as in the Arctic surface water), and with temperatures −1° to +2° its density is characterized by sigma values about 27. And because the Atlantic Deep Water in the salinity maximum layer beneath has formed by mixing processes, its sigma value is around 27·2–27·4. Hence, the layering is only just stable, and the added salt shed into the immediately underlying water when ice forms on the surface must soon increase the density sufficiently to start convection going and the deep water convected to the surface proceeds to melt the ice that has formed. WEYL calculates that with present conditions this sets a limit of about 1 m thickness to the seasonal formation of Antarctic ice. However, contrary to WEYL's reasoning, it would appear that any increase in the salinity of the Atlantic Deep Water (which is supplied largely by the mixing and cooling processes at the surface of the Labrador Sea and about Iceland) should increase the density difference between it and the surface layers of the Antarctic Ocean. Thus, convection should be increasingly inhibited there, the more saline the Atlantic Deep Water became, allowing more Antarctic ice to form. It is in the winter when the ice-forming processes are most intensely developed – but when oceanographic vertical sections extending right into the ice belt are hardest to get, and so far lacking – that the resulting convection should reach the greatest depths, penetrating in favourable circumstances right through the salinity maximum marked by the Atlantic Deep Water at the middle depths. This is therefore the time at which the Antarctic Bottom Water (which supplies the bottom water in all the oceans, as already noticed) is created. It would appear that an increase in the salinity of the Atlantic water could check the formation of Antarctic Bottom Water and possibly stop it altogether. If this occurred, the existing bottom water in all the oceans sealed off by the stable stratification above it would presumably stagnate until, very slowly, geothermal heat accumulated in it and raised its temperature enough to start the convection again.

It is clearly important to understand how the greater salinity of the Atlantic than other oceans is maintained and to watch the progress of any variations in this situation. The contrast is particularly marked between the North Atlantic and North Pacific Oceans, even where they are separated only by the isthmus of Panama (indeed the contrast is particularly strong there) though the climatic influences on the surface on either side might be expected to be very similar. Of course, the immense distance to be travelled before the water from one of these oceans can circulate into the other accounts for the possibility of maintaining the differences actually created by the climates of the various regions thereby involved.

The Atlantic possesses in the Mediterranean an extended area in the zone of maximum evaporation where water of particularly high salinity is created, but this water enters the Atlantic through the Straits of Gibraltar at sub-surface levels and spreads at intermediate depths. The main agency determining the difference of salinity, which is so marked in the surface waters, between North Atlantic and North Pacific seems likely to be the steady transport of moisture with the Trade Wind across Panama from the Atlantic to the Pacific. This is a steady drain of moisture from the surface of the one ocean to the other. It is not interrupted at any time of the year, and seems not to be balanced by any comparable reverse flow in the atmosphere (which would necessarily be in colder latitudes and where the winds are more variable). WEYL (*loc. cit.*) quotes computations by DEFFEYES which assess this moisture flux, as shown by the upper air observations at Panama, at $0{\cdot}1 \times 10^6$ m^3/sec (which is about one-tenth of the flow of all the world's rivers). Despite the modest seasonal variation, the observations show that secular changes do occur: over the years studied from 1956 to 1964 this moisture flux was declining – most of the decline took place in the 1960s – with the widespread evidence of declining vigour of the global atmospheric circulation at that time. By 1964 the decrease of the moisture flux amounted to 30%.

This seems to place the primary control elsewhere, in whatever determines the energy level of the atmospheric circulation. (The change was again most obvious, and first detected, in the 'Iceland' low.) The consequences must, however, be spread by the oceans over the world in the ways described in this section. The salinity of the surface waters of the North Atlantic and North Pacific might be expected to show a detectable response to such climatic changes within a few years or decades, but it probably takes hundreds of years for the effect to be carried by the deep water to the Antarctic.

(4) *Anomalies in the extent and position of ocean heat sources*

Shifts, extensions and constrictions of the areas in which the ocean gives off most heat to the atmosphere must affect the energy and steering of atmospheric disturbances. Two areas have been noted as particularly liable to such anomalies in the present era. The one, between 40° and 45°N, south of Newfoundland and Nova Scotia, concerns the limit of the warm Gulf Stream water. The other, between 30° and 45°N about 160–180°W in the Pacific, similarly concerns the limit of the warm Kuro Shiwo water. The effects of warm, or cold, anomalies in these areas upon the atmosphere seem to be strongest in producing respectively more cyclonic, or more anticyclonic, situations near 60°N 50 to 80° of longitude farther east (downstream in the sense of the meandering upper westerly winds). These effects receive attention in Chapter 10.

An impressive series of studies by BJERKNES (1966, 1969*a*, 1969*b*) has drawn attention to the importance of another area (see pp. 402–3, 409), that remarkable region of the equatorial zone in the eastern half of the Pacific, extending from the American seaboard sometimes as far as 170°W, where the water is 5–10°C cold for the latitude. Fig. 8.2 shows the geography: the zone narrows westwards, and its extent and intensity vary in different years. When this zone is cold, the heat source represented by the warmest water in the eastern and central Pacific is restricted and its axis lies near 10°N. This seems to place the jet stream, and hence the general atmospheric circulation development, over the eastern North Pacific farther north and make it weaker than in those years when the warm water extends to the equator (at least between 100° and 170°W). These situations undoubtedly affect the North Atlantic indirectly, in that a vigorous (and southward-shifted) circulation over the eastern Pacific is liable either to advect warm air over North America frustrating the development of the winter cold trough over the higher latitudes, or to establish the jet stream on a zonal (west–east) orientation in a lower than usual latitude across both sectors: the tendency is either to weaken the atmospheric circulation over the Atlantic sector or to displace it also to lower latitudes than normal, when the eastern equatorial Pacific is warm.

Some similar effect apparently operates also in the southern hemisphere, since

BJERKNES found over the years since 1957 for which data were available that atmospheric pressure tended to be generally above normal over the northern and southern polar regions when the eastern equatorial Pacific was warm.

The history of sea temperatures observed at the coast of Peru (see fig. 10.11) suggests that the eastern Pacific sector here considered may tend to warm and cool in anti-phase to the rest of the equatorial zone. Indeed, its changes within the twentieth century have been opposite in phase to the general experience of the rest of the world. The same may be true of the vigour, though not the latitude, of the atmospheric circulation development over the eastern North Pacific. What is important is to notice that, despite the global extent of the influence of sea-water anomalies in this key area, these anomalies appear to be no more than a link in the chain of cause and effect. The water spreading westwards along the equator in the Pacific from the coast of Peru tends to be cold just when the general atmospheric circulation over most of the world, and most particularly the southeasterly–easterly Trade Winds across South America and in the South Pacific, are strongest. This is because these are the conditions that promote upwelling (both at the coast of Peru–Ecuador and in the open ocean under the Trade Wind stream, near the equatorial low pressure trough).

(5) *Variations of world sea level*

The general level of the ocean surface stood about 100 m lower than now at the maximum development of the ice ages, and it should rise about another 50 m if all the present ice accumulated on land were melted. Apart from the addition, or removal, of water in this way, which must take thousands of years before the extreme state is attained, thermal expansion or contraction of the water columns when the ocean is generally warmed or cooled to some depth during some persistent climatic regime could affect world sea level appreciably. Warming or cooling of the entire ocean by 1°C throughout its depth would raise or lower the surface by 60 cm.

In some areas, particularly the Baltic and north Canadian seas and sounds, regionally amplified post-glacial rises of sea level, lasting some thousands of years after the melting of the great ice caps that had depressed the land surface, opened wide channels. But, contrary to earlier opinion, evidence now shows that the warmest post-glacial climates, even with these extra channels open between the Atlantic and the Arctic, presumably giving more interchange of water than can now occur, did not suffice to remove the pack-ice from the polar region. Later, isostatic recovery of the land lowered sea level again in these regions.

Changes of world sea level must affect climate in two ways, by increasing or decreasing the total area of land and by changing the amount of water that can be exchanged over the submarine ridges, or sills, at the entrances to the Arctic Ocean, the Baltic, the Mediterranean and so on. The history and geography of these changes will be considered in Volume 2.

The hypsographic curve (fig. 8.13), due to DIETRICH, indicates the total potentialities of

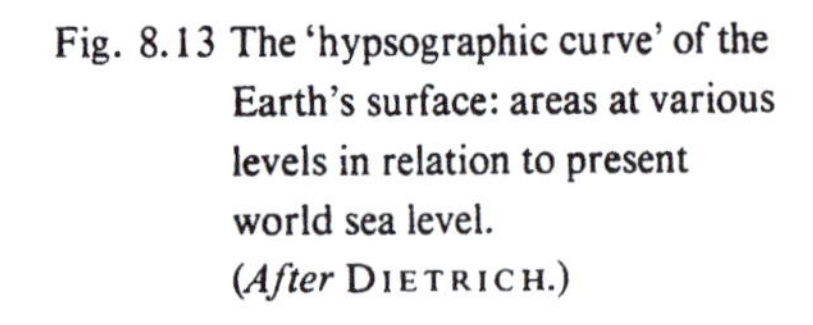
Fig. 8.13 The 'hypsographic curve' of the Earth's surface: areas at various levels in relation to present world sea level. (*After* DIETRICH.)

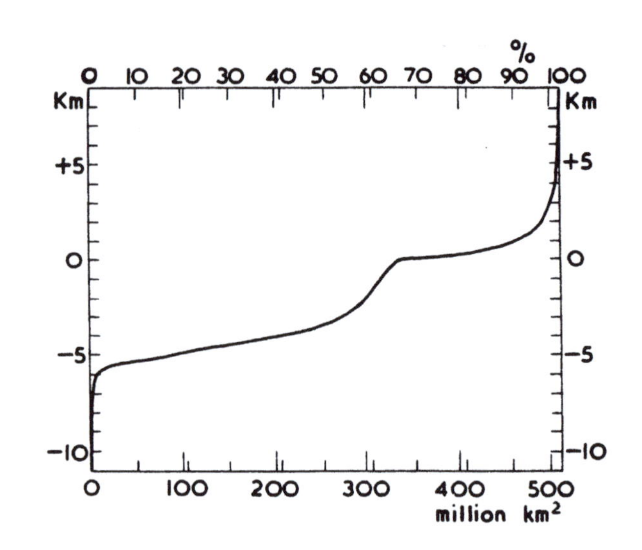

sea-level changes. It will be noticed that the curve indicates that most of the Earth's solid crust has its surface either 4000–6000 m below sea level (maximum frequency at 4950 m) or about 100 m above present sea level. The area at present submerged by less than 200 m, the continental shelf, amounts to 5·5% of the whole and more than half of this must have been dry land at the ice-age maxima. Melting of the remaining ice caps would submerge a bigger percentage of the land area than this.

The present rate of exchange of water through the Strait of Gibraltar is reckoned to be sufficient to replace the entire water of the Mediterranean in about 75 years, representing a supply to the Atlantic of over 3700 km^3 of highly saline water during that time. R. S. ANDERSON (1965) has made some calculations of the situation during the extreme stage of the Würm ice age, when the volume of the Mediterranean was probably 6% less (assuming no major tectonic alteration of this unstable area) and the Gibraltar Strait was halved in depth and narrowed by 70%. On the difficult and doubtful assumptions made about evaporation, precipitation and run-off in a colder, windier climate, with the Black Sea providing a fresh-water river into the Aegean and corresponding changes in the volume and freshness of the Nile and the European rivers, ANDERSON concluded that the exchange through the Strait of Gibraltar would proceed at 65% of its present rate, though, surely surprisingly, the salinity of the outflow (estimated at 37·9‰) would not be much different from today.

The Arctic Ocean is also a nearly land-locked sea. Its rate of exchange with the Atlantic probably varies considerably even with the minor climatic changes of modern times, due to the varying winds and without any great change of sea level, as indicated in our discussion of the ice drift, earlier in this chapter. The surface water from the rivers of the northern continents probably makes its way out in 3–30 years, depending on the variable tendency to go into repeated circuits of the inner Arctic gyre. The water and ice in the central parts of this gyre may linger there for 10–100 years. No estimates are known of the rate of renewal of the deep water in the Arctic at the present time.[1] The progress of the Atlantic water between 200 and 900 m depth – only a small offshoot of the Atlantic water circulating in the Norwegian Sea – can be traced (COACHMAN and BARNES 1963) in a cyclonic gyre – i.e. contrary to the movement of the surface water and ice – around the polar basin at a rate that would probably complete the circuit within about 10 years. Its temperature at the point of entry near west Spitsbergen is +3·5°C and gradually falls to about 0°, as heat is lost upwards by eddy conductivity: thereby, the level of the temperature maximum associated with this water gradually drops from 150 to 500 m in the remotest parts of the polar basin.

The greatest depths of water between Greenland and Iceland, over the Iceland–Faeroes–Scotland ridge, and between Scotland and Norway, the 'sills' over which Atlantic water has to pass to enter the Arctic, in all cases attain 200 m and in some areas much more. Despite the importance attached to this by EWING and DONN (1956) in their theory of the ice age–interglacial oscillations, the total reduction of the vertical cross-sectional area of these channels available for water movement at the lowest sea level in the ice age can hardly have exceeded 30% of today's.

1. There are two deep basins in the Arctic with general depths about 4000 m, one the Nansen Deep broadly between the Lomonosov ridge (that runs along the 140°E meridian to the pole and rises generally to less than 2000 m depth) and Spitsbergen–Severnaya Zemlya, and the other, broader and more isolated one between the Lomonosov ridge, the edge of the Siberian continental shelf near 80°N and near the coasts of Canada and Alaska. Both basins are internally partly divided by ridges and rises over which the depth is about 2000 m. Depths of over 3000 m, however, extend through the centre of the Greenland–Spitsbergen channel and south about the Greenwich meridian to 65°N.

The central, deepest part of the Greenland Sea about the Greenwich meridian, especially between 73° and 76°N, in the centre of the cyclonic gyre in the surface water (fig. 8.5), is occupied in depth by a vast dome of cold water mostly at temperature below 0°. Temperatures about, or below, −1°C are encountered at depths of 450–600 m between 74° and 75°N and extend over the entire breadth of the deep ocean bed in this sea. Convection from the surface downwards presumably occurs in the cooling season, starting near, but at the north and west side of, the centre of the gyre, where the stratification of the top layers is least stable, and spreading over a wider area later as the surface water becomes colder. AAGARD (1968) suggests that the amount of this convection during any one winter may be indicated by the accumulated number of degrees below −1°C shown by the monthly mean air temperatures at Jan Mayen. It appears that the water at 2500 m depth cooled by about 0·15°C as a result of the succession of cold winters in the area between 1962 and 1966; it had slowly warmed by twice this amount over the previous 50 years.

CHAPTER 9

The water cycle

Water is taken up as vapour into the atmosphere from the Earth's surface, from the seas, lakes, rivers, soils, and from plants and animals. It is then transported to various heights and sometimes over great distances with the air. After some time the vapour condenses again, forming ice crystals or water droplets, and thereby producing mists, fogs and all the different kinds of clouds (whose form depends on the air motion, the water content and the nature and sizes of the particles they carry). Sooner or later the H_2O molecules are put down once more on the Earth's surface: as rain, snow, hail, etc., or in a deposit of dew or ice. The average sojourn of a water molecule in the atmosphere is reckoned at about 10 days, though the figure covers an enormous range of variation. Some molecules get carried up into the stratosphere in which they are liable to stay for from 7 to 10 years. At the other extreme, some molecules evaporated from wet or sodden ground may be soon drawn into the lower layers of a thunderstorm cloud and precipitated again within the hour. The overall average itself probably varies with the vigour of the general atmospheric circulation. During their time in the atmosphere the water molecules are the stuff from which most kinds of weather are made. Once redeposited the water may spend weeks, months or years on its way to the sea, and as we saw in Chapter 8 some of it spends thousands of years in the deep ocean before being presented to the atmosphere again. Some water becomes trapped in the subsoil and in porous strata of rock for even longer periods: a radiocarbon age of 25 000 years has been measured in oases in the Sahara in water that has passed through the strata below the desert. And some of the ice in the depths of the Greenland and Antarctic ice sheets has been embedded for hundreds of thousands of years.

In this chapter we shall consider only those parts of this cycle that are of most direct concern to climate and its variations.

Evaporation

The quantity of water that the air can hold in the form of vapour (whether expressed by its mixing ratio in grammes per kilogramme, or per cubic metre, of air or by its vapour pressure) increases with temperature (see Table 9.1). The air's moisture content, or *absolute humidity*, may be measured either by the water vapour pressure or by the mixing ratio.

The quantity of vapour picked up in a given time depends on the temperature of the

TABLE 9.1
Saturation vapour pressure

Temp., °C	30	31	32	33	34	35	36	37	38	39
Vapour pressure, mb	42·5	44·9	47·5	50·3	53·2	56·2	59·4	62·8	67·3	69·9
Temp., °C	20	21	22	23	24	25	26	27	28	29
Vapour pressure, mb	23·4	24·9	26·4	28·1	29·8	31·7	33·6	35·7	37·8	40·1
Temp., °C	10	11	12	13	14	15	16	17	18	19
Vapour pressure, mb	12·3	13·1	14·0	15·0	16·0	17·1	18·2	19·4	20·6	22·0
Temp., °C	0	1	2	3	4	5	6	7	8	9
Vapour pressure, mb	6·1	6·6	7·1	7·6	8·1	8·7	9·3	10·0	10·7	11·5
Temp., °C	−10	−9	−8	−7	−6	−5	−4	−3	−2	−1
Saturation:										
Over water, mb	2·9	3·1	3·3	3·6	3·9	4·2	4·5	4·9	5·3	5·7
Over ice, mb	2·6	2·8	3·1	3·4	3·7	4·0	4·4	4·8	5·2	5·6

evaporating surface, the moisture gradient (i.e. lapse with height) in the air and the wind speed (which determines the eddy diffusivity). The rate of evaporation is given by the expression:

$$F = -\mu_e \cdot \frac{0{\cdot}621}{p} \cdot \frac{de}{dz}$$

where F is the amount of vapour passing through 1 cm²/sec of the water surface, μ_e is the eddy diffusivity, p is the atmospheric pressure and de/dz is the rate of change of vapour pressure with height. The minus sign expresses the fact that de/dz has negative values when evaporation is occurring.

SVERDRUP (1945) gives the following simple formula as adequate for deriving annual mean values of the amount of water evaporated E (in cm) from a sea surface, where winds of more than 5 m/sec prevail:

$$E = 3{\cdot}6(e_w - e_a)W$$

where e_w is the saturation vapour pressure at the mean temperature of the water surface, e_a is the average actual water vapour pressure measured in the air at the height of a ship's deck, say 8 m above the sea (both in mb), and W is the average wind speed at the same height (in m/sec). For W under 5 m/sec and a smooth sea, however, the evaporation is very much less and this expression cannot be used in such cases.

Table 9.1 provides a convenient series of values of the water vapour pressure that saturates air at different temperatures. From it we see the increasingly steep rise of possible water vapour content of the air as its temperature rises. This affects the evaporation from water surfaces as they become warmer, as may be seen through the effect upon e_w in the formula given above. Table 9.2 shows the implied changes of water vapour content of the air over the oceans in tropical latitudes as their temperature varies in different climatic eras, assuming that the air–sea temperature difference and prevailing wind speeds remain constant (*relative humidity* assumed to be at all times 85–90% saturation).

Another feature of Table 9.1 is the figures which show the smaller amount of water vapour that saturates the air in the presence of ice crystals as compared with the situation when only water or water droplets are present. (The difference increases the lower the temperature.) This has important consequences regarding the growth of ice crystals in cloud at temperatures below the freezing point.

TABLE 9.2

Apparent changes in the amount of water evaporated from the tropical oceans in different climatic eras

(Relative humidity and wind speed assumed unchanged)

ΔT = Assumed temperature change (°C) ΔE = Change of evaporation

	Pacific		*Atlantic*	
	ΔE	ΔT	ΔE	ΔT
	Departures from 1900 to 1920 values			
Ice Age climax	−10%	(−2°)	−25%	(−5°)
Post-glacial warmest climate	0	(0°)	+5%	(+1°)
Warmest decades of 20th century	Probably slight		+3%	(+0·6°)

Over land the amount of vapour taken up depends upon the availability of water to evaporate. A few special environments with dense vegetation may at times present a greater water surface than the flat surface of a lake. But many other environments, e.g. deserts and rock faces on mountains, as well as areas of parched vegetation after dry weather, provide little or no water to evaporate. The *potential evaporation* is determined by the condition of the air, but it may not be realized because of the state of the ground. PENMAN (1956) has given an expression for the potential evaporation E_d in mm/day in temperate latitudes, which may be written:

$$E_d = 0{\cdot}35(e_s - e_a)(0{\cdot}50 + 0{\cdot}53W_2)$$

where e_s is the saturation vapour pressure at the prevailing surface temperature, e_a is the average actual vapour pressure in the air and W_2 is the mean wind speed at a height of 2 m above the surface in m/sec. (e_s and e_a are again measured in millibars.) PENMAN's formula indicates, and direct measurement of evaporation from tanks confirms, that in present-day temperate climates, such as England's, potential evaporation exceeds rainfall for about 4 months in the summer half of the year. (In England the evaporation is greatest in June and July, the monthly totals averaging 75–80 mm at typical lowland sites in both months, but the excess of evaporation over rainfall is greatest in June (20–25 mm) and is greater in May than July.) Because of the part played by percolation and run-off, the surface of bare ground is usually dry (and the underlying soil at less than 'field capacity' [1]) for a good deal longer than the period during which the evaporation by itself exceeds the rainfall.

TURC (1958) has derived empirically formulae from which the evaporation and surplus water to be drained by run-off and the river system can be calculated for the widest possible range of climates and states of the soil and vegetation. For bare soil the evaporation over a 10-day period E_{10d} in mm is given by

$$E_{10d} = \frac{R + a}{\sqrt{1 + \left(\frac{R + a}{l}\right)^2}}$$

where R is the rainfall and a the amount of water over and above the rainfall which can be evaporated into the air in 10 days in the given situation. a takes values which are found to range from 1 for dry soil to 10 when the soil is at field capacity. l is given by

$$l = 0{\cdot}062(T + 2)\sqrt{I}$$

where T is the mean air temperature in °C and I is the total incoming radiation in cal/cm^2/day. When there is vegetation present, E_{10d} is given by

$$E_{10d} = \frac{R + a + V}{\sqrt{1 + \left(\frac{R + a}{l} + \frac{V}{2l}\right)^2}}$$

in which V represents the additional evaporation produced by the vegetation drawing moisture up from the soil. V depends on the height, density and nature of the plant growth. For a luxuriant plant cover which never lacks moisture $V = 70$.

From study of evaporation and run-off in drainage basins in many parts of the world, TURC derived the following expressions for calculating the probable yearly total evaporation E_y and the surplus water to be drained away yearly D_y in any climatic situation where the average yearly rainfall R_y and temperature T_y are known:

$$E_y = \frac{R_y}{\sqrt{0{\cdot}9 + \frac{R_y^2}{L^2}}}$$

$$\text{where} \quad L = 300 + 25T_y + 0{\cdot}05T_y^3$$

$$D_y = R_y - E_y$$

assuming no significant change in the quantity of water stored in the ground. E_y and R_y are measured in mm, T_y in °C. As the formula for E_y does not take account of the seasonal

1. *Field capacity* is a term used in soil moisture studies and agricultural meteorology. It describes the situation when the soil above the water table is holding as much moisture as it can against the pull of gravity, i.e. without drainage or flooding of the surface.

distribution of the rainfall or the temperature, it gives too much evaporation and too little run-off in climates such as Algeria's where the rainfall is concentrated in the cold season; the error goes the other way in the Ukraine, where the rainfall comes mainly in summer. From these formulae, however, LAMB *et al.* (1966) have found it possible to provide estimates of evaporation and run-off in England in various past climatic epochs, which appear reasonable in relation to botanical and geomorphological evidence of the times concerned.

Evaporation is a difficult item to measure, but figures which appear to be representative of what occurs over much larger water bodies can be obtained from an open tank as small as 2 m square by 60 cm deep, in which the water level is maintained daily, allowance being made for the measured rainfall received. Smaller tanks lose too much water, both because trapped incoming radiation warms the water, thereby increasing the evaporation, and through rain splashing out over the side. Knowledge of the distribution of evaporation over the world has therefore been extended by calculation using the empirical formulae established from a notably thin network of measuring sites. Fig. 9.1 maps the world distribution of evaporation

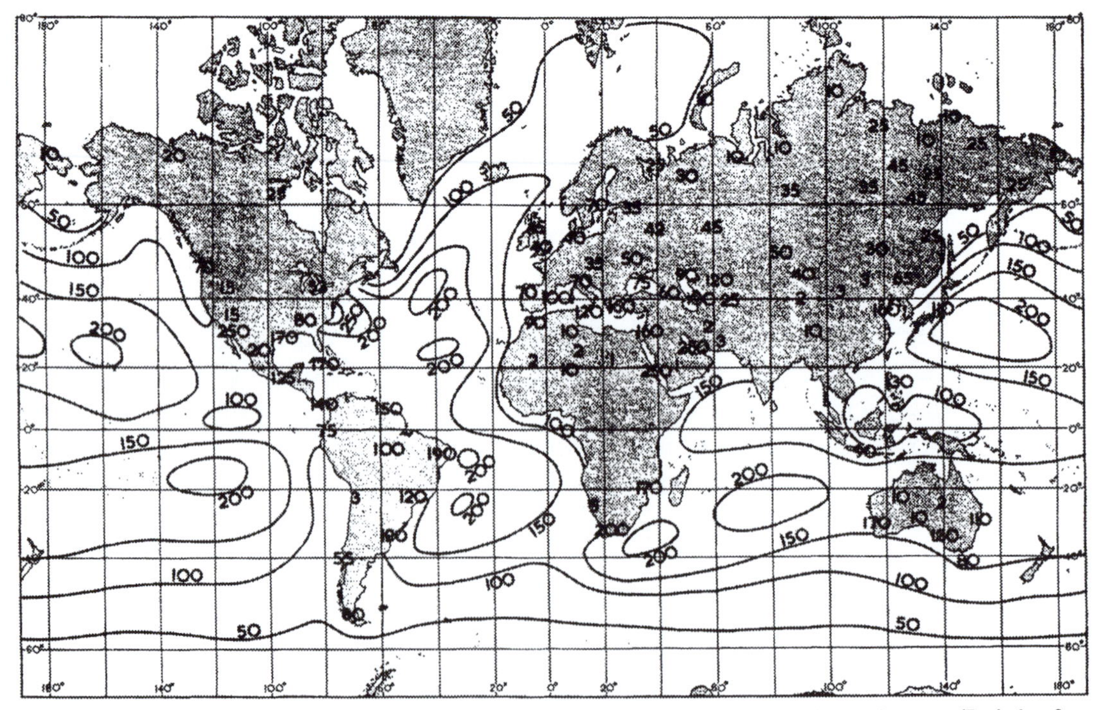

Fig. 9.1 Evaporation: average yearly amount of water evaporated from the surface in centimetres. (Periods of observations uncertain, believed generally about 1930–50.)
N.B. The *potential* evaporation rises to 150–250 cm/yr in the southernmost part of Soviet Asia (Qara Qum) east of the Caspian Sea, and totals of 250–400 cm/yr have been measured with dishes and tanks in the desert zones of the world: over 250 cm/yr also at Jerusalem and inland in southern and central Spain.

and shows that the greatest amounts are evaporated from the warm seas, particularly where the cold, dry airmasses from Canada and Siberia in winter most frequently come over the Gulf Stream and Kuro Shiwo waters. Where water of Gulf Stream origin penetrates the Arctic, in the Norwegian and Barents Seas, and cold northerly airstreams commonly blow over it, the evaporation is peculiarly high for the latitude. At the equator evaporation is somewhat less than in the region of the tropics because of the cloudiness of the equatorial rainbelt, and the somewhat lower temperatures resulting, and in the Pacific because of the notably cooler water there. Over land the evaporation varies too greatly with the topography for any detail to be shown on such a world map: the sample values marked indicate how it is reduced in the deserts by the scanty availability of water to evaporate.

Table 9.3 attempts a summary of the distribution of evaporation by latitude over the Earth in the present century. Table 9.4 gives a sample view of the seasonal variation for one area, England, and compares the evaporation month by month with the average rainfall.

TABLE 9.3
World distribution of evaporation
Average yearly totals (cm)
(Estimates by WÜST 1922)

Latitude	*Oceans*	*Continents*	*Overall average*
90–80°N	5	5	5
80–70	9	9	9
70–60	12	12	12
60–50	40	36	38
50–40	70	33	51
40–30	96	38	71
30–20	115	50	91
20–10	120	79	109
10–0	100	115	103
0–10°S	114	122	116
10–20	120	90	113
20–30	112	41	96
30–40	89	51	85
40–50	58	50	58
50–60	23	20	23
60–70	9	10	9
70–80	5	5	7
80–90	0	5	5
All latitudes average	84·2	50·4	74·3

TABLE 9.4

Monthly evaporation measured in England (cm) (London, Camden Square 1885–1919)
(After BILHAM 1938)

	J	F	M	A	M	J	J	A	S	O	N	D	*Yr*
35-year averages	0·25	0·63	1·68	3·83	6·17	7·39	7·54	5·92	3·51	1·57	0·63	0·23	38·23
Greatest	0·66	1·63	2·62	5·94	8·25	9·78	11·68	9·60	6·05	2·36	1·09	0·81	50·01
Least	−0·43	0·03	0·76	2·16	3·96	4·32	3·66	4·01	1·73	0·89	0·05	−0·36	32·00
35-year averages Rainfall	4·72	4·24	4·65	3·91	4·47	5·13	6·05	5·61	4·62	6·68	5·99	6·07	62·15

The minus signs in the above table indicate net condensation of water on the surface.

Over the oceans in middle and higher latitudes calculation indicates that evaporation is greatest in winter when the winds are strongest.

Transport

The lapse rate of temperature with height affects the vertical transport of water vapour from the air layers near the surface into the upper troposphere. Stable layers, especially inversions of temperature, check this transport by checking the convection; and in anticyclones, and with inversions over cold sea or polar ice, there may be much more moisture present in the bottom kilometre or so of the atmosphere than at higher levels. Any melting of the ice on the polar seas should therefore result in a marked increase of the water vapour content of the upper troposphere over the polar regions.

FLOHN (1963) has estimated the mean residence time of a water molecule in the atmosphere as 9 days over the equatorial zone and 11 days for the world average. These times, taken into consideration with the vertical transport and upper wind strengths prevailing, appear to correspond to mean distances travelled from the point of evaporation to the point of fall-out (in rain, snow, etc.) of about 900–1000 km in the equatorial zone and from 5000 to 11 000 km over other latitudes. The pattern and scale of this horizontal transport is, however, clearly subject to variation with the variations of the general atmospheric circulation that accompany the climatic differences between different years and decades, etc. KRAUS (1955, 1958) was able to point to a long period of generally reduced rainfall in the tropics and equatorial zone between about 1895 and 1940: this actually covered the time when the ocean surface temperature was generally highest and the input of water vapour into the atmosphere was presumably boosted by up to 3% as compared with the previous and subsequent decades (cf. Table 9.2). During that time it was in the middle and higher latitudes of both hemispheres (especially at all the places exposed to transport of moisture from the oceans with the prevailing westerly winds and on the high ice caps in Greenland and near the South Pole) that the downput of rain and/or snow was at its long-term maximum. With the reversion to weaker atmospheric circulation, and equatorward shifts of the main features of that circulation after 1960, the rainfall in the equatorial zone appears to have increased again, despite some lower sea surface temperatures (cf. fig. 10.11), whereas the downput has shown on the whole a decreasing tendency over the higher latitudes. These changes bear witness to some change in the prevailing scale of water vapour transport over the Earth, accompanying the changes of vigour and pattern of the global wind circulation.

Variations in the poleward transport of water vapour, as here discussed, must affect the shape of the curve of latitude distribution of precipitation illustrated in fig. 9.6 hereinafter. B. LETTAU (1969) has pointed out, further, that quantitative studies of the downput of snow (indicated by accumulation on the ice cap) in the interior of Antarctica indicate not a simple linear relationship to monthly mean circulation strength and temperature values but that the eddies which cause deviations from the mean play a vital part: monthly mean values underestimate the moisture transport, a great part of which evidently occurs during mild spells of shorter duration.

Condensation, clouds, etc.

When the temperature of the air falls (either by contact with a cold surface or by expansion) to the saturation point for the amount of water vapour in it, condensation occurs. In carefully purified air (up to four-fold) supersaturation is possible, but in the atmosphere there appear to be always enough suitable particles – salts and acids, especially *sea*-salt derived originally from evaporated spray – present to ensure that condensation occurs. Many of the condensation nuclei (notably the sea-salt particles) are hygroscopic and cause condensation to start somewhat before the saturation point is reached, so that the air begins to be misty at relative humidities rather below 100%. The first droplets or ice crystals formed are microscopic, a few microns in diameter, and the nuclei are so abundant that there are millions of them in one litre of air. The fall-speeds (terminal velocities) of such minute cloud particles are negligible – about 3 mm/sec, or 200 m/day, for droplets of 10 μm diameter – and they are borne aloft, rising and falling with the motions of the air itself, sometimes for days.

Growth of the minute cloud droplets into drizzle and rain demands concentration of the condensed water into fewer, but bigger, drops. The drops we call *drizzle* have diameters from 0·1 to 0·5 mm and fall-speeds of from 0·3 to 2 m/sec. *Rain drops* are bigger than this; diameters up to about 5 mm have been observed, but they begin to be deformed by the air resistance and shatter at sizes from 3 mm diameter upwards. The fall-speeds are about 4 m/sec for a 1-mm-diameter drop and may reach 10 m/sec for the biggest drops. Shattering of the large drops produces ionization and separation of electrical charges between drops and air, so that it is thought to be the process (or one of the processes) whereby the electrical charges in thunderclouds are built up.

Though there are always more than enough nuclei for cloud droplets to condense on, nuclei of the sorts and shapes on which ice crystals can form are much scarcer. As a result, supercooled clouds and fogs – consisting entirely of liquid droplets – are commonly observed at temperatures below the freezing point. When these droplets are carried by the wind against solid objects, whether trees and grasses, mountains or structures built by man, they freeze on impact, coating the objects with ice crystals. The rime deposit grows in the direction from which the droplets come, i.e. to windward. It is only when the temperature is as low as −40°C that condensation always produces ice crystals. Moreover, the particles in ice clouds are generally fewer than in water-droplet clouds. (The fibrous texture of the edges of ice-crystal clouds, in marked contrast to the sharp outlines of growing, typically cauliflower-shaped, cumulus clouds that still consist largely of minute water droplets, is partly due to the particles being sparser and, on the whole, larger than in water-drop clouds and partly because some of the particles are big enough to travel an appreciable distance through the drier air outside the cloud before evaporating.) These facts mean that situations may arise in nature in which a cloud of supercooled droplets comes in contact with ice crystals and the latter are liable to be present in much more limited numbers than the droplets. This is certainly one situation in which the cloud of microscopic droplets will be transformed into an ice, or snow, cloud consisting of much larger particles than before: since the air which is just saturated in the immediate vicinity of a water droplet is supersaturated in the presence of ice (cf. Table 9.1). The ice crystals must therefore grow, while evaporation of the liquid droplets sets in as the H_2O in the intervening air is transferred to the ice crystals. Snow may then fall, and when it reaches air layers with temperature above 0°C the flakes will gradually melt to give rain.

This sequence of events, as originally suggested by Bergeron (1935) and put forward in more detail by Findeisen (1938), was thought for many years to be essential to the formation of rain; and it may, indeed, be the means by which the drops grow into raindrops in the great majority of cases in middle and higher latitudes. In all these cooler climates the tops of the rain clouds nearly always reach well above the freezing level, and snow is usually present at least in the upper parts of the cloud. In shower clouds, when the whole height range of the cloud can be viewed from the side, transformation of the upper part of the cloud into an anvil-shaped structure with the fibrous appearance of an ice-crystal cloud can usually be seen to take place shortly before the first rain or snow falls. In these latitudes, therefore, most rain is melted snow.

There are no more efficient ice nuclei for condensation than minute ice particles themselves. Mason (e.g. 1969, pp. 454–5) has shown that tiny ice splinters are liable to be thrown off when supercooled water drops freeze as well as from fragile snow crystals falling through a cloud. For this reason, ice crystals are liable to be found in considerable concentrations at any temperature below 0°C. It is also noticed that the ice splinters appear to acquire a negative electric charge: so this process, too, may be relevant to thunderstorm development.

The ice-nucleus process cannot, however, be the only one capable of producing rain, because rain is often observed to fall in the tropics from clouds which do not reach the freezing level; and this happens sometimes in summer even in the temperate zone. It now appears, both from theoretical computations (Hocking 1959) and from laboratory experiment (WOODS and MASON 1964), that cloud droplets may grow into raindrops through collisions and coalescence provided there are enough (about one per litre of air) produced in the initial condensation that have a diameter of about 40μm or more. This condition is most likely to be satisfied in warm climates where the moisture content of the air at saturation point is high, though it may depend upon the occurrence of a certain number of giant nuclei. It is observed that, for a given liquid water content, typical cumulus clouds in maritime air will contain far fewer, but larger droplets, and will precipitate sooner, than continental clouds: for a liquid water content of 1 g/m^3 characteristic figures given by MASON (1969) are 50 droplets per cm^3 in a maritime cloud and 200 droplets per cm^3 in a cloud formed in continental air.

From these arguments it seems possible that a decrease in the availability of ice nuclei or of giant condensation nuclei of hygroscopic salts might bring about a climatic change in which clouds would persist longer without precipitating. A superabundance of ultra-minute nuclei produced, for example, by fires might also tend to maintain clouds with small droplet sizes and inhibit precipitation (SUTCLIFFE 1966, p. 59).

The patterns of air motion prevailing determine the type of cloud that will appear, though condensation affects the heat budget – and hence the vertical motion – of the air in which it occurs, through liberating latent heat. This increases the buoyancy of the air in which the cloud is forming, thereby tending to promote the vertical growth of the cloud and extend the moisture transport upwards. Clouds are chiefly of two types:

(1) *Heap clouds* (cumulus, cumulonimbus and some kinds of altocumulus)
These are the clouds in which the vertical component of motion is strongest. They occur therefore in air which has a lapse rate of temperature with height that is unstable for vertical motion or that is at least conditionally unstable (i.e. becomes unstable when the latent heat of condensation is liberated in rising air). They are associated with all the other phenomena of vertical convection, including in the more intense cases thunderstorms, tornadoes, and the greatest water contents that are ever found in the upper troposphere; occasionally a few of the highest towering cumulonimbi penetrate the tropopause, and some moisture is thus injected into the stratosphere.

(2) *Layer clouds* (stratus, stratocumulus and similar types)
These are the clouds that form in air that is stable in respect of vertical motion, especially below temperature inversions in anticyclones and at frontal surfaces with warm air above. In these cases, even the latent heat liberated by condensation does not impart sufficient buoyancy to carry the cloud-filled air through the layers above, which is sometimes very dry.

Smooth sheet clouds (stratus) may be distinguished from ripple clouds (stratocumulus and some altocumulus, etc.), the latter marking some turbulence in the air flow in and below the cloud layer, due either to thermal instability in the air below the stable layer or to shearing motion.

In some frontal zones smooth sheet clouds of great vertical thickness (nimbostratus, altostratus) develop through the slanting upward motion (upgliding) of moist air over the frontal surface, the vertical lapse rate of temperature within the warmer airmass concerned being stable.

The diagrams in figs. 3.22 and 3.25 illustrate some typical situations in which these different cloud types occur.

The prevalence in different situations and different parts of the world of heap or layer clouds has other important climatic connotations. High-reaching cumulus types occur where strong lapse rates of temperature prevail:

(i) over strongly heated lands, especially in low latitudes and in summer over most latitudes, though not in strongly developed anticyclones where subsiding air motion through the upper and middle troposphere checks such development;

(ii) in outbreaks of swiftly moving Arctic and Antarctic air that is strongly heated from below over much warmer seas;

(iii) in regions of marked horizontal convergence of the lower air in cyclonic situations (and even where the isobars are merely cyclonically curved), at fronts and in and near the equatorial zone of intertropical convergence. (A contributory factor in the last-named region is the heating of air from middle latitudes over some of the warmest seas on Earth.)

The total moisture content of these clouds can be gauged, at least roughly, by the amount that saturates air at the temperature of the cloud base. As the air rises in the cloud, more and more of its moisture is condensed as the temperature falls. The amount of liquid water and the amount remaining in the vapour form can be calculated, if the drawing in of drier air from the environment through which the cloud rises and losses by fall-out of rain be neglected. The results are illustrated in Table 9.5 from calculations by LUDLAM (1957) for cloud base at the 900-mb pressure level (typically about 1 km above the ground). The total water content in grammes per kilogramme of air is assumed constant, but the water content per cubic metre decreases with increasing height because of the expansion of the air rising in the cloud. In nature, some mixing with the drier air at the sides of the cloud takes place and observed water content in parts of the cloud affected by this is lower than the table would indicate; but in the heart of a growing cumulus or cumulonimbus cloud the table probably gives a realistic indication of the water content. Aircraft flying through such clouds have measured mean liquid water contents over the distance flown in 3–15 seconds of up to about 4 g/m^3 (ZAITSEV 1950). The peak values in the fastest-rising air columns are doubtless greater than this and probably close to the figures in the table. This suggests that the liquid water content may reach about 8 g/m^3 locally in the upper parts of tropical cumulonimbi and up to 5 g/m^3 in a temperate climate. It is possible, moreover, that, when the liquid drops continue to be supported and borne upwards by the strongest up-currents, until they are frozen into hailstones, the continuing supply of moisture through the cloud base may bring the total H_2O content in the hail region of the cloud to even greater values than the table indicates. The *average* liquid water content taken over the whole of an active tropical cumulus cloud, however, is probably seldom (if ever) more than 2 g/m³.

TABLE 9.5

Probable moisture content of a vertical up-current in cloud (in grammes H_2O/cubic metre) in liquid (L) and vapour (V) form

Height	*Temperature at cloud base*							
	25°		15°		5°		−5°	
	L	V	L	V	L	V	L	V
10 km	7·3	0·5						
7 km	7·7	4·3	5·8	1·2	3·6	0·2		
5 km	6·7	8·5	5·2	3·2	3·7	0·8	1·9	0·2
4 km	5·5	11·0	4·6	4·9	3·3	1·5	1·9	0·4
3 km	4·2	15·7	3·6	6·4	2·8	2·8	1·8	0·9
2 km	2·2	20·0	2·0	9·6	1·7	4·6	1·2	1·7
Cloud base (taken as 1 km)	0	24·0	0	13·2	0	6·8	0	3·2

From these figures it appears that by heights of 3–4 km in the middle troposphere a fifth to a third of the moisture entering the base of the cloud may have been condensed in the tropics and over half in colder climates.

The vertical transport of moisture in heap clouds, here considered, clearly makes possible heavier rainfalls on the mountains in climates where these clouds are common than could be produced by condensation of a similar proportion of the moisture normally present in the ambient air at, or about, the height of the mountain-tops. The prevailing amount of moisture present in the air at the heights concerned apart from that in the cumulus clouds must be less than the figures in the V columns of Table 9.5.

In layer clouds the liquid water condensed is generally very much less; though in deep layers of nimbostratus cloud produced by gliding uplift over frontal surfaces the quantity may sometimes approach that in cumulonimbus. In stratocumulus cloud up to 2·5 km thick measurements have indicated liquid water present in quantities from about 0·1 to 0·8 g/m³. In surface fogs in the temperate zone values of 0·1–0·2 g/m³ seem typical, though sometimes increasing to 0·4 g/m³ in the upper part of the fog (MASON 1957).

Layer clouds prevail where the vertical lapse rate of temperature is stable (discourages vertical motion into the overlying air): hence, they are common (*a*) in frontal zones, (*b*) beneath the temperature inversion at the lower limit of the subsiding air in anticyclones and (*c*) over cold surfaces. In winter anticyclones over land and in the polar regions the stability due to the strong inversion of temperature over the cold surface may enable anticyclonic subsidence to proceed to much lower levels than at any other time. There is then liable to be a very sharp change from cloudy, misty or fog-filled air below the inversion to air with much smaller H_2O content (and sometimes very low relative humidity – see Chapter 3, pp. 102–3) above the inversion. The upper levels of mountains, and occasionally even quite small hills[1] in winter, may experience this very dry air; and prolonged occurrences may parch the vegetation. FLOHN (1941) lists about ten cases in 20 years when relative humidity was continuously below 60% for a week or more on the tops of the Brocken, Feldberg and Schneekoppe, the extreme duration being 18 days (see also Table 9.9 on p. 384). Readily detected (smelt) ozone sometimes provides chemical evidence of the presence of subsided air in the dry, clear air immediately above the cloud and fog layer. The source of supply must ultimately be the ozone layer in the stratosphere, though 'leakage' through the tropopause folding and breaks near the jet stream maintain a fairly constant quantity of ozone in the upper troposphere. On contact with the ground ozone is very quickly reduced to ordinary oxygen, so its presence always means that some, or all, of the air concerned was recently brought down from higher levels. Ozone measurements in this *free air föhn* and in ordinary föhn winds (see later pp. 381–3) on the uplands in Germany show its concentration in the surface air increasing many-fold while the conditions last (WEICKMANN and UNGEHEUER 1952).

The frequency of layer clouds and fog in the lowest kilometre of the atmosphere over the European lowlands in winter gives an advantage to the mountains as regards winter sunshine. The Alps are well known for this, and the virtue for human health of a visit to the mountains in winter has been widely proclaimed. Mountain climates elsewhere in the temperate zone doubtless offer some of the same advantage to the inhabitants of the towns and industrial areas on the lower ground. Table 9.6 gives some comparative figures.

The places in Table 9.6 are grouped in and about the Alps, the Riesengebirge, the Harz and the Scottish Highlands. It is seen that some of the Alpine summits get more sunshine in winter than Trieste at the head of the Adriatic. The highest top in the Harz gets more winter sunshine than places on the surrounding north German plain, though Claustal on the middle levels of the same mountains gets rather less sun than either. In the more cyclonically disturbed climate of Scotland more cloud prevails at all seasons, and the highest levels get less sun than the lowlands; but, even there, in winter the sunshine figures at top and bottom of the mountain are about equal. This effect, like the occurrence of fog on the low ground (with which it is often associated), is at its maximum around the time of lowest sun: the situations which give more sun on the tops affect the statistics for November, December and January more than February or March. It is in quiet anticyclonic weather in winter, when the Earth's surface (particularly the low ground) is cold, that with both thermal and mechanical turbulence of the wind at a minimum subsidence can descend to the lowest levels it ever attains; and, under these conditions, the higher levels of the mountains commonly rise above the fogs, cloudiness and pollution concentrated in the lowest air.

1. The writer observed these conditions on a hill rising to only 170 m above sea level in Co. Limerick, Ireland, in an anticyclone in December 1939.

TABLE 9.6

Average sunshine (hours) on the mountains and neighbouring lowlands

Place	*Height a.s.l. (m)*	*Winter* DJF	*Spring* MAM	*Summer* JJA	*Autumn* SON	*Year*
Basle 47·5°N 7·6°E 1901–50	318	204	482	684	328	1698
Jungfraujoch 46·5°N 8·0°E 1932–47	3460	273	383	454	357	1467
Säntis 47·3°N 9·3°E 1901–50	2500	326	419	466	414	1625
Sonnblick 47·1°N 12·9°E 1901–50	3106	325	399	493	400	1617
Vienna 48·3°N 16·4°E 1901–50	203	179	546	754	360	1839
Trieste 45·7°N 13·7°E 1901–50	11	297	547	832	444	2120
Breslau (Wroclav) 51·1°N 17·1°E 1891–1930	147	138	502	696	324	1660
Schneekoppe 50·7°N 15·7°E 1901–30	1618	214	413	497	303	1427
Prague 50·1°N 14·4°E 1901–50	191	177	562	744	343	1826
Berlin, Potsdam 52·4°N 13·1°E 1891–1930	82	163	497	641	315	1616
Braunschweig 1891–1930	83	153	481	642	309	1585
Brocken (Harz) 51·8°N 10·6°E 1896–1930	1150	175	420	542	273	1410
Claustal-Zellerfeld (Harz) 51·8°N 10·3°E 1891–1930	585	144	460	579	288	1471
Fort William 56·8°N 5·1°W 1891–1903	66	86	415	416	202	1119
Ben Nevis 56·8°N 5·0°W 1884–1903	1343	83	251	270	132	736

Places at a coast, despite a certain incidence of sea fog (especially where the sea is relatively cold in spring and early summer), mostly get more sunshine than the hinterland through escaping the autumn and winter inland fogs and the convection cloud on summer days. This is illustrated in Table 9.7.

TABLE 9.7
Average sunshine (hours) at coasts and inland

Place	*Height a.s.l. (m)*	*Winter* DJF	*Spring* MAM	*Summer* JJA	*Autumn* SON	*Year*
Montrose 56·7°N 2·5°W 1921–50	13	170	434	502	294	1400
Perth 56·4°N 3·5°W 1921–50	27	141	419	498	265	1323
Scarborough 54·3°N 0·4°W 1921–50	33	143	428	533	287	1391
York 53·9°N 1·1°W 1921–50	38	123	411	517	259	1310
Cromer 52·9°N 1·3°E 1921–50	56	180	489	593	326	1588
Cambridge 52·2°N 0·1°E 1921–50	18	157	455	574	302	1488
Eastbourne 50·8°N 0·3°E 1921–50	37	201	561	707	359	1828
East Malling, Kent 51·3°N 0·4°E 1921–50	41	155	473	620	312	1560
London (Kew) 51·5°N 0·3°W 1921–50	19	141	451	585	283	1460
Westerland, Sylt 54·9°N 8·3°E 1891–1930	6	145	521	646	287	1599
Hamburg 53·5°N 10·0°E 1891–1930	29	109	459	545	267	1380

In this table, the places are grouped with one coastal observation point first in each section and then one or more places in the hinterland. The sunshine figures for places up to

30 km inland from the low-lying coasts of East Anglia and northwest Germany, as at Norwich and Oldenburg, are similar to those for the coast itself; but where a ridge of hills backs the coast, whether in eastern Scotland or Sussex, the cloudier and foggier inland regime, with its lower sunshine figures, begins immediately behind the hills.

The world distribution of cloudiness shows a strong association with the areas of frequent cyclonic activity over middle and high latitudes controlled by the jet stream. The distribution therefore varies with the variations of the jet stream, i.e. with the form and intensity at any given time of the circumpolar vortex and the course of the mainstream of the meandering upper westerly winds. The incidence of cloud cover is, of course, dependent on moisture supply and is therefore much greater on the windward side of mountain ranges, particularly those unprotected by other, intervening ranges from the moisture coming from the oceans. Some of the highest cloudiness figures in the world (80–90% cover, and even over 90% in some months) are concentrated on the windward sides of the Andes and parts of the Rocky Mountains in latitudes where there is a strong prevalence of either westerly or easterly surface winds. There is a sharp decrease of cloudiness to leeward of the ranges, noticeably producing regions of lower cloud amount even in eastern Scotland and east of the Norwegian mountains and the Urals. Great cloudiness (averages 80–90% or more in some months) also prevails on the windward slopes of the mountain ranges around the Bay of Bengal, in Indonesia and Japan. Cloudiness is on the whole greater over the oceans (overall average 58%) than over the landmasses (overall average 49%) largely because of the areas screened, or partly screened, by mountains from the sources supplying moisture to the prevailing winds.

The frequency of cyclonic activity affects the occurrence of cloudiness more than does the question of whether the sea is warm or cold for the latitude, except that fogs and very low cloud are common where winds from a warmer region blow over cold sea.

The all-longitudes average cloud distribution by latitude is shown in fig. 9.2. The bold curve presents the average for many years, as given by Brooks (1930). The fine line shows the average for the twelve months March 1962–February 1963, as estimated from Tiros satellite data (Clapp 1964). Note the minima of cloud in the desert zones and the imposing cloudiness averages for the Southern Ocean. It is not yet certain that the various sources of error peculiar to satellite observations (e.g. confusion of ice and snow with cloud) have been eliminated; but to some extent the large values of cloudiness for 1962 may be a true indication of the anomalous character of that year. Seasonal variation (not shown in the figure) should consist chiefly of a movement of the equatorial belt of clouds and rain to about 10°N in June–August and to near 10°S in December–February, according to the hitherto accepted normal values; but the Tiros satellite data show that the intertropical cloud maximum remained sharply defined near 5°N in all seasons in 1962–3; only in the southern summer a secondary maximum appeared about 10°S, affecting a broader belt of latitudes and with lower cloudiness values than the peak near 5°N. (This lack of seasonal movement of the equatorial rains seems to have been a remarkable feature of the 1960s and

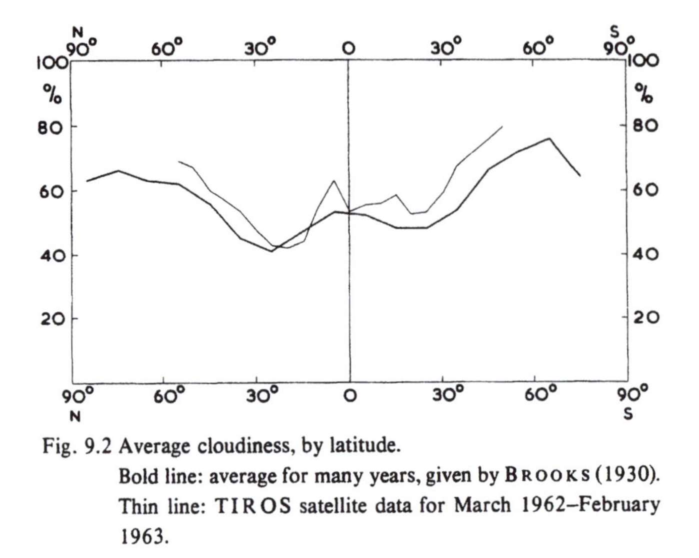

Fig. 9.2 Average cloudiness, by latitude.
Bold line: average for many years, given by BROOKS (1930).
Thin line: TIROS satellite data for March 1962–February 1963.

caused abnormally great rainfall totals in the zone over which the system lingered, while serious droughts occurred north and south of it.) The other principal seasonal change is the cloudiness of the Arctic in summer and its much clearer skies in winter (80–90°N July average 90%; January average 36%) attributable to the sealing-off of the moisture supply by ice and the remoteness of the main belt of cyclonicity, which moves to lower latitudes in winter than in the northern summer. The magnitude of the effect between latitudes 70° and 80° is similar in the Arctic and Antarctic (cloudiness averages: 70–80°N July–September 76%, January–March 56%; 70–80°S January–March 71%, June–August 52%). The distribution, and its variations, will be much better known in the future with the aid of satellite surveys.

Fog

Fog (apart from hill fog, where mountains and high ground are enveloped in cloud) occurs on average over 50 days a year only over areas of cold sea produced by the well-known regions of upwelling off the coasts of California, South America and southwest Africa, as well as over the cold Labrador Current water and over the polar seas and adjoining coasts. In a few of these areas the average exceeds 80 days/yr. The fogs over cold seas are commonest in spring and summer, when the temperature difference from the neighbouring warmer regions is greatest owing to the seasonal lag in the temperature of the sea. The frequency reaches 40 days/yr in some inland areas in the European lowlands and in eastern North and South America. Inland fogs are produced by night cooling, chiefly in autumn and winter, but also sometimes when the ground is wet at other seasons.

A classification of fogs should recognize the types produced by the following processes:

(1) Cooling of the air by:

(*a*) radiation cooling of the ground at night and in winter;
(*b*) advection of warm air over a cold surface (e.g. the cold seas mentioned in the previous paragraph);
(*c*) uplift of the air over mountains (hill fog);
(*d*) expansion due to pressure fall.

(2) Evaporation into cool air from:

(*a*) wet ground after rain;
(*b*) water bodies (seas, lakes, rivers) which are much warmer than the air (e.g. Arctic 'sea-smoke').

These fogs have the appearance of steam (and are sometimes called 'steam fogs'), often showing quite vigorous small-scale convection. They are seldom of great horizontal or vertical extent.

(3) Mixing of two nearly saturated air bodies of somewhat different temperature. (Also seldom extensive or of long duration.)

Precipitation

Figs. 9.3 (*a*) and (*b*) map the normal amount of water present in the atmosphere over all parts of the world in January and July. There is an obvious, strong relationship to the patterns of evaporation and transport by the winds, but the relationship to the world distribution of rainfall (fig. 9.4, p. 368) is a good deal less obvious and can be understood only by taking account of the processes that produce condensation and rainfall. In general, rainfall depends upon vertical motion in the atmosphere to produce the required chilling and condensation. Much more downput of moisture occurs in middle and subpolar latitudes, in the zones of cyclonic activity over the oceans and on the windward slopes of mountains, than might have been expected from the distribution of mean moisture content of the air. In lower latitudes, the desert zone receives even less – much less – rain than the moisture content of the air would indicate: the dryness is due to the prevailing anticyclonic subsidence motion, its warming of the air preventing condensation.

The world average water content of the atmosphere is about 25 mm, whereas the rainfall averages around 1000 mm a year. This means that on average the air holds only about 10 days' supply of water. However, the greatest falls of rain in 24 hours measured on the mountains in the tropics, and in the temperate zone, amount to many times the total water content of the air columns passing over the observation point at any given moment. This clearly demands a continued supply of moisture which is precipitated after a sojourn in the air that may be no more than an hour or two.

Precipitation may be classified according to the processes that produce it as follows:

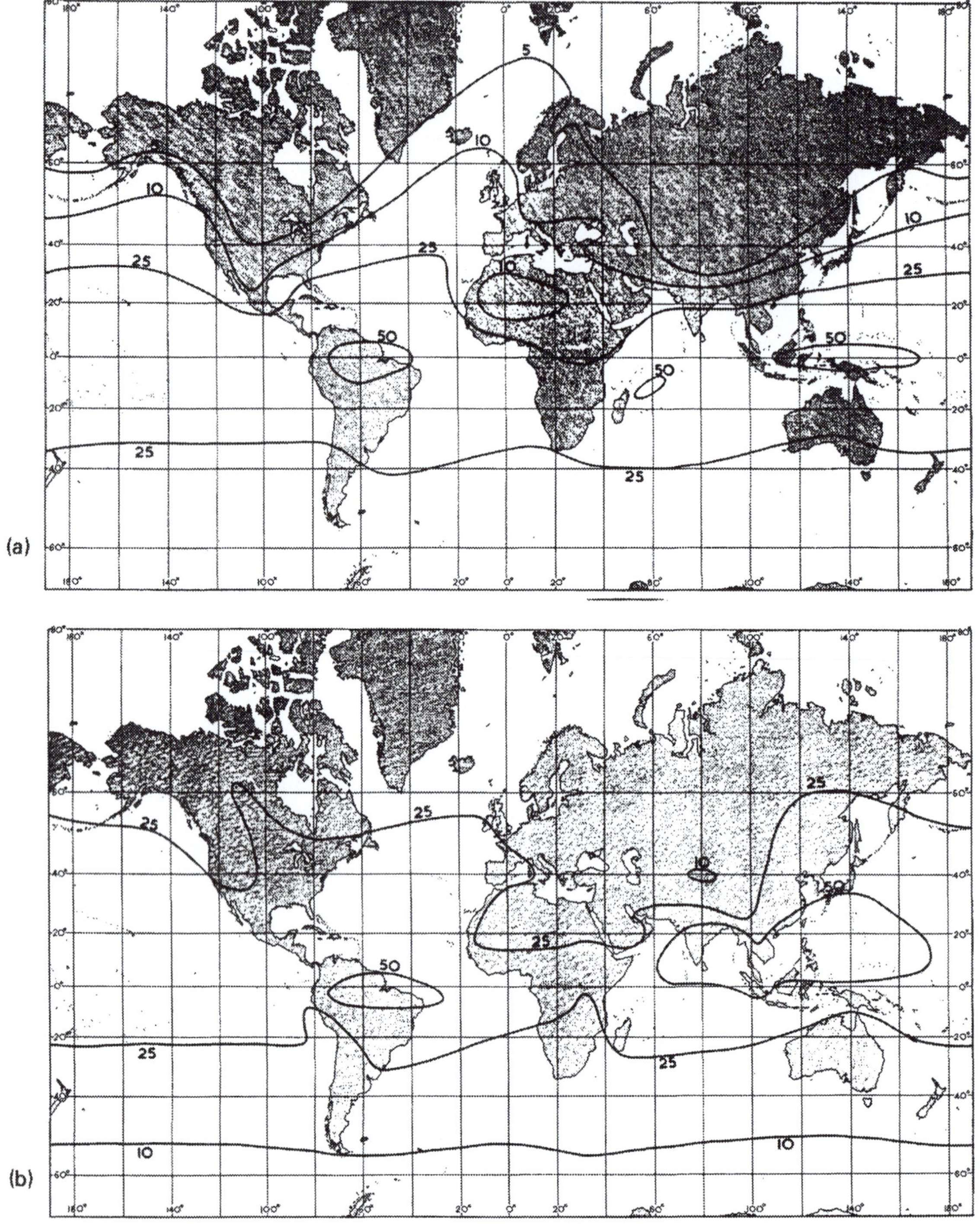

Fig. 9.3 Total moisture content of the atmosphere: average water content of the air columns in (*a*) January, (*b*) July, reduced to equivalent depth of water if it were all precipitated (mm).
(Periods of observations uncertain, believed generally about 1930–50.)

(1) *Orographic* – due to the forced uplift of moist airstreams blowing against mountain ranges. Prolonged rain or drizzle is characteristic on the windward side; dry weather is usual on the lee side and skies that are clear or which show only lenticular wave clouds in the undulating air motion downstream from the mountains.
(2) *Convective* – showers and downpours from convection clouds (cumulus and cumulonimbus types) formed in the up-currents in thermally unstable air. These are characteristically intense falls of rain, hail or snow; often sporadic in distribution, as are the cells of the thermal turbulence in air that is heated from below; and often of rather brief duration, as the cloud is carried past the observer by the wind. But, if the wind is light, the whole convection pattern may be nearly stationary and some places get prolonged heavy rain while other areas escape. In some situations, the shower cloud formation may be controlled and localized by hills or mountains and their day-time upslope (anabatic) breezes, giving prolonged downpours. These may be regarded as mixed orographic–convective rain situations.
(3) *Cyclonic* – due to the concentration of mainly ascending air motions in regions of convergent motion of the moist surface airstreams in a depression or where the isobars are cyclonically curved. If the air is also thermally unstable, mixed cyclonic–convective cloud development may produce wide regions of congested, dense cumulus-type cloud and moderate or heavy rainfall. This precipitation may be prolonged over many hours if the movement of the system is slow.
(4) *Frontal* – convergence of unlike airstreams at a front forces the less dense, warmer, and usually moister, air to rise over the frontal surface or over the line of convergence at the front. Gliding uplift of the warmer air over a frontal surface, as at advancing warm fronts, usually produces light to moderate rainfall, unless the uplifted airmass be thermally unstable in which case sporadic intense bursts of rain may occur. Intense, though brief, rainfall is also common with heap cloud development over the convergence line at an advancing cold front.

Statistical studies readily establish the general amounts and geographical extent of the rain, snow, etc., to be expected when each of these processes is at work. The yields in each case depend upon the moisture content of the air, the temperatures and temperature lapse with height prevailing, the vertical and horizontal velocities of the air motions involved, the size of the circulation system and its speed of translation (if it is mobile). The statistics therefore differ from one climate and one geographical situation to another and with the time of year. The size ranges of convection cloud and the larger circulation systems are indicated in Table App. I.12. It is impossible to indicate in these paragraphs the downputs typical of the great diversity of physical conditions mentioned above. In the case of orographic and convective rainfall all manner of geographical situations and local topographies introduce further variations.

A notable measure of the orographic effect was seen in March 1938 when persistent westerly winds blew across the mountains of Scotland all month. Though the situation was rather anticyclonic at times, the month's rainfall total at the head of Glen Barrisdale (at Loan 57·1°N 5·4°W 275 m a.s.l.) in the western Highlands was 1270 mm, a figure rarely exceeded in Britain. Just 140 km away across the main watershed of Scotland at Kettins (56·5°N 3·3°W 66 m a.s.l.) near Coupar Angus only 9 mm fell. The southwest monsoon blowing across the mountains of India and Burma produces effects of this magnitude each year, e.g. many-year average July rainfall at Akyab (20·1°N 92·9°E 9 m a.s.l.) on the coast of Burma 1400 mm and at Mandalay (22·0°N 96·1°E 77 m), across the Arakan Mts and in the north–south running Irrawaddy valley, only 69 mm.

SAWYER (1956) has studied the distribution of rainfall when a single depression crosses, or passes near to, the British Isles (using data for 1941–50): the figures of widest applicability are those giving the rainfall totals as a function of distance on either side of the track of the cyclone centre, particularly measurements over the eastern lowlands where the effect of the hills and mountains in the west and north is least. For each of the common cyclone tracks investigated there appeared to be an area close to, or just north of, the path of the centre where the average downput of rain was 15–16 mm except in the case of depressions passing just south of the country when the highest averages were about 13 mm along the south coast. The average falls exceeded 10 mm over a belt about 500 km wide in most cases. It appeared that the overall breadth of the precipitation belt was greatest with depressions of which the centres passed altogether north of Britain, these being characteristically the largest systems: they gave on average ⩾1·5 mm of rainfall everywhere in Britain and >2·5 mm everywhere north of a line from Exeter to York. The belt with over 5 mm average rainfall was clearly narrowest with those depressions which passed south of Britain, near the Channel; it was confined within about 250 km of the axis of the English Channel. These southern depressions tend to be smaller systems; though average rainfall exceeded 2·5 mm all up the eastern side of Britain, probably the effect of orographic exposure to east winds. There was, however, a wide range of variation of the rainfall given by a depression, the 'standard deviation' being 10–15 mm near the track of the depression. (The inverted commas draw attention to the fact that the distribution, as with most rainfall statistics, is skew, with a long tail towards large amounts.) Several slow-moving depressions in 1968 and 1969 passing east or northeast on tracks near, or over, the English Channel gave rainfall totals exceeding 100 mm in various lowland areas in southern England.

The rainfall with passing fronts was also examined in SAWYER's survey. With warm fronts which crossed the whole of Britain the average downput ranged from about 1·5 mm over an area from the Yorkshire plain to Norfolk and the south Midlands of England to 10 mm in the far west of Ireland and in the Hebrides: 60% of these warm fronts gave <0·1 mm near the coast of Norfolk and southeast England. With cold fronts which crossed the whole of Britain the average rainfall ranged from 1 to 2 mm over southern and eastern England to >10 mm everywhere north of a line from Galway and southwest Scotland to Wick and approached 15 mm in the far northwest of Scotland: over 40% gave <0·1 mm near the coast of Norfolk and southeast England. Orographic uplift and shelter clearly affect the figures in places: the average falls with warm fronts crossing the British Isles amount to ⩽2·5 mm over much of eastern Scotland; for cold fronts the average is about 3 mm near the east coast of Scotland and 2·5–3 mm along the south coast of Ireland. By contrast, the average exceeds 5 mm with either type of front in all the hill districts where the general height of the tops is over 800 m. And the tendency for stronger development of convection cloud probably accounts for locally greater average rainfall with cold fronts (about 2·5 mm) over inland parts of East Anglia and (2·5–4 mm) over the English Midlands.

It is when two or more of the rainfall processes in our classification are operating simultaneously that extreme falls can occur.

The world distribution of rainfall (depicted in fig. 9.4 for the present epoch) is the end-product of these processes. Because the operation of orographic uplift and convection is affected by every local change of slope and aspect, and convection by the nature of the surface besides, no map can display all the intricacies of rainfall distribution. Small-scale maps especially are drastically simplified representations and necessarily omit all the extremes. (Some examples of these will be referred to in what follows.) The broad-scale distribution of rainfall is itself quite inadequately known over the sea because of the difficulty of measurement on a ship, which constitutes an obstacle that diverts the path of the wind upwards (carrying the raindrops and snowflakes with it) and is itself tossed up and down by the sea. TUCKER (1961) has devised a method of estimating the rainfall at Ocean Weather Ships manned by trained meteorological staff reporting 'slight', 'moderate' or 'heavy' rain etc., according to standard practice.[1] The corresponding rates of rainfall are

1. The definitions given in the *Observer's Handbook* (London, H.M.S.O. 1956) are:

Slight rain	Up to 0·5 mm/hr
Moderate rain	0·5–4 mm/hr
Heavy rain	Over 4 mm/hr

Because convective processes are generally more vigorous than frontal or orographic upgliding, the rainfall from shower clouds is generally more vigorous and the following guidance is given to observers:

Slight shower	Up to about 2 mm/hr
Moderate shower	2–10 mm/hr
Heavy shower	10–50 mm/hr
Violent shower	Over about 50 mm/hr

The heavier rates of rainfall are not usually kept up for very long in the middle and higher latitudes investigated by TUCKER: hence statistical comparisons of observers' estimates with the measured amounts of water deposited in one complete hour of continuous rain in Britain gave the following averages:

Slight continuous drizzle	0·3 mm/hr
Moderate continuous drizzle	0·6 mm/hr
Slight continuous rain	0·7 mm/hr
Moderate continuous rain	1·6 mm/hr
Heavy continuous rain	2·4 mm/hr

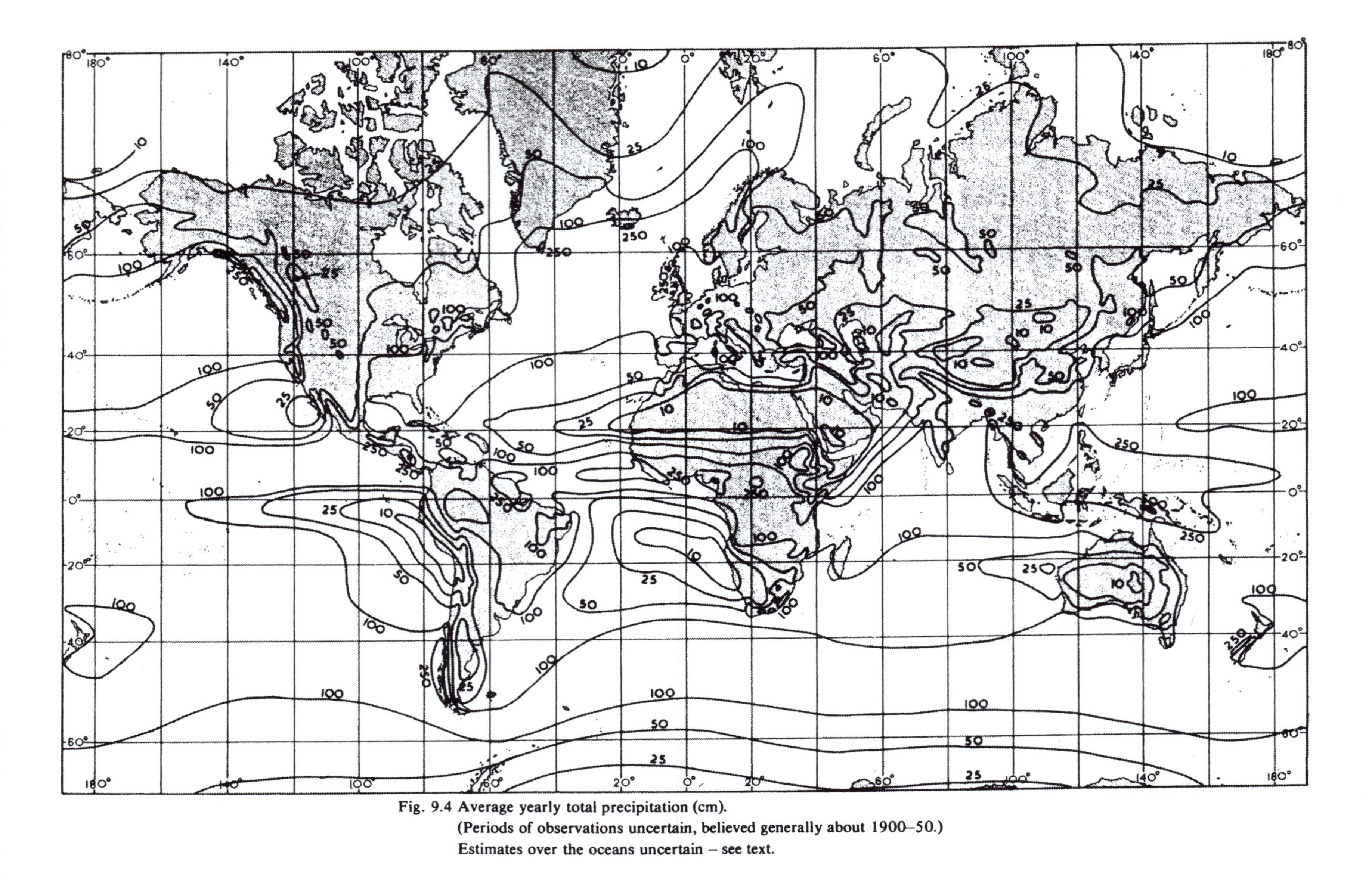

Fig. 9.4 Average yearly total precipitation (cm).
(Periods of observations uncertain, believed generally about 1900–50.)
Estimates over the oceans uncertain – see text.

quite well known from observatories on land. TUCKER's results for the North Atlantic 1953–7 average, sketched in fig. 9.5, differ materially from fig. 9.4 in showing much less rainfall over the region around the Azores and over Biscay and much greater totals north of 55 °N. The rainfall on islands and coasts may be no very good guide to the rainfall at sea because of orographic effects; from TUCKER's results it appears to be generally greater than the rainfall over the open sea. TUCKER's estimates may well be the best so far available, though the period of years covered is short. A similar poleward shift, and increase, of the greatest rainfalls shown in fig. 9.4 in association with the subpolar cyclonic activity may be needed for the North Pacific and Southern Oceans. Indeed, a similar result to TUCKER's has been found by ELLIOTT (1971), comparing rainfall measurements over the Pacific and on the Oregon coast.

The bold line in fig. 9.6 gives an overall view of the distribution of rainfall by latitude, as

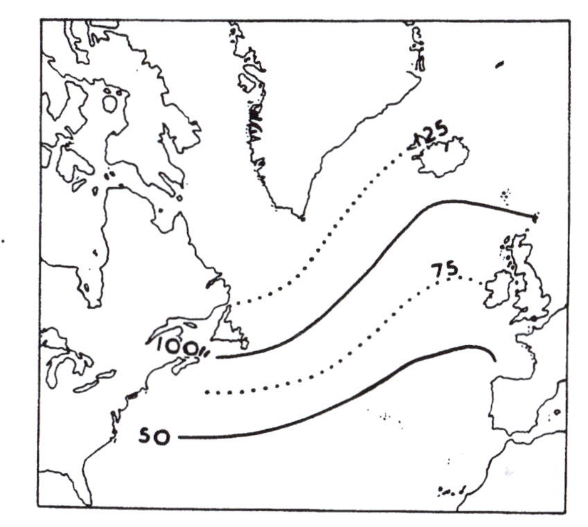

Fig. 9.5 Average yearly total precipitation over the North Atlantic, 1953–7. (*As derived by* TUCKER – see text.)

far as it is known, at the present epoch. The dotted line on the upper diagram indicates how the curve would be amended in the northern hemisphere portion if TUCKER's estimates of rainfall over the ocean were applied to both the Atlantic and the Pacific. Some amendment of this kind may also be required in the southern hemisphere. The thin line shows how the best available estimates of evaporation are distributed. Since this gives a world average evaporation of 74 cm/yr and the bold curve gives a world average downput of 100 cm/yr, the margin of uncertainty in present estimates becomes all too clear. However, it is certain that evaporation exceeds downput in the subtropical anticyclone zones and presumably also in much of the Trade Wind zones, whereas the downput of moisture is in excess in all other latitudes.

The great features of the world distribution are:

(i) the equatorial rainbelt, much broader over the continents than the oceans and including the monsoon rains over southern and eastern Asia;

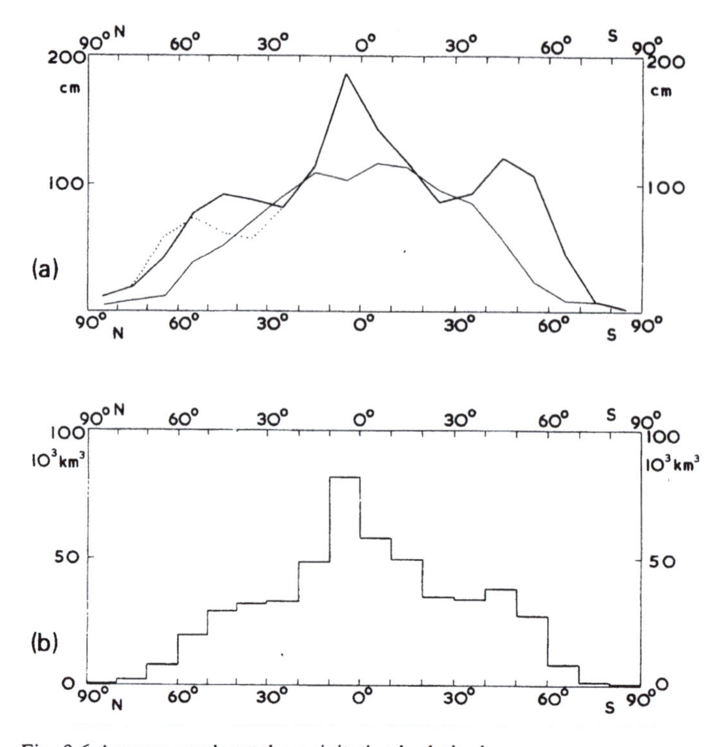

Fig. 9.6 Average yearly total precipitation by latitude.
Period of observations uncertain, believed generally about 1900–50.
(*a*) Averages in cm/yr.
Dotted line: amendment to correspond to TUCKER's amended derivation of precipitation over the ocean (observations 1953–7).
Thin line: estimated average evaporation.
(*As given by* GENTILLI *1958*.)
(*b*) Total precipitation over each latitude zone in thousands of cubic kilometres.
(*Adapted from* GENTILLI.)

(ii) the desert zones induced by the subtropical anticyclones;

(iii) the extensive precipitation belts, and the large amounts, associated with the temperate zone and subpolar frontal and cyclonic activity;

(iv) orographic intensification of the downput in (i) and (iii) on the windward slopes and crests of the mountain ranges;

(v) 'rain-shadow' (orographic diminution) in the lee of mountain ranges, producing deserts in a few extreme cases;

(vi) the cold deserts in the polar regions, where the total downput is small because of the small water content of the cold air.

The following details of the distribution of precipitation illustrate important points:

(1) The greatest regular rainfalls in the world occur where mountain ranges lie athwart moisture-laden winds and force them to rise.

The extreme example is the northeast corner of the Bay of Bengal where the curve of the Himalayas and the other ranges before them presents an open V-shaped trap towards the SW monsoon wind, which has no way out other than ascent. Thus, Cherrapunji, Assam (25·3°N 91·8°E, 1313 m a.s.l.) has an average yearly rainfall (1851–1950) of about 11 430 mm, 86% of it falling in the monsoon months of May to September. (In 1861, the wettest year at Cherrapunji, the total was 23 000 mm.) For similar reasons, the greatest annual rainfalls in Europe are reported from just near the main watershed of the mountains in Wales – again where the ranges open a V-shaped trap to the SW winds from the Atlantic – near Glaslyn (53·1°N 4·1°W) at a height of 762 m on Snowdon with an average of 5030 mm/yr and a point 1097 m a.s.l. on the southwest side of the mountains of Jugoslavia, above Kotor (42·4°N 18·5°E), near the Adriatic with 4630 mm/yr.

(2) The greatest rainfalls anywhere near sea level are in rather similar situations to (1), often near coasts athwart the Trade Winds near the equator. Examples are Ponape, Caroline Islands (7·0°N 158·2°E 12 m a.s.l.) with an average of 4875 mm/yr for 1931–60; Hinatuan on Mindanao, Philippine Islands (8·3°N 126·2°E 6 m a.s.l.) average 4305 mm/yr; Hilo, Hawaii (19·7°N 155·1°W 11 m) 3470 mm/yr; Cristobal, Canal Zone (9·3°N 79·9°W 11 m) 3361 mm/yr and Coco Beach, Gabon (1·0°N 9·6°E 13 m) 3509 mm/yr. The 1916–50 average 3342 mm/yr for Seathwaite, Cumberland, on the southwest side of the English Lake District (54·4°N 3·2°W) at only 128 m a.s.l., in the belt of prevailing westerlies coming from the warm Atlantic, is in the same class.

(3) The rainfalls near the east coasts of the U.S.A. and of Asia, and similarly near the east coasts of South America, South Africa and Australia, in subtropical latitudes are significantly increased by the warm ocean currents nearby. The reality of the connexion seems confirmed by computer-calculated models of the atmospheric and ocean circulations by MANABE and BRYAN (1969).

(4) Rainfall is greatly reduced over all the areas of cold sea produced by marked upwelling in the lower latitudes, e.g. off California, Peru and southwest Africa and over the cold-water region of the equatorial Pacific. The last-named is the zone of desert islands with a remarkably low average rainfall, notably Malden Island (4·1°S 155·0°W 1 m a.s.l.) 1890–1919 average 705 mm/yr, but with an extraordinary variability from year to year (already instanced in Chapter 6, p. 249), the wet years being associated with extension to the area from the North Pacific around 5°N of a normal equatorial sea surface temperature level and incursions of the intertropical frontal activity (also mainly from the North Pacific).

(5) Common positions of convergence lines in the flow of surface winds tend to locate precipitation maxima, e.g. the heaviest rainfall averages in the region of the great East African Lake Victoria are over the western half of the lake: the greatest averages approach 2300 mm/yr on the islands near 0·3°S 32·3°E, compared with figures about 1000–1200 mm/year around most of the edges of the lake, though the nearest part of the western shore has 2000 mm/yr (FLOHN and FRAEDRICH 1961). The convergence over the lake of the night land-breeze systems from the surrounding shores is displaced somewhat westwards by the prevailing upper E'ly winds over that zone. Another case is the maximum of summer rainfall about 40 km inland from the coast of Texas (EDDY 1966), similarly with the convergent sea breezes over the island of Malta (LAMB 1955*a*), and presumably the effect is quite general. Also the convergence of day-time upslope (anabatic) winds over mountain massifs, i.e. converging with the general wind or with each other, must determine local precipitation maxima near the ridges. (See, for example, FLOHN 1955, 1968.) This helps to produce particularly heavy rainfalls over massifs that are largely surrounded by the sea, where the combination of day-time sea breezes and upslope drift over the mountains, transports extra moisture: the mountains of south Norway, Scotland, the English Lake District and North Wales all provide examples.

Similarly, the additive effects of orographic uplift, and/or thermal convection intensified at some local wind convergence line, commonly locate the most intense falls of rain in a given cyclonic situation. Some of the greatest 24-hour totals, 1000–1250 mm, have been reported near the mountains of Luzon, Philippine Islands, as well as in Taiwan and Japan during the passage of typhoons. The world record was reported from Cilaos on Réunion (27·1°S 55·5°E 1200 m a.s.l.) in the tropical cyclone belt of the Indian Ocean on 10–11 September 1963 when 1870 mm fell. The greatest 24-hour fall recorded in Europe is believed to be 355 mm measured at Kirche Wang on the northern slopes of the Riesengebirge (Silesia) near Schneekoppe (50·7°N 15·8°E) on 29–30 July 1897, near the occlusion of a slow-moving depression (SCHERHAG 1948, p. 285). The greatest reported in England is 279 mm from a thunderstorm near the sea-breeze front in Dorset (Martinstown) on 18 July 1955.

In mountain regions it is often possible to point to a fairly general variation of the rainfall measurements with height above sea level, though the amounts are distinctively different on either side of the range, where the prevailing wind blows across it. HADER (1954) gives 1851–1950 average values of the rainfall over an area of the Alps in Austria, which increase with height from about 600 mm/yr on the lower ground near 200 m a.s.l. to 1000 mm/yr about 750 m a.s.l. and nearly 1900 mm/yr at 2100 m a.s.l. This increase with height is, however, attributable to the increasing uplift of the wind and steepness of the mountains, which, as we saw in a previous section of this chapter, can produce the greatest concentrations of condensation products in the upper part of cumuliform clouds. Similar examples can be found elsewhere in middle and subtropical latitudes. Many high mountain ranges, however, rise above a height zone of maximum precipitation to levels where the prevailing lower temperatures and lesser moisture content of the air appear as the main influence and induce a general decrease of precipitation at greater heights. This is observed on most high mountains in the tropics, often above about 1000 to 1500 m a.s.l., e.g. the Andes and Himalayas, also on the Altai in central Asia. A contributory factor, operative even where the mountains are steep, is the amount of moisture shed in precipitation on the lower slopes. Perhaps the clearest example of this variation with height is to be seen in the amounts of snow accumulating on the windward-facing south and west slopes of the ice-sheet in northwest Greenland, where the gradient of the ice surface is remarkably uniform (and slight): fig. 9.7 gives the results of measurements near 70° and 77°N, showing maximum accumulation near the 2000- and 1000-m levels respectively (BENSON 1962).

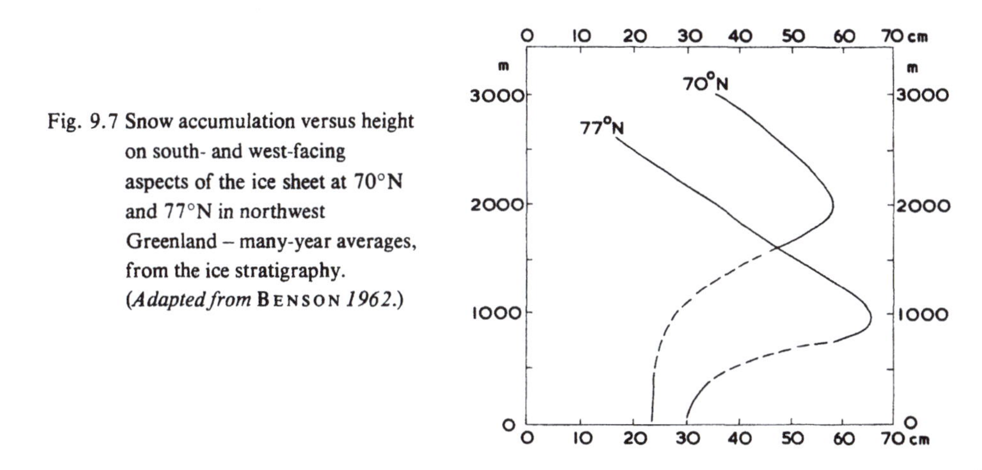

Fig. 9.7 Snow accumulation versus height on south- and west-facing aspects of the ice sheet at 70°N and 77°N in northwest Greenland – many-year averages, from the ice stratigraphy. (*Adapted from* BENSON *1962.*)

The study of year-to-year variations and shorter-term anomalies of rainfall is discussed elsewhere, in Chapters 6, 7 and 10. The normal patterns of diurnal and seasonal variation may be described as giving generally most rain over land at the warmest times and seasons, when thermal convection is strongest, and least around dawn and in late winter. Local departures from this pattern may be produced by some complex sequences of diurnal breeze development. Over the seas and oceans generally diurnal and seasonal variation is slight, except where dictated by a marked seasonal shift of cyclonic activity (e.g. Mediterranean climates with dry summers and wet, stormy winters). Higher water temperatures in summer, greater windiness and prevalence of cold, thermally unstable airstreams in winter produce some approach to similar amounts of rainfall. Over water bodies of more limited extent, from lakes to inland seas like Lake Victoria, convergence of the night breezes off the shores tends to produce the greatest rainfall at night.

Secular variation of rainfall totals seems likely to take various forms:

(1) An overall increase or decrease of the world average, as should, for instance, occur with changes in the evaporation and moisture content of the air due to:

(*a*) rise or fall of the general world temperature level;
(*b*) stronger or lighter winds. This may have a critical point in low latitudes (where the seas are warm and evaporation great (cf. fig. 9.1)) because of possible great changes in the extent of winds just strong enough to produce breaking waves: the mean strength of the Trade Winds is about this critical value.

(2) An increase or decrease over most of the middle and higher latitudes, due to:

(*a*) a change in the prevailing rate of wind transport from the lower latitudes where the greatest amounts of water vapour are taken up; or
(*b*) a change in the rate of evaporation from the seas in these latitudes owing to higher or lower temperatures, or stronger or lighter winds;
(*c*) a change in the abundance of ice nuclei, produced by a lowering or raising of the general temperature level in the upper air.

(3) A poleward or equatorward shift of the belts of cyclonic and frontal activity over either or both hemispheres.

(4) A longitude shift of the cyclonic activity generated near the eastern sides of cold troughs in the upper westerlies, due to a change of prevailing wave length.

The regions of high rainfall seen in fig. 9.4 over eastern North America and the western Siberian lowlands owe their origin partly to this relationship to upper cold troughs and seem subject to some longitude variations of this kind. Similar east–west secular shifts may be traceable, despite the orographic complications, also in east Asia and Japan as well as New Zealand and eastern Australia.

(5) More localized changes produced by possible long-term changes in the frequency of cyclonic development:

(*a*) over sea areas which in one epoch are notably warm for their surroundings and in another epoch not so;
(*b*) in areas which in one epoch commonly experience lee-trough or eddy effects of a mountain chain and in another epoch do not, because the prevailing windstream is otherwise directed.

(6) Changes in direction of the prevailing winds, which must alter the positions and intensity of topographic (hill or coast) uplift and shelter effects.

In later sections of this book (Volume 2) we shall notice observational evidence of some of these variations.

DROZDOV (1966) has examined statistically the changes in precipitation that occur over the rest of the northern hemisphere when the Arctic Ocean (north polar basin) is warmer than normal, and by reasoned extrapolation has presented maps in terms of percentage deviations, summer (April–October) and winter (November–March), for the simple case of a uniform warming by 5°C – as might happen if the floating pack-ice were removed. The maps show some intricate detail which is doubtless far from guaranteeable, but the main effects appear simple:

(i) a substantial increase in precipitation at most places north of 65°N, extending farther south over North America (to 40–50°N) and over Eurasia in winter (to about 55°N). The increase is attributable partly to a more northern position of the depression tracks and partly to increased moisture content of the air;

(ii) reduced precipitation over a belt in middle latitudes, which presumably marks a poleward shift of the influence of the subtropical anticyclones and reduced zonal flow of the winds. The lost precipitation in some places amounts to 30%. The zone affected appears to be 45–65°N across Europe and Asia in winter, and generally 40–50°N (though more devious and patchy, including parts of Britain and Scandinavia but not central Europe) in summer. Across the American sector it is shown at 40–50°N in winter, but 0–30°N in summer;

(iii) increased precipitation in latitudes between zone (ii) and the equator. Patchy areas of very big increase presumably indicate orographic influence, northward spread of monsoon rainfall and meridional wind-flow patterns.

Seasonal variation

Seasonal variations of rainfall have been noted in Chapter 4 and explained by the regular shifts of the main zones of the atmospheric circulation. These seasonal variations are nowhere more pronounced than near the fringes of the arid zones, either where the polar front depression rains bring a winter maximum and the summers are dry – as in the Mediterranean today – or where the equatorial and monsoon rains arrive in summer and the winters are dry – as at Freetown, Cherrapunji or Darwin – see the data table in Part II, Chapter 11 (also DEACON 1953). There are also places with two rainy seasons in the course of the year, as a given rainbelt passes first north and then south over the area – notably in equatorial East Africa and in parts of China – see Chapter 4 (and LAMB 1966).

Special forms of precipitation

Supercooled drizzle

Liable to form when large amounts of moisture are condensed in a cloud layer usually only a few hundred metres thick at temperatures between 0° and about –5°C.

Rime-like ice deposit on objects at the ground grows more rapidly than with fog; if very wet, glaze (a glassy coating of ice).

Freezing rain

Rain may fall when the surface air temperature is below 0° (even down to –5°C in some cases) when there is warm air overrunning a shallow freezing layer near the ground.

If the warm front is held stationary by the resistance of the flow of the dense cold air next to the surface, or by the stagnation (inertia) of the latter in a terrain sheltered by mountains, rain which freezes on impact with objects at the ground may continue for many hours. Heavy deposits of glazed ice soon form, doing great damage to electric wires, trees, etc.

These conditions, known in New England as 'ice-storms', are commonest in climates near the limit of a continental winter snow cover, with the threatened penetration of cyclonic activity developing over a warm ocean.

Snow

The prevailing form of precipitation at air temperatures below 0°C. The types of snow vary, largely according to the temperature at which they are formed, but are based on the hexagonal shape and characteristic angle (60°) of the individual ice crystal, however small it may be. Snow flakes grow by accretions to the six-pointed stars which are the commonest type formed between about −12° and −18°C. At very low temperatures any precipitation that falls consists of single ice crystals. The higher the temperature, the larger the snow flakes tend to become. Terminal fall velocities vary according to texture, but are typically about 1 m/sec for a flake of 5-mm diameter (against 8 m/sec for a raindrop of the same size) and reach about 2 m/sec for the largest snow flakes of ⩾20-mm diameter.

Such large flakes, especially when wet, and the more compact types of snow-grains which also have relatively great fall speeds for their size, survive longest in falling through air at temperatures above the freezing point. Wet snow, or sleet, may sometimes reach the ground therefore at air temperatures up to +2° or 3°C. Prolonged snowfall tends to cool the air, especially with the aid of evaporation cooling when the air is initially dry; extreme cases have been reported where long-continued precipitation through a deep isothermal layer, starting with the freezing level as high as 1500 m above the ground, ultimately turned from rain to snow and brought the freezing level right down. With big lapse rates of temperature with height, showers from convection clouds in Arctic air that has crossed part of the warm Atlantic, in spring, may bring large snowflakes to the ground when the surface temperature just before the shower was as high as +6°C.

Clearly, the probability of snow rather than rain reaching the ground depends greatly upon the height of the freezing level, the lapse rate of temperature with height and the fall-speeds of the flakes (possibly aided in squally showers by down-draughts of colder air). A convenient way of summarizing the position is in terms of 1000–500 mb thickness (LAMB 1955*b*). For places on the low ground in Britain and northern and central Europe snow or rain is on average about equally probable at a thickness of 5270 m. When there is already extensive snow lying snowfall is commoner than rain at all thicknesses below 5300, in some cases 5350 m: this is because of the frequency of light precipitation from cloud beneath an inversion. On the other hand, over the ocean and small island groups, such as

Shetland, with water temperatures about 10°C, rain is commoner than snow at all thicknesses greater than 5220 m.

The average 1000–500 mb thickness over the western and southern limit of the main snow cover of Eurasia in winter is also about 5270 m.

Snow begins to lie once the temperature at the very surface of dry ground is close to 0°. Where there is water it takes longer because of the latent heat given out. Thawing snow remains at 0°, and keeps the air near the surface within a few degrees of the freezing point (+3° is rarely exceeded at the usual height of thermometer screens as long as the snow cover remains complete): this is because of the latent heat absorbed. The maximum rates of disappearance of snow in middle Europe (50–55°N) appear to be about 10–15 cm in 24 hours due to rain from mild air, 70 cm in 10 days of spring (March–April) sunshine and (dry) warm air advection.

The density of lying snow varies with its texture and increases with age and bedding down. A layer of new-fallen snow holds much air trapped in the interstices: its specific gravity is oftenest about 0·1, i.e. 1 cm depth of new snow is equivalent to about 1 mm of rain. The range of variation of the S.G. of snow lying on the surface is from about 0·03 to 0·3. The older it is, in general, the denser it becomes; and when it is compressed into ice, as at all depths greater than 100 m in the Greenland and Antarctic ice caps, its density is about 0·9.

Recrystallization processes within the snow lying on glaciers and ice caps produce fairly regular seasonal differences of structure and density, which result in a stratigraphy of annual layers in which year-by-year differences of snowfall or accumulation and some traces of the temperatures prevailing can be recognized (Benson 1962, 1967). The air in the minute spaces within the snow is saturated with water vapour. At the end of each summer the ice surface begins to cool sharply by net loss of radiation to the sky, a strong gradient of temperature develops within the top layer of the snow and this creates an upward directed gradient of vapour pressure. Thus, in early September 1952 at 2150 m on the ice cap in northern Greenland at 77°N 46°W Benson measured −25°C at the snow surface and −15°C just 20 cm below it. The vapour pressure difference between these temperatures at saturation over ice is 1 mb. (It is worth noticing that this is a greater value than would occur at the lower temperatures prevailing in the winter.) Hence, there is a transport of water vapour upwards, some of it escaping into the atmosphere but much of it crystallizing within the surface layer of the snow where it encounters the lowest temperature. This transport is greatly assisted – indeed only takes place at a significant rate – when a wind, especially a gusty wind, blowing over the snow, accelerates it. The process produces wind-crust and regularly forms a thin layer, or layers, of relatively high density, and harder, ice as soon as the autumn cooling sets in: this is at the expense of the summer snow below, where substance is lost by sublimation. In consequence, the density becomes least in the summer layers of the stratigraphy. Langway (1967) found the following sequence of specific gravities in one year's stratum of the ice sheet in north Greenland, in a sample core taken in May:

Summer ice of the previous year	0·27
Autumn ice	A fluctuating increase of specific gravity, the earliest peak value 0·30 and the last peaks in late autumn, 0·35–0·36
Winter ice	0·35 in early winter, and gradual increase to 0·41
Spring ice/snow	0·30

At heights too great for any summer melting the upward transport of water vapour described in the previous paragraph is the only one, other than compaction, which redistributes the material within the annual layers. There is no corresponding downward transfer of vapour in summer when the surface is warmest, since this creates a very stable vertical stratification of the air within the snow, the lightest being on top. But where summer melting occurs, the water percolates downwards into the snow until it freezes again.

In order to describe the processes concerned in the formation and survival of a *glacier*, it is necessary to distinguish:

(i) the *firn line*, the upper limit of ablation (i.e. of net loss of substance by melting and run-off, evaporation and sublimation or removal by wind – the last-named known as 'deflation'). Snow on the surface of the glacier below this level disappears in summer;
(ii) the *dry snow line*, above which no melting occurs, even in summer, and the snow remains dry all the year through.

Between (i) and (ii) the snow becomes wet in summer and water percolates downward. *Firn* is the name given to snow more than one year old, which undergoes considerable transformation by percolation, recrystallization and granular texture development, and some compaction, but still contains some air and is not yet fully compacted into ice. Its specific gravity is from about 0·4 to 0·8.

The *snow line* on ice-free terrain is the level above which the snow survives the summer melting. It is therefore logically continuous with the firn line on glaciers. In reality it is far from being a continuous or horizontal line because it is subject to innumerable local topographic effects, as when snow blows off the ridges and tops, and drifts into sheltered hollows, clefts and crannies where, on the side shaded from the afternoon sun, it may survive the summer at elevations much lower than the general snow line. The general snow line itself is normally lower on the side of the hills away from the afternoon sun than it is on the sunny side. It also varies somewhat from year to year. Hence, any statement about the snow line can only be a generalization about the level (within a margin of some hundreds of metres) above which snow should be perennial (FLINT 1957). Fig. 9.8 shows how the height of the snow line varies with latitude in the present epoch. The greatest heights are in continental interiors, especially in the arid zones and particularly near the elevated tablelands of Tibet (where it reaches about 6000 m) and in subtropical North and South America (FLOHN 1959). The lowest snow lines for their latitudes occur in maritime climates with heavy precipitation.

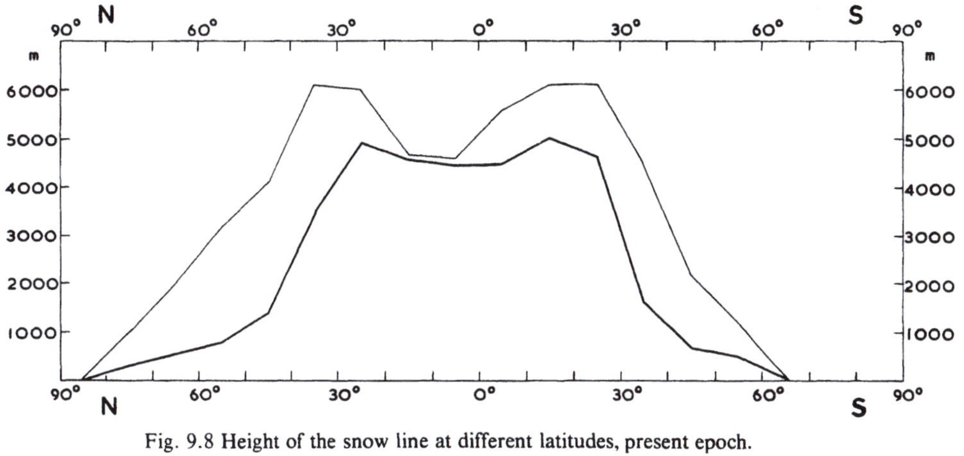

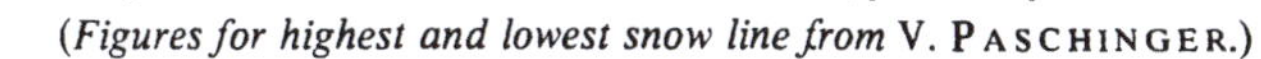
Fig. 9.8 Height of the snow line at different latitudes, present epoch.
(*Figures for highest and lowest snow line from* V. PASCHINGER.)

Soft hail (*Graupel*)

Formed from ice crystals or small frozen drops on collision with supercooled cloud droplets, which appear as a coating of rime on the central crystal or drop. The resulting shapes are mostly conical or irregular. The sizes are commonly 2–4 mm, occasionally up to 6 mm, in diameter; the corresponding fall speeds are generally 1·5–2·5 m/sec. This is a winter, or cold-climate, form of precipitation, mainly from shower clouds or rather turbulent layer clouds.

Hail

Formed in violent convection clouds, especially over continents in middle latitudes, where moisture is sufficiently abundant and great lapse rates of temperature with height are developed. It is therefore associated with thunderstorms, and in some parts of the U.S.A. it may occur in nearly every thunderstorm cloud. The formation process involves successive accretions of liquid water when the hailstone falls below the freezing level and of supercooled droplets when it is carried above that level again by the up-currents in the cloud. This process may be repeated as often as the growing hailstone falls into an air current strong enough to carry it up again. And the final structure of the resulting hailstone is of concentric shells of alternately clear and opaque ice. The commonest sizes of hailstones are 0·5 to 2 cm in diameter both in Europe and U.S.A. Large hailstones over 2·5 cm in diameter probably fall somewhere in England about once a year: the largest in about 10 years were 7–10 cm across, and weighed 200 g, at Horsham on 5 October 1958 from a storm that was also accompanied by a tornado. Giant hailstones up to about 1 kg in weight, and 10–15 cm across, have been reported in central Europe and in mid-western U.S.A. where the damage caused to buildings and crops sometimes amounts to millions of dollars from a single storm (FLORA 1956); the Canadian prairies are similarly affected, but less often and seldom so severely. Hailstones up to 2–4 kg in weight have been reported from Spain, India and China. The up-currents required to support giant hailstones, and their terminal velocities when falling, are up to 50 m/sec. (See LUDLAM 1961, where a clear diagram of the air motions and regions of hail formation within a cumulonimbus is given.)

Hail is most frequent at those times of day and in those months of the year when convection clouds are most strongly developed, i.e. in the afternoons and in spring–summer in the continental regions where it is commonest. In the U.S.A. the highest frequencies are over the Great Plains and eastern flank of the Rocky Mts, especially near the front/boundary of moist air from the Gulf of Mexico and a cold wave advancing across the mountains: the greatest reported frequency for any one locality averages 9 storms a year in part of Wyoming. Reported average frequencies in other parts of the world reach 13/yr in southwest Germany, 10 in Paris, 9 in Malta and 7 in Beirut. In the Mediterranean it is a winter phenomenon, also in southern Australia and Tasmania, where the frequency is not negligible. Hail is not unknown nearer the equator, and severe storms, with large hail, have been reported, though rarely, from northern Australia, Fiji, the Philippine Islands, and even in the Sahara desert.

Thunderstorms

The conditions for thunder are high moisture content of the air and a strong vertical lapse rate of temperature, both extending to great heights. The electrical charges are carried to different parts of the cloud through the splitting of the larger raindrops and shattering of ice crystals in the violent air currents developed. Thunderstorms therefore develop where and when air with a sufficient moisture supply is strongly heated from below, and especially where mountains and their local wind regimes cause added uplift. Coastal sea breezes also affect the development of thunderstorms, reducing the frequency just at the coast and in the zone of subsiding air that supplies the diurnal breeze, but increasing the frequency and severity of the storms in the zone of convergence with the general wind at some distance inland: in this the additional moisture supply from the sea also plays a part. Thunderstorms over land are therefore most frequent in the afternoon and in the warmer seasons of the year, but over the oceans and ocean coasts where thunderstorms occur in cold polar airstreams they are commonest in winter and (owing to radiation cooling of the cloud-tops) at night.

Fig. 9.9 shows the world distribution of thunder. This world survey is not complete, because observation standards and networks are not uniform – there is no 24-hour watch in many areas and the observation network hardly exists over the sea – but it is the best that can be provided at present, using 50–90 years of observation up to the 1950s wherever such long runs exist. (Some published references are unreliable in important areas through using only short runs of data since the political changes following the Second World War.) A more complete and uniform coverage may be possible in due course from the mapping of radio-fixed atmospherics instead of actual thunderstorm reports.

The average frequency is greatest, believed to exceed 200 days/yr, in parts of Java, in the central African highlands, near the coast of the Gulf of Guinea and on slopes near the Congo and Amazon basins. Thunder is rare in high latitudes except over warm seas and mountainous coasts: in these regions it is commonest in winter, when the air–sea temperature contrast is greatest. Over land in the temperate zones the highest frequencies are in summer and generally about the hottest part of the day. The thunderstorm frequency reaches rather over 50 days/yr in those parts of mid-western U.S.A. where hailstorms and tornadoes are most frequent and 20–30 days/yr in the corresponding areas of Europe.

Lightning causes on average 11 deaths a year in Britain and 230 in the United States, where the damage to property averages 10–15 million dollars a year. It is believed to be the cause of 70% of all forest fires in California and is estimated to start on average 10 000 forest fires yearly within the area of the United States (LANE 1968).

Effects of latent heat conversions upon local climates

A brief general survey of the importance of long-distance transport of latent heat in world climate distribution in the present epoch has been given in Chapter 2 and may be gleaned from figs. 2.16 and 2.17. Here we need only notice the effects of release of latent heat of condensation where they are most readily observed.

(i) *Föhn winds and föhn effect*

When a windstream is forced to rise over a range of hills or mountains, being reduced to its saturation temperature (dew point) and shedding moisture as drizzle and rain during the ascent over the windward side, the latent heat of condensation is converted into feelable heat and results in higher temperatures in the air descending to the same level on the leeward side. The effect is named after the warm south wind that descends from the Alps in Austria and Bavaria, but its operation is universal.

In general, the air rises (fig. 9.10) from low ground at height z_o on the windward side where the temperature is T_o to its condensation level at height z_c, its temperature falling to T_c at the dry adiabatic lapse rate α_d; it then continues its ascent to the height of the main ridge of the mountains z_r, its temperature falling at the saturated adiabatic rate α_s to T_r. Then $T_o - T_c = (z_c - z_o)\alpha_d$ and $T_c - T_r = (z_r - z_c)\alpha_s$. On the lee side the cloud typically ceases near the ridge and during descent the air's temperature increases at the dry adiabatic rate throughout, so that at any height z the temperature T is given by $T - T_r = (z_r - z)\alpha_d$. α_d is greater than α_s, so that height for height the temperatures on the lee side are greater than on the other side; the difference increases the greater the descent and the farther the cloud-base is below the ridge level on the windward side, i.e. the greater $(z_r - z_c)$ is.

The effect is made clear by two numerical examples in Table 9.8, chosen as typical values for two climates where it is very common; α_d is taken as 1°C/100 m and α_s as 0·6°C/100 m. In the diagram (fig. 9.10) z_p and T_p stand for the height of the plain, or valley bottoms, on the lee side. Typical values for eastern Scotland and for the northern Alpine valleys are

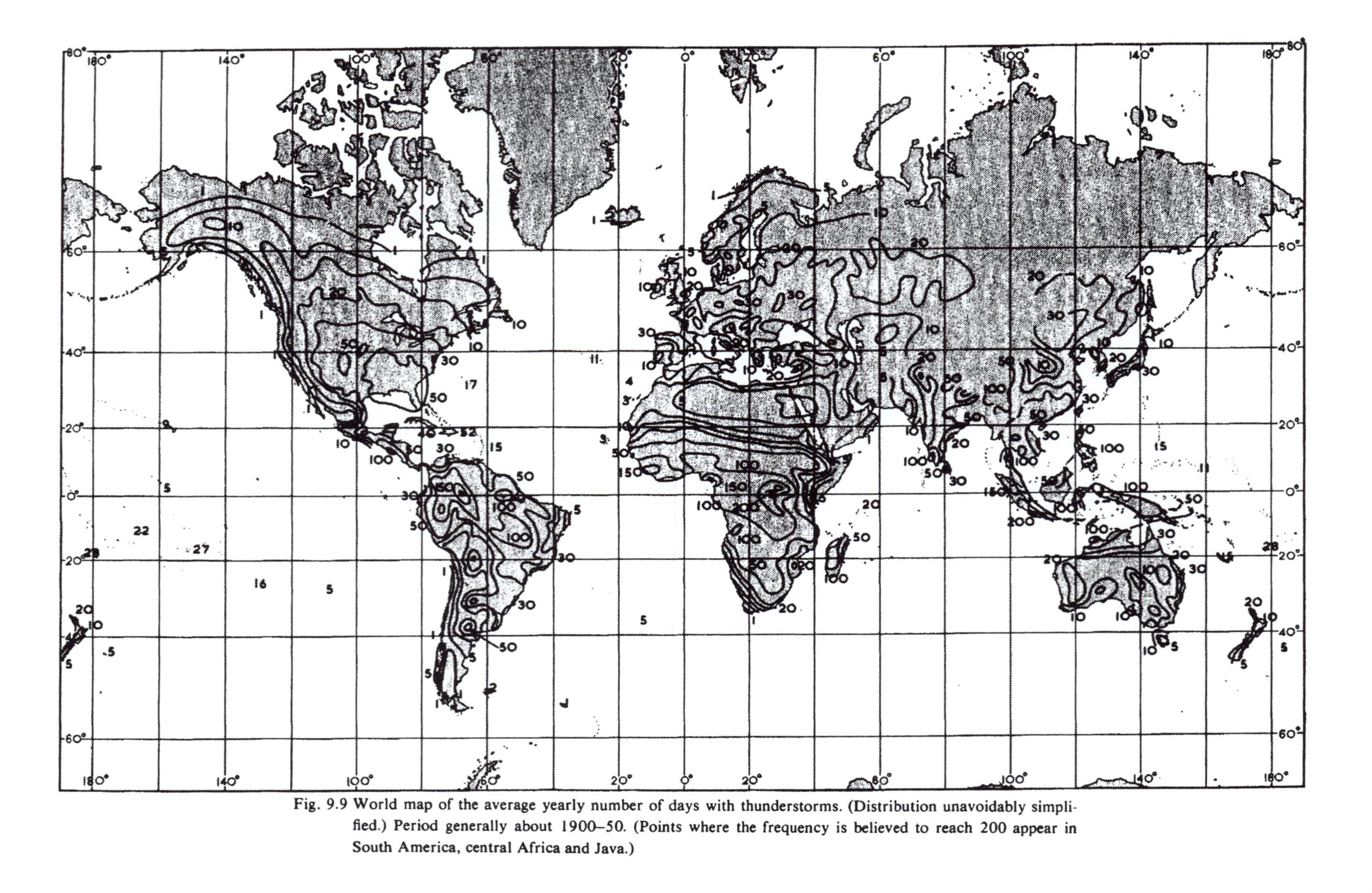

Fig. 9.9 World map of the average yearly number of days with thunderstorms. (Distribution unavoidably simplified.) Period generally about 1900–50. (Points where the frequency is believed to reach 200 appear in South America, central Africa and Java.)

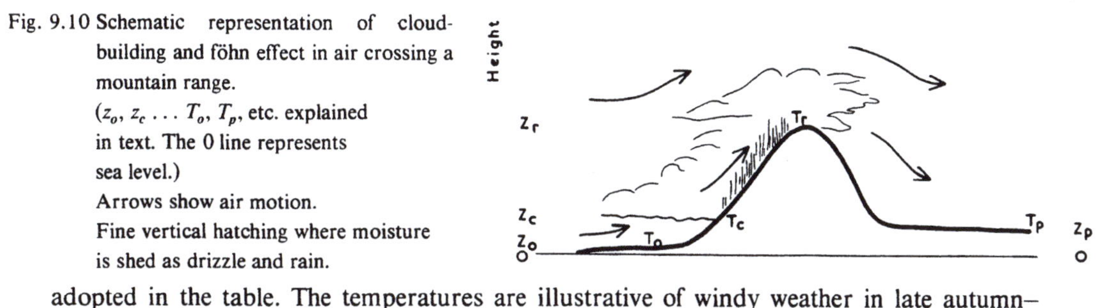

Fig. 9.10 Schematic representation of cloud-building and föhn effect in air crossing a mountain range.
(z_o, z_c . . . T_o, T_p, etc. explained in text. The 0 line represents sea level.)
Arrows show air motion.
Fine vertical hatching where moisture is shed as drizzle and rain.

adopted in the table. The temperatures are illustrative of windy weather in late autumn–early winter but could occur even in winter. The warmth so induced is sometimes a cause of avalanches in the Alps and of unseasonable warmth, accompanied by gusty winds, in the valleys. In the area between Munich and the foot of the Alps the föhn wind blows on average about 100 days/yr and makes the climate more genial than it could otherwise be. In eastern Scotland the effect of this extra warmth in the prevailing W'ly and SW'ly winds is registered in the rich countryside and fine old buildings, that are a legacy of the wealth of past generations, in inland districts (from Forfar to Elgin) east and northeast of the Grampians. A similar result is equally noticeable in southeast and central Norway (e.g. Gudbrandsdal) and Sweden. Even in southwest Greenland an area of open land, where the climate is relatively mild, exists between the ice cap and the sea, partly produced by the warmth of föhn effect in the SE winds that commonly blow there and cross the 2000-m-high crest of the inland ice on their way.

The adiabatic warming of the wind descending the lee sides is accompanied by largely clear skies – even frontal cloudbelts are commonly broken – and sunshine. The latter effects are commonly noticeable to some extent even when the wind crosses little hills 100 or 200 m high, and these can produce characteristic windward-intensification and lee-side rain-

TABLE 9.8
Examples of föhn effect

	Scotland W'ly wind				*Alps S'ly wind*		
	Temperatures °C				*Temperatures °C*		
Height a.s.l. (m)	*Windward side*		*Lee side*	*Height a.s.l. (m)*	*Windward side*		*Lee side*
1000		3·3°		2500		−2·3°	
300	7·5°			700	8·5°		
				500			17·7°
50	10°		12·8°	50	15°		
Assumed heights (m):							
(*a*) main ridge		1000				2500	
(*b*) cloud base	300		1000		700		2500
(*c*) low ground	50		50		50		500

shadow tendencies in the rainfall pattern (closely studied in Sweden by BERGERON (1968) in the 'Pluvius' research project).

(ii) *Free-air föhn*

In the anticyclonic development that commonly takes place at the rear (west side) of a cold trough in the upper westerlies, the warmth of the upper air (and increasing 1000–500 mb thickness) seen in the intensifying ridge immediately west of the upper trough is commonly associated with subsidence in air that has shed its moisture in ascending the warm front surface of the cyclonic disturbances still farther west. This may be regarded as a kind of föhn effect in which the subsiding warm air has blown over the crest of the surface cold airmass, and during the descent its temperature increases at the dry adiabatic rate: so we have both stages of the process that can be seen in air crossing the Alps (see fig. 9.10). It is for this reason that all subsiding air above the inversion in anticyclonic situations is often referred to in the German literature as *freier Föhn*. And, of course, the mountain-tops experience this kind of föhn oftener than the valleys. But the dynamics of the general wind circulation is involved in any anticyclogenetic situation, and the subsidence is induced by cross-isobaric flow and mass-convergence at and near the maximum wind level in the upper troposphere. Hence, in general, the upper surface of the cold airmass is itself being depressed by the general subsidence until it reaches the point where the sinking is checked by turbulence (thermal and mechanical) set up at the ground.

TABLE 9.9

Incidence of free air föhn

	J	F	M	A	M	J	J	A	S	O	N	D	*Year*
	Average number of days/yr with relative humidity $\leq 40\%$												
Zugspitze (47·4°N 11·0°E 2962 m) 1901–36	4·8	4·9	3·8	0·9	1·1	1·1	1·4	2·2	2·5	3·9	5·1	3·8	35·5
Schneekoppe (50·7°N 15·7°E 1618 m) 1901–36	5·9	4·9	3·3	1·3	1·1	0·7	0·4	0·7	1·8	3·6	5·0	5·2	33·9
Feldberg, Black Forest (47·9°N 8·0°E 1493 m) 1927–36	3·4	5·9	3·4	1·8	1·2	1·0	0·9	1·3	1·0	2·2	3·1	3·8	29·0
	Average number of days/yr with relative humidity $\leq 20\%$												
Zugspitze 1901–36	1·1	0·7	0·6	0·1	0·2	0·2	0·3	0·3	0·5	0·8	1·2	0·9	6·9
Brocken, Harz (51·8°N 10·6°E 1150 m) 1901–36	1·0	0·5	0·5	0·1	0·1	0·1	—	—	0·1	0·4	1·1	0·8	4·7
Achnagoichan, Speyside (57·1°N 3·8°W 305 m) 1956–64	—	0·2	0·1	0·2	0·3	0·1	0·1	—	—	—	—	—	1·0

Figures from FLOHN (1941) and GREEN (1965).

At Achnagoichan, which is far below the height of the neighbouring Cairngorm mountains (1300 m), it is possible that a few extreme cases of normal föhn wind enter into these figures.

CHAPTER 10

Some observed causes of climatic variation

In Chapter 7 we reviewed the spatial relationships of anomalies within the large-scale wind circulation over the Earth and tendencies for sequences (i.e. evolutions) of displaced or anomalous development which are known to affect Europe, in particular, and appear dependable enough to use in seasonal weather forecasting. In some of these sequences/evolutions, effects upon the thermal condition of the undersurface and the subsequent reactions (feedback) from this upon the atmosphere, were either known or suspected to be playing a part. In this chapter we examine cases where influences upon the atmosphere from the undersurface and from other extraneous circumstances can be either seen in operation or deduced.

Variations of sea ice

The responses of the Arctic sea ice to twentieth-century changes in the atmospheric circulation – the vigour and northward penetration of storminess from the Atlantic into the Arctic at first increasing over several decades and then decreasing – have been seen in Chapters 7 and 8. The retreat of the limit of 'permanent' pack-ice during the Arctic warming that accompanied the vigorous wind circulation, especially in the 1920s and 1930s, diminished the region over which sharp cooling of surface air beneath an inversion was liable to take place. Hence, in the region largely relinquished by the ice from Spitsbergen to Novaya Zemlya and Franz Josef Land the annual mean temperature for 1921–30 was 2·5°C higher than in the immediately preceding ten years, and by 1931–40 the total increase was over 3·5°C; the mean winter temperatures were up by 6°C in 1921–30 and by over 8°C in 1931–40 as against 1911–20 in the same region (KIRCH 1966). The re-advance of the sea ice in the 1960s reacted upon the atmosphere correspondingly by bringing at once a sharp lowering of surface temperatures prevailing in the regions directly affected by the new areas of ice, between Franz Josef Land,[1] Spitsbergen and Iceland, and affected temperatures in northerly winds as far south as the British Isles. Strong N'ly winds in February 1969 brought Arctic conditions of blowing snow on the night of the 7th–8th even to inner London (where they had not been seen for many years) and temperatures as low as –5 to –8°C with the blizzard in open country in the English Midlands and East Anglia. This

1. Average air temperatures at Franz Josef Land in the 1960s were about 2·4°C lower overall, and 6·3°C lower in the months December to March, than in the preceding decades.

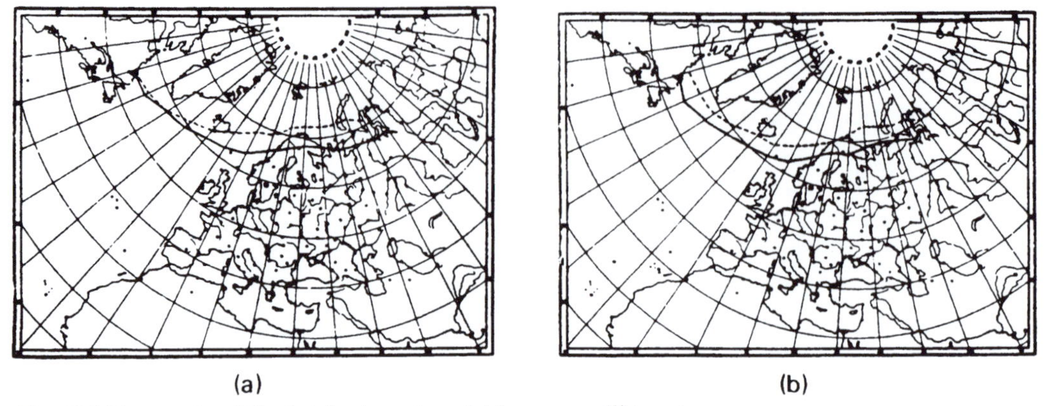

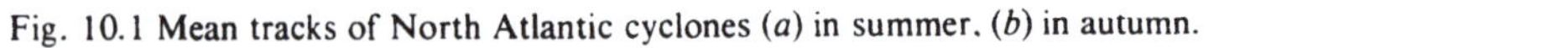

Fig. 10.1 Mean tracks of North Atlantic cyclones (*a*) in summer, (*b*) in autumn.
Continuous line: years with much ice in the East Greenland–Barents Sea in April–July (1881, 1882, 1887, 1891, 1895, 1896, 1906, 1911).
Broken line: years with little ice in these areas in April–July (1880, 1897, 1899, 1904, 1908).
(*After* WIESE *1924*.)

windstream began its passage across the warm waters of the Atlantic at Iceland instead of somewhere between 75° and 80°N as had been usual at least between 1920 and 1960. The ice north of Iceland had probably acquired a thickness comparable with that in the central polar basin, reducing heat flux from the water beneath to a negligible value. And the distance over open water traversed by the wind on its way to Britain was approximately halved as compared with the years which had come to be regarded as normal.

Since a change from ice to open ocean, or vice versa, has a strong effect upon the heating of the overlying atmosphere, prevailing 1000–500 mb thickness is changed and any strong thermal gradients generated near the ice margin tend to shift with it. Hence, the jet streams and the common depression tracks also tend to be displaced in parallel with the ice limit. This tendency must be particularly strong in the North Atlantic, where thermal gradients near the ice limit tend to be extra strong because the waters near Iceland (Irminger Sea) and in much of the Norwegian Sea are of warm (Gulf Stream) origin. The displacement of the most frequented cyclone tracks in summer and autumn as between years with much ice and years with little ice north of Iceland in spring–early summer is illustrated in fig. 10.1. That this is to be understood as a statistical result is shown in fig. 10.2, where the range of variability in the steering of individual cyclones is more readily gauged. The net shift south in heavy ice years is, however, in the sense long conjectured as marking the difference between warm eras and ice ages or between lesser warm and cold epochs.

We have noted evidence (Chapter 7, pp. 257–9, 301–2), however, that the slackness or vigour and steering of cyclonic storms in the sub-Arctic and sub-Antarctic belts can itself be responsible for extending the limit of ice or open water, particularly when the activity is maintained over several decades. Reverse relationships of the kind here demonstrated in figs.

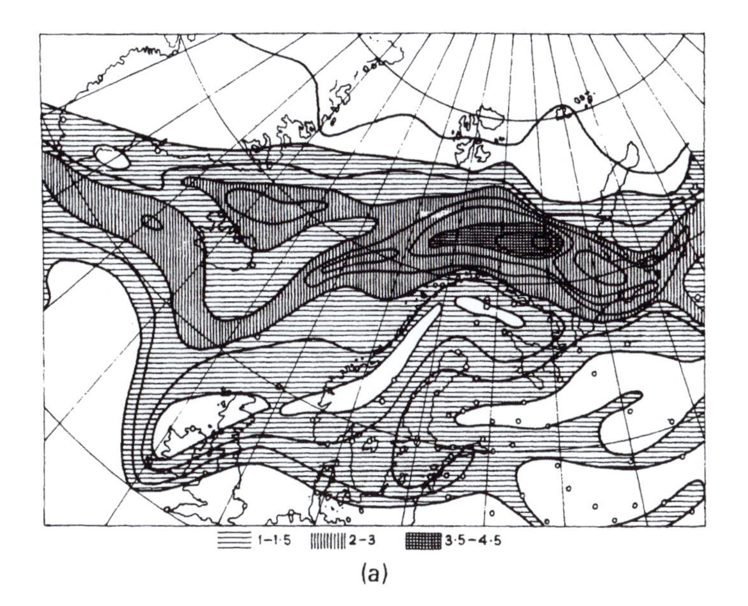

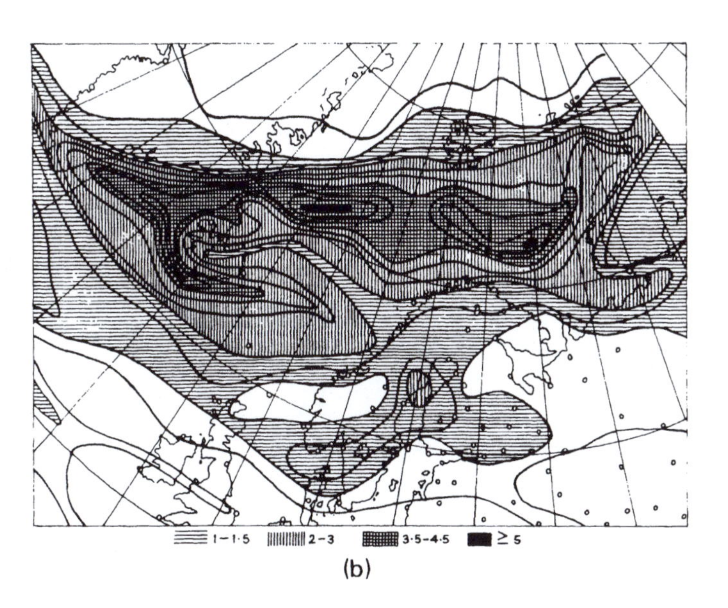

Fig. 10.2 Mean frequency of North Atlantic cyclones in autumn:
(*a*) Years with much ice in April–July (as for fig. 10.1).
(*b*) Years with little ice in April–July (as for fig. 10.1).
(*After* WIESE *1924*.)

10.1 and 10.2, where the extent of ice has some tendency to guide, or limit, the poleward penetration of the travelling storms, appear to have some forecasting validity (in terms of statistically significant lag associations) for 4–8 months ahead. Such hints as have been found of longer-lived associations, lasting some years, probably depend upon persistence of some more basic cause of climatic change. Transport of warmth by the winds and wind-driven ocean currents in the last millennia of the ice age is believed to have been able to melt most of the Atlantic sea ice while nearly all the North American inland ice was still present and the European ice dwindled only slowly.

A study by Brooks and Quennell (1928) of anomalies of seasonal mean surface pressures over the Atlantic, Europe and part of Siberia accompanying and following variations of the Arctic sea ice, principally the ice in the regions east of Greenland and about Iceland but also in the Barents and Kara Seas, showed a strong tendency for high pressure over Iceland and Greenland in the summers of the heavy ice years and to an only slightly less extent in the summers of the following year or two also. Apart from this, there were a number of other apparent relationships expressed by correlation coefficients which were near, or surpassed, the border-line of statistical significance. Among the most interesting were negative correlation coefficients between the amount of ice east of Greenland and on the Barents Sea in the (late) winter–spring and surface pressures prevailing (i) about the Faeroe and Shetland Isles in the following October–December ($r \cong -0{\cdot}55$ over two periods of about 30 years), (ii) over Ireland, Biscay and France in the following November–January ($r \cong -0{\cdot}35$ in each of two periods of about 30 years).[1] There was an appearance of negative correlation between the east Greenland–Barents Sea ice and atmospheric pressure over and between southern Ireland and France in the succeeding three winters and positive correlation between the ice and the pressure in spring over Europe between Ireland and Berlin for up to 3 years following.

Schell (1961) has shown that the centuries-long records of the ice at the coasts of Iceland are an index of the long-term variations of the European and North American climate, 50-year mean temperatures over the much wider area tending to be high when the ice is least, low when the ice is most extensive; but this begs the question of cause and effect. Speerschneider (1931), known for his compilation of data on the variations of Baltic ice from early times (1915), regards the ice extent – whatever its inertia and lag effects – as no more than a response to the winds: 'a severe ice year ought to be called a year with . . . abnormal wind conditions, as the wind is the source while the condition of the ice is the notable effect.'

Thermal condition of land surfaces

State of ground is only seen to have noteworthy effects upon the large-scale atmospheric circulation either where a sluggish response to heating is conditioned by vast tracts of

1. These studies have lately been carried further by I. I. Schell (Arctic ice . . . in the northeastern Atlantic and . . . seasonal foreshadowing. . . ., *Monthly Weather Rev.*, **98**, 833–50. Washington 1970).

sodden and partly flooded ground, as readily occurs with frozen subsoil, or where snow and ice cover is extensive and deep (cf. HOUGHTON 1958, LAMB 1955). The extent of the snow and ice must be comparable with that of a cold trough in the upper westerlies. These observations imply that the thermal inertia conferred by the large specific heat of water, and/or the latent heat of melting of snow and ice, must be involved. Dry, snowfree ground responds so readily to radiation gain or loss, and to the heat transported by the winds, that it adapts itself almost at once to whatever temperature regime the atmospheric circulation pattern favours.

A case, demonstrated by NAMIAS (1963*a*), where the influence of the thermal anomaly of a land surface was effective, was the general snow cover which persisted from mid February to mid March 1960 over the United States east of the Rockies to 36°N (i.e. far south of normal). The snow was maintained by several surface N'ly outbreaks separated by more anticyclonic periods. Since the normal surface intake of solar radiation in the region of the southern part of this snow is reckoned as some 300 cal/cm^2/day, NAMIAS considered that 80% of this must have been lost while the snow lay, i.e. about 240 cal/cm^2/day deficiency of the normal seasonal heating in that part of the hemisphere. The real loss may, however, have been somewhat less, when account is taken of the total albedo including the effect of the atmosphere with its clouds and haze in the normal year.

The great ice sheets of the ice ages are the supreme example of a state of ground which produces a persistent temperature anomaly in the air over it because of the great reduction in the normal intake of radiation throughout the year (by amounts of the same order as the figures discussed in the last paragraph). Even temporary incursions of warm air accompanied by strong winds can hardly raise the surface air temperature above +1° to +3°C at any point on the ice sheet, and in the interior much lower maximum values must have obtained, as they do today on the high ice plateaux in Greenland and Antarctica.

Anomalies of the surface water temperature of enclosed seas such as the Baltic and the North Sea have important effects on the surrounding lands. The wide reaches of the Baltic in February, ordinarily nowadays an open water surface of temperature +2° to +2·5°, and liable to be as high as +4° to +5° in very mild winters, are transformed in some winters into a thick ice surface, the temperature of the snow on which may fall many degrees below the freezing point. Most of such an ice surface will, moreover, persist into late April and much of it even into May in some cases. (In 1966 there was still some ice in the northern part of the Gulf of Bothnia at the beginning of June.) As long as such anomalies are present they must produce corresponding anomalies of surface temperature in the winds that have crossed the sea. Even the North Sea develops surface-water anomalies of up to 2–3°C above normal in mild winters and down to 4–6°C below normal in severe winters, when the temperature of its southern and eastern parts may be about 0°C and support patches of fresh-water ice derived from the river mouths. Such anomalies cannot be entirely eliminated in less than a month or two of the calmer, warm weather of spring and early summer; they doubtless account for the fact that the widest areas of significant month-to-month persistence tendency of air temperature anomaly in Europe, both in winter and summer, are centred about these seas and embrace just the immediately surrounding lands (see Chapter 7 and the references there cited). In summer greater anomalies of sea surface temperature are liable to be developed in the Baltic (up to $\pm 4°$) and in the seas between Greenland and Norway than in the North Sea, and it is about these areas, rather further north than in winter, that the greatest month-to-month persistence tendency of air temperatures is then developed. In other words, the apparent persistent warmth or cold peculiar to the land surfaces of Scandinavia in summer, and of England and the Baltic lands including southern Scandinavia in winter, really belongs to the thermal condition of the extensive water surfaces in the region. Other regions with some notable month-to-month persistence tendency are found close to the Mediterranean and Black Seas.

A more genuine case where the condition of a land surface itself seems to have been important was provided by the summer of 1968, when the extreme northern fringe and northeast of Europe, as well as a wider area of northwest Siberia, remained exceptionally cold. The previous winter's snow did not clear from the low ground in these areas until late June and even continued into July near the Gulf of Ob. The ground remained cold and sodden and, with the additional influence of the sea ice to the north having become thicker and more extensive in recent years, one season's quota of solar radiation and atmospheric heat transport was insufficient to warm out the anomaly. A sharp upper cold trough was maintained over the region between 50° and 90°E, and the cold surface itself doubtless contributed to this; the attendant cyclonic development, cloudiness and frequent cold rains in turn helped maintain the cold surface. There was renewed snowfall as far south as 66°N 60°E in late July and much farther into the northern Siberian lowlands in late August. Thus, autumn set in early with renewals and intensification of the upper cold trough over the same region, bringing a new snow cover on 11 September 1968 temporarily as far south as 53–55°N in places between 50° and 70°E. West of this abnormal cold regime, a blocking anticyclone lay over central Sweden and Norway through most of the summer, with a ridge reaching to Scotland, and fine warm weather prevailed over these regions. But over central Europe, southern England and France there was an exceptional prevalence of cold north-easterly winds and several periods of cyclonic weather with heavy rains. These rains were especially productive of flooding and landslides, because they came from stationary, or slow-moving, cyclonic systems associated with the southern part of the blocking pattern: i.e. stationary rain areas and long-continued downpours over the places affected.

Even in summer 1968, however, the extensive and strong ocean temperature anomalies in the cold waters south of Newfoundland and in the central Atlantic (see pp. 394–6), developed from late May onwards, were an important integral part of the regime and seem to have played a role in maintaining the atmospheric circulation pattern that governed it (MURRAY and RATCLIFFE 1969).

A somewhat similar, though less extreme, situation affected the North American sector in summer 1968 but did not persist into autumn. The regions about Hudson's Bay were primarily affected, though the summer was cold also in the United States far farther south.

It seems likely that the cold summers in the central United States in the 1830s, noted by WAHL (1968), were to a similar extent brought about by areas of persistent snow and cold sodden ground in the Canadian north and abnormally cold water, with some persistent ice, in Hudson's Bay. This could set up a pattern in that sector analogous to that affecting Europe and western Siberia in summer 1968, as described above. The same doubtless applies to some still more remarkable cold summers reported by early explorers of the American interior from the late 1500s onwards. Thus, FROBISHER reported bitter cold and heavy snowfall, one foot deep on his ship, in Hudson Strait on 5 August 1578 (New Style calendar);[1] and one of CHAMPLAIN's parties was able to walk over the ice to an island in Lake Superior in June probably of the year 1608. About the same time both Alaska and Finland seem, on the evidence of tree growth (researches by H. FRITTS and G. SIRÉN cited in *Aspen Conference* 1962), to have experienced a run of notably warm summers, particularly between 1550 and 1580; this was presumably under the influence of stationary Alaskan and Finnish–north Russian blocking anticyclones which stood in the same relation to the cold regime farther east, penetrating far south over the continent, as did the Scandinavian anticyclone in summer 1968. By 1580, however, the cold regime had become more widespread: the summers in Alaska and Finland also were much colder from then on. By analogy, and partly as a legacy from the more general cold climates of the seventeenth century, the warm anticyclonic summers which prevailed over western, central and northern Europe, from France to north Russia, between 1730 and 1810 (and in a few earlier years, e.g. 1718 and 1719) may have been associated with a persistent cold surface over much of northern and western Siberia, sustained by extensive remanent snow, permafrost and flooding in that region.

Study of the stratigraphy of the ice cap at 77°N 56°W suggests that Greenland underwent a similar experience to Alaska and Finland though not entirely in phase with them. Summers were apparently warm in northern Greenland (LANGWAY 1967) in the middle of the seventeenth century and on until about 1730, and at the same time the climate was dry, the average annual snow accumulation being no more than 70–80% of modern values. These observations fit the concept of extra frequent blocking anticyclones over Greenland between about 1650 and 1730; the N'ly winds originating on their eastern side must have played a part in the formidable increase of the East Greenland sea ice (which entirely surrounded Iceland in 1695) and in the coldness of the climate of Europe and North America during these same years, though it is also possible that the existence of a wide additional cold surface east of Greenland had much to do with the recurrent anticyclogenesis over northwest Greenland (i.e. a feed-back). The general warmth of the climate of the present century and in the early middle ages also affected northern Greenland: these warm centuries are marked in the ice stratigraphy by a number of hard ice layers, produced by melting in the warmest summers and subsequent refreezing, accompanied by a markedly

1. I.e. this date has been corrected to the Gregorian calendar now in use.

increased average annual accumulation of snow. This may be attributed partly to the increased vapour content of the warmer air and partly to the increased frequency of invasion of the region by cyclonic activity operating in high latitudes.

By contrast to all the cases cited hitherto in this section (where a cold surface persists owing to the presence of snow or ice or waterlogging), when an extensive continental land surface becomes well dried by a long, hot summer, it is as likely to cool down abruptly as to stay warm when the radiation conditions and weather regime change with the changing season. In other words, the dry land surface has very little temperature persistence tendency. Month-to-month persistence of sign of temperature anomalies over land is at a minimum in autumn. There is another minimum of persistence over dry land with the spring change of radiation regime. The spring–summer persistence tendency over the U.S. Great Plains is apparently exceptional, and NAMIAS (1963*a*, p. 354) believes it may be largely due to persistence in the development (i.e. latitude, strength and tendency to great or small extent) in any given year of the eastern Pacific and western Atlantic subtropical anticyclone cells over the oceans on either side of the narrow American continent. (Cf. fig. 10.3 which shows the reasonableness of this contention in terms of the distribution of spring–summer lag correlation of 700 mb heights.)

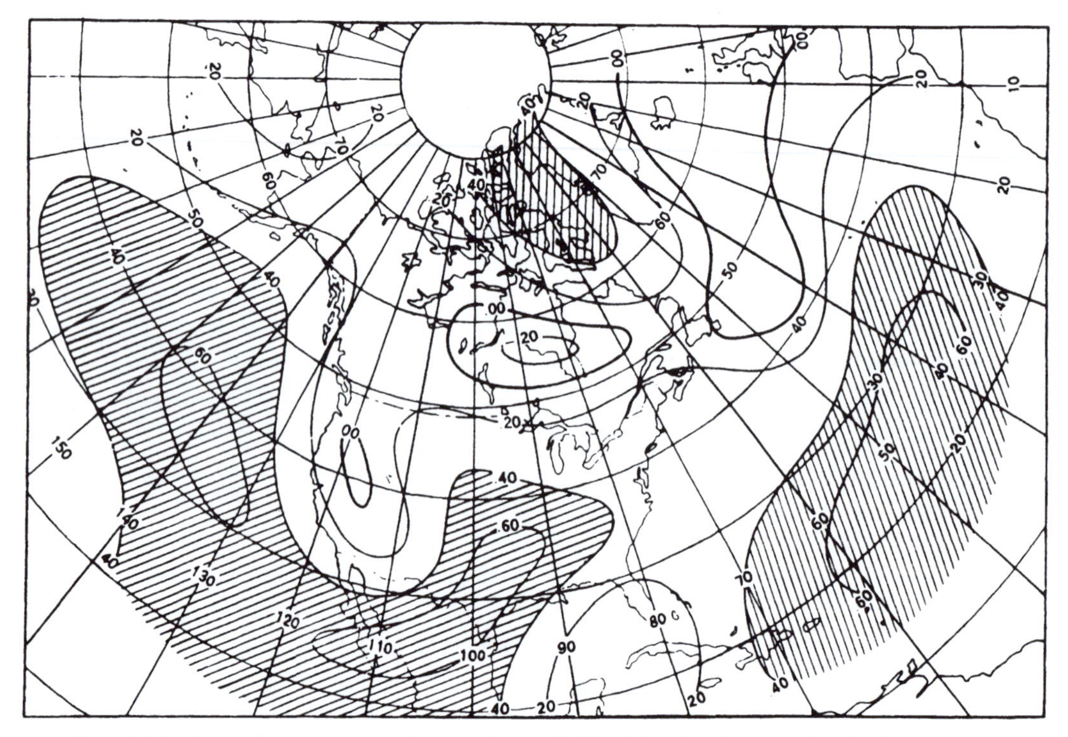

Fig. 10.3 Isopleths of spring–summer lag correlation of 700 mb heights, from data for 1933–58. Shaded areas: correlation coefficients exceed the 5% level of statistical significance. (*After* NAMIAS *1963a*.)

Variations of ocean temperature

Changes of annual mean sea surface temperature by as much as 1·5 °C over wide areas of the North Atlantic, and exceeding 3° in some areas, have been observed to accompany changes of the atmospheric circulation from groups of years with strongly developed westerly winds to years with frequent northern anticyclones. In the extreme case of summer 1968, alluded to in the previous section, much of the central North Atlantic and the area south of the Newfoundland Banks was generally 3–5 °C colder than the average of the previous 100 years.

BJERKNES (1963) has shown (see fig. 10.4) that from the 1890s to the 1920s the 5-year

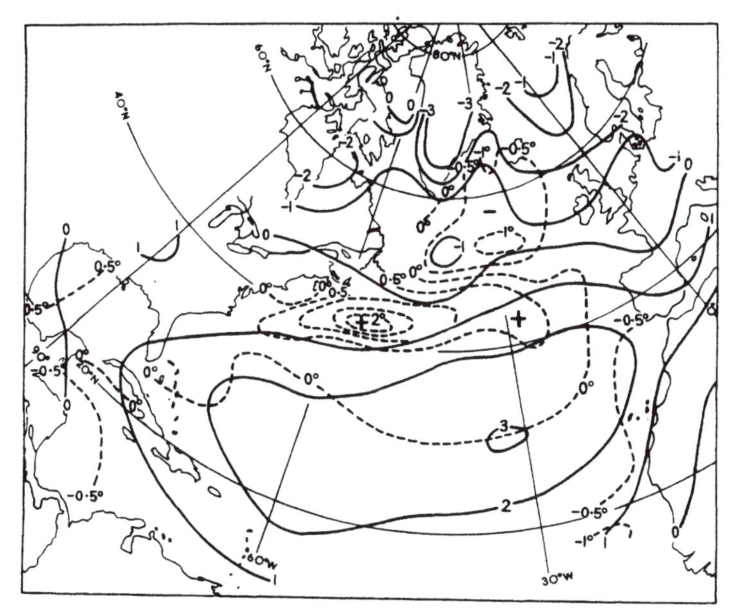

Fig. 10.4 Changes of annual mean barometric pressure (mb) (full lines) and of sea surface temperature (°C) (broken lines) from 1894–8 to 1920–4.
(*After* BJERKNES *1963.*)

mean values of sea surface temperature fell by more than 1 °C between 50° and 55 °N in mid Atlantic and rose by over 2 °C south of Newfoundland in the core (about 43 °N) of a zone of enhanced warmth between 35° and 50°N. Over the same years the mean atmospheric pressure rose by 2–3 mb over the whole subtropical anticyclone belt of the North Atlantic and fell by 2–3 mb over Iceland, Greenland and Labrador: these pressure changes indicated an intensification of both the 'Iceland' low and the 'Azores' high, and of the westerlies between them, as well as some northward displacement of the entire system.

The areas of greatest cooling and of greatest warming of the ocean surface were just where the surface W'ly winds respectively increased and decreased most. This points to changes in the rates of evaporation and of upwelling as fundamental. The upwelling is created by divergence of the surface water, whose drift tends to deviate to the right of the wind impelling it (the EKMAN drift, see Chapter 8). This causes a drift away of the surface water to all sides from cyclonic central areas, and outwards from regions of cyclonic curvature of the isobars; it also causes convergence of the surface waters in anticyclonic regions. Surface water temperatures are likely to be lowered in cyclonic central regions and wherever conditions become more cyclonic than before, and raised in anticyclonic regions, owing to the changes in the amount of evaporation and upwelling as well as through the differences of cloud cover and consequent changes of solar radiation supply to the sea surface. There is, however, an obvious feed-back from the ocean to the atmosphere in these changes (cf. BJERKNES *loc. cit.*, p. 316). The sea temperature situation evolved by the 1920s, with positive water-temperature anomalies concentrated in the zone 40–50°N and lowered water temperatures north of 50°N, provided an enhanced thermal gradient about 50°N, which must have been transmitted to the overlying atmosphere through the differential heating of the winds north and south of it. This must have produced a strengthened and northward-displaced flow of the upper westerlies, resulting in intensification of the subpolar low and of the Azores high-pressure belt and a northward shift of both. Hence, the situation tended to generate, and/or intensify, the very circulation features which appeared to cause it.

Particularly strong changes of prevailing sea surface temperature in the Atlantic in latitudes 50–60°N are illustrated in fig. 10.5 over years when the zonal atmospheric circulation was increasing and moving farther north and over other years in which the reverse changes were taking place.

Sea surface temperature anomalies are considered likely to be the most important single influence in causing long-term weather anomalies by SAWYER (1965), provided they cover a large area of the order of 1000 km or more across, are strong enough to raise or lower the total (feelable plus latent) heat input into the atmosphere by at least 50 cal/cm^2/day (of the order of one-tenth of the solar constant) and persist with little change through the period concerned.

Cases where the sea temperature anomaly distribution in the North Atlantic may serve as a predictor of atmospheric circulation behaviour for at least a month ahead have been studied by RATCLIFFE and MURRAY (1970). The largest numbers of similar cases available for study related to either positive or negative anomalies of sea surface temperature covering most of the ocean between 40° and 50°N, especially the western part, and with the greatest anomaly on average $\pm 1{\cdot}5$ to 2°C over and near the Newfoundland Banks, while the zero anomaly isopleth lay near 50° or 30°N (fig. 10.6). These are essentially situations in which the Gulf Stream–Labrador Current boundary lay respectively north or south of normal:

(1) When the ocean surface in the area described was warmer than normal by about this amount (or more), and was colder than normal in the fifties and (usually) south of 30°N,

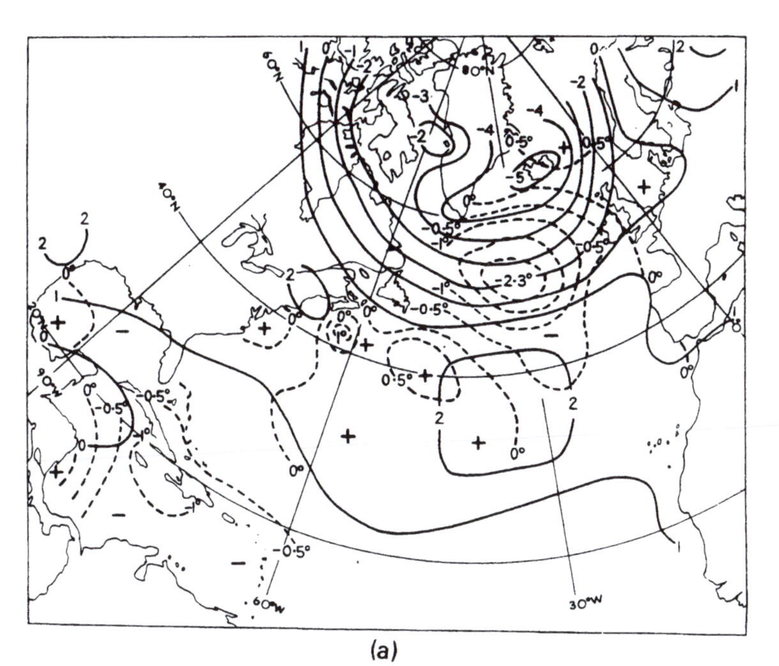

(a)

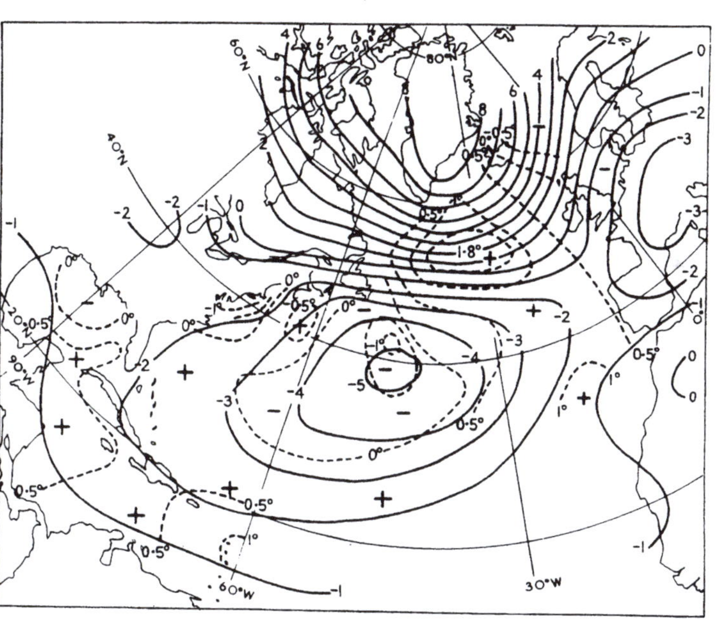

(b)

Fig. 10.5 Change of annual mean barometric pressure (mb) (full lines) and of sea surface temperature (°C) (broken lines):
(*a*) From 1902 to 1904.
(*b*) From 1913 to 1915.
(*After* BJERKNES *1963*.)

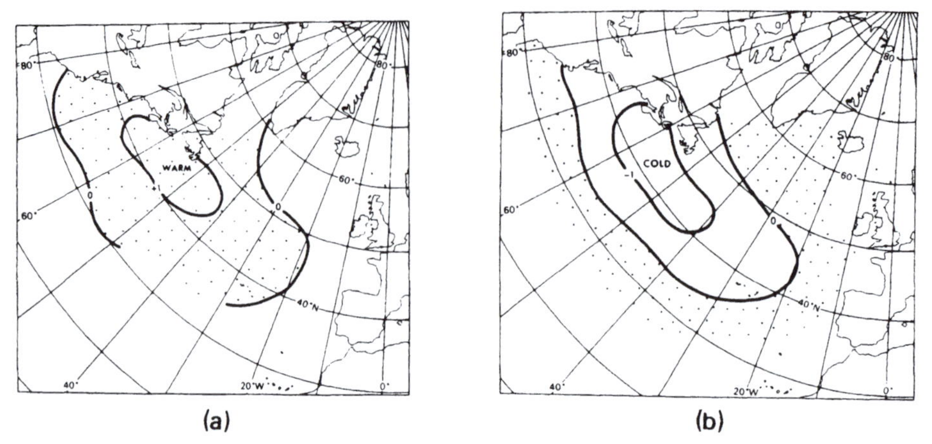

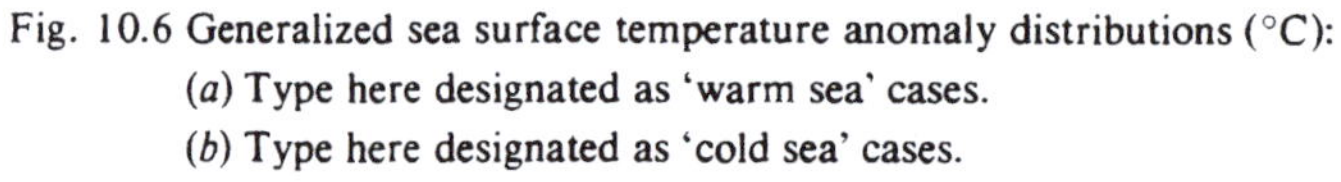
Fig. 10.6 Generalized sea surface temperature anomaly distributions (°C):
(*a*) Type here designated as 'warm sea' cases.
(*b*) Type here designated as 'cold sea' cases.

the monthly mean pressures at sea level over the following month tended to be below normal over a wide region of the northeastern Atlantic and Scandinavia. The region was centred about Iceland–Scotland–central Norway and Sweden; the greatest departures averaged −3·5 to −4 mb. The departures were bigger than this in the winter half-year. The pressure anomalies tended to be centred 10–20° longitude farther east in those months of the year when prevailing wave lengths in the upper westerlies are long (e.g. October and mid winter) and farther west in those periods (especially in late winter, May and November) when short wave lengths prevail. At the same time mean barometric pressure tended to be above normal over the Atlantic between the Azores and the Newfoundland Banks region (by about +1·5 mb on average) as well as between the Urals and the Gulf of Ob (by +2 to 2·5 mb).

(2) When the ocean surface in the area described was colder than normal by about the said amount (or more), and was (usually) warmer than normal near 30° and near 60°N, the monthly mean pressures at sea level over the following month tended to be above normal over a wide region of the northern and northeastern Atlantic and Scandinavia. This region was centred a little farther west than its counterpart in (1), the greatest average departures being +3·5 to +4 mb over east Greenland–Iceland–west Norway. Again the departures in the winter half-year were greater than the overall averages mentioned and tended to be centred farther east in the months when wave lengths are long, farther west in the months when wave lengths are short. The high-pressure regime also tended to sprawl far east from its centre to cover northern Europe in the later winter months, when the very low temperatures developed over a continental snow surface in lighter than normal winds produce great density of the surface air layers. Pressures tended to be below normal over the Atlantic between 30° and 45°N (departures −1·5 to

−2·0 mb on average over the region between 25° and 45°W) as well as in northwest Siberia (by −1·0 to −1·5 mb on average).

These patterns are illustrated in figs. 10.7, 10.8. The departures appear statistically significant at or beyond the 5% level (cf. RATCLIFFE and MURRAY *loc. cit.*) over some areas (stippled on the maps) of the size of the British Isles. The effects upon prevailing air temperatures, rainfall, sunshine, etc., are those appropriate to increased frequency of this or that wind direction, or of light winds, and anticyclonic or cyclonic conditions. (See, for example, fig. 10.9, which shows the temperature anomalies of the Februarys for which the pressure anomalies are illustrated in figs. 10.7, 10.8.)

Detailed study suggests that the positions of strongest surface pressure anomaly, developed essentially at the downstream end of the jet stream from the sector of boosted thermal gradient, are displaced west or east in parallel with any displacement west or east of the strongest sea surface temperature anomaly in the western Atlantic.

The general explanation proposed for the effects observed as characteristic of these two ocean temperature anomaly distributions upon the large-scale atmospheric circulation is that (1) gives a boosted thermal gradient across the Atlantic ocean near 50°N, puts more heat and moisture into the warm and cold airmasses in the frontal depressions traversing the ocean in the forties, and by these means deepens the depressions and strengthens the upper westerlies in a rather northern position. This also tends to increase the wave length and carry the travelling depressions farther east than in other circumstances. LAEVASTU (1965) has studied the heat transfer from ocean to atmosphere in terms of Q_H, the feelable heat transferred (which is proportional to wind speed and sea–air temperature difference), and Q_E, the latent heat transferred (which is proportional to wind speed and vapour pressure difference between sea and air). $Q = Q_H + Q_E$ is greater the warmer the sea is relative to the overlying air and is greatest of all behind cold fronts, on the west side of occluded cyclones and in the southeast sector of anticyclones (northern hemisphere case). By contrast, (2) gives a boosted thermal gradient across the Atlantic near 30°N and tends to weaken the gradient near 50°N. Hence, the cyclonic activity and the mainstream of the upper westerlies tend to be transferred to southern tracks; wave lengths probably tend to shorten, and there is room for polar anticyclones to spread over Greenland and the northern and northeastern Atlantic, towards Britain and south Scandinavia.

Some other ocean temperature anomaly distributions, though less frequently occurring, can also be shown to have characteristic effects upon the atmospheric circulation patterns while the given thermal pattern in the ocean lasts. Thus, a situation in which the entire western Atlantic is colder than normal and the eastern half of the ocean is warm tends to warm the airmasses over the eastern half while those in the west remain cool, thus turning the zone of main thermal gradient from the normal W–E (or WSW–ENE) alignment to a more SW–NE orientation. This tends to amplify upper cold troughs over the western Atlantic and upper warm ridges over the eastern Atlantic, so favouring SW–NE steering of

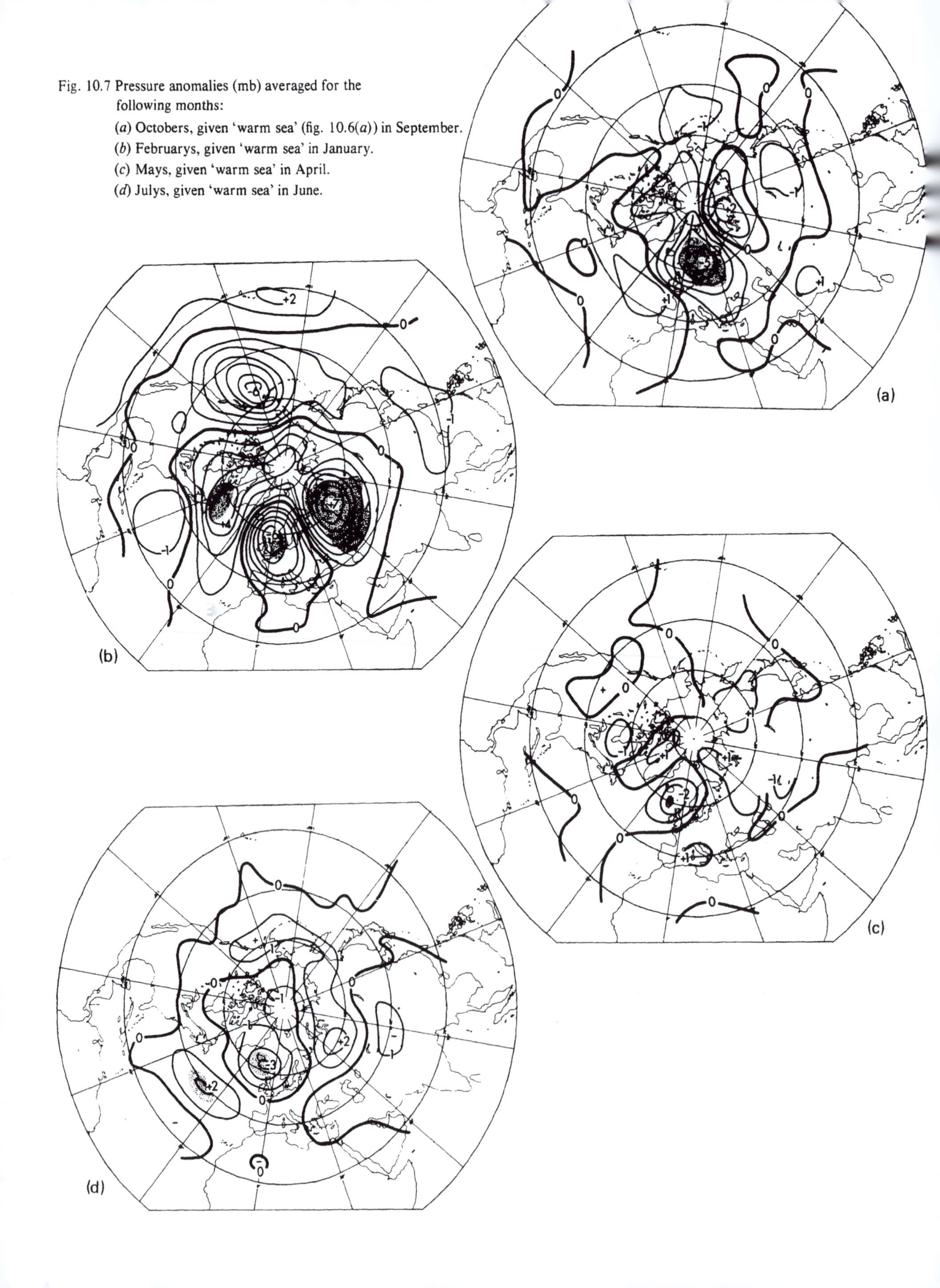

Fig. 10.7 Pressure anomalies (mb) averaged for the following months:

(*a*) Octobers, given 'warm sea' (fig. 10.6(*a*)) in September.
(*b*) Februarys, given 'warm sea' in January.
(*c*) Mays, given 'warm sea' in April.
(*d*) Julys, given 'warm sea' in June.

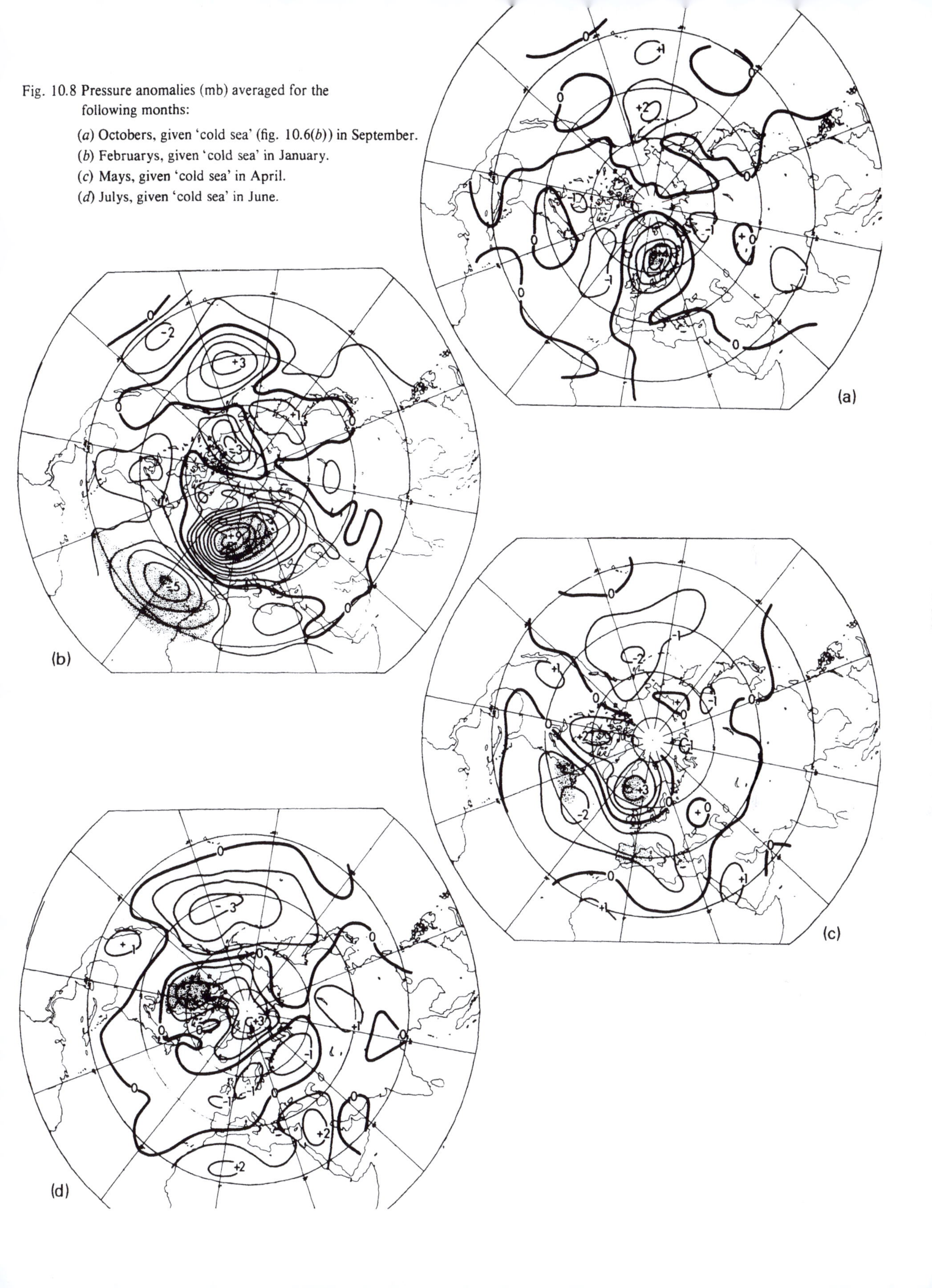

Fig. 10.8 Pressure anomalies (mb) averaged for the following months:

(*a*) Octobers, given 'cold sea' (fig. 10.6(*b*)) in September.
(*b*) Februarys, given 'cold sea' in January.
(*c*) Mays, given 'cold sea' in April.
(*d*) Julys, given 'cold sea' in June.

cyclonic activity far northeast into the East Greenland Sea and Norwegian Sea; these are situations associated with fine, warm weather in western and northwestern Europe with extension of the Azores anticyclone towards Britain and south Scandinavia. These conditions began to show themselves in April 1959; by midsummer the water temperatures were 2–3°C below normal in the western and central Atlantic and more than 1°C above normal

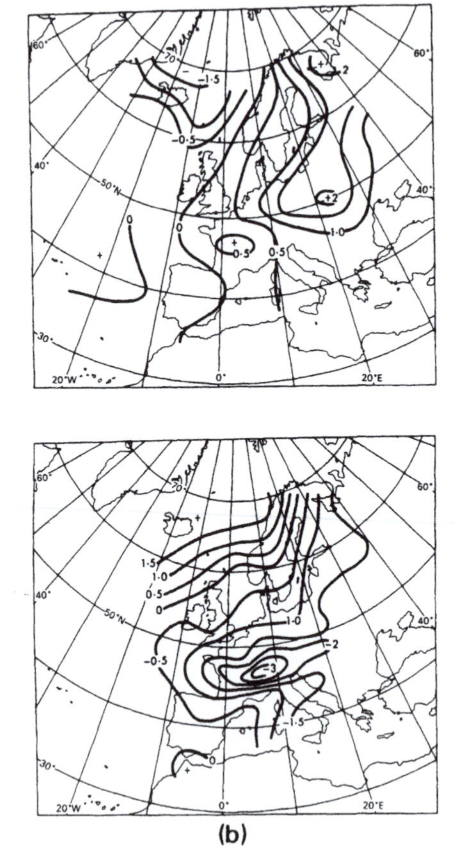

Fig. 10.9 Air temperature anomalies (°C) averaged for the Februarys of years with: (*a*) 'warm sea', (*b*) 'cold sea' in January.

east of about 20°W. The summer was marked by repeated extensions of the Azores anticyclone over Britain and the southern Baltic region, continuing into October. It was the longest summer, and one of the warmest and driest, in England for about 50 years. No month between April and September had more than 85% of the normal rain. By August–September the ground was parched, the grass and tree-leaves faded, over a wide area of northern Europe embracing Britain and the lands about the southern Baltic, while in southern France and northern Italy the grass was green, the soil moist and in places flooded.

When the western half of the Atlantic is warmer than normal and the eastern half is cold, the contrary tendencies prevail, with repeated upper cold troughs and cyclonic weather over Britain and northwest Europe.

In all these cases the thermal inertia of the ocean is basic to the forecast significance of the ocean temperature anomaly pattern, since such strong anomalies as we have been discussing can be expected to persist (at least as regards sign) through one or more following months. That there is a real influence from the ocean operating upon the atmosphere may doubtless be demonstrated quantitatively by calculating and comparing the heat fluxes from ocean to atmosphere in the unlike cases. It is also indicated qualitatively by the fact that a new sea temperature distribution, which sometimes comes into being quite quickly at the end of the autumn when the thermocline is destroyed by storminess and the last of the previous summer's particular areas of surface warmth disappear, promptly begins to show its own characteristic effects on the atmospheric pressure distribution and the steering of cyclonic activity.

Disappearance of the thermocline – i.e. development of near-isothermal conditions from the ocean surface downwards – in November–early December and the formation of a new thermocline and regions of warm surface water between late April and July in the next warm season, are the stages in the seasonal round when a radical change of ocean surface temperature anomaly distribution, and of corresponding steering patterns in the atmospheric circulation, are most liable to occur. When, however, an anomaly distribution is basically due to some more prolonged change in the energy supply, affecting the strength and heat content of the principal ocean currents over successive years, the given anomaly pattern will tend to reappear after any individual summer in which quite another distribution had been developed in the warm surface waters above the thermocline. Similar reversions are also observed after individual winters in which the pattern differed from the surrounding years but probably only through rather small lateral shifts of the strong thermal gradients at water-current boundaries.

Two Russian studies give additional statistical support to the general circulation tendencies over the North Atlantic noted by RATCLIFFE and MURRAY. From data for 1890–1960, VINOGRADOV (1967) found that the summers of years in which the sea temperatures in May were generally subnormal north of 50°N, and on average 0·5–1°C too cold about Denmark Strait and Iceland, but warm south of 50°N, tended to have an excess of the WANGENHEIM **W** type and subnormal air temperatures (by on average 1·0–1·5°C) prevailing over northern Europe. Years when the May water surface temperatures were generally above normal north of 50°N, and on average by ⩾1·0°C in the seas north of Iceland, generally gave an excess of the WANGENHEIM **E** type in summer, with above-normal air temperatures (by on average 2·0–2·9°C) over a region centred about an axis from Iceland and the Shetland Isles to Lapland and the extreme northern fringe of Europe. Finally, years when the May water temperatures were below normal in the western Atlantic and from there to the Denmark Strait and west Iceland, but above normal (by ⩾1·0°C) in Biscay and the neighbourhood of southern Britain, gave an excess of the WANGENHEIM **C** type in summer: the summer air temperatures averaged about 2·0–2·4°C above normal about an axis across the British Isles and south Sweden and were ⩾2·0°C below normal over the East Greenland–Iceland region. This last case was the pattern of 1959. SEMENOV (1967) has produced maps from data for 1889–1939 showing that the frequency of cyclone centres was substantially increased about Iceland and the Faeroe Isles in years when the water temperatures were below normal in the northern part of the North Atlantic and above normal in the southern part. The increase of cyclonic activity in such years as compared with years with the opposite sea temperature anomaly distribution amounted to 60–80% in winter and spring, 20–50% in summer and autumn, and was accompanied by 20–40% reduction of cyclonic activity south of about 55°N in all seasons.

An illuminating feature of SEMENOV's study is that in 70–75% of all cases in the years for which values were available, 1000–500 mb thickness and the heights of the 700 and 500 mb pressure levels showed anomalies of the same sign as the sea surface temperature anomaly at the same point. This indicates that the observed surface weather anomalies come about through amplification or diminution, as the case may be, of upper ridges and troughs passing over the warm or cold sea. It was noted that the crests of the upper warm ridges tended to move almost along the water surface isotherms as each ridge advanced, usually rather slowly, east across the Atlantic: in summer the amplitude increased, or remained great, when the ridges later moved over warm land, but in winter the amplitude of the ridges decreased as they moved over the European continent. The warm upper core of a blocking anticyclone tended to occur between 60°W and 40°E over whichever part of the surface below was warm.

Sea temperature data so far available for the North Pacific are hardly adequate for establishing whether similar effects are produced upon the atmospheric circulation by northward or southward displacements of the northern limit of the warm Kuro Shiwo current, comparable with those derived for the cases of unlike extension of the Gulf Stream water over the Newfoundland Banks. Preliminary study of the data which start about 1957 suggests, however, that given warmer than normal sea over an indicator area (first noticed by NAMIAS) extending roughly 30–45°N 155–175°W in the winter months December–February there is a tendency for greatly enhanced cyclonic activity over the Gulf of Alaska in about the same longitudes and above normal pressure over a broad belt across Arctic Canada and Greenland towards the British Isles. The mean anomalies appear greatest in these months and in December reach −5 mb near 40°N 160–170°W, +8 mb near 60°N 20–40°W. In March, August and September in cases of warm sea in the same area a similar distribution of pressure anomalies is recognizable but all shifted south by 5–10° latitude: in October and, perhaps, November a contrary shift of the whole pattern towards the north may be recognized.

A case study by BJERKNES (1966) of the El Niño phenomenon seems to suggest that abnormal warmth of the equatorial water in the eastern-central Pacific (where it is normally 25–27°C, and 19–20° at the South American coast, colder than in any other sector at that latitude) acts much like extra warmth in the Kuro Shiwo on the atmospheric circulation in that sector, boosting the thermal gradient between the equator and 30–40°N. There was a major El Niño which culminated in the northern winter of 1957–8 with sea surface temperatures near the equator 3–4°C above normal all the way from the Peruvian coast to 170–180°W, the observed values being over 28°C across most of this sector. BJERKNES' analysis interprets this as a southward displacement (from its normal position near 5°N), and intensification, of the main heat source in the eastern Pacific. As a result, the North Pacific westerlies were correspondingly displaced south and intensified in that sector, as may be seen from the comparison of the mean surface isobars of the 1957–8 winter with the two previous winters in fig. 10.10.[1] BJERKNES attributes this intensification of the eastern Pacific westerlies to greater energy, and hence more efficient northward flux of angular momentum, in the Hadley circulation. The intensified and southward displaced westerlies in the central and eastern North Pacific generated when the equatorial water in these sectors is warm seem to tend, especially in winter, to pass south of the Rocky Mountains and maintain the main flow of the circumpolar vortex in an abnormally low latitude around much of the northern hemisphere, commonly as far as the eastern Atlantic and occasionally farther east. Some similar tendency to generate the main flow in a low latitude may also operate in the southern hemisphere: for BJERKNES has found evidence of expanded polar anticyclones and higher pressure over the Arctic and Antarctic when the equatorial Pacific water is warm (though not without exception).

Intensification of the Hadley circulation should tend to occur in any sector when the equatorial ocean is warmer than normal, but it is probable that no other sector is liable to important anomalies of water surface temperature to be compared with those seen from time to time in the usually cold sector of the equatorial Pacific.

The history of prevailing water surface temperatures in various sectors near the equator and in other parts of the world's oceans where known, illustrated here in fig. 10.11, seems to indicate some general parallelism between rising water temperatures and the increasing vigour of the general atmospheric circulation from the beginning of our records up to some time between 1920 and 1950, followed by reversals of all these trends. The curves for the majority of the twelve different areas for which long records of surface water temperature are available suggest that a very general warming was in progress up to some time around the 1940s, followed by some decline. There are features which complicate the picture, but they seem to be either of lesser magnitude or rather localized importance. There was a secondary minimum of ocean temperature in many regions around 1920; the Peruvian coast station, which shows an odd history, may be representative primarily of the upwelling in that neighbourhood and should, perhaps, show trends of temperature inverse to those

1. The sector-to-sector relationships in extratropical latitudes between the east Pacific and Atlantic are not so regularly the same as in 1957–8 as BJERKNES supposed. They probably depend too much on other factors which also affect wave positions downstream. A later study by BJERKNES (1969), however, throws much light on sector-to-sector relationships along the equator in the Pacific and the working of the Southern Oscillation.

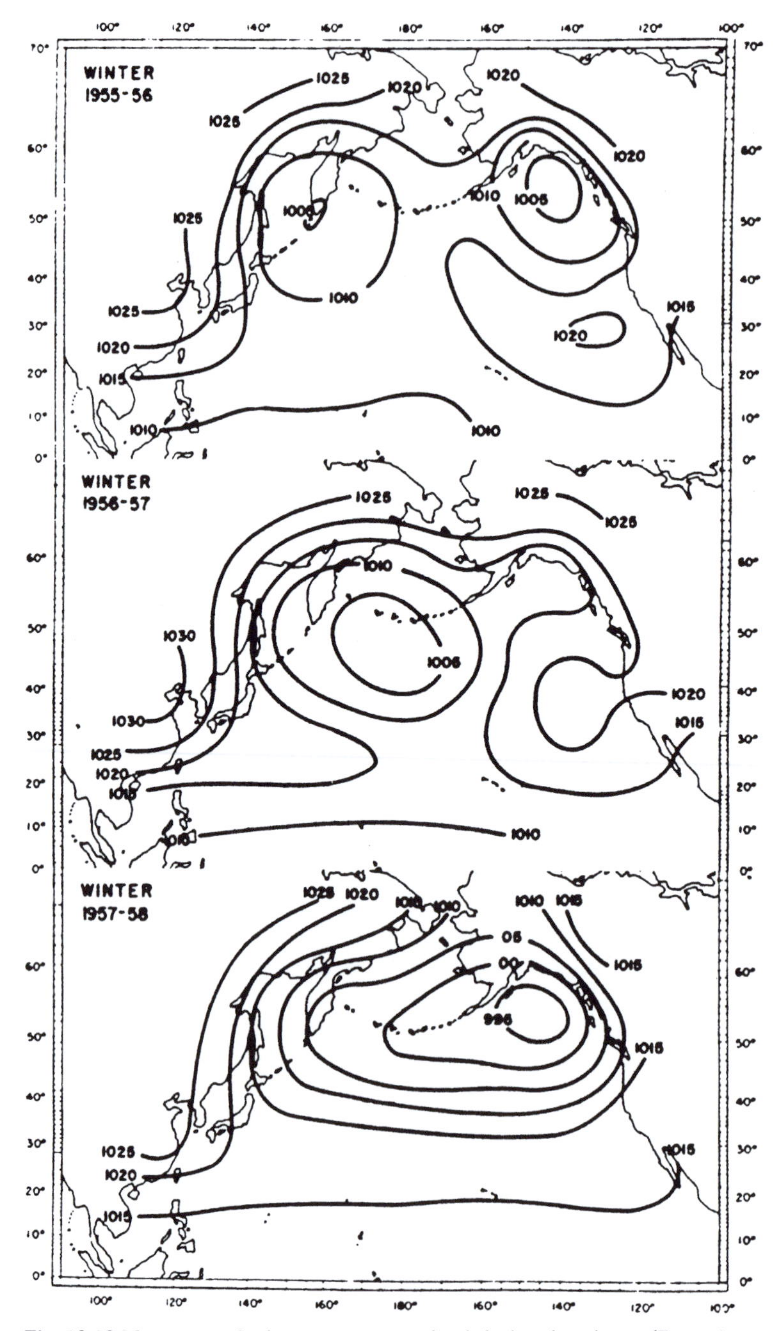

Fig. 10.10 Mean atmospheric pressure at sea level during the winters (December–February) of 1955–6 and 1956–7, before the 1958 El Niño warmth of the equatorial eastern Pacific, and 1957–8 at the height of that warmth. (*After* BJERKNES *1966*.)

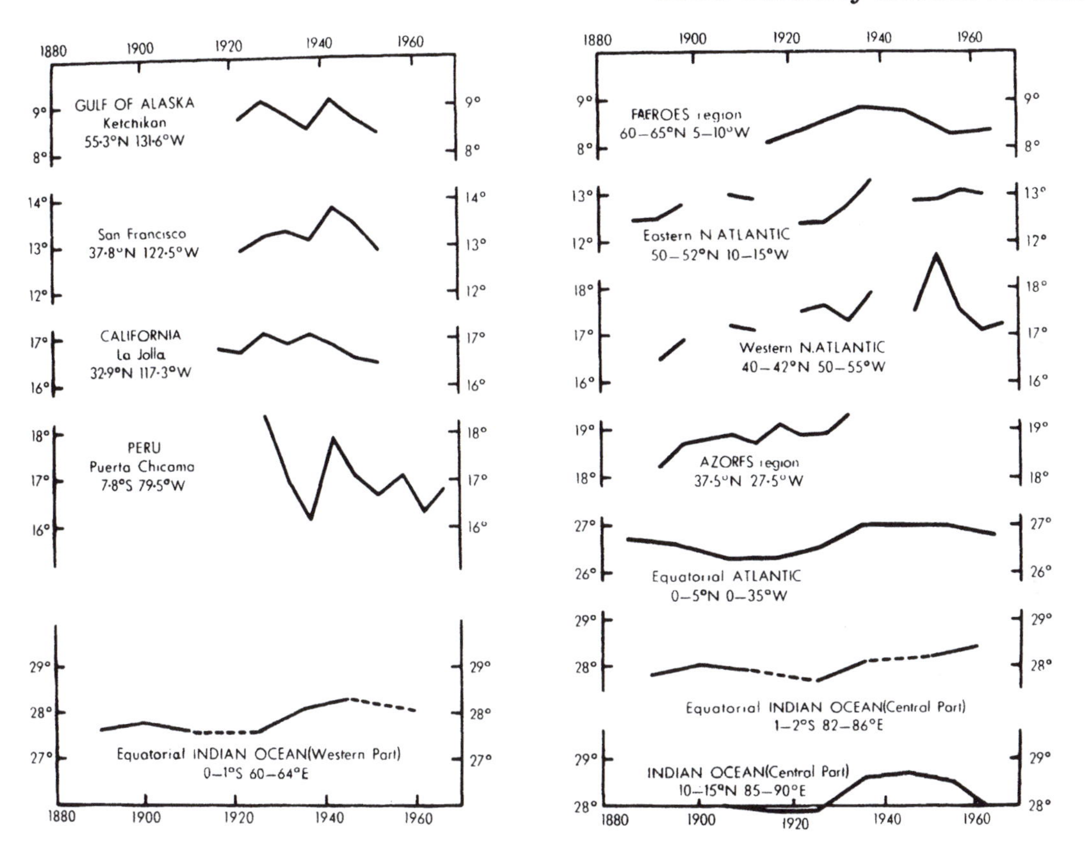

Fig. 10.11 Sea surface temperature history in various parts of the world's oceans.
Non-overlapping 5-year (in some curves 10-year) means plotted against the middle year of the period.
(Data from BJERKNES *1963*; BROWN *1963*; FROST *1966*; RODEWALD *1956, 1958a, b, 1964, 1968*; SMED *1946, 1948, 1965 and some unpublished Indian Ocean data kindly supplied by Captain* G. P. BRITTON.)

affecting most of the world's oceans. Off Peru we see cold water when the atmospheric circulation was most vigorous, and it may be that it is the Peru coast anomalies which tend to spread westwards across the equatorial Pacific. There is more yet to be understood about the sea temperatures in the Indian Ocean also, where two of the three areas studied show the same trends as we have found in most of our observation series in the oceans but the area in between does not. Even near the equator the highest water temperatures come *towards the end* of the epoch of strong atmospheric circulation and may be partly an effect of this; and both may be partly the effects of some common external cause. In any case, the parallelism affects the global atmospheric circulation more clearly than the local weather in equatorial regions. Rainfall in most of the equatorial zone seems actually to have decreased (KRAUS 1955, 1958) in the decades of highest water temperatures as compared with the nineteenth century or with the 1950s and 1960s, presumably owing to the more intense subtropical anticyclones and expansion of their area of influence in the warm decades despite the necessarily increased input of water vapour into the air. The time of low rainfall and low levels of the lakes in eastern equatorial Africa (LAMB 1966*b*) corresponded rather well with the period of strongest atmospheric circulation, between about 1900 and the 1950s, save that its beginning about 1896 and end about 1960 were sudden, as if depending on some critical threshold value of circulation strength. This must mean that the additional energy in the atmospheric circulation resulted in a larger-scale pattern of moisture transport, so that the condensation with its release of latent heat and rainfall was realized in other latitudes farther than before or since from the equatorial source. There is much evidence that the subtropical anticyclones were expanded poleward as well as equatorward in the decades of strongest general atmospheric circulation. The rainfall increases were in the zones of middle latitudes westerlies and were reflected also in increased downput of snow on the ice sheets in high latitudes (LAMB 1964, 1967*a*).

Fig. 10.12 portrays the average departures from early twentieth-century values of the sea surface temperatures observed by American ships all over the North Atlantic between 1780 and 1848, though weighted to the years 1840–8, in which 65% of the observations (collected by MAURY 1852) were made. The pattern and the values seem to verify the results derived (for the months of January and July only) for the period 1780–1820 from the first survey of British ships' observations (collected by RENNELL 1832), reproduced in fig. 8.9; they also indicate that much the same geographical distribution of departures from early twentieth-century values applied to the whole year. These 1780–1848 patterns all show a boosted thermal gradient in the ocean surface near 40°N, particularly in the sector south of the Newfoundland Banks. The southern positions of the North Atlantic atmospheric circulation features in the late eighteenth and early nineteenth centuries, indicated or implied in figs. 7.1 (*b*) and 7.22, may be largely explained along the lines examined earlier in this section (e.g. figs. 10.6–10.8) by these differences in the position and amount of the energy supply from the ocean. Prevailing atmospheric pressures at sea level, in summer and winter, appear to have been higher than now in the Arctic, especially Greenland, and also over parts of northern Europe.

There is one other way, which NAMIAS (1963*b*, 1965) has stressed, in which ocean surface temperature anomalies may exert a decisive influence on the large-scale atmospheric circulation pattern and development. This is through the large abnormal source of energy represented by any great extent of warmer than normal sea in an area commonly invaded by cold airmasses of Arctic origin. The whole northern Pacific, the Gulf of Alaska and those regions of the North Atlantic exposed to outbreaks of very cold air from northern Canada in winter or from the Arctic east of Greenland at most seasons are places where this influence can occur. The effect is seen in the rapid deepening of depressions which are often

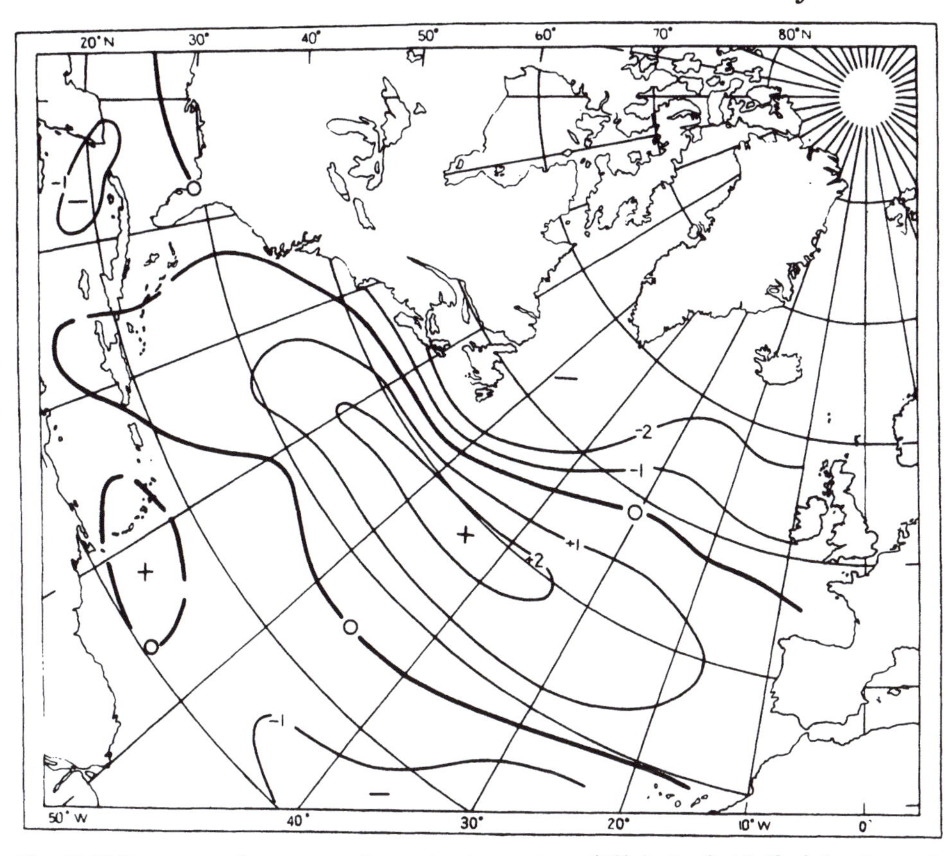

Fig. 10.12 Departures of average surface water temperatures (°C) in the first half of the nineteenth century from the values of the period 1887–99, 1921–38.
Using MAURY's collection of the observations from American ships (65% of the observations made in 1840–8, 18% 1830–9, 10% 1820–9, 7% between 1780 and 1819).

already deep, fully occluded and with their centres completely surrounded by cold air.[1] The resulting very deep systems are liable to become nearly stationary over the warm water region, where they continue to derive further energy. On their eastern sides persistent S'ly surface winds transport warm air polewards, and are liable to create a stationary warm ridge of considerable amplitude in the upper westerlies. This warm ridge may be something of a regional anomaly in a regime which favours a stationary upper cold trough over the continental sector downstream and which may determine a notably cold winter regime over most of the continent in question. A prime example of this, according to NAMIAS (1963*b*, 1965), was the influence of an extensive warm anomaly centred near 40°N 170°W in mid

1. First described, with a case-study and theoretical discussion by SUMNER (1951) in which there was vertical instability also in the warm airmass above the occlusion and the calculations of heat fluxes affecting the 1968–9 winter regime in the Pacific given by NAMIAS in a recent article (*J. Physical Oceanogr.*, **1** (2), 65–81. 1971).

Pacific upon the great North American and European winter of 1962–3. Repeated development of intense cyclonic activity in air drawn from the Arctic and northeast Siberia over the area of warm water in mid Pacific maintained a stationary large-amplitude upper warm ridge over, and off, the American Pacific seaboard and an equally stationary cold trough of great meridional extent over most of North America east of the Rockies. The circulation strength, latitude and wave length maintained the next downstream cold trough, also nearly stationary and of great meridional extent, over central Europe (where the winter was generally the coldest since 1830, in England the coldest since 1740). Further examples of the same effect could be seen in sudden deepenings of depressions over the warmer than normal eastern Atlantic in the 1968–9 winter: the consequence was interruptions of the cold weather prevailing in Britain in December and the maintenance of a very mild January all over western Europe, thanks to S'ly winds east of depressions stagnating off Ireland and in the Southwest Approaches. The periods concerned were markedly out of character with the rest of the 1968–9 winter in Britain and central Europe, which was cold with frequent E'ly and N'ly winds. The cyclone deepenings which interrupted this latter winter with S'ly winds over western Europe may be regarded as somewhat flukey, since the belt of warm water off the coast of Europe was not wide except just south of Iceland. The excessive deepenings of cyclones over the eastern Atlantic may also have owed something to the abnormal extent of ice on the seas immediately north of Iceland in 1968–9, presenting a source of cold airmasses much nearer than usual to the warm-water regions of the Atlantic.

In so far as the physical principles have been rightly identified, all the relationships between sea temperature and atmospheric circulation here demonstrated for the North Atlantic can be expected to work similarly over the North Pacific and the Southern Ocean. NAMIAS (e.g. 1969) has repeatedly stressed the role of warm water over a wide area of the central North Pacific, through what we may now call the NAMIAS–SUMNER effect, in supposedly controlling the large-scale circulation pattern of the atmosphere over much of the northern hemisphere by the extra deepening of cyclones produced by intense heating of the cold Arctic and Siberian airmasses. It appears, however, that the heart of the region concerned, between 30° and 45°N 155–175°W, which is liable to big anomalies of sea surface temperature, lies in the same relation to the boundary of the warm Kuro Shiwo ocean current and the cold water to the north as does the Newfoundland Banks key area in the North Atlantic to the Gulf Stream. In both cases, the area where extensive warm or cold anomalies of sea surface temperature are developed is near where the warm current weakens and fans out. In other words, therefore, the anomalies probably represent cases where the warm current is respectively *either* more strongly developed and pushes the current boundary farther north and northeast *or* is weak and fails to spread as far in these directions as usual. The effect upon the jet stream in the atmosphere may therefore work just as we have described for the North Atlantic (cf. p. 397). Indeed, RATCLIFFE's unpublished studies in the Meteorological Office suggest that in the autumn and winter months warm water in the mid-Pacific key area produces lower than normal pressure over the Gulf of Alaska and high atmospheric pressure over Canada and Greenland, whereas cold water produces the opposite tendency; in the spring and early summer months, when the atmospheric circulation is generally weaker and marked by shorter wave lengths and smaller-scale features, the pressure anomaly pattern is displaced west and southwest so far that warmer than usual water in mid Pacific tends to produce low pressure immediately north and northeast of it while above-normal pressure prevails over Alaska and neighbouring areas.

It is not surprising that the equatorial Pacific should be another such key region for the atmospheric circulation. It shares with the two regions just discussed, near the current boundaries in the North Atlantic and North Pacific, the property of being liable to peculiarly great sea temperature anomalies, amounting (as noted on p. 403) to +3 to 4°C in some cases. (The greatest negative anomalies may be smaller.) And the area covered by positive anomaly at one and the same time may stretch 10 000 km east and west, 3000 km north and south, exceeding the extent of the middle latitudes anomalies mentioned. Unpublished studies by P. R. ROWNTREE of the Meteorological Office, experimenting with a SMAGORINSKY mathematical model of the atmosphere, support BJERKNES' suggestion of an intensified Hadley circulation and development of an intensified jet stream in a southern position over the eastern North Pacific when the equatorial Pacific is abnormally warm.

There are probably not many situations in which the NAMIAS–SUMNER effect is operative as the cause of a climatic fluctuation lasting more than one to three or four years at most. This is because, unlike the case of the other sea temperature influences we have been discussing, the maintenance of cold air and cyclonic circulation over the region of anomalously warm sea must tend to eliminate the sea surface temperature anomaly concerned. Only near the eastern limits of each ocean in middle and higher latitudes the atmospheric circulation may (with the aid of the coastline to guide the water drift) maintain sufficient poleward drift of warm water to give rise to suitable situations for this effect to be repeated at intervals over many years.

Cause and effect, action and reaction between atmosphere and ocean, are intertwined at so many points in the variations that we have so far been considering that they can only be distinguished when we concentrate on this or that time-scale. It is both necessary and attractive therefore to turn our attention to things that may cause variations of the initial energy supply available to atmosphere and oceans.

In fig. 10.12 we find some areas of the ocean with surface temperature anomalies over ±2°, probably in some places over ±3°, characterizing an epoch that lasted more (and probably very much more) than half a century. Moreover, these ocean temperature anomalies exceeded the estimated world average air temperature anomaly (about −1°C for A.D. 1800 compared with 1900–50) of the time in question. Consideration of the distribution and magnitude of these sea surface temperature anomalies suggests that they can largely, but not wholly, be explained by a southward shift of the Labrador Current boundary and intensification of the Gulf Stream–North Atlantic Drift on a more southern and more nearly west to east track than in 1900–50. Averaged over the whole Atlantic between 55°N and 40°S, prevailing surface water temperatures around A.D. 1800 seem to have been a little cooler than early twentieth-century normal values: over the whole Atlantic Ocean between these latitudes a departure of −0·1°C around 1800 is indicated, though the probably more reliable figure for the North Atlantic between 55°N and the equator comes out at −0·2° to −0·3°C. If one were to assume that, as with the changing seasons, this secular change was caused entirely by a change in the radiation balance due to less intense insolation reaching the surface, a reduction in the average radiation supply by 1–3% would be implied.[1]

Volcanic activity and climate

Great volcanic eruptions of the explosive type sometimes create world-wide veils of finely divided dust which linger for some years in the high atmosphere and dim the sun and moon, though they brighten the background sky. Volcanoes are also a continuing source of injection of carbon dioxide, sulphur compounds (H_2S, SO_2, etc.) and water vapour into the atmosphere. None of these gases are, however, produced in sufficient quantity to have any important meteorological or climatic effect on time-scales less than many millions of years. Most assessments of the magnitudes of individual eruptions have concentrated on the scale of the local disaster in terms of topographic changes, quantity of lava, destruction of life and agricultural land, tidal waves, etc. (see, however, SAPPER 1917, 1927). These are irrelevant as regards meteorological effectiveness, which depends on the injection of sufficiently vast quantities of fine dust (or 'ash') into the stratosphere. Eruptions that produce great flows of lava, particularly those that exude the most fluid, dark, basaltic lavas, often as in Iceland from long line fissures in the surface crustal rocks, do not necessarily produce any explosion, commonly produce none sufficient to put material into the stratosphere and may not produce any ash at all. Equally, great ash eruptions do not always produce any lava. For study of the climatic effects, therefore, an assessment of eruptions in terms of the dust veil in the atmosphere, and its persistence, has to be used, as given by LAMB (1970) and in abbreviated form here below.

Injection and spread of volcanic dust in the atmosphere; duration of veils

Observations of the greatest heights reached by the columns of rising smoke and solid, or solidifying, fragments of matter over erupting volcanoes, as well as the heights of the dust clouds subsequently observed in the atmosphere, appear to fall into three groups:

(i) top of the dust at various heights in the troposphere, commonly 7–15 km but occasionally as low as 1–3 km;
(ii) top at 20–27 km, in the lower stratosphere;

1. The figure may be judged by comparison with the movement of the sea surface temperature isotherms in the open ocean (away from regions of coastal heating or upwelling and away from water-current boundaries) that accompanies the regular seasonal changes of radiation supply.

(iii) top in the region of 50 km, from where some of the dust and water vapour, etc., is liable to be carried up by convection within the mesosphere to the mesopause (where very thin, noctilucent clouds – also known as ultra-cirrus – are sometimes observed) at 80–82 km.

Dust which fails to enter the stratosphere will be washed back to Earth in rain (or snow) within a matter of days. Group (i) cases are clearly therefore of no climatic interest. Group (ii) is probably of the greatest importance, because the quantities of material reaching the lower stratosphere are sometimes very great and produce optically dense veils there. There seems, moreover, to be a permanent aerosol layer at heights around 22–25 km containing sulphate particles produced by the oxidation on encounter with atomic oxygen of sulphur compounds diffused up from the lower atmosphere. The concentration of these sulphate particles or globules is probably increased by volcanic eruptions, and a (sometimes fluid) coating of sulphate solution or of sulphuric acid is probably liable to increase the size and optical effectiveness of volcanic dust particles in this layer.

It is possible that some items among the particulate volcanic ejecta in the troposphere, which consist largely of minute fragments of silica (i.e. the volcanic dust or ash) plus hygroscopic sulphur compounds, particularly sulphuric acid, increase the abundance of condensation-nuclei (see, for instance, BERRY *et al.* 1945, p. 255). The effects of this must, however, be too localized in place and time to be of climatic importance, since they are limited to the few days or weeks in which concentrations of volcanic materials remain in the troposphere.

The sizes of the solid particles observed in persistent volcanic dust veils have been mainly from 0·5 to 2 μm[1]; though the addition of a liquid, or deliquescent, coating may mean that the effective size is most commonly 1–2 μm. There may also be quantities of still finer ash with particle diameters less than 0·5 μm that are not readily detected. Because their diameter is less than the wave length of most of the solar radiation, the effect of such particles, however, approximates to that of the air molecules themselves and hence is of no climatic importance. The terminal velocity w of fall of a particle of radius r and density ρ falling through still air (density ρ_a and viscosity M) is given by Stokes' law as

$$w = \frac{2}{9} g r^2 \left(\frac{\rho - \rho_a}{M} \right)$$

For very small particles this has to be corrected for slip associated with the mean free path of the gas molecules: this is achieved (HUMPHREYS 1940, p. 592) by multiplying the right hand side of the equation by a factor $\left(1 + \frac{B}{rp}\right)$ where B is a constant for any given temperature (B is about 0·005 to 0·007 at stratospheric temperatures of −80° to −30°C) and p (the atmospheric pressure) is measured in millibars. The calculated terminal velocities range from 0·005 cm/sec for a 0·5 μm particle diameter to 0·08 cm/sec for a 2 μm particle diameter at the 20-km level and to 1·0 cm/sec for the latter size particles at the 40-km level.

1. One micron (or one micro-metre), written 1μm (or 1μ in the older literature), is 10^{-6} m.

The corresponding expected residence times in the stratosphere for particles injected at different heights are given in the following table. The column headed (*a*) applies in latitudes between about 30° and the equator, where the tropopause is normally at a height of about 17 km; the column headed (*b*) applies in all other latitudes, where the lower tropopause at about 12 km prevails.

TABLE 10.1
Stratospheric residence times for volcanic dust
(Total times taken to fall through still air to tropopause at (*a*) 17 km, (*b*) 12 km)

Particle diameter	*Initial height (km)*	(*a*)	(*b*)
2 μm	40	25 weeks	41 weeks
	30	21 ,,	37 ,,
	25	16 ,,	31 ,,
	20	7 ,,	23 ,,
1 μm	40	1·9 years	3·1 years
	30	1·6 ,,	2·8 ,,
	25	1·3 ,,	2·4 ,,
	20	0·6 ,,	1·7 ,,
0·5 μm	40	7·8 years	12·5 years
	30	6·5 ,,	11·3 ,,
	25	5·0 ,,	9·7 ,,
	20	2·2 ,,	6·9 ,,

The dust is carried around the world by the upper winds. Survey of the fan-shaped deposits (e.g. fig 10.13 (*a*)) of measured depths of ash, cinders and larger blocks – the whole assortment collectively known as *tephra* – provides one possible measure of the yield of individual volcanic explosions. Such surveys of *buried* volcanic ash fans, some found in the subsoil, some as intrusions in sedimentary rocks and some in the present ocean bed, can, if extended to many eruptions, provide a fossil record of the upper winds prevailing at the time of the volcanic activity (e.g. EATON 1963, GILL 1961, p. 341, SCHWARZBACH 1961, pp. 70–2, THORARINSSON 1944, 1959). The total quantities of solid tephra of various sizes blown up into the atmosphere, as measured either by the matter lost from the site of the eruption or constituting the subsequent deposit, may be very large: estimates range from 50 to 150 km^3 for a few of the very greatest eruptions. Dust in the layers above the reach of rain- and snow-clouds soon spreads into a fairly uniform veil over the wind zone into which it was injected (see fig. 10.14). Spread laterally and fore and aft of the main concentration is helped by eddy motion and by such differences of wind speed and direction as exist within a dust layer several kilometres deep: this spread is commonly from 5 to 15% of the distance travelled by the main concentration. Plates 1*a* and *b* illustrate (for the first time with the aid of satellite photography) the beginnings of the spread of material in the atmosphere ejected

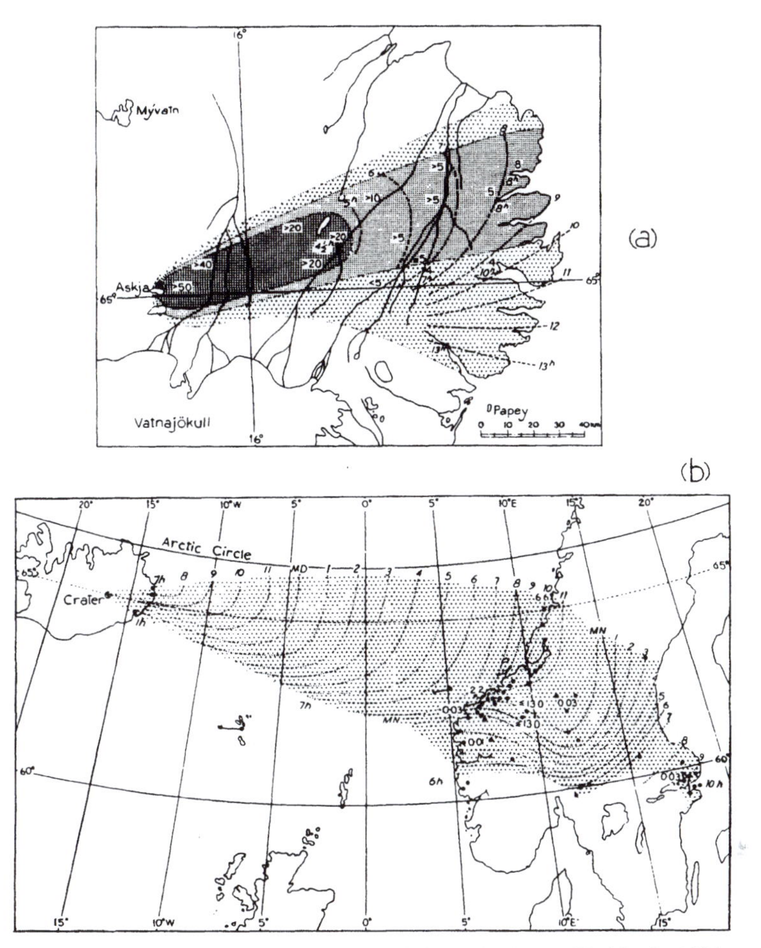

Fig. 10.13(*a*) Spread of dust over Iceland from Askja on 29 March 1875. Depths of deposit in centimetres. Dotted lines are isochrones, times GMT.
(*After* THORARINSSON *1944, p. 98.*)
(*b*) Spread of dust from the Askja volcano in Iceland to Scandinavia on 29–30 March 1875. Point depths of dust deposit in Scandinavia given in mm.
(*After* MOHN *1877.*)
Note: MD means midday, MN midnight.

by the 1970 eruption on the island of Jan Mayen (71°N 8°W). Often the dust layer consists in detail of overlapping veils at different heights resulting from a succession of explosive phases of the eruption, each layer extended downwards with time (by size-sorting and the range of terminal velocities represented). Typically, volcanic dust is likely to take

PLATE 1*a* (facing, above). *ESSA 8 Satellite photograph of the Earth, with latitude and longitude net superposed, as received by APT in the U.K. Meteorological Office, Bracknell.*

The picture shows the island of Jan Mayen (71°N 8°W) about 10h GMT on 21 September 1970 in an area of clear sky but with a trail of volcanic ash and steam apparently issuing from it, and extending southeast to where it becomes hidden by a larger, roughly rectangular area of denser 'cloud' extending farther along a similar path between about 68°N 2°W and 66–67°N 9°E. This denser cloud is also thought to have come from the Jan Mayen eruption, which had one or more explosive phases on 18 September: the cloud concerned appears to be disconnected with, and higher than, all the cloud systems associated with the weather patterns over the Norwegian Sea. Since the reflected light from the volcanic 'dust' cloud appears stronger, in spite of the slanting angle of incidence of the view from the satellite, than that from any other clouds in the picture except those in the polar front belt (and in a small wave depression on it near 57°N 25°E) over the Baltic States and northern Germany, this volcanic dust cloud was presumably dense enough to obscure the sun at places underneath it at this stage, 3 days after the main eruption. Study of the winds over those 3 days suggests that the dust cloud was travelling mainly at heights rather over 30 km in the stratosphere.

After the date of this photograph the dust cloud gradually became more spread out and diffuse, as it passed across central and southern Scandinavia on the 22nd and 23rd to the Baltic, where by the 25th it could no longer be distinguished from the other clouds seen in the satellite photographs of that date. Some of the thinning dust veil seems later to have spread around a high-reaching anticyclone over the North Sea and Norwegian Sea, the characteristic colours of volcanic dust sunsets and prolonged twilights being noticed in southern England in the last few days of September.

This photograph is believed to be the first clear record ever obtained by photography from an artificial satellite of the beginning of a volcanic dust veil. Unfortunately, because the dust deposit in this case mainly fell to the ocean bed, and the veil may have been largely confined to unusually high latitudes, it has so far been impossible to assess this eruption in terms of dust veil index (p. 421), though the value will probably turn out to be less than 50. (The eruption seems to have been on a comparable scale with the eruptions of Hekla in Iceland in 1947 and 1970.)

The author acknowledges the kind permission of the Controller of Her Majesty's Stationery Office, London to reproduce this photograph, which is Crown Copyright.

PLATE 1*b* (facing, below).

Air photograph taken at Jan Mayen, looking horizontally towards the north at a height of about 2000 m on 21 September 1970, with the main summit of Beerenberg (2340 m) in the foreground and showing the column of steam rising from the north end of the island. This was the origin of the narrow plume seen in the satellite photograph (Plate 1*a*) taken on the same day. The column of steam was reported to be originating at this time from the sea, at the point where the hot lava, streaming from the fissures opened on the northeast side of the mountain, reached the water. (Some days later a temperature of +39°C was measured in the water half a kilometre from the lava front, which was still enlarging the island, and many sea-birds were seen swimming in the hot water; the usual temperature of the sea in the area is +1°C.)

The photograph shows the column of steam on 21 September penetrating the flat layer of stratus clouds, which is so usual a feature of the Arctic, and rising to a height estimated as about 6 km: it is not surprising that the explosive phases of the eruption should have thrown dust and steam to very much greater heights.

The photograph, which was taken by R. C. Ulrichsen from a Hercules plane of the Royal Norwegian Air Force, is reproduced by kind permission and through the courtesy of *Aftenposten's A-Magasinet*, Oslo, in which it was first published.

20W

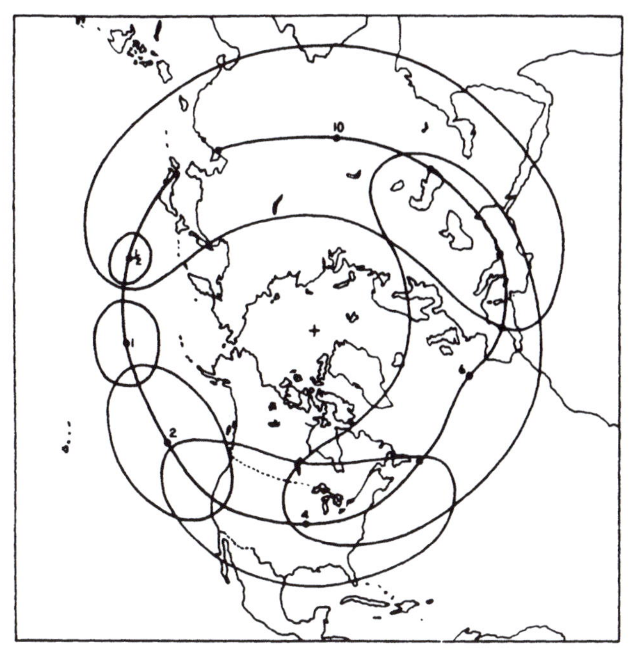

Fig. 10.14(*a*) Mean trajectory of constant-pressure-level balloon transosondes at the 300 mb level from Iwakuni, Japan in 1957–8 and areas (the ellipses) within which observations and computations indicate that 50% of the transosondes would be found a given number of days after release.
(*After* ANGELL *1960*.)

(*b*) Fourteen-day gradient wind trajectories at the 500 mb level from Perth, Australia; Christchurch, New Zealand; Port Stanley, Falkland Islands; and Cape Town for July 1957, January and July 1961; and from the South Pole for July 1957.
(*After* GILES *and* ANGELL *1963*.)

from 2 to 6 weeks to circuit the Earth in middle or lower latitudes and from 1 to 4 months to become a fairly uniform veil over the whole latitude zone swept by the wind system into which it was injected. In the lower stratosphere and tropopause layers mostly concerned, this usually means either (*a*) in the meandering upper westerlies over middle latitudes, (*b*) in the tropical easterlies or (*c*) in the intermittent equatorial-zone (approx. 10°N–10°S) stratospheric westerlies occurring in the course of the '26-month oscillation' (see pp. 205–7).

Spread of the dust to other latitudes outside the wind zone originally affected takes longer. It tends to occur in distinct pulses each autumn (probably also to some lesser extent in spring) associated with the breakdowns, including phases of meridionality in this or that sector, and migrations of the stratospheric circulation systems of the season that is ending and the poleward transfer of mass in the atmosphere with the onset of winter. Meridional components in the mean upper wind circulation elude direct observation because they are so much weaker than the zonal components. Knowledge of them has been derived partly from observation of the progress of tracer substances in the atmosphere, such as volcanic dust, nuclear fission products from atomic weapon tests, and ozone, partly also from theoretical considerations such as angular momentum studies. Fig. 5.3 (*b*) (p. 193) showed the directions derived for the mean motions in the lower atmosphere in the meridional and vertical plane; they were generally poleward between the equator and latitude 30° as well as over high latitudes but equatorward over the middle latitudes; they are, however, known to be poleward over all latitudes for a time in late summer and autumn and are equatorward over all latitudes at some stage in (late) spring. The eddy transfers in both directions (bold arrows in fig. 5.3 (*b*)) seem, however, to be, on the whole, of more effect than the mean circulation over latitudes 30–90°. Hence, volcanic dust put into the lower stratosphere over low latitudes is gradually spread over the whole Earth: by the mean meridional circulation as far as latitude 30° and thence by the eddies in the circumpolar westerlies and with some extra poleward breakthrough of the mean drift in the autumn. By contrast, volcanic dust originating in high latitudes hardly spreads in significant quantities beyond about latitude 30° in the hemisphere of origin; though small quantities may reach any part of the Earth, it effectively covers only latitudes 30–90° in the hemisphere of origin and tends to maintain its concentration mostly over latitudes 60–90°, i.e. the polar cap. Because of the shorter sojourn in the stratosphere that is normal for dust particles over the higher-level tropopause in latitudes between 30°N and 30°S (cf. Table 10.1), the final stage of a world-wide dust veil from an equatorial eruption is probably a concentration of the last remaining airborne dust over the two polar caps.

These inter-latitudinal (and the much slighter inter-hemispheric) exchanges are illustrated in fig. 10.15 by the concentrations at different latitudes from pole to pole of strontium 90 injected into the lower stratosphere by weapon tests near 75°N in 1961 and near 12°N in 1962–3. The half-life of Sr 90 is 27 years; so it is fall-out from the atmosphere that accounts for the decay of radioactivity with time registered in fig. 10.15.

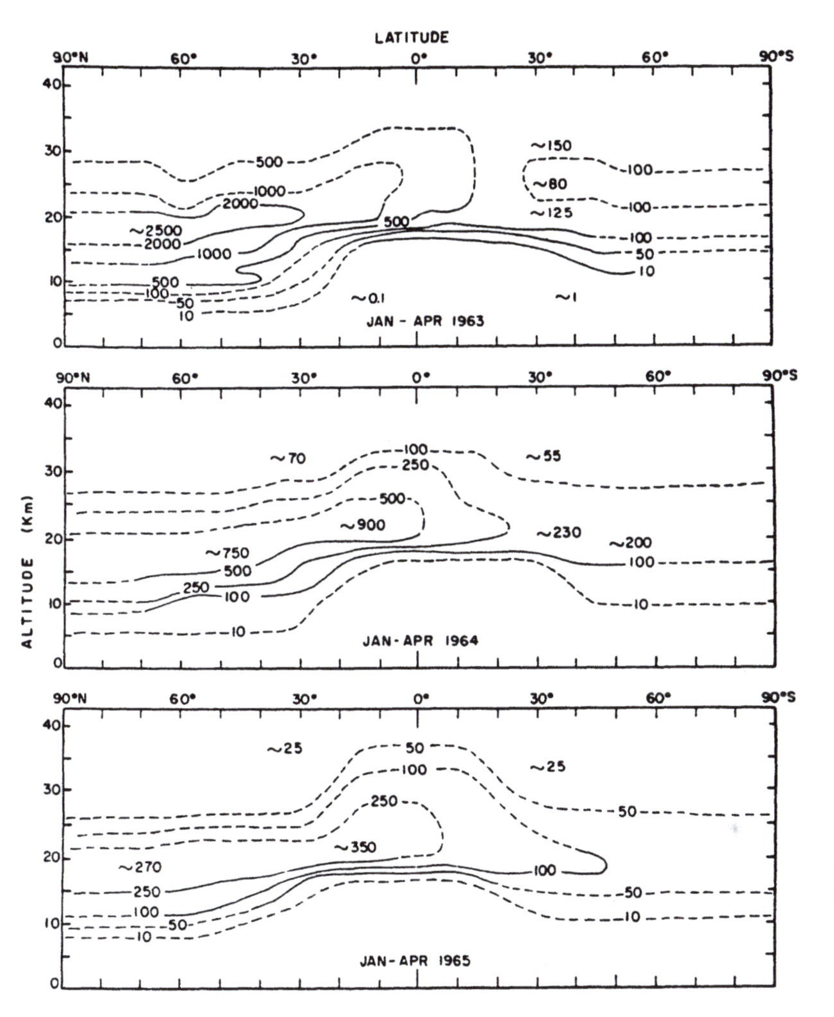

Fig. 10.15 Distribution of ^{90}Sr in the atmosphere by height and latitude in January to April, 1963, 1964 and 1965.
(*After* FEELY *et al. 1966.*)
Units: dpm per 1000 cu. feet of air at 1013 mb and 15°C.
Heights in km.
The ^{90}Sr particles have an estimated median residence time in the atmosphere of about 10 months.
The distributions here shown are attributed partly to Soviet tests near 75°N in 1961 and 1962 and partly to American tests near 12°N in 1962.

Effects of volcanic dust veils

The presence of the denser volcanic dust veils is readily apparent to the eye, through the dimming and reddening[1] of the sun and through the white glare from the sky near the sun, which becomes fiery red at sunrise and sunset, as well as from the hazy white or dull, leaden appearance of the rest of the sky. Reports of many such veils exist from before the era of instruments to measure the radiation effects. Their frequency indeed suggests something like a world-wide wave of exceptional volcanic activity between about 1750 and 1900; there was probably enhanced activity as compared with the immediately preceding times from about 1550 onwards and especially between 1660 and 1694. Among the most specific of these reports is that in June–July 1783, after great eruptions in Iceland, the sun was invisible until it had risen 17° above the horizon in southern France, remained copper-coloured in England at 20° elevation and even in Italy was so dim throughout the day that it could easily be looked at. No stars were seen below 40° elevation. Another great eruption took place in Japan in August of that year, and 1783 must certainly be accounted among the greatest volcanic dust years in the northern hemisphere of which we have report.

Records of the really great dust veils that covered the northern hemisphere are probably nearly complete from about A.D. 1500, if not earlier, though the eruptions which produced them are not always known. For Iceland, the Mediterranean and Japan the record of eruptions is reasonably complete from much earlier times. Great world-wide dust veils are not likely to pass unnoticed. In previous centuries they were liable to strike terror in those who beheld the optical effects. The writer has therefore attempted (LAMB 1970) a numerical assessment of volcanic dust in the atmosphere over the northern hemisphere from A.D. 1500 to date.

Radiation effect

Observations of the strength of the direct solar beam at observatories sited to minimize the effects of atmospheric haze and cloudiness are available from 1883. Fig. 10.16 displays the monthly values from 1883 to 1954 from observatories between 30° and 60°N. Note the great reduction of direct solar beam intensity, by 20–30% in some months, following the great volcanic eruptions in 1883, 1902 and 1912 as well as after the slightly lower order dust-producing eruptions in 1888 and 1907. (Erratically high values of the solar radiation measured in 1886–7 must be attributed to difficulty of eliminating the effects of weather, etc., in the early part of the record before 1892 when only one observatory was operating.) Fig. 10.17 presents the radiation measurements at the Goetz Observatory, Bulawayo (20°S 28½°E) comparing the monthly averages over the months January 1963–March 1964 (thin lines) with the overall means for the corresponding months in the previous 6 years (bold lines). The deficient strength of the direct solar beam after the great volcanic eruption in Bali, Indonesia (8½°S 115½°E) in March 1963 and the increase of diffuse radiation from the

1. On a few occasions, when unusually big dust particles are present, the sun and moon may appear white or even have a pale blue or greenish hue.

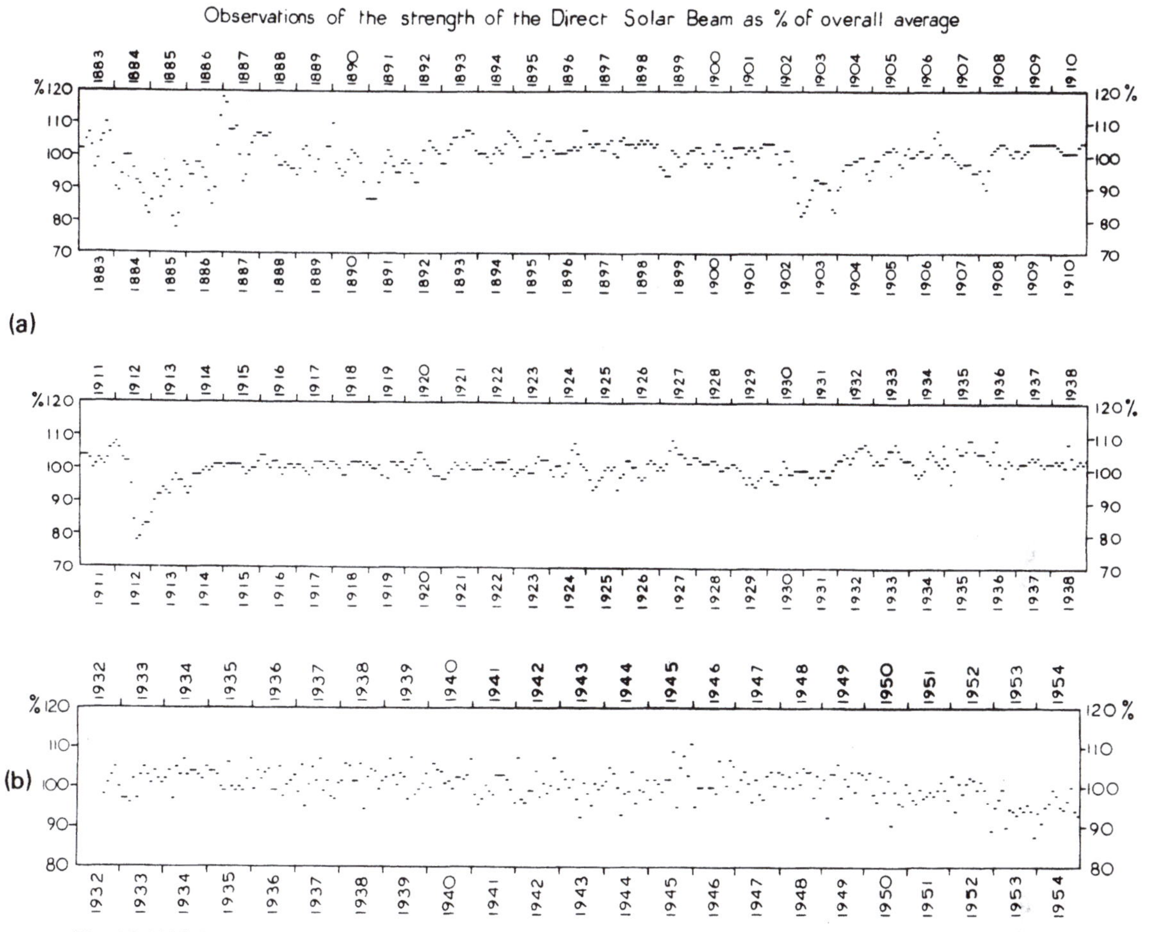

Fig. 10.16(*a*) Average monthly values of the strength of the direct solar radiation derived from observations at mountain observatories between 30° and 60°N in America, Europe, Africa and India from 1883 to 1938, as percentage of the overall mean.

(*b*) Average monthly values of the strength of the direct solar radiation derived from observations at Matsumoto (36° 15′N, 137°58′E) and Shimizu (32°47′N, 132°58′E) in southern Japan from 1932 to 1954, as percentage of the overall mean.

sky is plain. The fact that the resultant total radiation over the months in question approximately equalled the average of the previous 6 years must, however, be attributed to the finer than usual weather in 1963 and 1964, when the Rhodesian droughts of the 1960s were beginning: but for this, the total radiation would undoubtedly have shown a deficit. Similar observations from Melbourne, Australia (38°S 145°E) (DYER and HICKS 1965) show that although the direct radiation was cut by about 24% at one stage in 1963, the total radiation received in the same month was just 6% below normal. This indicates that the mainly forward scattering of energy which produces the enhanced diffuse radiation from the sky makes good about three-quarters of the diminution of the solar beam.

The intensity I_x of the solar beam after passing a distance of x cm through a layer of sparsely distributed dust particles of radius r cm, n particles occurring on average per cm^3, should (HUMPHREYS, 1940, p. 594) be

$$I_x = I \,.\, e^{-2n\pi r^2 x}$$

where I is the intensity of the incident solar beam before entering the atmosphere. HUMPHREYS shows that 10 km^3 of rock, split into spheres of 1 μm diameter and distributed in this way, should deplete the solar beam by several per cent. In reality most of the volcanic dust particles are probably thin plate-like fragments of shattered solidified bubbles;

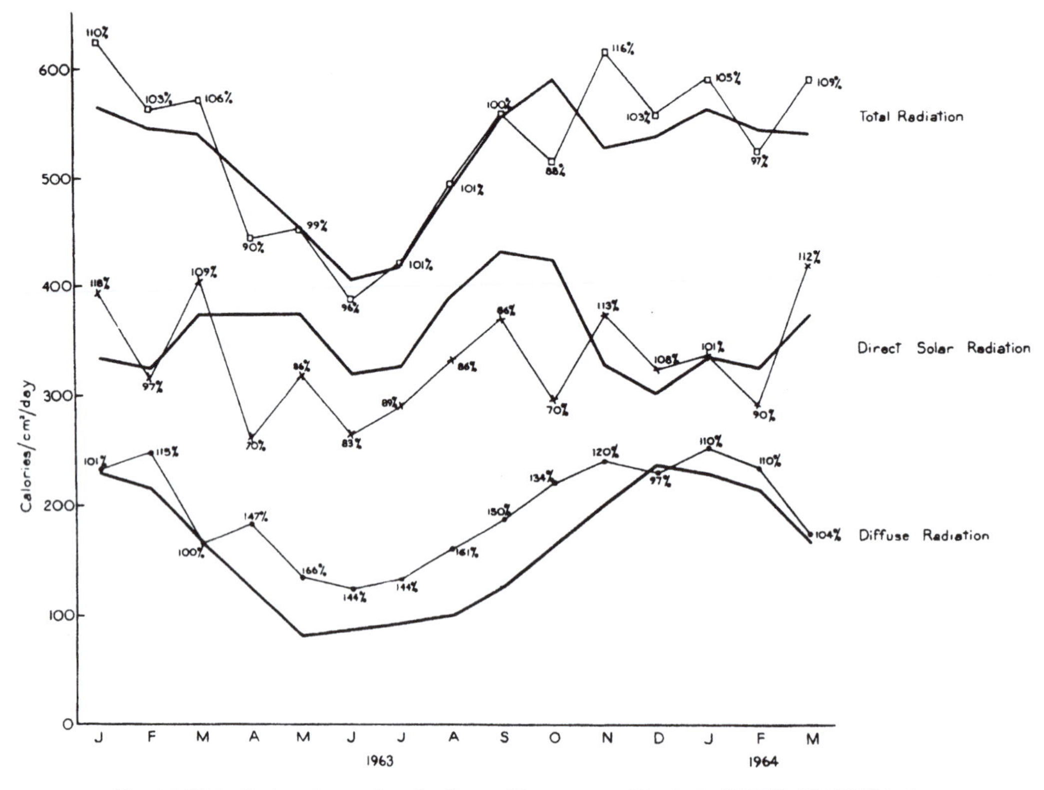

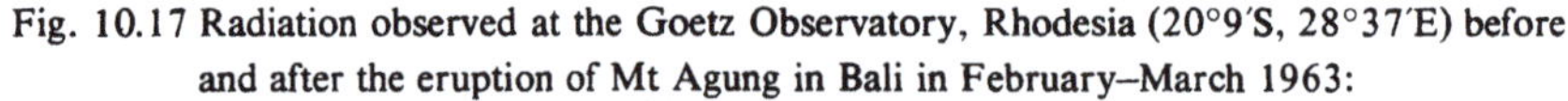

Fig. 10.17 Radiation observed at the Goetz Observatory, Rhodesia (20°9′S, 28°37′E) before and after the eruption of Mt Agung in Bali in February–March 1963:
(*a*) Total radiation, monthly average values.
(*b*) Direct solar beam, monthly average values.
(*c*) Diffuse radiation, monthly average values.
Bold lines, averages for 1957–62; thin lines, 1963–4, monthly percentages of the 1957–62 average being indicated.

though they may be encased in spherical droplets of sulphurous fluid, and calculations based on spherical shape probably give a reasonable first approximation to their effect. This theoretical reduction of the solar beam depends on the particle diameters being bigger than the wave lengths of most of the incident solar energy, over three-fourths of which is at wave lengths less than 1 μm. Since the sizes concerned are, however, much smaller than the wave lengths of most of the outgoing terrestrial radiation, most of the intensity of which is at wave lengths between about 5 and 50 μm, the radiation loss from the Earth is not much diminished. Humphreys gives the intensity E_y of the terrestrial radiation after passing a length of y cm through the dust layer as

$$E_y = E \cdot e^{-6000 n r^6 y/\lambda^4}$$

where E is the initial intensity of the radiation emitted from the Earth and λ is its wave length. He concluded that passing a given distance through a volcanic dust layer, with characteristic particle diameters around 1·85 μm (derived from optical effects of the 1883 Krakatau eruption), should diminish the solar beam intensity about thirty times as much as it would the radiation escaping from the Earth.

The effect is clearly likely to be greatest in high latitudes, where the always low angle of incidence of the solar beam implies long paths through any dust layers, and where, as we have seen, the presence of the dust may be most prolonged. Hence, production of more ice on the Arctic seas should be expected to be a common consequence of great volcanic dust veils, and the time taken to melt such excess ice afterwards may further prolong the surface temperature reduction in high latitudes by a few years.

Effect on prevailing surface temperatures

Available observations of average yearly temperature from 1811 to 1910, mainly at continental stations in Europe, North America, Africa and India, are displayed in fig. 10.18. Here, the marked lowering of prevailing temperatures after the great volcanic eruptions in 1811–12, 1815 and 1835, allegedly also in 1878, stand out, as well as the effects of the later eruptions already seen in fig. 10.16. The temperature lowering in middle latitudes is commonly 0·5–1·0°C for the year after a great eruption, even a high-latitude eruption in the hemisphere concerned; an estimated lowering of 1·3°C occurred after the two great eruptions in one year in 1783. Longer duration of the effect in middle latitudes than in the tropical zone is also apparent in fig. 10.18 about 1820, 1840 and from the 1880s into the 1890s; still later continuance of after-effects may be presumed in high latitudes.

Numerical assessment of dust veil magnitudes

The available types of observation of the effects of volcanic explosions afford three alternative ways of arriving at a numerical assessment of dust veil magnitude, using the following formulae for a Dust Veil Index (**DVI**) in which the coefficients are adjusted to give **DVI** = 1000 for the 1883 eruption of Krakatau (6°S 105½°E). The most reliable assess-

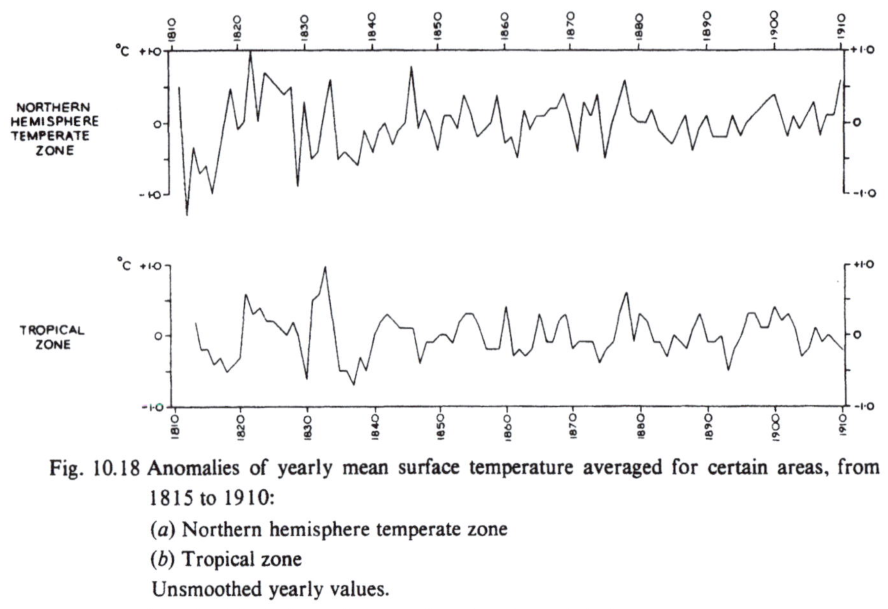

Fig. 10.18 Anomalies of yearly mean surface temperature averaged for certain areas, from 1815 to 1910:

(*a*) Northern hemisphere temperate zone

(*b*) Tropical zone

Unsmoothed yearly values.

Standard deviations of the yearly values here plotted:

Northern temperate zone (land stations) 0·38°C.

Tropical zone (land stations) 0·29°C.

(*After* KÖPPEN *1873, 1914.*)

ments are obtained by using as many of these formulae as can be applied to a given eruption and taking a round figure for the arithmetic mean as the index of order of magnitude.

$$\mathbf{DVI} = 0{\cdot}97 \,.\, R_{d\,\max} \,.\, E_{\max} \,.\, t \tag{1}$$

$$\mathbf{DVI} = 52{\cdot}5 \,.\, T_{d\,\max} \,.\, E_{\max} \,.\, t \tag{2}$$

$$\mathbf{DVI} = 4{\cdot}4 \,.\, q \,.\, E_{\max} \,.\, t \tag{3}$$

In these expressions:

$R_{d\,\max}$ is the greatest percentage depletion of the direct solar beam shown by any monthly average for middle latitudes of the hemisphere concerned after the eruption.

$T_{d\,\max}$ is the estimated lowering of average temperature for the most affected following year in middle latitudes of the hemisphere concerned.

q is the estimated volume in km^3 of solid matter dispersed as dust in the atmosphere.

$E_{\max}$ is the greatest proportion of the Earth at some time covered by the dust veil (taken as 1 for eruptions between 20°N and 20°S, 0·7 for eruptions between latitudes 20 and 35°, 0·5 for latitudes 35–42°, 0·3 for latitudes greater than 42°.

t is the total time in months elapsed after the eruption to the last readily observed effects in middle latitudes (optical effects cease and temperature and radiation measurements return to their former level).

No more refined expressions, using exponential decay rates and the like, would be appropriate to the quality of the data available for most eruptions in the past. Even in the 1960s it is hardly possible to produce a refined world average value of the radiation reaching the Earth's surface, owing to the still sparse and irregular network of observation points. Moreover, these formulae are sufficient to rank eruptions in orders of magnitude, any with **DVIs** ⩾ 100 being worthy of consideration for possible climatic effects, especially if they occur in close enough succession to be cumulative. As regards likely effects on the differential heating of the atmosphere in different latitudes, and hence on the strength and pattern of the mainstreams of the atmospheric circulation, eruptions clearly ought to be separated according to magnitude and latitude of occurrence.

Geographical distribution and frequency of great eruptions

Table 10.2 on p. 424 surveys the geographical distribution of great dust-producing eruptions by frequency per century over as much of the time from A.D. 1500 to the present day as available information covers. The volcanoes of the Pacific Ocean perimeter bulk large, but those of the Pacific Islands in and near the equatorial zone together with those in the West Indies stand out as the main dust veil producers. This is partly because of the frequency of explosive eruptions in these areas and partly because eruptions in low latitudes spread their dust veils over more of the Earth. The world total frequency over the period surveyed can be taken as averaging five eruptions per century of magnitude greater than or equal to the Krakatau 1883 eruption (i.e. giving **DVI** ⩾ 1000). The numbers of dust veils observed, and eruptions of this apparent magnitude known to have occurred, in each century since A.D. 1500, without correction for increasing likelihood of reports in the last 100–150 years, give the appearance of a world-wide wave of volcanic activity: 1500–99, 4; 1550–1649, 6; 1600–99, 6–7; 1650–1749, 4; 1700–99, 5; 1750–1849, 10; 1800–99, 7–9; 1850–1949, 2–4; 1900–65, 0.

Observed climatic effects in recent times

Effects observable in the patterns of the large-scale mean atmospheric circulation in January and July in the neighbourhood of the British Isles or over the Atlantic have been explored in the case of forty eruptions since A.D. 1680 which gave **DVI** ⩾ 100 over the northern hemisphere alone. Characteristic results as regards differences of mean pressure across the zone of prevailing westerlies, averaged over all months of the year, for the eruptions between 1875 and 1912, are illustrated in fig. 10.19. Only the weakening in the year of a great volcanic eruption in low latitudes appears statistically significant (at about the 2–3% level); but there is a fairly general tendency for this to be followed by above-average values of this circulation index in the following year, and in the case of high-latitude eruptions this tendency for enhancement of the strength of the westerlies is the only effect apparent in the diagram. The January circulation behaviour over the longer run of eruptions since 1680, strictly numerical values of the circulation index being available from 1750, showed the

TABLE 10.2

Volcanic dust ejected in the main producing regions

Region	*First reported eruption*	*Average number of eruptions per century since 1500 (or later start of information)* **DVI/E_{max}*** ⩾100	⩾250	⩾1000
Iceland	A.D. 1104	2·6	2·6	0·6
Mediterranean	1500 B.C. approx.	1·9	1·3	0
Azores–Canary Islands	A.D. 1558	1·0	0·7	0
Mexico–West Indies–Andes (north of 20°S)	1539	4·0	3·8	0·7
Andes (south of 20°S)	1520 (1800 taken as datum, reports thought incomplete earlier)	1·8	0·6	0
Antarctic and Sub-Antarctic (south of 60°S)	1821	(1–4?)	?	?
New Zealand and Southwest Pacific Islands south of 15°S	1846	2·5	2·5	0·8
Indonesia, Moluccas, Celebes, Philippine Is., etc.	1500	8·8	8·8	2·6
Japan	*c.* A.D. 800	1·5	1·5	0
Kamchatka, Aleutians, Alaska	1737	3·9	3·5	0

* **DVI/E_{max}** is used in this table to compare the eruptions in different regions as dust-producers without regard to the extent of spread of the dust veils over the Earth.

very same trends in the 3–4 years following. In the month of July the one statistically significant tendency found in the area investigated was for a southward displacement, on average by 3–6° latitude, of the subpolar low-pressure zone in the longitude of Greenwich in the first July after a great eruption (the larger displacements, significant at the 1% level, for high-latitude eruptions and somewhat smaller southward displacements, significant at the 5% level, for low-latitude eruptions). The southward anomaly tendency appears to continue for 3 or 4 years after a high-latitude eruption.

These findings are in line with the observation that many of the coldest, wettest summers in Britain – and it has been suggested in eastern North America and Japan also – occur in volcanic dust years. The summers of 1912, 1903, 1879, the 1840s, 1816, the 1760s, 1725, 1695 in England are outstanding examples, and it may be that volcanic dust has played a part in all the very worst summers of these centuries. It seems clear that there is also a tendency for cold winters in Europe after some volcanic eruptions, evidently those in low latitudes, that produce really world-wide dust veils. There is evident logic in the apparent expectation that the following two or three winters have a tendency to enhanced atmospheric circulation, more prevalent westerlies and hence mildness, and that this tendency is set up immediately after high-latitude eruptions: for these are the situations in which the chilling effect of volcanic dust is likely to be concentrated over the higher latitudes, with an enhanced thermal gradient between the region so affected and the lower latitudes that are either unaffected or becoming free of the volcanic dust veil. As always, these are no more than statistical tendencies; since the initial thermal state of the atmosphere, oceans and land surfaces at the time of the eruption must affect the patterns developing.

The evolution over the 3–4 years after a great eruption, as here described and illustrated, seems to represent a preferred time scale, but after some eruptions may work itself out more rapidly, after others more slowly. The radiation and surface temperature curves (e.g. figs. 10.16, 10.18) seem to indicate, not surprisingly, that the evolution and recovery to normal values take longer to complete – 5 to 7 years being typical in middle latitudes – after the very greatest dust-veil-producing eruptions, as in 1835, 1815 and 1783.

In illustration of the effects, figs. 10.20 and 10.21 display the monthly mean pressures for six Januarys and six Julys chosen to represent about the maximum effect of five great world-wide dust veils and one veil from a high-latitude eruption (in Alaska in 1912) within 6–18 months of their formation. All the Januarys with the exception of 1903 show more than usual blocking tendency somewhere in the north European region, and they all appear to have been colder than normal winter situations for northeastern North America. All the Julys are more cyclonic than normal over the British Isles and northeastern North America, most being also cold weather situations.

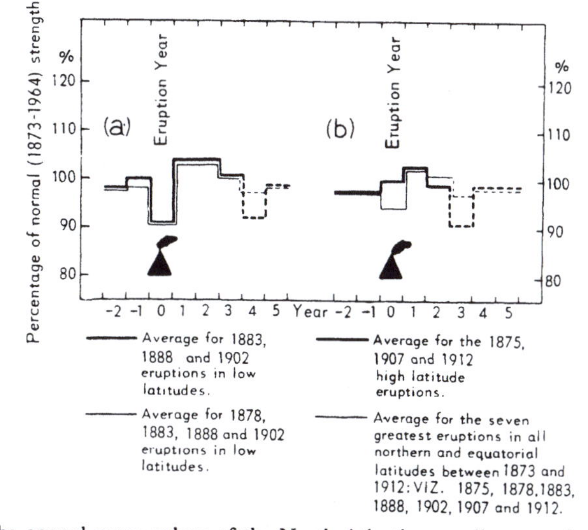

Fig. 10.19 Course of the annual mean values of the North Atlantic overall range of monthly mean pressure (between regions of highest and lowest mean pressure), an index of circulation strength: from 2 years before to 5 years after years of great volcanic eruptions in various latitudes. (The standard deviation of the individual yearly values 1873–1964 is 7·74% of the overall average value of the index.)

(*a*) Averages for eruptions in low latitudes.

(*b*) Averages for (bold line) eruptions in high northern latitudes and (thin line) for great eruptions in all northern and equatorial latitudes. Broken lines indicate portions of the curves complicated by possible overlapping effects of eruptions in different years.

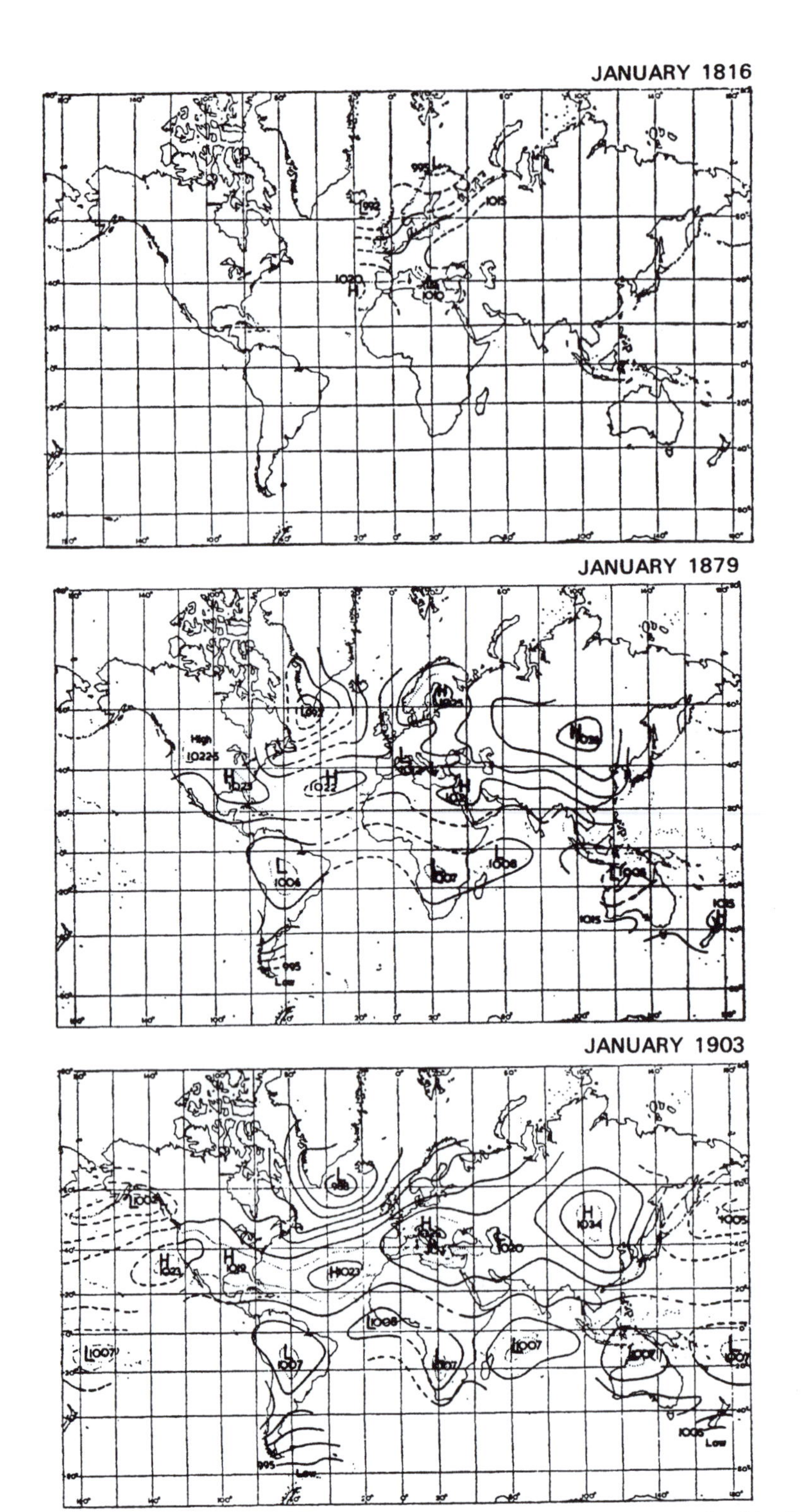

Fig. 10.20 Average barometric pressure at sea level in six Januarys,

JANUARY 1837

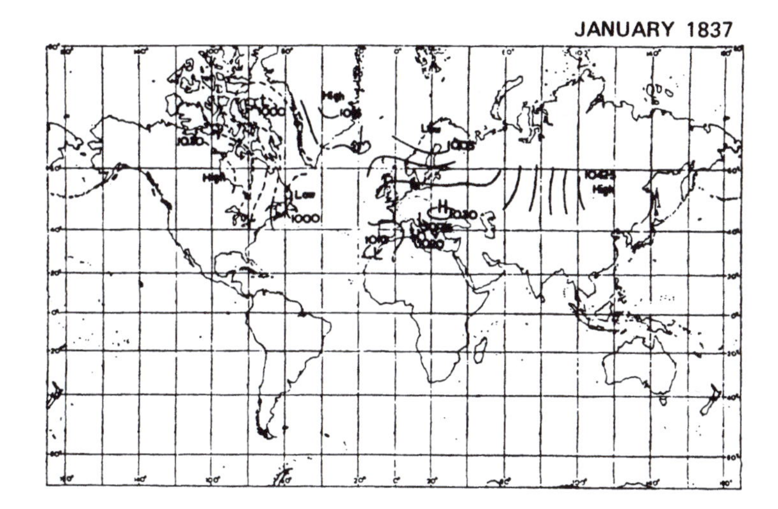

JANUARY 1885

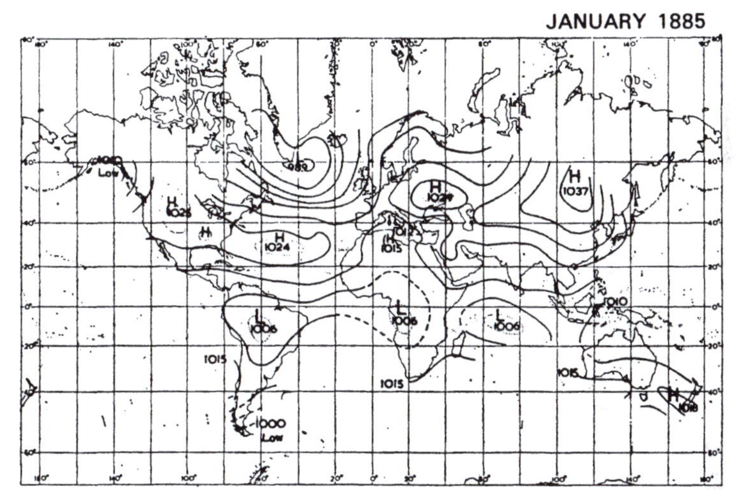

JANUARY 1913

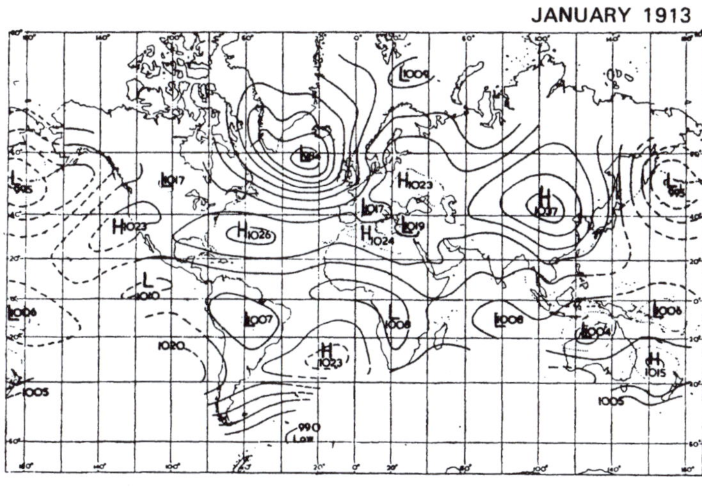

each about one year (6–18 months) after a great volcanic eruption.

JULY 1816

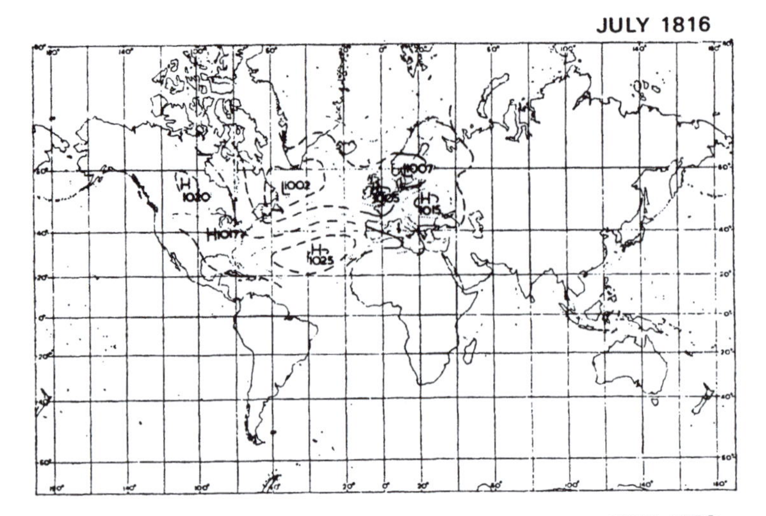

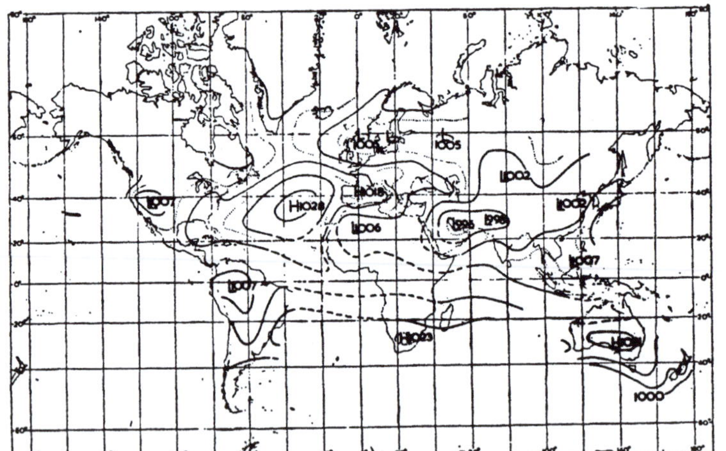

JULY 1903

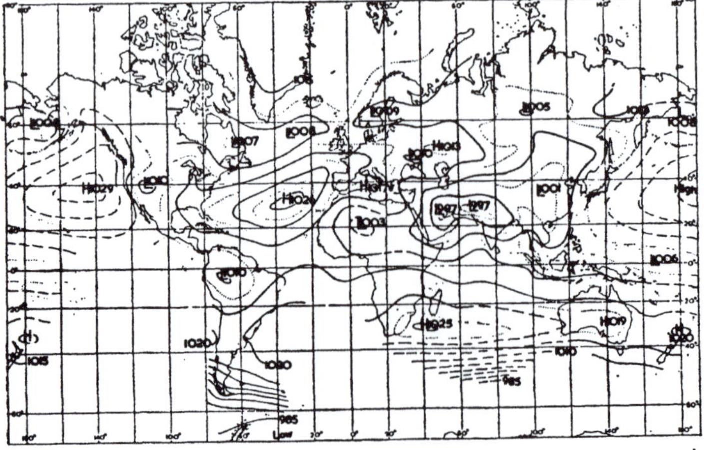

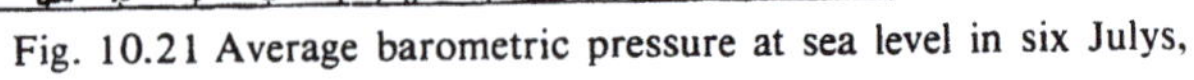
Fig. 10.21 Average barometric pressure at sea level in six Julys,

latitudes.[1] They also indicate that cumulative effects at times when volcanic explosions are abnormally frequent can influence the prevailing temperature level and climatic pattern of a century or more. Nevertheless volcanic dust is not the only, and probably not the main, cause of climatic variations within the period surveyed.

Waves of more than usually frequent volcanic activity in the past can be detected from various types of field evidence and sometimes seem to be of world-wide extent. One such, which we have also traced in the data cited in this section, occurred between about A.D. 1550 and 1900, principally between 1750 and 1850. Other earlier bouts of enhanced frequency of volcanic explosions have been traced by AUER (1958, pp. 83, 229; 1965, pp. 11–21) in radiocarbon-dated thick layers of volcanic ash in the bogs of Patagonia, the dates being 0–500 B.C., 2000–3000 B.C., and around 7000 and 9000 B.C. AUER and others have believed it possible to trace parallel variations in activity in Kamchatka, northwestern North America, Iceland and the Mediterranean. Indeed, the last bout of volcanic activity in Europe north of the Alps, in the German Eifel (Laacher See) also appears to have been about 9000 B.C. Although the occurrence of frequent dust veils at such times must have had climatic effects in the years and centuries concerned similar to those observed in the recent period which we have discussed, there seems to be a tendency for the dates in each case to be about the end of a long and fairly homogeneous climatic era. This suggests that something about the preceding climatic regime, most likely the gradual change of world sea level brought about by it (e.g. by progressive melting of glaciers), may build up strains in the Earth's crust which are ultimately released in volcanic outbreaks. In this sense the climate may cause the volcanic activity, though the latter probably then upsets the previously existing radiation and heat balance regime, so that after the volcanic wave a somewhat different climatic regime emerges.

The suggestion put forward by W. J. HUMPHREYS many years ago that volcanic dust may have caused the Quaternary ice ages, which interested WEXLER (1956) who derived a climatic pattern with increased meridional extension of an upper cold trough over central longitudes of North America for a hypothetical epoch in which available incoming radiation was cut by 20%, has been taken up in a new form by BUDYKO (1968). Starting with the assumptions that the frequency of great volcanic eruptions in the last 100 years (which he takes as 4) is a reasonable approximation to the overall average frequency in the last 10^6 years, and that eruptions are mutually unconnected and subject to purely random variations of frequency, BUDYKO estimates the statistical probabilities (expectations) of various, more or less briefly sustained, high frequencies. He concludes that 40 great eruptions in one century should be expected about once in 10 000 years, 130 great eruptions within a

1. The author has presented elsewhere (LAMB 1967*b*) the case for believing that observed effects on prevailing temperatures and ice in latitudes south of 45°S between 1896 and 1907, resembling the aftermath of a great volcanic eruption, provide the most reasonable explanation of the disappearance of Thompson Island, an island 2 km across and rising to about 500 m above sea level near 54°S 5°E on the mid-ocean ridge, the existence of which had been confirmed as late as 1893.

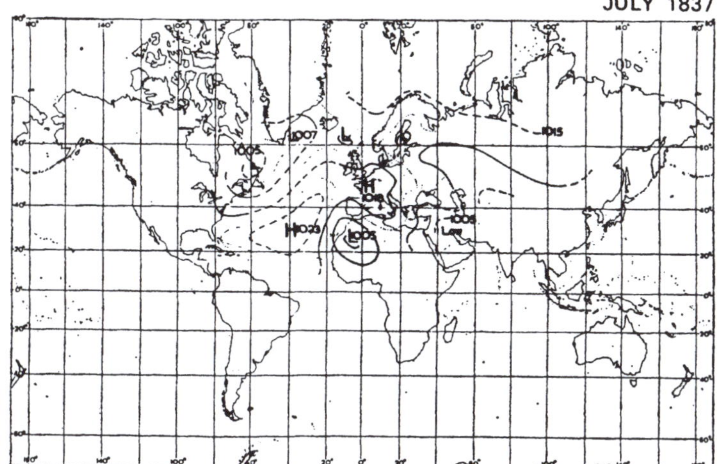

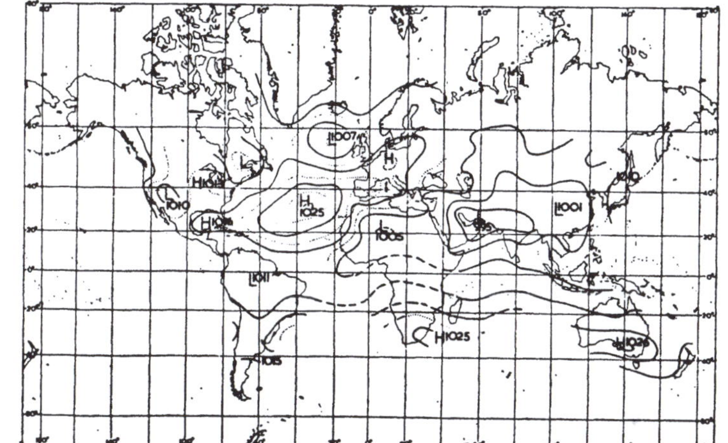

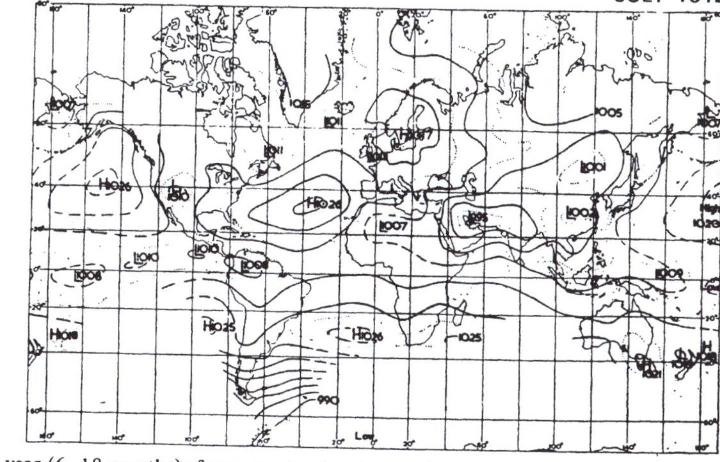

each about a year (6–18 months) after a great volcanic eruption.

A further complication in attempting to single out the characteristic evolution of the effects of a great volcanic dust veil on radiation received, surface temperature measurements and large-scale wind circulation development, is the frequency (especially in the eighteenth and nineteenth centuries) with which new eruptions reinforced and restored an atmospheric dust veil already in existence. Clearly this should be expected to prolong or renew the effects characteristic of the young (i.e. full-extent) stage of a dust veil. Figs. 10.22 and 10.23 illustrate the apparent associations between cumulative effects of volcanic activity and miscellaneous indicators of world climate. Our longest comparison, showing the known incidence of volcanic eruptions in Iceland over a thousand years, shows such parallelism with the amount of Arctic sea ice observed that one is tempted to suppose that the frequency of volcanic outbreaks in Iceland must also parallel the world-wide frequency: at the least, it repeats, and tends to verify, the appearance of a wave of abnormal frequency between about

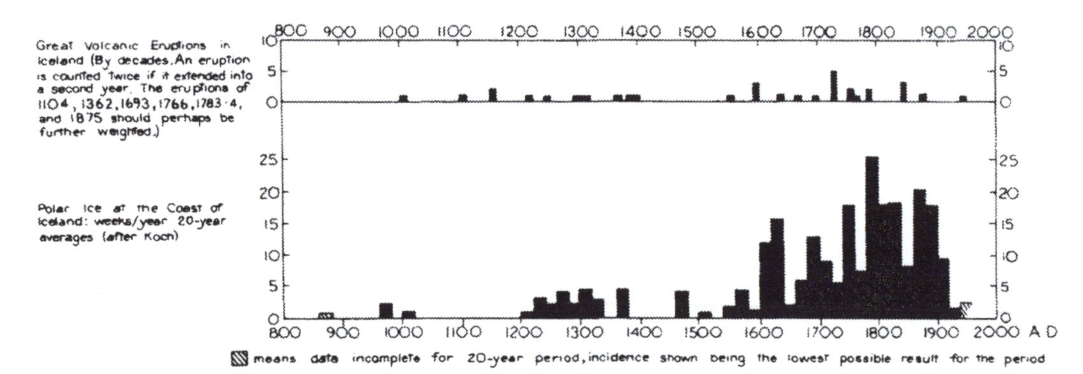

Fig. 10.22 Great volcanic eruptions in Iceland and incidence of Arctic sea ice in that area since A.D. 870.

1600 and 1900. The cumulative indices of volcanic dust in fig. 10.23 clearly show, however, that the maximum incidence of volcanic dust came almost 200 years after the time of generally coldest climate and can hardly be held responsible for more than delaying a recovery of world temperature that began around the year 1700.

Similarly, though it has often been pointed out (DEFANT 1924, WAGNER 1940) that the increasing vigour of the atmospheric circulation and warmth in high latitudes in the early twentieth century coincided with increasing atmospheric transparency, as the dust of the volcanic eruptions in the nineteenth century settled out, the renewed weakening of the circulation and cooling of the Arctic since about 1940 began when there had been no noteworthy volcanic eruptions since 1912 and proceeded for 15–20 years before any significant increase of volcanic dust could be alleged.

Wider implications: waves of volcanic activity and long-term climatic effects

These studies establish that great volcanic dust veils do have significant effects over the first 2–7 years following the eruption, possibly for as long as 10–15 years in high

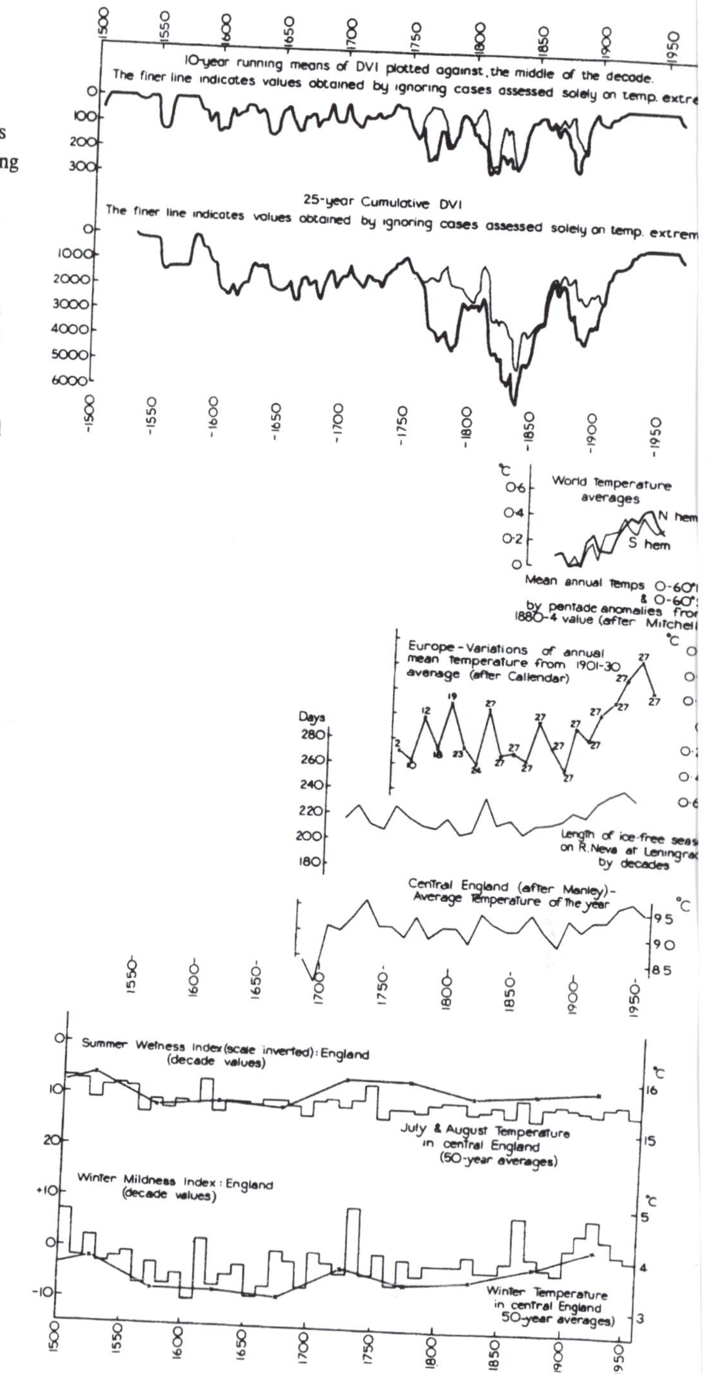

Fig. 10.23 Volcanic dust over the northern hemisphere and various climatic indicators. In the top two sections of the diagram the finer line indicates **DVI** values obtained by ignoring cases of dust veils assessed solely on evidence of temperature anomaly. There is some reason to suppose that if the **D V I** values for the 1680s and 1690s could be computed from year to year variations of temperature, the values for those decades should be much greater. The computations of world temperature averages for the northern and southern hemispheres (bold line northern hemisphere) are 5-year means due to MITCHELL (1961); those of the variations of annual mean temperature over Europe are decade means for stations in the area 40–60°N 0–30°E due to G. S. CALLENDAR (*personal communication*), the numbers of stations used being shown against each point; the average temperatures of the year in England are taken from MANLEY's (1958, 1964) series for central England; the summer wetness index and winter mildness index decade values for England come from LAMB (1966, pp. 97–8, 217–221) and the corresponding 50-year average temperature values for central England, derived by regression equations from the index values for periods before the beginning of MANLEY's series of homogenized thermometric observations, come from fig. 6.6 (p. 236) in this book.

century about once in 100 000 years and 100 great eruptions within 5 years about once in a million years. The last-named frequency, it is supposed, should reduce the total incoming radiation by 50% in low latitudes and by more than 80% in high latitudes. Unfortunately, our observation that rather over half the last 100 years were within the latter part of a 400-years-long wave of enhanced volcanic activity undermines BUDYKO's first premise: his estimate of the long-term average frequency is likely to be falsely high. Moreover neither the recurrent ice ages of the Quaternary era nor world volcanic activity appear to be so nearly random in their incidence as BUDYKO requires.

Others have proposed that volcanic dust could upset the world climatic balance in other ways. For example, BLOCH (1965) has suggested that volcanic dust blackening ice caps at times of exceptional volcanic activity should cause sufficent abnormal melting to raise world sea level and could be responsible for ending an ice age. The data of the survey of volcanic eruptions here discussed, however, indicate that no single eruption is likely to have covered more than 1–2% of the surface of Antarctica or of the former great northern ice sheets to a depth of more than 1 cm. Moreover, a darkened area of this size would surely be covered by new snow within a few days or weeks, before any further eruption occurred (see LAMB 1968).

A view closer to our own conclusions above was arrived at by FUCHS (1947) from the geological evidence of variations of volcanic activity over the last million years. FUCHS found that there had been three major, long periods of enormously greater volcanic activity in Africa than the present. These were in some sense associated with crustal movements in the great Rift Valley. The timing coincided at least roughly with the onset of the three main pluvial periods in East Africa, in the Lower, Middle and Upper Pleistocene respectively. (The activity in the Middle Pleistocene was apparently the greatest of all.) In other words, a new climatic regime, and probably a colder one that favoured glaciation in high latitudes, emerged after the pre-existing climatic balance had been disturbed by the volcanic dust. It may not be fair to go on to conclude that the cold climate regime was maintained over tens of thousands of years afterwards by continued volcanic dust veils. FUCHS drew attention to the appearance of parallelism, at least on the coarse geological time-scale, of these three gigantic bursts of volcanic activity culminating in the early parts of the Lower, Middle and Upper Pleistocene with the known course of volcanic activity in Java and suggested world-wide rhythmic variations of volcanic activity. The large-amplitude, very long-term rhythm was supposed to have its origin in tectonic movements, which would nowadays probably be attributed to convection within the Earth's sub-crustal mantle.

Dust veil assessments and chronology of major eruptions cited

The Dust Veil Index (**DVI**) assessments of the eruptions in recent centuries covered in this section, are listed in Table 10.3.

TABLE 10.3

Dust veil index assessments and chronology of major eruptions cited

Year	*Volcano*	*Situation*	**DVI**
1680	Krakatau	6°S 105½°E	400
1680	Tongkoko, Celebes	1½°N 125°E	1000
1693	Hekla, Iceland	64°N 19½°W	100
1693	Serua, Molucca Is.	6½°S 130°E	500
1694	Amboina, Molucca Is.	4°S 128°E	≮250
1694	'Celebes'	1°N–6°S 119°–125°E	≮250
1694	Gunung Api, Molucca Is.	4½°S 130°E	400

NOTE: If the low temperatures prevailing in England, as well as Iceland and a wide surrounding region, over the years 1694–8 were representative of a world-wide anomaly of about the same amount, and provided their departure from the temperatures prevailing in the immediately preceding and following years were entirely due to volcanic dust, the total **DVI** for 1694–8 should be 3000–3500.

Year	*Volcano*	*Situation*	**DVI**
1707	Vesuvius	41°N 14°E	150
1707	Santorin	36½°N 25½°E	250
1707	Fujiyama, Japan	35°N 139°E	350
1712	Miyakeyama, Japan	34°N 139½°E	200
1717	Vesuvius	41°N 14°E	100
1717	Kirishima Yama, Japan	32°N 131°E	200
1721	Katla, Iceland	63½°N 19°W	250
1730	Roung, Java	8°S 114°E	300
1744	Cotopaxi, Ecuador	1°S 78°W	300
1752	Little Sunda Is., possibly Tambora	8°S 118°E	1000
1754	Taal, Luzon, Philippines	14°N 121°E	300
1755	Katla, Iceland	63½°N 19°W	400
1759	Jorullo, Mexico	19°N 102°W	300
1760	Makjan, Moluccas	½°N 127½°E	250
1763	'Molucca Is.'	2°N–3°S 125–131°E	600 (?)
1766	Hekla, Iceland	64°N 19½°W	200
1766	Mayon, Luzon, Philippines	13½°N 123½°E	2300 (?)
1768	Cotopaxi, Ecuador	1°S 78°W	900
1772	Gunung Papandayan, Java	7½°S 108°E	250
1775	Pacaya, Guatemala	14°N 91°W	1000 (?)
1779	Sakurashima, Japan	31½°N 131°E	450
1783	Eldeyjar, off Iceland	63½°N 23°W }	700
	Laki and Skaptar Jökull, Iceland	64°N 18°W }	
1783	Asama, Japan	36½°N 138½°E	300
	Total veil, 1783:		1000
1786	Pavlov, Alaska	55½°N 162°W	150
1795	Pogrumnoy, Umanak Is., Aleutians	55°N 165°W	300
1796	Bogoslov, Aleutians	54°N 168°W	100
1799	Fuego, Guatemala	14½°N 91°W	600
1803	Cotopaxi, Ecuador	1°S 78°W	1100 (?)
1807–10	Various, including Gunung		
	Merapi, Java	7½°N 110½°E	(?)
	and São Jorge, Azores	38½°N 28½°W	(?)
	Total veil, 1807–10:		1500 (?)
1811	Sabrina, Azores	38°N 25°W	200
1812	Soufrière, St Vincent	13½°N 61°W	300
1812	Awu, Great Sangihe, Celebes	3½°N 125½°E	300
1813	Vesuvius	41°N 14°E	100

TABLE 10.3 – *continued*

Year	*Volcano*	*Situation*	**DVI**
1814	Mayon, Luzon	13½°N 123½°E	300
1815	Tambora, Sumbawa	8°S 118°E	3000
	Total veil, 1811–18:		4400
1821	Eyjafjallajökull, Iceland	63½°N 19½°W	100
1822	Galunggung, Java	7°S 108°E	500
1826	Kelud, Java	8°S 112½°E	300
1831	Giulia or Graham's Island	37°N 12–13°E	200
1831	Pichincha, Ecuador	0°S 78½°W	?
1831	Babuyan, Philippine Is.	19°N 122°E	300
1831	Barbados	13°N 60°W	?
	Total veil, 1831–3:		about 1000
1835	Coseguina, Nicaragua	13°N 87½°W	4000
1845	Hekla, Iceland	64°N 19½°W	250
1846	Armagora, S. Pacific	18°S 174°W	1000
1852	Gunung Api, Banda, Moluccas	4½°S 130°E	200
1856	Cotopaxi, Ecuador	1°S 78°W	700
1861	Makjan, Moluccas	½°N 127½°E	800
1875	Askja, Iceland	65°N 17°W	300
1878	Ghaie, New Ireland, Bismarck Archipelago	4°S 152°E	possibly 1250
1883	Krakatau	6°S 105½E	1000
1888	Bandai San, Japan	38°N 140°E	250
1888	Ritter Is., Bismarck Archipelago	5½°S 148°E	250
	Total veil 1883–90:		about 1500
1902	Mont Pelée, Martinique	15°N 61°W	100
1902	Soufrière, St Vincent	13½°N 61°W	300
1902	Santa Maria, Guatemala	14½N 92°W	600
	Total veil, 1902:		about 1000
1907	Shtyubelya Sopka/Ksudatch, Kamchatka	52°N 157½°E	150
1912	Katmai, Alaska	58°N 155°W	150
1963	Mt Agung (Gunung Agung), Bali	8½°S 115½°E	800
1966	Awu, Great Sangihe, Celebes	3½°N 125½E	150–200
1968	Fernandina, Galapagos	½°S 92°W	50–100
	Total veil, 1963–8:		about 1100
1970	Deception Is.	63°S 60½°W	(200)

NOTE: Sources of continuing information on volcanic eruptions at the present time, from which dust-veil assessments may be attempted, are:

(1) The yearly *Bulletins of Volcanic Eruptions* issued since 1961 by the Volcanological Society of Japan on behalf of the International Association of Volcanology, I.U.G.G.

(2) The *Smithsonian Institution Center for Short-Lived Phenomena*, Cambridge, Mass., which exchanges up-to-date reports by airmail postcard with on-the-spot observers and interested institutions.

Other types of dust and pollution carried by the atmosphere

In these cases the effects of each type in isolation in modifying climate have not yet been adequately studied, and certainly cannot be said to be seen in operation; though some guide lines may be derived from volcanic dust and other specialized studies. An exception is the local toxic effects on human and other life, and the effects on sunshine, daylight and visibility, of local concentrations of certain pollutants (e.g. automobile-exhaust gases, carbon monoxide, sulphur oxides, ozone, suspended particles or globules of tarry substances, etc.) trapped in the surface layer of the atmosphere beneath an inversion; this happens particularly in smog situations (e.g. surface radiation cold in still, anticyclonic weather in winter and at some Arctic camps, townships and airfields over a longer season) and lesser, though long-continued, concentrations in sheltered industrial valleys and at special sites (e.g. Los Angeles) beside a cold-water coast.

Attention has also been drawn (SCHAEFER 1966) to the possible importance of the abundant lead oxide, emitted in automobile exhausts owing to the use of tetra-ethyl-lead in petrol, in increasing the concentration of nuclei for ice-crystal formation in the atmosphere. Lead iodide is apparently formed instantly on exposure of the lead oxide to iodine, which occurs in minute amounts (0·05–0·5 $\mu g/m^3$) as a free vapour in the atmosphere from the burning of fossil fuels and the effluent of growing plants. This substance acts like silver iodide particles as an ice nucleus. Concentrations of 10^4–10^5/litre have been observed in rural air near roads, and over 10^6/l in fields near eastern New York, and seem to have increased tenfold in 13 years. The rapidity with which supercooled droplet clouds at −15°C are changed to ice-crystal clouds (in a few seconds) has been demonstrated in the laboratory. This presumably speeds up the precipitation of atmospheric moisture, and may be liable to decrease cloudiness downwind in areas remote from moisture sources, but any general climatic effect is far from having been demonstrated. It is conceivable that, through re-evaporation of the precipitated water, the main effects should be rather some increase of precipitation, especially near moisture sources, and a general speeding up of the precipitation–evaporation cycle.

Sea spray, and the salt therefrom, may damage vegetation near windy coasts and may play a part in shifting the vegetation species boundaries near coasts, when a shift of the large-scale climatic pattern increases windiness there.

(a) *Other types of dust*

Wind-blown Saharan sand has been deposited in readily seen amounts on ships' decks as far away as 46°W, and has been collected in measurable quantities on nets specially exposed (DELANEY, PARKIN *et al.* 1967) to the Trade Wind at the Bahamas (near 25°N 75–78°W) for scientific study. It has fallen, sometimes in orange-coloured mud rain, in northern Europe from Britain to Scandinavia: there were examples in southern Scandinavia and Finland on 30–31 January 1947 (when it was brought by S'ly winds aloft a few hours before a different windstream brought brownish black tephra (ash) from a volcanic eruption in Iceland) and over most of Britain on 1–2 July 1968 after a one-day heat wave in a S'ly wind which brought the highest air temperatures (32°C near London) for 12 years. On the latter occasion the dust clouds carried by the upper wind over England were dense enough to turn the sun white and pale by 6 p.m., some hours before the cold-front rain which produced the mud deposit. Thus clouds of dust blown up from the surface are sometimes dense enough to affect radiation receipt after travelling about a tenth of the way round the world. Smoke-trails from industrial areas and great population centres, as well as from forest and bush fires, can also be traced over great distances. In September 1950, for several days the sun was dimmed and appeared blue or bluish-white[1] over England, seen through the smoke-cloud in the upper troposphere from exceptional forest fires in western Canada.

Blowing dry soil, especially the light loess soils in open situations in Eurasia (and perhaps parts of North America) south of the ice margin during the ice ages, was probably also capable of dimming the sun at times. BRYSON (1967) has lately drawn attention to the density of the dust pall over India, especially between heights of 3 and 12 km where it is less rapidly washed out by rain than lower down; this dust is continually blown up from the dry soils of the desert and semi-desert areas between Arabia, Persia and Rajasthan in northwest India. WEXLER (1936) studied the atmospheric turbidity produced in this way in the famous dry, 'dust-bowl' years of the early 1930s in the U.S. Middle West when wind erosion of soil was serious to agriculture: concentrations (particle counts) were greatest in the surface air in autumn and winter, but despite clearer conditions at the surface total turbidity in the air columns was greatest in spring, when the greater thermal instability associated with surface heating of cold air, carried (and dispersed) the dust up to great heights in the troposphere. But the radiation measurements in fig. 10.16 (*a*), which in the 1930s were largely from United States observatories, indicate no general deficit over the particular years affected. Chemical analysis of dust found in the atmosphere or in sediments suggests that the turbidity of the northern hemisphere atmosphere north of the tropic nearly all arises from Asia, that of the air nearer the equator is largely Saharan and that over the southern hemisphere is of Australian origin. None of these agencies can, however, raise dust in any significant quantity into the stratosphere. Hence any that does not settle out earlier must generally be brought down out of the atmosphere again within about 10 days by rain. Comparability with a volcanic dust veil can only have arisen when the supply of dust was sustained over long periods, as was perhaps possible in the ice-age loess situations but hardly in the other cases. More important is that in all these cases, unlike volcanic dust in the stratosphere, the incoming solar radiation penetrates to the troposphere; in most of them indeed the dust etc. that partially intercepts the solar radiation before reaching the surface is in the lowest 1–2 km of the atmosphere. It is not clear, therefore, that there should be any significant decrease in the heat that ultimately reaches the surface air except in so far as the reflectivity of a dust haze increases the albedo. As with low clouds, it seems that there must be a reduction of the diurnal range of temperature, with lower day maxima and higher night minima. Similar considerations apply to smoke trails from industrial areas.

Meteoric or extra-terrestrial dust is probably always present in the atmosphere as particles of sizes up to a few hundred microns, and is traceable in deposits of all past ages back

1. The optical observation implies particle diameters in the range 1–5 μm.

to at least the Ordovician, but is probably negligible in connexion with variability of climate, unless (or except when) it is itself subject to some great variation. The global total which falls to Earth is estimated as normally about one million tons yearly. For comparison, the estimated quantities of solid matter put into the stratosphere as dust after great volcanic explosions (**DVI** $\geqslant$ 100), and constituting the persistent dust veils observed at such times, range from one million to perhaps 100 million tons – the latter figure in the case of the very greatest eruptions of recent times (e.g. Tambora 1815, Coseguina 1835). These figures are about a thousandth part of the estimated quantities of rock lost from the volcano site or of the measured deposits of tephra around, and downwind from, the volcano.

There have been many indications of increasing turbidity of the lower atmosphere during the twentieth century. Thus, DAVITAJA (1965) found by careful measurements that the amount of dust contaminating the glaciers in the high Caucasus, which had remained almost constant from the 1790s to around 1930 at 10–20 milligrams/litre (mg/l), apart from a peak of about 30 mg/l in the 1870s, rose to an average of 50 mg/l in the 1930s and 1940s (temporarily to over 150 when the battles of the Second World War were near) and to an average of about 220 mg/l in the 1950s. DAVITAJA attributed the continued net melting of the glaciers in the Caucasus and central Asia in the years since 1950 when the climate was becoming colder to the darkening of the glacier surface by dust. Similarly, MCCORMICK and LUDWIG (1967) report that direct measurements of atmospheric turbidity in Washington, D.C., show an increase of the annual mean value[1] from 0·098 in 1903–7 to 0·154 in 1962, and at Davos at a height of about 1600 m in the Swiss Alps from 0·024 for 1914–26 to 0·043 for 1957–9. BRYSON, BUDYKO, DAVITAJA and others tentatively attribute to this the gradually declining strength of the direct solar beam measurements after about 1945, seen here in fig. 10.16 (*b*) in the Japanese observations, and also reported by most other observatories, e.g. Tashkent (DAVITAJA *loc. cit.*) where it appears as a steady decline from at least 1930. It may, however, be due to a variation of the sun's output at source (cf. pp. 22–3). In all cases the decline seems to mount by 1954–60 to several per cent.

(b) *Pollution of the upper atmosphere by exhaust gases and vapours from rockets and high flying aircraft*

This seems likelier than the items considered in (*a*) to affect the thermal climate and the energy of the wind circulation in the lower atmosphere, because here the radiation balance is liable to be altered outside (above) the lower atmosphere (troposphere). This could happen both through absorption of solar and terrestrial radiation by the foreign substances intro-

1. Figures quoted for atmospheric turbidity are β in the expression

$$P_\lambda = e^{-\beta/\lambda^\alpha}$$

where P_λ is the solar transmission factor (values from 0 to 1)
λ is the wave length
and α is a so-called wave-length factor dependent on particle size.

duced into the upper atmosphere and through chemical action of these substances in disturbing the ozone amount in the stratosphere (STUBBS 1963) as well as by their increasing the Earth's albedo. Moreover, calculations by PRESSMAN *et al.* (1963) indicate that, after release normally between 90 and 130 km above the Earth, the rocket-exhaust materials would take perhaps half a year to fall through the rarified layers to the top of the stratosphere but would take some hundreds of years to fall through the stratosphere from the 50-km to the 12- to 20-km level. The main exhaust gases are water vapour and carbon dioxide, which despite their effectiveness as absorbers of solar and terrestrial radiation (see Chapter 2) are probably not of much importance because their molecules would be broken down (dissociated) by the solar ultra-violet radiation (STAGG 1964). Potentially far more important are the chemically active metals, nitric oxides and fluorine contained in the exhausts of some fuels that are, or may be, used.

The extra water vapour injected by high-flying aircraft increasingly since about 1940 in the tropopause layer, where it is not exposed to dissociation, may have facilitated the formation of more cirrus cloud than formerly. Where the condensation trails broaden out into persistent cloud, the Earth's albedo must be affected.

Since a mere 25 kg is the total atmospheric content of sodium which is responsible for the normal night airglow, and the temperature of the stratosphere is accounted for by the presence of ozone amounting to just 10–20 parts per million, it is reasonable to suggest that the exhausts of rockets which carry some 2000 tons of fuel may be enough to affect the balance of the atmosphere at levels where the total mass is of the order of 10^7 (at 130 km) to 10^9 tons.

In this section it should also be mentioned that the use of thermonuclear bombs in war would pollute the stratosphere with dust on a massive scale, since each ground burst bomb appears to be equivalent in this respect to a volcanic eruption of **DVI** probably about 50–150 depending upon the latitude (cf. BUDYKO's estimates of the reduction of total incoming radiation at the Earth's surface which could result from 100 great volcanic eruptions in 5 years, pp. 432–3). Testing of atomic weapons by bursts in the air or below ground creates no dust and presumably has no such effect.

(c) *Carbon dioxide*

The atmosphere's content of CO_2 was increased by about 10% from the nineteenth-century average of 290 parts per million to 320–30 parts per million by 1967 due to the burning of fossil fuels; half this increase took place after 1945 as the pace of the industrial revolution increased. This should have a warming effect on the lower atmosphere since, like the glass of a greenhouse, this gas is more transparent to the incoming solar short-wave radiation than it is to the long-wave radiation sent out from the Earth. The magnitude of the effect has been discussed in Chapter 2 (pp. 45–6). Since observation shows that the average surface temperature over the Earth has nevertheless gone down by a few tenths of a degree between 1950 and the 1960s, it is clear that other agencies have had a more powerful effect than the carbon dioxide.

Solar disturbance

Historical introduction

The possibility that climatic variations might be explained by variations of the sun's output at source has long claimed attention. The idea goes back almost to the invention of the thermometer and to the earliest years of careful observation of the sun. The Italian Jesuit priest RICCIOLI suggested in 1651 that the temperature of the Earth falls with increasing spottedness of the sun (HELLAND-HANSEN and NANSEN 1917, p. 139), though in 1801 WILLIAM HERSCHEL came to the opposite view from an examination of wheat prices in England. Nearer our own times, the subject was extensively investigated by KÖPPEN (1873, 1914), WALKER (1915), BROOKS (e.g. 1923, 1934) and others. Since the sun is the ultimate source of all the energy in (natural) climatic processes, since its enormous size and heat output could be presumed to entail its seething with convection on many time- and space-scales (before the means existed to observe the surface traces of convection patterns), and since the occurrence of variable amounts of darkening of the sun's face in spots had long been observed, the thought that this should be the origin of climatic variations (some of which might be cyclic) was an obvious one.

The quest for evidence and understanding has, however, proved extremely difficult. Sunspots are but one of several types of solar disturbance, the one for which we have by far the longest record, fairly reliable as regards comparative yearly sunspot numbers back to 1700, and as regards the times of maxima and minima of the '11-year' sunspot cycles back to 1610, and capable of being extended on the basis of aurorae much farther back in time (see Appendix I). Sunspots occur in so-called 'active regions' of the sun, within which other types of disturbance also must be distinguished, particularly the brightened areas or *faculae* (also in the photosphere) round about the sunspots and the *flares* that occur in the chromosphere above (see Chapter 1); also *prominences* above the flares, and the bright hydrogen and calcium clouds, known as *flocculi* or *plages*, in association with which the short-lived (20 minutes to a few hours) solar flares or chromospheric eruptions take place. At least the faculae and flares are as likely to have effects upon weather and climate as the sunspots, yet we only have sufficiently complete records of faculae since 1874 and of flares in the sunspot cycles since perhaps 1930–50. The sunspot record therefore gives but a partial view of the history of solar variation as it may have affected solar output and weather, unless some systematic relationships between sunspots and the associated areas of intensified solar output (either in the course of the circa 11-year or the much longer sunspot cycles) can be established (see also pp. 18–30).

At least one (non-linear) relationship between atmospheric behaviour and faculae rather than sunspots has been reported. SUDA (1963) found that tropopause height over Japan on average lowers over the 2–3 days following a peak area of faculae and the more so the greater the area of faculae. Tropopause height at Leopoldville near the equator (4·4°S 15·0°E), however, averaged over 13 months, rose by over 1 km (from 15·2 to 16·4 km), increasing almost linearly with sunspot number from the 1954 solar minimum to the 1957 maximum, while the temperature at the tropopause remained constant (STRANZ 1959). If this should prove to be a general experience, it may be attributable to a variation in the production of ozone and the heating of the ozone layer in the stratosphere. It evidently has effects in the troposphere as it reduces the vertical gradient of temperature, especially in the upper troposphere, when the tropopause is high. This observation is suggestive but refers to only a fraction of one (exceptionally intense) sunspot cycle.

It is hardly surprising that WALKER's world-wide survey of correlation coefficients connecting sunspot number alone with atmospheric pressure, temperature and rainfall produced mostly low values, between 0 and ±0·2. Actually, the prospects of finding useful coefficients were reduced by WALKER's ignoring in this instance all monsoonal/seasonal changes and working only with annual values. Despite this, correlation coefficients between sunspot number and annual mean (surface) pressure in a few of the areas with 40 years of data ranged up to −0·30 to −0·45 (India, Ceylon, Cape Town and south Australia) and up to +0·20 to +0·35 (Greenland, southern U.S.A., across mid European Russia and across South America near 35°S). From 20 years of data the coefficients were +0·20 and +0·25 at the Azores and Hawaii respectively. There was a suggestion here that the world's high- and low-pressure systems tended to be intensified at times of much sunspot activity, a conception that has been carried over in other forms into much modern Russian thinking and writing on the subject under the name of the 'Law of accentuation of pressure systems'.

One particular experience dispelled the early optimism about finding valid associations between solar disturbance and weather. It was noticed that the level of the great Lake Victoria in eastern central Africa (a sheet of water eight-tenths of the size of Ireland) rose and fell by about one metre, its maximum and minimum years between 1893–4 and 1927 coinciding with those of the sunspot cycle. BROOKS (1923) found a correlation coefficient of +0·87 (statistically significant beyond the 0·1% level) connecting yearly sunspot numbers and the level of the lake over 20 years within this period, which however amounted to barely two solar cycles. Despite the strength of this association over the preceding years, prediction of a low level of the lake with the next sunspot minimum in 1933 was immediately falsified. We can now see, in retrospect, that after 1927 peak levels of the lake occurred twice in each sunspot cycle, in 1932, 1937, 1942, 1947, 1952, 1957 and 1964, in or about sunspot minimum as well as sunspot maximum years; the range of variation of the lake level also became smaller as the time for development of each stage was halved. (It is now known, moreover, that bigger and more lasting changes of Lake Victoria's water level have occurred on two occasions apparently unconnected with solar phenomena: a decline of 2·4 m between 1876 and 1898, believed to have occurred mainly between 1893 and 1898, and a rise of 1·5–2 m in 1961.) The conclusion was perhaps too readily drawn that solar variation had nothing to do with the changes of the world's wind circulation and rainfall distribution involved. Research in this field was discouraged or dropped, particularly among British meteorologists, for more than a generation. The subject was further discredited by the 1950s through its continuing pursuit mainly by over-optimistic or naïve amateurs working in isolation without adequate criticism of either data or results. A review by TUCKER (1964) stressing the difficulties that had been so often ignored by proponents of solar influence, nevertheless was not entirely negative in its conclusions.

Doubt has also been thrown on KÖPPEN's (1914) inverse relationship between sunspot number and prevailing surface temperatures at tropical stations, which had given a correlation coefficient $r = -0{\cdot}48$ ($P < 0{\cdot}1\%$) from 1813 to 1910, because of an apparent phase-change in the relationship, the correlation coefficient being positive from the 1920s to 1950 (TROUP 1962). SCHERHAG (1950) has also reported phase changes about 1748, 1824, 1853 and 1897 in the supposed relationship of winter temperatures at Berlin to sunspot number. Such phase changes could indicate that the temperatures have a natural periodicity close to, but not identical with, and unconnected with, that of sunspots.

If there is any influence of solar variation upon weather and climate, it is clearly necessary to establish:

(i) by how much the so-called 'solar constant', i.e. the total energy in the solar beam, varies;

(ii) any characteristic course or evolution of this variation;

(iii) any mechanisms by which the known variations of the sun's output of (*a*) ultra-violet waves, (*b*) corpuscle streams, could affect the lower atmosphere and any regularly observed effects.

For reasons explained in Chapter 2, it has been extraordinarily difficult to demonstrate any variation of the solar constant on the basis of observations made at the bottom of the Earth's atmosphere (or, so far, from artificial satellites). Nevertheless, a systematic variation, amounting to about a half of one per cent, in the form of a double wave within the '11-year' sunspot cycle, has been tentatively derived by BAUR from the averages of several cycles between 1920 and 1957 and is reproduced as fig. 2.1 in this book. Maximum values of the solar constant appear to occur not at either extreme of the sunspot cycle but about two years after sunspot maximum and around the middle of the rising phase of activity after sunspot minimum. It may be too early to regard this solar output variation, as shown in fig. 2.1, as a firmly established feature of the sun's behaviour until it has been observed over many more sunspot cycles and, preferably, by instruments mounted on vehicles outside the Earth's atmosphere. In the light of the findings for 1920–57, however, it seems that WALKER's search for a straight correlation between sunspot number and the weather elements was probably not well chosen to show the strongest solar effects.

Research has continued, in recent years on an increasing scale, in institutions in Germany, Soviet Russia and the U.S.A. and has by now produced an impressive body of apparently reliably observed associations between solar disturbances, weather and climate on various time-scales. In some cases, the first, still tentative, indications of the chain of physical causation may be glimpsed. (Among the more useful reviews and collections of results the reader is referred to BAUR 1949, 1956, 1958, BELINSKI 1957, FLOHN 1951, JULIAN *et al.* 1957, MIRONOVITCH 1960, MITCHELL 1965, MUSTEL 1967, ROBERTS 1963, SAZONOV 1964.)

We shall here survey results which seem to show some statistically significant association between solar events and climate, weather or the behaviour of the Earth's atmosphere over many repeats of the same process.

Apparent climatic relationships to long solar cycles (23–400 years)

BRAY (1967) has defined a Solar Activity Index based on a count which is a weighted combination of the frequency of auroral reports and of sunspots, taken from the independent collections of data by SCHOVE (1955, 1962) and KANDA. (See also Appendix I.)

Mean values of this index can be given over groups of successive 11-year cycles back to 527 B.C., the values being designed to be comparable with the average of the highest (smoothed) monthly mean Zürich sunspot numbers of the 11-year cycles in each group. It was noticed that all through the record groups of three or more cycles of high activity (index values averaging 107–137 for the group) alternated with intervals in which nearly all the index values were below 100 (averaging 60–89 for the group): this was the basis on which the cycles were grouped. Reliability of BRAY's index over the whole series was indicated by a number of tests of which the most telling were:

(i) A statistically highly significant ($P < 1\%$) negative correlation coefficient connecting (*a*) mean yearly sunspot number and length of the approximately 11-year cycle between A.D. 1699 and 1964, (*b*) the solar activity index and sunspot cycle length between 527 B.C. and A.D. 1964. (Average cycle lengths have varied from 10·2 years from 1914 to 1964, and apparently 10·0 years between A.D. 800 and 900, to between 12 and 14 years in several periods of low activity, e.g. A.D. 1656–1723, 1799–1833, 1870–1913.)

(ii) A statistically highly significant ($P < 1\%$) negative correlation coefficient connecting atmospheric radiocarbon (^{14}C) activity with the above-mentioned groups of active and weak sunspot cycles from 129 B.C. to A.D. 1964. With the changes from active to weak sunspot cycles and vice versa, the amount of ^{14}C in the atmosphere underwent an inverse change in twenty-two out of twenty-four cases and in one of the remaining cases there was no change. (The χ^2 value indicates a less than one in a thousand probability of this result being due to chance.)

The equivalence of BRAY's sunspot-auroral index and the amount of ^{14}C in the atmosphere as indicators of solar activity can be seen in the following table, in which the ^{14}C activity is expressed as mean percentage departure from the modern (1850) level.

The apparent discrepancy in the bottom line of the table could be attributed to the small number of observations, but may be real since we shall have occasion to notice other discrepant associations with times of extremely high sunspot number.

TABLE 10.4

Radiocarbon activity relative to BRAY's sunspot–auroral index of solar activity since 527 B.C.

Sunspot-auroral index	*Carbon-14 anomaly*	
	No. of observations	*Mean ± S.D.*
21–40	7	+1·2 ± 1·0
41–60	38	+0·8 ± 1·0
61–80	78	+0·2 ± 1·2
81–100	85	−0·3 ± 1·0
101–120	80	−0·4 ± 1·0
121–140	53	−0·7 ± 1·2
141+	8	+0·1 ± 0·6

BRAY (1965, 1968) has compared these indicators of the prevailing degree of solar activity with (*a*) tree growth near the forest limit in British Columbia since A.D. 1656, (*b*) the records of sea ice at the coasts of Iceland since A.D. 870, (*c*) advances and retreats of glaciers in all parts of the world since 700 B.C. The extension back to 700 B.C. was included because the ^{14}C data suggested the same association of low solar activity with a great period of advancing glaciers about that time as between A.D. 1400 and 1850. All these climatic indicators appear to show statistically significant associations with the general level of solar activity prevailing over periods not less than a century in length, in the sense that greater solar activity appears to correspond to warmer climates.

The details of BRAY's solar activity index since 527 B.C. and the correspondence between this and ^{14}C measurements on the one hand and available knowledge of the climatic indices referred to are set forth in Tables 10.5 and 10.6 below.

One cannot easily derive information about individual '11-year' sunspot cycles or the weather prevailing in them from either ^{14}C measurements or glaciers because of the lags in their response. The ^{14}C in the atmosphere is partly accounted for by the level of solar activity in previous decades and the slow attainment of equilibrium between the atmospheric $^{14}CO_2$ and that in the bicarbonates dissolved in the oceans. Following a sharp change in the level of solar activity a lag of up to 40 years in the change of atmospheric ^{14}C seems typical. Every glacier responds individually to changes of prevailing temperature and precipitation according to its catchment basin and the terrain down to the glacier snout; lags of 10–30 years are very common, and some surges forward of a single glacier snout have more to do with conditions at the bedrock face and waves passing down the glacier than with climate. Hence, it may not be possible by means of these indicators to descend to much more detail than is given in these tables.

BRAY's study of forest growth, not included in these tables, was based on forty-nine Engelmann spruce trees selected because they were all over 300 years old and near the

TABLE 10.5

BRAY's solar activity index, sunspot cycle lengths and atmospheric ^{14}C activity prevailing in 31 periods of alternately lower and higher solar activity from 527 B.C. to A.D. 1964

Period	BRAY'S sunspot-auroral index	Average sunspot cycle length (years)	Carbon-14 activity		
			No. of values	Av. departure (%) from 1850 level	S.D.
B.C.					
527–456	113	10·2	4	+0·6*	±0·6
455–399	<100	11·2	2	+0·1*	
398–334	<121	10·7	1	+0·6*	
333–255	(low)	11·4	2	−0·6	
254–187	126	11·4	4	−0·9	±1·0
186–130	81	11·4	1	+0·4	
129–21	111	10·8	8	−1·3	±1·2
20 B.C.–A.D. 169	79	11·2	21	−0·7	±0·9
A.D.					
170–202	107	11·0	5	−1·7	±0·7
203–283	(low)	11·6	3	−1·2	±1·1
284–316	114	11·3	6	−1·3	±0·2
317–348	76	10·3	2	−0·1	
349–380	126	10·0	2	−0·7	
381–426	86	12·2	4	−0·7	±0·5
427–591	107	10·9	19	−1·0	±0·7
592–707	87	11·4	18	−0·4	±1·1
708–880	108	10·8	21	−1·1	±0·9
881–957	84	11·1	11	−0·4	±1·3
958–1010	118	10·6	7	−0·9	±1·0
1011–1083	84	12·5	8	−0·1	±0·6
1084–1211	116	10·3	17	−1·2	±0·6
1212–1358	83	11·6	17	−0·4	±0·6
1359–1387	137	9·7	6	−0·3	±1·0
1388–1524	69	11·4	26	+1·0	±1·3
1525–1586	135	10·3	10	+0·7	±0·5
1587–1723	68	11·4	40	+1·1	±0·9
1724–1798	110	10·0	16	+0·5	±0·9
1799–1834	60	13·9	14	+0·8	±0·5
1835–1879	113	10·2	42	−0·1	±0·5
1880–1914	74	12·1	19	+0·1	±0·4
1915–1964	130	10·2	3	−2·0	

* These consistently positive values of ^{14}C departure at the beginning of the sunspot auroral index series are taken to be a legacy from low solar activity in the preceding centuries (when the ^{14}C departures are also known to have been positive).

When account is taken of the S.D. of the ^{14}C measurements, the appearance of reliability and the interest lies in the long runs of consistently negative and consistently positive results.

upper forest limit in two neighbouring valleys (the Yoho and Little Yoho Valleys, both glacier-headed and near the Rocky Mountains divide) in British Columbia. Mean basal area increase of the boles of these trees was obtained for each individual sunspot cycle from 1699 to 1964 and gave a correlation coefficient of +0·76 ($P < 0{\cdot}1\%$) with the average yearly sunspot number for each '11-year' cycle. The mean basal area growth was 9·6 cm²/yr for 1934–64 and 9·4 over the groups of active sunspot cycles between 1723 and 1799; it was 9·7 over the four active cycles between 1834 and 1879; but it averaged only 8·8 cm²/yr for 1699–1723, 8·7 for 1799–1823 and 8·9 for 1879–1923, all weak sunspot cycles. Similarly the average growth rate of Formosan cypress trees (at a height of 1000 m a.s.l. 60 km south of Taipeh) was found by OUTI (1961, 1962) to give a correlation coefficient of +0·75 with Zürich sunspot number for the cycles 1749–1950.

TABLE 10.6

BRAY's solar activity index, behaviour of glaciers and an index of the Arctic sea ice

Period	*Sunspot–auroral index*	*Numbers of reported maximum glacier advances and major readvances*	*Arctic ice at Iceland coast*	*^{14}C anomaly (%)*
B.C.				
700–528	(low?)	8		+0·6
527–334	<112	2		+0·4
333–21	<106	1		−1·0
20 B.C.–A.D. 202	84	2		−0·9
A.D.				
203–380	<105	2		−1·0
381–591	103	2		−0·9
592–880	100	3		−0·7
881–1010	101	0	0·4	−0·6
1011–1211	105	3	0·5	−0·8
1212–1387	95	2	6·9	−0·3
1388–1586	91	6	3·2	+0·9
1587–1798	81	36	22·6	+0·8
1799–1879	91	20	40·7	+0·1
1880–1964	109	5	13·9	−0·2

Indications pointing in the same direction, albeit weakly and from data of inadequate length, were obtained by BRIER (1952) using an approach of interest because of its implications regarding the manner of the reaction of the general atmospheric circulation to solar activity. BRIER used the monthly mean pressure values at four points in the northern hemisphere (Gulf of Alaska, the Irminger Sea off Iceland, the U.S. Rockies and Lake Baikal in Siberia) where winter–summer differences are especially big, to define an 'index of wintriness'. Over all the months of the year in the 40 years, 1899–1939, investigated, BRIER found more summer-like pressure distribution the higher the sunspot number (the correlation coefficient being −0·26, significant at about the 5% level, between the index of wintriness and sunspot number).

A harmonic analysis by ARAI (1958) of 5-day mean and monthly mean 500-mb charts of the northern hemisphere over the years 1946–56 for long waves in the upper westerlies, studying particularly the ratio of the amplitudes to normal in wave numbers 1–5 (i.e. the first 5 harmonics) indicated that:

(i) amplitude in wave number 1 was negatively correlated with Zürich sunspot number, rather weakly in summer and spring but significantly in autumn and winter (highly so in winter). In other words, the higher the sunspot number the less the eccentricity of the circumpolar vortex;
(ii) amplitude in wave number 2 appeared significantly positively correlated with sunspot number in winter but not in other seasons; indeed there was a nearly significant negative correlation in autumn;
(iii) amplitude in wave number 3 appeared to go more with the magnitude of the cyclonic shear on the poleward side of the jet stream than with solar activity, but there was some tendency to positive correlation with the latter in spring;
(iv) no stable relationships with wave numbers 4 and 5 were found.

TETRODE (1952) noticed that for those places in Europe and North America for which reliable series of temperature measurements go back far enough, the two warmest periods of 3-years' duration (*any* 36 consecutive months) coincided with the highest sunspot maxima thus far observed and the 2 years following, i.e. 1778–80 and 1947–9. The next following sunspot maxima, in 1787 (almost as high) and in 1957 (still higher sunspot number, but known that the spot areas outweighed the areas of brightened sun), which were the last high maxima preceding a decline of solar activity, were, however, accompanied by already slightly lower temperatures.

The climatic effects of long-term increases of solar activity, if the association be real, seem to operate through some general increase of energy supply to the Earth. These effects appear to be cumulative when centuries of high or low solar activity occur in succession, presumably owing to thermal inertia associated with heat storage in the oceans. The 'glacier advances' column of Table 10.6 is doubtless affected by the growth of our knowledge of the facts towards recent times: nevertheless, it seems that there was rather little glacier advance in the period A.D. 1388–1524 (mean solar index 69 preceded by many centuries with index values > 100) by comparison with the widespread glacier advances between 1587 and 1723 and 1799–1834 (solar index averages 60 and 69, preceded by centuries of predominantly weak solar activity). Correspondingly, there was apparently little glacier recession in the rather brief interval of great solar activity 1525–86 (mean solar index 135) which was preceded by 150 years of quiescent sun.[1]

1. BRAY (1968) believes it possible to state that 75–80% of all known major glacier advance events and other indicators of cold climate in Late Glacial and post-glacial time occurred during intervals of weak solar activity

That these variations may show some approach to cyclic repetitions is suggested by several features:

(i) SUESS reports (from the results shown in fig. 2.4 (*b*) in this book and subsequent additional measurements) that ^{14}C measurements on wood of known age indicate a cyclic fluctuation in the atmospheric ^{14}C content of period length about 200 years back to 0–300 B.C. and 400 years before that to the earliest measurements around 5000 B.C. The correspondence with sunspot variations and the severity of winters in Europe, especially eastern Europe, is seen in fig. 10.24. The evidence that the response to solar changes is

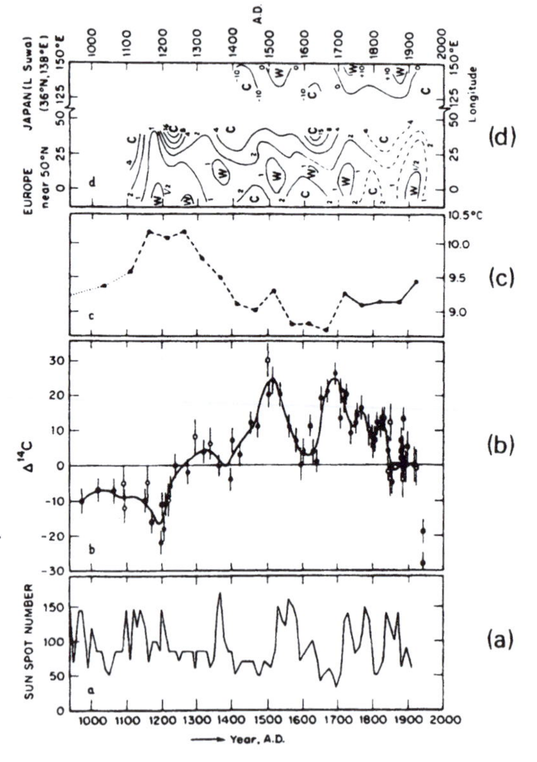

Fig. 10.24(*a*) (*Bottom diagram*) Zürich sunspot numbers (highest smoothed monthly mean) at the maximum of each cycle since A.D. 1749 and estimates by SCHOVE (1955) for earlier cycles (see Appendix I).
(*b*) Variations per thousand in the atmosphere's radioactive carbon content from measurements on wo·d of various known ages (SUESS 1965).
(*c*) Mean temperature over the year in central England, 50-year averages (LAMB 1965) (as in fig. 6.6).
(*d*) Winter severity indices for different longitudes in Europe near 50°N and Japan (LAMB 1961, 1966). '**W**' marks predominance of warmth, '**C**' predominance of cold, in the available reports.
The three investigators worked without knowledge of each other's results. Positive anomalies of ^{14}C (diagram (*b*)) are seen to accumulate during the longer periods with low sunspot numbers, negative anomalies are prominent only when solar activity had been high for many centuries previously. A correlation coefficient $r = -0{\cdot}80$ was found between the atmospheric radiocarbon values and 50-year mean temperature in central England about 150 years later. The strongest correlation between the radiocarbon values and winter severity, in eastern Europe (about 35°E in the top diagram), appears to involve little lag. (*This entire diagram is reproduced from* SUESS *1968 by kind permission.*)

and a similar percentage of glacier recession and warm climate indicators occurred with high solar activity, though his inclusion of several millennia in the Late Glacial and early post-glacial periods before the existence of firmly dated radiocarbon anomalies (i.e. before 5000 B.C.) involves a dangerous extrapolation. Of more certain value is his observation that these climatic fluctuations, which appear to be associated with solar variations on time-scales from 200 to 2000 years or rather more, are superposed on a much longer-term underlying trend (possibly a MILANKOVITCH cycle, though affected by a 5000-year lag as noticed in Chapter 2), whereby the length and severity of the glacial advance periods successively decrease to around 3000 B.C. and have been increasing again since.

quickest near the heart of the great Eurasian continent is what should be expected if the radiation supply changes.

(ii) OUTI (1961, 1962) finds traces of quasi-periodic repetitions in the Formosan tree-growth series on time-scales of about 100, 20–22, 11 and 6 years.

(iii) LINK (1958, 1964), LINK and LINKOVA (1959), by an analysis of Chinese astronomical discoveries over more than 4000 years, found evidence of an approximately 400-year periodicity, repeated through six cycles since 2300 B.C. in (*a*) average length and activity of the '11-year' solar cycles, (*b*) average cloudiness of the nights in China.

(iv) ABBOT (1963) has stressed the importance of the (approximately) 23-year solar period and its multiples, particularly about 46 and 91 years, in association with recurrences of prolonged droughts over the Great Plains and prairies of the U.S.A. and Canada, affecting also the measured levels of the Great Lakes since A.D. 1860. (A quasi-periodicity of roughly 23 years appears to be substantiated from much longer runs of data of various kinds instanced in Chapter 6, including data related to climate from the geological past. But how far, or in what way, the climatic effects registered are associated with solar disturbances remains uncertain. This spacing was obvious between the severe winters in Europe in 1895, 1917, 1940 and 1963, although the basic sunspot cycle averaged 10·2 years between 1917 and 1963: moreover, this particular series of outstanding winters cannot readily be traced back before 1895 when the double sunspot cycle was closer to 23 years.)

In the 2500 years long history covered by Table 10.5 (due to BRAY), successive spans of about 400 years (370–440 years) are made up of just two or three alternations of groups of active and weak solar cycles. The subperiods, each comprising one well-marked alternation, have been close to 200 years, or occasionally 80–100 years, in the last millennium: but in earlier times this is not apparent either because the data are too sparse or because the period may really be subject to variation. SUESS's measurements suggest (e.g. to some extent discernible in fig. 2.4 (*b*)) more regularity of a 400-year (and sometimes 200-year) fluctuation of atmospheric ^{14}C superposed on still longer-term trends, than has so far been identified in any directly solar data. LINK's studies (*loc. cit.*) seem to verify this solar fluctuation and one associated climatic item over 4400 years past.

There is a hang-together of the trends of various indices of the strength and frequency of the prevailing westerly winds of middle latitudes, north and south of the equator, of the Trade Winds and the Asian monsoon currents, and in particular of the frequency of SW'ly surface winds in England, that makes it reasonable to take variations of the latter (as in fig. 6.8) as an indicator of the varying energy level of the global wind circulation. (This is illustrated by a correlation coefficient of $+0{\cdot}75$ ($P < 0{\cdot}1\%$) between the decade averages of the SW wind at London and of the amount of snow accumulation at the South Pole from A.D. 1770 to the 1950s.) The curve shows an apparent 200-year fluctuation of the westerlies since A.D. 1340 which keeps a constant phase relationship to the solar variations discussed in this section, the maxima of the SW wind coming in each case about 30–40 years before the main sunspot maximum. By the time the maximum solar activity occurs, the strength of the global wind circulation as indicated by the frequency of the westerlies is already falling off sharply. A decrease of energy at that stage might be explained by the final great maximum of sunspot area outstripping (or occurring at the expense of) any corresponding increase in the brightened areas of faculae round about, as was demonstrable in the 1950s. This is indicated by the values of BAUR's solar index (see Chapter 2) over the years 1874–1957. Moreover, the closely parallel course of its cumulative anomaly to the variation of indices of the westerlies and general circulation strength over these years (fig. 10.25) is impressive. The cumulative departures of BAUR's solar index from its 1874–1957 average also appear (MIRONOVITCH 1960) to show strong relationships with:

(i) yearly values of Azores–Iceland pressure difference (in-phase relationship);
(ii) yearly numbers of depressions near the Azores (inverse relationship);
(iii) continentality (July–January temperature difference) of central Europe (inverse relationship);
(iv) yearly numbers of typhoons noted by the Shanghai Observatory (inverse relationship).

The apparently lowered energy level of the general wind circulation during the periods, long or short, of much weaker solar activity that follow the great maxima are presumably associated with a lower energy output, all types of solar disturbance being reduced (quiescent sun).

In a series of papers on solar variability and climate WILLETT (e.g. 1949*a*, *b*, 1964) has stressed the *circa* 80- to 100-year and 20- to 24-year sunspot cycles and expansion and contraction of the circumpolar vortex. The longer of these cycles is divided into quarters, starting with the sudden break from very high to very low level of prevailing solar activity (which may be more regularly marked at 170- to 200-year intervals than at 80–100 years): the first quarter is allegedly dominated by low latitude of the mainstream of the zonal wind circulation, cold climates and glacier growth in middle latitudes; during the second and third quarters the zonal circulation, at first slowly and then faster, shifts poleward until a pronounced high-latitude zonal pattern predominates with warm climates in middle and high latitudes; in the fourth quarter the atmospheric circulation tends to break down into meridional patterns, with subpolar continental-sector anticyclones giving warm dry summers and cold dry winters in middle latitudes. This analysis has points of similarity with the findings of the Russian investigators DZERDZEEVSKI (1961) and GIRS (1956, 1967) with regard to the last 70–80 years, which have also been used as a basis for forecasting the prevailing atmospheric circulation and climatic character of the next 20–40 years. All these diagnoses rely excessively, if not solely, however, on the course of the last 80- to 100-year cycle and cannot be regarded as statistically established. WILLETT's assertions, sum-

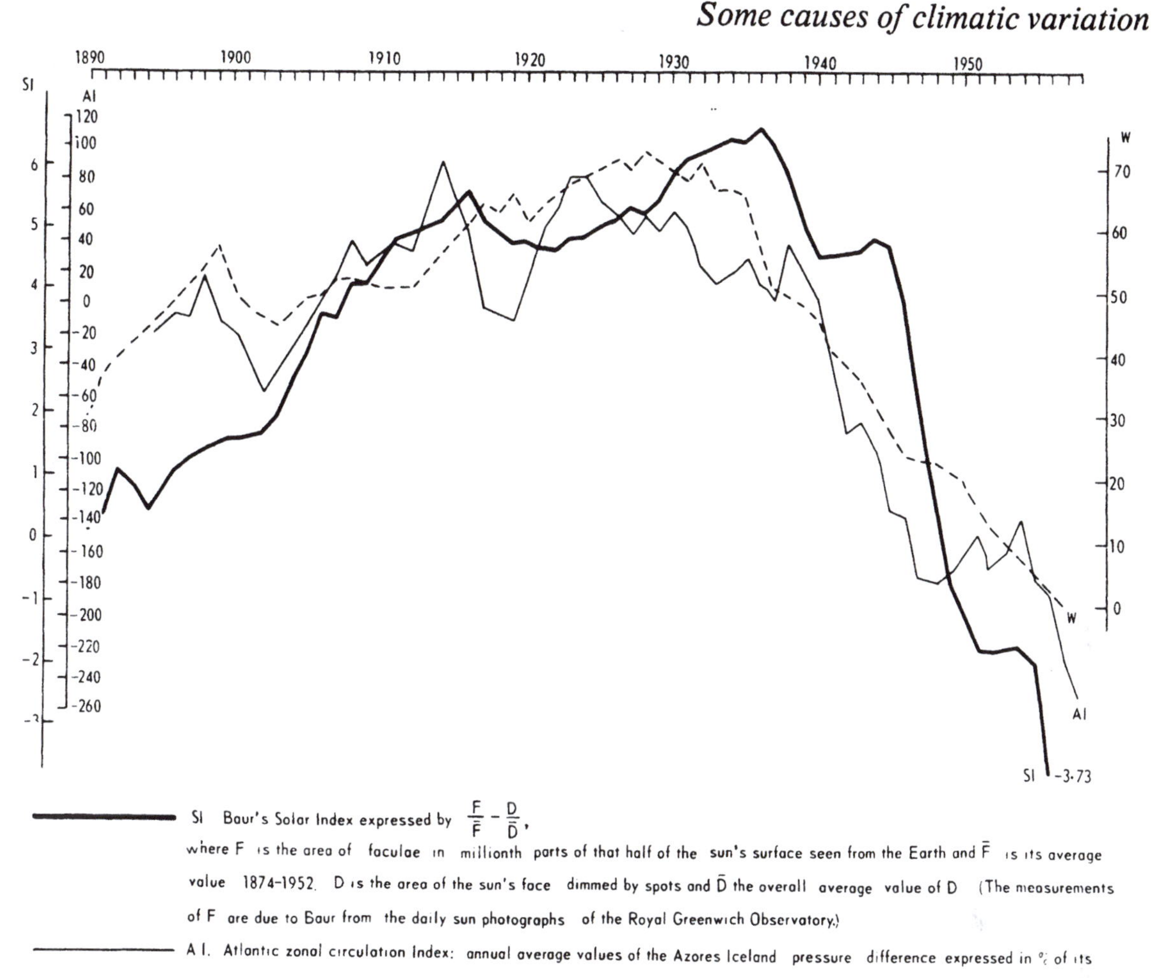

Fig. 10.25 Cumulative anomalies of solar energy (indicated by faculae minus spot areas) and strength and prevalence of the middle latitudes westerlies over the northern hemisphere. (*After* MIRONOVITCH *1960.*)

marized above, agree plausibly enough with the history of glacier variation in middle latitudes over recent centuries, although it seems likely that the characteristics he suggests for the third and fourth quarters would have to be very much weaker in some long cycles than in others. Comparison with long series of monthly mean pressure maps for the months of January and July going back to 1750 (i.e. over 100 years earlier than any charts available to WILLETT), constructed by LAMB and JOHNSON (1966), showed a latitude variation of the 'Iceland' low and Azores high in July in good agreement with WILLETT's description through two long cycles since about 1800 and for the South Atlantic subtropical anticyclone since about 1890, though it seems that the return towards lower latitudes began already in the fourth quarter of each long cycle. The latitude variations in January appeared to be different in phase, a drift towards lower latitudes setting in as early as the third quarter, though differences of timing between the curve for the Iceland pressure minimum and the Azores maximum suggest that the latitude values may be affected in some complicated way by the positions of blocking anticyclones and partial blocking in winter, making this result less reliable than that derived from the July maps (see curves published by LAMB and JOHNSON, 1959, 1961, figs. 7 (*a*) and 27 (*a*)).

Significant time patterns of climatic variation within these longer-term solar variations of about 90 and 170–200 years also appear to be indicated by some other parameters for which long series of observations have been investigated, viz.:

(i) The Bai-Ü rains at Seoul, Korea ($37\frac{1}{2}$°N 127°E), as measured by the May + June total rainfall, averaged for each '11-year' sunspot cycle from 1780 to 1954, yielded a correlation coefficient $r = +0{\cdot}46$ (P approx. 5%) with the mean sunspot number. As measured by the number of rainy days in May and June, averaged for each '11-year' cycle, the Bai-Ü rains broadly paralleled the mean sunspot number at least since 1740 (YAMAMOTO 1967).

(ii) Annual rainfall in central Europe, as represented by Prague–Klementinum since 1804 and by north German stations since 1850, has followed an inverse course to that of the average yearly number of sunspots for each cycle (KRIVSKY 1953).

(iii) Sea level, measured around the Atlantic Ocean, which should be high when ice sheets and glaciers melt, also the salinity of the Baltic Sea and its water exchange with the Atlantic, have been found to show a positive correlation with long-term solar activity, whereas the amount of ice on the Barents Sea has a negative correlation with it, and all these items have therewith an 80- to 90-year cycle (RUBASHEV 1964).

For their significance to be regarded as established, however, relationships of this kind need to be seen to operate over many more than the one or two long-term fluctuations of solar activity covered by actual rainfall (or other instrumental) observations. (Hence the potential value of tree-ring studies at places where the varying ring widths can be mainly attributed to just one climatic variable, be it rainfall or temperature.)

Apparent climatic relationships to the '11-year' solar cycle

Patterns of significant relationship to the course of the approximately 11-year sunspot cycle appear to be well established in at least the following cases:

(i) Severe winters (December–February) in central Europe, defined by the lowest sextile (17%) among the temperatures averaged over De Bilt (Holland), Berlin and Vienna from 1755 to 1950, were clustered about both sunspot extremes (BAUR 1948, 1956). Of the

31 winters with mean temperature more than 1·8°C below average 22 were between 0·4 year before and 1·7 year after either sunspot maximum or sunspot minimum.

(ii) Winters in central Europe centred between 1·1 year after a sunspot maximum and 3·6 years before the following sunspot minimum were almost never very cold (data 1755–1950): only one winter out of the 43 was severe, whereas 28 were milder than average (BAUR 1956).

Despite the indications given in (i) and (ii) of a double wave in the European winter temperatures (most mild winters being attributable to strength and prevalence of W'ly winds in the East Atlantic–European sector), BODURTHA (1952) found a well-marked single wave in the frequency of anticyclogenesis (pressure rise 7 mb or more within 24 hours) over Alaska in the winter months from 1899 to 1938: 67 cases in the 12 winters with highest sunspot number, 30 cases in the 12 winters with lowest sunspot number.

(iii) Summer (June–August) rainfall in central Europe, measured by the average of 10 stations (data in most cases from 1804 to 1956)[1] showed significantly lower rainfall than

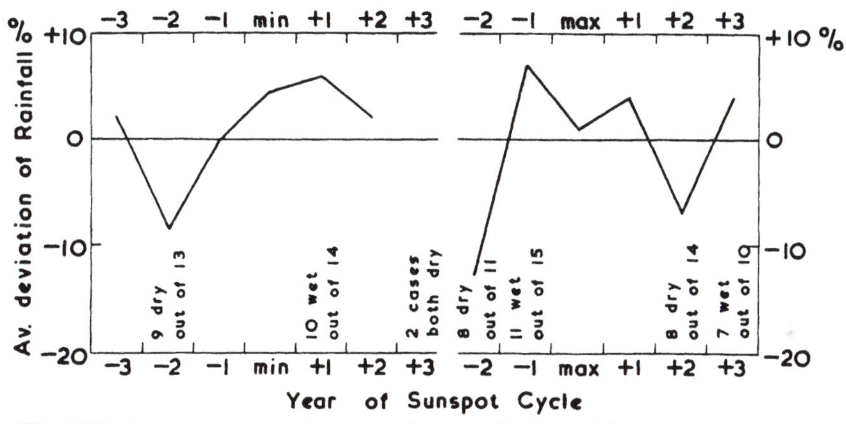

Fig. 10.26 Average percentage departures of summer (J J A) rainfall in central Europe from the 1804–1956 mean at different stages of the '11-year' sunspot cycle. (*Adapted from* BAUR *1959*.)

normal at three points of the '11-year' sunspot cycle (see fig. 10.26), namely 2 years before either sunspot extreme and the second summer after sunspot maximum (BAUR 1959). (The frequency of dryness was greater for the second *summer* than for the summer of the second *year* after sunspot maximum.)

(iv) Average summer (June–August) temperatures at New Haven, Connecticut (data 1781–1966) have shown (LANDSBERG 1967) a smooth fluctuation with consistently

1. The stations (and their average summer rainfall in mm) were Zwanenburg-Hoofdorp (217 mm), Bremen (223), Frankfurt-am-Main (201), Stuttgart (235), Basle (274), Bayreuth (215), Dresden (230), Prague (179), Warsaw (220), Kremsmünster (384) – overall mean 238 mm.

high values 2 years before sunspot minimum and low values one year before sunspot maximum (difference between the respective means of 17 cases 1·3°C). Prevailing summer temperatures had a second peak 2 years after sunspot maximum.

BAUR attributes these effects to greater energy received from the sun at the intermediate phases of the sunspot cycle, resulting in increased vigour of the atmospheric circulation and a slight poleward displacement of the well-developed subtropical anticyclone belt, than at either extreme of the sunspot cycle (cf. fig. 2.1). To test BAUR's hypothesis against actual data on the atmospheric circulation which were not available to him, LAMB and JOHNSON's (1966) long series of monthly mean pressure maps for each January were arranged according to phase of the '11-year' sunspot cycles from 1780. The three Januarys centred about (*a*) each sunspot maximum, (*b*) the middle of each waning phase of solar activity and (*c*) each sunspot minimum were combined into three groups (comprising respectively 39, 43 and 44 Januarys) for which the overall average pressure distributions were computed.[1] The average pressures in key areas in the subpolar and subtropical zones differed slightly, but apparently significantly, as shown in Table 10.7, in a sense agreeing with BAUR's supposition. Though the differences of (*circa* 40-year) mean pressure are small, the effects on pressure gradient are big and they are of the same order of magnitude as the differences between the 40-year average pressure around A.D. 1800 when cold winters were common in Europe and between 1900 and 1939 when they were rare.

TABLE 10.7

Average pressure (mb) in January (at m.s.l.) and solar cycle phase

Solar cycle phase	*65°N 25°W near west Iceland*	*60°N 10°E Oslo*	*40°N 30°W northwest of Azores*
About sunspot max.	999·8	1013·7	1018·1
Middle of waning	996·9	1009·4	1019·0
About sunspot min.	1001·3	1012·9	1017·6
Av. pressure 1900–39	999	1011	1018

The changes of mean pressure at 55°N 5°W over the British Isles were similar to those at Oslo.

Since the foregoing paragraphs were written our knowledge of the matter has been extended by a study carried out by Mr B. N. PARKER of the Meteorological Office, resulting in maps of the world-wide distribution of anomalous atmospheric circulation tendencies at different phases of the '11-year' sunspot cycles which are illustrated in Appendix IV.

1. I am indebted to Miss J. R. EASTWOOD for this work.

Other seasons of the year have also been found to show some apparently significant relationships to phase of the sunspot cycle. Thus, DROGAJČEV (1968) found the distribution of rainfall in May over the agricultural belt of the Ukraine–Kazakhstan shown in the following table (m = sunspot minimum, M = sunspot maximum).

TABLE 10.8

May rainfall in the Russian grainlands 1891–1965 and the solar cycle

Year	*m – 1 to m + 2*	*M – 1 to M + 1*	*All other years*	*Totals*
Wet	13	3	9	25
Normal	10	5	10	25
Dry	4	13	8	25

Chi-squared value 12·90, significant at the 1–2% level. Over central Europe (average of 16 stations between Emden, Tilsit and Vienna) BAUR (1956) found that springs (March, April, May) centred 1·0–2·1 years after a sunspot maximum (data 1801–1950) gave without exception more than the long-term average rainfall.

The reader may also refer in this connexion to the works by WEXLER, GASJUKOV and MAKSIMOV cited in Chapter 6.

KHRABROV (1958) has drawn attention to the exceptionally high frequency of large changes (over 6°C) of mean temperature from one month to the next in European Russia in 1956 in the period when solar activity was abruptly increasing to the 1957 maximum. An exceptional number of 3 and 4 σ temperature anomalies occurred in the Soviet Union at that time and again during the sharp fluctuations of solar activity in 1967–9.

Relationships have been found in different parts of the world between the frequency of thunderstorms (number of days in the year with thunder) and solar activity. The nature of the relationship clearly differs in different regions, being for example strong and in-phase ($r = +0{\cdot}88$ ($P < 0{\cdot}1\%$)) in Siberia from the average of numerous stations' data (43 to 71°N 59 to 150°E) over the years 1888–1924 (SEPTER 1926) and likewise in China, but strongly out-of-phase in many other regions (North America, Africa); hence, generalizations about the overall global effect are as yet unjustified, and different writers (BROOKS 1934, KLEJMENOVA 1967) have differed according to the weighting given to different regions. The most reliable and detailed results appear to be still those derived for central Europe by VON BEZOLD (1884) from the yearly number of lightning strikes on buildings in Bavaria through four solar cycles from 1833 to 1882 and substantiated by MYRBACH (1935) as applying equally to numbers of thunderstorm days in Vienna 1878–1934 and Kremsmünster (Upper Austria, 48°N 14°E) over the years 1810–1934. In this region (see fig. 10.27) there is a double wave of thunderstorm frequency within the '11-year' sunspot cycle, with sharp peaks 2–3 years before and 1–2 years after sunspot minimum, separated by one sharp minimum near sunspot minimum and another about 2 years before sunspot maximum, the latter followed by several years of rather low thunder frequency. This complex wave of thunderstorm activity in Europe appears to bear the logical inverse relationship to BAUR's findings about the tendency to dry summers: hence we may suppose that the thundery years are years when the subtropical anticyclone largely fails to move north in summer in this sector and the continent is oftener dominated by cyclonic activity.

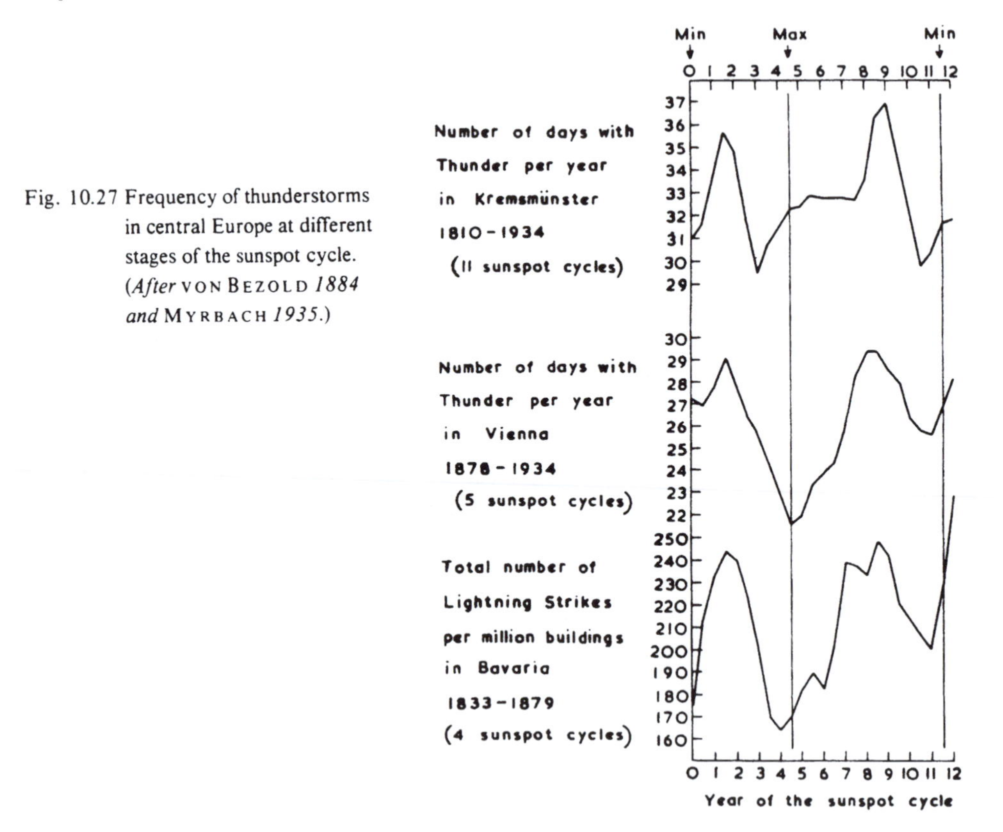

Fig. 10.27 Frequency of thunderstorms in central Europe at different stages of the sunspot cycle. (*After* VON BEZOLD *1884 and* MYRBACH *1935*.)

At Athens the number of days from May to October with Etesian winds – i.e. N'ly wind strong enough to exclude the afternoon sea breeze – was found by CARAPIPERIS (1960) to have a simple relationship to yearly sunspot number, expressed by a correlation coefficient $r = +0{\cdot}63$ for the years 1893–1937. The average was 65 days at sunspot maximum, 40–45 days at sunspot minimum. This is a wind system related to the Asian summer monsoon rather than to the westerlies over northern and central Europe. The finding may therefore be related to an observation (RAGHAVAN 1961) that over the years 1923–59 the number of monsoon storms striking the Midnapur–Bengal coast near Calcutta during the monsoon season showed an allegedly simple, inverse relationship to sunspot number.[1] Upper winds at 4 km and above over Calcutta indicated that the decline of storm frequency was accompanied by a southward expansion, or shift, of the warm Tibetan summer upper anticyclone in the years of maximum solar activity. There was also a secular variation over the years studied such that the storm frequency was greatest about 1940–5.

Irregular variations of solar activity within a year

Turning next to short-term variations of solar activity over periods of a few months to a year, we find several relationships which suggest that more energy is generally fed into the atmospheric circulation following periods of rising solar disturbance. The contrary seems to occur over the next few months when solar activity flags:

(i) When the 3-month mean Zürich sunspot number rose by 5 or more from the period about July (J J A) to the period about November (O N D), of which there were 30 cases between 1851 and 1960, in 27 cases (90%) the mean temperature of January and February in central Europe was above average (statistically significant beyond the 1% level) (DINIES 1957, 1965).

(ii) When the 2-month mean Zürich sunspot number fell by 5 or more from August and September (A S) to October and November (O N), the following January and February in central Europe showed a heightened tendency to be either very mild or very cold. This, at first sight strange, result was clearly due to increased frequency of blocking which can give either S'ly or N'ly surface winds over Europe: pressure at the North Pole was above normal in 8 of the 9 Januarys concerned for which this could be verified (since 1899) and was below normal at 10–40°N.

(iii) BAUR (1956) found that when the 5-month mean Zürich sunspot number rose by more than 4 from the period about February to the period about July and then the 2-month mean number fell from A S to O N by more than 4, subject to two further conditions about the atmospheric circulation in November (Atlantic sector not very blocked) and early December (not very mild in Berlin), the following winter (D J F) was cold in central Europe in all 12 cases (data used 1865–1952).

(iv) B. N. PARKER (unpublished) reports that over nine years investigated (1960–8) the difference of monthly mean pressure 40–45°S 172°E (index of New Zealand westerlies) tended to increase after rises of solar activity, the strongest correlation coefficient with Zürich sunspot number being $r = +0{\cdot}38$ ($P < 0{\cdot}1\%$) at 10 months' lag.

SCHUURMANS (1969), having noticed in recent years a pattern of 500-mb height changes over the 24 hours after a solar flare such as to produce, on the average of some 80

1. It is not so certain that this is a simple single-wave variation of storm frequency within the '11-year' sunspot cycle. As 5-year moving means were employed in the investigation of the Bengal storms, any double oscillation of about this period length within the 11-year cycle would be suppressed by the method.

cases examined in the northern hemisphere and some 60 cases in the southern hemisphere, a significant swing towards (or intensification of) meridional circulation in all sectors in middle latitudes, proposes to use year-to-year variability of sunspot number as an index of the probable incidence of flare activity much farther back in the past. Comparison of the results with both summer and winter temperatures in Europe since 1749 suggested a general association between this activity and meridional or blocked circulation patterns. There is some discussion of the probable mechanism of such solar flare-induced changes in SAZONOV and KASOGLEDOVA (1968).

Processes by which solar activity may influence weather and climate

There is probably most hope of establishing the existence and nature of a chain of causation linking physical processes on the sun, and between the sun and the Earth, with events in the lower atmosphere in the case of the briefest types of intense solar disturbance. This applies especially to solar flares or chromospheric eruptions that characteristically rise within a few minutes to maximum intensity and die away over the ensuing hours. Many studies have shown promising results (e.g. DUELL and DUELL 1948, MUSTEL 1967, PALMER 1953, ROBERTS 1963, SAZONOV and ZORINA 1967, STAGG 1931, VITELS 1949). Zero day is variously taken as marked by either:

(*a*) an index of the occurrence of the flare and its intensity or of flocculi (plage);

(*b*) passage of the flare-producing sunspot area across the central solar meridian as seen from the Earth and at a solar latitude $< 15°$;

(*c*) the greatest associated disturbance in the ionosphere;

(*d*) the greatest associated geomagnetic disturbance (which usually has a 'sudden commencement' corresponding to the sharp build-up of the flare); or

(*e*) the greatest auroral activity.

The last three, (*c*), (*d*) and (*e*), mark the arrival at the Earth of an invasion of solar corpuscles. For clarity of interpretation attention is often concentrated on periods of rather little solar activity so that successive events are well spaced out and their effects not too likely to overlap. The fullest account has been given by SAZONOV (1964).

WARD (1960) and LONDON *et al.* (1959), investigating temperature and circulation changes at the 100-mb level, i.e. in the lower stratosphere not far above the tropopause, found statistically significant variations over the Arctic and over the United States about 3–5 days *before* the greatest geomagnetic storms. This can only imply that the link is with some form of solar energy propagated at extremely high speeds. Streams of solar corpuscles, mostly particles (protons and electrons) with energy of 10^3–10^4 electron volts,[1] take from 17 to 20 hours up to several days to reach the Earth and cannot penetrate the Earth's atmosphere except within about 20° of the geomagnetic poles, towards which they are

1. The *electron volt* is a unit defined as the energy acquired by a charged electron passing down through a potential difference of one volt in an electric field. One eV = $1\cdot602 \times 10^{-12}$ ergs.

directed by the Earth's magnetic field, save during magnetic disturbances when such particles may reach latitudes 50–55°. Higher-energy solar protons and alpha particles with energies above 10^6 eV (many with 10^7–10^8 eV), which SAZONOV calls 'solar cosmic rays' or 'hard particles', behave differently and can penetrate the atmosphere in most geomagnetic latitudes (everywhere poleward of 30°); they are much less deflected by the Earth's magnetic field. Solar flares produce particles with energies up to a few times 10^9 eV. The energy spectrum is fairly flat up to 10^7 eV.

The solar particle streams ('solar wind') are accompanied by strong magnetic fields emanating from the sun, which disturb the lines of force of the Earth's magnetic field when they approach it. Solar corpuscular streams widen geometrically and also by thermal expansion as they proceed along their path. The Earth's magnetosphere extends to 8–12 Earth-radii on the side towards the sun, but many times farther out on the side away from the sun, forming the so-called 'tail of the magnetosphere'. This means that the Earth's magnetic field is more diffuse, weaker and more easily disturbed on the night side of the Earth and therefore particularly in winter in middle to high latitudes (SAZONOV and ZORINA 1967): and so it is in these situations that most solar and galactic particles arrive.

Galactic cosmic-ray particles, which approach the Earth from all directions in space, are most abundant in the energy range 10^8–10^{10} eV. They are liable to be trapped, concentrated ('focused') and directed (rendered 'anisotropic') when they come into the magnetic fields of the solar wind. It is no doubt due to this gathering effect of the magnetic fields thrown out from the sun with the corpuscular streams, whenever and in whatever direction these occur, that fewer cosmic-ray particles reach the Earth from the galaxy at times of active solar disturbance (the reduction being greatest for the lowest energy particles from the galaxy), resulting in less ionization of the atmosphere (FORBUSH effect) – particularly at high levels, e.g. 10 mb, where the measurements about the 1957 sunspot maximum were only 50% of those about the 1954 sunspot minimum – and less production of radioactive carbon. The effect is also measurable in the troposphere and, through reducing the atmosphere's electrical conductivity around sunspot maximum, might be expected to affect the frequency of thunderstorms (see NEY 1959). It is only in years of low solar activity that the galactic cosmic rays reach the Earth largely unaffected. Together with other high-energy particles trapped in the 'VAN ALLEN radiation belts' they may be projected into the Earth's atmosphere when and where the geomagnetic field is weakened and distorted, particularly by solar disturbance. At most stages of the '11-year' sunspot cycle the energy brought into the atmosphere appears to be mostly accounted for by particles of energy level about $2{\cdot}5 \times 10^9$ eV. Escape of the trapped particles is easiest from the lobes or 'horns' of the VAN ALLEN radiation belts which are at geomagnetic latitudes about 30° (inner belt) and 50° (outer belt). This and the curved path of the solar wind, which comes in towards the Earth at an angle of 50–60° to the Earth–sun line, tends to bring in the various extraterrestrial particles at high geomagnetic latitudes (greater than 55°). There is always, however, a prism-like tendency of the Earth's magnetic field so that the lower the geomagnetic latitude the smaller the proportion of low-energy particles in the invasion.

SAZONOV (1966, 1967) has discussed how far this energy penetrates into the Earth's atmosphere. The bombardment by particles of energy level $\leqslant 10^6$ eV (typical solar corpuscles) leads only to some intensification of the chaotic atomic motions mostly at ionospheric levels; it may be regarded as the pouring in of a little hot gas into the uppermost reaches of the atmosphere, i.e. in the auroral zone and above. The higher-energy (solar and galactic cosmic-ray) particles, and the secondary particles (protons, neutrons, alpha-particles and pi-mesons) resulting from their impact upon the upper atmospheric particles, mostly continue with direction unaltered by their collisions; most of their energy is apparently spent at or near the deepest level of the atmosphere to which they penetrate. This level is about 30–35 km for particles with energy 10^7–10^8 eV, but as low as 15 km (i.e. near the tropopause) for 10^9–10^{10} eV.

It may be important that the direction of impact is slanting and so can give an impulse to the wind. The total energy involved in such invasions, up to several ergs/cm^2/sec in the case of solar cosmic rays, can, however, be no more than an additional impulse or retardation, a potential trigger mechanism helping to unleash or to direct energy that has been mainly built up within the terrestrial atmospheric processes. Despite this disparity of the total energy there is no lack of apparently associated effects. Moreover, the general concordance between the structure of the Earth's magnetic field (see fig. 10.28) and the shape of the main

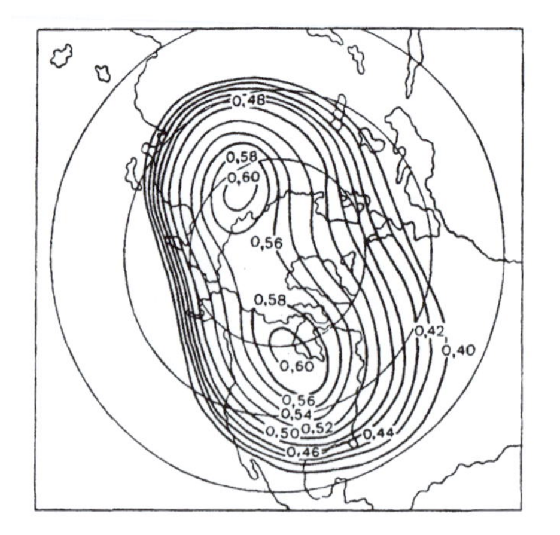

Fig. 10.28 Vertical component of the Earth's magnetic field (strength in oersteds). N.B. An *oersted* is the magnetic field strength produced at the centre of a plane circular coil of wire of one turn, radius 1 cm, carrying a current of $\frac{1}{2}\pi$ abampères. (One abampère = 10 ampères.) (*After* SAZONOV *1964*.)

circumpolar vortex in the atmosphere (cf. FLOHN 1952, 1959) indicates that, in the upper troposphere, cyclones prevail where the horizontal component of the magnetic field is weak, anticyclones where it is strong (SAZONOV 1967).

MUSTEL (1967) found a general fall of surface pressure (averaging 2–4 mb) over the general areas of France–Italy, Novaya Zemlya and northwest Siberia, and in Arctic Canada and northwest Greenland, as well as at Vostok (78·4°S 106·9°E) near the south magnetic pole, culminating about 4 days after a solar flare. At the same time pressure rose by a

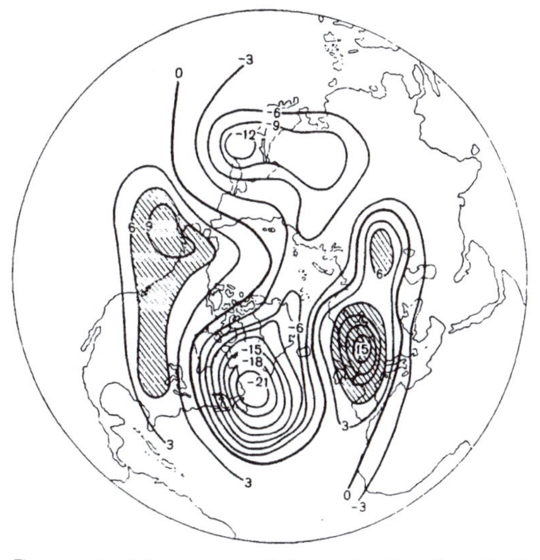

Fig. 10.29 Average change of 5-day mean height (in tens of feet) of the 500 mb surface 10 days after a geomagnetic storm. (*After* SAZONOV *1964*, *adapted from* ASAKURA *and* KATAYAMA *1958*.)

similar amount over regions extending from Greece to Moscow and from the Persian Gulf to central Siberia. The map given by SAZONOV (here reproduced as fig. 10.29) for the average change (data for 10 winters 1946–56) of 500-mb height over the northern hemisphere over about 10 days after a geomagnetic storm – i.e. rather longer after a chromospheric outburst – suggests that the changes of surface pressure in the regions mentioned above tend to be accompanied, or followed, by blocking anticyclones over northern Europe–western Siberia.

SAZONOV's own investigations (1964) of the geographical distribution of intense high-level (500-mb) anticyclones, and of the longitude of highest pressure along each 5° latitude circle in the upper troposphere, in the winters and summers of the years 1949–62, revealed a ring-shaped 'zone of anticyclogenesis' (figs. 10.30 (*a*), (*b*)) roughly paralleling the auroral zone. This was a feature both of the positions of maximum 500-mb height at each latitude and of the distribution of centres of intense 500-mb level anticyclones.

This SAZONOV ring is markedly eccentric to the geographic pole: it roughly parallels the auroral zone (fig. 10.31), more closely in summer and with increasing height towards the tropopause. Its centre is in the general region of the geomagnetic pole over Arctic Canada–Greenland (at present 78·5°N 69°W). The ring therefore passes across, or near, Scandinavia, the Siberian Arctic coast and Alaska, and it merges with the subtropical anticyclone zone over America and the Atlantic. The similar shape and course of the zone of maximum 1000–500 mb thickness (fig. 10.32) ignores even the strongest winter-time thermal differences in the undersurface, and hence points to some extraterrestrial cause. This suggestion is further strengthened by high correlation coefficients obtained by SAZONOV (*loc. cit.*) between sunspot number and the occurrence of very intense atmospheric pressure systems.

Finally, the close similarity between the course of the SAZONOV ring of anticyclogenesis

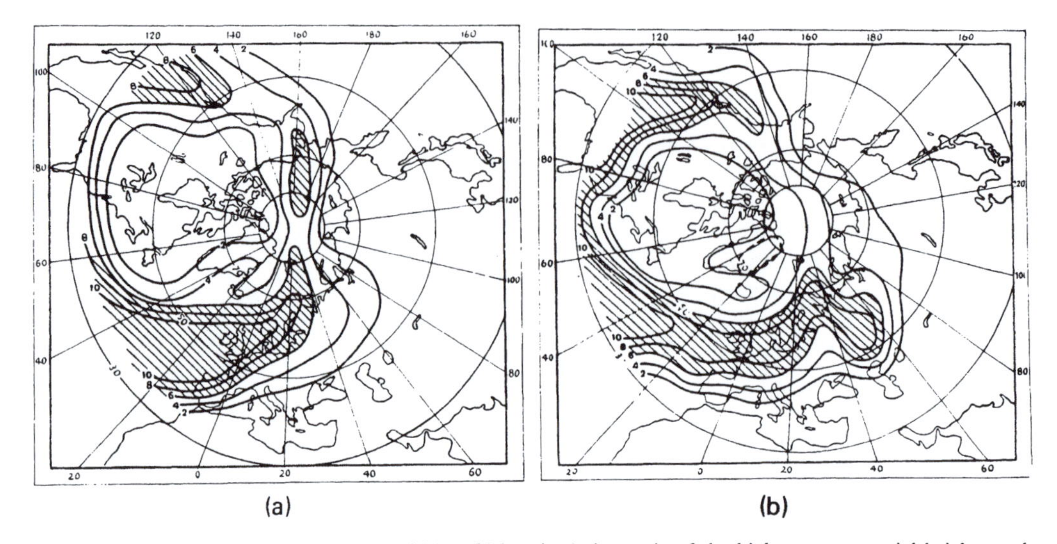

Fig. 10.30 Frequency (%) of occurrence within 10°-longitude intervals of the highest geopotential height on the 500 mb chart at different latitudes. (Data 1949–62.)
(*a*) December to March (*b*) June to September
(*After* SAZONOV *1964*.)

and the highest standard deviations of monthly mean surface pressure in the northern hemisphere (as illustrated in figs. 7.11 (*a*) and (*c*) and for all other months of the year except July) cannot escape notice. The arc of highest standard deviations of monthly mean pressure over the southern hemisphere (fig. 7.12) shows some semblance of a similar relationship to the southern geomagnetic pole[1] (78·5°S 111°E): its course also ignores gross differences in the thermal nature of the undersurface.

The climatic implications of more, or less, anticyclonic activity along the SAZONOV ring over periods, decades or centuries long need to be considered. Logic, as well as experience of the 1930s in the northern hemisphere, suggests that more anticyclones on this ring mean increased heat and moisture transport into the Arctic and a warming of climates in middle and higher latitudes. The probable shape of the southern hemisphere ring (deduced from fig. 7.12) indicates that this should apply equally, or even more strongly, to the Antarctic with the present position of the geomagnetic pole there. Over longer periods, however, the Earth's magnetic poles wander: the northern magnetic dip pole can be traced from a

1. The southern magnetic pole (the actual *dip* pole as opposed to the *geomagnetic* poles that correspond to the idealized dipole most nearly equivalent to the Earth's magnetism) lies near the coast of Antarctica (67°S 140°E) south of Australia, much farther from the geographical pole than the northern one. Hence, the auroral zone (and the magnetic lines of force, like those of constant magnetic dip (PIGGOTT and SHAPLEY 1962), that parallel it) over the southern hemisphere is highly eccentric to the geographical pole. Passing roughly over the ring of maximum variability of monthly mean surface pressure, it indicates an anomalous situation over west Antarctica and the Weddell Sea–South Atlantic, 'the South Atlantic anomaly' (where observatories far from man-made magnetic interference are effectively in a rather low geomagnetic latitude).

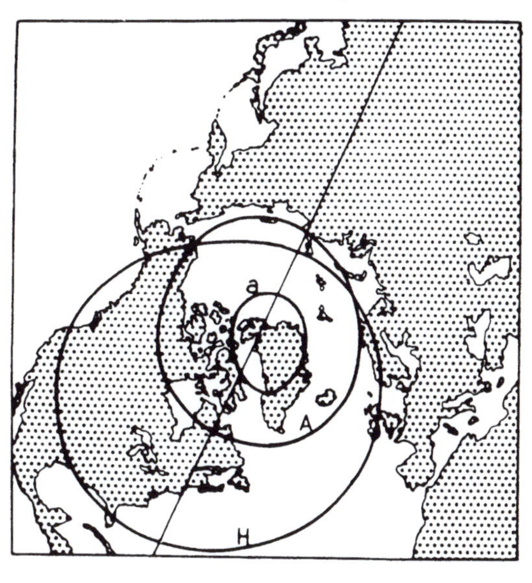

Fig. 10.31 The auroral ring and the axis of the ring-shaped zone of frequent anticyclogenesis identified by SAZONOV:
(*a*) Inner limit } of maximum frequency
(*A*) Outer limit } of aurora
(*H*) Axis of highest 500 mb values for each latitude.
(*After* SAZONOV *1964*.)

position about 82°N 125°W in A.D. 1600 to near 70°N 95°W from 1830 to 1920 and 74°N 100°W in 1948, and some parallel movement of the geomagnetic poles seems probable. Hence, the maxima of blocking anticyclone activity accompanying solar disturbance may move to different sectors in different millennia. At any period when the geomagnetic pole lies close to the geographical pole these questions do not arise in the same form: the additional anticyclonic activity would, however, probably be in a higher geographical latitude than the subtropical belt. In view of the secondary maximum of variable blocking anticyclone activity at the surface, and of S.D. of monthly mean pressure, over Greenland (see figs. 3.16 (*c*) and (*d*), 7.11) in the present century, there is need to investigate whether

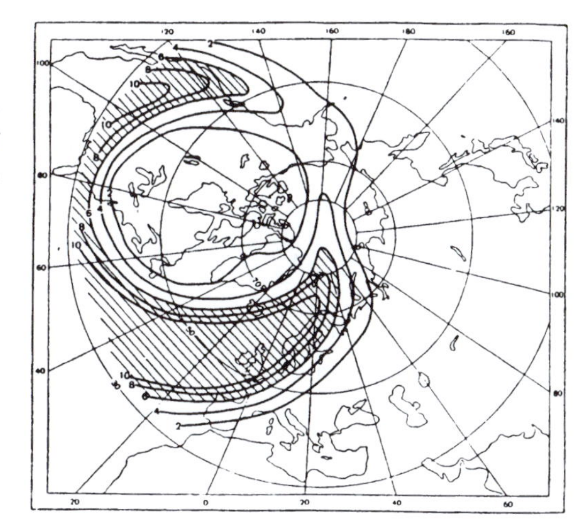

Fig. 10.32 Frequency (%) of occurrence within 10°-longitude intervals of the greatest value of 1000–500 mb thickness for each latitude, December to March. (Data 1949–62.)
(*After* SAZONOV *1964*.)

the SAZONOV ring may also undergo expansion and contraction, possibly lying in rather lower geomagnetic latitudes at any times when very high energy particles are exceptionally abundant.

Monitoring solar activity: data for further studies

The standard sources of continuing information on current solar activity are the following:
(1) *Preliminary Reports and Forecasts of Solar Geophysical Activity*, issued weekly by the Space Environment Services Center, NOAA, Boulder, Colorado.

This is the source of most up-to-date (airmail) information: reproductions of solar photographs, data on flares, plages, prominences, spots and spot areas, cosmic ray effects, bursts of radio noise on various wave lengths, geomagnetic storms and geomagnetic A and K indices.
(2) *Sunspot Bulletins*, issued monthly by the Swiss Federal Observatory (Eidgenössische Sternwarte) Zürich.

This gives provisional values of the Zürich relative sunspot number for each day of the past month, and the monthly mean, as well as forecast monthly mean values for 6 months ahead.

A final bulletin giving the definitive relative sunspot numbers for each day and each month of the past year is issued yearly.
(3) The International Astronomical Union's *Quarterly Bulletins on Solar Activity*, issued by the Eidgenössische Sternwarte, Zürich, with UNESCO support, give definitive data on sunspot numbers, flares, coronal activity and solar radio-wave emission.
(4) *Photoheliographic Results*, issued yearly since 1955, some years in arrears, by the Royal Greenwich Observatory, Herstmonceux, Sussex.

This publication also gives positions and sizes of sunspots and is the only source that gives areas of faculae.

The standard sources for data on solar activity in the past are:

BAUR, F. *Meteorologische Beziehungen zu solaren Vorgängen*: **I** (1964) – Neufestsetzung der Epochen der Minima und Maxima der Sonnenflecken; **II** (1967) – Meteorologischer Nachweis von Strahlungsschwankungen der Sonne, *Met. Abhandl.*, **50** (3 and 4). Berlin (Inst. f. Met. Geophys. der Freien Univ.).

WALDMEIER, M., 1961, *Sunspot activity in the years 1610–1960.* Zürich (Schulthess).

Selected data derived from these sources are printed here in Appendix I as well as the attempted extensions of the series back to much earlier times, referred to in the text.

Appendixes to Part I

APPENDIX I

Supplementary tables

1 Solar Variation

2 Earth Dimensions

3 Dimensions of Atmospheric Systems

APPENDIX I

Supplementary tables

1 Solar Variation

2 Earth Dimensions

3 Dimensions of Atmospheric Systems

1 Solar Variations

TABLE App. I.1

BAUR's Solar Index: monthly and yearly average values

Derived from measurements made on the Royal Greenwich Observatory's daily sun photographs (see text p. 22).
SI = $100\,(F/\bar{F} - D/\bar{D})$ F and $\bar{D}$ used in calculating these values of the index are the respective averages for 1900–1949

	Jan.	*Feb.*	*Mar.*	*Apr.*	*May*	*Jun.*	*Jul.*	*Aug.*	*Sep.*	*Oct.*	*Nov.*	*Dec.*	*Year*
1874				−67	−80	−61	−30	5	64	23	−10	−41	−22
1875	42	−6	−11	−16	1	−24	3	−5	22	3	−13	1	0
1876	−11	−18	−29	15	11	7	−8	−10	−1	−13	−7	−14	−6
1877	−40	−9	−12	−16	−29	−17	−6	−5	−17	−8	−14	2	−14
1878	−4	−10	−9	2	−5	−1	2	1	−3	3	−4	0	−2
1879	3	3	5	−1	4	−9	−1	−13	−6	−5	−11	−3	−3
1880	−1	7	12	10	−3	−19	48	−34	−58	−4	−1	11	−3
1881	21	79	56	34	11	12	3	23	34	−10	−14	42	23
1882	0	−13	12	−4	46	30	100	47	29	24	−1	36	25
1883	10	35	14	−68	18	−64	−22	22	−5	8	−2	21	−3
1884	28	51	−23	−16	4	−40	4	−28	−3	42	72	35	10
1885	44	−23	3	−23	−23	−56	−2	11	−5	21	−4	9	−4
1886	−5	−7	−56	−41	−7	−13	−31	−3	−3	4	23	−8	−12
1887	0	−3	14	0	−16	−10	−26	−22	4	5	3	−21	−6
1888	−3	6	8	7	2	5	8	7	0	3	−7	−5	3
1889	5	−14	−5	−4	0	−4	−6	−25	−1	4	4	−1	−4
1890	3	4	5	1	0	4	−2	−5	1	26	19	24	6
1891	17	5	28	14	−1	10	42	43	6	25	41	59	24
1892	50	54	121	97	98	54	62	12	90	138	88	51	76
1893	12	4	13	−17	15	17	9	−50	11	−4	−57	−64	−9
1894	−68	−48	−15	−44	−61	−77	−44	−26	−11	20	−4	4	−31
1895	22	9	21	18	40	30	59	28	44	8	31	−4	25
1896	37	−40	−4	1	9	−17	5	31	12	44	22	35	11
1897	39	66	27	33	26	22	5	33	−2	27	31	5	26
1898	16	8	−7	34	13	6	12	−13	−6	5	0	13	7
1899	−7	0	−4	−3	2	−10	7	11	−6	−14	6	−4	−2
1900	−7	−2	0	−11	−2	−16	−5	−7	−9	−20	−4	1	−7
1901	0	−3	−7	0	−17	−9	2	−2	−1	−5	−1	0	−4
1902	53	1	−13	9	7	11	2	−2	−4	−13	7	20	2
1903	−2	12	3	13	37	15	34	36	36	−12	7	20	17
1904	10	21	17	24	58	26	36	26	67	25	40	76	35
1905	40	9	49	81	56	63	40	105	79	22	−3	70	51
1906	100	98	47	73	49	55	−5	68	7	62	28	19	50

1907	7	−58	58	36	21	15	1	−10	−26	11	16	58	11
1908	86	102	66	20	32	27	61	−25	−17	131	32	30	46
1909	−21	40	2	30	19	24	18	32	−4	−33	−31	−31	4
1910	18	6	29	35	24	44	49	36	21	3	51	34	29
1911	33	22	29	14	25	20	19	24	17	13	12	15	20
1912	15	14	2	4	9	17	9	6	−3	9	7	−1	7
1913	11	3	7	4	4	4	2	1	2	−2	4	4	4
1914	7	2	7	−4	20	16	17	23	22	16	10	4	12
1915	17	−11	7	6	5	−56	−23	5	37	17	48	37	7
1916	21	15	6	28	−35	−40	7	26	12	−12	1	48	10
1917	−1	3	−42	−17	−81	−50	−29	−76	−57	−42	−113	−160	−56
1918	−78	−23	−63	−55	−24	−3	−99	−47	−32	−20	−16	24	−36
1919	24	−24	−6	−4	−65	−100	52	−39	5	1	5	10	−12
1920	−32	13	−38	45	7	8	19	16	−18	−4	46	−6	5
1921	14	9	2	0	13	−10	−21	−6	7	−5	0	2	0
1922	12	−21	−48	28	19	10	−2	2	−1	−3	4	−19	−2
1923	19	12	3	0	7	−8	11	14	−13	−8	−3	9	3
1924	13	−7	7	−13	−12	−16	1	17	10	15	15	33	5
1925	41	27	18	24	−10	19	39	41	40	34	75	15	30
1926	125	74	79	118	68	39	32	9	0	−33	14	24	46
1927	22	33	32	−1	4	30	6	−2	−6	−3	−16	31	11
1928	−39	−21	−12	13	26	5	16	37	27	75	64	35	19
1929	4	23	42	47	55	52	56	58	113	35	12	29	44
1930	99	114	85	51	38	73	23	5	20	−15	17	21	40
1931	27	−29	10	14	13	38	29	18	6	9	1	5	12
1932	19	−5	7	−4	−7	10	18	3	−1	6	1	9	5
1933	19	3	18	18	12	3	7	3	−3	−1	6	2	7
1934	−2	5	2	−3	5	17	19	12	17	6	−1	0	7
1935	22	13	−3	20	−6	−23	1	24	20	−6	−32	−10	2
1936	37	18	15	21	49	−11	41	−17	22	26	−52	−9	12
1937	−18	20	49	−29	−30	−33	−24	21	52	−28	95	9	7
1938	−35	−14	74	42	−16	3	−92	20	48	−8	−72	1	−4
1939	25	7	−1	−47	−64	−19	−38	−68	−79	−13	9	40	−21
1940	23	−17	−81	−15	−39	−72	−40	−97	−11	−16	−14	−36	−35
1941	−15	−10	−21	10	41	−25	−37	−17	−36	−13	−5	21	−9
1942	13	−38	−30	−45	4	31	4	4	3	5	−25	−8	−7
1943	4	−22	−13	0	41	32	10	3	11	2	6	−6	6
1944	10	12	−1	14	−1	7	12	8	−1	4	2	−18	4
1945	12	17	−1	−10	4	0	−14	3	−15	−63	0	26	−3
1946	−9	−42	0	−11	−35	−6	−72	−83	−44	−34	−86	−56	−40
1947	−33	−101	−103	−137	−197	−115	−91	−142	−137	−139	−112	−57	−114

TABLE App. I.1—*continued*

	Jan.	*Feb.*	*Mar.*	*Apr.*	*May*	*Jun.*	*Jul.*	*Aug.*	*Sep.*	*Oct.*	*Nov.*	*Dec.*	*Year*
1948	−48	−40	−51	−204	−154	−162	−105	−168	−136	−136	−77	−171	−121
1949	−94	−192	−154	−135	−51	−54	−65	−88	−107	−91	−107	−62	−100
1950	−52	−39	−66	−86	−65	−39	−61	−73	−13	−43	−55	−33	−52
1951	−55	−43	−32	−99	−127	−96	8	−20	−65	−8	−25	−19	−48
1952	−13	7	−1	−6	0	−20	−19	−61	−8	−13	−17	−39	−16
1953	−15	19	9	−35	−4	−20	2	−24	−11	4	9	−1	−6
1954	3	3	−14	1	1	1	−7	−4	7	12	1	0	0
1955	24	−7	12	7	−25	−19	4	−30	−9	−41	−99	−60	−24
1956	−31	−143	−93	−46	−123	−109	−119	−197	−204	−144	−250	−236	−142
1957	−150	−109	−194	−203	−201	−264	−245	−194	−356	−351	−263	−310	−237
1958	−255	−162	−248	−244	−200	−188	−255	−265	−256	−207	−151	−253	−224
1959	−255	−96	−199	−167	−170	−146	−106	−211	−105	−62	−73	−18	−134
1960	−72	−6	−35	−82	−81	−54	−77	−96	−94	−20	−18	−29	−55
1961	14	41	17	−9	19	−37	−21	11	−24	8	16	12	4
1962	10	−6	12	−7	15	30	68	25	−40	−21	18	2	8
1963	10	10	23	−5	−28	6	34	−3	−24	−11	24	35	6
1964	18	12	24	39	23	17	22	5	7	12	7	−7	15

TABLE App. I. 2

Solar Faculae: yearly average values in millionths of the area of the visible surface of the sun, corrected for foreshortening

As derived from measurements of the daily sun photographs of the Royal Greenwich Observatory. (See *Sunspot and Geomagnetic Storm Data . . . 1874–1954*, published for the Royal Greenwich Observatory by H.M.S.O., London 1955, and the *Monthly Notices of the Royal Astronomical Society*, London, for subsequent years.)

Main maxima in **bold print**			Main minima in *italics*		
1874	1011	1905	**2612**	1935	1100
		1906	2320	1936	2545
1875	505	1907	1999	1937	**3505**
1876	257	1908	2098	1938	3205
1877	162	1909	1355	1939	2349
1878	*67*				
1879	136	1910	971	1940	1523
		1911	459	1941	1288
1880	923	1912	210	1942	809
1881	1951	1913	*95*	1943	568
1882	**2154**	1914	454	1944	*344*
1883	1864				
1884	2034	1915	1521	1945	940
		1916	1785	1946	2188
1885	1487	1917	**2305**	1947	**2894**
1886	579	1918	1882	1948	2331
1887	304	1919	1729	1949	2597
1888	239				
1889	*131*	1920	1219	1950	1754
		1921	739	1951	1379
1890	304	1922	415	1952	711
1891	1415	1923	*222*	1953	331
1892	**3267**	1924	575	1954	*137*
1893	2404				
1894	1877	1925	1750	1955	793
		1926	2556	1956	2195
1895	2278	1927	2212	1957	2270
1896	1410	1928	**2589**	1958	2225
1897	1149	1929	2567	1959	**2697**
1898	891				
1899	337	1930	1630	1960	2494
		1931	801	1961	1617
1900	180	1932	400	1962	1210
1901	*30*	1933	*267*	1963	907
1902	178	1934	354	1964	502
1903	970				
1904	1768				

Mean of monthly values 1900–1949: 1431

TABLE App. I. 3

Sunspot extremes since . . 1610 (Zürich data)

Year of minimum	Lowest (smoothed) monthly average Relative Sunspot Number	Year of maximum	Highest (smoothed) monthly average Relative Sunspot Number	Duration of Rise (years)	Duration of Fall (years)	Conventional number of cycle
1610·8		1615·5		4·7	3·5	
1619·0		1626·0		7·0	8·0	
1634·0		1639·5		5·5	5·5	
1645·0		1649·0		4·0	6·0	
1655·0		1660·0		5·0	6·0	
1666·0		1675·0		9·0	4·5	
1679·5		1685·0		5·5	4·5	
1689·5		1693·0		3·5	5·0	
1698·0		1705·5		7·5	6·5	
1712·0		1718·2		6·2	5·3	
1723·5		1727·5		4·0	6·5	
1734·0		1738·7		4·7	6·3	
1745·0		1750·3	92·6	5·3	4·9	
1755·2	8·4	1761·5	86·5	6·3	5·0	1
1766·5	11·2	1769·7	115·8	3·2	5·8	2
1775·5	7·2	1778·4	158·5	2·9	6·3	3
1784·7	9·5	1788·1	141·2	3·4	10·2	4
1798·3	3·2	1805·2	49·2	6·9	5·4	5
1810·6	0·0	1816·4	48·7	5·8	6·9	6
1823·3	0·1	1829·9	71·7	6·6	4·0	7
1833·9	7·3	1837·2	146·9	3·3	6·3	8
1843·5	10·5	1848·1	131·6	4·6	7·9	9
1856·0	3·2	1860·1	97·9	4·1	7·1	10
1867·2	5·2	1870·6	140·5	3·4	8·3	11
1878·9	2·2	1883·9	74·6	5·0	5·7	12
1889·6	5·0	1894·1	87·9	4·5	7·6	13
1901·7	2·6	1907·0	64·2	5·3	6·6	14
1913·6	1·5	1917·6	105·4	4·0	6·0	15
1923·6	5·6	1928·4	78·1	4·8	5·4	16
1933·8	3·4	1937·4	119·2	3·6	6·8	17
1944·2	7·7	1947·5	151·8	3·3	6·8	18
1954·3	3·4	1957·9	201·3	3·6	6·6	19
1964·5	9·6	1968·9	112·2	4·3		20

NOTE: The Smoothed value of Relative Sunspot Number W for month i is taken as:

$$W_i = \tfrac{1}{2}\left(\frac{W_{i-6} + W_{i-5} + \ldots + W_{i+5}}{12} + \frac{W_{i-5} + W_{i-4} + \ldots + W_{i+6}}{12}\right)$$

Dates of sunspot minimum and maximum are often expressed, as in Table App. I.3, in years and decimals. Thus, 1745·0 means the beginning of 1745.

BAUR (1964) has published tables of successive monthly values from 1749 to 1962 of the 9- and 17-month average sunspot numbers derived from the Zürich data. These are very nearly averages over 10 and 19 rotations of the sun respectively; they correspond more nearly to whole rotations than the conventional smoothings (e.g. 5- and 13-month averages). The dates of the various sunspot maxima and minima since 1749 appear slightly shifted by this calculation, but differ on average only by 0·1 year from those given here in Table App. I.3. BAUR gives (*loc. cit.*) a table of the dates derived of sunspot maxima and minima and of the divisions of each cycle since 1749 into septiles of the rising and falling phases, for convenience in relation to significant associations noted with the behaviour of the large-scale circulation of the atmosphere.

Sunspot activity does not necessarily fall to nothing at the minima, because the activity of the new cycle generally begins (in the higher solar latitudes) before the activity of the old cycle has died out near the sun's equator. The overlap is sometimes more than 2 years.

Regarding each cycle as a burst of activity, there may be more sense in considering the cycle spacing, or interval between one maximum and the next, than the cycle length (which is conventionally measured from minimum to minimum).

Vigorous cycles produce early maxima, weak cycles reach their maximum later in the cycle. This can be seen from Table App. I.3 by the ratios of the length of the rising to that of the declining phase of each cycle, the ratio being small in the most active cycles (brisk rise, long-drawn-out fall) and sometimes exceeding unity in the weak cycles.

The cycle intervals tend to be shorter when the maxima are high, averaging more nearly 10 than 11 years in some centuries.

GNEVYSHEV (1967) has revealed that, at least for the 8 solar cycles from 1874 to 1962, there are in reality two waves of activity per cycle observable on the sun. These waves of activity overlap in time, but culminate in separate maxima and at different solar latitudes. The two maxima appear to affect all forms of solar disturbance. The first maximum is the well-known one (doubtless because the widest range of solar latitudes is then affected), the corona being bright all round the sun with greatest intensity at solar latitudes around 25°. The second wave of activity is concentrated in the lower solar latitudes with greatest intensity about 13–15° solar latitude. The maximum of the second or low-latitude wave occurs 2–2½ years after the first maximum. The suggestion that the zone of maximum disturbance drifts towards the solar equator is held to be an illusion due to the partial superposition of these two waves of activity around different, but in each case constant, latitudes on the sun.

The approximate dates of sunspot maxima (and minima) can be discerned over many centuries before the period covered in Table App. I.3 by studying the frequency of polar aurorae. The results for the period 700 B.C. to A.D. 1610, and the degree of consistency among six different investigators, may be seen in fig. App. I.1 below (reproduced from LINK 1964 by kind permission of the author and publishers).

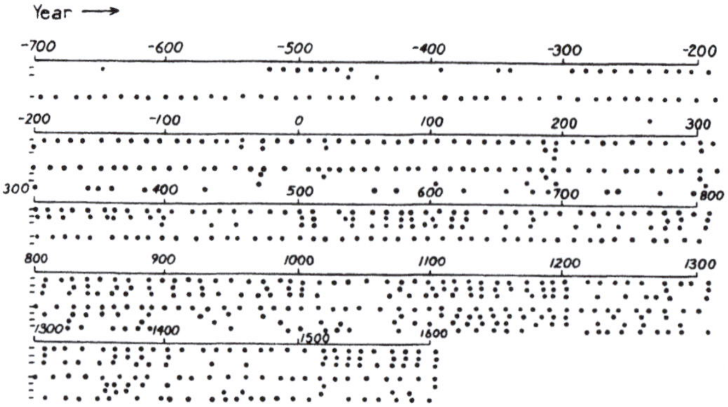

Fig. App. I.1 Dates of maxima of the '11-year' sunspot cycles since 700 B.C. according to different estimates.
The dots indicate the dates as estimated in the following sources – in the order of the rows, reading from the top:
SCHOVE, D. J. (1955) *J. Geophys. Res.*, **60**, 127.
NICOLINI, T. (1942) *Contr. Astr. Obs. Napoli*, **2**, No. 7.
FRITZ, H. (1893) *Vjsht. Naturf. Ges. Zürich*, **38**, 89.
SCHOSTAKOWITSCH, W. B. (1928 and 1931) *Beitr. Geophys.*, **29** Nr. 2 and **30**, 281.
KANDA, S. (1933) *Proc. Imp. Acad. Japan*, **9**, 293 (5th row from sunspot reports, 6th row from aurorae).
LINK, F. (1964) *Planetary and Space Sci.*, **12**, 339.

SCHOVE (1955) has thought it possible also to estimate the relative magnitudes of the sunspot maxima in most cycles since 649 B.C. His estimates, which have been widely quoted and apparently found to have some general reliability, are listed here in Table App. I.4. His assumption, however, that there are always just 9 maxima in a century risks forcing the evidence and seems to account for some differences from other workers' findings (cf. SCHOSTAKOWITSCH in fig. App. I.1 between A.D. 0 and 100 and in the fourth and twelfth centuries. In the twelfth century SCHOVE along with KANDA and probably most investigators, finds evidence of $9\frac{1}{2}$—10 cycles).

The foregoing table may be taken as reliable in its recording of various extreme cases for which there is evidence and in its indication of some longer runs of strong or weak cycles.

It has often been suggested that sun-spottedness is subject to an irregular 80- to 90-year cycle (e.g. GLEISSBERG 1958, VITINSKII 1965), and to a 170- to 200-year and various longer-term quasi-periodicities, superimposed on the sequence of approximately 11-year cycles and affecting their intensity and length. This is not very clearly seen in Tables App. I.3 and App. I.4 but the maxima which stand out are those of A.D. 1372, possibly also 1572, 1778 and 1957 (see fig App. I.2). There seems to have been more generally higher sunspot activity for a long time before A.D. 1200.

BRAY (1965) has produced a different way of obtaining an indication of the vigour of

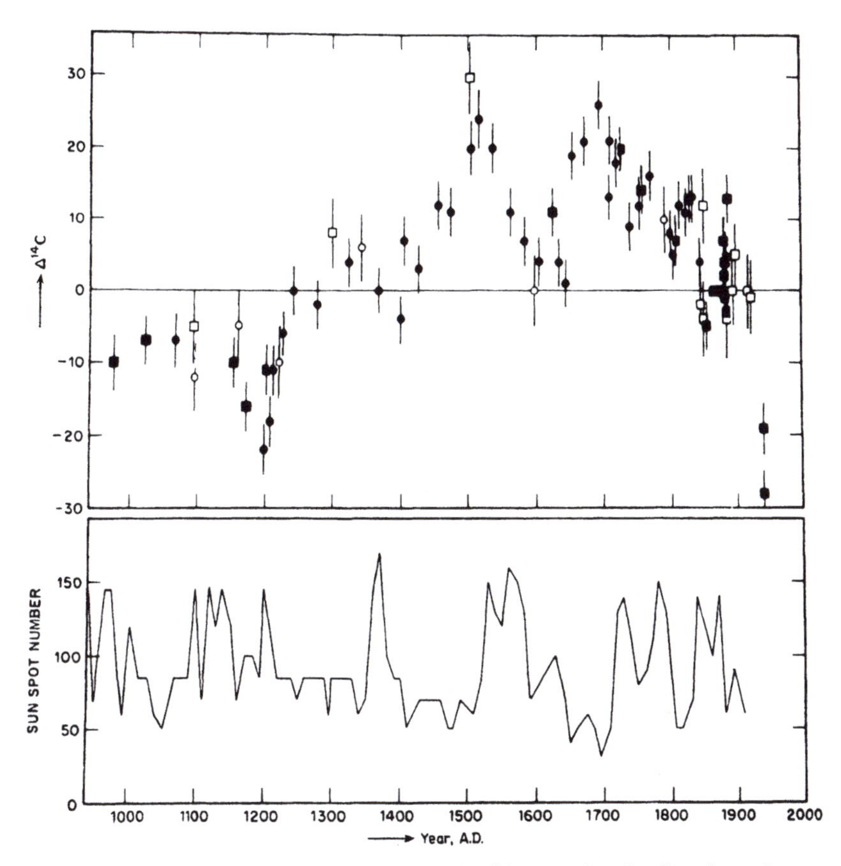

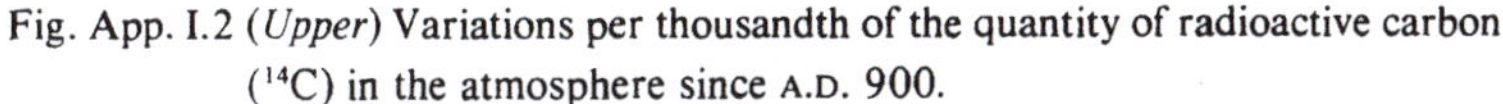

Fig. App. I.2 (*Upper*) Variations per thousandth of the quantity of radioactive carbon (^{14}C) in the atmosphere since A.D. 900.
(*After* H. E. SUESS *1965, collected results from various laboratories.*)
(*Lower*) Average sunspot numbers as estimated by SCHOVE (1955). Despite the uncertainty of these estimates before 1750, they suggest a negative correlation with ^{14}C whereby low values of ^{14}C occur during, or a little after, periods of especially active solar disturbance.
The rough 170- to 200-year quasi-periodicity alleged to affect solar activity may be traced in the high maxima in the lower figure and in the dips of the ^{14}C concentration (upper diagram).

sunspot activity over various periods in the past before the start of observation by telescope in 1610. He noticed the tendency for 3 or more cycles of highly active solar disturbance to be followed by runs of 3 or more low-activity cycles, and vice versa, and that the cycle length and cycle interval both tended to be shorter when the disturbance was great. From the Zürich data for the cycles between the 1698 and 1964 minima (i.e. over 24 cycles), he found a correlation coefficient of $-0{\cdot}64$ between the cycle interval (the time elapsed from the preceding maximum to the maximum of a given sunspot cycle) and the mean yearly

TABLE App. I.4

Sunspot maxima since 649 B.C. Dates and estimates of strength given by SCHOVE (1955)

Pure interpolations in brackets ( )

SCHOVE'S *suggested equivalents*

	Annual mean sunspot number in maximum year
SSS = extremely strong	probably >160
SS = very strong	probably 140–150
S = strong	probably 110–130
MS = moderate to strong	about 100
M = moderate	about 80–90
MW = moderate to weak	about 70
W = weak	about 60
WW = very weak	about 50
WWW = extremely weak	probably <45
X = unknown	

Year of max.	–648....		....–522	(–512)	–501	(–491)	–481	
Character	S?		S?	X	S?	MW	S	
–471	–461....			....–393....		....–349	–340....	
S	S			S		S	S	
(–293)	(–283)	–272	(–261)	(–249)	–236	–223	–214	–205
X	X	X	X	X	X	X	S	S
–192	–182	–172	–163	(–149)	–135	–125	–113	–104
S	MW	MW	S	W	M	MS	S	S
–91	(–82)	(–72)	–62	–53	–42	–27	–16	(–5)
SS	M	MW	SS	SS	MS	S	MW	MW
A.D. (8)	20	(31)	42	53	65	(76)	(86)	(96)
M–W	S	M–W	M–W	S	S	M–W	M–W	M–W
105	(118)	(130)	(141)	(152)	(163)	175	186	196
M–S	M–W	M–W	M–W	M–W	M–W	M–S	S	S
(208)	(219)	(230)	(240)	(252)	(265)	(277)	290	302
X	X	X	X	X	X	X	M–S	SS
311	321	330	342	354	362	372	387	396
M	M	W	W	S	S	SS	M	M
410	(421)	430	441	452	465	479	490	501
W	W	M–S	M–S	S	W	M–S	M	SS
511	522	531	542	557	567	578	585	597
S	M–W	S	M	M	SS	M	S	M–W
(607)	618	628	642	654	665	677	(689)	(699)
W	M	M–W	M	M–S	M	S	W	W

TABLE App. I.4 – *continued*

Schove's *suggested equivalents*								
714 S	724 S	735 W	745 SS	754 M	765 SS	(776) MS	(787) W	(798) MS
809 S	821 W	829 S	840 SS	850 MS	862 M	872 S	887 M	898 W
907 W	917 M	926 SS	938 MS	(950) MW	963 SS	974 SS	986 M	(994) W
1003 S	1016 M	1027 M	1038 W	(1052) WW	1067 M	1078 M	1088 M	1098 SS
(1110) MW	1118 SS	1129 S	1138 SS	1151 S	1160 MW	1173 MS	1185 MS	1193 M
1202 SS	1219 M	1228 M	1239 M	1249 MW	1259 M	1276 M	1288 M	1296 W
1308 M	1316 M	1324 M	1337 W	1353 MW	1362 SS	1372 SSS	1382 MS	1391 M
1402 M	(1413) WW	(1429) MW	(1439) MW	1449 MW	1461 MW	(1472) WW	(1480) WW	1497 MW
1505 W	1519 M	1528 SS	1539 S	1548 S	1558 SS	1572 SS	1581 S	1591 MW
1604·5 MW	1615·5 M	1626·0 MS	1639·5 MW	1649·0 WW	1660·0 WW	1675·0 W	1685·0 WW	1693·0 WWW
1705·5 WW	1718·2 S	1727·5 SS	1738·7 S					

sunspot number within the cycle: this is statistically significant, having a less than 1 in 1000 probability of occurring by chance. Bray quotes Kanda's estimates of the dates of sunspot maxima and mean cycle intervals for earlier centuries as given in Table App. I.5: from these figures some indication of the activity of the cycles can be derived (cf. Table App. I.6).

Bray further notes the appearance of a correlation with climate whereby the periods of high solar disturbance and rapid cycles in the ninth to thirteenth, eighteenth, and twentieth centuries coincide with evidence of warmth from forest growth near the thermal limit of forest in northwestern North America, with widespread glacier recession (the eighteenth century hardly fits as regards this item), or with observed high temperatures prevailing.

TABLE App. I.5

Sunspot maxima and cycle intervals in those early centuries for which KANDA's summaries of observations* in the Far East are adequate

A.D. 302·0 (or from aurorae 306·0)	311·3 (or from aurorae 313·9)	322·1	(missed)	342·2	354·2	359·9	372·6	388·9	398·4
			9 cycles, average interval 10·7 years						
807·9	(missed)	830·0 (or from aurorae 827·3)	840·0	851·9	865·1	875·6	(from aurorae 882·6)	888·0	
			8 cycles, average interval 10·0 years						
(from aurorae 1100·4)	1108·7	1120·5 (or from aurorae 1117·1)	1130·2 (or from aurorae 1129·2)	1137·7 (or from aurorae 1140·0)	1148·4 (or from aurorae 1150·7)	1160·5 (or from aurorae 1158·7)	1171·9	1185·3 (or from aurorae 1181·2)	1193·9 (or from aurorae 1193·0)
			9 cycles, average interval 10·3 years						
1202·8 (or from aurorae 1204·3)	(missed)	(from aurorae 1227·0)	1239·0	(from aurorae 1247·6)	1257·8 (or from aurorae 1261·1)	(missed)	1277·4	(from aurorae 1288·9)	
			8 cycles, average interval 10·8 years						

* According to BRAY (*loc. cit.* 1965) the dates in this table are derived from actual sunspot reports unless otherwise stated.

TABLE App. I.6

Mean yearly sunspot number per cycle, average cycle length and cycle interval (maximum to maximum) from Zürich data

(Adapted from BRAY 1965)

Period	*Number of cycles*	*Average cycle length (minimum to minimum)*	*Average cycle interval*	*Mean yearly sunspot number*
1914–1964	5	10·2	10·2	62
1755–1798	4	10·8	9·5	57
1834–1879	4	11·3	10·2	56
1723–1755	3	10·6	10·7	48
1879–1914	3	11·6	12·1	35
1698–1723	2	12·7	12·6	22
1798–1823	2	12·5	13·9	26
1611–1655	4	11·3	11·2	
1655–1698	4	10·7	11·0	

2 Earth Dimensions

TABLE App. I.7(*a*)

Areas of different 5° latitude zones, proportions of ocean and overall totals

Latitude	*Proportion* of the total area of the hemisphere that lies poleward, %*	*Total area of the zone between two latitudes 10^6 km²*	*Northern hemisphere Proportion of ocean 5° zones, %*	*Northern hemisphere Proportion of ocean 10° zones, %*	*Southern hemisphere Proportion of ocean 5° zones, %*	*Southern hemisphere Proportion of ocean 10° zones, %*	*Latitude*
90°	0						90°
		0·98	100·0		0		
85°	0·4			90		0	85°
		2·93	86·9		0		
80°	1·5						80°
		4·85	77·1		10·7		
75°	3·4			70		27	75°
		6·74	65·5		38·6		
70°	6·1						70°
		8·57	28·7		79·5		
65°	9·5			30		91	65°
		10·33	31·2		99·7		
60°	13·5						60°
		12·01	45·0		99·9		
55°	18·2			43		99	55°
		13·59	40·7		98·5		
50°	23·5						50°
		15·07	43·8		97·5		
45°	29·4			48		97	45°
		16·43	51·2		96·4		
40°	35·9						40°
		17·66	56·8		93·4		
35°	42·9			57		89	35°
		18·75	57·7		84·2		
30°	50·1						30°
		19·70	59·6		78·4		
25°	57·9			63		77	25°
		20·50	65·2		75·4		
20°	65·9						20°
		21·15	70·8		76·4		
15°	74·4			74		78	15°
		21·63	76·5		79·6		
10°	82·7						10°
		21·96	75·7		76·9		
5°	91·3			77		76	5°
		22·12	78·6		75·9		
0°	100						0°
Totals		254·97	60·7		80·9		

* The proportion of the total area of a hemisphere that lies poleward of any latitude ϕ on a truly spherical Earth would be $100\,(1-\sin\phi)\%$. Departures from this spherical form amounting to more than 0·1% by area only occur within 13° latitude of the poles.

60·1% lies poleward of the present position of the tropics.

8·3% lies poleward of the present position of the polar circles.

Total area of the Earth's surface 510×10^6 km². Total area of ocean (present epoch) 361×10^6 km². Total area of land (present epoch) 149×10^6 km².

(Figures mainly taken from Smithsonian Meteorological Tables (6th revised edition) in *Smithsonian Miscellaneous Collections*, **114**. Washington (1958).)

TABLE App. I.7(*b*)

Other dimensions

Mean radius	6371 km	(3960 English statute miles)
Equatorial radius	6378 km	
Polar radius	6357 km	
Mass	5975×10^{24} kg	

TABLE App. I.8

Sizes of the main landmasses

	10^6 km^2
Asia	44·68
Africa	29·84
North America	20·02
Central and South America	18·46
Europe	9·71
Australia	7·63
Greenland	2·18
Antarctica, continent with shelf ice*	14·1
Antarctic ice surface†:	
Summer	*c.* 14·2
Winter	*c.* 45

* Ross Ice Shelf 0·54, Filchner Ice Shelf 0·33, Amery (Lambert Glacier) Ice Shelf 0·05 million km^2.
† Present total extent of ice-free (mountain and oasis) areas in Antarctica estimated 0·60 million km^2.

TABLE App. I.9

Sizes of water bodies (present epoch)

(*a*) Oceans and seas

	Area, 10^6 km^2	*Mean depth, m*
Pacific Ocean	165·25	4282
Atlantic Ocean	82·44	3926
Indian Ocean	73·44	3963
Oceans, excluding adjoining seas	321 approx	4120 approx.
Oceans, including adjoining seas	361	3850
Arctic Ocean including Norwegian Sea and Davis Strait–Baffin Bay	14·09	1205
Larger seas adjacent to Asia (includes South China Sea between China and East Indies 2·50 million km^2)	8·14	1212
Mediterranean and Black Sea	2·97	1429
Baltic Sea	0·42	55
North Sea	0·57	94
Irish Sea	0·10	60
English Channel	0·07	54
Caribbean Sea	2·30	2216 (Caribbean Sea and Gulf of Mexico)
Gulf of Mexico	1·86	
Bering Sea	2·27	1437
Hudson's Bay	1·23	128
Gulf of St Lawrence	0·24	128
Okhotsk Sea	1·53	838
East China Sea	1·25	188
Japan Sea	1·01	1350
Andaman Sea	0·80	870
Persian Gulf	0·24	25
Red Sea	0·44	491
Weddell Sea*	2·5–3·0 (measurement of natural basin out as far as the South Orkney Is. and 70°S near 4°W, but 2·0 × 10^6 km^2 to a limit from Joinville Is., Grahamland to Cape Norvegia)	
Ross Sea*	1·0	

Figures largely derived from Kossinna (1921).
* Excluding the 'permanent' ice shelf.

TABLE App. I.9 – *continued*

(*b*) Inland seas and lakes

	Area, 10^6 km^2	Mean depth, m
Caspian Sea (decreased $0{\cdot}03 \times 10^6$ km^2 between 1930 and 1960)	0·44	
Black Sea	0·43	
Aral Sea	0·063	16
L. Baikal	0·030	
L. Balkash	0·018	
L. Urmia	0·0045	
Dead Sea	0·0008	
L. Victoria	0·068	80 (max. depth)
L. Nyasa	0·037	750 (max. depth)
L. Tanganyika	0·033	600 (max. depth)
L. Chad	0·007–0·020 Seasonal and longer term variations	av. 3 (ranges from about 1–8 m)
L. Ladoga	0·018	90
L. Onega	0·0098	60
L. Vänern	0·0056	33
L. Vättern	0·0019	39
L. Mälaren	0·0017	
L. Geneva	0·0006	154
L. Superior	0·081	145
L. Huron	0·062	76
L. Michigan	0·058	99
L. Erie	0·026	21
L. Ontario	0·019	91
Great Salt Lake, Utah	0·0051	
L. Titicaca, Bolivia–Peru	0·0083	106
L. Taupo, New Zealand	0·0006	112

(*c*) The Earth's total water inventory

(Estimates by Professor H. HOINKES, Innsbrück)

Total available water (i.e. not chemically fixed in the rocks) of which	1360×10^6 km^3
c. 2·5% is Fresh water	35×10^6 km^3
most of this being Ice (chiefly on Antarctica)	26×10^6 km^3
Only *c*. 0·02% is in the Lakes, Rivers, etc.	$0{\cdot}3 \times 10^6$ km^3
and *c*. 0·6% is in the Soil, Subsoil, porous strata of the Rocks etc.	8×10^6 km^3
and a minute proportion *c*. 10^{-5} is in the Atmosphere	*c*. 13×10^3 km^3
most of this being changed 30–40 times a year.	

Notice that the amount 'locked up' as ice in Antarctica today and additionally in other ice sheets at the maximum development of ice ages is enough to have a measurable effect on the saltiness of the sea and its content of the oxygen isotope 0^{18}, etc.

TABLE App. I.10

(*a*) Estimated total extent of ice and snow surfaces, various epochs

	Winter maximum			*Summer minimum*		
	% of total area of the hemisphere	Difference from now	*Equivalent latitude of the limit*	*% of total area of the hemisphere*	Difference from now	*Equivalent latitude of the limit*
A.D. *1900–1940 approx.*						
Northern hemisphere	25	—	48·8	5·5	—	71·2
Southern hemisphere	13–14	—	60	7·7	—	68
About A.D. *1800*						
Northern hemisphere	26–27	+1–2	47·3	6·1	+0·6	70·2
Southern hemisphere	13	0 to −1	60–61	7·7		68
Maximum phase of Pleistocene ice ages						
Northern hemisphere	34–35	+9–10	41·0	16–17	+10–12	(57·0
Southern hemisphere	24	+10–11	50	(21)	+13–14	(53)

NOTE: The equivalent latitude of the ice limit means the position it would occupy if the entire ice-covered area were circular and centred on the pole.

(*b*) Estimated quantity of ice on land at the present time and about the maxima of the Quaternary ice ages

(i) Present epoch

	Area, 10^6 km²	*Volume, 10^6 km³*	*Mean thickness m*
Antarctica	13·0	30·0	2300
Greenland (main ice sheet)	1·73	2·6	1500
Other glaciers	0·45	0·24	
Whole Earth	15·2	32·8	

Sources of the above estimates:

BAUER, A., 1955, Über die in der heutigen Vergletscherung der Erde als Eis gebundene Wassermasse, *Eiszeit und Gegenwart*, **6**, 60–70. Öhringen/Württ.

HOLLIN, J. T., 1962, On the glacial history of Antarctica, *J. Glaciol.*, **4**, 173–95. Cambridge.

Compare also detailed list of estimates in:

FLINT, R. F., 1957, *Glacial and Pleistocene Geology*, 51. New Work (Wiley),

but note that estimates of the total quantity of ice on Antarctica have been increased by surveys made in and since the International Geophysical Year. The volume of the Antarctic ice sheet was estimated by BAUER (1955) at 18 900 000 km³ and FLINT (1957) gave the world total of ice on land as about 24 000 000 km³. The latter estimate indicates that world sea level would rise by about 60 m if all the ice were melted, whereas the estimate of 32 000 000 km³ of ice now adopted would produce a sea-level rise of about 82 m if it all melted.

(ii) Glacial Maxima estimates for the last, i.e. Würm, ice age in *italics*

	Area, $10^6 \times km^2$	*Volume, $10^6 \times km^3$*	*Mean thickness, m*
Antarctica	13·2	25–34	2300 ± 300 approx.
Greenland	2·2	(3–3·3)	(1350–1500)
Laurentide ice sheet, North America	13·1/*12·5*	(16–20)	(1200–1500)
Cordilleran ice sheet, North America	2·5/*2·2*	(2–4)	(800–1500)
Scandinavian ice	5·5/*4·25*	(10–12)	(*c.* 2000)
British Isles ice	0·46/*0·37*	(0·3–0·5)	(700–1000)
Northwest Siberian and Novaya Zemlyan ice	4·2/*2·17*	(4–5)	(1000–1200)
Northeast Siberian ice	1·14/*0·93*		
Central Asian ice, including Himalaya	1·14/*0·87*		
Southern South America (main ice sheet)	0·75/*0·68*		
South America, other glaciers	0·19/*0·15*		
Other glaciers elsewhere	0·6/*0·5*		
Whole Earth	45/*40*	70–73	

The area estimates in the above list are taken from FLINT (*loc. cit.* p. 53).

The change of total volume of ice on land is derived from the general estimate of the lowering of world sea level at the greatest of the Quaternary glaciations as about 100 m below the present level. FLINT (*loc. cit.* p. 270) estimates this lowest Quaternary sea level as 120 m below the highest stand in the last warm interglacial period.

The ice volume and mean thickness estimates in brackets in the above columns are related to various factors, including:

(i) estimates of the heights of the ice summits above sea level derived from the evidence of past loading of the Earth's crust. In particular, in a map printed in WOLDSTEDT (1958, p. 121), ice plateau summits above 3000 m are indicated over the northern Baltic, Spitsbergen, Novaya Zemlya and northwest Siberia;

(ii) assuming the existence of some relationship between the total mass of ice and the time taken to melt it when warmer climates returned;

(iii) the limitations imposed by the known increases of ice-covered area in the various regions and the total extra volume of water withdrawn from the oceans.

It has been suggested by FLOHN (1963), supported by WOLDSTEDT (1965, pp. 274–5), that the evidence of a formerly 300-m greater thickness than now of the ice sheet covering Antarctica relates not to the times of maximum glaciation in other parts of the world but to the warm interglacial periods. The thickness of the Antarctic ice sheet at the times of glacial maxima elsewhere may have been less than now, possibly by more than 300 m, even if it had a slightly greater spread owing to the lowering of sea level. This suggestion that the history of the great Antarctic ice sheet runs in opposite phase to the variations of the glaciers elsewhere is in agreement with the increased accumulation observed on the ice cap near the South Pole as world climates warmed up from A.D. 1760 to about 1940 and the reversal of both trends since (GIOVINETTO and SCHWERDTFEGER 1966, LAMB 1967), and there is evidence of similar trends in the Arctic north of 70°N, in Greenland and Jan Mayen (LAMB *et al.* 1962). Unfortunately, the estimates of the volumes of the former great ice sheets listed in the above table are not close nor reliable enough to derive the ice age volume of the Antarctic ice sheet as that needed to make up the known world total volume of ice on land.

The following estimated area changes of certain smaller ice masses are from FLINT (*loc. cit.* pp. 51, 53).

(iii) Areas (km^2) of other glaciers now and at glacial maxima*

	Present epoch	*Quaternary max.*	*Würm max.*
Alps	3600	38 850	33 670
Caucasus	1970	12 950	11 000
Iceland and Jan Mayen	12 600	142 450	116 550
Spitsbergen	58 000	200 000	155 000
Australasia, mainly New Zealand	1000	66 500	58 000
Sub-Antarctic Islands	3000	10 000	10 000

* For additional sources referred to in connexion with these estimates, see References to Appendix I.

TABLE App. I.11

Earth's rotation speeds at different latitudes and the effect on air transferred from one latitude to another

Latitude	*Circumference (length) of the parallel (360° long.)*	*Rotational speed of a point on the surface, eastwards about the Earth's axis*	*Change of speed for 10° change of latitude = west wind speed acquired by previously still air on being transferred 10° poleward*	*Cumulative change of speed* *Equivalent west wind speed acquired by still air from the equator*
0°	40 000 km (21 600 naut. miles)	465 m/sec (900 knots)		
			7 m/sec (14 knots)	
10°	39 400 km (21 250 naut. miles)	458 m/sec (886 knots)		7 m/sec (14 knots)
			21 m/sec (41 knots)	
20°	37 600 km (20 300 naut. miles)	437 m/sec (845 knots)		28 m/sec (55 knots)
			34 m/sec (66 knots)	
30°	34 600 km (18 700 naut. miles)	403 m/sec (779 knots)		62 m/sec (121 knots)
			46 m/sec (89 knots)	
40°	30 600 km (16 550 naut. miles)	357 m/sec (690 knots)		108 m/sec (210 knots)
			57 m/sec (111 knots)	
50°	25 700 km (13 880 naut. miles)	300 m/sec (579 knots)		165 m/sec (321 knots)
			67 m/sec (129 knots)	
60°	20 000 km (10 800 naut. miles)	232 m/sec (450 knots)		232 m/sec (450 knots)
			74 m/sec (143 knots)	
70°	13 680 km (7400 naut. miles)	158 m/sec (307 knots)		306 m/sec (593 knots)
			77 m/sec (151 knots)	
80°	6950 km (3750 naut. miles)	81 m/sec (156 knots)		384 m/sec (744 knots)
			81 m/sec (156 knots)	
90°	0	0		465 m/sec (900 knots)

3 Dimensions of Atmospheric Systems

TABLE App. I.12
Wind circulation features: normal sizes and typical values of atmospheric pressure and wind velocity (at full development)

System	*Cross measurement*	*Vertical extent*	*Atmospheric pressure at m.s.l. in central region*	*Winds* *Horizontal component*	*Winds* *Vertical component*
Waves in the upper westerlies:					
(*a*) Stationary waves	5000–10 000 km west to east (one complete wave length through ridge and trough)	*c*. 10–15 km (from a base around 2 km up to 15–20 km a.s.l.)	—	Generally 20–30 m/sec in the flow near max. wind level but more in jet stream cores	
	1000–4000 km (*c*. 10–40° lat.) north to south (full amplitude range, between ridge and trough)				
(*b*) Mobile waves (such as the warm ridge accompanying the warm sector of a single frontal cyclone and the rear-side cold trough)	1000–2500 km west to east (wave length through ridge and trough)	As for (*a*)	See under Cyclones	See under Jet Stream	
	500–2000 km (*c*. 5–20° lat.) north to south (ridge-trough full amplitude range)				
Jet Stream	500–5000 km long	1–5 km (centred at heights about 10–15 km a.s.l.)		Commonly 30–70 m/sec Extremes >100 m/sec in many parts of the world outside the tropics, up to 150 m/sec over Japan and southern Indian Ocean near 35–40°S, possibly also at same latitude near Atlantic seabord of North America	
	50–500 km broad				

TABLE App. I.12 – *continued*

System	*Cross measurement*	*Vertical extent*	*Atmospheric pressure at m.s.l. in central region*	*Winds* *Horizontal component*	*Winds* *Vertical component*
Anticyclone:					
(*a*) In the subtropical belt (or extension thereof)	3000–4000 km long 750–1500 km broad	Throughout troposphere (12–20 km): leans towards equator (or towards the warmest side) with increasing height	5–20 mb above the average for the latitude or region	Mostly <10 m/sec at sea level, but stronger near the periphery	Subsidence rates of order of 1 cm/sec
(*b*) Blocking anticyclone (warm cut-off anticyclone) in higher latitudes	1500–3000 km (occas. 4000 km) major axis (in various orientations) 750–1500 km minor axis	As for (*a*)	20–50 mb above the average for the latitude or region Extreme values 1050 to 1080 mb approx. over cold surfaces in the northern hemisphere	Mostly 0–10 m/sec but stronger near periphery	As for (*a*)
(*c*) Cold, polar anticyclones and ridges	1000–3000 km (roughly circular). Occas. linked with neighbouring warm blocking anticyclones into a very large system	Commonly 1–5 km	10–30 mb above the average for the latitude or region	Mostly 0–10 m/sec (occas., locally and at periphery, up to 20 m/sec)	As for (*a*)
Wave on the polar front (incipient frontal cyclone)	200–500 km long 50–200 km broad	Commonly 3–6 km	2–5 mb below surrounding region	Commonly 5–10 m/sec at sea level	
Polar front cyclone	1000–2500 km, occas. 3000 km over oceans in seasons of maximum vigour of circulation	From m.s.l. up to 5–15 km as a complete whirl. Commonly throughout the troposphere, up to 15 km at occluded (cold centre) stage, as a stationary cyclone in high latitudes	Commonly 20–50 mb below the average for the latitude over the ocean in winter, 10–30 mb below the average for the region over land in summer. Extreme values about 925 mb North Atlantic, and 910 mb Antarctic Ocean	10–25 m/sec at surface in strongest windstreams, occas. up to about 50 m/sec in gusts	Rates of order of 5 cm/sec in upgliding near frontal surface, but much greater in cumulonimbus clouds *q.v.*

Cold cut-off cyclone	500–1500 km	From m.s.l. commonly up to 10–15 km	20–30 mb below the average for the latitude or region	7–20 m/sec at sea level	Strongest in cumulonimbus clouds *q.v.*
Tropical cyclone	300–500 km	5–15 km	50 to 150 mb below the average for the latitude. Extreme values about 850–870 mb in North Pacific and South Indian Oceans	25–50 m/sec common at sea level	As in cumulonimbus clouds elsewhere *q.v.*
Tornado (accompanied by waterspout if over water or dust whirl over land)	Up to 200 m in max. velocity ring: diameter varies rapidly, 1–10 m commonest	Of order of 5 km	Commonly over 100 mb below surroundings. Extreme pressure values probably 500–700 mb (from observed heights of water columns raised from ponds and seas)	10–100 m/sec at surface in max. velocity ring, but over a breadth of the order of (≤) 1 m	Extremes possibly >50 m/sec
Cumulus convection cells, cumulonimbus clouds	5–50 km (occas. neighbouring cells arranged in line along a front or line squall, giving congested cells merging together along a line up to 500–1000 km long but only 20–50 km broad) (*cont. in next col.*)	Commonly 1–10 km. Extremes *c.* 15 km in Europe, 20 km in tropics (*cont. from previous col.*) Up to 400 km across in congestions of multiple-cell cumulonimbus in afternoons over some islands in the tropics	—	5–50 m/sec in gusts and squalls	Mostly 1–5 m/sec but 10–50 m/sec in active cumulonimbus clouds with squalls and thunderstorms
Land and sea breezes	Commonly 10–20 km on either side of the coast. Occas. linked with other regional circulations (e.g. mountain winds) into one wind system 100–200 km across	0·5–1·5 km	—	Mostly 1–5 m/sec in the temperate zone, up to 10 m/sec in warmer climates or where linked into a bigger circulation	

TABLE App. I.12 – *continued*

System	*Cross measurement*	*Vertical extent*	*Atmospheric pressure at m.s.l. in central region*	*Winds*	
				Horizontal component	*Vertical component*
Mountain and valley winds (anabatic, up-slope and katabatic, down-slope breezes)	Seldom more than 10–50 km from the mts and valleys concerned, often only within a few metres of the slope	Very restricted if air thermally stable; if unstable up to max. cumulonimbus heights (but strong winds only among the topography, especially in narrow valley channels, at mountain-tops and along the edge of the mts.; also in very localized lee-wave concentrations down-wind).		Very various. Anabatic breezes up open mountain slopes commonly 1–2 m/sec. Katabatic winds 'funnelled' in Greenland fjords up to 50–60 m/sec (also very strong where the Mistral blows through the Rhone gap). Strongest winds just over the mountain-tops and along mountain sides often occurring just when thermally stable stratification checks wider deflection of the wind.	Commonly $\leqslant 1$ m/sec. Cumulonimbus rates apply to extreme cases of ascent where such clouds form over the ridges. Extreme rates of descent vary widely

APPENDIX II

The production and decay of radioactive carbon ^{14}C in the atmosphere

When cosmic ray primary particles, mostly protons, enter the Earth's atmosphere, they produce showers of secondary radiation, which includes neutrons, in the upper reaches of the atmosphere. Most of the neutrons collide with nitrogen atoms, ^{14}N, and are captured by the atomic nucleus; few survive to reach ground level. In the collisions the dominant reaction produces radiocarbon, ^{14}C, a proton being ejected from the nucleus and one electron shed with it:

$$^{1}_{0}n + {}^{14}_{7}N \rightarrow {}^{14}_{6}C + {}^{1}_{1}H$$

The radiocarbon is readily oxidized to appear as a minute contribution to the atmospheric carbon dioxide which is taken up by living organisms and by sea water. Another reaction occurs in the case of about 1% of the neutrons captured:

$$^{1}_{0}n + {}^{14}_{7}N \rightarrow {}^{12}_{6}C + {}^{3}_{1}H$$

This transmutes the nitrogen atom nucleus into ordinary carbon and releases radioactive hydrogen, tritium.

The rate of decay of any radioactive atoms is always proportional to the number present

$$-\mathrm{d}q/\mathrm{d}t = \lambda q$$

where q is the quantity of atoms present at any time t. λ, the constant of proportionality, is called the 'decay constant'. If the number of atoms originally present was q_0, the decay follows the exponential curve expressed by

$$q = q_0 e^{-\lambda t}$$

The half-life $T_{\frac{1}{2}}$ of a radioactive species is given by putting $q = \frac{1}{2}q_0$ in this equation, which leads to

$$T_{\frac{1}{2}} = \frac{1}{\lambda}\log_e 2$$

The decay products of a ^{14}C atom are nitrogen $^{14}_{7}N$, a β^- particle and a neutrino. The decay of a ^{14}C atom is detected by making use of the ionization produced in a gas along the path of the ejected β^- particle.

APPENDIX III

Orbit data

The Earth's orbit

The elements of the Earth's orbit and the variations calculated by MILANKOVITCH (1930) can be seen in fig. App. III.1.

The orbit is shown in perspective as the ellipse *PQAR*, where *P* marks perihelion (the nearest approach to the sun) and *A* marks aphelion (where the Earth is farthest from the sun). This ellipse also defines the plane of the ecliptic. The sun is at *S*, one focus of the ellipse. The centre of the ellipse is at *O*. The direction of the Earth's movement round its orbit is shown by the arrow near the Earth *E*. The lengths of the major and minor axes of the ellipse, *PA* and *QR*, are designated as $2a$ and $2b$.

The **eccentricity** (e) of the orbit is given by OS/OP.

VS is drawn vertical at the sun *S* to the plane of the ecliptic, and *NS* is a line through

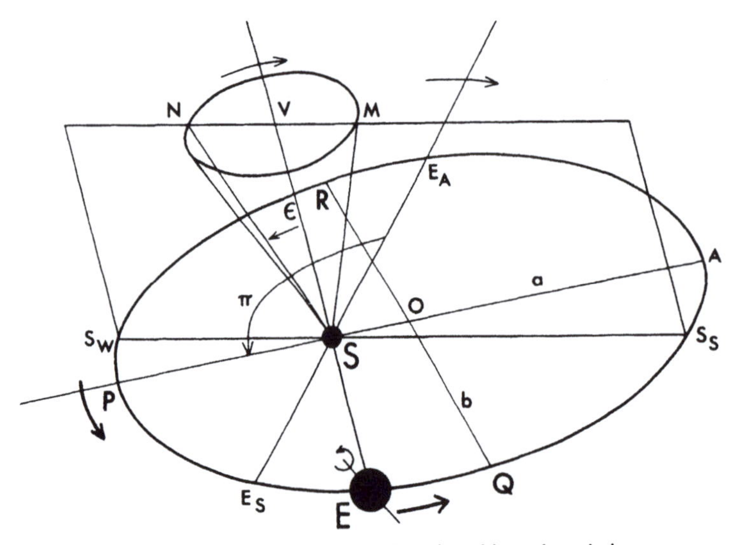

Fig. App.III.1 The elements of the Earth's orbit and variations to which they are subject. (Explained in text.)

the sun drawn parallel to the Earth's axis of rotation in the present epoch. The angle NSV is what is called the **obliquity of the ecliptic** (ϵ).

S_W and S_S are the present positions of the Earth at the (northern) winter and summer **solstices.** E_S and E_A are the positions of the (northern) spring and autumn **equinoxes.** The line E_S E_A cuts S_W S_S at right angles.

The plane NS_W SS_SM and the line of the equinoxes E_S E_A, at right angles to it, undergo a clockwise movement (as indicated by the arrows beyond E_A and beyond the conic section NVM) completing one revolution in 26 000 years: in this time the Earth's polar axis describes a cone in space similar to the cone SNM. The displacement amounts to 50″ of arc a year. All the time, however, the major axis PA of the elliptical orbit moves slowly in the opposite direction (as indicated by the arrow beside P), completing one revolution in space after an irregular interval that averages about 96 600 years. As a result, perihelion returns to the same time of year, conventionally measured by the angle PSE_A which is known as the longitude of perihelion (π) in less than 26 000 years, actually after an irregular period which averages about 20 600 years. This resultant cyclic progress of the Earth's seasonal attitude relative to its position in its orbit is known as the **precession of the equinoxes.**

The eccentricity (e) of the ellipse $PQAR$ also varies as PA revolves in space. This variation therefore has the same mean period of 96 600 years.

The obliquity of the ecliptic (ϵ) varies with a period of about 40 000 years.

The characteristics of these variations are analysed further by BERNARD (1962).

The following terms are used in discussing orbital characteristics and behaviour:

The **Apsides** are the points on an orbit at which the attracting force is greatest and least.
The **Apsidal line** of the Earth's orbit (fig. App. III.1) is the line PA, joining perihelion and aphelion, i.e. the major axis of the ellipse. In the case of the moon's orbit it is the line joining perigee and apogee.
The **Nodes of an Orbit** are the two points at which the orbit intersects some reference plane, e.g. the moon's orbit cutting the ecliptic.

The Moon's orbit

The moon's orbit (fig. App. III.2) lies in the plane of the ecliptic once every third year: at present this occurs some time about mid September. Its nodal points may then be taken as perigee and apogee, a situation described as nodal apsides. PETTERSSON's (1930) account of this in relation to the long-term variations of the tidal force, from which fig. App. III.2 is adapted, proceeds from the case where it is the apogee apex (α) of the lunar orbit which approaches the plane of the ecliptic from the north. It did so about the winter solstice in the year A.D. 1433 and is represented by the position α_0 in fig. App. III.2. At that moment the apsidal lines of the Earth's and the moon's orbits nearly coincided in the same plane.

After an interval of 2·9985 years there is again a nodal apside situation, but this time α

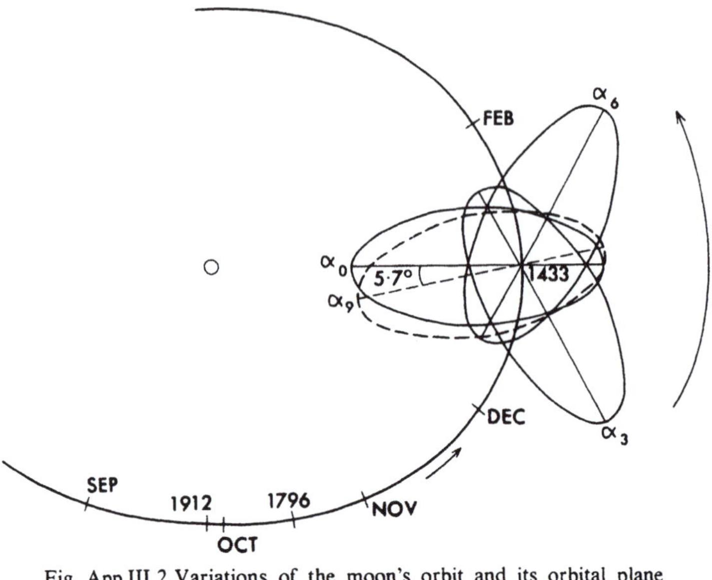

Fig. App.III.2 Variations of the moon's orbit and its orbital plane relative to the ecliptic. (Explained in text.)
(*After* PETTERSSON *1930*.)
(Size of the moon's orbit greatly exaggerated relative to that of the Earth and the position of the sun.)

approaches the plane of the ecliptic from the southern side and during that interval of nearly 3 years the moon has always attained its apogee on the southern side of the ecliptic. The position of the moon's apsidal line this time as it passes through the plane of the ecliptic is indicated by α_3 in fig. App. III.2; it makes an angle of 121·9° of arc with the previous alignment α_0.

Nearly 3 years later again there is another nodal apside situation, this time indicated by α_6 in the diagram and making an angle of 243·8° with the first occurrence. α again approaches the plane of the ecliptic from the northern side, where the apogee positions have all been for nearly 3 years.

The last occurrence illustrated in the diagram is just under 9 years after the first. Instead of being exactly in the original alignment, however, the apogee position of the moon falls at α_9, making an angle of 5·7° with the direction of α_0. And this time apogee approaches the plane of the ecliptic from the southern side. This was the situation in the year 1442.

Fig. App. III.2 also indicates the time of year and position of the Earth in its orbit at the moment when the moon's apogee lay in the plane of the ecliptic and with its apsidal line directed nearly towards the sun in 1796 and 1912.

Appendix III

TABLE App. III.1

Planetary years and other details

Planet	*Sidereal year*	*Mass (satellites excluded)*	*Solar distance in astronomical units*	*Eccentricity of orbit*	*Inclination of orbit to the (Earth's) ecliptic*
	(days)	(relative to Earth = 1)	(half the major axis of the ellipse relative to Earth = 1)		
Mercury	87·969	0·053	0·39	0·206	7°0′
Venus	224·701	0·815	0·72	0·007	3°24′
Earth	365·256	1	1	0·017	—
Mars	686·980	0·107	1·52	0·093	1°51′
Asteroids			2·9		
	(years)				
Jupiter	11·862	318·00	5·20	0·048	1°19′
Saturn	29·458	95·22	9·55	0·055	2°30′
Uranus	84·018	14·55	19·2	0·047	0°46′
Neptune	164·78	17·23	30·1	0·009	1°47′
Pluto	248·4	0·9	39·5	0·247	17°9′

From C. W. ALLEN, 1955, *Astrophysical quantities*, Univ. London Athlone Press (pp. 155–7), where further details are given. The variable quantities in the last two columns are here the A.D. 1900 values.

APPENDIX IV

Global distribution of departures from the long-term average of monthly mean pressure at sea level associated with different phases of the sunspot cycle

Fig. App.IV.1 January pressure anomalies (mb):
(*a*) Average for the years around sunspot maxima (years X − 1, X, X + 1).

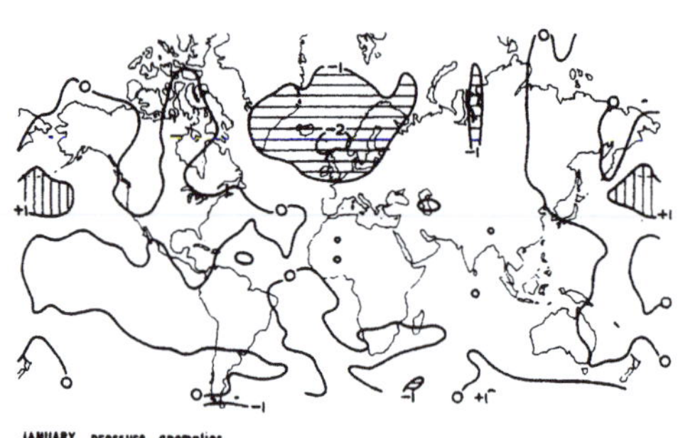

(*b*) Average for the years in the descending phase of the sunspot cycle (years X + 2, X + 3, X + 4, N − 2).

(*c*) Average for the years around sunspot minima (years N − 1, N, N + 1).

(*d*) Average for the years in the ascending phase of the sunspot cycle (years N + 3, X − 2).

Figs. App. IV.1 and 2 are composite maps, from unpublished work by B. N. PARKER of the Meteorological Office, which present averages for the greatest number of years for which reliable maps can be drawn in different parts of the world. The averages here mapped cover all 19 sunspot cycles between 1750 and 1958 (i.e. they mostly represent 57 years for each sunspot cycle phase) over the North Atlantic and Europe, for which the basic series of monthly mean pressure maps for each year produced by LAMB and JOHNSON (1966) have been shown to contain only random errors, with the result that averages for 40 years or more are believed to have a very small margin of error. Over the southern hemisphere the maps here reproduced comprise far fewer cases.

Among the most interesting general features of these maps are the indications of increased middle-latitudes westerlies chiefly in the long descending phase of the sunspot cycles in winter over the North Atlantic–Europe and over the South Pacific–southwest Atlantic. At the same time the westerlies seem to be either shifted poleward or reduced in strength over the opposite sector of both hemispheres. Boosted strength of the westerlies appears in the North Pacific in winter around sunspot minima and south of Australia in summer at that

Fig. App.IV.2 July pressure anomalies (mb):
(*a*) Average for the years around sunspot maxima (years X − 1, X, X + 1).

(*b*) Average for the years in the descending phase of the sunspot cycle (years X + 2, X + 3, X + 4, N − 2).

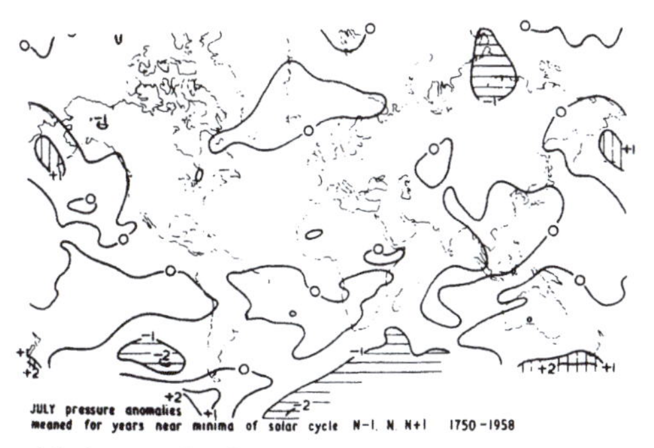

(*c*) Average for the years around sunspot minima (years N − 1, N, N + 1).

(*d*) Average for the years in the ascending phase of the sunspot cycle (years N + 3, X − 2).

time. Blocking seems to affect Europe and South America particularly in the winters around both sunspot extremes, but the patterns are distinctive – i.e. different positions most affected by the block and differently orientated windstreams – around sunspot maximum and minimum.

Preliminary investigation of statistical significance, by Mr PARKER, of the northern hemisphere (except Pacific) portions of these maps suggests that most of the strong features shaded are probably to be treated as significant, viz:

Januarys

About Sunspot Maxima

Urals over	+2 mb	5% level
Mid Atlantic	−1mb	1% level

Descending Phase

Faeroes region	−2 mb	5% level

About Sunspot Minima

Central Canada	−1 mb	5%

Ascending Phase

Northeast Asia	−4 mb	1%
Central U.S.A.	−1 mb	5%

Julys

About Sunspot Maxima

Labrador	−1 mb	1%

Possibly also Siberian sector of Arctic Ocean +1 mb (a very constant feature)

Descending Phase

No significant anomalies apparent in the northern hemisphere

About Sunspot Minima

North Scandinavia (a very constant feature)	0 to −1 mb	1%

Ascending Phase

Greenland to Novaya Zemlya	below −1 mb	5%

It seems possible that more significant associations between the summer pressure patterns and the solar cycle might be obtained if the years were further sorted according to the phase of the quasi-biennial oscillation (see Chapters 5 and 6), which introduces some rather striking differences over the northeast Atlantic and Europe between alternate years.

Part II · Climate now

CHAPTER 11

Observational reference data and classification of climates

Reference data on present climates are given here in three forms:

(1) World maps of the distribution of various items of climatic importance for their effects upon man. The selection here is of items not often found in atlases and not given elsewhere in the text of this work.
(2) A world-wide classification of climates in terms of things critical for plants and the associated animal life, growth of crops, etc. The classification here presented is the one which is in widest use, due to KÖPPEN.
(3) A climatic table of data on characteristic temperature and rainfall values for each month of the year, and averaged for the year as a whole, for a network of places all over the world.

(1) The world maps

The maps which follow provide averages for the years 1900–50 approximately.

Maps of mean atmospheric pressure at m.s.l., prevailing surface winds, average temperatures of the air and of the sea, mean yearly rainfall (or its equivalent in the downput of snow) and mean yearly evaporation (as affected by lack of available water in regions of dry surface) will be found in the text of Part I.

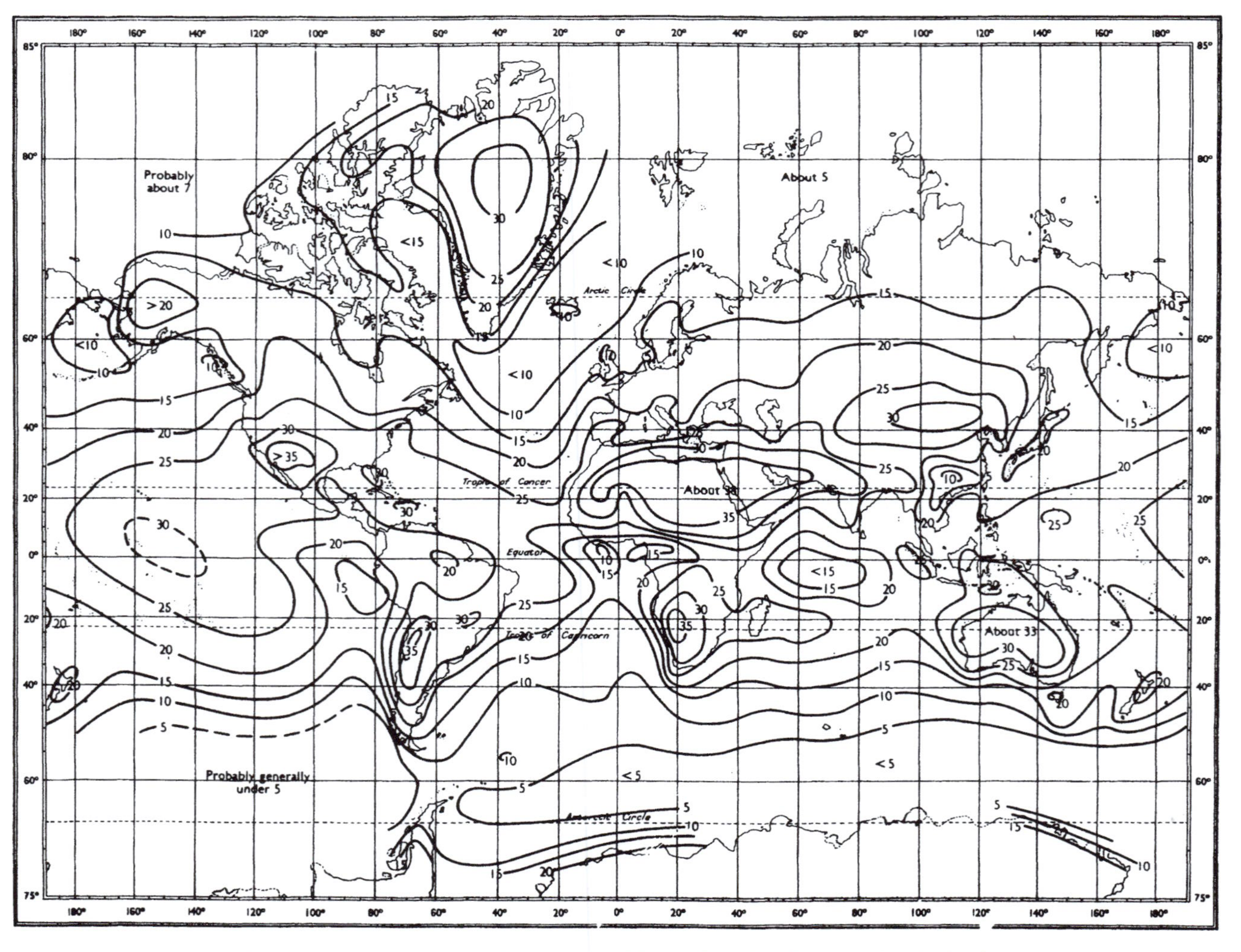

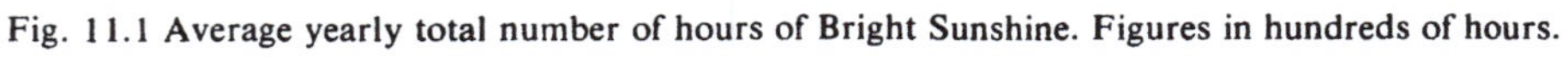

Fig. 11.1 Average yearly total number of hours of Bright Sunshine. Figures in hundreds of hours.

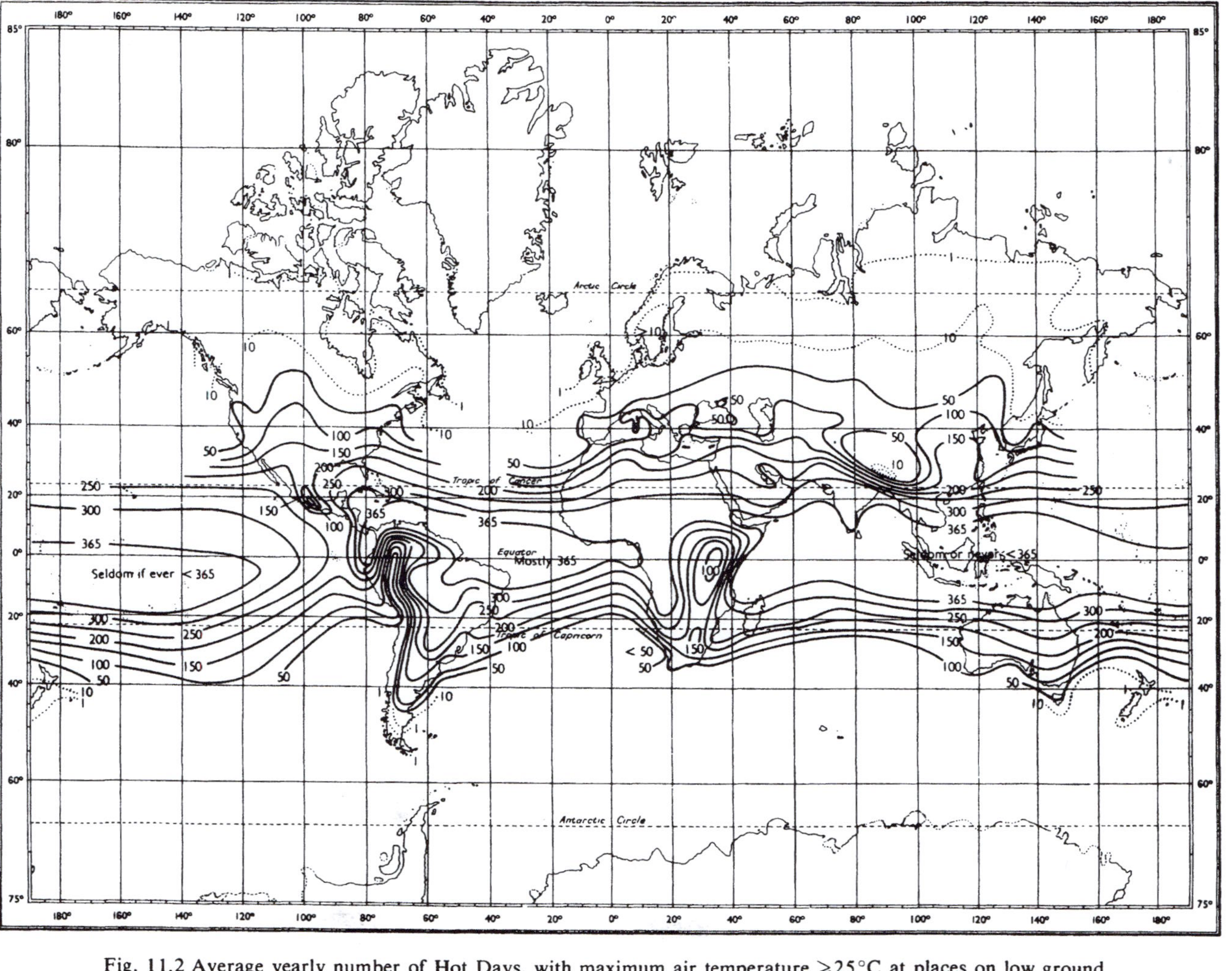

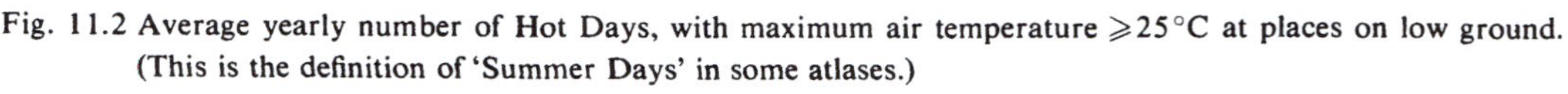

Fig. 11.2 Average yearly number of Hot Days, with maximum air temperature ⩾25 °C at places on low ground. (This is the definition of 'Summer Days' in some atlases.)

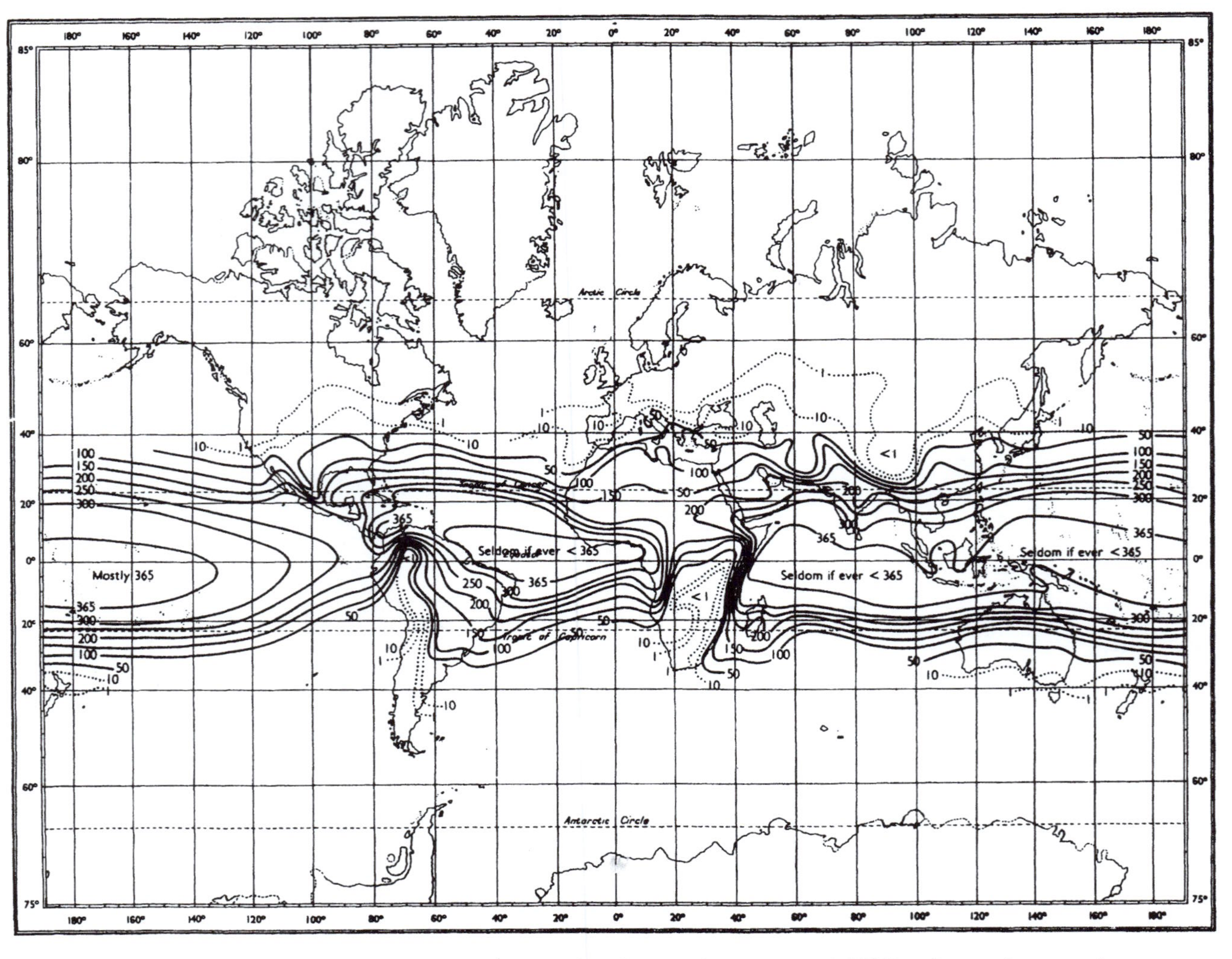

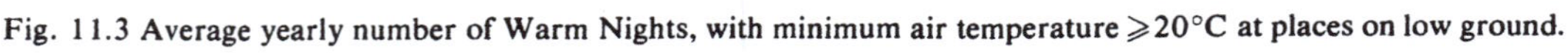
Fig. 11.3 Average yearly number of Warm Nights, with minimum air temperature $\geqslant 20°C$ at places on low ground.

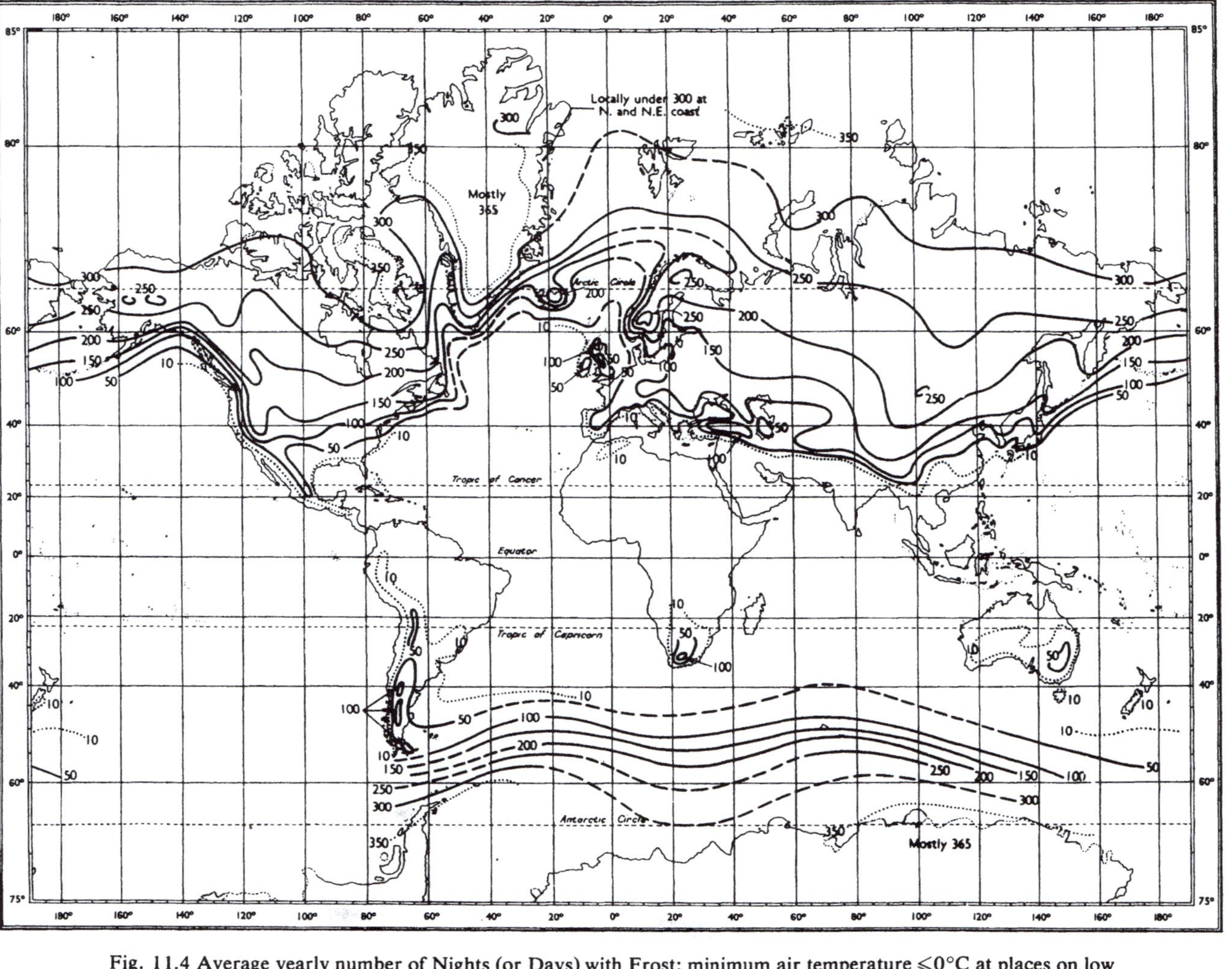

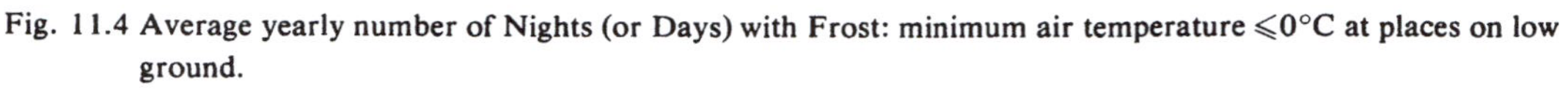
Fig. 11.4 Average yearly number of Nights (or Days) with Frost: minimum air temperature ⩽0°C at places on low ground.

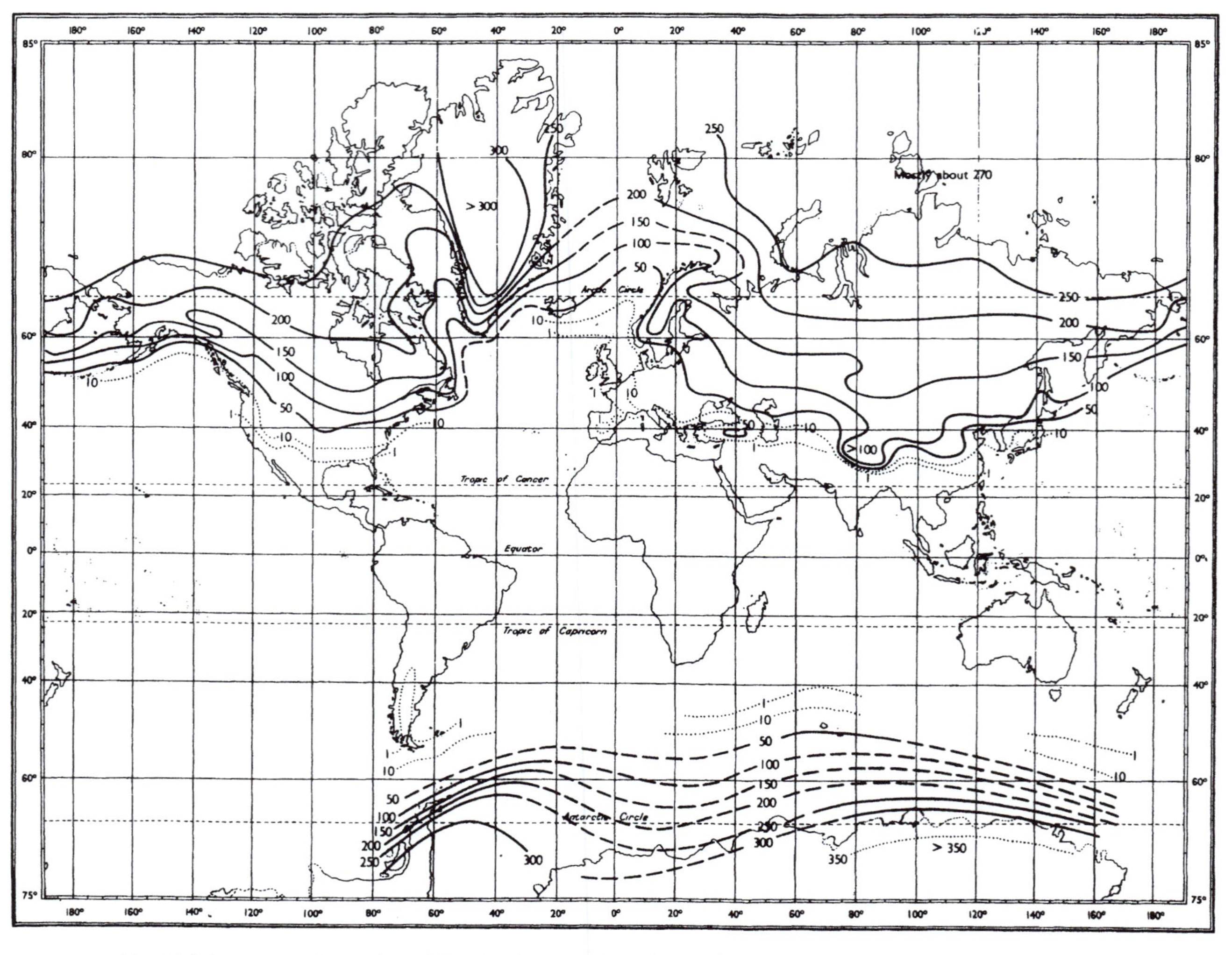

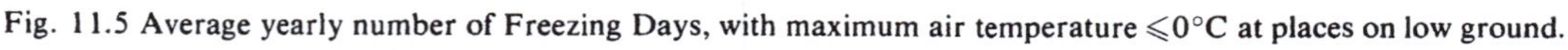
Fig. 11.5 Average yearly number of Freezing Days, with maximum air temperature ⩽0°C at places on low ground.

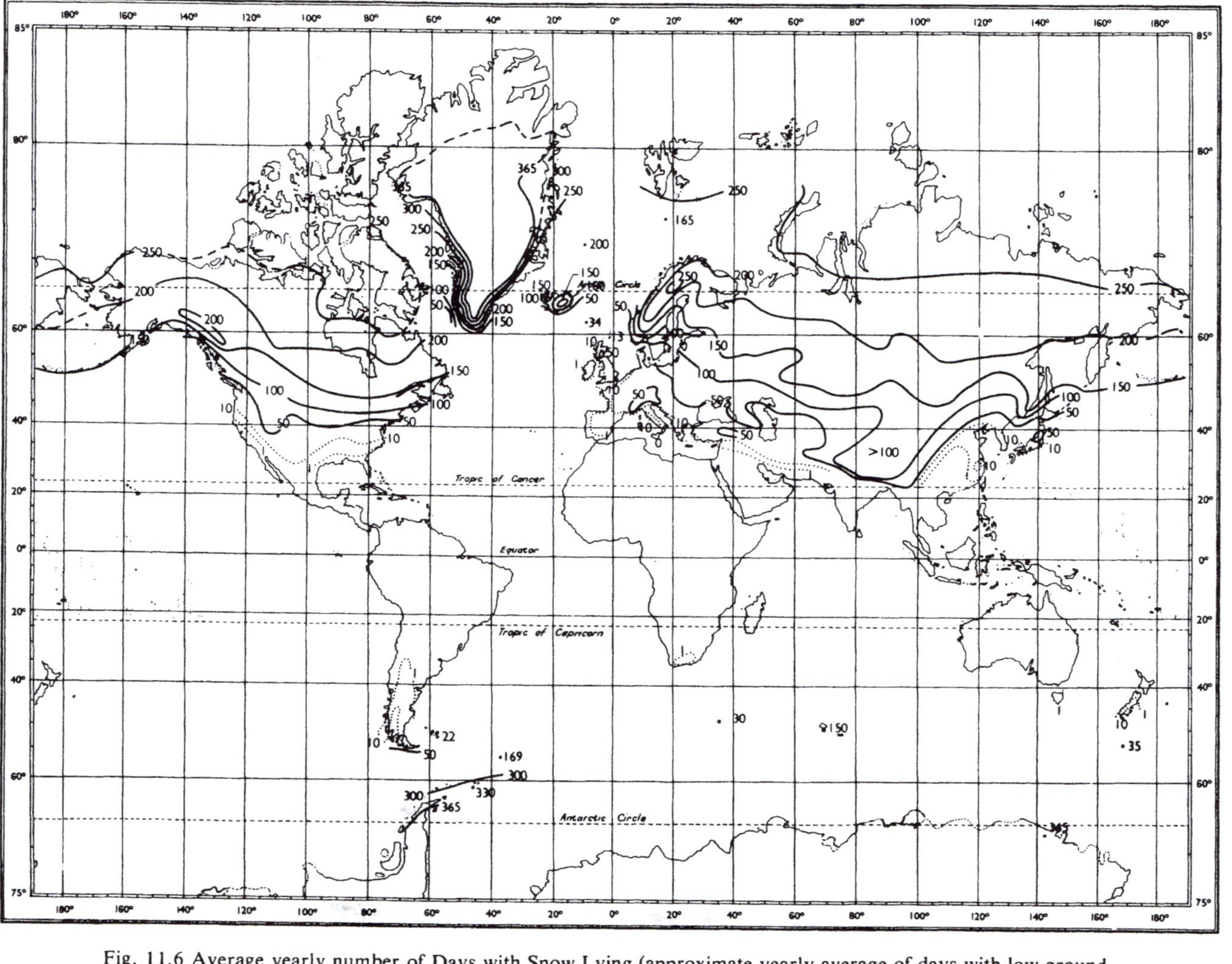

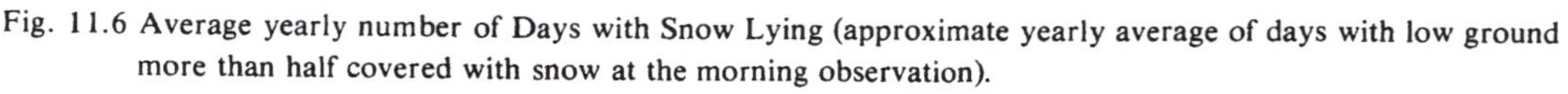

Fig. 11.6 Average yearly number of Days with Snow Lying (approximate yearly average of days with low ground more than half covered with snow at the morning observation).

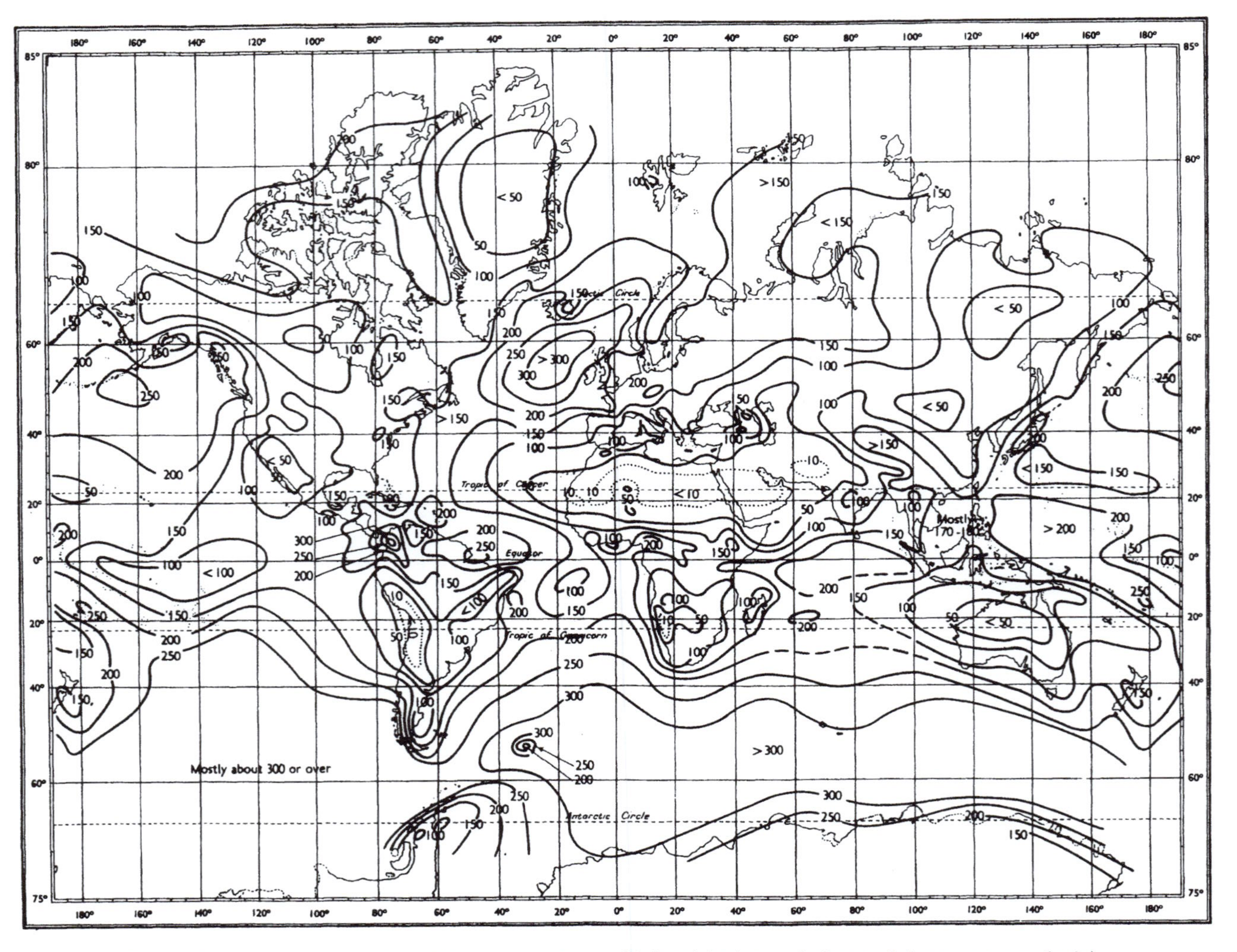

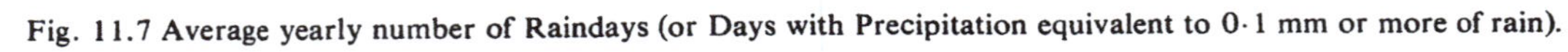

Fig. 11.7 Average yearly number of Raindays (or Days with Precipitation equivalent to 0·1 mm or more of rain).

(2) Classification of climates [1]

The simplest classifications are *genetic* ones, which describe the distribution of climatic regimes in terms of the conditions of radiation, surface heating and the prevailing winds which produce them. A simple list of the types of climate occurring in different parts of the world as related to the general wind circulation is given in Chapter 3 of this book.

Classifications that delimit climates in terms of those threshold values of temperature, rainfall, etc., which determine their *effects* upon vegetation and animal life also have obvious usefulness. That due to KÖPPEN (1923, 1931), elaborated below, is by far the most widely used. The first drafts of this system were worked out by KÖPPEN as early as 1900 and 1918, using still earlier approaches by biologists concerned with the effects of the climate on vegetation and animal life. The history and argumentation of the system were conveniently summarized by KÖPPEN (1936). Other classifications of interest, and the history of their development, starting with SUPAN's (1879) definition of climatic zones of warmth in terms of annual mean temperature and KÖPPEN's modification of this scheme only a few years later (1884) to take account of the duration of various temperature levels, are discussed by BLÜTHGEN (1966, pp. 444–546) and more briefly by FLOHN (1957) and GENTILLI (1958). The more important contributions and classifications are included in the references to Chapter 11.

Köppen's world-wide classification system

Climates are ranged in the following main groups:

A Tropical rain climates
B Arid climates
C (Warm) temperate rain climates
D Boreal forest and snow climates
E Cold snow climates (treeless)

A, **C** and **D** are all essentially tree climates; **BS** is the steppe grassland climate; **BW** defines the (warm) deserts (**W** standing for desert-waste/*Wüste*); **E** climates are those of the tundra and polar desert.

Subdivisions are designated as follows:

s Summer dry season
In group **A** climates confined to exceptional areas such as lee-sides of islands in the Trade Wind zone.
In groups **C** and **D** the wettest month of the year must have $\geqslant$ 3 times the average rainfall of the driest month.

1. For bibliography on climatic classification, see References to Chapter 11, pp. 594–5.

w Winter dry season
Group **A** climates are only designated **Aw** when at least one month normally has <60 mm rainfall.
In groups **C** and **D** the wettest month of the year is required to have ⩾10 times the average rainfall of the driest month.

s′, w′ Special cases of summer or winter dry season climates, where the chief rains come in late summer–autumn.

x Climates where the early summer is wet, late summer fine.

s″, w″ Special cases of summer or winter dry season climates, where the rainy season is split – i.e. has two peaks, with another, short dry season in between.

f Moist climates, with no marked dry season.

m Monsoon climates, with only a short dry season in the winter half-year.

On this system the chief types of climate appearing on the world map (fig. 11.8) are defined as follows:

Tropical rain climates

Mean temperature of the coldest month in all cases >+18°C.

Af Driest month has on average ⩾60 mm rain.

Am Driest month <60 mm but rainfall in the rainy season compensates this enough to allow forest.

Driest month R	60	40	20	0 mm
Required yearly R	1000	1500	2000	2500 mm

(**Af** and **Am** are both characteristically accompanied by tropical evergreen rain-forest.)

Aw Driest month <60 mm; annual rainfall insufficient to compensate.
(**Aw** produces Savanna.)

Arid climates

Defined by (i) $R < 2T + 28$ where summer is the season with most rain.
or (ii) $R < 2T + 14$ where there is no season with more liability to rainfall than another.
or (iii) $R < 2T$ where winter is the season with most rain.

In these formulae R is the average rainfall for the year in centimetres; T is the average yearly temperature (°C).

BS $R \geqslant T + 14$ for summer rain areas
$R \geqslant T + 7$ where there is no season with more liability to rain than another
$R \geqslant T$ for winter rain areas
(In **BS** climates the vegetation ranges from bush to grassland.)

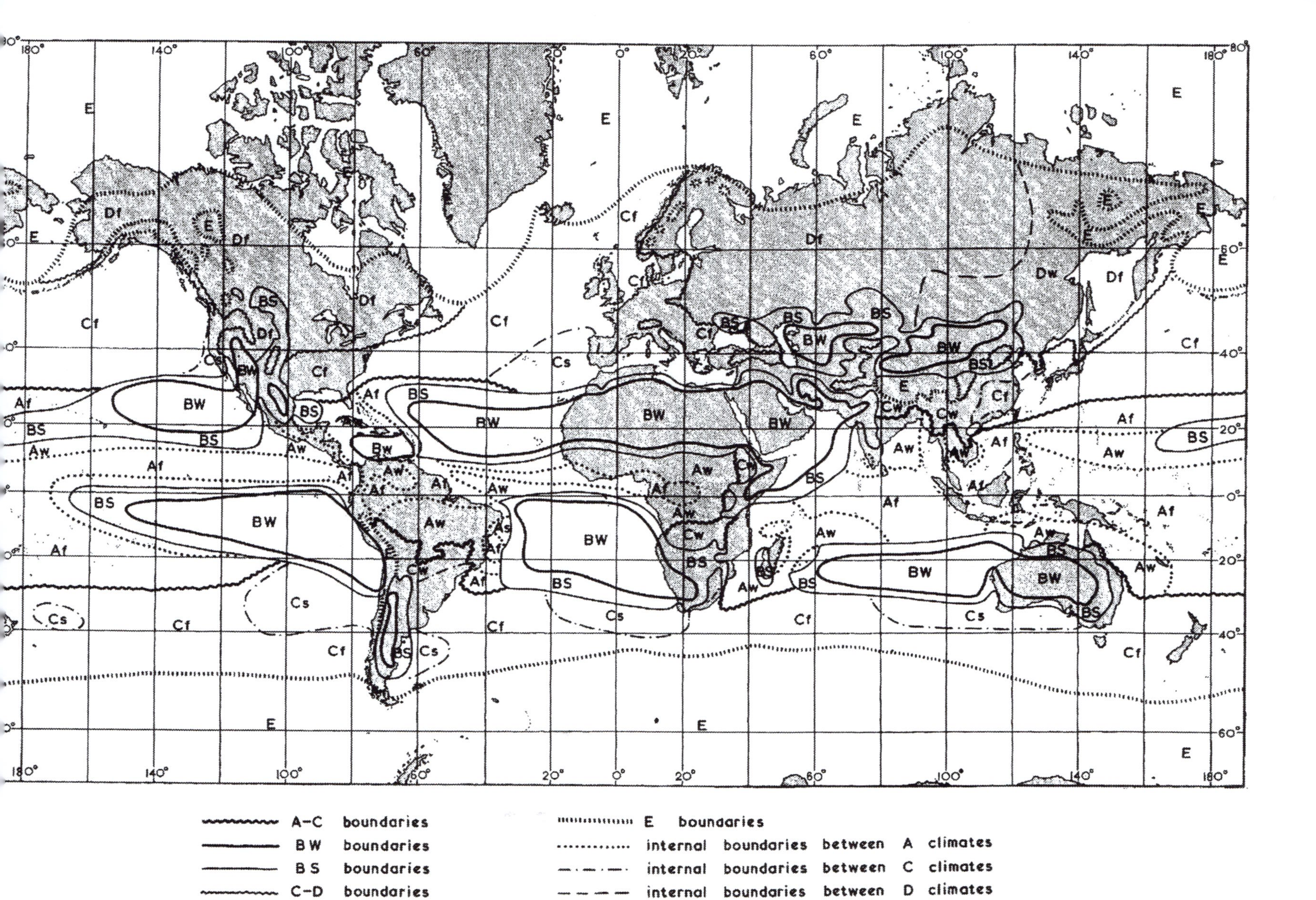

Fig. 11.8 KÖPPEN's world classification of climates.

BW $R < T + 14$ for summer rain areas
$R < T + 7$ where there is no season with more liability to rainfall than another
$R < T$ for winter rain areas
(**BW** produces desert.)

Temperate rain climates

Mean temperature of the coldest month between +18°C and −3°C.

Cs Wettest (winter) month of the year has ⩾3 times the average rainfall of the driest (summer) month.
(**Cs** is typically accompanied by evergreen broad-leafed forest.)

Cf No marked dry season.
Sub-types:
Cfa Hot summers: warmest month > + 22°C
Cfb Warm summers: at least 4 months > + 10°C
Cfc Cool summers: only 1–3 months > + 10°C.
(**Cfa** and **Cfb** vegetation generally deciduous broad-leafed forest, but in areas with very mild winters evergreen broad-leafed forest.
Cfc vegetation needle-tree forest.)

Cw Wettest (summer) month has ⩾10 times the rainfall of the driest winter month.
(**Cw**, occurring mainly on mountain heights above **Aw** climates, produces evergreen forest vegetation.)

Boreal forest and snow climates

Mean temperature of the warmest month > + 10°C, coldest month < − 3°C.

Dw Wettest (summer) month has ⩾10 times the average precipitation of the driest (winter) month.

Df Less seasonal difference in the average downput of rain and snow.
Sub-types:
Dfa and **Dwa** Hot summers: warmest month > + 22°C
Dfb and **Dwb** Warm summers: at least 4 months > + 10°C
Dfc and **Dwc** Cool summers: only 1–3 months > + 10°C.
(**Dfa, Dfb, Dwa, Dwb** climates go with deciduous (broad-leafed) forest.
Dfc, Dwc go with needle-tree forest.)

D climates and their vegetation are only found in the northern hemisphere, where the great continents provide the requisite conditions.

Cold snow climates

Mean temperature of the warmest month < + 10°C.

ET Mean temperature of the warmest month > 0°C.
(**ET** vegetation is tundra, dwarf tree species and mosses.)

EF (sometimes called simply **F**).
Mean temperature of the warmest month <0°C.
Brief thaws and rain can occur, but have little or no effect on the long-term condition of the surface.
(No vegetation in **EF** climates: polar desert.)

Further letters are sometimes used to distinguish particular climates and subdivisions of the KÖPPEN types:

d coldest month < − 38°C
g Ganges type climate, warmest month before the summer solstice, height of summer rainy
h hot: yearly mean temperature > + 18°C
i isothermal: difference between warmest and coldest months <5°C
k cold-winter climate: yearly mean temperature < + 18°C, but warmest month > + 18°C
l lukewarm, even warmth: all months between +10 and +22°C
n foggy
n′ moist, but not foggy: with mean summer temperatures over +24°C designated **n″**, over +28°C **n‴**
t′ Cape Verde Islands type: warmest in autumn
t″ Sudan type: coolest month just after the summer solstice.

It may be held that the types of climate covered by group **C** embrace too wide a range, and that the cooler temperate climates where birch forest (or mixed birch and pine) constitutes a large element should be distinguished from those of the domain of oak forest where the average temperature of the warmest month is generally about +15°C or over.

The proportions of the Earth's surface classified as falling within KÖPPEN's various types at the present day are estimated as follows:

	Portion of land surface	*Portion of total surface*
A	20%	36%
B	26%	11%
C	15½%	27%
D	21%	7%
E	17%	19%

KÖPPEN's classification can be criticized because it makes no mention of the level of incoming radiation and allows only indirectly and crudely for differences in the rate of evaporation. But these criticisms apply equally to most other classifications. The most widely used of these, that by THORNTHWAITE (1933), makes use of a complex formula for *precipitation effectiveness* (*PE*) where

$$PE = 1{\cdot}65\left(\frac{T + 12{\cdot}2}{R}\right)^{10/9}$$

where T is monthly mean temperature (°C) and R is monthly rainfall (mm). Arid climates occur where the 12-month total of $PE < 16$, semi-arid climates where $16 < PE < 31$, sub-humid climates where $31 < PE < 63$, humid climates where $63 < PE < 127$, per-humid climates where $PE > 127$. A modified version of KÖPPEN's system by TREWARTHA (1954) has also attained some recognition: it is said to be useful in defining the climatic regions of the Earth in relation to the general wind circulation and following at least some of the vegetation boundaries quite closely. BUDYKO (1955) has defined an *aridity index* $\mathbf{R}/L.R$ in which $\mathbf{R}$ stands for the average value over the year of the net radiation gain ('radiation balance') at the Earth's surface, R is the year's total rainfall and L is the latent heat of evaporation. The values of this aridity index have been mapped over the territory of the Soviet Union (GRIGOR'EV and BUDYKO 1959), and more recently over the whole world (GIESE 1969): they appear to lend themselves to a useful demarcation of different climates by means of threshold values of the index significant for the vegetation.

The chief virtues of KÖPPEN's classification are its simplicity and the fact that the threshold values used were arrived at by its author after trial and error and improvement over a long period of years, so that they probably express better than any other simple values could the critical boundaries in the vegetation. Because of its wide use (and availability in the form of wall maps), this classification is also known in nearly all countries.

(3) The world climatic table

Data are given, where possible, for the 100-year period 1851–1950. It is not yet possible to give figures for this same (or any other) long period everywhere. Indeed, for some places, chiefly in high latitudes, only a very short (usually recent) period can be covered. At the date of writing, the 30-year averages, or so-called 'normals', for 1931–60 published by the World Meteorological Organization are available for more places in more parts of the world than for any other period of comparable length. To give some bearing upon the representativeness of the figures for 1931–60 (which in many places covered the warmest years of record[1] and might be considered highly abnormal) and of shorter, more recent periods which sometimes have to be quoted, averages at a few places are given for 1851–1950, 1931–60 and 1951–60 or 1960–9. (Brackets around any dates quoted imply doubt as to whether the data covered the whole period.)

1. The period 1931–60 was in many parts of the world the warmest since observations began, but not quite everywhere (e.g. in much of Australia).

The data given include some, or all, of the following items:

T_m average air temperature, measured at a height of 1–2 m above the ground in the shade (usually in a ventilated white screen).

σT_m the year-to-year variability, measured by the standard deviation, of the monthly mean air temperature of the given month.

T_x the absolute maximum air temperature observed.

T_n the absolute minimum air temperature observed.

drT the mean daily range of temperature, taken as the difference between the average daily maximum and average daily minimum temperatures.
(The figures for *drT* have in most cases had to be limited to a crude indication of magnitude to the nearest 0·5 °C.)

R_m the average total rainfall or total downput of rain and snow when melted.

$R\ Q_5\geqslant$ these figures mark the boundary of the top quintile of precipitation, i.e. the downput which is exceeded in just one-fifth of the years – the wettest 20%.

$R\ Q_1\leqslant$ these figures mark the boundary of the bottom quintile, i.e. the downput which is not exceeded in just one-fifth of the years – the driest 20%.

Rdx the maximum downput observed in a day (in the form of rain, or snow when melted).

N_p the average number of days with precipitation.

N_r the average number of days with rain.

It has not been possible to include both T_x and T_n and *drT* for more than a very limited selection of places, but these may suffice to illustrate the relationships between average daily range and extremes of temperature in different types of climate.

The places incorporated in the table have been chosen to illustrate the average conditions prevailing in all the main climatic regimes in the world today, and the data make it possible to spot the ranges of variation – diurnal range, year-to-year variability, variation with height (and in some cases with aspect), maritime to continental and so on – that characterize them. A number of places have been included because they are the ones for which unusually long records exist or where comparisons can be made with some evidence of conditions in much earlier times, before the invention of meteorological instruments.

An asterisk after a date indicates a change of observation site great enough to make comparisons between the measurements before and after the change meaningless unless carefully determined adjustments are made. This difficulty arises chiefly through the effects of *urbanization* in recent years at most sites in cities, including those where observations have been made over the longest periods.[1] Cities are generally warmer than the surrounding

1. For a bibliography on urbanization effects, see References to Chapter 11 at the end of the book.

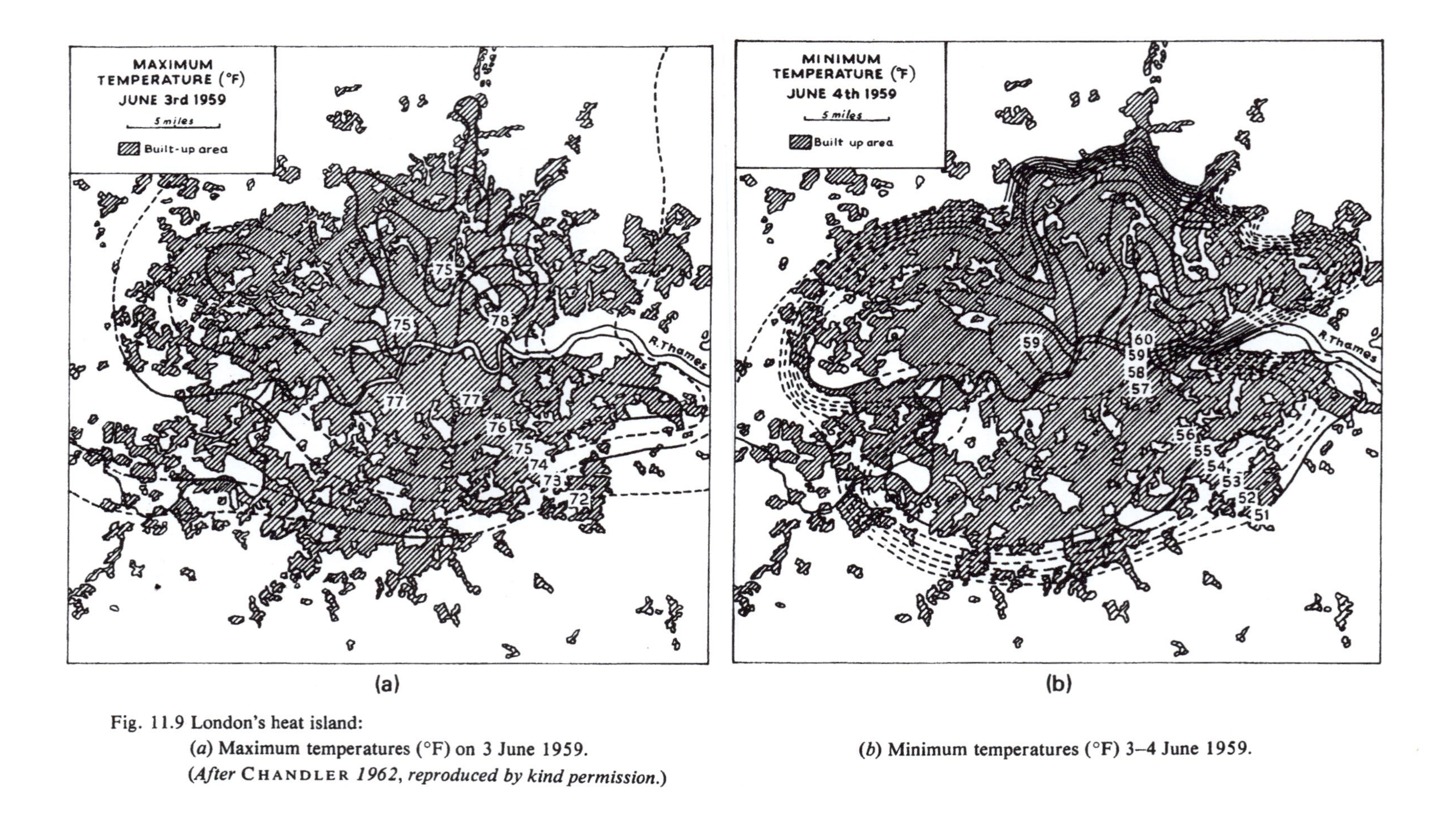

Fig. 11.9 London's heat island:
(*a*) Maximum temperatures (°F) on 3 June 1959.
(*b*) Minimum temperatures (°F) 3–4 June 1959.
(*After* CHANDLER *1962, reproduced by kind permission.*)

countryside for a number of reasons: (*a*) the larger amounts of incoming radiation absorbed by the paved area and walls and roofs of buildings than by the natural vegetation and soil surfaces (moreover, the buildings and pavements continue to re-radiate their excess heat throughout the night), (*b*) the artificial, rapid drainage of water from the paved surfaces so that less heat is used up in evaporation, (*c*) the artificial heat 'leaking' out from the buildings, etc. (and the body-heat of the people and animals). The extra warmth of the city, the *heat-island*, attains its greatest magnitude in terms of difference of temperature from the surrounding countryside in still, clear weather that goes with strong sunshine in summer and greatest radiation heat-loss in winter; in the latter case the artificial heat of the city often accumulates (like the pollution) beneath a temperature inversion which is partly due to anticyclonic subsidence and is strong enough to allow a considerable surface temperature rise without any deep convection being started. Rainfall tends to be increased over cities, both because of the extra convective uplift over the heat-island during much of the year and because of the up-currents forced by the buildings and the additional friction on the wind.

Figs. 11.9 (*a*) and (*b*) illustrate the intensity, and the geography, of London's heat-island during fine summer weather. Notice particularly the strong temperature gradient near the limits of the built-up area. On the occasion shown the temperature difference between the inner city area and the surrounding country amounted to +6 to 7°F (about +4°C) in the afternoon and to +9 to 10°F (about +5°C) in the following night. The strongest development of the heat-island is displaced a little away from the centre of the built-up area by the prevailing light breeze; there is a similar displacement of its average position over the year as a whole by the prevailing (SW'ly) wind, and the precipitation maximum is found over about the same area as London's thermal maximum. Nevertheless, the temperature gradient within the town is flat by comparison with that at its fringe. A similar map of Vienna on a night of radiation frost in May showed air temperatures as low as −2° and −3°C in the low-lying meadows on the eastern and southeastern outskirts, while in the heart of the city the temperature remained as high as +4° to 5°C.

The mean temperature difference between the inner parts of any town with more than 50 000–100 000 inhabitants and unbuilt or country areas has to be allowed for in using any climatic table. Averaged over the year, this difference has now attained the values listed in Table 11.1.

The increasing warmth of all large and growing towns relative to a green countryside accounts for at least part of the climatic trend apparent at urban sites towards ever-increasing warmth. (There is no reason, however, to expect this increase of warmth to affect cities in the arid zones where the surrounding landscape has characteristics approximating to those of the paved city area.) DRONIA (1967) has pointed out that this, unhappily, makes the values reported by most of the longest-established climatic stations in the world unrepresentative in recent decades, and may have falsified the usual estimates of the magnitude and duration of the early twentieth-century warming of world climates. The effect is certainly

TABLE 11.1

Average temperature difference (observations mainly 1950s and 1960s) between inner city sites and surrounding country

London (Kew Gardens *minus* Harpenden) The figure varies from 1·2°C in summer to 1·9°C in winter, having roughly doubled since 1880. It is almost certainly greater still (*c.* 2·0°C?) if *central* London be compared with the country outside.	1·5°C
Berlin (centre *minus* northwest outskirts)	1·7°C
(Dahlem *minus* northwest outskirts)	0·8°C
Kiel	0·7°C
Toronto (centre *minus* Malton airport, west of city)	1·2°C
New York	1·1°C
Los Angeles	0·7°C
Buenos Aires	1·0°C
Moscow	0·7°C
Barnaul, Siberia	0·4°C

seen in the more recent observations in cities in our world climatic table. Most North American cities' observation sites included in *World Weather Records* have, presumably partly for this reason, been moved to the airports since about 1950, though any such move also entails doubts about the comparability of the records before and after the change. The values for the long-established station in central Toronto and those for the modern airport station have both been included in the table to illustrate this: the overall difference between the sites was 0·9°C in the 1940s and 1·2°C in the 1950s. The very rapid increase of the temperatures observed in Mexico City, where the 1940s were apparently 0·8°C warmer overall than the 1930s and 1·3° above the average for 1878–1950, was presumably also largely attributable to growth of the city: since 1950 the observation site has been moved to a western suburb (Tacubaya). Also because of city growth the temperatures observed in Rome in the 1950s seem likely to be at least 0·5°C above what they would be without the city effect. (A collection of studies of many aspects of urban climates has lately been published by the World Meteorological Organization, 1970.)

The increases of precipitation averaged over the year over urban centres are commonly in the range 5–10% as compared with the surrounding countryside.

A few other points about the climates illustrated by places in our world climatic table deserve special mention.

Various cities in, or near, the arid zone – Cape Town, Cairo (Helwan) and Cordoba, Argentina – show some net cooling in the 1950s despite increased warmth in summer: at these places the effect of a global cooling trend may be nearly balanced by that of urbanization. At Sydney, Australia the balance is just the other way with the mean for 1951–60 0·2°C above that for 1931–60. Over most of the Australian region, and as far as Cocos Island in the equatorial Indian Ocean, the global cooling trend in the 1950s is dominant. At

Srinagar in Kashmir the 1950s average was 0·1°C below that for 1931–60, and the increased difference between there and Bombay and Calcutta probably indicates that the latter cities in the monsoon zone now return temperatures not less than 0·5°C too high because of the urban effects on temperature.

Few long-term comparisons are satisfactory in Africa and India because of frequent changes of observing site. The increase of the regular equatorial rains over East Africa (e.g. at Entebbe) since 1960, as compared with the previous 60 years, is however plainly very significant, the last decade averages for October and November falling within the wettest quintile of the previous period. The climatic change at the Azores, with the increased frequency of blocking anticyclones in the higher latitudes of the northern hemisphere since 1940, does not quite reach the same level of significance but is shown by a 36% increase of rainfall nevertheless. At Easter Island, in the usual realm of the South Pacific anticyclone, the wetness of some months hints at similar possibilities of substantial secular change.

The enormous month-to-month changes of average temperature in central Asia, and the normal diurnal ranges exceeding 15°C there and in the deserts of North Africa, Australia, and inland in the lower latitudes of North and South America, reaching 20–24°C in a few places at the higher levels in these continental interiors, indicate that these regions should respond quickly (and probably in a simple, direct way) to any changes in the incoming radiation supply. By contrast, climates with cloudy skies are marked by small diurnal ranges, and small seasonal changes, for their latitude. Secular, or longer-term, changes may well, however, be complicated by changes in the amount of cloud.

The difference of temperature with height in the free air has some effect on the temperatures observed at the few mountain stations in the table, but comparisons with neighbouring low-level stations are affected in some cases (e.g. in Cyprus and Agadir–Marrakech) by daytime sea breezes cooling the low-level place while the higher levels inland (like plateaux in any sunny climate) become very warm by day and show an enhanced diurnal range (cf. Lhasa, Tibet, and Kandahar, Afghanistan). In Ecuador (compare Guayaquil and Quito) the diurnal range is reduced, and the prevailing temperatures moderated, by the cloudiness of the coast due to the cool off-shore waters.

Notice also the generally smaller daily and seasonal ranges of temperature at oceanic and island stations than at continental and inland places at the same latitude.

Rainfall variation with height and proximity to the nearest ocean from which the prevailing wind blows is amply illustrated, but details of orography and slope towards (or away from) the prevailing wind produce very big local differences which cannot be covered in a world table. Seasonal passages, and switches of orientation, of the prevailing tracks of cyclonic activity stand out in the average rainfall figures – e.g. the two seasonal maxima in equatorial (especially East) Africa, and to some extent in China and Japan, and the great increase of rainfall from August to September in the Faeroes, Iceland and north Norway, while rainfall decreases in eastern England and Scotland and throughout central Europe.

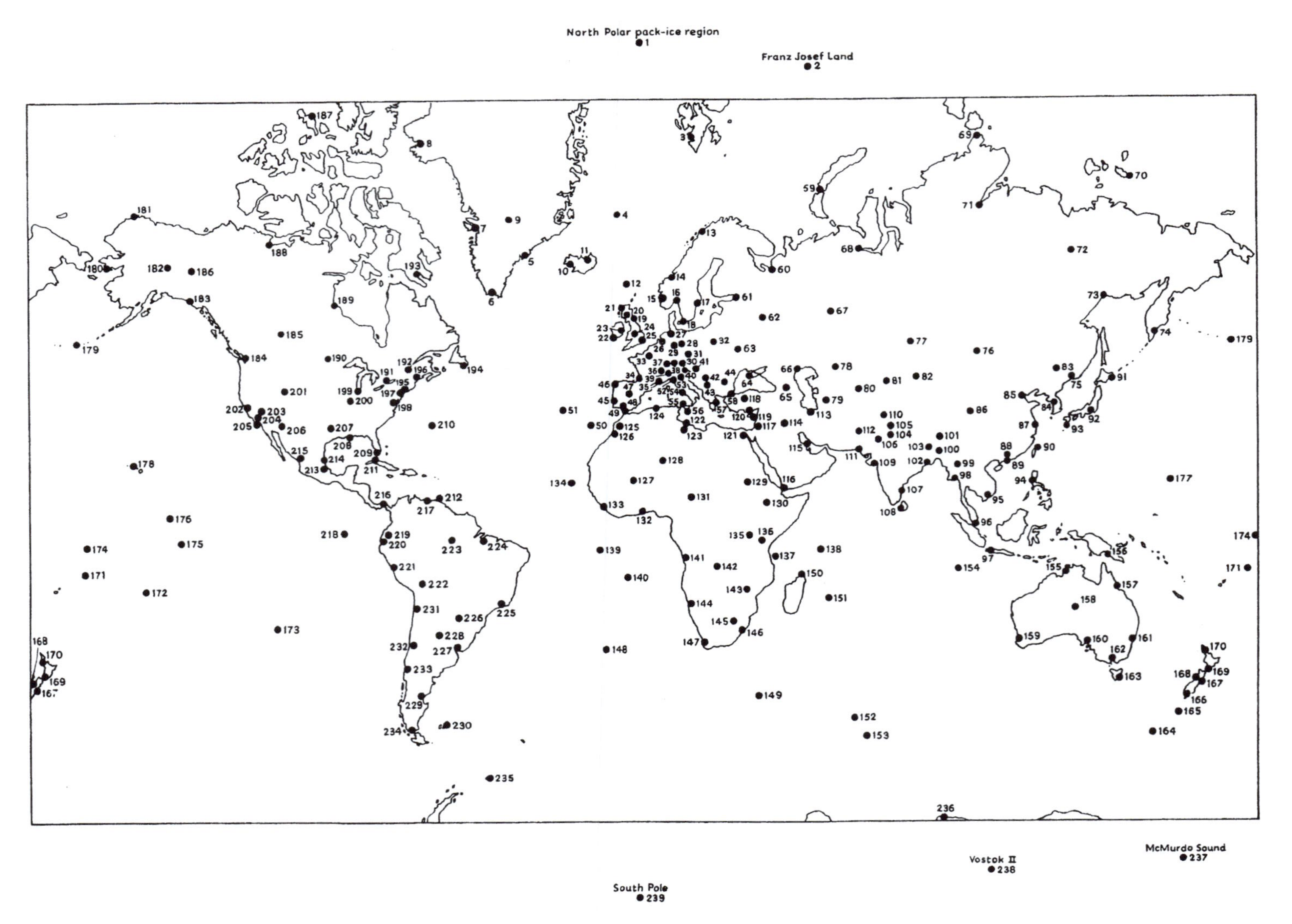

Fig. 11.10 Map of places in the World Climatic Table.

Principal sources used in compiling the world climatic table

World Weather Records, published by the Smithsonian Institution, Washington. *Smithsonian Misc. Collections*, **79** (1927), **90** (1944), **105** (1947).

World Weather Records, published by U.S. Department of Commerce, Weather Bureau, Washington (data for 1941–50, published in one vol., 1959; data for 1951–60, in 6 vols, published 1965–8).

Monthly Climatic Data for the World, published monthly by U.S. Department of Commerce, Weather Bureau (now the National Oceanic and Atmospheric Administration), Washington (used for data for the most recent years).

Climatic normals (CLINO) for CLIMAT and CLIMATSHIP stations for the period 1931–1960, published by the World Met. Org., Geneva (WMO – No. 117. TP. 52) 1962.

Tables of Temperature, Relative Humidity and Precipitation for the World, published by the Met. Office (H.M.S.O.), London (in 6 parts: M.O. 617 a–f) 1958.

CRADDOCK, J. M. (1964) The interannual variability of monthly mean temperature over the northern hemisphere, *Sci. Paper No. 20.* London (Met. Office).

NAGAO, T. (1951) Standard deviation of monthly mean temperature, *J. Met. Res.*, **3**, 434–448. Tokyo (Central Met. Observatory).

WORLD CLIMATIC TABLE

(with some comparative values for different epochs)

Abbreviations used:

T_m	average air temp. (°C), usually ½ (average max. + average min.)	$R : Q_5 \geqslant$	lower boundary of top quintile of monthly rainfall
σT_m	standard deviation of monthly mean air temp.	$R : Q_1 \leqslant$	upper boundary of lowest quintile of monthly rainfall
T_x	absolute max. air temp. observed	R_{dx}	greatest observed 24-hour rainfall
T_n	absolute min. air temp. observed	N_p	average number of days with precip.
drT	mean diurnal range of temp. (difference between mean max. and mean min.)		
R_m	average rainfall (mm)		* indicates change of observation site about the date marked.

Place, position and height above sea level	*Item*	*Years of observation covered*	J	F	M	A	M	J	J	A	S	O	N	D	*Year*
1 North Polar pack-ice region (Observations on the floating ice surface far from land 75–90°N)	T_m	1893–6	−35·6	−35·8	−30·3	−22·8	−11·0	−1·8	0·1	−1·7	−9·0	−21·7	−28·7	−32·4	−19·2
		1922–4, 1937–8	−30·4	−29·8	−30·9	−21·9	−12·2	−2·0	0·0	−0·7	−8·2	−16·4	−24·5	−28·0	−17·1
		1950–60	−33·7	−34·0	−32·2	−25·1	−10·7	−1·8	−0·2	−1·6	−9·2	−17·1	−26·9	−31·4	−18·7
	T_x	1893–6, 1937–8, 1952–5	−6·9	−5·4	−3·0	−3·6	2·2	4·0	3·5	2·9	1·0	−1·0	−5·0	−14·0	
	T_n		−50·3	−50·0	−52·0	−41·1	−28·4	−10·9	−3·4	−9·0	−31·0	−36·5	−44·1	−45·6	
	drT		0·7	0·2	1·6	3·5	2·0	1·3	0·8	1·0	0·9	0·5	0·7	0·4	
	R_m (equiv.)	1922–4, 1959–60	3–5	7–8	4–7	2–7	7–8	2–16	13–19	14–26	5–17	4–19	1–6	5–8	*c.* 100
	N_p	1893–6, 1954–5	10	11	13	13	19	18	20	20	19	12	8	8	171
2 Franz Josef Land (Bukhta Tikhaya) 82°20′N, 52°48′E (6 m)	T_m	1929–35	−18·0	−18·3	−21·9	−17·5	−8·3	−0·8	1·4	0·8	−2·2	−9·2	−14·4	−17·2	−10·5
	drT		8·5	9·5	7·5	7·0	4·5	4·0	4·0	4·0	3·5	5·0	6·5	7·5	
	R_m (equiv.)		8	8	5	5	5	8	20	28	18	8	8	5	126
3 Spitsbergen, west coast (Isfjord Radio) 78°04′N, 13°38′E (5 m)	T_m	1931–41, 1946–50	−9·2	−11·3	−11·8	−9·7	−3·3	2·0	5·2	4·7	1·4	−3·2	−6·1	−7·5	−4·0
		1960–8	−13·7	−14·3	−9·8	−9·5	−4·0	1·2	4·4	4·1	0·6	−4·9	−8·0	−11·3	−5·4
	σT_m		6·2	5·8	5·4	4·1	2·0	0·9	1·3	1·1	1·5	2·5	4·1	5·3	
	R_m (equiv.)	1931–41, 1946–60	26	26	24	15	20	19	25	42	35	39	36	35	342

4 Jan Mayen	T_m	1931–60	−4·0	−5·2	−4·8	−3·4	−0·5	2·4	5·2	5·5	3·8	0·9	−1·2	−2·9	−0·1
71°01′N, 8°25′W (12 m)	R_m (equiv.)	1951–60	79	54	63	58	23	28	36	61	83	93	82	75	735
5 Angmagssalik, Greenland	T_m	1931–59*	−6·8	−7·4	−5·5	−2·6	2·3	5·6	7·4	6·7	4·3	0·1	−3·0	−5·1	−0·3
65°37′N, 37°33′W (32 m)	R_m (equiv.)	1931–50	57	81	57	55	52	45	28	70	72	96	87	75	775
	R_{dx}	1898–1936	86	51	43	56	127	43	43	81	69	99	81	79	
6 Ivigtut, Greenland	T_m	1931–60	−5·3	−4·5	−2·8	0·4	5·2	8·4	9·8	8·8	6·0	1·6	−1·7	−4·1	1·8
61°12′N, 48°10′W (5 m)	drT		6·5	8·0	8·5	8·0	8·0	8·5	8·5	8·0	6·0	6·0	5·5	6·0	
	R_m (equiv.)	1931–50	86	125	86	81	99	100	75	109	172	187	144	77	1340
7 Jakobshavn, Greenland	T_m	1873–1950	−16·2	−17·3	−14·4	−8·7	0·1	5·1	7·9	6·7	2·3	−3·6	−8·3	−11·9	−5·1
69°13′N, 51°02′W (39 m)		1931–50	−13·3	−13·5	−11·5	−7·1	0·5	6·0	8·0	6·5	2·5	−3·3	−7·7	−9·9	−3·6
	σT_m	1873–1950	5·0	6·0	5·0	3·6	1·9	1·6	1·1	1·2	1·5	2·3	2·4	3·6	
8 Thule, Greenland	T_m	1951–60	−24·2	−25·1	−26·4	−17·1	−4·9	2·8	5·7	4·4	−2·2	−10·4	−17·6	−23·3	−11·5
76°31′N, 68°50′W (11 m)	R_m (equiv.)	1951–60	7	9	5	3	5	6	17	14	15	17	13	5	117
9 Eismitte, Greenland	T_m	1930–1	−41·7	−47·2	−40·0	−31·9	−21·1	−16·7	−12·2	−18·3	−22·2	−35·8	−42·8	−38·3	−30·6
70°53′N, 40°42′W (3000 m)	T_x	1930–1	−15·0	−22·8	−15·6	−12·2	−8·9	−5·0	−2·8	−5·6	−8·3	−13·3	−18·9	−19·4	
	T_n		−64·4	−64·4	−65·0	−58·9	−45·6	−30·0	−28·3	−35·0	−38·9	−56·1	−58·3	−56·7	
10 Stykkisholmur, Iceland	T_m	1851–1950	−1·7	−2·1	−1·7	0·8	4·7	8·2	10·0	9·4	6·9	3·4	0·7	0·7	3·0
65°05′N, 22°46′W (25 m)		1921–50	−0·7	−0·5	0·5	2·4	6·3	9·5	11·3	10·9	8·0	4·2	1·6	0·4	4·7
		1960–8	−0·4	−0·3	−0·7	1·8	5·0	8·4	10·1	9·5	7·4	4·7	1·0	−1·1	3·8
	σT_m	1873–1950	2·3	2·3	2·6	1·9	1·7	1·0	0·9	1·1	1·3	1·7	1·5	1·9	
	drT	1901–20	5·5	5·0	5·5	5·0	5·5	6·0	5·5	5·0	4·5	4·5	4·5	5·0	
	T_x	1901–20	9·4	9·4	8·9	13·3	20·0	18·3	21·7	20·6	17·8	13·3	10·6	10·6	
	T_n		−29·4	−19·4	−19·4	−17·2	−6·7	−1·7	2·8	−1·1	−2·8	−9·4	−13·3	−19·4	
	R_m (equiv.)	1856–1950	78	69	55	40	35	39	38	45	73	78	72	71	693
		1921–50	92	78	74	47	33	41	40	52	82	81	88	87	796
		1960–8	66	71	45	49	27	36	33	38	53	71	72	60	621
	R_{dx}	1901–20	38	36	18	28	15	33	20	25	28	51	25	25	
11 Akureyri, Iceland	T_m	1924–54	−1·1	−1·4	−0·6	1·7	6·1	9·4	11·1	10·3	7·8	3·6	0·8	0·3	4·0
65°41′N, 18°05′W (23 m)	T_x	1924–54	13·9	13·3	16·1	16·1	21·7	28·3	23·3	25·0	22·2	17·8	15·6	13·3	
	T_n		−20·0	−20·0	−22·2	−16·7	−8·9	−3·9	0·6	−1·1	−8·3	−11·7	−18·3	−20·7	
	R_m (equiv.)	1931–60	45	42	42	32	16	23	35	38	46	57	45	54	475
	$R : Q_5 \geq$	1931–60	63	58	60	45	20	36	46	57	71	85	69	73	
	$Q_1 \leq$		28	20	19	17	8	10	18	16	26	18	19	37	
	R_{dx}	1924–54	20	20	28	15	23	20	28	53	91	30	28	28	
12 Thorshavn, Faeroes	T_m	1931–60	3·9	3·7	4·6	5·4	7·3	9·2	11·0	11·1	10·0	7·9	6·1	5·0	7·0
62°03′N, 6°45′W (24 m)	R_m	1931–60	149	136	114	106	67	74	79	96	132	157	156	167	1433

World Climatic Table—*continued*

Place, position and height above sea level	Item	Years of observation covered	J	F	M	A	M	J	J	A	S	O	N	D	Year
13 Tromsø, Norway 69°39′N, 18°57′E (102 m)	T_m	1931–60	−2·7	−3·3	−2·0	1·0	4·6	8·7	12·0	11·1	7·7	3·7	0·5	−1·3	3·3
	drT	1875–1950	4·5	4·5	5·0	5·5	5·5	6·0	6·0	6·0	5·0	4·0	4·0	4·0	
	T_x	1875–1950	7·2	8·3	8·3	13·9	20·6	27·8	28·3	26·7	22·2	15·6	11·1	8·3	
	T_n	1875–1950	−17·2	−18·3	−16·1	−14·4	−8·9	−4·4	1·1	−1·1	−4·4	−13·3	−16·1	−16·7	
	R_m (equiv.)	1931–60	118	94	113	75	65	57	56	83	115	131	97	115	1119
14 Trondheim, Norway 63°25′N, 10°27′E (133 m)	T_m	1851–1946	−2·4	−2·1	−0·4	3·6	8·0	11·9	14·4	13·2	9·6	5·0	1·0	−1·5	5·0
		1762–1850	−3·9	−2·7	−0·9	3·5	8·4	12·4	14·8	13·6	9·9	4·1	0·2	−2·7	4·8
	σT_m	1851–1946	2·7	2·8	2·1	1·4	1·6	1·8	1·7	1·6	1·1	1·6	1·9	2·7	
15 Bergen, Norway 60°24′N, 5°19′E (43 m)	T_m	1851–1950	1·5	1·3	2·5	5·7	9·6	12·7	14·6	14·0	11·3	7·6	4·2	2·3	7·3
	σT_m	1851–1950	2·0	2·1	1·6	1·2	1·5	1·6	1·4	1·4	1·2	1·4	1·6	2·0	
	R_m	1931–60	179	139	109	140	83	126	141	167	228	236	207	203	1958
	R_{dx}	(1821–1950)	94	94	117	97	61	66	79	89	119	89	132	119	
16 Oslo/Christiania 59°55′N, 10°43′E (25 m)	T_m	1851–1950	−4·0	−3·6	−0·6	4·7	10·5	15·3	17·4	15·7	11·3	5·8	0·5	−2·9	5·8
		1931–40*	−2·5	−2·1	0·3	5·3	11·4	15·7	17·8	16·8	11·6	6·3	2·4	−1·5	6·8
	drT	(1816–1950)	5·5	6·5	8·5	9·0	10·5	10·0	9·5	9·0	8·5	6·5	4·5	4·0	
	T_x	(1816–1950)	11·7	13·9	17·2	23·9	28·3	33·9	32·8	31·1	25·0	21·1	13·9	11·7	
	T_n	(1816–1950)	−29·4	−27·8	−23·3	−15·0	−3·3	0·6	5·6	2·8	−3·3	−11·1	−16·7	−23·3	
	R_m	1866–1920	32	28	31	32	38	48	68	82	60	61	44	39	563
		1931–40	53	32	30	44	39	58	92	70	78	79	64	69	707
17 Stockholm 59°21′N, 18°04′E (45 m)	T_m	1851–1950	−2·9	−3·2	−1·1	3·6	9·1	14·3	17·2	15·7	11·6	6·5	1·7	−1·3	5·9
		1756–1850	−4·4	−3·9	−1·9	3·3	9·0	14·8	17·4	16·3	11·9	6·6	1·3	−2·2	5·6
		1931–60	−2·9	−3·1	−0·7	4·4	10·1	14·9	17·8	16·6	12·2	7·1	2·8	0·1	6·6
		1960–8	−3·8	−2·7	−0·1	4·7	10·1	15·9	17·3	15·5	12·3	7·9	2·5	−1·5	6·5
	T_x	(1756–1936)	10·6	12·2	15·0	25·0	28·9	32·8	36·1	32·8	28·9	20·0	13·9	11·1	
	T_n	(1756–1936)	−32·2	−30·0	−25·6	−22·2	−6·7	0·0	4·4	2·2	−3·3	−8·9	−17·8	−22·8	
	R_m	1931–60	43	30	26	31	34	45	61	76	60	48	53	48	555
		1960–8	38	28	18	28	37	45	91	87	57	53	46	53	581
	R_{dx}	(1756–1936)	18	23	20	38	45	33	69	69	43	36	41	28	
18 Copenhagen 55°41′N, 12°36′E (27 m)	T_m	1851–1950	−0·1	−0·3	1·5	5·9	11·0	15·2	17·1	16·4	13·1	8·5	4·1	1·3	7·8
	T_x	1910–30	9·4	10·0	16·7	27·8	27·8	31·1	32·8	31·7	28·9	23·3	13·3	11·7	
	T_n	1910–30	−14·4	−19·4	−10·0	−6·1	−1·1	2·8	5·6	4·4	−1·7	−4·4	−10·6	−16·1	
	R_m	1821–1950	40	34	35	37	41	50	62	67	55	58	50	44	573
		1931–60	49	39	32	38	42	47	71	66	62	59	48	49	602
	$R: Q_5 \geqslant$	1931–60	66	54	47	55	58	64	99	104	83	84	68	66	
	$Q_1 \leqslant$	1931–60	24	17	16	22	27	27	43	35	28	34	31	26	
	R_{dx}	1910–30	15	20	18	20	28	30	41	41	33	28	28	23	

Station	Element	Period													
19 Edinburgh 55°55′N, 3°11′W (76 m)	T_m	1851–1950	3·5	3·8	4·8	7·1	9·8	13·0	14·7	14·3	12·2	8·8	5·7	4·1	8·5
	σT_m	1851–1950	1·5	1·7	1·5	1·1	1·1	1·0	1·0	0·9	1·0	1·1	1·3	1·6	
	drT	1921–50	4·5	4·5	6·0	6·0	6·5	8·0	7·0	6·5	6·5	5·0	4·5	4·5	
	T_x	1921–50	13·9	13·9	20·0	22·2	24·4	28·3	28·3	27·8	26·1	21·7	19·4	13·9	
	T_n	1921–50	−8·3	−9·4	−6·1	−3·9	−1·1	2·8	5·6	4·4	0·6	−2·2	−4·4	−6·7	
	R_m	1851–1950	55	42	44	39	52	53	71	78	60	67	58	57	676
	R_{dx}	1921–50	25	28	43	33	38	33	48	71	53	38	38	25	
20 Ben Nevis, Scotland 56°48′N, 5°00′W (1343 m)	T_m	1884–1903	−4·4	−4·6	−4·4	−2·4	0·6	4·3	5·1	4·7	3·3	−0·3	−1·7	−3·8	−0·3
	T_x		9·1	7·8	7·8	11·4	13·3	19·1	17·8	17·5	17·0	14·0	11·1	7·2	
	T_n		−17·4	−16·8	−14·7	−11·5	−9·8	−5·1	−3·2	−2·8	−7·6	−9·8	−12·1	−13·8	
	R_m	1884–1903	466	344	387	215	201	192	274	339	400	392	390	484	4084
21 Stornoway, Outer Hebrides 58°11′N, 6°21′W (10 m)	T_m	1901–50	4·7	4·7	5·6	6·7	9·2	11·4	13·3	13·3	11·4	9·2	6·7	5·3	8·5
	R_m	1931–60	107	75	63	65	52	68	87	88	97	118	111	110	1041
	$R: Q_5 \geqslant$	1931–60	160	115	97	99	67	84	110	115	137	156	146	166	
	$Q_1 \leqslant$	1931–60	91	59	43	45	30	41	56	48	70	85	83	84	
22 Valentia, Ireland 51°56′N, 10°15′W (9 m)	T_m	1869–1950	7·2	7·2	7·6	9·1	11·4	13·8	14·9	15·1	13·7	11·2	8·8	7·7	10·6
		1921–50	7·5	7·4	8·2	9·3	11·6	13·9	15·2	15·1	13·9	11·7	9·1	7·9	10·9
	R_m	1871–1950	165	123	104	89	82	85	102	120	114	144	144	164	1436
23 Dublin 53°22′N, 6°21′W (47 m)	T_m	1911–50	5·0	5·0	6·4	7·8	10·3	13·6	15·0	15·0	12·8	10·0	6·9	5·3	9·4
	R_m	1911–50	69	56	51	48	58	51	71	76	71	69	69	66	755
	R_{dx}	1911–50	48	46	38	28	38	28	46	38	69	56	38	48	
24a England and Wales (Av. of 7 districts)	R_m	1851–1950	87	65	62	56	61	63	77	84	77	99	90	91	912
		1751–1850	62	61	57	51	56	73	79	86	82	97	83	81	868
		1916–50	94	68	58	61	65	57	81	83	78	94	97	90	926
		1960–9	81	61	61	65	72	62	78	87	90	91	99	95	942
24b Central England (lowland sites)	T_m	1851–1950	3·7	4·1	5·4	8·1	11·1	14·2	15·9	15·5	13·3	9·6	6·0	4·2	9·3
		1680–1700	2·3	2·6	4·3	7·0	10·5	13·6	15·3	14·7	12·2	9·1	5·4	3·7	8·4
⅓ (Stonyhurst or Nelson, Lancs. + Ross-on-Wye + Cambridge)		1921–50	3·9	4·2	5·9	8·3	11·3	14·4	16·3	15·8	13·6	10·0	6·4	4·4	9·6
		1960–9	3·3	3·6	5·6	8·3	11·3	14·5	15·4	15·3	13·5	10·9	6·2	3·4	9·3
	σT_m	1851–1950	1·8	1·9	1·4	1·1	1·1	1·0	1·2	1·1	1·1	1·2	1·4	1·8	
25 London (Greenwich) 51°29′N, 0°00′ (45 m)	T_m	1921–50	4·2	4·4	6·7	8·9	12·2	15·6	17·8	17·2	15·0	10·6	6·7	4·7	10·3
	drT	1921–50	5·0	5·5	8·0	9·0	10·0	10·0	10·0	10·0	9·0	8·0	5·5	5·0	
	T_x	1921–50	15·0	16·7	22·2	28·9	32·8	33·9	35·0	37·2	31·7	28·9	20·0	15·6	
	T_n	1921–50	−11·7	−12·8	−6·1	−3·9	−2·2	1·7	6·7	3·9	1·1	−5·0	−5·0	−7·8	
	R_m	1921–50	51	38	36	46	46	41	51	56	46	58	63	51	583
	R_{dx}	1921–50	41	20	28	18	25	41	36	56	33	41	36	28	

Place, position and height above sea level	Item	Years of observation covered	J	F	M	A	M	J	J	A	S	O	N	D	Year
26 Utrecht (De Bilt), Netherlands 52°06′N, 5°11′E (3 m)	T_m	1851–1950	1·8	2·5	4·7	8·1	12·1	15·3	16·9	16·5	14·0	9·7	5·2	2·6	9·1
	σT_m	1851–1950	2·4	2·6	1·7	1·4	1·4	1·4	1·3	1·3	1·2	1·3	1·6	2·3	
	R_m	1931–60	68	52	45	49	52	58	77	88	71	72	70	63	765
	$R: Q_5 \geqslant$	1931–60	97	75	64	67	67	83	111	128	97	105	117	87	
	$Q_1 \leqslant$		40	27	24	28	29	37	41	45	37	27	39	44	
27 Hamburg 53°38′N, 10°00′E (16 m)	T_m	1931–60	0·0	0·4	3·3	7·6	12·2	15·6	17·3	16·8	13·6	9·1	4·9	1·8	8·6
	R_m	1931–60	57	48	39	52	53	64	84	83	63	59	59	59	720
28 Berlin (inner) 52°30′N, 13°20′E (35 m)	T_m	1851–1950*	−0·1	0·8	3·9	8·6	13·7	17·3	19·0	18·1	14·6	9·4	4·1	1·0	9·2
		1751–1850	−2·2	0·3	3·0	8·5	13·8	17·2	18·7	18·0	14·3	8·9	3·6	0·1	8·7
		1931–60	−0·5	0·2	3·9	9·0	14·3	17·7	19·4	18·8	15·0	9·6	4·7	1·2	9·5
	σT_m	1851–1950	2·9	3·0	2·1	1·7	1·9	1·6	1·3	1·3	1·3	1·5	1·7	2·2	
	T_x	1881–1930	12·8	16·7	22·8	28·3	33·3	33·9	35·6	34·4	34·4	25·0	17·8	14·4	
	T_n		−17·2	−26·1	−13·9	−6·7	−2·2	1·7	6·1	6·1	−0·6	−9·4	−13·3	−19·4	
	R_m	1851–1950*	43	37	38	40	48	60	76	61	44	46	44	46	583
		1921–50	48	40	29	46	51	61	79	65	44	51	49	39	602
	R_{dx}	1881–1930	28	15	13	25	33	38	66	58	41	18	48	13	
29 Brocken (Harz), Germany 51°48′N, 10°37′E (1142 m)	T_m	1931–60	−4·6	−4·7	−2·0	1·2	5·7	9·1	10·8	10·7	7·9	3·6	−0·3	−3·0	2·9
	drT	1851–1930	5·0	5·0	4·5	5·5	6·5	6·0	6·0	6·0	5·5	4·5	4·5	4·0	
	T_x	1851–1930	10·6	14·4	15·0	20·6	24·4	24·4	26·7	25·0	24·4	20·0	15·0	10·6	
	T_n		−24·4	−26·1	−17·2	−13·9	−7·2	−2·8	0·0	0·0	−2·2	−11·1	−16·7	−20·6	
	R_m	1931–60	158	126	94	105	96	115	143	117	105	122	115	126	1422
30 Munich/München 48°08′N, 11°42′E (529 m)	T_m	1931–60	−2·2	−1·0	3·3	7·9	12·5	15·9	17·7	16·9	13·7	8·2	3·1	−0·7	7·9
		1960–8	−2·5	0·0	3·1	8·8	11·9	16·2	17·3	16·6	14·0	9·3	3·7	−1·5	8·1
	R_m	1931–60	59	55	51	62	107	125	140	104	87	67	57	50	964
	$R: Q_5 \geqslant$	1931–60	81	88	71	85	129	160	173	141	131	100	77	69	
	$Q_1 \leqslant$		33	27	31	37	63	98	89	75	55	29	29	32	
	R_{dx}	1851–1933	15	18	28	63	53	43	58	71	30	41	25	13	
31 Prague (Klementinum) 50°05′N, 14°25′E (202 m)	T_m	1851–1950	−0·8	0·4	4·0	9·0	14·2	17·7	19·5	18·7	15·0	9·5	3·9	0·4	9·3
	R_m	1851–1950	20	20	25	36	54	61	61	57	38	31	27	24	454
32 Warsaw 52°13′N, 21°01′E (120 m)	T_m	1851–1950	−3·3	−2·3	1·4	7·5	13·6	17·2	18·8	17·7	13·6	8·0	2·3	−1·7	7·7
	drT	(1803–1933)	5·0	5·0	7·0	9·0	10·5	10·5	10·5	10·0	9·5	7·0	4·5	4·0	
	T_x		10·0	12·2	20·6	23·9	33·9	32·2	35·0	36·7	31·1	25·0	15·6	10·6	
	T_n		−30·0	−22·2	−20·0	−4·4	−1·1	2·8	6·7	5·0	0·0	−8·3	−17·2	−20·6	
	R_m	1851–1950	32	29	33	39	51	65	80	73	45	42	39	36	564
	R_{dx}	(1803–1933)	30	15	33	38	38	51	86	74	46	36	33	25	

33 Paris (Parc St Maur)	T_m	1851–1950	2·8	3·9	6·4	10·0	13·5	16·7	18·5	17·9	15·0	10·4	6·0	3·3	10·4
48°48′N, 2°30′E (50 m)	T_x	(1821–1950)	15·0	17·8	23·9	31·1	33·3	36·7	40·0	36·1	34·4	28·3	20·6	16·7	
	T_n	(1821–1950)	−13·9	−15·6	−6·7	−2·2	0·6	4·4	6·1	6·1	1·7	−3·9	−9·4	−13·3	
	R_m	1821–1950	38	33	38	43	51	53	53	51	51	56	51	48	566
	R_{dx}	(1821–1950)	23	25	28	28	41	38	36	33	30	51	28	33	
34 Bordeaux (Merignac)	T_m	1931–60	5·2	5·9	9·3	11·7	14·7	18·0	19·6	19·5	17·1	12·7	8·4	5·7	12·3
44°50′N, 0°42′W (49 m)	drT	1891–1950	7·0	9·0	10·0	10·5	11·0	11·5	12·0	12·5	11·5	10·5	8·0	6·5	
	T_x	1891–1950	18·9	26·1	28·9	31·1	35·0	38·3	38·3	38·9	37·8	31·1	25·0	20·0	
	T_n	1891–1950	−12·2	−11·7	−6·1	−5·6	0·6	2·8	5·6	0·6	−1·7	−5·6	−7·2	−12·8	
	R_m	1931–60	90	75	63	48	61	65	56	70	84	83	96	109	900
	R_{dx}	(1891–1950)	36	38	41	48	33	53	41	48	38	38	48	28	
35 Marseilles	T_m	1851–1950*	6·7	7·6	9·8	12·8	16·4	20·0	22·5	22·1	19·5	15·2	10·5	7·3	14·2
43°27′N, 5°23′E (75 m)		1931–60	5·5	6·6	10·0	13·0	16·8	20·8	23·3	22·8	19·9	15·0	10·2	6·9	14·2
	σT_m	1851–1950	1·8	1·8	1·4	1·0	1·2	1·1	1·2	1·1	1·3	1·5	1·5	1·9	
	T_x	1823–1949	20·6	21·7	26·1	28·3	33·9	37·2	37·8	38·3	33·3	30·0	24·4	21·1	
	T_n	1823–1949	−10·0	−12·8	−6·7	−2·2	0·0	5·0	8·3	8·3	1·1	−2·8	−6·1	−11·7	
	R_m	1823–1949	48	38	46	51	48	25	15	23	66	94	79	56	589
	R_{dx}	(1823–1949)	56	58	51	48	74	46	104	74	211	221	150	91	
36 Geneva	T_m	1851–1950	0·6	1·9	5·3	9·5	13·7	17·4	19·5	18·7	15·2	9·9	5·0	1·4	9·8
46°12′N, 6°09′E (405 m)	R_m	1901–40	50	54	72	73	71	83	77	102	89	87	79	79	916
37 Basle (Binningen)	T_m	1851–1950	−0·2	1·4	4·7	8·9	13·0	16·5	18·3	17·6	14·2	9·1	4·0	0·6	9·0
47°33′N, 7°35′E (317 m)	σT_m	1851–1950	2·7	2·5	1·8	1·5	1·6	1·4	1·6	1·4	1·5	1·5	1·7	2·7	
	drT	(1864–1950)	6·0	8·0	10·0	11·0	11·5	11·5	12·0	11·5	10·5	9·0	6·0	5·5	
	T_x	(1864–1950)	18·9	19·4	22·8	30·6	33·3	38·3	38·9	38·9	35·0	29·4	21·7	18·3	
	T_n	(1864–1950)	−23·9	−23·9	−14·4	−5·6	−2·8	2·8	5·0	3·3	−1·1	−5·6	−10·6	−21·1	
	R_m	1871–1950	43	41	50	64	82	93	87	84	77	71	59	50	801
	R_{dx}	(1864–1950)	28	46	30	25	61	46	43	48	53	43	51	36	
38 Zürich	T_m	1871–1950	−1·6	0·0	3·8	7·9	12·1	15·6	17·3	16·6	13·3	8·1	3·1	−0·3	8·0
47°23′N, 8°34′E (569 m)	R_m	1901–40	64	55	74	94	114	140	142	132	104	87	67	76	1149
39 St Gotthard	T_m	1871–1950	−7·4	−7·2	−5·2	−2·0	1·9	5·4	8·1	8·0	5·2	0·7	−3·6	−6·6	−0·2
46°33′N, 8°34′E (2095 m)		1781–90	−7·4	−8·3	−7·3	−3·1	2·4	5·4	7·4	7·4	5·0	−0·2	−5·1	−7·1	−0·9
		1951–60	−7·5	−7·5	−4·6	−2·1	2·2	5·7	8·4	7·8	5·7	1·1	−3·1	−5·3	0·1
	R_m	1901–40	185	186	178	209	211	172	167	206	202	232	210	169	2327
40 Sonnblick	T_m	1878–1950	−12·9	−13·1	−11·7	−8·8	−4·1	−1·1	1·1	0·9	−1·2	−4·7	−8·9	−11·8	−6·3
47°03′N, 12°57′E (3106 m)		1931–60	−13·2	−13·0	−11·2	−8·2	−3·8	−0·6	1·6	1·4	−0·5	−4·3	−8·3	−11·4	−6·0
	drT	1878–1950	5·0	5·0	5·0	5·0	4·0	4·5	4·5	4·5	4·0	3·5	4·0	4·5	
	T_x	1878–1950	1·1	3·3	3·3	3·9	9·4	12·8	13·9	12·2	10·0	8·3	5·6	1·1	
	T_n	1878–1950	−37·2	−36·7	−34·4	−26·7	−20·0	−15·6	−10·6	−10·0	−16·1	−20·6	−28·3	−32·8	
	R_m	1878–1950	109	122	132	152	150	124	137	122	104	117	107	119	1485
		1931–60	115	109	112	153	136	142	154	134	104	118	108	111	1495

Place, position and height above sea level	Item	Years of observation covered	J	F	M	A	M	J	J	A	S	O	N	D	Year
41 Vienna/Wien (Hohe Warte) 48°15′N, 16°22′E (203 m)	T_m	1851–1950	−1·3	0·3	4·3	9·5	14·2	17·5	19·4	18·6	15·0	9·7	3·9	0·1	9·3
		1931–60	−1·4	0·4	4·7	10·3	14·8	18·1	19·9	19·3	15·6	9·8	4·8	1·0	9·8
	σT_m	1851–1950	2·8	2·8	2·1	1·7	1·7	1·3	1·3	1·2	1·4	1·6	1·9	2·5	
	T_x	1851–1950	16·7	19·4	23·3	28·3	33·3	36·1	36·7	36·1	31·1	27·8	21·7	18·9	
	T_n	1851–1950	−22·2	−25·6	−16·1	−7·8	−2·8	3·9	7·2	5·6	−0·6	−8·9	−14·4	−20·0	
	R_m	1851–1950	38	36	45	52	71	69	77	69	50	52	48	46	653
		1931–60	40	43	45	45	70	67	83	72	41	56	53	45	660
	$R: Q_5 \geq$	1931–60	50	61	59	75	90	89	110	104	51	93	81	62	
	$Q_1 \leq$	1931–60	25	16	24	20	28	30	50	40	14	13	26	17	
	R_{dx}	(1851–1950)	23	61	46	48	76	66	74	43	86	69	48	51	
42 Budapest 47°17′N, 19°01′E (118 m)	T_m	1851–1950	−1·1	0·8	5·6	11·5	16·7	20·0	22·0	21·1	16·8	11·3	4·9	0·5	10·8
	drT	(1871–1950)	5·0	6·5	8·5	10·0	11·0	11·5	11·5	12·0	11·5	9·0	5·5	4·0	
	R_m	1851–1950	40	35	44	53	71	69	52	50	46	55	57	50	622
43 Belgrade 44°48′N, 20°27′E (132 m)	T_m	1931–60	−0·2	1·6	6·2	12·2	17·1	20·5	22·6	22·0	18·3	12·5	6·8	2·5	11·8
	T_x	1920–35	16·7	20·0	26·7	31·1	33·3	36·7	39·4	41·7	35·6	34·4	29·4	20·6	
	T_n	1920–35	−18·9	−25·6	−14·4	−6·1	−1·7	5·0	9·4	7·2	1·7	−12·8	−11·1	−19·4	
	R_m	1931–60	48	46	46	54	75	96	60	55	50	55	61	55	701
44 Bucharest 44°25′N, 26°06′E (91 m)	T_m	1857–1950	−3·2	−0·8	4·9	11·4	16·9	20·5	22·9	22·3	17·7	11·8	4·9	−0·5	10·7
	σT_m	1857–1950	3·2	3·2	2·7	1·7	1·7	1·2	1·2	1·4	1·7	1·9	2·6	2·7	
	T_x	1889–1938	16·1	22·2	26·7	32·8	35·6	40·6	40·6	40·6	36·1	31·1	29·4	20·6	
	T_n	1889–1938	−27·8	−24·4	−14·4	−5·6	0·0	4·4	7·8	6·7	−1·7	−10·6	−17·8	−26·7	
	R_m	1865–1924	34	28	42	44	63	88	68	51	40	43	48	41	588
45 Lisbon 38°43′N, 9°08′W (95 m)	T_m	1856–1950	10·4	11·2	12·8	14·4	16·5	19·4	21·3	21·8	20·3	17·2	13·7	11·1	15·9
	drT	1856–1941	5·5	6·0	6·5	6·5	7·0	8·5	9·0	9·0	8·0	6·5	5·5	5·5	
	T_x	1856–1941	18·9	22·8	28·3	30·6	34·4	37·2	39·4	37·8	37·2	31·7	25·0	18·9	
	T_n	1856–1941	−1·1	−1·7	1·1	4·4	5·6	9·4	11·1	12·8	10·6	6·1	1·1	−0·6	
	R_m	1856–1950	89	83	87	64	42	17	5	4	36	75	101	98	701
	R_{dx}	1856–1941	81	51	61	53	43	36	38	46	76	56	91	112	
46 La Coruña, Spain 43°22′N, 8°24′W (58 m)	T_m	1881–1950	9·5	9·8	10·8	11·9	14·0	16·2	17·8	18·2	17·3	14·8	12·0	10·3	13·5
	R_m	1931–60	121	80	95	70	60	46	29	47	71	92	125	139	975
47 Madrid 40°25′N, 3°41′W (667 m)	T_m	1931–60	4·9	6·5	10·0	13·0	15·7	20·6	24·2	23·6	19·8	14·0	8·9	5·6	13·9
	T_x	1913–47	17·8	21·1	25·6	29·4	32·8	36·7	38·3	38·9	35·6	30·0	22·2	17·8	
	T_n	1913–47	−10·0	−7·2	−3·9	−1·7	0·0	4·4	8·3	9·4	3·9	−1·1	−3·9	−8·9	
	R_m	1931–60	38	34	45	44	44	27	11	14	31	53	47	48	436

48 Seville, Spain 37°29′N, 5°59′W (39 m)	T_m	(1913–47)	10·0	11·7	14·2	16·7	20·3	24·4	27·5	28·1	24·7	19·7	14·4	11·1	18·6
	drT	(1913–47)	10·0	10·0	10·5	12·0	13·0	14·5	16·0	16·0	14·0	11·5	10·0	9·0	
	T_x	(1913–47)	22·8	26·7	30·0	32·8	38·9	45·0	45·6	47·2	42·8	38·9	27·8	22·2	
	T_n		−2·8	−1·7	0·0	3·9	2·8	5·0	11·1	12·2	8·9	6·1	0·0	−2·8	
	R_m	(1913–47)	56	74	84	58	33	23	3	3	28	66	94	71	593
49 Gibraltar 36°06′N, 5°21′W (3 m)	T_m	1851–1950	12·7	13·2	14·1	15·8	18·0	20·7	23·0	23·7	22·0	18·8	15·7	13·3	17·6
	σT_m	1851–1950	1·0	1·0	0·9	0·9	1·1	1·1	0·9	0·9	1·0	1·0	1·0	1·0	
	T_x	1945–1957	23·3	23·9	26·7	27·8	30·6	32·8	38·3	37·2	33·3	33·3	28·9	23·9	
	T_n		2·2	0·6	3·3	8·3	8·3	13·9	14·4	13·9	13·9	10·0	7·8	2·2	
	R_m	1852–1930*	119	114	119	67	41	14	1	2	33	83	165	136	894
50 Madeira (Funchal) 32°38′N, 16°54′W (25 m)	T_m	1880–1950*	15·6	15·4	15·7	16·5	17·7	19·7	21·4	22·3	22·0	20·6	18·4	16·5	18·5
		1921–50*	16·0	15·9	16·2	17·0	18·0	19·8	21·5	22·3	22·0	20·9	18·8	16·9	18·8
	drT	1901–50	5·5	5·0	5·5	5·0	5·0	5·0	5·0	5·0	5·0	5·0	5·5	5·0	
	R_m	1880–1950	77	85	83	43	24	8	2	2	29	91	107	84	635
		1921–50	63	74	78	34	19	6	1	2	26	78	89	84	554
		1960–9	118	129	73	22	17	21	0·2	3	22	90	97	78	670
51 Ponta Delgada, São Miguel Is., Azores 37°45′N, 25°40′W (36 m)	T_m	1894–1950	14·5	14·2	14·3	15·2	16·5	18·8	20·9	21·9	20·9	18·8	16·8	15·5	17·4
	σT_m	1894–1950	0·7	0·6	0·7	0·6	0·6	0·7	0·7	0·6	0·7	0·7	0·7	0·7	
	R_m	1894–1940	77	77	63	57	54	35	21	36	66	84	85	85	740
		1941–60	132	108	123	65	71	39	29	27	78	103	135	97	1007
	$R: Q_5 \geqslant$	1931–60	165	142	145	101	91	51	47	47	117	130	172	130	
	$Q_1 \leqslant$		65	55	61	32	25	19	9	14	39	49	69	72	
52 Milan 45°27′N, 9°11′E (147 m)	T_m	1841–1960	1·4	3·9	8·3	13·2	17·7	21·8	24·3	23·3	19·4	13·4	6·9	2·7	13·0
		1931–60	1·8	4·1	8·7	13·7	18·1	22·4	25·0	24·0	20·0	13·6	7·5	3·2	13·5
	T_x	(1901–55)	17·8	20·6	24·4	30·0	31·7	34·4	35·6	35·6	31·1	26·7	20·6	15·6	
	T_n		−15·0	−15·0	−7·2	0·0	4·4	5·6	10·6	9·4	5·6	−0·6	−5·0	−13·3	
	R_m	1851–1950	57	56	80	90	104	82	68	71	86	118	103	77	992
		1921–50	60	56	66	83	101	75	51	68	72	91	98	70	891
	R_{dx}	(1901–55)	43	61	58	53	63	63	66	91	76	71	104	56	
53 Padua/Padova, Italy 45°24′N, 11°51′E (13 m)	R_m	1851–1950	50	47	69	74	88	81	56	59	74	91	81	59	829
		1751–1850	61	52	57	77	86	88	69	68	81	96	87	65	887
54 Rome 41°54′N, 12°29′E (18 m)	T_m	1851–1950	6·8	8·0	10·4	13·8	17·9	21·9	24·7	24·4	21·2	16·6	11·5	8·0	15·4
		1951–60	7·6	8·6	10·7	13·9	18·2	22·5	25·0	24·7	21·6	16·7	12·0	9·4	15·9
	σT_m	1851–1950	1·5	1·5	1·3	0·9	1·3	1·1	1·1	1·0	1·2	1·2	1·4	1·5	
	drT	1946–55	8·5	9·5	11·0	12·0	10·5	12·0	13·5	13·5	12·0	11·0	9·5	8·5	
	R_m	1851–1950	79	69	72	67	57	39	15	25	69	126	117	99	834
55 Palermo, Sicily 38°07′N, 13°19′E (105 m)	T_m	1851–1950	10·7	11·2	12·7	15·2	18·5	22·1	24·9	25·2	23·2	19·8	15·7	12·2	17·6
	T_x	(1880–1955)	21·7	25·6	33·8	30·0	35·0	39·4	40·6	45·0	41·1	32·2	28·9	26·1	
	T_n		1·7	0·0	−0·6	5·6	8·9	11·1	15·0	17·2	10·6	8·9	5·0	3·7	
	R_m	1880–1955	127	92	66	43	20	5	3	40	31	93	99	89	708
	R_{dx}	1880–1955	58	61	43	46	51	38	18	38	61	66	91	99	

Place, position and height above sea level	Item	Years of observation covered	J	F	M	A	M	J	J	A	S	O	N	D	Year
56a Malta (Valetta)	T_m	1853–1950*	12·8	12·8	13·9	15·8	18·8	22·7	25·7	26·2	24·5	21·6	17·8	14·4	18·9
35°54′N, 14°31′E (71 m)		1921–50	12·5	12·7	13·7	15·9	18·9	22·8	25·8	26·3	24·7	21·7	18·1	14·2	18·9
	drT	1853–1947	4·5	4·5	5·5	5·5	5·5	6·5	6·5	6·5	5·5	5·5	5·0	4·5	
56b Malta (central plateau)	R_m	1854–1953	85	53	43	23	10	3	4	5	28	76	97	103	530
35°52′N, 14°23′E (200 m)		1928–42	112	87	46	23	12	1	2	4	35	65	115	122	624
57 Athens (Observatory)	T_m	1858–1950	9·0	9·6	11·8	15·5	20·1	24·5	27·5	27·3	23·9	19·5	14·7	11·1	17·9
37°58′N, 23°43′E (107 m)	T_x	1858–1938	20·6	22·8	28·3	32·8	38·3	42·8	41·1	41·7	39·4	35·0	30·6	21·7	
	T_n	1858–1938	−6·7	−6·1	−6·7	1·7	5·6	12·2	14·4	15·0	8·9	7·2	−1·1	−4·4	
	R_m	1861–1950	57	40	35	21	21	15	6	9	16	44	65	70	399
	R_{dx}	1858–1938	53	41	43	41	38	43	53	43	53	84	150	74	
58 Istambul (Goztepe)	T_m	1929–50	4·6	5·4	6·4	11·2	16·2	20·6	23·0	23·1	19·5	14·9	11·6	7·6	13·7
40°58′N, 29°05′E (40 m)	drT	1934–51	5·0	5·5	7·0	9·0	8·5	9·5	9·0	8·5	8·0	7·0	6·0	5·5	
	R_m	1929–50	94	74	60	37	30	26	27	15	48	72	83	83	648
59 Malye Karmakuly,	T_m	(1882–1920)	−16·7	−16·3	−15·9	−10·3	−4·6	1·0	6·1	6·0	2·0	−3·3	−11·2	−13·9	−6·4
Novaya Zemlya		1943–50	−11·2	−14·9	−13·3	−7·7	−3·6	2·0	6·0	6·5	3·9	−1·1	−5·6	−9·2	−4·0
73°23′N, 52°42′E (16 m)	R_m	1896–1910	11	15	13	9	27	21	35	39	44	40	13	12	279
60 Archangel	T_m	1851–1950*	−12·9	−12·4	−8·0	−1·0	5·4	12·3	15·8	13·5	8·1	1·4	−5·1	−10·5	0·5
64°35′N, 40°36′E (7 m)	drT	1876–1911	4·0	6·5	8·0	7·0	6·0	6·5	7·0	6·0	4·5	3·5	3·5	4·0	
	T_x	1876–1911	2·8	2·8	7·8	19·4	26·7	31·1	32·8	31·1	23·3	16·1	7·2	2·8	
	T_n	1876–1911	−45·0	−40·6	−37·2	−27·2	−13·3	−3·3	1·1	0·0	−6·7	−20·6	−26·1	−35·6	
	R_m	1881–1916	23	19	21	18	31	47	64	63	55	42	30	24	437
61 Leningrad/St Petersburg	T_m	1851–1950	−7·9	−8·1	−4·2	2·6	−9·2	14·8	17·8	15·9	10·9	4·8	−0·9	−5·5	4·1
(town) 59°56′N, 30°16′E		1751–1850	−9·7	8·5	−4·6	2·1	8·9	14·9	17·7	16·2	10·7	4·3	−1·5	−6·9	3·6
(4 m)		1931–60	−7·6	−7·9	−4·3	3·3	9·9	15·4	18·4	16·8	11·2	5·1	−0·2	−4·4	4·6
	σT_m	1871–1950	3·7	3·7	2·6	2·1	2·3	1·7	1·7	1·5	1·5	2·0	2·3	3·3	
	T_x		6·1	5·6	13·3	22·8	30·0	32·8	32·2	32·2	27·2	21·1	12·2	7·2	
	T_n		−36·1	−34·4	−31·7	−21·7	−6·7	0·0	6·1	2·8	−1·7	−12·8	−22·2	−37·8	
	R_m	1881–1950	28	27	25	33	45	57	61	82	61	48	40	32	539
		1931–60	36	32	25	34	41	54	69	77	58	52	45	36	559
62 Moscow/Moskva	T_m	1851–1950	−10·3	−9·7	−4·6	3·8	11·8	16·1	18·3	16·3	10·7	4·3	−2·2	−7·9	3·9
55°50′N, 37°33′E (164 m)		1960–8	−10·4	−8·2	−3·6	5·4	12·8	16·8	18·6	16·7	11·0	5·5	−1·7	−5·6	4·8
	drT	1896–1910	6·5	7·0	8·5	9·0	11·5	12·0	11·5	11·0	10·0	6·5	4·5	5·5	
	T_x	1896–1910	3·9	4·4	11·7	26·7	30·0	34·4	35·0	35·6	30·0	23·3	8·3	5·6	
	T_n	1896–1910	−32·8	−27·2	−28·9	−13·3	−3·9	−0·6	6·1	2·2	−3·3	−8·3	−20·6	−32·8	
	R_m	1931–60	31	28	33	35	52	67	74	74	58	51	36	36	575
		1960–8	49	42	39	35	57	51	90	81	57	46	38	57	642
	R_{dx}	1896–1910	13	13	23	25	33	51	46	30	28	41	23	18	

Station	Element	Period													Year
63 Kiev	T_m	1851–1950	−6·0	−5·3	−0·6	7·2	14·5	17·6	19·5	18·5	13·8	7·5	1·0	−3·9	7·0
50°24′N, 30°27′E (179 m)	R_m	1931–60	43	39	35	46	56	66	70	72	47	47	53	41	615
64 Simferopol, Crimea	T_m	1931–60	−0·4	0·0	2·7	9·1	15·0	19·0	21·6	20·7	15·5	10·5	5·7	1·8	10·1
45°01′N, 33°59′E (205 m)	R_m	1931–60	44	40	35	36	49	75	58	34	31	41	44	41	528
65 Tiflis/Tblisi	T_m	1851–1950	0·6	2·3	6·6	11·8	17·4	21·2	24·4	24·2	19·5	13·8	7·7	2·8	12·7
41°41′N, 44°57′E (490 m)		1931–60	1·3	3·1	6·0	12·1	17·5	21·4	24·6	24·4	19·8	13·7	7·8	2·9	12·9
	R_m	1931–60	20	21	36	43	87	69	50	37	42	46	37	20	508
	R_{dx}	1875–1910	13	18	36	33	61	66	38	66	48	33	43	46	
66 Astrakhan	T_m	1851–1950	−6·8	−6·0	0·2	9·4	18·0	22·8	25·3	23·4	17·3	9·8	2·8	−3·2	9·4
46°22′N, 48°03′E (−14 m)	drT	1876–1912	5·0	5·5	8·5	9·5	9·5	8·5	9·0	9·5	9·0	9·0	5·5	4·5	
	R_m	1876–1912	13	12	10	15	15	18	13	10	15	10	15	15	161
67 Sverdlovsk/Ekaterinburg	T_m	1846–1950	−16·1	−13·7	−7·5	2·1	9·8	15·2	17·4	15·0	9·0	1·1	−6·8	−13·7	1·0
56°48′N, 60°38′E (237 m)		1931–60	−14·6	−13·4	−7·5	3·3	10·3	16·4	17·8	15·8	9·4	1·9	−7·1	−13·0	1·6
	drT	1882–1911	6·0	7·0	8·5	9·0	10·0	9·0	9·0	8·5	7·0	5·0	5·0	5·5	
	R_m	1890–1915	16	13	12	21	47	66	73	67	39	28	30	23	435
		1931–60	15	17	17	22	40	59	80	82	49	29	25	27	462
	R_{dx}	1882–1911	15	10	15	25	33	66	76	79	33	20	28	15	
68 Salekhard/Obdorsk	T_m	1882–1915*	−25·6	−21·8	−18·0	−10·5	−2·2	7·1	13·8	11·1	5·1	−4·9	−16·7	−21·9	−7·0
66°31′N, 66°35′E (25 m)		1931–60	−22·0	−22·3	−18·8	−7·8	−0·6	8·8	14·1	12·0	5·6	−3·0	−13·5	−20·6	−5·7
		1960–8	−24·7	−23·2	−18·4	−8·5	−1·5	7·8	14·7	10·7	4·9	−5·0	−16·6	−21·8	−6·8
	σT_m		3·9	4·3	4·4	3·3	2·6	2·3	1·9	2·0	1·6	2·6	3·5	4·0	
	T_x	1883–1911	0·0	1·1	3·3	8·9	21·1	27·8	29·4	27·2	20·0	13·9	5·6	1·7	
	T_n	1883–1911	−52·8	−53·9	−47·2	−33·3	−25·6	−6·7	0·5	−1·1	−9·4	−27·2	−47·2	−51·7	
	drT	1883–1911	4·5	6·5	8·5	8·0	5·5	5·0	6·5	6·0	4·0	3·5	4·0	4·5	
	R_m	1882–1915*	7	8	6	6	17	35	49	50	38	18	12	10	256
		1931–59	24	20	24	32	39	51	57	57	54	46	31	29	464
		1960–8	19	26	24	25	28	53	81	58	58	36	22	20	450
67 Cape/Mys Chelyuskin	T_m	1932–60	−27·3	−27·4	−28·0	−21·0	−9·7	−1·0	1·5	0·8	−2·2	−10·3	−20·1	−25·0	−14·1
77°43′N, 104°17′E (13 m)	R_m	1932–60	12	14	18	21	24	25	27	28	24	21	16	14	244
70 Wrangel Island	T_m	1927–38	−23·6	−25·6	−23·3	−17·2	−8·6	0·3	3·1	1·9	−1·9	−8·6	−17·2	−20·8	−11·8
70°58′N, 178°33′W (3 m)	R_m	1927–38	5	5	5	5	5	10	15	23	13	10	3	5	104
71 Khatanga	T_m	1933–60	−31·7	−30·6	−28·6	−17·9	−6·3	5·7	12·5	8·8	1·7	−11·3	−25·8	−29·4	−4·4
71°59′N, 102°28′E (24 m)	R_m	1933–60	17	13	12	12	16	27	38	48	39	31	24	19	296

World Climatic Table—*continued*

Place, position and height above sea level	Item	Years of observation covered	J	F	M	A	M	J	J	A	S	O	N	D	Year
72 Verkhoyansk	T_m	1884–1935*	−50·3	−44·7	−32·2	−15·3	0·5	12·2	13·6	9·4	1·7	−15·3	−37·5	−47·8	−17·1
67°33′N, 133°23′E (137 m)		1931–60*	−46·8	−43·1	−30·2	−13·5	2·7	12·9	15·7	11·4	2·7	−14·3	−35·7	−44·5	−15·2
	T_x	1884–1935*	−16·7	−10·0	3·3	11·1	26·1	34·4	36·7	33·3	25·0	12·8	1·1	−10·6	
	T_n		−67·2	−67·8	−60·6	−54·4	−28·3	−7·2	−1·7	−7·8	−16·7	−44·4	−56·7	−64·4	
	R_m	1884–1935*	5	5	3	5	8	23	28	25	13	8	8	5	136
		1931–60*	7	5	5	4	5	25	33	30	13	11	10	7	155
73 Okhotsk	T_m	1931–60	−22·4	−19·2	−14·2	−5·4	1·4	6·4	11·9	13·1	8·7	−2·1	−13·5	−20·1	−4·6
59°22′N, 143°12′E (6 m)	R_m	1931–60	11	6	14	17	38	44	65	55	54	39	25	10	378
74 Petropavlovsk	T_m	1890–1915	−11·0	−11·2	−7·2	−2·0	2·2	6·7	10·6	11·9	9·2	3·9	−2·5	−7·6	0·2
52°58′N, 158°43′E (102 m)	R_m	1890–1915	87	58	59	66	(33)	51	60	66	72	75	67	80	772
75 Vladivostok	T_m	1873–1930	−14·2	−10·0	−3·3	4·4	9·4	14·2	18·6	20·8	16·4	8·9	−1·1	−10·0	4·4
43°07′N, 131°55′E (29 m)	σT_m		2·5	2·0	1·6	1·0	1·0	1·0	1·5	1·2	1·2	1·2	1·5	2·2	
	T_x	1873–1930	2·8	7·8	13·3	18·9	23·3	31·1	33·3	32·2	28·9	22·8	17·2	10·6	
	T_n		−30·0	−28·9	−21·7	−8·3	−0·6	3·9	8·3	10·0	3·9	−8·3	−17·8	−26·1	
	drT	1873–1930	7·0	9·0	8·0	6·5	6·5	6·0	6·0	6·0	7·0	8·0	6·5	6·5	
	R_m	1873–1930	8	10	18	30	53	74	84	119	109	48	30	15	598
	R_{dx}	1873–1930	23	28	28	41	56	53	58	145	109	76	119	36	
76 Irkutsk	T_m	1830–1915	−20·7	−17·5	−9·3	1·4	8·6	14·9	18·0	15·5	8·7	0·3	−10·7	−17·8	−0·7
52°16′N, 104°19′E (467 m)		1931–60	−20·8	−17·8	−9·3	1·6	8·8	15·4	17·9	15·1	8·2	1·1	−10·8	−18·5	−0·8
	σT_m		3·7	2·9	2·7	1·8	1·2	1·0	1·3	1·1	1·0	1·9	2·6	3·8	
	T_x	1830–1913	2·2	5·0	14·4	29·4	31·1	35·0	36·7	33·3	28·9	22·8	13·3	2·8	
	T_n		−50·0	−43·9	−36·7	−31·1	−14·4	−4·4	0·6	−2·8	−10·0	−30·5	−39·4	−46·1	
	drT	1830–1913	10·0	13·0	13·0	12·0	13·0	13·5	11·0	11·0	12·0	11·0	10·0	9·0	
	R_m	1830–1915	18	12	(9)	17	27	67	75	63	45	22	18	20	393
		1931–60	12	8	9	15	29	83	102	99	49	20	17	15	458
	R_{dx}	1830–1913	10	5	7	13	15	48	41	61	28	13	10	30	
77 Barnaul	T_m	1838–1950	−18·3	−16·7	−10·1	1·4	10·9	17·3	19·7	16·8	10·3	2·1	−8·4	−15·6	0·8
53°21′N, 83°49′E (160 m)		1931–60	−17·7	−16·3	−9·1	2·8	11·8	18·0	20·0	17·3	10·8	3·3	−8·6	−15·3	1·4
	σT_m		3·7	3·7	2·9	2·5	1·8	1·6	1·6	1·4	1·4	2·2	3·2	3·5	
	R_m	1838–1950	21	15	15	18	34	44	57	47	33	33	30	26	373
		1921–50	24	20	21	26	40	46	74	53	43	42	40	31	460
		1960–8	24	21	25	25	33	45	58	36	32	53	48	36	436
78 Kasalinsk	T_m	1881–1920	−11·2	−9·9	−2·2	9·9	18·9	23·9	26·0	23·8	16·9	7·9	−0·7	−7·5	8·0
45°46′N, 62°07′E (67 m)	R_m	1881–1915	10	10	12	11	12	9	6	7	8	13	12	13	122
79 Ashkhabad	T_m	1931–60	2·1	4·7	8·8	16·3	23·3	28·6	31·2	29·3	23·5	15·9	7·7	2·8	16·2
37°57′N, 58°23′E (227 m)	T_x	1892–1911	22·2	28·9	33·3	36·1	43·3	43·3	45·0	43·3	41·7	38·9	32·2	26·1	
	T_n		−22·8	−25·6	−13·9	−3·9	5·6	7·8	9·4	8·9	2·8	−3·9	−15·0	−17·2	
	R_m	1931–60	22	21	44	38	28	6	2	1	3	11	15	19	210
	R_{dx}	1892–1911	15	18	28	23	25	25	25	8	10	13	20	23	

80 Tashken 41°20′N, 69°18′E (479 m)	T_m	1881–1950	−1·1	1·9	7·8	14·3	20·1	24·8	27·0	24·9	19·2	12·5	6·5	1·7	13·3
		1931–60	−0·2	2·7	7·3	14·5	20·1	24·8	27·1	24·8	19·1	12·6	5·4	0·9	13·3
	σT_m		3·5	3·9	2·7	2·6	1·8	1·3	1·0	1·5	1·6	2·2	2·4	3·0	
	T_x	1892–1911	18·9	24·4	30·0	32·8	39·4	41·1	41·1	38·9	35·6	35·0	27·2	22·2	
	T_n	1892–1911	−28·3	−25·6	−19·4	−5·0	0·6	6·1	8·9	7·8	0·6	−6·1	−21·7	−24·4	
	drT	1892–1911	9·0	9·5	9·5	10·0	12·0	14·0	15·5	16·0	15·5	13·5	10·0	8·5	
	R_m	1881–1950	46	46	64	56	30	11	3	1	4	24	36	49	370
		1931–60	49	51	81	58	32	12	4	3	3	23	44	57	417
81 Alma Ata 43°14′N, 76°56′E (848 m)	T_m	1931–60	−6·7	5·1	1·6	10·8	16·0	20·4	23·3	22·3	17·4	10·0	−0·1	−5·4	8·7
	R_m	1931–60	26	32	64	89	99	59	35	23	25	46	48	35	581
82 Urumchi 43°47′N, 87°37′E (913 m)	T_m	(1907–33)	−16·1	−13·9	−5·8	8·9	15·3	18·9	21·1	20·0	14·4	4·7	−5·8	−10·8	4·3
	R_m	(1907–31)	9	15	15	33	25	33	16	35	15	47	22	11	276
83 Harbin 45°45′N, 126° 38′E (143 m)	T_m	(1909–52)	−20·1	−15·8	−6·0	5·8	14·0	19·8	23·3	21·6	14·3	5·7	−6·6	−16·7	3·3
	drT	(1909–52)	12·0	13·5	12·5	13·0	13·5	12·5	10·5	10·5	12·0	12·5	10·5	11·0	
	R_m	(1909–52)	4	6	17	23	44	92	167	119	52	36	12	5	577
	R_{dx}	(1909–52)	5	11	17	31	62	79	147	113	54	28	15	8	
84 Seoul, Korea 37°34′N, 126°58′E (85 m)	T_m	1931–60	−4·9	−1·9	3·6	10·5	16·3	20·8	24·5	25·4	20·3	13·4	6·3	−1·2	11·1
	drT	1931–60	9·2	9·3	9·8	11·3	11·6	10·2	8·2	8·8	10·4	11·9	10·4	8·6	
	R_m	1931–60	17	21	56	68	86	169	358	224	142	49	36	32	1259
85 Peking 39°57′N, 116°19′E (52 m)	T_m	(1915–52)	−4·7	−1·9	4·8	13·7	20·1	24·7	26·1	24·9	19·9	12·8	3·8	−2·7	11·8
	σT_m		2·1	2·0	1·5	1·5	1·2	1·1	0·8	0·8	0·9	1·3	1·6	1·7	
	T_x	(1915–52)	14·2	18·5	28·1	35·8	38·1	42·6	40·5	38·3	34·3	31·1	24·2	13·5	
	T_n	(1915–52)	−22·8	−17·8	−13·8	−3·3	3·4	10·1	14·9	11·3	1·6	−4·7	−13·5	−19·6	
	drT	(1915–52)	11·4	11·7	12·7	13·9	13·8	13·2	10·2	10·0	12·1	13·8	11·5	10·9	
	R_m	1841–1952	4	5	8	17	35	78	243	141	58	16	11	3	623
	R_{dx}	(1841–1952)	14	11	25	67	109	203	225	144	96	37	55	11	
86 Lanchow 36°01′N, 103°59′E (1508 m)	T_m	1932–52	−6·5	−1·7	5·4	12·1	17·4	20·9	22·8	21·4	16·3	10·1	1·7	−5·3	9·5
	drT	1932–52	15·4	15·3	14·3	14·8	15·0	14·1	12·9	12·5	11·5	12·6	13·4	14·3	
	R_m	1932–52	1	3	8	14	34	40	66	92	55	18	4	2	338
87 Shanghai (Zi-ka-wei) 31°12′N, 121°26′E (5 m)	T_m	1873–1953	3·4	4·3	8·2	13·7	18·9	23·1	27·1	27·2	23·0	17·7	11·6	5·9	15·3
		1921–48	3·5	4·5	8·7	14·2	19·3	23·3	27·5	27·7	23·4	17·8	12·2	6·3	15·7
	T_x	1873–1953	23·3	28·5	32·0	34·8	35·7	39·3	40·2	40·0	37·8	33·6	29·8	24·1	
	T_n	1873–1953	−12·1	−10·3	−5·8	−1·3	3·0	10·5	15·9	10·1	6·8	1·1	−5·1	−10·2	
	R_m	1873–1953	47	58	85	91	96	177	149	139	132	74	53	38	1139
		1921–50	46	66	77	84	107	168	139	132	160	62	53	41	1135
	R_{dx}	1873–1953	55	39	65	57	90	161	148	155	195	108	74	45	

World Climatic Table—*continued*

Place, position and height above sea level	Item	Years of observation covered	J	F	M	A	M	J	J	A	S	O	N	D	Year
88 Canton	T_m	(1912–52)	13·6	14·2	17·2	21·6	25·6	27·3	28·8	28·2	27·2	24·0	19·7	15·7	21·9
23°00′N, 113°13′E	T_x	1912–52	28·0	29·2	30·6	33·0	35·7	36·7	37·2	37·7	37·6	36·0	32·0	29·0	
(18 m)	T_n	1912–52	0·0	0·0	0·0	8·9	10·6	16·7	21·0	20·4	13·7	10·0	1·1	−0·3	
	R_m	1912–52	27	65	101	185	256	291	264	249	149	49	51	34	1720
89 Hongkong	T_m	1884–1950	15·7	15·2	17·4	21·3	25·1	27·3	27·8	27·7	27·2	24·6	20·9	17·3	22·3
22°18′N, 114°10′E		1951–60	15·4	15·8	18·2	21·8	25·6	27·5	28·4	27·9	27·3	24·7	21·2	17·4	22·6
(33 m)	σT_m		1·6	1·7	1·3	0·9	0·8	0·8	0·4	0·5	0·5	0·8	0·6	1·1	
	T_x	1884–1934	26·1	26·1	28·3	31·7	32·8	34·4	34·4	36·1	34·4	34·4	30·0	27·8	
	T_n	1884–1934	0·0	3·3	7·2	11·1	15·6	19·5	22·2	22·2	18·3	13·9	6·7	5·0	
	drT	1884–1934	4·4	4·4	3·9	4·4	4·4	3·9	5·0	5·0	4·4	4·4	5·0	5·0	
	R_m	1884–1950	33	47	76	140	280	402	393	363	264	105	42	27	2172
		1951–60	30	60	70	133	332	479	286	415	364	33	46	17	2265
	$R: Q_5 \geq$	1951–60	51	93	108	184	497	637	437	546	567	474	91	19	
	$Q_1 \leq$	1951–60	9	12	40	59	137	269	154	307	152	96	1	5	
	R_{dx}	1884–1934	99	56	97	157	509	320	533	282	203	292	150	91	
90 Taipei, Taiwan (Formosa)	T_m	1931–60	15·2	15·4	17·5	20·9	24·5	26·8	28·4	28·3	26·9	23·3	20·5	17·2	22·1
25°02′N, 121°31′E (8 m)	drT	1897–1940	6·8	6·6	7·0	7·7	7·9	8·5	8·6	8·7	8·4	7·4	6·9	6·8	
	R_m	1889–1940	91	147	164	182	205	322	269	266	189	117	71	77	2100
	$R: Q_5 \geq$	1931–60	120	233	234	266	316	423	365	375	258	179	116	112	
	$Q_1 \leq$	1931–60	45	67	92	86	87	211	163	143	113	48	32	36	
91 Nemuro, Hokkaido	T_m	1886–1950	−5·3	−5·8	−2·5	3·1	6·7	10·3	14·7	17·8	16·1	10·8	4·4	−1·7	5·7
43°20′N, 145°35′E		1931–60	−4·8	−5·6	−2·2	2·8	6·8	10·0	14·3	17·5	15·5	10·8	4·7	−1·3	5·7
(26 m)	drT	1886–1950	7·0	7·0	7·0	7·0	8·0	7·0	7·0	6·5	8·0	7·5	6·5	6·5	
	R_m	1886–1950	38	30	61	76	94	94	99	104	147	104	89	58	994
		1931–60	49	40	77	77	99	97	104	106	152	124	92	63	1081
	R_{dx}	1886–1950	51	43	119	63	89	117	109	124	135	97	74	112	
92 Tokyo	T_m	1886–1945	3·3	4·2	7·2	12·5	16·9	20·8	24·7	26·1	22·5	16·7	10·8	5·8	14·3
35°41′N, 139°46′E (4 m)		1931–60	3·7	4·3	7·6	13·1	17·6	21·1	25·1	26·4	22·8	16·7	11·3	6·1	14·7
	σT_m		1·0	1·1	1·2	0·8	0·7	1·0	1·1	0·9	1·0	0·7	1·0	1·0	
	T_x	1931–60	21·3	24·9	25·2	27·2	31·4	34·7	37·0	38·4	36·4	32·3	27·3	22·7	
	T_n	1931–60	−9·2	−7·9	−5·6	−3·1	2·2	8·5	13·0	15·4	10·5	−0·5	−3·1	−6·8	
	R_m	1886–1945	48	74	107	135	147	165	142	152	234	208	97	56	1565
		1931–60	48	73	101	135	131	182	146	147	217	220	101	61	1563
	$R: Q_5 \geq$	1931–60	71	98	141	171	167	239	207	234	257	303	136	89	
	$Q_1 \leq$	1931–60	25	47	66	90	85	90	60	55	154	135	58	23	
	R_{dx}	1931–60	48	91	87	81	121	278	151	171	393	164	169	85	

Station	Element	Period													Year
93 Kagoshima	T_m	1931–60	6·6	7·7	10·8	15·1	19·0	22·6	26·8	27·1	24·4	18·9	14·0	9·0	16·8
31°34′N, 130°33′E (5 m)	T_x	1931–60	23·9	24·1	25·8	27·7	31·1	34·1	36·6	37·0	34·2	32·2	28·6	24·4	
	T_n		−5·7	−6·7	−3·9	−1·0	3·9	9·0	15·9	16·8	9·8	2·6	−1·5	−5·5	
	drT	1931–60	10·2	10·4	10·7	10·7	9·8	8·0	7·7	8·3	9·1	10·7	11·3	11·0	
	R_m	1931–60	75	116	149	228	249	454	343	220	213	120	89	79	2337
	R_{dx}	1931–60	113	90	94	145	156	306	234	217	174	167	98	169	
94 Manila	T_m	1887–1940	24·7	25·4	26·6	28·1	28·5	27·8	27·0	27·0	26·8	26·6	25·8	25·1	26·6
14°35′N, 120°59′E (1 m)	drT		9·5	10·5	11·0	11·0	10·0	9·0	7·5	6·5	7·0	7·5	8·5	9·0	
	R_m	1887–1940	20	12	19	31	140	264	437	436	357	180	150	75	2121
95 Saigon	T_m	1951–60	25·8	26·3	27·8	28·8	28·2	27·4	27·1	27·1	26·7	26·5	26·1	25·7	27·0
10°49′N, 106°40′E (9 m)	R_m	1951–60	6	13	12	65	196	285	242	277	292	259	122	37	1808
96 Singapore	T_m	(1869–1932)	26·4	26·9	27·5	27·5	27·8	27·5	27·5	27·2	27·2	26·9	26·9	26·9	27·2
1°18′N, 103°50′E (10 m)	σT_m		0·5	0·5	0·4	0·5	0·5	0·4	0·2	0·3	0·3	0·2	0·5	0·4	
	T_x	(1869–1932)	33·9	34·4	34·4	35·0	36·1	35·0	33·9	33·9	33·9	33·9	33·3	33·9	
	T_n		20·0	18·9	19·4	21·1	21·1	21·1	21·1	20·5	20·5	20·5	20·5	20·5	
	drT	(1869–1932)	7·0	8·5	7·0	7·0	8·0	7·0	7·0	6·5	6·5	7·0	7·0	7·0	
	R_m	1869–1932	251	173	193	188	173	173	170	196	178	208	254	257	2414
	R_{dx}	1869–1932	218	140	168	135	216	99	185	107	99	152	183	236	
97 Jakarta/Batavia	T_m	1864–1945	26·1	26·1	26·7	27·2	27·2	26·9	26·7	26·7	27·2	26·9	26·7	26·4	26·7
6°11′S, 106°50′E (7 m)		1931–60	26·2	26·3	27·1	27·2	27·3	27·0	26·7	27·0	27·4	27·4	26·9	26·6	26·9
	T_x	1864–1945	33·9	33·3	33·3	34·4	33·9	33·9	33·3	34·4	35·5	36·7	35·5	33·9	
	T_n		20·5	20·5	20·5	20·5	21·1	19·5	19·5	19·5	18·9	20·5	20·0	19·5	
	R_m	1864–1945	300	300	211	147	114	97	63	43	66	112	142	203	1798
		1931–60	335	241	201	141	116	97	61	50	78	91	151	193	1755
	R_{dx}	1864–1945	66	71	53	48	43	38	30	20	28	41	43	53	
98 Rangoon	T_m	1878–1940*	25·0	26·4	28·6	30·3	29·2	27·2	26·9	26·9	27·2	27·8	26·9	25·3	27·3
16°46′N, 96°10′E (20 m)	drT	1878–1940	13·5	14·0	14·0	11·5	8·5	5·5	5·0	5·0	5·5	6·5	8·5	13·5	
	R_m	1878–1940*	3	5	8	51	307	480	582	528	394	180	69	10	2617
		1951–60	8	5	6	17	260	524	492	574	398	208	34	3	3964
99 Mandalay	T_m	1920–39	20·3	23·1	27·5	31·7	31·4	29·7	29·7	29·2	28·6	27·2	24·2	20·3	26·9
21°59′N, 96°06′E (75 m)	T_x	1920–39	32·8	37·2	42·2	43·3	43·9	41·7	41·1	38·3	39·4	38·9	36·7	32·2	
	T_n		7·2	8·3	12·2	17·8	20·6	20·0	22·2	21·7	20·5	16·7	13·3	6·7	
	drT	1920–39	15·0	16·0	17·0	13·5	10·5	8·5	8·5	8·5	8·5	9·0	10·5	13·0	
	R_m	1887–1940	12	4	5	32	148	130	85	116	152	135	53	11	883
		1920–39	3	3	5	30	147	160	69	104	137	109	51	10	828
	R_{dx}	1920–39	18	18	20	58	132	107	132	89	79	99	99	38	

World Climatic Table—*continued*

Place, position and height above sea level	Item	Years of observation covered	J	F	M	A	M	J	J	A	S	O	N	D	Year
100 Cherrapunji	T_m	(1903–50)	11·8	12·9	16·4	17·9	19·2	20·1	20·4	20·4	20·5	18·9	16·1	12·8	17·3
25°15′N, 91°44′E	drT	1906–40	8·0	7·0	8·0	6·5	6·0	4·5	4·0	4·5	4·5	6·0	7·0	8·0	
(1313 m)	R_m	1851–1950	19	54	237	765	1341	2692	2620	1955	1245	432	56	13	11 429
	$R: Q_5 \geq$	1931–60	34	71	301	800	2363	3685	3227	2520	1684	648	84	5	
	$Q_1 \leq$	1931–60	0	7	31	312	1111	2102	1701	1202	705	110	0	0	
	R_{dx}	1906–40	86	91	307	462	813	925	838	683	630	592	333	191	
101 Lhasa, Tibet	T_m	1941–8	−1·7	1·1	4·7	8·1	12·2	16·7	16·4	15·6	14·2	8·9	3·9	0·0	8·4
29°40′N, 91°07′E	T_x	1941–8	16·1	22·2	20·6	24·4	26·1	31·7	28·9	27·2	25·6	23·3	20·6	16·1	
(3685 m)	T_n	1941–8	−16·1	−15·0	−10·0	−7·8	−2·8	2·2	1·7	2·8	0·0	−7·8	−12·2	−15·0	
	drT	1941–8	16·5	15·5	8·5	9·5	9·0	9·0	8·5	8·0	8·5	15·5	17·5	17·5	
	R_m	1941–8	1	13	8	5	25	63	122	89	66	13	3	0	408
	R_{dx}	1941–8	2	38	13	8	20	28	28	28	23	15	23	0	
102 Calcutta (Alipore)	T_m	1878–1950	19·5	22·1	27·2	30·0	30·4	29·8	28·8	28·6	28·7	26·8	23·3	20·4	26·3
22°32′N, 88°20′E (6 m)		1931–60	20·2	23·0	27·9	30·1	31·1	30·4	29·1	29·1	29·2	27·9	24·0	20·6	26·8
	σT_m		0·9	1·0	1·0	1·1	0·8	0·6	0·6	0·4	0·4	0·6	0·7	0·9	
	T_x	1881–1940	31·7	36·7	40·0	41·7	42·2	43·9	36·7	35·5	36·1	35·6	33·3	30·6	
	T_n	1881–1940	6·7	7·8	10·0	16·1	18·3	21·1	22·8	23·3	22·2	17·2	10·6	7·2	
	drT	1881–1940	14·0	14·0	13·5	12·0	10·5	7·0	5·5	6·0	6·5	8·5	11·0	13·5	
	R_m	1881–1940	10	30	36	43	140	297	325	328	251	114	20	5	1599
	R_{dx}	1881–1940	43	81	69	107	155	302	183	254	368	173	84	53	
103 Darjeeling	T_m	1931–60	6·4	7·7	11·2	14·3	15·7	17·0	17·5	17·5	17·2	15·1	11·3	8·1	13·3
27°03′N, 88°16′E (2127 m)	R_m	1931–60	22	27	53	109	187	522	713	573	419	116	14	5	2760
104 New Delhi	T_m	1941–60	14·3	17·3	22·9	29·1	33·5	34·5	31·2	29·9	29·3	25·9	20·2	15·7	25·3
28°35′N, 77°12′E (220 m)	drT	(1934–43)	14·5	14·5	16·0	16·0	14·5	10·5	8·5	8·0	10·0	15·5	17·5	15·0	
	R_m	1931–60	25	22	17	7	8	65	211	173	150	31	1	5	715
	$R: Q_5 \geq$	1931–60	44	37	23	12	11	75	310	246	247	49	1	9	
	$Q_1 \leq$	1931–60	3	0·3	1	0	0·5	13	114	77	42	0	0	0	
	R_{dx}	(1934–43)	38	104	15	13	15	236	130	178	165	56	3	33	
105 Simla	T_m	1931–60	5·1	6·4	10·4	15·0	18·9	20·1	18·1	17·3	16·4	14·1	10·9	7·7	13·4
31°06′N, 77°10′E (2202 m)	R_m	1931–60	65	71	58	38	54	147	414	386	195	45	7	24	1504
106 Jodhpur	T_m	1931–60	17·1	19·9	25·2	30·3	34·4	34·3	31·3	29·2	29·4	27·7	22·7	18·7	26·7
26°18′N, 73°01′E (217 m)	T_x	(1891–1943)	32·8	37·8	41·1	45·0	48·9	46·1	44·4	40·6	42·2	41·1	36·7	32·2	
	T_n	(1891–1943)	−1·1	2·8	9·4	17·2	18·9	22·2	22·2	22·8	20·5	13·9	8·9	3·9	
	drT	(1891–1943)	15·5	15·5	16·0	16·0	14·5	12·0	9·5	9·0	10·5	16·5	18·5	16·0	
	R_m	1931–60	8	5	2	2	6	31	122	146	47	7	3	1	380
	$R: Q_5 \geq$	1931–60	8	10	2	1·4	11	47	178	212	102	7	3	0·5	
	$Q_1 \leq$	1931–60	0	0	0	0	0	9	40	35	1·1	0	0	0	

107 Madras	T_m	1870–1950*	24·4	25·6	27·5	30·2	32·9	32·5	30·8	30·2	29·6	28·0	25·9	24·7	28·5
13°04′N, 80°15′E (7 m)	drT	1870–1940	10·0	11·0	10·5	9·5	10·5	10·5	9·5	9·5	9·5	8·5	7·0	8·5	
	R_m	1870–1950*	35	10	11	16	31	50	90	115	123	300	352	145	1278
	R_{dx}	1870–1940	213	124	63	84	132	58	117	79	99	234	236	262	
108 Colombo	T_m	1931–60	26·2	26·4	27·2	27·7	28·0	27·4	27·1	27·2	27·2	26·6	26·2	26·1	26·9
6°54′N, 79°52′E (7 m)	drT	1911–52	8·0	8·5	8·0	6·5	5·0	4·5	4·5	4·5	4·5	5·5	6·5	7·0	
	R_m	1931–60	88	96	118	260	353	212	140	124	153	354	324	175	2397
	$R: Q_5 \geq$	1931–60	152	171	178	361	475	279	183	185	217	428	429	239	
	$Q_1 \leq$	1931–60	26	39	54	133	223	149	59	35	56	229	205	65	
	R_{dx}	1911–52	79	132	97	183	290	94	183	127	152	257	211	114	
109 Bombay (Colaba)	T_m	1873–1940	23·9	23·9	26·1	28·1	29·7	28·9	27·2	26·9	26·9	28·1	27·2	25·6	26·9
18°54′N, 72°49′E (11 m)		1931–60	24·3	24·9	26·9	28·7	29·9	29·1	27·5	27·1	27·4	28·3	27·5	25·9	27·3
	σT_m		0·8	0·9	0·9	0·8	0·6	0·6	0·4	0·4	0·4	0·6	0·9	0·9	
	drT	1873–1940	9·0	9·0	8·0	7·0	6·0	5·5	4·5	5·0	5·0	7·0	9·0	10·0	
	R_m	1846–1950	3	1	1	1	16	498	646	356	285	54	15	2	1878
		1951–60	0	1	0	2	16	639	814	541	246	102	7	0	2368
	$R: Q_5 \geq$	1931–60	0·4	0	0·1	0·4	29	764	866	541	408	136	27	0·3	
	$Q_1 \leq$	1931–60	0	0	0	0	0	251	517	193	115	19	0	0	
	R_{dx}	1873–1940	48	41	33	28	127	409	305	287	549	150	122	25	
110 Srinagar, Kashmir	T_m	1931–60	1·1	3·5	8·5	13·4	17·9	21·7	24·7	23·9	20·5	14·1	7·7	3·5	13·4
34°05′N, 74°50′E		1951–60	0·9	3·9	8·0	13·2	17·2	21·5	24·3	23·6	20·9	14·0	7·7	4·0	13·3
(1585 m)	T_x	1891–1942	13·3	20·6	25·6	31·1	35·6	37·2	37·2	36·1	35·0	33·9	23·3	17·2	
	T_n	1891–1942	−13·3	−14·4	−5·0	0·6	2·8	7·2	11·1	10·6	4·4	−1·7	−7·8	−10·6	
	R_m	1931–60	73	72	104	78	63	36	61	63	31	28	20	36	665
		1951–60	77	61	111	87	76	29	56	79	36	43	27	42	725
	$R: Q_5 \geq$	1931–60	102	118	139	112	91	57	96	89	47	39	39	57	
	$Q_1 \leq$	1931–60	38	30	64	46	33	10	22	25	7	5	0·5	7	
	R_{dx}	1891–1942	147	61	71	58	61	46	66	69	102	61	28	48	
111 Karachi (Manora)	T_m	(1875–1940)	18·9	20·3	24·4	27·5	30·0	30·8	30·0	28·6	28·1	27·5	24·2	20·3	25·9
24°48′N, 66°59′E (2 m)		1931–60	18·9	21·2	24·3	26·9	29·2	30·4	29·3	28·2	27·6	27·1	24·9	21·3	25·8
	drT	(1875–1940)	12·0	11·5	10·0	9·5	8·0	6·0	5·5	5·0	6·0	10·5	13·0	13·0	
	R_m	1931–60	7	11	6	2	0	7	96	50	15	2	2	6	204
	R_{dx}	(1875–1940)	41	28	53	104	30	183	201	137	206	13	23	46	
112 Kandahar, Afghanistan	T_m	1940–7	6·4	9·4	13·9	19·2	23·6	26·9	28·9	27·2	22·2	18·1	12·5	7·2	18·0
31°36′N, 65°40′ E	drT	1940–7	14·0	14·5	16·5	18·5	19·5	20·5	20·0	20·0	23·5	23·0	20·5	15·5	
(1055 m)	R_m	1940–7	79	43	20	8	5	2	3	1	0	1	2	20	184
113 Teheran (Mehrabad)	T_m	1943–60	3·5	5·2	10·2	15·4	21·2	26·1	29·5	28·4	24·6	18·3	10·6	4·9	16·5
35°41′N, 51°19′E	drT	1883–1935	10·0	10·0	11·0	12·0	13·5	15·0	15·0	14·5	14·5	12·5	11·0	10·0	
(1200 m)	R_m	1943–60	37	23	36	31	14	2	1	1	1	5	29	27	208

Place, position and height above sea level	Item	Years of observation covered	J	F	M	A	M	J	J	A	S	O	N	D	Year
114 Baghdad	T_m	1937–60	9·9	12·2	15·8	22·2	28·4	32·9	34·8	34·5	30·7	24·7	17·2	11·2	22·9
33°20′N, 44°24′E (34 m)	T_x	(1888–1952)	25·0	30·0	32·2	40·0	44·4	48·3	49·4	48·9	46·7	41·7	34·4	26·1	
	T_n	(1888–1952)	−7·8	−5·0	−2·8	2·8	10·6	14·4	16·7	17·8	10·6	3·9	−1·7	−6·7	
	drT	(1888–1952)	11·5	12·0	12·5	15·5	16·5	17·5	19·0	19·0	19·0	17·0	14·5	12·0	
	R_m	1937–61	26	28	28	17	7	0	0	0	0	3	21	26	156
	$R: Q_5 \geqslant$	1941–60	34	38	52	29	10	0	0	0	0	4	34	35	
	$Q_1 \leqslant$	1941–60	5	10	13	1	0	0	0	0	0	0	1	8	
	R_{dx}	(1888–1952)	36	33	56	20	15	2	1	1	1	15	36	33	
115 Bahrein (Muharraq)	T_m	1951–60	17·4	18·3	21·2	25·6	29·6	32·0	33·8	34·3	32·5	29·0	24·5	19·2	26·4
26°16′N, 50°37′E (2 m)	drT	(1928–43)	6·0	6·0	6·5	8·0	8·0	8·0	8·0	8·5	8·5	8·5	7·0	6·0	
	R_m	1931–60	16	15	11	6	1	0	0	0	0	0	9	18	76
	$R: Q_5 \geqslant$	1931–60	27	22	18	7	1·5	0	0	0	0	0	10	36	
	$Q_1 \leqslant$	1931–60	0·5	0	1	0	0	0	0	0	0	0	0	0·3	
116 Aden (Khormaksar)	T_m	1951–60	25·5	25·6	27·2	28·7	30·7	32·8	32·2	31·6	31·7	28·9	26·6	26·0	28·9
12°50′N, 45°02′E (3 m)	drT	1947–53	5·5	5·5	5·5	6·5	6·5	8·0	8·0	8·0	7·5	8·5	7·0	5·5	
	R_m	1941–60	7	3	6	0	1	0	3	2	7	1	3	6	39
	R_{dx}	1947–53	10	3	25	1	1	1	8	8	2	8	1	8	
117 Jerusalem (Old City)	T_m	1918–38	8·9	9·4	13·1	16·4	20·6	22·5	23·9	24·2	23·1	21·1	16·4	11·1	17·6
31°47′N, 35°14′E (760 m)	σT_m		1·0	1·5	1·2	0·9	1·0	0·9	0·9	1·0	0·9	0·9	1·1	1·2	
	drT	1918–38	8·0	8·0	10·5	13·0	13·5	14·0	13·5	13·0	13·0	12·0	9·5	8·0	
	R_m	1846–1954													560
		1921–50	128	135	61	24	3	0	0	0	1	7	54	83	496
	R_{dx}	1918–38	99	86	36	38	13	3	0	0	10	23	56	76	
118 Ankara	T_m	1916–49	−0·3	1·1	5·0	10·8	16·1	18·6	22·5	22·8	18·3	13·6	8·3	2·2	11·6
39°57′N, 32°53′E (861 m)	drT	1916–49	8·5	9·0	11·0	13·0	13·5	14·0	15·0	15·5	14·5	14·0	11·0	8·0	
	R_m	1916–49	33	30	33	33	48	25	13	10	18	23	30	48	344
119 Famagusta, Cyprus	T_m	1907–39	11·4	11·9	13·9	16·9	21·1	25·3	28·1	28·3	25·8	21·9	17·8	13·6	19·9
35°07′N, 33°57′E (23 m)	drT	1907–39	10·5	11·5	12·0	12·5	13·5	14·0	14·0	13·5	14·0	14·0	12·0	10·5	
	R_m	1907–39	97	66	33	20	15	5	1	1	5	33	56	109	441
	R_{dx}	1907–39	79	43	38	33	28	33	18	25	20	61	102	122	
120 Nikosia, Cyprus	T_m	(1888–1953)	10·0	10·3	12·5	16·7	21·9	25·6	27·8	27·8	25·6	20·8	16·4	11·9	18·9
35°09′N, 33°17′E (220 m)	drT	(1888–1953)	9·0	9·5	11·5	13·0	12·5	14·5	15·5	15·5	14·5	12·5	11·5	9·5	
	$R_{...}$	1888–1953	74	51	33	20	28	10	1	1	5	23	43	76	365

121 Helwan (near Cairo) 29°52′N, 31°20′E (116 m)	T_m	1904–50	13·2	14·3	17·2	21·1	25·0	27·3	28·1	28·0	25·9	23·8	19·7	14·8	21·5
		1921–40	13·3	14·4	17·5	21·5	25·4	27·7	28·3	28·1	26·2	24·1	19·8	15·1	21·8
		1941–50*	13·4	14·4	16·9	20·8	25·5	27·4	28·3	28·2	25·9	23·7	20·0	15·2	21·6
	σT_m		0·9	1·0	0·9	0·9	0·9	0·9	0·9	1·0	0·9	0·8	1·0	1·0	
	T_x	1904–45	31·1	33·3	38·3	45·0	46·7	47·2	42·8	42·8	42·2	42·8	37·8	30·6	
	T_n	1904–45	1·7	1·7	3·3	5·6	9·4	12·8	16·1	17·2	14·4	10·6	5·6	1·1	
	drT	1904–45	10·0	11·5	13·0	14·5	15·5	15·0	14·5	13·5	12·0	11·5	11·0	10·0	
	R_m	1904–50	7	4	5	3	2	0	0	0	0	1	4	5	31
	R_{dx}	1904–45	25	23	25	38	28	3	0	0	1	23	13	28	
122 Tripoli, Libya 32°54′N, 13°11′E (22m)	T_m	1879–1939	12·2	13·3	15·3	18·1	20·3	23·3	25·6	26·1	25·6	22·5	18·3	13·6	19·5
	T_x	1879–1939	28·3	32·8	38·3	40·6	42·8	44·4	45·6	44·4	45·0	41·1	35·6	30·0	
	T_n	1879–1939	1·1	2·8	3·9	6·1	6·1	10·0	15·6	16·7	15·0	10·0	5·6	0·5	
	drT	1879–1939	8·0	8·0	8·5	8·5	8·5	8·0	8·0	8·0	8·0	8·0	9·0	8·5	
	R_m	1879–1939	81	46	28	10	5	3	1	1	10	41	66	94	386
	R_{dx}	(1879–1939)	51	124	30	51	20	10	5	28	79	66	130	71	
123 El Azizia 32°32′N, 13°01′E (112 m)	T_m	1913–40, 1947–51	11·4	13·3	15·6	19·2	23·1	27·2	28·9	28·9	27·8	23·6	18·1	12·8	20·8
	T_x	1913–40	30·0	33·3	44·4	48·3	49·4	52·8	50·6	56·1	57·7	48·9	37·8	30·6	
	T_n	1913–40	−3·3	−2·2	−0·6	1·1	5·6	8·3	12·8	10·6	12·8	5·6	1·1	−0·6	
	R_m	1913–40	48	30	23	8	8	1	1	1	5	15	25	61	226
124 Algiers (Dar el Beida – Maison Blanche) 36°43′N, 3°15′E (25 m)	T_m	1931–60	10·3	10·8	13·0	15·2	18·0	21·8	24·4	25·1	23·1	18·9	14·9	11·7	17·3
	drT	(1885–1937)	5·5	6·5	6·0	7·0	8·0	7·0	7·0	8·0	6·5	6·0	5·5	5·0	
	R_m	1931–60	116	76	57	65	36	14	2	4	27	84	93	117	691
125 Marrakech 31°37′N, 8°02′W (468 m)	T_m	1931–60	11·5	13·4	16·1	18·6	21·3	24·8	28·7	28·8	25·4	21·2	16·5	12·5	19·9
	drT	1921–55	14·0	14·0	14·5	15·0	15·0	16·5	19·0	17·5	16·0	14·5	13·5	13·5	
	R_m	1931–60	28	29	32	31	17	7	2	3	10	21	28	33	241
126 Agadir 30°23′N, 9°34′W (25 m)	T_m	1931–60	13·8	15·0	16·7	18·0	19·2	20·8	22·1	22·6	21·9	20·5	18·1	14·6	18·6
	drT	1934–55	14·0	13·5	12·0	11·0	10·0	9·0	9·0	9·0	9·5	11·0	11·5	13·0	
	R_m	1931–60	48	32	24	16	5	0	0	1	6	22	29	41	224
127 Timbuktu 16°43′N, 3°00′W (263 m)	T_m	1951–60	22·6	25·4	28·5	31·8	34·3	33·9	31·5	29·1	31·2	31·7	28·3	22·8	29·3
	T_x	1896–1912, 1949–55	38·9	41·7	46·1	47·8	47·8	48·3	48·3	43·9	46·1	45·0	42·8	38·9	
	T_n	1896–1912, 1949–55	5·0	5·6	8·9	13·9	18·9	20·0	18·9	19·4	20·0	17·2	8·3	5·6	
	R_m	1931–60	0	0	0	1	3	19	65	95	37	5	0	0	225
	$R : Q_5 \geqslant$		0	0	0	0	6	29	87	133	61	13	0	0	
	$Q_1 \leqslant$		0	0	0	0	0	5	35	61	18	0	0	0	
	R_{dx}	1896–1912, 1949–55	1	3	5	2	10	20	33	38	28	18	2	1	
128 Tamanrasset 22°47′N, 5°31′E (1366 m)	T_m	1931–60	12·3	14·7	18·3	22·5	26·0	29·0	28·9	28·3	26·7	23·0	18·2	13·6	21·8
	drT	(1925–50)	15·5	16·0	16·5	16·5	16·0	14·0	13·5	13·5	14·0	14·5	15·0	15·0	
	R_m	1951–60	2	0·5	1	5	6	12	3	17	17	3	4	4	75
	$R : Q_5 \geqslant$	1931–60	2	1	1	2	5	13	5	17	13	2	1	1	
	$Q_1 \leqslant$	1931–60	0	0	0	0	0	0	0	0	0	0	0	0	

Place, position and height above sea level	Item	Years of observation covered	J	F	M	A	M	J	J	A	S	O	N	D	Year
129 Khartum 15°36′N, 32°33′E (328 m)	T_m	1931–60	22·5	23·8	27·2	30·7	33·1	33·3	30·8	29·4	30·9	31·4	27·5	23·7	28·7
	σT_m		1·5	1·7	1·5	1·1	0·7	0·8	1·0	1·2	1·1	0·7	0·9	1·1	
	drT	1900–45	17·0	18·0	19·0	18·5	16·5	15·0	13·5	12·0	14·0	16·0	16·0	16·5	
	R_m	1931–60	0	0	0	1	5	7	48	72	27	4	0	0	164
	$R: Q_5\geq$	1931–60	0	0	0	0	7	10	97	109	46	7	0	0	
	$Q_1\geq$		0	0	0	0	0·1	0·1	10	36	5	0	0	0	
	R_{dx}	1900–45	1	1	15	25	18	33	86	74	41	46	3	0	
130 Addis Ababa 9°02′N, 38°45′E (2408 m)	T_m	1951–60	17·2	18·1	19·0	18·6	19·3	17·4	15·3	14·9	16·5	16·2	16·8	17·0	17·2
	drT		18·0	16·0	15·5	15·0	15·0	14·0	10·5	10·5	13·0	16·5	16·5	18·0	
	R_m	1951–60	24	25	67	93	53	105	239	266	174	43	3	17	1109
	R_{dx}		30	38	43	53	79	53	71	74	74	38	36	28	
131 Fort Lamy, Chad 12°08′N, 15°02′E (294 m)	T_m	1951–60	23·5	25·9	30·1	32·7	32·3	30·5	27·5	26·2	27·1	28·6	27·1	24·1	27·9
	R_m	1931–60	0	0	0	5	36	66	156	257	104	23	1	0	646
132 Accra (Airfield) 5°36′N, 0°10′W (65 m)	T_m	1941–60	27·3	27·7	27·8	27·7	27·0	25·7	24·6	24·3	25·3	26·1	27·1	27·3	26·5
	drT		8·0	7·0	6·5	6·5	6·5	5·5	4·5	5·0	4·5	6·0	6·5	7·0	
	R_m	1931–60	16	37	73	82	145	193	49	16	40	80	38	18	787
		1951–60	16	31	73	97	161	228	31	19	37	117	35	19	866
	$R: Q_5\geq$	1931–60	28	53	105	111	185	317	92	27	63	133	55	32	
	$Q_1\leq$		0	5	39	47	83	88	4	2	11	29	16	3	
	R_{dx}	(1888–1955)	89	107	109	137	150	302	104	94	114	140	94	76	
133 Freetown (Lungi) 8°37′N, 13°12′W (25 m)	T_m	1951–60	26·6	27·2	27·4	27·7	27·0	26·1	25·1	25·1	25·8	25·9	26·5	26·5	26·4
	R_m	1951–60	17	8	32	65	226	389	730	800	528	301	171	54	3321
134 Porto da Praia, Santiago, Cape Verde Islands 14°54′N, 23°31′W (34 m)	T_m	1904–30	22·5	22·2	22·8	23·3	24·2	25·0	26·1	26·7	26·9	26·9	25·6	23·9	24·7
	drT	1904–30	5·0	5·5	5·5	5·5	6·0	5·5	4·5	4·5	4·0	5·0	4·5	4·5	
	R_m	1904–30	3	2	1	1	0	1	5	97	114	30	8	3	265
135 Entebbe (Airport) 0°03′N, 32°27′E (1155 m)	T_m	(1931–60)	22·0	22·1	22·2	21·8	21·6	21·1	20·6	20·7	21·2	21·7	21·8	21·6	21·5
	drT	(1900–51)	9·0	9·0	8·0	7·5	7·0	7·5	8·0	8·5	9·0	9·0	8·5	9·0	
	R_m	1900–50	72	88	162	261	259	121	80	76	74	94	123	115	1525
		1960–9	69	120	193	269	248	99	67	79	73	159	198	106	1680
	$R: Q_5\geq$	1931–60	121	128	212	356	407	144	104	113	107	119	197	165	
	$Q_5\leq$		28	44	119	225	189	67	42	37	37	48	66	65	

136 Nairobi (Kabete)	T_m	1930–55*	18·3	19·1	19·1	18·4	17·3	16·2	15·4	15·4	16·9	18·0	17·6	17·5	17·4
1°16′S, 36°45′E (1820 m)	T_x	(1930–55)	28·9	30·6	30·0	27·8	27·8	26·7	26·1	26·7	27·8	30·0	27·8	27·8	
	T_n		8·3	8·9	9·4	11·1	8·9	7·2	6·1	6·7	5·0	7·2	6·1	8·3	
	drT	(1930–55)	13·0	13·5	11·0	9·5	9·0	9·5	10·0	10·0	13·0	11·5	10·0	10·5	
	R_m	1930–50	36	50	112	204	133	47	15	29	28	55	98	85	892
	R_{dx}	(1930–55)	69	53	79	112	76	56	13	63	36	41	63	53	
137 Zanzibar (Chukwani)	T_m	1931–51	28·3	28·6	28·9	27·5	26·4	25·8	25·0	25·3	25·6	25·8	27·8	28·1	26·9
6°15′S, 39°13′E (19 m)	drT	1931–51	8·0	8·5	8·0	6·0	5·0	5·0	5·5	6·0	6·5	7·0	8·0	7·5	
	R_m	1931–51	58	66	147	320	290	53	28	30	41	66	170	140	1409
138 Port Victoria, Mahé,	T_m	1887–1954	26·4	26·9	27·2	27·5	27·2	26·7	25·6	25·6	26·1	26·1	26·4	26·1	26·5
Seychelles	drT	1887–1954	4·0	4·0	4·5	5·0	4·5	4·5	3·5	3·5	3·5	4·5	5·0	4·5	
4°37′S, 55°27′E (1 m)	R_m	1887–1954	386	267	234	183	170	102	84	69	130	155	231	340	2351
		1931–60	358	242	204	170	161	85	83	78	116	167	206	332	2203
	R_{dx}	1887–1954	257	216	155	165	163	114	69	94	262	140	132	155	
139 Ascension Island	T_m	(1889–1954)	26·1	26·9	27·5	27·5	26·9	26·1	25·6	25·0	24·7	25·0	25·0	25·6	26·0
(Georgetown)	R_m	1899–1954	5	10	18	28	13	13	13	10	8	8	5	3	134
7°56′S, 14 25′W (17 m)	R_{dx}	1899–1954	5	41	46	211	38	41	41	10	8	5	3	5	
140 St Helena (Jamestown)	T_m	1853–62	23·6	24·2	24·7	24·2	21·9	20·8	19·7	19·7	19·7	20·3	20·8	21·7	21·8
15°55′S, 5°43′W (12 m)	R_m	1853–62	8	10	20	10	18	18	8	10	5	3	0	3	113
141 Luanda, Angola	T_m	1931–60	25·9	26·6	26·9	26·4	25·0	22·2	20·2	20·3	21·8	24·0	25·2	25·6	24·2
8°51′S, 13°14′E (70 m)	drT	1914–40	5·0	5·5	6·0	5·5	5·0	5·0	5·0	5·5	5·0	4·5	5·0	5·0	
	R_m	1931–60	26	35	97	124	19	0	0	1	2	6	34	23	367
142 Lulumbashi/Elisabethville,	T_m	1951–60	20·5	20·3	20·8	20·8	19·1	17·0	16·9	19·4	22·3	23·1	21·6	20·5	20·2
Congo	T_x	1919–49	32·8	32·2	33·9	32·2	31·7	30·0	31·7	34·4	37·2	36·7	36·1	33·9	
11°40′S, 27°29′E	T_n		10·0	12·2	7·8	5·0	3·3	1·1	0·6	0·6	2·8	7·2	10·0	12·2	
(1298 m)	drT	(1919–49)	11·5	11·0	11·5	14·0	17·0	19·5	20·0	20·5	20·5	18·5	14·5	11·0	
	R_m	1931–60	248	263	204	57	4	0	0	1	3	27	164	258	1229
	R_{dx}	(1919–49)	74	86	112	94	5	0	0	1	15	56	89	69	
143 Salisbury, Rhodesia	T_m	1931–60	20·0	19·8	19·4	18·7	15·9	13·6	13·6	15·6	19·0	21·3	20·8	20·4	18·2
(Kutsaga Observatory)	drT		10·0	10·0	10·5	13·0	14·0	14·5	14·5	15·0	14·5	14·0	11·5	10·5	
17°56′S, 31°06′E	R_m	1931–60	216	172	99	36	11	4	1	3	5	30	100	186	863
(1478 m)		1960–9	170	145	61	55	7	5	0	3	7	39	93	171	756
	$R: Q_5 \geq$	1931–60	301	231	144	62	25	10	1	5	9	55	147	233	
	$Q_1 \leq$		147	103	41	10	0·5	0	0	0	0	2	59	114	
144 Walvis Bay	T_m	1916–50	18·9	19·4	19·2	18·3	17·2	16·1	14·7	13·9	13·9	15·0	16·9	18·1	16·8
22°56′S, 14°30′E (7 m)	drT	1916–50	8·0	8·0	8·5	11·0	13·0	14·5	13·0	12·0	10·0	9·0	9·5	8·5	
	R_m	1916–50	2	5	8	3	3	0	1	3	1	1	2	1	30

World Climatic Table—*continued*

Place, position and height above sea level	Item	Years of observation covered	J	F	M	A	M	J	J	A	S	O	N	D	Year
145 Johannesburg (Germiston)	T_m	1932–50	20·0	19·7	18·3	16·1	12·5	10·3	10·6	13·1	15·8	18·3	18·9	19·7	16·1
26°14′S, 28°09′E	T_x	1932–50	32·8	32·8	31·1	29·4	25·6	24·4	23·3	26·1	30·0	32·2	33·9	33·3	
(1665 m)	T_n		5·6	7·2	5·0	−1·1	−5·6	−7·2	−7·2	−6·7	−2·8	0·0	1·7	5·6	
	drT	1932–50	11·0	10·5	11·0	12·0	12·5	13·0	13·5	14·0	14·0	13·5	12·0	11·5	
	R_m	1932–50	114	109	89	38	25	8	7	8	23	56	107	124	708
	R_{dx}	1932–50	91	102	76	48	74	18	28	20	48	36	74	89	
146 Durban	T_m	(1873–1950)	23·9	23·9	23·3	21·7	18·9	17·2	16·7	17·5	18·9	20·3	21·7	22·8	20·6
29°50′S, 31°02′E (5 m)	T_x	(1873–1950)	33·3	31·7	32·2	37·2	35·0	32·2	33·3	31·7	41·7	30·6	38·9	32·2	
	T_n		13·9	15·0	14·4	10·6	6·7	5·0	3·9	5·0	7·8	10·0	10·6	13·3	
	R_m	1873–1950	109	122	130	76	51	33	28	38	71	109	122	119	1008
	R_{dx}	(1873–1950)	43	150	74	145	160	239	46	63	53	56	135	69	
147 Cape Town (Royal Observatory)	T_m	1857–1950	21·2	21·4	20·7	17·5	15·1	13·3	12·7	13·2	14·4	16·6	18·3	20·1	17·0
33°56′S, 18°29′E (12 m)		1921–50	21·4	21·8	20·6	18·0	15·4	13·6	12·9	13·5	14·6	16·8	18·9	20·4	17·3
		1951–60	21·7	21·5	20·3	17·3	15·0	13·6	12·8	13·4	14·9	16·3	18·9	20·5	17·2
	T_x	1932–50	37·2	37·8	39·4	38·9	35·0	29·4	28·9	31·7	33·9	32·2	33·9	37·8	
	T_n		6·7	5·0	5·6	3·3	−0·6	−1·7	−2·2	−0·6	0·6	1·1	4·4	5·0	
	drT	1932–50	10·0	10·5	10·5	10·5	10·0	10·5	10·0	10·0	9·0	10·0	10·0	10·0	
	R_m	1841–1950	17	15	22	49	93	111	91	83	57	40	26	20	624
		1921–50	15	15	18	43	82	99	93	76	57	35	22	16	571
		1951–60	9	23	19	76	123	92	96	97	41	39	21	14	650
	R_{dx}	1932–50	25	30	30	43	61	41	48	38	38	48	18	18	
148 Tristan da Cunha	T_m	1956–60	17·4	18·2	17·0	15·7	14·3	13·2	12·0	11·8	11·8	12·8	14·2	16·3	14·6
37°03′S, 12°19′W (23 m)	drT	1943–7	4·0	4·5	4·5	4·0	4·0	4·0	4·0	4·0	4·0	4·5	4·0	4·0	
	R_m	1956–60	103	88	88	153	144	187	147	193	129	139	117	167	1655
149 Marion Island	T_m	1951–60	6·8	7·3	7·2	5·9	4·7	4·1	3·6	3·3	3·2	4·4	5·1	5·7	5·1
46°53′S, 37°52′E (26 m)	R_m	1951–60	194	185	216	221	235	221	228	178	190	167	190	226	2451
150 Diego Suarez, Madagascar	T_m	1941–60	26·9	26·8	27·2	27·1	26·3	25·0	24·2	24·2	24·6	25·5	26·6	27·2	26·0
12°21′S, 49°18′E (114 m)	drT		7·0	7·5	7·0	7·0	7·5	7·5	8·0	8·0	7·5	7·5	7·5	8·0	
	R_m	1931–60	277	211	187	56	8	8	7	7	5	11	28	111	915
	R_{dx}		74	66	508	97	10	15	8	5	5	18	48	53	
151 Mauritius (Royal Alfred Observatory/	T_m	1875–1950*	26·4	26·1	25·6	24·4	22·5	20·8	20·3	20·	21·0	22·2	23·9	25·5	23·2
Pamplemousses)	drT	(1875–1928)	7·0	6·5	6·5	6·5	7·0	7·0	7·0	7·5	8·0	8·0	8·0	7·5	
20°06′S, 57°32′E (55 m)	R_m	1875–1950*	207	195	218	141	97	67	57	59	37	39	51	115	1283
	R_{dx}	1875–1928	175	361	193	124	206	150	76	81	48	140	127	89	

152 Kerguelen (Port aux Français) 49°20′S, 70°13′E (14 m)	T_m	1951–60	7·3	8·0	7·2	6·1	3·8	2·1	2·2	2·1	2·5	3·5	4·6	6·2	4·6
	R_m	1951–60	75	50	66	90	129	117	113	100	107	86	94	93	1121
153 Heard Island 53°01′S, 73°23′E (5 m)	T_m	1948–53	3·3	3·3	2·8	2·2	1·1	−0·3	−0·8	−0·8	−1·4	−0·3	0·6	2·2	1·0
	drT	1948–53	3·5	3·5	3·5	3·5	3·5	4·0	4·0	4·0	4·0	4·0	3·5	3·5	
	R_m	1948–53	147	147	145	155	147	99	91	56	63	94	102	130	1376
	R_{dx}	1948–53	38	46	33	30	41	28	30	23	25	36	46	33	
154 Cocos (Keeling) Island 12°05′S, 96°53′E (5 m)	T_m	1905–42	27·5	27·8	27·8	27·5	26·9	26·4	26·1	26·1	26·1	26·7	26·9	27·2	26·9
	R_m	1905–42	137	196	216	264	201	229	221	122	94	84	107	117	1988
155 Darwin 12°28′S, 130°51′E (30 m)	T_m	1882–1924	28·8	28·6	28·9	28·9	27·7	26·1	25·2	26·3	28·1	29·6	29·9	29·5	28·1
	R_m	1870–1924	405	328	272	104	18	4	2	2	12	55	123	262	1587
156 Port Moresby, Papua 9°26′S, 147°13′E (28 m)	T_m	1941–60	27·6	27·3	27·3	26·9	25·4	26·2	25·8	26·1	26·5	27·2	27·5	27·7	26·8
	R_m	1951–60	150	194	170	173	49	31	119	37	53	21	97	164	1150
157 Townsville 19°16′S, 146°46′E (3 m)	T_m	1941–60	27·3	27·0	26·3	24·6	22·1	19·6	19·2	20·0	21·9	24·6	26·5	27·4	23·9
	drT	(1870–1937)	6·0	6·5	7·0	8·0	9·0	9·0	9·0	9·0	8·0	6·5	7·0	6·0	
	R_m	1941–60	332	364	275	83	34	26	22	10	10	21	55	102	1333
	R_{dx}	(1870–1937)	221	262	130	114	56	109	89	38	66	231	74	137	
158 Alice Springs 23°38′S, 133°37′E (580 m)	T_m	1879–1940	28·6	28·2	24·8	20·0	15·5	12·4	11·4	14·4	18·4	23·1	26·0	27·8	20·9
		1921–40*	28·3	27·7	24·7	19·5	15·4	12·1	11·4	14·6	17·9	22·8	25·4	27·2	20·6
	T_x	1879–1940	43·9	42·8	43·3	37·2	35·6	30·0	30·0	35·6	37·8	41·1	42·2	43·9	
	T_n	1879–1940	10·6	8·9	7·2	2·2	−1·7	−5·6	−7·2	−3·9	−0·6	3·9	5·6	10·0	
	drT	1879–1940	15·0	14·5	15·0	15·0	15·0	14·5	15·5	16·5	18·0	16·5	16·0	15·5	
	R_m	1879–1940	42	40	31	17	17	16	9	9	9	19	26	37	272
		1921–40*	37	33	35	11	15	17	6	6	7	18	25	31	241
	R_{dx}	1879–1940	99	81	147	51	38	51	51	28	30	58	69	119	
159 Perth, Western Australia 31°57′S, 115°51′E (64 m)	T_m	1875–1942	23·3	23·3	21·7	19·2	16·1	13·9	13·1	13·3	14·7	16·4	19·2	21·7	18·0
	drT	1875–1942	12·0	12·0	11·0	10·5	9·0	8·0	8·5	9·0	9·5	9·5	10·5	11·0	
	R_m	1875–1942	8	10	20	43	130	180	170	145	86	56	20	13	881
		1921–40	8	9	26	50	153	195	185	151	85	59	19	12	952
		1941–60	6	15	19	53	121	197	189	128	69	53	27	18	895
	R_{dx}	1875–1942	43	41	76	63	76	99	76	71	46	43	36	43	
160 Adelaide 34°56′S, 138°35′E (43 m)	T_m	1857–1950	23·0	23·2	21·0	17·7	14·5	11·9	11·1	12·2	14·0	16·5	19·3	21·6	17·2
		1931–60	22·6	21·0	20·9	17·2	14·6	12·1	11·2	12·0	13·4	16·0	18·5	20·7	16·7
	T_x	1857–1948	47·8	45·6	43·9	37·2	31·7	24·4	23·3	29·4	32·8	39·4	45·0	46·1	
	T_n	1857–1948	7·2	7·2	6·7	4·4	2·8	0·6	0·0	0·0	0·6	2·2	5·0	6·1	
	drT	1857–1948	14·0	13·5	12·0	10·0	9·0	8·0	8·0	9·0	10·0	12·0	13·5	13·5	
	R_m	1839–1950	20	19	25	43	69	76	66	63	53	45	30	27	536
		1931–60	23	23	21	50	66	61	61	59	49	47	36	27	523
	R_{dx}	1839–1948	58	142	89	79	69	53	43	56	41	56	53	61	

World Climatic Table—*continued*

Place, position and height above sea level	Item	Years of observation covered	J	F	M	A	M	J	J	A	S	O	N	D	Year
161 Sydney 33°52′S, 151°12′E (42 m)	T_m	1859–1945	21·9	21·9	20·8	18·1	15·0	12·5	11·7	13·1	15·0	17·5	19·4	21·1	17·3
		1931–60	21·9	21·9	21·2	18·3	15·7	13·1	12·3	13·4	15·3	17·6	19·4	21·0	17·6
		1951–60	21·9	22·0	21·6	19·2	15·7	13·6	12·6	13·7	15·1	17·5	19·6	21·1	17·8
	T_x	1859–1945	45·6	42·2	39·4	32·8	30·0	26·7	25·6	27·8	33·3	37·2	39·4	41·7	
	T_n	1859–1945	10·6	9·4	9·4	7·2	4·4	2·2	2·2	2·8	5·0	5·6	7·8	8·9	
	R_m	1859–1945	89	102	127	135	127	117	117	76	74	71	74	74	1183
		1931–60	104	125	129	101	115	141	94	83	72	80	77	86	1205
	R_{dx}	1859–1945	180	226	282	190	213	132	198	135	145	163	107	119	
162 Melbourne 37°49′S, 144°58′E (38 m)	T_m	1931–60	19·9	19·7	18·4	15·1	12·5	10·2	9·6	10·5	12·4	14·3	16·2	18·4	14·8
	drT	1856–1948	11·5	11·5	11·0	9·5	8·5	7·0	8·5	9·0	9·5	10·5	11·0	11·5	
	R_m	1931–60	45	59	50	69	54	52	54	50	58	74	70	58	691
163 Hobart 42°53′S, 147°20′E (55 m)	T_m	1843–1942	16·7	16·7	15·3	13·1	10·6	8·3	7·8	8·9	10·6	12·5	13·9	15·6	12·5
		1931–60	16·3	16·1	15·1	12·4	10·5	8·3	7·8	8·8	10·6	11·8	13·6	15·1	12·2
	T_x	1843–1942	40·6	40·0	34·4	28·9	23·9	18·9	18·9	20·0	23·9	31·7	32·8	36·1	
	T_n	1843–1942	4·4	3·9	1·7	0·6	−1·7	−1·7	−2·2	−1·1	−1·1	0·0	1·7	3·3	
	drT	1843–1942	10·0	10·0	9·5	8·5	7·5	6·5	6·5	7·5	9·0	9·5	10·0	10·0	
	R_m	1843–1942	48	38	46	48	46	56	53	48	53	58	61	53	608
		1931–60	42	47	52	63	51	66	47	53	53	72	58	64	668
	R_{dx}	1843–1942	43	30	71	104	38	36	56	48	36	30	43	69	
164 Macquarie Island 54°30′S, 158°57′E (5 m)	T_m	1951–60	6·7	6·6	6·0	5·2	3·9	2·9	2·9	3·2	3·3	3·8	4·5	6·1	4·6
	R_m	1951–60	99	81	104	94	81	81	68	68	68	73	64	72	952
165 Campbell Island 52°35′S, 169°07′E (19 m)	T_m	1941–60	9·3	9·3	8·5	7·2	6·1	4·6	4·5	4·8	5·5	6·2	7·2	8·5	6·8
	R_m	1941–60	121	111	138	120	140	127	103	115	112	116	120	119	1442
166 Invercargill 46°26′S, 168°21′E (4 m)	T_m	(1890–1940)	13·9	13·9	12·8	10·6	8·1	6·1	5·3	6·9	8·6	10·6	11·4	13·1	10·1
	R_m	1890–1940	107	84	102	104	112	91	81	81	81	104	107	102	1156
167 Christchurch, New Zealand 43°32′S, 172°37′E (8 m)	T_m	(1863–1940)	16·4	16·1	14·4	11·9	8·9	6·4	5·8	6·7	9·2	11·7	13·6	15·6	11·4
	T_x	(1863–1940)	35·6	34·4	32·2	27·8	25·6	20·6	21·1	21·1	27·2	31·1	32·2	33·3	
	T_n	(1863–1940)	1·1	1·7	−1·1	−3·3	−6·1	−5·6	−5·0	−5·0	−5·0	−3·3	−0·6	0·6	
	drT	(1863–1940)	9·5	9·0	9·0	9·5	9·0	8·5	8·5	9·0	9·5	10·0	10·5	10·0	
	R_m	(1863–1940)	56	43	48	48	66	66	71	48	46	43	48	56	639
168 Hokitika, New Zealand 42°43′S, 170°58′E (3 m)	T_m	(1866–1940)	15·3	15·6	14·4	12·2	9·7	7·5	7·2	7·8	9·4	11·4	12·5	14·2	11·4
	R_m	(1866–1940)	262	191	239	236	244	231	218	239	226	292	267	262	2907
	R_{dx}	(1866–1940)	140	234	193	208	140	114	152	135	117	137	117	112	

169 Wellington, New Zealand 41°16′S, 174°46′E (126 m)	T_m	1862–1940	16·9	16·9	15·8	13·9	11·4	9·7	8·6	9·2	10·8	12·2	13·6	15·9	13·2
	T_x	1862–1940	29·4	31·1	27·2	23·3	21·7	20·6	19·4	19·4	20·6	23·9	27·2	28·3	
	T_n	1862–1940	3·9	5·0	3·9	2·2	0·0	−1·1	−1·7	−1·7	−0·6	1·1	2·2	3·3	
	drT		7·0	7·0	7·0	6·5	6·0	6·0	6·0	6·0	6·0	6·5	7·0	7·0	
	R_m	1862–1940	81	81	81	97	117	117	137	117	97	102	89	89	1203
	$R_m: Q_5 \geqslant$	1931–60	99	157	125	124	181	173	188	163	124	168	111	121	
	$Q_1 \leqslant$	1931–60	39	49	35	63	76	70	70	64	49	71	41	57	
	R_{dx}	1862–1940	114	160	145	127	145	76	76	97	97	89	66	152	
170 Auckland (Albert Park) 36°51′S, 174°46′E (41 m)	T_m	1921–50*	19·2	19·2	18·3	16·4	13·8	11·6	10·7	11·3	12·5	14·2	15·9	17·8	15·1
	T_x	1921–50	32·2	32·2	30·0	27·2	22·8	21·1	19·4	19·4	21·7	23·9	27·2	31·7	
	T_n	1921–50	7·2	8·3	5·6	3·9	2·2	1·7	0·6	1·1	1·1	1·7	5·0	6·1	
	R_m	1921–50*	85	104	71	108	123	139	139	108	98	106	89	78	1248
171 Apia, Samoa 13°48′S, 171°47′W (2 m)	T_m	1941–60	26·9	26·8	26·9	27·1	26·7	26·5	25·9	26·1	26·2	26·4	26·6	26·7	26·6
	T_x	1890–1937	32·8	33·3	32·8	32·8	32·2	32·2	32·8	32·2	32·2	33·9	33·3	32·8	
	T_n	1890–1937	20·6	21·1	21·1	20·6	19·4	19·4	17·2	18·3	18·3	18·9	20·6	21·1	
	drT	1890–1937	6·0	5·0	6·5	6·0	6·0	6·0	6·0	5·0	5·5	5·5	6·5	6·0	
	R_m	1931–60	424	364	352	214	186	130	115	111	147	221	279	385	2928
	$R_m: Q_5 \geqslant$	1931–60	528	487	462	281	272	208	161	181	216	287	399	515	
	$Q_1 \leqslant$	1931–60	252	262	238	137	106	45	60	35	61	128	148	223	
	R_{dx}	1890–1937	86	132	94	89	114	74	48	206	99	109	86	91	
172 Papeete, Tahiti 17°32′S, 149°35′W (2 m)	T_m	1951–60	26·0	26·2	26·5	26·3	25·5	24·6	24·1	23·9	24·3	24·8	25·6	26·0	25 3
	drT	1880–1940	9·5	9·5	9·5	9·5	9·5	9·5	10·0	10·0	9·5	9·5	9·5	9·0	
	R_m	(1880–1940)	251	244	429	142	102	76	53	43	53	89	150	249	1881
173 Easter Island 27°10′S, 109°26′W (30 m)	T_m	1911–13	22·8	23·3	23·1	21·4	20·6	18·3	17·8	17·8	18·1	18·6	19·7	21·7	20·3
	drT	1911–13	6·5	8·0	6·0	6·0	6·5	5·5	5·5	5·5	6·0	7·0	6·0	6·5	
	R_m	1911–13	132	41	229	130	99	241	81	66	86	56	127	74	1362
174 Canton Island 2°46′S, 171°43′W (3 m)	T_m	1931–60	28·4	28·3	28·4	28·7	28·8	28·8	28·7	28·6	28·7	28·7	28·7	28·4	28·6
	drT		6·0	5·5	6·5	6·5	7·0	6·5	6·5	7·0	6·5	6·5	6·5	5·5	
	R_m	1931–60	66	54	63	92	110	67	66	64	31	28	41	65	748
	$R_m: Q_5 \geqslant$	1931–60	109	83	99	139	171	97	99	100	51	43	65	99	
	$Q_1 \leqslant$	1931–60	0	0	13	40	27	31	28	23	9	4	0	0	
175 Malden Island 4°03′S, 155°01′W (8 m)	T_m	1890–1926	27·8	28·1	28·1	28·3	28·3	28·3	28·1	28·3	28·1	28·1	28·1	27·8	28·1
	R_m	1890–1926	89	48	114	114	109	53	48	41	20	23	18	18	695
	R_{dx}	1890–1926	107	74	99	97	102	84	91	107	36	114	155	46	
176 Fanning Island 3°54′N, 159°23′W (5 m)	T_m	1903–28*	27·5	27·5	27·8	27·8	27·8	28·1	28·1	28·3	28·1	28·3	28·3	27·8	27·9
	R_m	1903–28*	274	267	272	358	320	254	208	112	81	91	74	203	2514

World Climatic Table—*continued*

Place, position and height above sea level	Item	Years of observation covered	J	F	M	A	M	J	J	A	S	O	N	D	Year
177 Wake Island	T_m	1931–60	25·2	25·1	25·4	25·8	26·5	27·4	27·8	28·0	28·1	27·6	26·9	26·1	26·7
19°17′N, 166°39′E (3 m)	R_m	1931–60	29	34	37	47	52	48	117	180	133	134	78	46	936
178 Honolulu, Hawaii	T_m	1906–50	21·6	21·6	22·2	23·0	24·2	25·2	25·9	26·2	25·9	25·1	23·8	22·6	23·9
(Oahu, Observatory)		1951–60	21·6	21·6	21·9	23·0	23·9	24·8	25·5	25·9	25·4	24·9	23·9	22·5	23·7
21°18′N, 158°06′W (3 m)	drT		4·0	5·0	5·5	5·5	5·5	5·0	5·0	5·0	5·0	5·5	5·5	5·0	
	R_m	1901–50	91	71	57	33	19	12	8	14	19	33	48	90	495
		1921–50	90	62	38	40	18	9	10	12	18	45	44	84	470
		1951–60	93	77	101	13	14	1	6	15	8	40	65	79	512
	R_{dx}		114	163	343	203	119	76	30	53	152	117	140	155	
179 Adak, Aleutian Islands	T_m	1942–60	0·6	0·6	1·3	2·8	4·5	6·7	9·3	10·9	8·9	5·7	2·7	1·1	4·6
51°53′N, 176°39′W (5 m)	R_m	1942–60	169	136	151	107	117	79	75	111	139	176	185	194	1639
		1951–60	196	151	195	130	167	115	81	99	153	207	208	209	1911
180 Nome, Alaska	T_m	1931–60	−15·3	−14·7	−13·4	−6·0	1·7	7·7	9·7	9·4	5·5	−1·3	−8·6	−14·3	−3·3
64°30′N, 165°26′W (4 m)	drT	1917–53	8·0	9·0	9·5	9·0	7·0	8·5	6·0	6·5	6·5	6·0	6·5	7·0	
	R_m	1931–60	26	24	22	20	18	24	58	97	68	43	29	25	454
181 Barrow Point, Alaska	T_m	1921–53	−26·4	−28·1	−26·1	−18·1	−7·5	1·1	4·2	3·6	−0·8	−8·3	−17·2	−23·6	−12·3
71°18′N, 156°47′W (9 m)		1960–8	−25·5	−28·7	−26·3	−19·1	−6·7	0·8	3·6	2·8	−1·5	−10·0	−17·9	−24·3	−12·7
	σT_m		4·8	4·0	2·5	2·0	1·8	1·4	1·1	1·0	1·8	2·7	3·2	2·9	
	T_x	1921–53	0·6	−0·6	−1·1	5·6	7·2	21·1	25·6	22·8	15·0	4·4	3·9	1·1	
	T_n	1921–53	−47·2	−48·9	−46·7	−41·1	−27·8	−13·3	−5·6	−6·7	−15·6	−28·3	−40·0	−48·3	
	drT	1921–53	7·0	7·5	8·0	8·5	6·0	5·5	7·0	6·0	4·0	5·5	6·5	7·0	
	R_m	1921–53	5	3	3	3	3	8	23	20	13	13	8	7	109
	R_{dx}	1921–53	18	8	8	5	8	10	20	13	13	25	10	8	
182 Fairbanks, Alaska	T_m	1900–47*	−23·9	−17·5	−12·5	−1·4	8·3	14·7	15·6	12·8	6·4	−3·1	−15·8	−21·9	−3·2
64°51′N, 147°43′W	T_x	1900–47	5·6	10·0	13·3	20·6	30·0	35·0	37·2	32·2	26·7	19·4	12·2	14·4	
(134 m)	T_n	1900–47	−54·4	−50·0	−48·9	−35·6	−17·8	−2·2	−1·1	−7·2	−9·4	−33·3	−47·8	−50·6	
	drT	1900–47	10·0	11·5	15·0	14·0	13·5	14·0	13·5	12·0	11·5	9·5	9·5	9·5	
	R_m	(1900–47)*	23	13	18	8	15	33	48	53	33	20	18	15	297
	R_{dx}	(1900–47)	46	46	20	10	20	43	58	99	38	30	23	25	
183 Yakutat, Alaska	T_m	1931–60	−2·6	−1·9	−0·3	2·5	7·0	10·3	12·3	12·1	9·6	5·5	1·0	−2·2	4·4
59°31′N, 139°40′W (9 m)	R_m	1931–60	276	208	221	184	203	129	214	277	420	498	407	312	3348
	$R: Q_5 \geqslant$	1931–60	384	274	302	253	288	194	301	386	535	635	565	422	
	$Q_1 \leqslant$	1931–60	164	137	134	107	116	56	117	153	307	357	219	183	

Station	Element	Period													
184 Vancouver (Airport)	T_m	1941–60	2·3	4·2	5·8	9·1	12·6	15·2	17·6	17·0	14·3	10·1	6·0	3·9	9·8
49°11′N, 123°10′W (3 m)	σT_m		2·5	2·4	1·2	1·2	1·0	0·8	1·1	0·9	0·8	1·1	1·3	1·7	
	T_x		15·0	16·1	20·0	26·1	28·3	33·3	32·8	33·3	29·4	25·0	23·3	15·6	
	T_n		−16·7	−13·3	−9·4	−2·8	0·6	1·7	4·4	3·9	−1·1	−6·1	−12·2	−13·3	
	R_m	1941–60	139	121	96	60	48	51	26	36	56	117	142	156	1048
	R_{dx}		71	71	91	41	41	41	43	61	79	84	112	89	
185 Edmonton	T_m	1931–60	−14·1	−11·6	−5·5	4·2	11·2	14·3	17·3	15·6	10·8	5·1	−4·2	−10·4	2·7
53°34′N, 113°31′W	σT_m		4·0	5·0	2·7	2·0	1·6	1·4	1·4	1·0	2·0	2·4	3·0	3·5	
(671 m)	T_x		13·9	16·7	22·2	31·1	34·4	37·2	36·7	35·6	32·2	28·3	23·3	16·1	
	T_n		−49·4	−49·4	−40·0	−26·1	−12·2	−3·9	−1·7	−3·3	−11·1	−26·1	−42·2	−43·3	
	drT		10·5	13·0	12·0	13·5	14·5	14·0	14·0	14·0	13·5	12·0	10·0	9·0	
	R_m	1931–60	24	20	21	28	47	80	85	65	34	23	22	25	474
	$R: Q_5 \geq$	1931–60	34	27	33	40	74	123	97	98	50	39	29	33	
	$Q_1 \leq$		13	9	8	10	19	37	62	36	15	6	12	12	
186 Dawson	T_m	1931–60	−27·6	−23·9	−14·6	−1·4	8·1	13·8	15·4	12·5	6·4	−3·1	−16·4	−24·9	−4·6
64°04′N, 139°26′W	T_x		1·7	8·9	11·1	20·6	29·4	32·8	35·0	31·1	26·1	20·0	11·1	12·8	
(323 m)	T_n		−55·6	−58·3	−45·6	−36·1	−12·8	−3·9	−1·7	−8·3	−13·3	−30·6	−45·0	−52·8	
	R_m	1931–60	21	14	13	8	25	33	50	49	31	28	27	24	323
187 Isachsen	T_m	1951–60	−34·6	−36·6	−35·1	−24·2	−11·5	−0·2	3·7	1·4	−8·4	−18·8	−28·2	−32·2	−18·7
78°47′N, 103°32′W	R_m	1951–60	2	2	1	4	8	3	22	23	18	10	4	2	99
(25 m)															
188 Coppermine	T_m	1931–60	−28·6	−30·1	−25·8	−17·2	−5·6	3·4	9·3	8·4	2·6	−6·9	−19·9	−26·3	−11·4
67°49′N, 115°05′W (7 m)	drT		8·0	8·5	8·5	9·5	8·5	7·0	9·0	6·5	6·0	6·0	8·0	7·5	
	R_m	1931–60	13	8	15	14	12	20	34	44	29	27	17	13	246
189 Churchill	T_m	1951–60	−27·6	−25·6	−19·5	−10·1	−2·2	6·2	12·2	11·7	5·6	−1·4	−11·9	−21·6	−7·0
58°45′N, 94°04′W (30 m)	R_m	1951–60	17	17	21	36	40	48	39	60	52	42	46	25	443
190 Winnipeg (Airport)	T_m	1931–60	−17·7	−15·5	−7·9	3·3	11·3	16·5	20·2	18·9	12·8	6·2	−4·8	−12·9	2·5
49°54′N, 97°14′W		1951–60	−18·1	−14·1	−8·2	3·7	11·4	16·6	19·9	18·9	12·4	6·3	−4·3	−12·4	2·7
(240 m)	T_x		7·8	8·3	23·3	32·2	37·8	38·3	42·2	39·4	37·2	30·0	21·7	11·7	
	T_n		−44·4	−43·9	−38·9	−27·8	−11·7	−6·1	1·7	−1·1	−8·3	−20·6	−36·7	−47·8	
	drT		11·0	11·5	12·0	11·5	14·5	13·5	13·5	14·0	12·0	11·0	9·5	10·0	
	R_m	1931–60	26	21	27	30	50	81	69	70	55	37	29	22	517
		1951–60	25	21	23	31	52	97	75	70	50	41	34	19	538
	R_{dx}		18	15	38	23	48	48	69	66	66	74	18	25	
191a Toronto (City)	T_m	1836–1951	−5·0	−5·3	1·1	5·6	11·9	17·5	20·6	19·7	15·6	8·9	2·8	2·8	7·4
43°40′N, 79°24′W		1931–60	−3·8	−3·7	0·2	6·9	13·2	19·0	21·8	20·9	16·4	10·6	4·3	−1·7	8·7
(116 m)		1951–60	−4·0	−2·6	0·4	8·2	13·4	19·2	22·1	21·3	17·1	11·1	4·9	−1·3	9·2
	drT	1836–1951	8·0	8·5	8·0	9·0	10·5	10·5	11·0	10·5	10·0	9·0	6·5	6·5	
	R_m	1836–1951	69	61	66	63	73	69	74	68	74	61	71	66	815
		1931–60	67	59	67	66	70	63	74	61	65	60	63	61	776
	R_{dx}	1836–1951	63	43	43	58	69	61	99	94	89	97	78	48	

World Climatic Table—*continued*

Place, position and height above sea level	Item	Years of observation covered	J	F	M	A	M	J	J	A	S	O	N	D	Year
191b Toronto (Malton Airport) 43°41′N, 79°38′W (176 m)	T_m	1951–60	−5·7	−4·5	−1·0	7·2	12·6	18·5	21·3	20·5	16·2	9·8	3·6	−2·8	8·0
	R_m	1951–60	52	50	61	70	69	56	59	75	58	62	55	56	723
192 Montreal (Dorval Airport) 45°28′N, 73°45′W (30 m)	T_m	1941–60	−9·6	−8·3	−2·2	5·9	13·1	18·6	21·4	20·3	15·4	9·4	2·5	−6·4	6·7
	drT		8·5	8·5	8·0	9·5	9·5	9·5	9·5	9·0	9·0	8·0	6·5	7·0	
	R_m	1941–60	83	81	78	72	72	85	89	77	82	78	85	89	971
193 Frobisher Bay, Baffin Land 63°45′N, 68°33′W (21 m)	T_m	1942–60	−26·1	−24·8	−21·2	−13·2	−2·2	3·9	8·1	7·0	2·7	−5·0	−13·8	−21·7	−8·8
	R_m	1942–60	32	55	24	26	21	42	70	58	38	43	40	28	457
194 St Johns, Newfoundland 47°34′N, 52°42′W (74 m)	T_m	1872–1940	−4·7	−5·6	−2·5	1·9	5·8	11·4	15·3	16·1	12·5	8·1	2·8	−1·7	4·9
		1960–8	−3·3	−4·2	−2·2	1·6	6·2	11·2	16·1	15·8	12·4	8·3	4·3	−0·4	5·5
	σT_m		1·7	2·0	2·1	1·1	0·9	1·3	1·6	1·2	0·9	0·8	1·2	1·8	
	T_x	1872–1940	15·0	13·3	19·4	22·2	27·2	30·6	32·2	33·9	28·9	30·6	20·0	15·6	
	T_n	1872–1940	−28·3	−29·4	−25·6	−18·3	−6·7	−2·8	0·6	0·0	−1·7	−5·6	−14·4	−20·0	
	drT	1872–1940	6·0	6·5	6·0	6·0	8·5	9·5	9·5	9·0	8·5	7·0	5·5	5·5	
	R_m	(1872–1940)	135	124	117	107	91	89	89	94	97	135	150	140	1368
		1931–40	113	101	107	78	80	85	67	98	122	115	113	147	1226
		1960–8	160	156	145	101	91	70	61	86	74	121	135	148	1348
	R_{dx}	(1872–1940)	41	53	61	46	48	58	41	69	58	74	33	56	
195 New Haven, Conn. 41°18′N, 72°56′W (32 m)	T_m	1851–1950*	−2·0	−1·8	2·3	8·3	14·2	19·3	22·3	21·2	17·6	11·8	5·6	−0·3	9·9
		1780–1850	−3·1	−2·3	2·3	8·3	14·4	19·4	22·2	21·4	17·0	10·6	4·5	−0·8	9·5
		1921–50*	−1·1	−1·1	3·4	8·7	14·7	19·8	22·6	21·7	18·2	12·4	6·7	0·5	10·5
		1960–8	−2·0	−1·5	2·7	8·3	13·6	19·3	22·1	21·3	17·9	12·4	6·7	−0·3	10·0
	σT_m	1851–1950	2·6	2·3	2·2	1·3	1·3	1·2	1·1	1·2	1·3	1·5	1·7	2·2	
	R_m	1804–1920	97	101	103	90	99	80	109	113	94	94	91	93	1164
		1921–50*	101	86	107	99	99	98	95	108	91	76	99	100	1159
196 Mt Washington, New Hampshire 44°16′N, 71°18′W (1909 m)	T_m	1933–60	−14·2	−14·7	−11·6	−5·0	1·7	7·1	9·6	8·7	5·0	−0·6	−6·5	−12·9	−2·7
	drT		10·0	9·0	8·0	6·5	6·5	6·5	6·5	6·0	6·0	6·5	6·5	7·0	
	R_m	1933–60	135	134	147	150	149	164	162	168	172	154	169	163	1867
	R_{dx}		53	109	97	69	69	104	71	97	114	94	61	61	
197 New York (City) 40°47′N, 73°58′W (132 m)	T_m	(1869–1949)	−0·8	−0·6	3·1	9·7	15·8	20·3	23·3	22·8	20·8	15·0	6·7	1·7	11·5
		{1789–95, 1804–15}	−1·8	−0·7	3·7	9·6	15·3	20·7	23·2	22·7	18·4	11·8	5·9	0·6	10·8
	T_x	1869–1949	20·0	22·8	28·9	32·8	35·0	36·1	38·9	38·9	37·8	32·2	23·9	20·6	
	T_n	1869–1949	−21·1	−25·6	−16·1	−11·1	1·1	6·7	12·2	10·6	3·9	−2·8	−13·9	−25·0	
	drT	(1869–1949)	7·0	8·0	8·5	8·5	8·5	9·5	9·0	8·0	10·5	11·0	8·0	6·5	
	R_m	(1869–1949)	94	97	91	81	81	84	107	109	86	89	76	91	1086

Station	Element	Period	J	F	M	A	M	J	J	A	S	O	N	D	Year
198 Washington, DC	T_m	1851–1950*	1·3	1·9	6·3	12·1	17·9	22·6	24·9	23·8	20·2	13·9	7·8	2·5	12·9
38°54′N, 77°03′W (22 m)	σT_m	1851–1950	2·9	2·5	2·5	1·7	1·5	1·3	1·1	1·1	1·5	1·7	1·6	2·1	
	T_x	1870–1949	25·0	28·9	33·9	35·0	36·1	38·9	41·1	41·1	40·0	35·6	28·3	23·3	
	T_n	1870–1949	−25·6	−26·1	−15·6	−9·4	0·6	6·1	11·1	9·4	2·2	−3·3	−11·7	−25·0	
	drT	1870–1949*	8·5	9·0	10·0	11·0	11·5	11·0	10·5	10·0	10·5	10·5	9·5	9·0	
	R_m	1870–1949*	86	76	91	84	94	99	112	109	94	74	66	79	1064
	R_{dx}	1870–1949	76	58	71	81	89	107	147	185	145	102	81	69	
199 Chicago	T_m	1870–1949	−3·9	−2·8	2·2	8·6	14·2	19·7	23·1	22·2	18·6	12·2	4·7	−1·4	9·8
41°53′N, 87°38′W	T_x	1870–1949	18·3	20·0	27·8	32·8	36·7	38·9	40·6	38·9	37·8	31·1	25·6	20·0	
(251 m)	T_n	1870–1949	−28·9	−29·4	−24·4	−8·3	−2·8	1·7	9·4	8·3	−1·7	−10·0	−18·9	−26·1	
	R_m	1870–1949	51	51	66	71	86	89	84	81	79	66	61	51	836
	R_{dx}	1870–1949	48	51	84	104	76	86	104	157	86	63	86	69	
200 St Louis (City), Missouri	T_m	1851–1950*	−0·1	1·5	6·7	13·2	18·9	23·9	26·4	25·3	21·2	14·6	7·1	1·7	13·4
38°38′N, 90°12′W	σT_m	1851–1950	3·2	2·8	2·8	1·9	1·7	1·6	1·4	1·5	1·8	2·1	2·1	2·8	
(173 m)	drT	1870–1949	9·0	9·5	10·0	10·0	11·0	10·0	9·5	10·0	10·0	10·0	9·0	8·5	
	R_m	1837–1923	58	65	89	96	114	116	92	88	81	72	73	64	1008
		1857–1950*	58	63	89	97	114	114	89	86	81	74	71	63	999
		1921–50*	59	48	92	102	104	97	74	96	86	74	69	62	963
	$R: Q_5 \geq$	1931–60	73	71	109	128	133	162	127	119	105	104	94	72	
	$Q_1 \leq$	1931–60	21	24	42	55	45	47	34	31	31	37	31	22	
	R_{dx}	1870–1949	91	112	99	160	86	122	175	224	107	102	91	76	
201 Salt Lake City, Utah	T_m	1931–60	−2·1	0·6	4·7	9·9	14·7	19·4	24·7	23·6	18·3	11·5	3·4	−0·2	10·7
40°47′N, 111°58′W	drT	1874–1949	10·0	9·5	11·0	13·5	14·5	16·5	17·0	16·5	16·5	14·5	10·5	10·0	
(1287 m)	R_m	1931–60	34	30	40	45	36	25	15	22	13	29	33	31	353
	$R: Q_5 \geq$	1931–60	48	42	58	62	55	38	23	33	22	46	53	43	
	$Q_1 \leq$	1931–60	17	15	18	27	10	6	3	4	3	12	13	18	
	R_{dx}	1874–1949	33	33	46	36	69	51	28	36	46	43	41	36	
202 San Francisco	T_m	1851–1950*	9·7	10·9	11·9	12·7	13·4	14·3	14·3	14·5	15·6	15·2	13·2	10·5	13·0
37°47′N, 122°25′W		1921–50*	9·8	11·4	12·5	13·0	13·8	14·7	14·6	14·9	16·2	15·7	13·7	10·9	13·4
(47 m)	σT_m	1851–1950	1·3	1·3	1·3	1·1	1·0	1·0	0·8	0·8	0·9	1·0	1·0	1·1	
	T_x	1850–1948	25·6	26·7	30·0	31·7	36·1	37·8	37·2	33·3	38·3	35·6	28·3	23·3	
	T_n	1850–1948	−1·7	0·6	0·6	4·4	5·6	7·8	8·3	7·8	8·3	6·1	3·3	−2·8	
	drT	(1850–1948)	5·5	6·5	7·0	7·0	6·5	8·0	6·5	6·5	8·0	8·0	6·5	5·5	
	R_m	1850–1948	119	97	79	38	18	3	0	1	8	25	63	112	563
		1921–50*	102	99	71	38	15	4	0·3	0·5	3	27	57	103	520
	R_{dx}	1850–1948	119	91	94	61	36	30	5	8	91	51	102	86	
203 Death Valley (Greenland Ranch), California	T_m	1911–52	11·1	14·4	18·9	23·9	28·9	34·2	38·6	37·2	31·9	23·9	16·1	11·4	24·2
36°28′N, 116°51′W	T_x	1911–52	29·4	33·3	37·8	42·8	48·9	51·1	56·7	52·8	49·4	43·3	33·9	30·0	
(−54 m)	T_n	1911–52	−9·4	−6·1	−1·1	1·7	5·6	9·4	16·7	18·3	5·0	0·0	−4·4	−7·2	
	drT	1911–52	15·5	15·5	16·5	16·5	16·5	17·0	16·0	16·5	18·5	18·0	16·5	15·0	
	R_m	(1911–52)	3	1	2	3	5	3	8	8	5	2	3	2	45

World Climatic Table—*continued*

Place, position and height above sea level	Item	Years of observation covered	J	F	M	A	M	J	J	A	S	O	N	D	Year
204 Mt Wilson, California 34°14′N, 118°04′W (1783 m)	T_m		5·8	6·4	7·2	9·7	13·6	19·4	22·8	22·2	19·2	14·2	10·3	7·2	13·2
	R_m		160	170	155	66	30	5	1	3	13	28	48	112	791
205 Los Angeles 34°03′N, 118°15′W (95 m)	T_m	1878–1949	13·1	13·6	14·2	15·6	16·9	18·9	21·4	21·7	20·8	18·3	16·4	13·9	17·1
	T_x	1878–1949	32·2	33·3	37·2	37·8	39·4	40·6	42·8	41·1	42·2	38·9	35·6	33·3	
	T_n	1878–1949	−2·2	−2·2	−0·6	2·2	4·4	7·8	9·4	9·4	6·7	4·4	1·1	−1·1	
	R_m	(1878–1949)	79	76	71	25	10	3	1	1	5	15	30	66	382
206 Phoenix, Arizona 33°26′N, 112°01′W (339 m)	T_m	1931–60	10·4	12·5	15·8	20·4	25·0	29·8	32·9	31·7	29·1	22·3	15·1	11·4	21·4
	T_x	1896–1949	28·9	33·3	35·0	39·4	45·6	47·8	47·8	46·1	45·0	40·6	35·6	28·9	
	T_n	1896–1949	−8·9	−4·4	−1·1	1·7	3·9	9·4	17·2	14·4	9·4	2·2	−2·8	−5·6	
	R_m	1931–60	19	22	17	8	3	2	20	28	19	12	12	22	183
	$R: Q_5\geqslant$	1931–60	32	34	28	15	7	5	29	45	34	21	24	43	
	$Q_1\leqslant$	1931–60	4	5	0	0	0	0	3	7	0	0	0	3	
207 Dallas, Texas 32°46′N, 96°47′W (156 m)	T_m	1892–1949	7·5	10·0	13·6	18·3	22·5	26·9	29·2	28·9	25·6	19·7	13·6	8·6	18·7
	drT	1892–1949	10·5	11·0	11·5	11·0	10·5	10·5	10·5	11·0	11·0	11·5	10·5	10·5	
	R_m	1892–1949	63	61	84	107	114	97	71	76	69	71	69	64	946
	R_{dx}	1892–1949	130	84	135	155	157	114	137	234	157	99	127	104	
208 New Orleans 29°57′N. 90°04′W (17 m)	T_m	1931–60	12·3	13·4	15·8	19·4	23·3	26·4	27·3	27·4	25·4	21·1	15·3	12·7	20·0
	drT	1871–1949	8·5	8·5	9·0	9·0	8·5	8·0	8·0	7·5	7·0	8·5	8·5	8·0	
	R_m	1931–60	98	101	136	116	111	113	171	136	128	72	85	104	1369
	$R: Q_5\geqslant$	1931–60	131	143	212	161	162	163	222	182	175	125	129	144	
	$Q_1\leqslant$	1931–60	58	49	51	60	51	54	116	87	65	8	26	61	
209 Key West (City), Florida 24°33′N, 81°48′W (7 m)	T_m	1851–1950*	20·9	21·5	22·8	24·5	26·3	28·0	28·7	28·8	28·2	26·2	23·6	21·6	25·1
	σT_m	1851–1950	1·7	1·6	1·3	1·0	0·7	0·8	0·6	0·6	0·5	0·7	1·1	1·5	
	R_m	(1871–1949)*	51	33	36	33	89	107	84	114	170	152	56	43	968
	R_{dx}	1871–1949	102	76	114	201	147	140	191	211	302	343	226	99	
210 Bermuda (Hamilton) 32°17′N, 64°46′W (46 m)	T_m	(1852–1940)	17·2	16·9	17·0	18·3	21·1	23·9	26·1	26·7	25·6	23·3	20·3	18·3	21·2
	drT	(1852–1940)	5·5	6·0	6·0	6·5	6·5	6·5	6·5	6·5	6·5	5·5	6·0	5·5	
	R_m	(1852–1940)	112	119	122	104	117	112	114	137	132	147	127	119	1462
	R_{dx}	(1852–1940)	76	137	130	290	140	203	112	150	155	145	152	117	
211 Havanna, Cuba 23°08′N, 82°21′W (24 m)	T_m	(1873–1945)	22·2	22·2	23·3	24·7	26·1	27·2	27·8	27·8	27·5	26·1	23·9	22·8	25·1
	T_x	(1873–1945)	31·7	32·8	33·3	32·8	34·4	35·0	35·0	34·4	35·0	33·9	32·8	31·7	
	T_n	(1873–1945)	10·0	10·0	11·7	12·8	15·0	18·9	18·9	20·0	19·4	17·2	12·8	10·6	
	drT	(1873–1945)	8·0	8·0	8·0	8·5	8·0	8·0	8·0	8·0	7·0	6·0	6·0	6·0	
	R_m	1873–1945	71	46	46	58	119	165	124	135	150	173	79	58	1224

Station	Element	Period	J	F	M	A	M	J	J	A	S	O	N	D	Year
212 Trinidad (Piarco Airport) 10°37′N, 61°21′W (12 m)	T_m	1951–60	24·5	24·7	25·4	26·3	26·6	26·1	25·9	26·1	26·2	25·9	25·4	24·8	25·7
	drT	1946–58	10·0	9·5	10·5	10·0	9·0	8·0	8·0	9·0	9·0	9·5	9·5	9·0	
	R_m	1951–60	77	61	27	71	129	269	243	213	144	151	212	153	1750
213 Mexico City (Central) 19°26′N, 99°08′W (2259 m)	T_m	1878–1950*	13·0	14·6	16·6	18·1	18·7	18·0	17·1	17·1	16·7	15·6	14·2	13·1	16·1
		1942–50*	14·7	16·1	18·1	19·4	20·1	19·1	18·4	18·4	18·1	16·7	15·4	14·2	17·4
	T_x	(7 years)	23·3	27·2	28·9	32·2	31·7	30·6	28·3	27·2	25·6	25·6	25·0	22·8	
	T_n	(7 years)	−2·8	−1·7	1·1	0·6	6·1	9·4	8·3	9·4	1·1	1·7	2·2	0·0	
	drT		13·5	14·5	15·5	14·5	13·5	11·5	11·0	10·5	11·5	11·0	12·0	13·0	
	R_m	1878–1950*	4	5	11	18	46	100	116	114	102	37	13	5	571
		1941–50*	2	2	10	18	29	94	91	108	91	24	13	2	485
	R_{dx}	(12 years)	33	13	18	36	38	53	48	58	51	63	28	28	
214 Tampico, Mexico 22°12′N, 97°51′W (20 m)	T_m	1921–60	18·9	20·3	22·0	24·7	26·8	28·0	28·0	28·3	26·5	25·6	22·0	19·7	24·2
	R_m	1921–60	38	19	13	19	49	143	151	130	297	146	48	30	1083
215 Mazatlan, Mexico 23°11′N, 106°25′W (78 m)	T_m	1921–60	19·8	19·7	20·2	21·8	24·4	26·9	28·0	28·0	27·8	27·0	24·0	21·2	24·1
	T_x	1932–42	25·0	26·1	27·8	30·6	30·6	32·8	32·8	33·3	31·7	31·7	30·6	27·8	
	T_n	1932–42	11·7	11·1	12·8	13·9	15·0	21·1	20·6	20·0	20·0	20·6	16·7	12·2	
	drT	1932–42	5·5	6·0	5·5	6·0	5·5	4·5	5·0	5·0	4·5	5·0	5·0	5·5	
	R_m	1921–60	12	8	3	0	1	34	174	215	250	63	17	27	805
	R_{dx}	1932–42	10	13	5	0	3	79	147	213	145	46	36	30	
216 Cristobal (Colon), Panama Canal Zone 9°23′N, 79°54′W (11 m)	T_m	1908–25	26·7	26·7	26·9	27·2	27·1	26·8	26·8	26·8	26·9	26·6	26·3	26·7	26·8
	drT		4·5	5·0	4·5	5·0	5·5	5·5	5·0	5·0	6·0	6·0	5·0	5·0	
	R_m	1863–1923	95	41	40	109	314	337	406	375	317	384	526	289	3233
	R_{dx}	(37 years)	86	201	89	137	191	216	173	175	191	279	249	262	
217 Caracas (Cagigal Observatory), Venezuela 10°30′N, 66°55′W (1042 m)	T_m	1951–60	19·2	19·7	20·7	21·7	22·0	21·5	21·1	21·6	21·8	21·5	20·8	19·9	21·0
	drT		10·5	11·5	11·5	11·5	10·0	9·0	9·5	10·0	10·5	10·0	9·5	11·0	
	R_m	1951–60	22	28	11	43	92	121	107	105	107	110	73	44	863
218 San Cristobal, Galapagos Islands 0°54′S, 89°37′W (6 m)	T_m	1951–60	25·1	25·9	26·1	25·7	25·2	23·8	22·7	21·7	21·5	22·0	22·6	23·8	23·8
	R_m	1951–60	39	155	124	106	27	4	12	7	5	7	6	14	506
219 Quito, Ecuador 0°08′S, 78°29′W (2811 m)	T_m	1931–60	13·0	13·0	12·9	13·0	13·1	13·0	12·9	13·1	13·2	12·9	12·8	13·0	13·0
	drT		14·5	13·5	13·5	13·0	13·0	14·5	15·5	15·5	15·5	14·5	15·0	14·5	
	R_m	1931–60	119	131	154	185	130	54	20	25	81	134	96	104	1233
	$R: Q_5 \geq$	1931–60	166	169	195	225	167	87	32	43	109	179	151	139	
	$Q_1 \leq$	1931–60	59	87	111	131	91	25	5	6	41	97	47	75	
220 Guayaquil 2°12′S, 79°53′W (7 m)	T_m	1931–60	25·5	26·0	26·4	26·3	25·6	24·4	23·5	23·2	23·8	24·0	24·6	25·4	24·9
	drT		10·0	9·0	9·0	10·0	11·0	10·5	9·5	11·5	11·5	10·0	11·0	10·0	
	R_m	1931–60	212	289	292	207	54	11	4	0	0	1	2	28	1000

World Climatic Table—*continued*

Place, position and height above sea level	Item	Years of observation covered	J	F	M	A	M	J	J	A	S	O	N	D	Year
221 Lima (Campo del Marte),	T_m	1931–60	21·5	22·3	21·9	20·1	17·8	16·0	15·3	15·1	15·4	16·3	17·7	19·4	18·2
Peru 12°04′S, 77°02′W	R_m	1951–60	1	0·4	0·6	0·1	1	3	3	4	4	2	1	0·5	21
(137 m)															
222 La Paz (El Alto),	T_m	1951–60	9·1	8·8	9·3	9·0	8·4	7·4	7·4	8·1	8·6	9·9	10·1	10·0	8·8
Bolivia 16°30′S,	drT	1891–1948	11·0	11·0	12·0	14·0	15·0	15·5	16·0	15·5	14·5	14·5	14·0	12·5	
68°10′W (4103 m)	R_m	1951–60	139	108	56	22	9	4	4	18	34	37	48	76	555
	R_{dx}	1891–1948	48	41	28	33	18	10	20	25	33	25	43	38	
223 Manaus, Brazil	T_m	1951–60	26·0	25·8	25·7	25·8	26·5	26·7	26·8	27·6	28·1	27·7	27·1	26·6	26·7
3°08′S, 60°01′W (44 m)	drT		7·0	7·0	7·0	6·5	7·0	7·0	8·0	9·0	9·5	9·0	8·5	8·5	
	R_m	1931–60	278	278	300	287	193	99	61	41	62	112	165	220	2095
224 Belem	T_m	(1932–60?)	25·6	25·4	25·4	25·7	26·0	26·0	25·9	26·0	26·0	26·2	26·5	26·2	25·9
1°27′S, 48°29′W (24 m)	R_m	1931–60	317	413	436	382	265	164	160	113	119	106	94	201	2770
	R_{dx}	(1882–1918)	81	127	61	66	91	20	30	40	71	40	18	43	
225 Rio de Janeiro	T_m	1871–1950	25·4	25·7	25·1	23·6	22·0	20·8	20·1	20·7	21·0	21·6	22·8	24·5	22·8
22°54′S, 43°10′W (27 m)		1931–60	26·0	26·1	25·5	23·9	22·3	21·3	20·8	21·1	21·5	22·3	23·1	24·4	23·2
		1951–60	26·6	26·5	25·8	24·3	22·4	21·5	21·2	21·6	22·3	23·2	23·7	24·8	23·7
	T_x	(1851–1935)	38·9	36·7	36·1	35·0	34·4	32·2	32·8	33·9	37·8	38·9	37·8	38·9	
	T_n		15·6	17·2	17·8	15·6	13·3	11·1	11·1	11·7	10·0	13·9	15·0	13·3	
	drT		6·0	6·5	6·0	6·0	6·0	6·5	6·5	6·5	5·5	6·0	6·0	6·0	
	R_m	1851–1950	133	118	130	106	75	52	42	42	62	80	98	136	1074
		1931–40	107	115	80	94	63	30	35	41	47	68	100	99	879
		1941–60	150	148	161	127	78	50	47	44	55	77	86	139	1162
	$R: Q_5\geq$	1931–60	216	219	194	162	107	61	62	78	74	94	136	164	
	$Q_1\geq$		71	52	69	67	39	18	17	14	30	42	62	83	
	R_{dx}	1851–1935	97	122	145	224	63	206	48	51	71	51	99	170	
226 Asuncion, Paraguay	T_m	1951–60	29·2	28·8	27·2	23·1	20·6	18·8	16·3	20·2	22·1	24·5	27·2	29·0	24·1
25°16′S, 57°38′W (64 m)	drT	(1893–1930)	13·5	13·0	13·0	10·5	10·5	10·5	11·5	11·5	13·0	13·5	14·0	13·5	
	R_m	1951–60	177	182	140	154	171	80	55	38	96	164	122	153	1532
	R_{dx}	(1893–1930)	221	124	125	152	145	79	97	61	81	175	175	99	
227 Buenos Aires (City	T_m	1856–1950	23·2	22·7	20·5	16·5	13·0	10·1	9·7	10·8	13·0	15·6	18·9	21·7	16·3
Observatory) 34°35′S,		1931–60	23·7	23·0	20·7	16·6	13·7	11·1	10·5	11·5	13·6	16·5	19·5	22·1	16·9
58°29′W (55 m)	T_x	(1856–1941)	40·0	39·4	37·2	36·1	28·9	25·0	28·9	30·6	30·0	32·8	35·0	38·9	
	T_n		6·1	4·4	3·9	−2·2	−3·9	−5·0	−5·6	−2·8	−2·2	−2·2	2·2	3·9	
	R_m	1861–1950	81	74	113	89	74	62	56	65	80	88	81	98	961
		1931–60	104	82	122	90	79	68	61	68	80	100	90	83	1027
	R_{dx}	(1856–1941)	157	160	124	122	165	86	94	112	127	178	97	107	

Station	Element	Period	J	F	M	A	M	J	J	A	S	O	N	D	Year
228 Cordoba, Argentina	T_m	1873–1950	23·6	22·9	20·5	16·9	13·3	10·2	10·3	12·0	14·8	17·5	20·4	22·7	17·1
31°24′S, 64°11′W		1931–60	24·2	23·2	20·7	16·8	13·6	11·0	10·6	12·3	15·1	17·9	20·8	23·1	17·4
(425 m)		1951–60	23·9	22·8	20·9	16·3	13·9	10·8	10·8	12·7	15·3	17·7	20·8	22·2	17·3
	T_x		45·6	43·9	37·2	34·4	33·3	31·7	35·0	36·7	37·8	40·6	40·6	42·8	
	T_n		5·6	3·3	0·6	−0·6	−6·7	−8·3	−10·6	−7·2	−6·1	−1·1	2·2	3·9	
	drT		9·5	9·0	8·0	8·0	8·5	9·0	9·5	10·5	10·0	9·0	9·0	9·0	
	R_m	1873–1950	103	100	90	45	29	9	10	12	25	67	94	113	697
		1931–60	101	88	92	39	25	10	8	15	29	77	88	108	680
		1951–60	121	92	94	36	24	13	5	21	24	82	104	116	732
	R_{dx}		114	145	84	63	91	15	18	38	56	69	71	94	
229 Sarmiento, Argentina	T_m	1931–60	17·3	16·9	14·3	10·8	7·0	3·9	4·0	5·5	8·0	11·6	14·3	16·4	10·8
45°35′S, 69°08′W	drT	1916–44	14·5	14·5	13·0	11·0	10·0	8·5	9·0	10·0	11·5	14·0	13·5	14·0	
(266 m)	R_m	1931–60	10	8	11	15	24	16	17	14	10	6	12	9	152
		1951–60	12	3	9	14	21	16	15	9	10	5	7	7	128
	$R: Q_5 \geq$	1931–60	16	13	15	26	37	25	21	25	17	7	19	18	
	$Q_1 \leq$		2	0	1	2	5	7	5	4	3	1	1	1	
230 Stanley, Falkland Islands	T_m	1941–60	9·2	9·5	8·5	6·2	3·8	2·6	2·3	2·6	3·9	5·8	7·4	8·5	5·9
51°42′S, 57°52′W (53 m)		1951–60	9·1	9·4	8·6	6·0	3·9	2·4	2·3	2·7	3·8	5·7	7·6	8·2	5·8
	T_x	(1874–1941)	24·4	23·3	21·1	17·2	14·4	10·6	10·0	11·1	15·0	17·8	21·7	21·7	
	T_n		−1·1	−1·1	−2·8	−6·1	−6·7	−11·1	−8·9	−11·1	−10·6	−5·6	−3·3	−1·7	
	drT	(1874–1941)	8·0	8·0	7·0	6·5	5·5	5·5	5·0	5·5	6·5	7·0	8·5	8·5	
	R_m	1941–60	73	59	51	52	55	50	47	47	43	35	42	71	625
231 Antofagasta, Chile	T_m	1951–60	20·0	20·3	18·9	16·5	15·1	13·6	13·4	13·5	14·4	15·3	17·0	18·2	16·4
23°28′S, 70°26′W	R_m	1951–60	0	0	0	0	0	0	0·3	0·1	0	0	0	0	0·4
(122 m)															
232 Santiago, Chile	T_m	1860–1950	19·9	19·1	16·9	13·6	10·5	8·0	7·9	9·1	11·3	13·7	16·4	18·9	13·8
33°27′S, 70°42′W		1951–60	20·8	20·4	18·1	14·2	11·1	8·6	8·2	9·7	11·8	14·2	18·1	20·0	14·6
(520 m)	T_x		35·6	36·7	34·4	31·1	30·6	26·7	27·2	29·4	31·1	33·3	36·1	37·2	
	T_n		6·1	6·1	3·3	0·6	−2·8	−3·3	−4·4	−3·3	−0·6	0·0	2·8	2·2	
	drT		17·5	17·5	17·0	16·0	13·5	11·5	12·0	13·0	13·5	15·0	16·5	17·5	
	R_m	1867–1950	2	3	4	14	65	86	76	57	29	15	6	4	361
		1921–50	3	5	4	13	64	88	66	60	25	17	6	3	355
		1951–60	1	0·4	6	21	74	65	57	55	26	8	2	2	317
	R_{dx}		23	15	13	33	74	104	63	66	28	25	30	38	
233 Valdivia, Chile	T_m	1951–60	16·5	16·2	14·6	11·6	9·7	7·9	7·6	8·0	9·1	11·6	14·0	16·2	11·9
39°48′S, 73°14′W (13 m)	drT	(1853–1942)	11·5	12·0	11·0	9·0	7·5	5·5	6·0	8·0	9·5	10·5	10·5	10·5	
	R_m	1852–1921	67	73	140	236	402	437	415	348	213	139	126	112	2708

World Climatic Table—*continued*

Place, position and height above sea level	Item	Years of observation covered	J	F	M	A	M	J	J	A	S	O	N	D	Year
234 Punta Arenas (Magellanes), Chile 53°10′S, 70°54′W (28 m)	T_m	1888–1950	11·0	10·6	8·9	6·7	4·1	2·5	1·9	2·7	4·5	6·8	8·4	10·1	6·5
	T_x	(1888–1939)	30·0	26·1	23·9	20·6	17·2	11·1	11·7	12·8	16·1	19·4	24·4	23·9	
	T_n	(1888–1939)	−3·3	−2·2	−4·4	−5·0	−8·9	−11·7	−11·1	−9·4	−7·2	−3·9	−5·0	−5·0	
	drT	(1888–1939)	7·0	7·5	7·0	6·0	5·5	4·5	5·0	5·0	6·0	7·0	8·0	8·0	
	R_m	1888–1950	34	28	46	42	43	35	33	31	30	24	28	33	407
		1921–50	33	27	42	44	45	41	34	32	33	28	28	32	419
		1951–60	37	30	45	43	44	25	37	52	25	23	39	41	441
235 Laurie Island (Orcadas del Sud), S. Orkney 60°44′S, 44°44′W (4 m)	T_m	1903–50	0·1	0·2	−0·6	−3·3	−7·1	−10·4	−10·9	−10·2	−6·9	−3·9	−2·3	−0·8	−4·7
		1951–60	0·5	1·0	0·2	−1·8	−6·2	−8·7	−10·5	−8·4	−4·9	−2·6	−1·3	−0·4	−3·6
	σT_m	1903–50	0·5	0·6	0·8	1·7	2·6	2·9	2·8	3·1	2·1	1·5	1·1	0·6	
	T_x	1903–50	12·2	9·0	10·8	7·6	9·2	5·9	7·8	8·2	6·5	.8·7	8·8	9·6	
	T_n	1903–50	−7·0	−9·8	−15·1	−31·5	−31·9	−38·3	−36·9	−40·1	−32·6	−31·2	−20·4	−13·2	
	drT	1903–50	3·2	3·3	4·0	5·2	7·1	8·5	9·3	9·2	7·8	6·0	4·2	3·4	
	R_m	1903–50	35	39	48	41	32	26	32	32	29	29	32	27	402
		1951–60	34	46	62	45	43	33	29	33	38	31	37	26	457
236 Mirny, Antarctica 66°33′S, 93°01′E (30 m)	T_m	1956–9	−1·9	−5·2	−9·2	−11·8	−13·2	−16·5	−16·4	−17·9	−18·0	−12·8	−7·6	−2·0	−11·0
	R_m	1956–9	4	9	32	33	105	66	106	78	103	49	13	28	626
	N_p	1956–9	13	9	10	10	13	11	17	14	14	14	8	13	146
237 McMurdo Sound 77°51′S, 166°37′E (*c*. 20m)	T_m	1902–4, 1911–12, 1956–8	−4·7	−8·4	−16·2	−22·6	−23·0	−24·4	−26·4	−26·9	−22·3	−19·1	−8·7	−3·8	−17·2
	T_x	1902–4, 1911–12, 1956–8	4·4	4·4	−2·8	−7·2	−8·3	−6·1	−8·9	−7·8	−8·9	−4·4	1·1	5·6	
	T_n		−15·6	−22·8	−29·5	−41·1	−50·1	−44·0	−47·4	−46·9	−50·7	−41·3	−21·7	−15·6	
	N_p	1902–4, 1911–12, 1956–8	11	10	13	12	12	11	11	12	10	13	9	8	132
238 Vostok II 78°27′S, 106°52′E (3420 m)	T_m	1958–60	−33·6	−44·0	−54·6	−63·1	−63·4	−66·7	−66·9	−70·6	−67·2	−59·0	−43·8	−32·2	−55·4
	T_x	1958–60	−22·4	−25·1	−37·7	−43·3	−42·7	−47·3	−44·1	−52·1	−44·7	−40·5	−33·5	−25·8	
	T_n	1958–60	−47·9	−64·0	−75·0	−73·5	−78·3	−81·1	−82·5	−88·3	−82·3	−75·5	−59·0	−48·0	
	R_m	1958–9	0·6	1	7	4	9	12	6	5	5	2	0·6	0·8	53
	N_p	1958	23	12	21	29	23	27	31	31	22	30	23	15	287
239 South Pole 2880 m	T_m	1957–9	−27·0	−37·5	−53·8	−57·8	−56·4	−59·2	−58·5	−59·9	−60·1	−51·1	−37·6	−28·2	−48·9
	T_x	1957–9	−24·7	−21·1	−37·6	−48·8	−46·1	−39·3	−34·9	−48·7	−37·6	−33·7	−25·6	−23·3	
	T_n	1957–9	−31·1	−46·7	−67·4	−72·9	−67·9	−74·3	−71·6	−73·4	−74·8	−63·3	−49·4	−38·0	
	R_m	1957–9	2	3	0·5	0·5	(54)	0·5	0·5	(20)	0·5	0·2	(50)	(47)	(179)

NOTE: Precipitation totals given for places in Antarctica include snow blown into the gauge.

References

Chapter 1

The Meteorological Glossary (1963) compiled by I. D. MCINTOSH. London (H.M.S.O. for Met. Office) 4th edn.

NEWTON, H. W. (1958) *The Face of the Sun.* London (Pelican Books).

VITINSKII, YU. I. (1965) *Solar Activity Forecasting.* Jerusalem (Israel Program for Scientific Translations). Originally published under the title *Prognozy Solnechnoi Aktivnosti*, Leningrad (Izdatel'stvo Akademii Nauk SSSR) 1962.

Chapter 2

ALISSOW, B. P.; DROSDOW, O. A. and RUBINSTEIN, E. S. (1956) *Lehrbuch der Klimatologie*, Berlin (Deutscher Verlag). (German edn; original in Russian, Leningrad 1952.)

ALLEN, C. W. (1958) Solar radiation, *Quart. J. R. Met. Soc.*, **84**, 307–18. London.

ANDERSON, R. Y. (1961) Solar-terrestrial climatic patterns in varved sediments, *Symposium on Solar Variations . . ., Ann. N. Y. Acad. Sci.*, **95**, Art. 1, 424–39.

ARAKAWA, H. and TSUTSUMI, K. (1956) A decrease in the normal incidence radiation values for 1953 and 1954 and its possible cause, *Geophys. Mag.*, **27**, 205–8. Tokyo (Japanese Met. Agency).

BAUR, F. (1949) Beziehungen des Grosswetters zu kosmischen Vorgängen, in HANN-SÜRING *Lehrbuch der Meteorologie* (5. Auflage) Teil 8: Die Erscheinungen des Grosswetters, 8. Kapitel. Band II (1951). Leipzig (Hirzel).

BAUR, F. (1956) *Physikalisch-statistische Regeln als Grundlagen für Wetter- und Witterungsvorhersagen*, **I**. Frankfurt-am-Main (Akademische Verlagsgesellschaft MBH). See also *ibid.*, **II** (1958).

BAUR, F. (1959) *Die Sommerniederschläge Mitteleuropas in den letzten $1\frac{1}{2}$-Jahrhunderten und ihre Beziehungen zum Sonnenfleckenzyklus.* Leipzig (Akad. Verlag. Geest & Portig K.-G.).

BAUR, F. (1963) Beziehungen irdischer Erscheinungen zu Vorgängen auf der Sonne, *Sterne u. Weltraum*, **2**, 155–8. Mannheim.

References

BAUR, F. (1964) Ist die sogenannte Solarkonstante wirklich konstant? (Bericht über die III. meteorologische Fortbildungstagung für Grosswetterkunde und langfristige Witterungsvorhersage) *Met. Rundschau*, **17**, 19–25. Berlin.

BERLAGE, H. P. (1957) Fluctuations of the general circulation of the atmosphere of periods more than one year, their nature and prognostic value, *Mededelingen en Verhandlingen*, **69**, De Bilt (Koninklijk Nederlands Met. Inst.).

BERNARD, E. A. (1962) Théorie astronomique des pluviaux et interpluviaux du Quaternaire africain, *Acad. Royale des Sciences d'Outre-Mer, Classe des Sciences Naturelles et Médicales*, (*Nouvelle série*) **XII**, fasc. 1. Brussels.

BERNHARDT, F. and PHILIPPS, H. (1958) Die räumliche und zeitliche Verteilung der Einstrahlung, der Ausstrahlung und der Strahlungsbilanz im Meeresniveau. Teil I, Einstrahlung. *Abhandl. des Met. und Hydr. Dienstes der DDR*, **45**, Berlin, Potsdam (Akad. Verlag).

BLACK, J. N. (1956) The distribution of solar radiation over the Earth's surface, *Archiv für Meteorologie, Geophysik und Bioklimatologie*, **B**, **7**, 165–89. Vienna.

BORISOV, A. A. (1959) *Klimaty SSSR*. Moscow (Gosudarst. Ucebno-Pedagog. Izdat. Min. Prosvesc. RSFSR). English edition, *Climates of the USSR*, translator R. A. LEDWARD, published Edinburgh and London (Oliver & Boyd) 1965.

BRAY, J. R. (1965) Forest growth and glacier chronology in northwest North America in relation to solar activity, *Nature*, **205**, 440–3 (30 January 1965). London.

BRAY, J. R. (1966) Atmospheric carbon-14 content during the past three millennia in relation to temperature and solar activity, *Nature*, **209**, 1065–7 (12 March 1966). London.

BRAY, J. R. (1967) Variation in atmospheric carbon-14 activity relative to a sunspot–auroral solar index, *Science*, **156**, 640–2 (5 May 1967). Washington.

BROOKS, C. E. P. (1949) *Climate through the Ages*. London (Ernest Benn) 2nd edn.

BROUWER, D. and VAN WOERKOM, A. J. J. (1950) *Astron. Papers Amer. Ephemeris*, **13** (part 2). Washington (Nautical Almanac Office, U.S. Naval Observatory).

BUDYKO, M. I. (1955) *Atlas Teplovogo Balansa* (Atlas of the heat balance). Leningrad (Glav. Geofiz. Observatoria Voeikova).

CALLENDAR, G. S. (1960) In discussion of a paper by E. B. KRAUS on 'Synoptic and dynamic aspects of climatic change', *Quart. J., R. Met. Soc.*, **86**, 572–3. London.

DAMON, P. (1968) Radiocarbon and climate, 151–4 in 'Causes of climatic change', *Met. Monographs*, **8**, No. 30 (ed. J. M. MITCHELL). Boston, Mass. (Amer. Met. Soc.).

EMILIANI, C. (1961) Cenozoic climatic changes, as indicated by the stratigraphy and chronology of deep sea cores . . ., *N. Y. Acad. Sci.*, **95**, Art. 1, 521–36. New York.

FAIRBRIDGE, R. W. (1961) Climatic change and ice ages, *Symposium on Solar Variations . . ., Ann. N. Y. Acad. Sci.*, **95**, Art. 1, 542–79. New York.

FLOHN, H. (1958) Bemerkungen zum Problem der globalen Klimaschwankungen, *Archiv f. Met., Geophys. und Biokl.* **B**, **9**, 1–13. Vienna.

FLOHN, H. (1963) *Klimaschwankungen und grossräumige Klimabeeinflussung*. Köln und Opladen (Westdeutscher Verlag).

FRITZ, S. (1949) The albedo of the planet Earth and of clouds, *J. Met.*, **6**, 277–82. Lancaster, Pa.

HOUGHTON, D. M. (1958) Heat sources and sinks at the Earth's surface, *Met. Mag.*, **67**, 132–43. London.

HOUGHTON, H.G. (1954) On the annual heat balance of the northern hemisphere, *J. Met.*, **11**, 1–9. Lancaster, Pa.

JACOBS, W. C. (1951) The energy exchange between the sea and the atmosphere and some of its consequences, *Bulletin of the Scripps Institute of Oceanography*, **6** (No. 2) 27–122. La Jolla (University of California).

KONDRATYEV, K. YA. and NIKOLSKY, G. A. (1970) Solar radiation and solar activity, *Quart. J., R. Met. Soc.*, **96**, 509–22. London.

KÖPPEN, W. and WEGENER, A. (1924) *Die Klimate der geologischen Vorzeit*. Berlin (Bornträger).

KUIPER, G. P. (1953) *The Sun*, Vol. **1**, edited by G. P. KUIPER, in a 4-volume series entitled *The Solar System*. Chicago (University Press).

LAMB, H. H. (1963) On the nature of certain climatic epochs which differed from the modern (1900–1939) normal, *Proc. of the WMO–UNESCO Rome 1961 Symposium on Changes of Climate*. Paris (UNESCO, *Arid Zone Research Series XX*).

LAMB, H. H. (1965a) The early medieval warm epoch and its sequel, *Palaeogeography, Palaeoclimatology and Palaeoecology*, **1**, 13–37. Amsterdam (Elsevier).

LAMB, H. H. (1965b) Trees and climatic history in Scotland (Discussion), *Quart. J., R. Met. Soc.*, **91**, 546–7. London.

LAMB, H. H.; LEWIS, R. P. W. and WOODROFFE, A. (1966) Atmospheric circulation and the main climatic variables between 8000 and 0 B.C.: meteorological evidence, *Quart. J., R. Met. Soc.*, special issue: *Conference on World Climates 8000–0* B.C. London.

LANDSBERG, H. E. (1962) Biennial pulses in the atmosphere, *Beiträge zur Physik der Atmosphäre*, **35**, 184–94. Frankfurt/Main (Akad. Verlag.).

LINK, F. (1964) Manifestations de l'activité solaire dans le passé historique, *Planetary and Space Science*, **12**, 333–48. London (Pergamon Press).

LINKE, F. (1962) *Met. Taschenbuch*, Neue Ausgabe, 2 Auflage. Neubearbeitet von F. BAUR. **I**. Leipzig (Akad. Verlag. Geest & Portig K.-G.).

MANABE, S. and WETHERALD, R. T. (1967) Thermal equilibrium of the atmosphere with a given distribution of relative humidity, *J. Atmos. Sci.*, **24** (3), 241–59. Lancaster, Pa.

MATTHEWS, M. A. (1959) The Earth's carbon cycle, *New Scientist*, **6**, 644–6 (8 October 1959). London.

MILANKOVITCH, M. (1930) Mathematische Klimalehre und astronomische Theorie der Klimaschwankungen, in *Handbuch der Klimatologie*, **I**, Teil A. Berlin (Köppen & Geiger).

References

MILANKOVITCH, M. (1938) Astronomische Mittel zur Erforschung der erdgeschichtlichen Klimate, in *Handbuch der Geophysik*, **9**, 593–698. Berlin.

MILLER, A. (1966) *Meteorology* (p. 34). Columbus, Ohio (Merrill Physical Science Series).

MIRONOVITCH, V. (1960) Sur l'évolution de l'activité solaire et ses liaisons avec la circulation atmosphérique générale, *Met. Abhandl., Inst. f. Met. und Geophys.*, **IX**/Heft 3. Berlin (Free University).

MITCHELL, J. M. (1965) The solar inconstant, *Proc. of the seminar on possible responses of weather phenomena to variable extra-terrestrial influences. NCAR Tech. Note, TN–8* (ed. J. E. KUTZBACH). Boulder, Colorado.

NEWTON, H. W. (1958) *The Face of the Sun.* London (Harmondsworth, Pelican).

ÖPIK, E. J. (1958) Solar variability and palaeoclimatic changes, *Irish Astronomical Journal*, **5**, 97–109. Armagh, N. Ireland.

ÖPIK, E. J. (1964) Ice ages. Pp. 152–73 (Chapter 10) in *The Planet Earth* (ed. D. R. BATES). Oxford (Pergamon).

PIVAROVA, Z. I. (1968) The long-term variation of solar radiation intensity according to observations at actinometric stations, *Trudy*, **233**, 17–37. (In Russian.) Leningrad (Glav. Geofiz. Obs.).

PLASS, G. N. (1954) Some problems in atmospheric radiation, *Proc. Toronto Met. Conf. 1953*, 53–9. London (R. Met. Soc. and Amer. Met. Soc.).

PLASS, G. N. (1956) The carbon dioxide theory of climatic change, *Tellus*, **8**, 140–54. Stockholm.

RALPH, E. K. and MICHAEL, H. N. (1967) Problems of the radiocarbon calendar, *Archaeometry*, **10**, 3–11. Oxford.

RASOOL, S. I. (1964) Global distribution of the net energy balance of the atmosphere from TIROS radiation data, *Science*, **143** (No. 3606), 567–9 (7 February 1964). Washington.

SAWYER, J. S. (1965) Notes on the possible physical causes of long-term weather anomalies, *W.M.O.-I.U.G.G. symposium on research and development aspects of long-range forecasting, Boulder, Colorado 1964. W.M.O. Tech. Note No. 66.* Geneva (W.M.O., No. 162, T.P. 79).

SCHOVE, D. J. (1955) The sunspot cycle, 649 B.C. to A.D. 2000, *J. Geophys. Res.*, **60**, 127–46. Washington.

SCHWARZBACH, M. (1961) *Das Klima der Vorzeit.* Stuttgart (Ferdinand Enke Verlag), 2. Auflage.

SEVERNY, A. (1959) *The Sun* (translated from the Russian by G. YANKOVSKY). London (Lawrence and Wishart).

SHAW, N. (1930) *Manual of Meteorology*, **3**, 155–6. Cambridge (University Press).

SIMPSON, G. C. (1928) Further studies in terrestrial radiation, *Mem. R. Met. Soc.*, **III**, No. 21. London.

SIMPSON, G. C. (1929) The distribution of terrestrial radiation, *Mem. R. Met. Soc.*, **III**, No. 23. London.

SIMPSON, G. C. (1940) Possible causes of change in climate and their limitations, *Proc. of the Linnean Soc. of London, Session 152, 1939–40*, Part 2, 190–219. London.

STUIVER, M. (1961) Variations in radiocarbon concentration and sunspot activity, *J. Geophys. Res.*, **66** (No. 1) 273–6. Washington.

SUESS, H. E. (1965) Secular variations of the cosmic ray-produced carbon-14 in the atmosphere and their interpretations, *J. Geophys. Res.*, **70** (No. 23) 5937–52. Washington.

SUESS, H. E. (1968) Climatic changes, solar activity, and the cosmic-ray production rate of natural radiocarbon, *Met. Monographs*, **8**, No. 30, 146–50. Boston, Mass (Amer. Met. Soc.).

WALDMEIER, M. (1961) *Sunspot Activity in the Years 1610–1960.* Zürich (Schulthess).

WILLIS, E. H.; TAUBER, H. and MÜNNICH, K. O. (1960) Variations in the radiocarbon concentration over the past 1300 years, *Amer. J. Sci., Radiocarbon Supplement*, **2**, 1–4. New Haven, Conn.

WINSTON, J. S. (1955) Physical aspects of rapid cyclogenesis over the Gulf of Alaska, *Tellus*, **7**, 481–500. Stockholm.

WOERKOM, A. J. J. VAN (1953) The astronomical theory of climate changes, in H. SHAPLEY, *Climatic Change*, Chapter 11, 147–57. Cambridge, Mass. (Harvard University Press).

ZEUNER, F. E. (1958) *Dating the Past.* London (Methuen) 4th edn.

ZEUNER, F. E. (1959) *The Pleistocene Period.* London (Hutchinson).

Chapter 3

Academia Sinica, Staff Members of the Met. Inst., Peking (1957) On the general circulation over Eastern Asia, I, *Tellus*, **9**, 432–46. Stockholm.

Academia Sinica, Staff Members of the Met. Inst., Peking (1958) *Ibid.*, II, III, *Tellus*, **10**, 58–75, 299–312. Stockholm.

ÅNGSTRÖM, A. (1935) Teleconnections of climatic changes in present time, *Geografiska Annaler*, **17**, 242–58. Stockholm.

ÅNGSTRÖM, A. (1939) The change of temperature climate in present time, *Geogr. Ann.*, **21**, 119–31. Stockholm.

ÅNGSTRÖM, A. (1949) Atmospheric circulation, climatic variations and continentality of climate, *Geogr. Ann.*, **31**, 316–20. Stockholm.

BJERKNES, J. (1937) Theorie der aussertropischen Zyklonenbildung, *Met. Zeitschrift*, **54**, 462–6. Braunschweig.

BOYDEN, C. J. (1963) Development of the jet stream and cut-off circulations, *Met. Mag.*, **92**, 287–99. London.

BRUNT, D. (1941) *Physical and Dynamical Meteorology.* Cambridge (University Press).

References

BÜDEL, J. (1954) Die Klimazonen des Eiszeitalters in *Eiszeitalter und Gegenwart*, **I**, *Allgemeinen Erscheinungen des Eiszeitalters* (ed. P. WOLDSTEDT). Stuttgart (Enke Verlag).

BUSHBY, F. H. and WHITELAM, C. J. (1961) A three-parameter model of the atmosphere suitable for numerical integration, *Quart. J., R. Met. Soc.*, **87**, 374–92. London.

CHARNEY, J. G. and ELIASSEN, A. (1949) A numerical method for predicting the perturbations of the middle latitudes westerlies, *Tellus*, **I**, 38–54. Stockholm.

DEFANT, A. (1921) Die Zirkulation der Atmosphäre in den gemässigten Breiten der Erde, *Geogr. Ann.*, **3**, 209–65. Stockholm.

EXNER, F. M. (1917) *Dynamische Meteorologie*. Leipzig and Berlin (Teubner).

FAUST, H. (1953) Die Nullschicht, der Sitz des troposphärischen Windmaximums, *Met. Rundschau*, **6**, 6–15. Berlin.

FAUST, H. (1955) Der Nullschichteffekt als Funktion der Schärfe des Maximums in der vertikalen Windverteilung, *Met. Rundschau*, **8**, 48–50. Berlin.

FLOHN, H. (1955) Zur vergleichenden Meteorologie der Hochgebirge, *Archiv. f. Met., Geophys. und Biokl.*, **B**, **6**, 193–206. Vienna.

FLOHN, H. (1959) Die heutige Vergletscherung und Schneegrenze in Hochasien, *Abhandl. der math.-naturw. Klasse Jahrg.*, Nr. 14, 309–31. Mainz (Akad. der. Wiss. und Lit.).

FLOHN, H. (1964) Grundfragen der Paläoklimatologie im Lichte einer theoretischen Klimatologie, *Geol. Rundschau*, **54**, 504–15. Stuttgart.

FLOHN, H. (1967) Thermische Unterschiede zwischen Arktis und Antarktis, *Met. Rundschau*, **20**, 147–9. Berlin.

FLOHN, H. and FRAEDRICH, K. (1966) Tagesperiodische Zirkulation und Niederschlagsverteilung am Victoria-See (Ostafrika), *Met. Rundschau*, **19**, 157–65. Berlin.

FULTZ, D. (1961) Developments in controlled experiments on larger scale geophysical problems, *Advances in Geophys.*, **7**, 1–103. New York and London (Academic Press).

GABITES, J. F. (1963) Historical survey of tropical cyclones, in *Proc. of the Inter-Regional W.M.O. Seminar on Tropical Cyclones in Tokyo, 1962*, 1–5. Tokyo (Japanese Met. Agency).

HADLEY, G. (1735) Concerning the cause of the general Trade Wind, *Phil. Trans., Roy. Soc.*, **39**, 58–73. London.

HALLEY, E. (1686) An Historical Account of the Trade Winds and Monsoons, observable in the Seas between and near the Tropicks, with an Attempt to assign the Phisical Cause of the said Winds, *Phil. Trans., Roy. Soc.*, **16**, No. 181. London.

HOLLMANN, G. (1954) Zur Frage des Nullschichteffekts und der Strahlströme, *Met. Rundschau*, **7**, 166–70. Berlin.

KNIGHTING, E. (1956) Progress in numerical weather prediction (Meteorological Office Discussion), *Met. Mag.*, **85**, 176–9. London.

LAMB, H. H. (1955*a*) Malta's sea breezes, *Weather*, **10**, 256–64. London.

LAMB, H. H. (1955*b*) Two-way relationships between the snow or ice limit and 1000–500 mb thicknesses in the overlying atmosphere, *Quart. J., R. Met. Soc.*, **81**, 172–89. London.

LAMB, H. H. (1957) Fronts in the intertropical convergence zone, *Met. Mag.*, **86**, 76–84. London.

LAMB, H. H. (1959) The southern westerlies, *Quart. J., R. Met. Soc.*, **85**, 1–23. London.

LAMB, H. H. *et al.* (1957) Jet streams over North Africa and the central Mediterranean in January and February 1954, *Met. Mag.*, **86**, 97–111. London.

MINTZ, Y. (1965) Very long-term global integration of the primitive equations of atmospheric motion. *W.M.O.–I.U.G.G. Symposium on research and development aspects of long-range forecasting, Boulder, Colorado 1964. W.M.O. Tech. Note No. 66*. Geneva (W.M.O. No. 162, T.P. 79).

MURRAY, R. and DANIELS, S. M. (1953) Transverse flow at entrance and exit to jet streams, *Quart. J., R. Met. Soc.*, **79**, 236–41. London.

PALMÉN, E. (1951) The role of atmospheric disturbances in the general circulation, *Quart. J., R. Met. Soc.*, **77**, 337–54. London.

PETTERSSEN, S. (1950) Some aspects of the general circulation of the atmosphere, *Cent. Proc., R. Met. Soc.*, 120–55 (maps pp. 142–4). London.

PRIESTLEY, C. H. B. (1949) Heat transport and zonal stress between latitudes, *Quart. J., R. Met. Soc.*, **75**, 28–40. London.

REINEKE, I. (1950) Untersuchungen über die Abweichungen vom Gradientwind in der oberen Troposphäre, *Met. Abhandl., Inst. f. Met. und Geophys.*, **I**, Heft 1. Berlin (Free University).

RIEHL, H. (1954) *Tropical Meteorology*, Chapter 9. New York (McGraw-Hill).

ROSSBY, C.-G. (1939) Relation between variations in the intensity of the zonal circulation of the atmosphere and the displacements of the semi-permanent centres of action, *J. Marine Res.*, **2**, 38–55. New Haven, Conn.

ROSSBY, C.-G. (1941) The scientific basis of modern meteorology, *U.S. Yearbook of Agriculture, 1941*, 599–655. Washington.

SAWYER, J. S. (1963) Notes on the response of the general circulation to changes in the solar constant, *Proc. of the WMO–UNESCO Rome 1961 Symposium on Changes of Climate*. Paris (UNESCO, *Arid Zone Research Series XX*).

SAWYER, J. S. (1966) Possible variations of the general circulation of the atmosphere, *Quart. J., R. Met. Soc.*, special volume (*Proc. of the International Conference on World Climate 8000–0 B.C.*) London.

SCHERHAG, R. (1934) Zur Theorie der Hoch- und Tiefdruckgebiete: Die Bedeutung der Divergenz in Druckfeldern, *Met. Zeit.*, **51**, 129–38. Braunschweig.

SCHERHAG, R. (1948) *Neue Methoden der Wetteranalyse und Wetterprognose*. Berlin (Springer).

SONECKIN, D. M. (1963) On the correlation between the wind and pressure fields in the jet stream zone, *Trudy*, **121**, 14–17. (In Russian.) Moscow (Central Institute of Forecasting).

SUTCLIFFE, R. C. and FORSDYKE, A. G. (1950) The theory and use of upper air thickness patterns in forecasting, *Quart. J., R. Met. Soc.*, **76**, 189–217. London.

SUTTON, SIR GRAHAM (1960) Theories of the circulation of the Earth's atmosphere (The Halley Lecture for 1960). *Observatory*, **80**, 169–90. London.

TANNEHILL, I. R. (1956) *Hurricanes*. Princeton (University Press).

Chapter 4

Academia Sinica, Staff Members of the Institute of Geophys. and Met., Peking (1957) On the general circulation over Eastern Asia, *Tellus*, **9**, 432–46; continued in *Tellus*, **10** (1958), 58–75. Stockholm.

BANERJI, S. K. (1950) Methods of foreshadowing monsoon and winter rainfall in India, *Indian J. Met. and Geophys.*, **1**, 4–14. Poona.

BAUR, F. (1947) *Musterbeispiele europäischer Grosswetterlagen*. Wiesbaden (Dietrich).

BAUR, F. (1964) Die Singularitäten des Winters 1963/64, *Berliner Wetterkarte 21.5.1964, Beilage 55/64 (SO 24/64)*. Berlin, Free Univ. (Inst. f. Met.).

BAYER, K. (1959) Witterungssingularitäten und allgemeine Zirkulation der Erdatmosphäre, *Práce Geofysik. Ustav.*, **125**, 521–634. Prague (Česk. Akad. Ved.).

BELINSKY, N. A. (1957) *Ispol'zovaniye nekotorykh osobennotsey atmosfernykh protsessov dlya dolgosrochnykh prognozov* (*Use of certain features of atmospheric processes for long-range forecasts*). Leningrad (Gidromet. Izdat.).

BERLAGE, H. P. (1957) Fluctuations of the general atmospheric circulation of more than one year . . ., *Med. en Verhandl.*, **69**. De Bilt (K. Ned. Met. Inst.).

BHULLAR, G. S. (1952) Onset of monsoon over Delhi, *Indian J. Met. Geophys.*, **3** (No. 1), 25–30. New Delhi.

BLÜTHGEN, J. (1966) *Allgemeine Klimageographie*. Berlin (de Gruyter), 2. Auflage.

BRADKA, J. (1957) Der Jahresverlauf der zyklonalen und antizyklonalen Aktivität auf der Nordhemisphäre, *Studia Geophysica et Geodaetica*, **1**, 342–71. Prague (Česk. Akad. Ved.).

BUDYKO, M. I. (1963) *Atlas teplovogo balansa zemnogo shara* (*Atlas of the Earth's heat balance*). Moscow (Akad. Nauk.).

CLAPP, P. F. (1964) Global cloud cover for seasons using TIROS nephanalyses, *Monthly Weather Rev.*, **92**, 492–507. Washington.

DAVIS, N. E. (1968) An optimum summer weather index, *Weather*, **23**, 305–17. London.

FLOHN, H. (1943) Kontinentalität und Ozeanizität in der freien Atmosphäre, *Met. Zeit.*, **60**, 325–31. Braunschweig.

FLOHN, H. (1947) Stratosphärische Wellenvorgänge als Ursache der Witterungs-singularitäten, *Experimentia*, **III/8**, 1–12. Basel.

4. Seasonal changes

FLOHN, H. (1948) Zur Kenntnis des jährlichen Ablaufs der Witterung im Mittelmeergebiet, *Geofisica pura e applicata*, **13**, Fasc. 5–6, 1–24. Milan.

FLOHN, H. (1949) Klima und Witterungsablauf in Zürich im 16. Jahrhundert, *Vierteljahrsheft der Naturf. Gesell. in Zürich*, **95**, 28–41. Zürich.

FLOHN, H. (1950) Ablauf und Struktur des ostasiatischen Sommermonsuns. **II** in 'Studien zur allgemeinen Zirkulation der Atmosphäre', *Berichte des dt. Wetterdienstes in der U.S. Zone*, Nr. 18, 21–33. Bad Kissingen.

FLOHN, H. (1954) *Witterung und Klima in Mitteleuropa*. Zürich (Hirzel).

FLOHN, H. (1960) Recent investigations on the mechanism of the 'Summer Monsoon' of southern and eastern Asia, in *Monsoons of the World*, 75–88. Poona (India Met. Dept., Proc. 1958 symposium).

FLOHN, H. (1963) Comments on a synoptic climatology of southern Asia, in *Meteorology and the Desert Locust*, 245–52. *W.M.O. Tech. Note No. 69*. Geneva (W.M.O., No. 171, T.P. 85).

FLOHN, H. and HESS, P. (1949) Grosswettersingularitäten im jährlichen Witterungsverlauf Mitteleuropas, *Met. Rundschau*, **2**, 258–63. Berlin.

FULTZ, D. (1956) A survey of certain thermally and mechanically driven systems of meteorological interest, *Fluid models in geophysics, Proc. 1st symposium on the use of models in geophysical fluid dynamics, Baltimore 1953*, 27–63. (See also 'Controlled experiments on larger scale geophysical problems', in *Advances in Geophysics*, **7**, 1–103. New York (Academic Press) 1961.)

GORDON, A. H. (1953) Seasonal changes in the mean pressure distribution over the world and some inferences about the general circulation, *Bull. Amer. Met. Soc.*, **34**, 357–67. Lancaster, Pa.

KOTESWARAM, P. (1958) The easterly jet stream in the tropics, *Tellus*, **10**, 43–57. Stockholm.

KOTESWARAM, P. (1963) Movement of tropical storms over the Indian Ocean, *Tech. Rep., No. 21*, 173–83. Tokyo (Japanese Met. Agency).

LAMB, H. H. (1950) Types and spells of weather around the year in the British Isles . . ., *Quart. J., R. Met. Soc.*, **76**, 393–438. London.

LAMB, H. H. (1953) British weather around the year, *Weather*, **8**, 131–6, 176–82. London.

LAMB, H. H. (1958) Differences in the meteorology of the northern and southern polar regions, *Met. Mag.*, **87**, 364–79. London.

LAMB, H. H. (1959) The southern westerlies: a preliminary survey . . ., *Quart. J., R. Met. Soc.*, **85**, 1–23. London.

LAMB, H. H. (1964) *The English Climate*. London (English Universities Press).

LAMB, H. H. and JOHNSON, A. I. (1959) Climatic variation and observed changes in the general circulation, I and II, *Geogr. Ann.*, **41**, 94–134. Stockholm.

LAMB, H. H. and JOHNSON, A. I. (1961) Ibid., III, *Geogr. Ann.*, **43**, 363–400. Stockholm.

References

LOON, H. VAN (1961) Charts of average 500 mb absolute topography and sea level pressure in the southern hemisphere, *NOTOS*, **10**, 105–12. Pretoria.

LOON, H. VAN (1965) A climatological study of the atmospheric circulation in the southern hemisphere during the IGY. Part I: 1 July 1957–31 March 1958, *J. Appl. Met.*, **4**, 479–91. Lancaster, Pa.

LOON, H. VAN (1966) On the annual temperature range over the Southern Oceans, *Geogr. Rev.*, **56** (No. 4), 497–515. New York.

MEINARDUS, W. (1929) Die Luftdruckverhältnisse und ihre Wandlungen südlich von 30° südl. Breite, *Met. Zeit.*, **46**, 41–9, 86–96. Braunschweig.

NAMIAS, J. (1950) The index cycle and its role in the general circulation, *J. Met.*, **7**, 130–9. Lancaster, Pa.

PETTERSSEN, SV. (1950) Some aspects of the general circulation of the atmosphere, *R. Met. Soc., Cent. Proc.*, 120–55. London.

RAMAGE, C. S. (1952) Relationship of general circulation to normal weather over southern Asia and the western Pacific during the cool season, *J. Met.*, **9**, 403–8. Lancaster, Pa.

RAMASWAMY, C. (1956) On the subtropical jet stream and its role in the development of large-scale convection, *Tellus*, **8**, 26–60. Stockholm.

RAMASWAMY, C. (1958) A preliminary study of the . . . Indian southwest monsoon in relation to the westerly jet stream, *Geophysica*, **6** (Nos. 3–4), 455–7. Helsinki.

RAMASWAMY, C. (1962) Breaks in the Indian summer monsoon as a phenomenon of interaction between the easterly and subtropical westerly jet streams, *Tellus*, **14**, 337–349. Stockholm.

RAMDAS, L. A.; JAGANNATHAN, P. and GOPAL RAO, S. (1954) Prediction of the date of establishment of southwest monsoon along the west coast of India, *Indian J. Met. Geophys.*, **5** (No. 4), 305–14. New Delhi.

RAO, K. N. and JAGANNATHAN, P. (1960) Trends in monsoon/annual rainfall of India, in *Monsoons of the World*, 225–8. Poona (India Met. Dept., Proc. 1958 symposium).

REITER, E. R. and HEUBERGER, H. (1960) A synoptic example of the retreat of the Indian summer monsoon, *Geogr. Ann.*, **42**, 17–35. Stockholm.

RIEHL, H. (1954) *Tropical Meteorology*, Chapter 9. New York (McGraw-Hill).

ROSSBY, C.-G. and WILLETT, H. C. (1948) The circulation of the upper troposphere and lower stratosphere, *Science*, **108**, 643–52. Washington.

SCHERHAG, R. (1943) Die Verwendung der Hohenkarten im Wetterdienst, *Forschungs- und Erfahrungsberichte des Reichswetterdienstes*, **B**, No. **11**. Berlin.

SCHWERDTFEGER, W. (1960) The seasonal variation of the strength of the southern circumpolar vortex, *Monthly Weather Rev.*, **88**, 203–8. Washington.

SCHWERDTFEGER, W. and PROHASKA, F. (1956) Der Jahresgang des Luftdrucks auf der Erde und sein halbjährige Komponente, *Met. Rundschau*, **9**, 33–43. Berlin.

SETH, S. K. (1963) Changes of climate in India during the protohistorical and historical periods. Pp. 443–54 in *Changes of Climate, Proc. WMO UNESCO Rome 1961 symposium.* Paris (UNESCO – *Arid Zone Research Series*, XX).

SIMPSON, G. C. (1919) *British Antarctic Expedition 1910–1913, Meteorology*, **I** (Discussion). Calcutta.

STEHNOVSKY, D. I. (1962) *The Earth's Barometric Pressure Field.* (In Russian.) Moscow (Cent. Inst. Prog., Gidrometeoizdat).

STEHNOVSKY, D. I. (1966) Features of the circulation of the atmosphere over the northern and southern hemispheres with large pressure anomalies in 1955–1959, *Akad. Nauk. Mežd. Geofiz. Projekt.* (*Met. Researches*), Article No. 11, 126–57. (In Russian.) Moscow.

TEICH, M. (1955) Beitrag zum Problem der allgemeinen Zirkulation, insbesondere der mitteltroposphärischen Hochdruckgebiete der nördlichen Hemisphäre, *Abhandl. des Met. und Hydr. Dienstes der D.D.R.*, **V**, No. 36. Berlin, Potsdam (Akad. Verlag).

TROUP, A. J. (1965) The southern oscillation, *Quart. J., R. Met. Soc.*, **91**, 490–506. London.

WAHL, E. (1952) The January thaw in New England, *Bull. Amer. Met. Soc.*, **33**, 380–6. Lancaster, Pa.

WAHL, E. (1953) Changes in the general circulation reflected in the occurrence and intensity of weather singularities, *Geophys. Res. Papers, No. 24*, 129–41. Cambridge, Mass. (Air Force Cambridge Research Center).

WALKER, G. T. (1924) World weather II, *Memoir No. 24*, 275 ff. Poona (India Met. Dept.).

WEXLER, H. (1959) Seasonal and other temperature changes in the Antarctic atmosphere, *Quart. J., R. Met. Soc.*, **85**, 196–208. London.

WRIGHT, P. B. (1967) Changes in 200-mb circulation patterns related to the development of the Indian southwest monsoon, *Met. Mag.*, **96**, 302–15. London.

Chapter 5

ANGELL, J. K. and KORSHOVER, J. (1962) The biennial wind and temperature oscillations of the stratosphere and their possible extension to higher latitudes, *Monthly Weather Rev.*, **90**, 127–32. Washington.

BREWER, A. W. (1949) Evidence for a world circulation provided by measurements of helium and water vapour distribution in the stratosphere, *Quart. J., R. Met. Soc.*, **75**, 351–63. London.

COURT, A. (1942) Tropopause disappearance during the Antarctic winter, *Bull. Amer. Met. Soc.*, **23**, 220–38. Lancaster, Pa.

DAVIS, N. E. (1967) The summers of northwest Europe, *Met. Mag.*, **96**, 178–87. London.

DOBSON, G. M. B.; HARRISON, D. N. and LAWRENCE, J. (1929) Measurements of the amount of ozone in the Earth's atmosphere and its relation to other geophysical conditions, Part II, *Proc. Roy. Soc.*, **A**, **114**, 521–41. London.

DYER, A. J. and HICKS, B. B. (1968) Global spread of volcanic dust from the Bali eruption of 1963, *Quart. J., R. Met. Soc.*, **94**, 545–54. London.

References

EBDON, R. A. (1961) Some notes on the stratospheric winds at Canton Island and Christmas Island, *Quart. J., R. Met. Soc.*, **87**, 322–31. London.

EBDON, R. A. (1963) The tropical stratospheric wind fluctuation, *Weather*, **18**, 2–7. London (R. Met. Soc.).

EBDON, R. A. (1967) Possible effects of volcanic dust on stratospheric temperatures and winds, *Weather*, **22**, 245–9. London (R. Met. Soc.).

FAUST, H. (1967) Interaction between the different layers of the homosphere, *Archiv. f. Met., Geophys. und Biokl.*, **A**, **16**, 12–30. Vienna.

FAUST, H. and ATTMANNSPACHER, W. (1961) Die allgemeine Zirkulation der aussertropischen Breiten bis 60 km Höhe auf der Basis der Nullschichtkonzeption, *Met. Rundschau*, **14**, 6–10. Heidelberg. (The Nullschicht concept is explained by FAUST, and three different types of Nullschicht occurring at different heights in the atmosphere are defined, in an article entitled 'Nullschichteffekt und allgemeine Zirkulation', *Geofisica pura e applicata*, **44**, 257–64. Milan 1959.)

GAIGEROV, S. S. (1967) On stratospheric warmings in the Antarctic and Arctic, in *Polar Meteorology*, 407–15. *W.M.O. Tech. Note No. 87.* Geneva (W.M.O., No. 211, T.P. 111).

GEB, M. (1966) Synoptisch-statistische Untersuchungen zur Einleitung blockierender Hochdrucklagen über dem Nordatlantik und Europa, *Met. Abhandl.*, **LXIX** (Heft 1). Berlin (Inst. f. Met. und Geophys., der Freien Universität).

GOLDSMITH, P. and BROWN, F. (1961) World-wide circulation of air within the stratosphere, *Nature*, **191**, 1033–7 (9 September 1961). London.

GOODY, R. M. (1954) *The Physics of the Stratosphere.* (Cambridge Monographs on Physics.) Cambridge (University Press).

HARE, F. K. (1962) The stratosphere, *Geogr. Rev.*, **52** (4), 527–47. New York (Amer. Geogr. Soc.).

HARE, F. K. and BOVILLE, B. W. (1965) The polar circulations, in *The Circulation in the Stratosphere, Mesosphere and Lower Thermosphere*, Chapter III, 43–78. *W.M.O. Tech. Note No. 70.* Geneva (W.M.O., No. 176, T.P. 87).

JUNGE, Ch. E.; CHAGNON, C. W. and MANSON, J. E. (1961) Stratospheric aerosols, *J. Met.*, **18**, 81–108. Lancaster, Pa.

KRAUS, E. B. (1960) Synoptic and dynamic aspects of climatic change, *Quart. J., R. Met. Soc.*, **86**, 1–15. London.

KULKARNI, R. N. (1962) Comparison of ozone variations and of its distribution with height over middle latitudes of the two hemispheres, *Quart. J., R. Met. Soc.*, **88**, 522–34. London.

KULKARNI, R. N. (1966) The vertical distributions of atmospheric ozone and possible transport mechanisms in the stratosphere of the southern hemisphere, *Quart. J., R. Met. Soc.*, **92**, 363–73. London.

LABITZKE, K. (1962) Beiträge zur Synoptik der Hochstratosphäre, *Met. Abhandl.*, **XXVIII** (Heft 1). Berlin (Inst. f. Met. und Geophys. der Freien Universität).

LABITZKE, K. (1965) On the mutual relation between stratosphere and troposphere during periods of stratospheric warmings in winter, *J. Appl. Met.*, **4**, 91–9. Lancaster, Pa.

LADURIE, E. Le R. (1967) *Histoire du climat depuis l'an mil.* Paris (Flammarion).

LAMB, H. H. (1960) In, A discussion on the results of the Royal Society's expedition to Halley Bay, Antarctica during the International Geophysical Year, *Proc. Roy. Soc.*, A, **256**, 193–7. London.

LAMB, H. H. (1970) Volcanic dust in the atmosphere . . ., *Phil. Trans., Roy. Soc.*, **A**, **266** (No. 1178), 425–533. London.

LANDSBERG, H. (1962) Biennial pulses in the atmosphere, *Beiträge zur Physik der Atmosphäre*, **35**, 184–94. Frankfurt/Main.

LANDSBERG, H. E.; MITCHELL, J. M., CRUTCHER, H. L. and QUINLAN, F. T. (1963) Surface signs of the biennial atmospheric pulse, *Monthly Weather Rev.*, **91**, 549–56. Washington.

MURGATROYD, R. J. (1957) Winds and temperatures between 20 km and 100 km – a review, *Quart. J., R. Met. Soc.*, **83**, 417–58. London.

MURGATROYD, R. J. (1965) (*a*) The 26-month oscillation, Chapter VI, 123–39; (*b*) General climatology . . ., Chapter VIII, 170–99, in *The Circulation in the Stratosphere, Mesosphere and Lower Thermosphere. W.M.O. Tech. Note No. 70.* Geneva (W.M.O., No. 176, T.P. 87).

MURGATROYD, R. J. (1969) A note on the contributions of mean and eddy terms to the momentum and heat balances of the troposphere and lower stratosphere. *Quart. J., R. Met. Soc.*, **95**, 194–202. London.

MURGATROYD, R. J. (1970) The structure and dynamics of the stratosphere. Pp. 159–95 in *The Global Circulation of the Atmosphere* (ed. G. A. CORBY). London (R. Met. Soc.).

PČELKO, I. G. (1967) On the evolution of the polar stratospheric vortex and the interrelationship between processes in troposphere and stratosphere. *Glav. Uprav. Gidromet. Služb., Met. Gidr.*, No. 12, 21–32. (In Russian.) Leningrad.

PROBERT-JONES, J. R. (1964) An analysis of the fluctuations in the tropical stratospheric wind, *Quart. J., R. Met. Soc.*, **90**, 15–26. London.

RAMANATHAN, K. R. and KULKARNI, R. N. (1960) Mean meridional distributions of ozone in different seasons calculated from *Umkehr* observations and probable vertical transport mechanisms, *Quart. J., R. Met. Soc.*, **86**, 144–55. London.

REED, R. J. (1963) On the cause of the 26-month periodicity in the equatorial stratospheric winds, *Met. Abhandl.*, **XXXVI**, 245–57. Berlin (Inst. f. Met. und Geophys. der Freien Universität).

RIETSCHEL, E. (1929) Die 3–3½ jährige und die 2 jährige Temperaturschwankung, *Veröffentlichungen der Geophys. Inst. Leipzig*, Serie 2, **4** (Nr. 1). Leipzig.

SAZONOV, B. I. and ZORINA, V. P. (1967) On the nature of stratospheric warming, *Trudy*, **211**, 31–45. (In Russian.) Leningrad (Glav. Geofiz. Obs.).

SCHERHAG, R. (1958) The role of tropospheric cold air poles and of stratospheric high-pressure centres in the Arctic weather, in *Polar Atmosphere Symposium*, Part I, *Meteorology Section*, 101–17. London (Pergamon).

SCHERHAG, R. (1959) Über die Luftdruck-, Temperatur- und Windschwankungen in der Stratosphäre, *Abhandl. math.-naturw*, Klasse 1959, **15**. Weisbaden. (Akad. der Wiss. und. Lit. in Mainz).

SCHERHAG, R. (1963) Synoptik der Hochatmosphäre, *Geofisica pura e applicata*, **54**, 166–81. Milan.

SHAPIRO, R. and WARD, F. (1962) A neglected cycle in sunspot numbers?, *J. Atmos. Sci.*, **19**, 506–8. Lancaster, Pa.

SUN CHU CHING; CHENG LUNG HSUN and HSIEH TU CHENG (1964) Interaction between stratospheric and tropospheric circulation. Chapter I, in TAO SHIH YEN *et al.*, *Studies of tropospheric circulation effects dependent on solar activity and stratospheric circulation.* Peking (Chinese Geophysical Institute). (In Chinese: translation available in Met. Office Library, Bracknell.)

TEWELES, S. (1964) The energy balance of the stratosphere. Chapter V, 107–22 in *W.M.O. Tech. Note No. 70.* Geneva (W.M.O., No. 176, T.P. 87).

VERYARD, R. G. and EBDON, R. A. (1961) Fluctuations in tropical stratospheric winds, *Met. Mag.*, **90**, 125–43. London.

VINCENT, D. G. (1968) Mean meridional circulations in the northern hemisphere lower stratosphere during 1964 and 1965, *Quart. J., R. Met. Soc.*, **94**, 333–49. London.

WRIGHT, P. B. (1968) Wine harvests in Luxemburg and the biennial oscillation in European summers, *Weather*, **23**, 300–4. London.

Chapter 6

ALLEN, C. W. (1955) *Astrophysical Quantities.* University of London (Athlone Press).

ALTER, D. (1922) A rainfall period equal to one-ninth of the sunspot period, *Kansas Univ. Sci. Bull.*, **13**, No. 11.

ALTER, D. (1933) Correlation periodogram investigation of English rainfall, *Monthly Weather Rev.*, **61**, 345–50. Washington.

ALTER, D. (1937) A simple form of periodogram, *Ann. Math. Statist.*, **8**, 121. Ann Arbor, Michigan.

ANDERSON, R. Y. (1961) Solar–terrestrial climatic patterns in varved sediments, *Ann. N.Y. Acad. Sci.*, **95**, Art. 1, 424–39. New York.

BAUR, F. (1956) *Physikalisch-statistische Regeln als Grundlagen für Wetter- und Witterungsvorhersagen*, **I**. Frankfurt/Main (Akad. Verlag).

6. Cyclic and quasi-periodic phenomena

BAUR, F. (1959) *Die Sommerniederschläge Mitteleuropas in den letzten 1½-Jahrhunderten und ihre Beziehungen zum Sonnenfleckenzyklus.* Leipzig (Akad. Verlag. Geest & Portig K.-G.).

BAYER, K. and BAYEROVA, V. (1969) Langfristige Schwankungen der troposphärischen Zirkulation über dem Nordatlantik im Zeitraum 1949–1966, *Ann. der Met.*, N.F., No. **4**, 221–6. Offenbach.

BERLAGE, H. P. (1957) Fluctuations of the general atmospheric circulation of more than one year, their nature and prognostic value, *Med. en Verhandl.*, **69**. De Bilt (K. Ned. Met. Inst.).

BERLAGE, H. P. (1961) Variations in the atmospheric and hydrospheric circulation of periods of a few years duration affected by variations of solar activity, *Ann. N.Y. Acad. Sci.*, **95** (Art. 1) 354–67. New York.

BERLAGE, H. P. (1966) The Southern Oscillation and world weather, *Med. en Verhandl.*, **88**. De Bilt (K. Ned. Met. Inst.).

BERLAGE, H. P. and DE BOER, H. J. (1960) On the Southern Oscillation, its way of operation and how it affects pressure patterns in the higher latitudes, *Geofisica pura e applicata*, **46**, 329–51. Milan.

BERNARD, E.-A. (1962*a*) Théorie astronomique des pluviaux et interpluviaux du Quaternaire africain, *Acad. Royale des Sciences d'Outre-Mer, Classe des Sciences Naturelles et Médicales* (*Nouvelle série*) **XII**, fasc. 1. Brussels.

BERNARD, E.-A. (1962*b*) Le caractère tropical des paléoclimats à cycles conjoints de 11 et 21 000 ans et ses causes: migration des pôles ou dérive des continents, *Acad. Royale des Sciences d'Outre-Mer, Classe des Sciences Naturelles et Médicales* (*Nouvelle série*) **XIII**, fasc. 6. Brussels.

BERNARD, E.-A. (1964) The laws of physical palaeoclimatology and the logical significance of palaeoclimatic data, in *Problems in Palaeoclimatology*, 309–21 (ed. A. E. M. NAIRN). New York (Wiley).

BERRY, W. B. N. and BARKER, R. M. (1968) Fossil bivalve shells indicate longer month and year in Cretaceous than present, *Nature*, **217**, 938–9 (9 March 1968). London.

BEZOLD, W. VON (1884) Über zündende Blitze im Königreich Bayern während des Zeitraums 1833 bis 1882, *Abhandl. des. kgl. Bayer. Akad. d. Wiss.*, II Klasse, **XV**, Abtheilung I. Munich.

BLACKMAN, R. B. and TUKEY, J. W. (1959) *The Measurement of Power Spectra.* New York (Dover).

BÖHME, W. (1967) Eine 26-monatige Schwankung der Häufigkeit meridionaler Zirkulationsformen über Europa, *Zeit.f. Met.*, **19**, 113–15. Berlin.

BOLLINGER, C. J. (1964) *Atlas of Planetary Solar Climate.* **4**: *Planetary Periodicities in Sun-tide Cycles and Climate Variation.* Norman, Oklahoma.

BOLLINGER, C. J. (1968) Sun tides: an unexplored astronomical approach to climatic cycles and trends, *Tellus*, **20** (No. 3), 412–16. Stockholm.

References

BRIER, G. W. (1961) Some aspects of long-term fluctuations in solar and atmospheric phenomena, *Ann. N.Y. Acad. Sci.*, **95**, Art. I, 173–87. New York.

BROOKS, C. E. P. (1934) The variation of the annual frequency of thunderstorms in relation to sunspots, *Quart, J., R. Met. Soc.*, **60**, 153–66. London.

BROOKS, C. E. P. (1949) *Climate through the Ages*. London (Benn) 2nd edn.

BRÜCKNER, E. (1890) *Klimaschwankungen seit 1700, nebst Bemerkungen über die Klimaschwankungen der Diluvialzeit*. Vienna (Hölzel).

BRUNT, D. (1925) Periodicities in European weather, *Phil. Trans., Roy. Soc.*, **A**, **225**, 247–302. London.

BRYSON, R. A. (1948) On a lunar bi-fortnightly tide in the atmosphere, *Trans. Amer. Geophys. Union*, **29**, 473–5. Washington.

CHAPMAN, S. (1919) The lunar tide in the Earth's atmosphere, *Quart, J., R. Met. Soc.*, **45**, 113–39. London.

CRADDOCK, J. M. (1965) The analysis of meteorological time series for use in forecasting, *The Statistician*, **15** (2), 167–90. London.

CRADDOCK, J. M. (1968*a*) Some meteorological applications of Sherman's statistic. Unpublished memorandum, Met. Office.

CRADDOCK, J. M. (1968*b*) *Statistics in the Computer Age*. London (English Universities Press).

DOBERITZ, R. (1967*a*) Statistical investigations of the climatic anomalies of the equatorial Pacific, *Bonner Met. Abhandl.*, Heft 7. Bonn (Met. Inst. der Univ.).

DOBERITZ, R. (1967*b*) Teleconnections and phase relations of rainfall at the tropical Pacific Ocean, *U.S. Army European Research Office, Contract No. DA-91-591-EUC-3983, Final Technical Report*. Bonn.

DOBERITZ, R. (1968) Cross spectrum analysis of rainfall and sea temperature at the equatorial Pacific Ocean, *Bonner Met. Abhandl.*, Heft 8. Bonn (Met. Inst. der Univ.).

EASTON, C. (1928) *Les hivers dans l'Europe occidentale*. Leyden (E. J. Brill).

EICKERMANN, W. and FLOHN, H. (1962) Witterungszusammenhänge über dem äquatorialen Südatlantik, *Bonner Met. Abhandl.*, Heft I. Bonn (Met. Inst. der Univ.).

FAIRBRIDGE, R. W. (1961) Climatic change and ice ages, convergence of evidence, *Ann. N.Y. Acad. Sci.*, **95**, Art. I, 543–79. New York.

FRENZEL, B. (1966) Climatic change in the Atlantic/sub-Boreal transition on the northern hemisphere: botanical evidence, *Proc. Internat. Symposium on World Climate 8000 to 0 B.C.*, 99–123. London (R. Met. Soc.).

GASJUKOV, P. S. and SMIRNOV, N. P. (1967) Pressure-field oscillations over the northern hemisphere within the 11-year cycle of solar activity, *Doklady*, **173**, 567–9. (In Russian.) Moscow (Akad. Nauk.).

HEDIN, SV. (1940) *The Wandering Lake*. London (Routledge).

HELLAND-HANSEN, B. and NANSEN, F. (1917) *Temperatur-Schwankungen des Nordatlantischen Ozeans und in der Atmosphäre*. Kristiania (Dybwad).

JEFFREYS, H. (1954) Dynamics of the Earth–Moon system, in *The Earth as a Planet*, Chapter 2, 42–56 (ed. G. P. KUIPER). Chicago (Univ. Press).

JOHNSEN, S. J.; DANSGAARD, W. and CLAUSEN, H. B. (1970) Climatic oscillations 1200–2000 A.D., *Nature*, **277**, 482–3. London.

KLEJMENOVA, E. P. (1967) On the variation of thunderstorm activity in the solar cycle, *Glav. Uprav. Gidromet. Služb., Met. Gidr.*, **8**, 64–8. (In Russian.) Leningrad.

KUTZBACH, J. E.; BRYSON, R. A. and SHEN, W. C. (1968) An evaluation of the thermal Rossby number in the Pleistocene, *Met. Monographs*, **8**, No. 30, 134–8. Lancaster, Pa. (Amer. Met. Soc.).

LAMB, H. H. (1963) On the nature of certain climatic epochs which differed from the modern (1900–1939) normal, *Proc. WMO–UNESCO Rome 1961 Symposium on Changes of Climate*, 125–50. Paris (UNESCO, *Arid Zone Research Series XX*).

LAMB, H. H. (1965) The early medieval warm epoch and its sequel, *Palaeogeogr., Palaeoclim., Palaeoecol.*, **1**, 13–37. Amsterdam (Elsevier).

LAMB, H. H. (1967) Britain's changing climate, *Geogr. J.*, **133**, 445–68. London.

LAMB, H. H. and JOHNSON, A. I. (1966) Secular variations of the atmospheric circulation since 1750, *Geophys. Mem.*, **110**. London (Met. Office).

LAMB, H. H.; LEWIS, R. P. W. and WOODROFFE, A. (1966) Atmospheric circulation and the main climatic variables between 8000 and 0 B.C.: meteorological evidence, *Proc. Internat. Symposium on World Climate 8000–0 B.C.* London (R. Met. Soc.).

LANDSBERG, H. E. (1962) Biennial pulses in the atmosphere, *Beiträge zur Physik der Atmosphäre*, **35**, 184–94. Frankfurt Main.

LILJEQUIST, G. H. (1949) On fluctuations of the summer mean temperature in Sweden, *Geogr. Ann.*, **31**, 159–78. Stockholm.

LINK, F. (1958) Kometen, Sonnentätigkeit und Klimaschwankungen, *Die Sterne*, **34**, 129–40. Leipzig (Barth).

LINK, F. (1964) Manifestations de l'activité solaire dans le passé historique, *Planet. Space Sci.*, **12**, 333–48. London (Pergamon).

MAKSIMOV, I. V. (1952) On the 80-year cycle of climatic variations, *Dok. Akad. Nauk. SSSR*, **86** (5), 917–20. (In Russian.) Moscow.

MAKSIMOV, I. V. (1954) Secular variations in ice formation in northern parts of the North Atlantic, *Trudy*, **8**. (In Russian.) Moscow (USSR Inst. Okeanologiya).

MAKSIMOV, I. V. (1958) Nutation phenomena in the atmosphere of the high latitudes of the Earth and their climate-forming role, *Problemy severa*, No. 1, *Geofiz.*, 97–115. (In Russian.) Moscow (Akad. Nauk. SSSR).

MAKSIMOV, I. V. and ABRAMOV, R. V. (1966) On the nutational migration of the Iceland low, *Problems of the Arctic and Antarctic*, **23**, 14–19. (In Russian.) Leningrad (Arctic and Antarctic Institute).

References

MAKSIMOV, I. V. and SLEPTSOV, B. A. (1963) Study of eleven-year variations in atmospheric pressure in the Antarctic, *Soviet Antarctic Expedition*, No. 43. (In English.) Moscow.

MAKSIMOV, I. V. and SMIRNOV, N. P. (1965) A contribution to the study of the causes of long-period variations in the activity of the Gulf Stream, *Oceanology*, **5**, (2). (Amer. Geophys. Union translation from Russian.) Moscow.

MAKSIMOV, I. V. and others (1967) Nutational migration of the Iceland low pressure, *Doklady*, **177**, 88–91. (In Russian.) Moscow (Akad. Nauk).

MAKSIMOV, I. V.; SARUHANJAN, E. I. and SMIRNOV, N. P. (1970) On the relation between deformation force and the movements of the atmosphere's centres of action, *Doklady*, **190**, 1095–7. (In Russian.) Moscow (Akad. Nauk).

MITCHELL, J. M. (1965) The solar inconstant, *Proc. Seminar on Possible Responses of Weather phenomena to Variable Extraterrestrial Influences. NCAR Tech. Note*, **TN-8**. Boulder, Colorado.

MITCHELL, J. M. *et al.* (1966) Climatic change, *W.M.O. Tech. Note No. 79*. Geneva (W.M.O., No. 195, T.P. 100).

MYRBACH, O. (1935) Sonnenfleckenzyklus und Gewitterhäufigkeit in Wien, Kremsmünster und Bayern, *Met. Zeit.*, **52**, 225–7. Braunschweig.

PALMER, C. E. and PYLE, R. L. (1966) The climatological setting of the Galapagos, in *The Galapagos*, Chapter 12, 93–9, ed. R. L. BOWMAN. (*Proceedings of the Symposium of the Galapagos International Scientific Project.*) Los Angeles (Univ. Cal. Press).

PANOFSKY, H. A. and BRIER, G. W. (1958) *Some Applications of Statistics to Meteorology*. Pennsylvania (State Univ. Press).

PETTERSSON, O. (1914) Climatic variations in historic and prehistoric time, *Svenska Hydrografisk-Biologiska Kommissionens Skrifter*, Häft 5. Göteborg.

PETTERSSON, O. (1930) The tidal force, *Geogr. Ann.*, **12**, 261–322. Stockholm.

RALPH, E. K. and MICHAEL, H. N. (1967) Problems of the radiocarbon calendar, *Archaeometry*, **10**, 3–11. Oxford.

RAWSON, H. E. (1907) Anticyclones as aids to long-distance forecasts, *Quart. J., R. Met. Soc.*, **33**, 309–10. London.

RAWSON, H. E. (1908) The anticyclonic belt of the southern hemisphere, *Quart. J., R. Met. Soc.*, **34**, 165–88. London.

RAWSON, H. E. (1909) The anticyclonic belt of the northern hemisphere, *Quart. J., R. Met. Soc.*, **35**, 233–48. London.

RICHTER-BERNBURG, G. (1964) Solar cycle and other climatic periods in varvitic evaporites, in *Problems in Palaeoclimatology* (ed. A. E. M. NAIRN), 510–19, New York (Wiley).

SCHELL, I. I. (1965) The origin and possible prediction of the fluctuations in the Peru Current and upwelling, *J. Geophys. Res.*, **70**, 5529–40. Chicago.

SCHERHAG, R. (1967) Bermerkungen zur Welt-Wetterlage im meteorologischen Jahr 1966–67, *Berliner Wetterkarte, Beilage 186/67* (19 December 1967). Berlin, Free Univ. (Inst. of Met.).

SHAPIRO, R. and WARD, F. (1962) A neglected cycle in sunspot numbers?, *J. Atmos. Sci.*, **19**, 506–8. Lancaster, Pa.

SHAW, N. (1936) *Manual of Meterology:* **2** – *Comparative Meteorology.* Cambridge (Univ. Press).

SHERMAN, B. (1950) A random variable related to the spacing of sample values, *Ann. Math. Statistics,* **21**, 339–61. Ann Arbor, Michigan.

SHOSTAKOVICH, V. B. (1931) Die Bedeutung der Untersuchung der Bodenablagerungen der Seen für einige Fragen der Geophysik, *Internat. Ver. theor. angew. Limnol., Verhandl.*, **5**, 307–17. Stuttgart.

SHOSTAKOVICH, V. B. (1944) An experiment on geochronological analysis of mud deposits of Malinovoie Lake in connexion with the uplift of the shore of the White Sea, *Geograficheskoe obschestvo SSSR izvestia,* **76**, 203–6. (In Russian.) Moscow.

SIRÉN, G. (1961) Skogsgränstallen som indikator för klimatfluktuationerna i norra Fennoskandien under historisk tid, *Communicationes Inst. Forest. Fenniae*, **54** (2). Helsingfors.

ŠNITNIKOV, A. V. (1949) General features of the cyclic fluctuations in the level of lakes and the wetness of the territory of Eurasia in connexion with solar activity, *Bull. of the* [USSR] *Commission on Investigation of the Sun*, Nos. 3, 4. (In Russian.)

ŠNITNIKOV, A. V. (1957) The variation of the water regime of the continents of the northern hemisphere, *Zap. Geogr. O-va SSSR*, **16** (New Series). (In Russian.) Leningrad (Akad. Nauk. Izdat).

STUIVER, M. (1961) Variations in radiocarbon concentration and sunspot activity, *J. Geophys. Res.*, **66**, 273–6. Washington.

SUESS, H. E. (1956) Absolute chronology of the last glaciation, *Science*, **123** (No. 3192), 355–7 (2 March 1956). Washington.

TAKAHASHI, K. (1957) Tidal oscillations on the sun as the cause of periodicities in meteorological phenomena, *Papers in Met. and Geophys.*, **7**, No. 4. Tokyo (Met. Res. Inst.).

TROUP, A. J. (1965) The southern oscillation, *Quart. J., R. Met. Soc.*, **91**, 490–506. London.

VITINSKII, YU. J. (1965) *Solar Activity Forecasting.* Jerusalem (Israel Program for Scientific Translations).

VOLTZINGER, N. E. *et al.* (1966) The internal structure of natural climatic series exposed by a periodogram analysis technique, *Arkt. i. Antarkt. Inst. Issled.*, **277**, 147–57. (In Russian.) Leningrad.

VOROBEVA, E. V. (1967) Cyclic changes in the intensity of the zonal circulation in the middle troposphere, *Trudy,* **211**, 56–67. (In Russian.) Leningrad (Glav. Geofiz. Obs.).

WALKER, G. T. (1923) A preliminary study of world weather–correlation in seasonal variations of weather, *Mem. India Met. Dept.*, **24**, Part 4. Calcutta.

WALKER, G. T. (1924) A further study of world weather, *Mem. India Met. Dept.*, **24**, Part 9. Calcutta.

WALKER, G. T. (1927) World weather III, *Mem. R. Met. Soc.*, **2**, No. 17. London.

WALKER, G. T. (1928) World weather – an address, *Quart. J., R. Met. Soc.*, **54**, 79–87. London.

WALKER, G. T. (1929) World weather IV – some applications to seasonal foreshadowing, *Mem. R. Met. Soc.*, **3**, No. 24. London.

WALKER, G. T. and BLISS, E. W. (1933) World weather V, *Mem. R. Met. Soc.*, **4**, No. 36. London.

WEXLER, H. (1956) Variations in insolation, general circulation and climate, *Tellus*, **8**, 480–94. Stockholm.

WOOD, S. M. (1946) The planetary cycles, *The Illinois Engineer*, **22**, No. 2 (February 1946). Illinois.

WOOD, S. M. (1949) Long-term weather cycles and their causes, *The Illinois Engineer*, **25**, No. 3 (March 1949). Illinois.

ZAYTSEV, G. N. (1965) Features of long-period variations of level and salinity in the northern Caspian Sea. (Amer. Geophys. Union, transl. from Russian.) *Oceanology*, **5** (2). Moscow.

ZEUNER, F. E. (1958) *Dating the Past.* London (Methuen), 4th edn, 145.

ZEUNER, F. E. (1959) *The Pleistocene Period.* London (Hutchinson), especially 173–219.

Chapter 7

ASTAPENKO, P. D. (1960) Atmospheric processes in high latitudes of the southern hemisphere. *Seria resultaty MGG* (*Results of the IGY*). (In Russian.) Moscow.

BAHR, H. (1911) Die interdiurne Veränderlichkeit des Luftdruckes, *Met. Zeit.*, **28**, 497–502. Braunschweig.

BAUR, F. (1947) *Musterbeispiele europäischer Grosswetterlagen.* Wiesbaden (Dietrich).

BAUR, F. (1956) *Physikalisch-statistische Regeln als Grundlagen für Wetter- und Witterungs-Vorhersagen*, **I**. Frankfurt/Main (Akad. Verlag).

BAUR, F. (1958) Ibid., **II**.

BAYER, K. (1965) Dlouhodobe variace v tlakovém poli nad Evropu (Langfristige Variationen im Druckfeld in Europa), *Met. Zpravy*, **18**, 167–9. Prague.

BENDER, J. A. and GOW, A. J. (1961) Deep drilling in Antarctica, *Ass. Hydr. Sci., Proc. Verb., U.G.G.I. Helsinki Meeting 1961*, 132–41. Gentbrugge, Belgium.

BERGER, E. (1961) Siebenjährige Mittelkarten der interdiurnen Veränderlichkeit des Luftdruckes am Boden . . . 1951–1957, *Met. Abhandl.*, **XIV** (Heft 5). Berlin (Inst. f. Met. Geophys. der Freien Univ.).

Berlin, Free Univ. (1953) Normalwerte des Luftdruckes auf der Nordhemisphäre . . . 1900–1939, *Met. Abhandl.*, **II** (Heft 1). Berlin (Inst. f. Met. Geophys. der Freien Univ.).

BERRY, F. A.; BOLLAY, E. and BEERS, N. R. (1945) *Handbook of Meteorology*. New York (McGraw-Hill).

BIEL, E. (1929) Die Veränderlichkeit der Jahressummen des Niederschlags auf der Erde, *Geogr. Jahresber. Österreich*, **14/15**, 151–80. Vienna.

BIRKELAND, B. J. (1949) Old meteorological observations at Trondheim, *Geofys. Publ.*, **XV**, No. 4. Oslo.

BORISOVA, L. G. and RUDIČEVA, L. M. (1968) The use of special features of natural synoptic seasons in making monthly weather forecasts, *Trudy*, **12**, 12–18. Leningrad (Gidromet. Nauč-Issled. Cent. SSSR).

BREZOWSKY, H.; FLOHN, H. and HESS, P. (1951) Some remarks on the climatology of blocking action, *Tellus*, **3**, 191–4. Stockholm.

BROOKS, C. E. P. (1925) The problem of mild polar climates, *Quart. J., R. Met. Soc.*, **51**, 83–94. London.

BROOKS, C. E. P. (1949) *Climate through the Ages*. London (Ernest Benn) 2nd edn.

BROWNE, I. M. and CRARY, A. P. (1959) The movement of ice in the Arctic Ocean. 'Scientific studies at Fletcher's Ice Island, T-3 (1952–5)', **I**, 37, *AGARD Geophys. Res. Papers No. 63*. Bedford, Mass.

BURMISTROVA, V. D. (1965) On the coupling of anticyclogenesis in the subtropical belts of the North and South Pacific, *Trudy*, **78**, 5–7. (In Russian.) Moscow (Akad. Nauk., Inst. Okeanologii).

CRADDOCK, J. M. (1964) The interannual variability of monthly mean air temperatures over the northern hemisphere, *Sci. Paper*, No. 20. London (Met. Office).

CRADDOCK, J. M. and FLOOD, C. R. (1969) Eigenvectors for representing the 500 mb geopotential surface over the northern hemisphere, *Quart. J., R. Met. Soc.*, **95**, 576–93. London.

CRADDOCK, J. M. and WARD, R. (1962) Some statistical relationships between the temperature anomalies in neighbouring months in Europe and western Siberia, *Sci. Paper*, No. 12. London (Met. Office).

CURRY, L. (1962) Climatic change as a random series, *Ann. Assoc. of American Geographers*, **52** (1), 21–31. Minneapolis.

DAVIDOVA, N. G. (1967) Types of synoptic processes and associated wind fields in oceanic regions of the southern hemisphere, pp. 263–91 in *Polar Meteorology, Tech. Note No. 87*. Geneva (W.M.O., No. 211, T.P. 111).

DZERDZEEVSKI, B. L. (1966) Some aspects of dynamic climatology, *Tellus*, **18**, 751–60. Stockholm.

DZERDZEEVSKI, B. L. (1968) *Circulation Mechanisms in the Atmosphere of the Northern Hemisphere in the 20th Century*. (In Russian.) Moscow (Akad. Nauk., Inst. Geogr.).

References

ELLIOTT, R. D. and SMITH, T. B. (1949) A study of the effects of large blocking highs on the general circulation in the northern hemisphere westerlies, *J. Met.*, **6**, 67–85. Lancaster, Pa.

ESSENWANGER, O. (1953) Statistische Untersuchungen über die Zirkulation der Westdrift in 55°N. Breite, *Berichte des dt. Wetterdienst., U.S. Zone*, **7**, 3–23. Bad Kissingen.

EVJEN, S. (1954) The number of cyclones and anticyclones in northwest and middle Europe, *Met. Ann.*, **3**, Nr. 16. Oslo.

GALLÉ, P. H. (1916) On the relation between fluctuations in the strength of the Trade Winds of the North Atlantic Ocean in summer and departures from normal of the winter temperature in Europe, *Kon. Akad. Wet. te Amsterdam*, **18**, 1435–48. Amsterdam.

GEDEONOV, A. D. (1967) Areas of large anomalies of monthly mean temperature . . ., *Trudy*, **211**, 68–74. (In Russian.) Leningrad (Glav. Geofiz. Obs.).

GIOVINETTO, M. B. and SCHWERDTFEGER, W. (1966) Analysis of a 200-year snow accumulation series from the South Pole, *Archiv f. Met., Geophys. und Biokl.*, **A**, **15**, 227–50. Vienna.

GLASSPOOLE, J. (1925) The relation between annual rainfall over Europe and that at Oxford and Greenwich, *British Rainfall (1925)*, 254–69. London (H.M.S.O.).

HANN, J. (1932) *Handbuch der Klimatologie*, **I** (4te Auflage: revised by K. KNOCH). Stuttgart (Engelhorn).

HAY, R. F. M. (1966) Spring temperature anomalies and summer rainfall, *Weather*, **21**, 219–27. London.

HAY, R. F. M. (1967) The association between autumn and winter circulations near Britain, *Met. Mag.*, **96**, 167–78. London.

HAY, R. F. M. (1968) Relations between summer and September temperature anomalies in central England, *Met. Mag.*, **97**, 76–90. London.

HAY, R. F. M. (1970) October daily pressures . . . near Iceland related to temperature quintiles of the following winter in central England, *Met. Mag.*, **99**, 49–55. London.

HESS, P. and BREZOWSKY, H. (1952) Katalog der Grosswetterlagen Europas, *Berichte des dt. Wetterdienst., U.S. Zone*, **33**. Bad Kissingen.

HOFMEYR, W. L. (1957) Atmospheric sea-level pressure over the Antarctic, in *Meteorology of the Antarctic*, Chapter 4, 51–70 (Ed. M. P. VAN ROOY). Pretoria (Weather Bureau).

JOHANSEN, H. (1958) On continental and oceanic influences in the atmosphere, *Met. Ann.*, **4**, Nr. 8. Oslo.

KATZ (or KAČ), A. L. (1960) *Seasonal Changes in the General Atmospheric Circulation and Long-range Forecasts.* (In Russian.) Leningrad (Gidrometeoizdat).

KIDSON, J. W. and NEWELL, R. E. (1969) Exchange of atmospheric angular momentum between the hemispheres, *Nature*, **221**, 352–3 (25 January 1969). London.

KLEIN, W. H. (1957) Principal tracks and mean frequencies of cyclones and anticyclones in the northern hemisphere, *Res. Paper No. 40.* Washington (U.S. Weather Bureau).

KOCH, L. (1945) The East Greenland ice, *Medd. om Grønland*, **130**, Nr. 3. Copenhagen.

KRAUS, E. B. (1955) Secular changes of tropical rainfall regimes, *Quart. J., R. Met. Soc.*, **81**, 198–210. London.

LAMB, H. H. (1955) Two-way relationship between the snow or ice limit and 1000–500 mb thickness in the overlying atmosphere, *Quart. J., R. Met. Soc.*, **81**, 172–89. London.

LAMB, H. H. (1956) Meteorological results of the *Balaena* expedition, *Geophys. Mem.*, **94**. London (Met. Office).

LAMB, H. H. (1957) On the frequency of gales in the Arctic and Antarctic, *Geogr. J.*, **123**, 287–97. London.

LAMB, H. H. (1959) The southern westerlies, *Quart. J., R. Met. Soc.*, **85**, 1–23. London.

LAMB, H. H. (1966) Climate in the 1960s, *Geogr. J.*, **132** (2), 183–212. London.

LAMB, H. H. (1967) On climatic variations affecting the Far South. Pp. 428–53 in *Polar Meteorology, W.M.O. Tech. Note No. 87.* Geneva (W.M.O., No. 211, T.P. 111).

LAMB, H. H. (in press) British Isles weather types, *Geophys. Mem.*, **116**. London (Met. Office).

LAMB, H. H. and BRITTON, G. P. (1955) General atmospheric circulation and weather variations in the Antarctic, *Geogr. J.*, **121**, 334–49. London.

LAMB, H. H. and JOHNSON, A. I. (1959) Climatic variation and observed changes in the general circulation. (Parts I and II.) *Geogr. Ann.*, **41**, 94–134. Stockholm.

LAMB, H. H. and JOHNSON, A. I. (1961) Ibid, Part III. *Geogr. Ann.*, **43**, 363–400.

LOWNDES, C. A. S. (1962) Wet spells at London, *Met. Mag.*, **91**, 98–104. London.

LOWNDES, C. A. S. (1964) Dry spells of 3 days or more at London, November to April, *Met. Mag.*, **93**, 231–7. London.

LOWNDES, C. A. S. (1965) Forecasting dry spells of 3 days or more in southeast England, May to October, *Met. Mag.*, **94**, 214–51. London.

MILES, M. K. (1961) Factors associated with the formation and persistence of anticyclones over Scandinavia in the winter half year, *Sci. Paper No. 8.* London (Met. Office).

MURRAY, R. (1967*a*) Persistence in monthly mean temperature in central England, *Met. Mag.*, **96**, 356–63. London.

MURRAY, R. (1967*b*) Sequences in monthly rainfall over England and Wales, *Met. Mag.*, **96**, 129–35. London.

MURRAY, R. (1968*a*) Some predictive relationships concerning seasonal rainfall over England and Wales and seasonal temperature in Central England, *Met. Mag.*, **97**, 303–10. London.

MURRAY, R. (1968*b*) Sequences in monthly rainfall over Scotland, *Met. Mag.*, **97**, 181–3. London.

NAMIAS, J. (1952) The annual course of month-to-month persistence in climatic anomalies, *Bull. Amer. Met. Soc.*, **33** (7), 279–85. Lancaster, Pa.

References

NAMIAS, J. (1954) Further aspects of month-to-month persistence in the mid-troposphere, *Bull. Amer. Met. Soc.*, **35** (3), 112–17. Lancaster, Pa.

NAMIAS, J. (1963) Surface–atmosphere interactions as fundamental causes of drought and other climatic fluctuations. Pp. 345–59 in *Changes of Climate – Proc. Rome WMO–UNESCO Symposium.* Paris (UNESCO – *Arid Zone Research Series* **XX**).

PEPPER, J. (1954) *Meteorology of the Falkland Islands and Dependencies 1944–1950.* London (Crown Agents).

PETTERSSEN, S. (1949) Changes in the general circulation associated with the recent climatic variation, *Geogr. Ann.*, **31**, 212–21. Stockholm.

RATCLIFFE, R. A. S. (1965) New criteria concerning fine spells in southeast England, May to October, *Met. Mag.*, **94**, 129–37. London.

RATCLIFFE, R. A. S. (1968) Forecasting monthly rainfall for England and Wales, *Met. Mag.*, **97**, 258–70. London.

RATCLIFFE, R. A. S. and COLLISON, P. (1969) Forecasting rainfall for the summer season in England and Wales, *Met. Mag.*, **98**, 33–9. London.

RATCLIFFE, R. A. S. and PARKER, A. E. (1968) Forecasting wet spells in southeast England, *Met. Mag.*, **97**, 1–12. London.

REX, D. F. (1950) Blocking action in the middle troposphere and its effect upon regional climate, *Tellus*, **2**, 196–211 (Part I), 275–301 (Part II). Stockholm.

REX, D. F. (1951) *Ibid., Tellus*, **3** (Nr. 2), 3–16 (Part III). Stockholm.

RÖDER, W. (1966) Synoptisch-statistische Untersuchungen über den Zerfall blockierender Hochdruckgebiete über den Nordostatlantik und Europa, *Met. Abhandl.*, **LXIX** (Heft 2). Berlin (Inst. f. Met. Geophys. der Freien Univ.). (Cf. GEB ref. in Chapter 5.)

SABINE, E. (1846) On the cause of remarkably mild winters which occasionally occur in England, *Phil. Mag. and J. Sci.*, **28**, 317 ff. London, Edinburgh and Dublin.

SCHELL, I. I. (1956) Interrelations of Arctic ice with the atmosphere and ocean in the North Atlantic–Arctic and adjacent areas, *J. Met.*, **13**, 46–58. Lancaster, Pa.

SHEPPARD, P. A. (1964) Basic ideas on the general circulation of the atmosphere. Pp. 322–31, 361 in *Problems in Palaeoclimatology* (ed. by A. E. M. NAIRN). New York (Wiley).

SORKINA, A. I. (1966) On the coupling of synoptic processes over the northern Atlantic and Pacific Oceans, *Trudy*, **198**, 53–88. (In Russian.) Leningrad (Glav. Geofiz. Obs.).

SUESS, H. E. (1968) Climatic changes, solar activity and the cosmic-ray production rate of natural radiocarbon. Pp. 146–50 in 'Causes of climatic change', *Met. Monographs*, **8**, No. 30. Boston, Mass. (Amer. Met. Soc.).

SUMNER, E. J. (1959) Blocking anticyclones in the Atlantic–European sector of the northern hemisphere, *Met. Mag.*, **88**, 300–11. London.

TAUBER, G. M. (1964) The centres of action of the atmosphere over the southern hemisphere oceans, *Akad. Nauk, Mežd. Geofiz.*, **II** and **X**, *IGY Programme No. 7 (Meteorology) and No. 12 (Oceanology)*. (In Russian.) Moscow.

TAUBER, G. M. (1967) Interconnexions of zonal and meridional circulation and their anomalies in the northern and southern hemispheres, *Gidromet. Nauč-Issled. Cent. SSSR*, **5**, 78–103. (In Russian.) Leningrad.

TAULIS, E. (1934) De la distribution des pluies au Chili, *Materiaux pour l'étude des calamités*, **33**, 3–20. Geneva (Soc. Geogr.).

U.S. Weather Bureau (1952) Normal weather charts for the northern hemisphere, *Tech. Paper No. 21*. Washington.

U.S. Weather Bureau (1965) Climatic relationships between temperatures of adjacent months in the United States. Washington. (Unpublished memorandum.)

WAHL, E. W. (1968) A comparison of the climate of the eastern United States during the 1830s with the current normals, *Monthly Weather Rev.*, **96**, 73–82. Washington.

WANGENHEIM (or VANGENGEJM), G. Ja. (1964) *Catalogue of Macroscopic Processes, according to the Classification of G. Ja. Wangenheim, 1891–1962* (ed. by M. Š. BOLOTINSKAYA and L. Ju. RYZAKOV. (In Russian.) Leningrad (Arkt. i Antarkt. Nauč-Issled. Inst.).

WEYL, P. K. (1968) The role of the oceans in climatic change: a theory of the ice ages. Pp. 37–62 in Causes of climatic change, *Met. Monographs*, **8**, No. 30. Boston, Mass. (Amer. Met. Soc.).

WIESE, W. (1925) Studien über die Erhaltungstendenz der mittleren monatlichen Temperaturanomalien, *Met. Zeit.*, **42** (6), 217–25. Braunschweig.

Chapter 8

AAGAARD, K. (1968) Temperature variations in the Greenland Sea deep water, *Deep Sea Research*, **15**, 281–96. London (Pergamon).

AAGAARD, K. (unpublished) The wind-driven circulation of the Greenland and Norwegian Seas and its variability, *I.C.E.S. Dublin 1969 Symposium on Physical variability in the N. Atlantic*. Copenhagen (I.C.E.S.).

ANDERSON, R. S. (1965) Palaeo-oceanography of the Mediterranean Sea: some consequences of the Würm glaciation. Unpublished thesis of the U.S. Naval Postgraduate School, Monterey, Calif. (Supplied by courtesy of the author and Prof. WARREN THOMPSON.)

BIEN, G. S.; RAKESTRAW, N. W. and SUESS, H. E. (1963) Radiocarbon dating of deep water of the Pacific and Indian Oceans, in *Radioactive Dating*, 159–73. Vienna (Internat. Atomic Energy Agency).

BJERKNES, J. (1966) A possible response of the atmospheric Hadley circulation to equatorial anomalies of ocean temperature, *Tellus*, **18** (4), 820–8. Stockholm.

BJERKNES, J. (1969*a*) Atmospheric teleconnexions from the equatorial Pacific, *Monthly Weather Rev.*, **97** (3), 163–72. Washington.

BJERKNES, J. (1969*b*) Examples of equatorial influence on Southern California rainfall, with appendix on additional atmospheric teleconnexions from the equatorial Pacific,

California Water Resources Center & National Science Foundation, Research Report (by courtesy of the author).

BOYUM, G. (1966) The energy exchange between sea and atmosphere at ocean weather stations M, I and A, *Geofys. Publ.*, **27**, Nr. 7. Oslo.

BROOKS, C. E. P. (1926) *Climate through the Ages.* London (Benn).

BROWNE, I. M. and CRARY, A. P. (1959) The movement of the ice in the Arctic Ocean. Pp. 37–49 in *Scientific studies at Fletcher's Ice Island, T-3 (1952–1955)*, **I** (ed. V. BUSHNELL). *Geophys. Res. Papers No. 63.* Bedford, Mass. (Air Force Cambridge Research Center).

COACHMAN, L. K. and BARNES, C. A. (1963) The movement of Atlantic water in the Arctic Ocean, *Arctic*, **16** (1), 9–16. Montreal.

DEACON, G. E. R. (1945) Water circulation and surface boundaries in the oceans, *Quart. J., R. Met. Soc.*, **71**, 1–10. London.

DEFANT, A. (1941) Die absolute Topographie des physikalischen Meeresniveaus und der Druckflächen sowie die Wasserbewegungen im Atlantischen Ozean, *Wiss. Ergebnisse Deutsch. Atlant. Exped. 'Meteor' 1925–27*, **6** (2). Berlin.

DEFANT, A. (1961) *Physical Oceanography* (2 vols). London (Pergamon).

DIETRICH, G. and KALLE, K. (1957) *Allgemeine Meereskunde.* Berlin (Borntraeger).

DORONIN, JU. P. (1968) On the problem of eradicating the Arctic ice, *Probl. Arkt. Antarkt.*, **28**, 21–8. Leningrad (Arkt. i Antarkt. Nauč.-Issled. Inst.).

EKMAN, V. W. (1905) On the influence of the Earth's rotation on ocean currents, *Ark. f. Mat. Astr. och Fysik*, **2** (No. 11). Stockholm (Kgl. Sv. Vet. Akad.).

ELLET, D. J. (1963) Surface temperatures in the southern North Sea, Jan.–March 1963, *I.C.E.S. Hydrogr. Com., C.M. 1963*, No. 115. Copenhagen.

EWING, M. and DONN, W. L. (1956) A theory of ice ages, *Science*, **123**, 1061–6. New York.

FLETCHER, J. O. (1969) Ice extent on the Southern Ocean and its relation to world climate, *Memorandum RM-5793-NSF.* Santa Monica, Calif. (Rand Corporation).

HUNKINS, K. and KUTSCHALE, H. (1965) Quaternary sedimentation in the Arctic Ocean, *Proc. VII INQUA Congress.* Boulder, Colorado.

KOCH, L. (1945) The East Greenland ice, *Medd. om Grønland*, **130**, Nr. 3. Copenhagen.

KU, T. L. and BROECKER, W. S. (1965) Rates of sedimentation in the Arctic Ocean, *Proc. VII INQUA Congress.* Boulder, Colorado.

LAMB, H. H. (1964) *The English Climate.* London (English Univ. Press).

LAMB, H. H. (1967) On climatic variations affecting the Far South. Pp. 428–37 in *Polar Meteorology. W.M.O. Tech. Note No. 87.* Geneva (W.M.O., No. 211, T.P. 111).

LAMB, H. H. and JOHNSON, A. I. (1966) Secular variations of the atmospheric circulation since 1750, *Geophys. Mem.*, **110**. London (Met. Office).

LEE, A.; RODEWALD, M. and others (1967) Selected papers . . . on . . . fluctuations in sea and air temperature . . . since 1950, *Internat. Com. for the NW. Atlantic Fisheries, Redbook 1967*, Part IV. Dartmouth, N.S.

LIESE, R. (1963) Beitrag zur Ermittlung der Höhe kommender Sturmfluten, *Deutsche gewässerkundl. Mitteilungen*, **7**, 11–23, 35–9. Koblenz.

LIGHTHILL, M. J. (1969) Unsteady wind-driven ocean currents, *Quart. J., R. Met. Soc.*, **93**, 675–88. London.

MAGAARD, L. (1964) Der Cromwell Strom, *Umschau*, **21**, 658–60. Frankfurt/Main.

MALMBERG, S. A. and STEFÁNSSON, U. (unpublished) Recent changes in the water-masses of the East Icelandic Current, *I.C.E.S. Dublin 1969 Symposium on Physical Variability in the N. Atlantic.* Copenhagen (I.C.E.S.).

MODEL, F. (1950) Warmwasserheizung Europas, *Berichte des dt. Wetterdienst. U.S. Zone*, **12**, 51–60. Bad Kissingen.

NEUMANN, G. (1968) *Ocean Currents.* Amsterdam (Elsevier).

PIVAROVA, Z. I. (1968) The long-term variation of solar radiation according to the observations of actinometric stations, *Trudy*, **233**, 17–37. Leningrad (Glav. Geofiz. Obs.).

REID, J. L. (1961) On the geostrophic flow at the surface of the Pacific Ocean with respect to the 1000-decibar surface, *Tellus*, **13** (4), 489–502. Stockholm.

RENNELL, J. (1832) *An Investigation of the Currents of the Atlantic Ocean.* London (Rivington).

ROSSBY, C.-G. (1936) On the frictional force between air and water . . ., *Papers in Phys. Oceanogr. and Met.*, **4** (No. 3). Woods Hole.

SABINE, E. (1846) On the cause of the remarkably mild winters which occasionally occur in England, *London, Edin. & Dublin Phil. Mag. & J. Sci.* (April 1846).

SOKALSKI, JU. M. (1959) *Okeanografiya.* (In Russian.) Leningrad (Gidrometeoizdat), 2nd edn.

STOMMEL, H. (1958) *The Gulf Stream.* Berkeley (Calif. Univ. Press) and Cambridge (Univ. Press).

SVERDRUP, H. U. (1945) *Oceanography for Meteorologists.* London (Allen & Unwin).

SVERDRUP, H. U. (1955) Discussions on the relationship between meteorology and oceanography, *J. Marine Res.*, **14**, 501–3. New Haven, Conn.

SVERDRUP, H. U.; JOHNSON, M. W. and FLEMING, R. H. (1942) *The Oceans.* New York (Prentice-Hall).

TRESHNIKOV, A. F.; MAKSIMOV, I. V. and GINDYSH, B. V. (1966) The great eastward drift of the Antarctic Ocean, *Probl. Arkt. i Antarkt.*, **22**, 18–34. Leningrad (Arkt. i Antarkt. Nauč.-Issled. Inst.).

WELANDER, P. (1959) On the vertically integrated mass transport in the oceans. Pp. 95–101 in *The Atmosphere and the Sea in Motion* (*Rossby Memorial Volume*) (Ed B. BOLIN). New York (Oxford and Rockefeller Inst. Press).

WEYL, P. K. (1968) The role of the oceans in climatic change: a theory of the ice ages. Pp. 37–62 in 'Causes of climatic change', *Met. Monographs*, **8**, No. 30. Boston, Mass. (Amer. Met. Soc.).

References

WÜST, G. (1957) Stromgeschwindigkeiten und Strommengen in den Tiefen des Atlantischen Ozeans, *Wiss. Ergeb. Deutsch. Atlant. Exped. 'Meteor' 1925–27*, **6** (2), 261–420. Berlin.

Chapter 9

BENSON, C. S. (1962) Stratigraphic studies in the snow and firn of the Greenland ice sheet, *Res. Report 70*. Hanover, New Hampshire (U.S. Army Cold Regions Res. and Eng. Lab.).

BENSON, C. S. (1967) Polar regions snow cover. Pp. 1039–63 in *Physics of Snow and Ice*, **I** (2) (Ed. H. OURA). *Proc. Internat. Conf. Low Temp. Sci. 1966*. Sapporo (Hokkaido Univ., Japan).

BERGERON, T. (1935) On the physics of cloud and precipitation, *Mém. Ass. Météor. UGGI. II Lisbon 1933*, 156–78. Paris.

BERGERON, T. (1968) Studies of the oreigenic effects on the areal fine structure of rainfall distribution, *Reports No. 6*. Uppsala (Met. Inst. Kungl. Univ.).

BILHAM, E. G. (1938) *Climate of the British Isles*. London (Macmillan).

BROOKS, C. E. P. (1930) The mean cloudiness over the Earth, *Mem. R. Met. Soc.*, **I**, No. 10, 127–38. London.

CLAPP, P. F. (1964) Global cloud cover for seasons using TIROS nephanalyses, *Monthly Weather Rev.*, **92**, 495–507. Washington.

DEACON, E. L. (1953) Climatic change in Australia since 1880, *Australian J. Phys.*, **6**, 209–18. Melbourne.

DROZDOV, O. A. (1966) On the variation of precipitation over the northern hemisphere with variation of the temperature of the polar basin, *Trudy*, **138**, 3–16. (In Russian.) Leningrad (Glav. Geofiz. Obs.).

EDDY, A. (1966) The Texas coast sea breeze, *Weather*, **21**, 162–70. London.

ELLIOTT, W. P.; EGAMI, R. and ROSSKNECHT, G. (1971) Rainfall at sea, *Nature*, **229**, 108–9 (8 January 1971). London.

FINDEISEN, W. (1938) Der Aufbau der Regenwolken, *Z. f. angewandte Met.*, **55**, 208–25. Berlin.

FLINT, R. F. (1957) *Glacial and Pleistocene Geology*. New York (Wiley).

FLOHN, H. (1941) Häufigkeit, Andauer und Eigenschaften des freien Föhns auf deutschen Bergstationen, *Beiträge Phys. freien Atm.*, **27**, 110–24. Leipzig.

FLOHN, H. (1955) Zur vergleichenden Meteorologie der Hochgebirge, *Archiv. f. Met. Geophys. und Biokl.*, **B, 6**, 193–206. Vienna.

FLOHN, H. (1959) Die heutige Vergletscherung und Schneegrenze in Hochasien, *Abhandl. math.-naturw. Klasse*, **Nr. 14**. Mainz (Akad. der Wiss. und Lit.).

FLOHN, H. (1963) *Klimaschwankungen und grossräumige Klimabeeinflussung*. Köln and Opladen (Westdeutscher Verlag).

FLOHN, H. (1968) Ein Klimaprofil durch die Sierra Nevada de Meridia (Venezuela), *Wetter und Leben*, **20**, 181–91. Vienna.

FLOHN, H. and FRAEDRICH, K. (1961) Tagesperiodische Zirkulation und Niederschlags verteilung am Victoria-See, *Met. Rundschau*, **6**, 157–65. Heidelberg.

FLORA, S. D. (1956) *Hailstorms of the United States*. Norman (Oklahoma Univ. Press).

GENTILLI, J. (1958) *A Geography of Climate*. Perth (Univ. Western Australia Press).

GREEN, F. H. W. (1965) The incidence of low relative humidity in the British Isles, *Met. Mag.*, **94**, 81–8. London.

HADER, F. (1954) Nordostalpine Seehöhenmittel der Niederschlagsmenge, *Archiv. f. Met., Geophys. und Biokl.*, **B, 5**, 331–43. Vienna.

HOCKING, L. M. (1959) The collison efficiencies of small drops, *Quart. J., R. Met. Soc.*, **85**, 44–50. London.

KRAUS, E. B. (1955) Secular changes in tropical rainfall regimes, *Quart. J., R. Met. Soc.*, **81**, 198–210. London.

KRAUS, E. B. (1958) Recent climatic changes, *Nature*, **181**, 666–8 (8 March 1958). London.

LAMB, H. H. (1955*a*) Malta's sea breezes, *Weather*, **10**, 256–64. London.

LAMB, H. H. (1955*b*) Two-way relationships between the snow or ice limit and 1000–500 mb thickness in the overlying atmosphere, *Quart. J., R. Met. Soc.*, **81**, 172–89. London.

LAMB, H. H. (1966) Climate in the 1960s . . ., *Geogr. J.*, **132** (2), 183–212. London.

LAMB, H. H.; LEWIS, R. P. W. and WOODROFFE, A. (1966) Atmospheric circulation and the main climatic variables between 8000 and 0 B.C.: meteorological evidence. Pp. 174–217 in *World Climate 8000–0 B.C.* London (Roy. Met. Soc.).

LANGWAY, C. C. (1967) Stratigraphic analysis of a deep ice core from Greenland, *Research Report 77*. Hanover, New Hampshire (U.S. Army Cold Regions Research and Engineering Laboratory).

LANE, F. W. (1968) *The Elements Rage*, Vol. 1. London (Sphere Books).

LETTAU, B. (1969) The transport of moisture into the Antarctic interior, *Tellus*, **21**, 331–40. Stockholm.

LUDLAM, F. H. (1957) The large-scale physics of clouds, in *The Physics of Clouds*, by B. J. MASON. Oxford (Clarendon Press).

LUDLAM, F. H. (1961) The hailstorm, *Weather*, **16**, 152–62. London.

MANABE, S. and BRYAN, K. (1969) Climate calculations with a combined ocean-atmosphere model, *J. Atmos. Sci.*, **26**, 786–9. Lancaster, Pa.

MASON, B. J. (1957) *The Physics of Clouds*. Oxford (Clarendon Press).

MASON, B. J. (1969) Some outstanding problems in cloud physics, *Quart. J., R. Met. Soc.*, **95**, 449–85. London.

PENMAN, H. L. (1956) Estimating evaporation, *Trans. Amer. Geophys. Union*, **37** (1), 43–6. Washington.

References

SAWYER, J. S. (1956) Rainfall of depressions which pass eastward over or near the British Isles, *Prof. Note No. 118*. London (H.M.S.O. for Met. Office).

SCHERHAG, R. (1948) *Neue Methoden der Wetteranalyse und Wetterprognose*. Berlin (Springer).

SUTCLIFFE, R. C. (1966) *Weather and Climate*. London (Weidenfeld and Nicolson).

SVERDRUP, H. U. (1945) *Oceanography for Meteorologists*. London (Allen & Unwin).

TUCKER, G. B. (1961) Precipitation over the North Atlantic Ocean, *Quart. J., R. Met. Soc.*, **87**, 147–58. London.

TURC, L. (1958) Le bilan d'eau des sols: rélations entre les précipitations, l'évaporation et l'écoulement, *Soc. hydrotech. de France: Compte rendu des troisièmes journées de l'hydraulique, Algiers*, 36–44. Paris.

WEICHMANN, L. and UNGEHEUER, H. (1952) *Klima, Wetter, Mensch*, 57–8. Heidelberg (Quelle & Meyer).

WOODS, J. D. and MASON, B. J. (1964) Experimental determination of collection efficiencies for small water droplets in air, *J. Atmos. Sci.*, **23**, 404–11. Lancaster, Pa.

WÜST, G. (1922) Verdunstung und Niederschlag auf der Erde, *Z. der Gesell. f. Erdkunde* (57th year), Nr. 1–2, 35–43. Berlin.

ZAITSEV, V. A. (1950) Liquid water content and distribution of drops in cumulus clouds, *Trudy*, **81** (No. 19), 122–32. Leningrad (Glav. Geofiz. Obs.). (*Nat. Res. Council of Canada, Tech. Trans. TT-395, Ottawa 1953.*)

Chapter 10

ABBOT, C. G. (1963) Solar variation and weather, *Smithsonian Misc. Coll.*, **146**, No. 3. Washington.

ANGELL, J. K. (1960) An analysis of operational 300 mb transo-sonde flights from Japan in 1957–8, *J. Meteor.*, **17**, 20–35. Lancaster, Pa.

ARAI, Y. (1958) Characteristics of long waves in westerlies related to solar activity, *J. Met. Soc. Japan*, **36**, 46–54. Tokyo.

ASAKURA, T. and KATAYAMA, A. (1958) On the relation between solar activity and the general circulation of the atmosphere, *Papers in Met., Geophys.*, **9** (No. 1) 15–23. Tokyo (Met. Res. Inst.).

Aspen Conference (1962) Proceedings of the conference on the climate of the 11th and 16th centuries, *NCAR Tech. Notes*, **63–1**, 21–2. Boulder, Colorado.

AUER, V. (1958) The Pleistocene of Fuego-Patagonia, Part II, *Suom. Tiedakat. Toimituksia, Series A, III*, **50**. Helsinki.

AUER, V. (1965) *Ibid.*, Part IV. *Ibid., Series A, III*, **80.**

BAUR, F. (1948) *Einführung in die Grosswetterkunde*. Wiesbaden (Dietrich).

BAUR, F. (1949) Zurückführung des Grosswetters auf solare Erscheinungen, *Archiv. f. Met., Geophys. und Biokl.*, **A**, **1**, 358–74. Vienna.

BAUR, F. (1956, 1958) *Physikalisch-statistische Regeln als Grundlagen für Wetter- und Witterungsvorhersagen*, **I**, 1956, **II**, 1958. Frankfurt/Main (Akad. Verlag).

BAUR, F. (1959) *Die Sommerniederschläge Mitteleuropas in den letzten* $1\frac{1}{2}$*-Jahrhunderten und ihre Beziehungen zum Sonnenfleckenzyklus*. Leipzig (Akad. Verlag. Geest and Portig K.-G.).

BELINSKI, N. A. (1957) *Ispol'zovaniye nekotorykh osobennostey atmosfernykh protsessov dlya dologosrochnykh prognozov* (Use of certain features of atmospheric processes for long-range forecasts). (In Russian, English translation Ohio U.S. Air Force base, 1959.) Leningrad (Gidromet. Izdat.).

BERRY, F. A.; BOLLAY, E. and BEERS, N. R. (1945) *Handbook of Meteorology*. New York (McGraw-Hill).

BEZOLD, W. VON (1884) Über zündende Blitze im Königreich Bayern während des Zeitraumes 1833 bis 1882, *Abhandl. des kgl. Bayer. Akad. d. Wiss.*, II Kl., **XV**, I Abt. Munich.

BJERKNES, J. (1963) Climatic change as an ocean-atmosphere problem. Pp. 297–321 in *Changes of Climate – Proc. Rome WMO–UNESCO Symposium*. Paris (UNESCO, *Arid Zone Research Series* **XX**).

BJERKNES, J. (1966) A possible response of the atmospheric Hadley circulation to equatorial anomalies of ocean temperature, *Tellus*, **18** (4), 820–8. Stockholm.

BJERKNES, J. (1969) Atmospheric teleconnections from the equatorial Pacific, *Monthly Weather Rev.*, **97** (3), 163–72. Washington.

BLOCH, M. R. (1965) A hypothesis for the change of ocean levels depending on the albedo of polar ice caps, *Palaeogeogr., Palaeoclim., Palaeoecol.*, **1**, 127–42. Amsterdam (Elsevier).

BODURTHA, F. T. (1952) An investigation of anticyclogenesis in Alaska, *J. Met.*, **9**, 118–25. Lancaster, Pa.

BRAY, J. R. (1965) Forest growth and glacier chronology in northwest North America in relation to solar activity, *Nature*, **205** (4970), 440–3 (30 January 1965). London.

BRAY, J. R. (1967) Variation in atmospheric carbon-14 activity relative to a sunspot-auroral solar index, *Science*, **156** (3775), 640–2 (5 May 1967). New York.

BRAY, J. R. (1968) Glaciation and solar activity since the fifth century B.C. and the solar cycle, *Nature*, **220** (5168), 672–4 (16 November 1968). London.

BRIER, G. W. (1952) Forty-year sea level pressure and sunspots, *Tellus*, **4**, 262–9. Stockholm.

BROOKS, C. E. P. (1923) Variations in the level of the central African lakes, *Geophys. Mem.*, **20**. London (Met. Office).

BROOKS, C. E. P. (1934) Variation of the annual frequency of thunderstorms in relation to sunspots, *Quart. J., R. Met. Soc.*, **60**, 153–65. London.

BROOKS, C. E. P. and QUENNELL, W. A. (1928) The influence of Arctic ice on the subsequent distribution of pressure over the eastern North Atlantic and western Europe, *Geophys. Mem.*, **41**. London (Met. Office).

References

BROWN, P. R. (1963) Climatic fluctuation over the oceans and in the tropical Atlantic. Pp. 109–23 in *Changes of climate – Proc. Rome WMO–UNESCO Symposium*. Paris (UNESCO, *Arid Zone Research Series* **XX**).

BRYSON, R. A. and BAERREIS, D. A. (1967) Possibilities of major climatic modification and their implications: northwest India a case for study, *Bull. Amer. Met. Soc.*, **48**, 136–42. Lancaster, Pa.

BUDYKO, M. (1968) On the causes of climatic variations, *Sveriges Met. och Hydr. Inst. Meddelanden*, Serie B, **28**, 6–13. Stockholm.

CARAPIPERIS, L. N. (1960) On the variation of the Etesian wind within the sunspot cycle, *Geofisica pura e applicata*, **46**, 190–2. Milan.

DAVITAJA, F. F. (1965) On the possible influence of the dust content of the atmosphere on the diminution of glaciers and the warming of climate, *Izvestia, Ser. Geogr.*, **2**, 3–22. (In Russian.) Moscow (Akad. Nauk.).

DEFANT, A. (1924) Die Schwankungen der atmosphärischen Zirkulation über dem nordatlantischen Ozean im 25-jährigen Zeitraum 1881–1905, *Geogr. Ann.*, **6**, 13–41. Stockholm.

DELANEY, A. C.; PARKIN, D. W. and others (1967) Airborne dust collected at Barbados, *Geochimica et Cosmochimica Acta*, **31**, 885–909. London (Pergamon).

DINIES, E. (1957) Ein Beitrag zur Vorhersage der Temperatur des Hochwinters auf Grund der Sonnenfleckentätigkeit in den Vormonaten, *Met. Rundschau*, **10**, 45–7. Berlin.

DINIES, E. (1965) Die solare und terrestrische Vorbereitung kalter Winter in Mitteleuropa, *Met. Rundschau*, **18**, 59–61. Berlin.

DROGAJČEV, D. A. (1968) Long-range forecasting of precipitation in May, *Gidromet. Nauč.-Issled. Cent. SSSR*, **12**, 19–44. (In Russian.) Leningrad.

DUELL, B. and DUELL, G. (1948) The behaviour of barometric pressure during and after solar particle invasions and solar ultraviolet invasions, *Smithsonian Misc. Coll.*, **110**, No. 8. Washington.

DYER, A. J. and HICKS, B. B. (1965) Stratospheric transport of volcanic dust inferred from solar radiation measurements, *Nature*, **208** (5,006) 131–3 (9 October 1965). London.

DZERDZEEVSKI, B. L. (1961) The general circulation of the atmosphere as a necessary link in the sun–climatic variations chain, *Ann. N.Y. Acad. Sci.*, **95** (Art. 1), 188–99. New York.

EATON, G. P. (1963) Volcanic ash deposits as a guide to atmospheric circulation in the geologic past, *J. Geophys. Res.*, **68**, 521–8. Washington.

FEELY, H. W.; SEITZ, H.; LAGOMARSINO, R. J. and BISCAYE, P. E. (1966) Transport and fall-out of stratospheric radioactive debris, *Tellus*, **18**, 316–28. Stockholm.

FLOHN, H. (1951) Solare Vorgänge im Wettergeschehen, *Archiv. f. Met., Geophys. und Biokl.*, **A**, **3**, 303–29. Vienna.

FLOHN, H. (1952) Atmosphärische Zirkulation und erdmagnetisches Feld, *Berichte des dt. Wetterdienstes, U.S. Zone*, **38**, 46–51. Bad Kissingen.

FLOHN, H. (1959) Kontinental-Verschiebungen, Polwanderungen und Vorzeitklimate im Lichte paläomagnetischer Messergebnisse, *Naturwiss. Rund.*, **10**, 375–84. Stuttgart.

FROST, R. (1966) Major storms in west Pakistan in September in relation to the Mangla Dam project, *Met. Mag.*, **95**, 57–63. London.

FUCHS, V. E. (1947) The volcanics of East Africa and pluvial periods, in V. E. FUCHS and T. T. PATERSON on The relation of volcanicity and orogeny to climatic change, *Geol. Mag.*, **84**, 321–33. London.

GILES, K. C. and ANGELL, J. K. (1963) A southern hemisphere horizontal [wind] sounding system: a preliminary study, *Bull. Amer. Met. Soc.*, **44**, 687–96. Lancaster, Pa.

GILL, E. D. (1961) The climates of Gondwanaland in Kainozoic times. Pp. 332–51 in *Descriptive Palaeoclimatology* (ed. A. E. M. NAIRN). New York (Interscience Publishers).

GIRS, A. A. (1956) Long period changes of the atmospheric circulation types and variations of solar activity, *Met. i Gidr.*, **10**. (In Russian.) Leningrad.

GIRS, A. A. (1967) On peculiarities of the Arctic meteorological regime in different stages of the circulation epoch of 1949–1964. Pp. 454–77 in *Polar Meteorology, W.M.O. Tech. Note No. 87*. Geneva (W.M.O., No. 211, T.P. 111).

HELLAND-HANSEN, B. and NANSEN, F. (1917) *Temperatur-Schwankungen des nordatlantischen Ozeans und in der Atmosphäre*. Kristiania (Dybwad).

HOUGHTON, D. M. (1958) Heat sources and sinks at the Earth's surface, *Met. Mag.*, **67**, 132–43. London.

HUMPHREYS, W. J. (1940) *Physics of the Air*. New York and London (McGraw-Hill).

JULIAN, P.; RUFF, I. and NORDØ, J. (1957) Possible responses of terrestrial atmospheric circulation to changes in solar activity, *Tech. Rep. No. 1*. Boulder (Univ. Colorado High Altitude Observatory, Institute for Solar-Terrestrial Research).

KHRABROV, JU. B. (1958) Some problems in taking account of solar activity in the making of monthly weather forecasts, *Trudy*, **71**, 11–16. (In Russian.) Moscow (Centr. Inst. Prog.).

KIRCH, R. (1966) Temperaturverhältnisse in der Arktis während der letzten 50 Jahre, *Met. Abhandl.*, **69** (3), Berlin (Free Univ., Inst. f. Met. Geophys.).

KLEJMENOVA, E. P. (1967) On the variation of thunderstorm activity in the solar cycle, *Glav. Uprav. Gidromet. Služb., Met. Gidr.*, **8**, 64–8. (In Russian.) Leningrad.

KOCH, L. (1945) The East Greenland ice, *Medd. om Grønland*, **130**, Nr. 3. Copenhagen.

KÖPPEN, W. (1873) Über mehrjährige Perioden der Witterung, insbesondere über die 11-jährige Periode der Temperatur, *Z. f. Met.*, **8**, 241–8, 257–67. Vienna.

KÖPPEN, W. (1914) Lufttemperaturen, Sonnenflecke und Vulkanausbrüche, *Met. Zeit.*, **31**, 305–28. Braunschweig.

KRAUS, E. B. (1955) Secular changes in tropical rainfall regimes, *Quart. J., R. Met. Soc.*, **81**, 198–210. London.

KRAUS, E. B. (1958) Recent climatic changes, *Nature*, **181**, 666–8 (8 March 1958). London.

References

KRIVSKY, L. (1953) The long-range variability of annual precipitation in Prague-Klementinum 1805–1951 and in relation to solar activity, *Publications of the Astrophys. Obsy. of the Czechoslovak Akad. Sci.*, **23** (by Z. GREGOR and L. KRIVSKY), 37–72. Prague.

LAEVASTU, T. (1965) Daily heat exchange in the North Pacific: its effects on the ocean and its relation to weather, *I.C.N.A.F. Special publication No. 6. Environmental symposium, Rome 1964*. Dartmouth, Nova Scotia.

LAMB, H. H. (1955) Two-way relationships between the snow or ice limit and 1000–500 mb thicknesses in the overlying atmosphere, *Quart. J., R. Met. Soc.*, **81**, 171–89. London.

LAMB, H. H. (1961) Climatic change within historical time as seen in circulation maps and diagrams, *Ann. N. Y. Acad. Sci.*, **95** (Art. 1), 124–61. New York.

LAMB, H. H. (1964) The role of atmosphere and oceans in relation to climatic changes and the growth of ice-sheets on land. Pp. 332–48, 361–2 in A. E. M. NAIRN, *Problems in Palaeoclimatology*. New York (Wiley).
Also in H. H. LAMB *The Changing Climate*. London (Methuen) 1966.

LAMB, H. H. (1966*a*) *The Changing Climate*. London (Methuen).

LAMB, H. H. (1966*b*) Climate in the 1960s, *Geogr. J.*, **132**, 183–212. London.

LAMB, H. H. (1967*a*) On climatic variations affecting the Far South. Pp. 428–53 in *Polar Meteorology. W.M.O. Tech. Note No. 87*, Geneva (W.M.O., No. 211, T.P. 111).

LAMB, H. H. (1967*b*) The problem of 'Thompson Island': volcanic eruptions and meteorological evidence, *Brit. Antarct. Surv. Bull.*, **13**, 85–8. London.

LAMB, H. H. (1970) Volcanic dust in the atmosphere, *Phil. Trans., Roy. Soc.*, **A**, **266** (No. 1178), 426–533. London.

LAMB, H. H. (1970) Volcanic dust in the atmosphere, *Phil. Trans., Roy. Soc.*, **A**, **266** (No. 1178), 426–533. London.

LAMB, H. H. and JOHNSON, A. I. (1959, 1961) Climatic variation and observed changes in the general circulation, *Geogr. Ann.*, **41**, 94–134; **43**, 393–400. Stockholm.

LAMB, H. H. and JOHNSON, A. I. (1966) Secular variations of the atmospheric circulation since 1750, *Geophys. Mem.*, **110**. London (Met. Office).

LANDSBERG, H. E. (1967) Two centuries of New England climate, *Weatherwise*, **20**, 52–7. Boston. (See also J. M. MITCHELL and H. E. LANDSBERG, *J. Geophys. Res.*, **71**, 5844–9.)

LANGWAY, C. C. (1967) Stratigraphic analysis of a deep ice core from Greenland, *Res. Report 77, U.S. Army C.R.R.E.L.* Hanover, New Hampshire.

LINK, F. (1958) Kometen, Sonnentätigkeit und Klimaschwankungen, *Die Sterne*, **34** (7–8), 129–40. Leipzig (Barth).

LINK, F. (1964) Manifestations de l'activité solaire dans le passé historique, *Planet. Space Sci.*, **12**, 333–48. London (Pergamon).

LINK, F. and LINKOVA, Z. (1959) Méthodes astronomiques dans le climatologie historique: période solaire et climatique de 400 ans, *Studia Geoph. et Geod.*, **3**, 43–61. Prague.

LONDON, J.; RUFF, I. and TICK, L. J. (1959) The relationship between geomagnetic variations and the circulation at 100 mb, *J. Geophys. Res.*, **64** (11), 1827–33. Washington.

MCCORMICK, R. A. and LUDWIG, J. L. (1967) Climate modification by atmospheric aerosols, *Science*, **156** (3870), 1358–9 (9 June 1967). New York.

MANLEY, G. (1958) Temperature trends in England, 1698–1957, *Archiv. f. Met., Geophys. und Biokl.*, **B**, **9**, 413–33. Vienna.

MANLEY, G. (1964) The evolution of the climatic environment. Chapter 9 in *The British Isles: a systematic geography* (Ed. J. W. WATSON and J. B. SISSONS). London (R. Geogr. Soc. for I.G.U., **XX**).

MAURY, M. F. (1852) *Wind and Current Charts of the North Atlantic.* Washington (Bureau of Ordnance and Hydrography, U.S. Navy).

MIRONOVITCH, V. (1960) Sur l'évolution séculaire de l'activité solaire et ses liaisons avec la circulation atmosphérique générale, *Met. Abhandl.*, **9** (3). Berlin (Inst. f. Met. Geophys. der Freien Universität).

MITCHELL, J. M. (1961) Recent secular changes of global temperature, *Ann. N.Y. Acad. Sci.*, **95** (1), 235–50. New York.

MITCHELL, J. M. (1965) The solar inconstant, *Proc. Seminar on Possible Responses of Weather Phenomena to variable Extraterrestrial Influences. NCAR Tech. Note*, **TN-8**. Boulder, Colorado.

MOHN, H. (1877) Askeregnen den 29–30 marts 1875, *Forh. Vid. Selsk. Christiania*, No. 10. Christiania.

MURRAY, R. and RATCLIFFE, R. A. S. (1969) The summer weather of 1968: related atmospheric circulation and sea temperature patterns, *Met. Mag.*, **98**, 201–19. London.

MUSTEL, E. R. (1967(?)) *Manifestation of Solar Activity in the Troposphere and the Stratosphere.* Moscow (Astron. Council of the USSR Acad. Sci. and USSR Hydromet. Centre – in English).

MYRBACH, O. (1935) Sonnenfleckenzyklus und Gewittertätigkeit in Wien, Kremsmünster und Bayern, *Met. Zeit.*, **52**, 225–7. Braunschweig.

NAMIAS, J. (1963*a*) Surface-atmosphere interactions as fundamental causes of drought and other climatic fluctuations. Pp. 345–59 in *Changes of Climate – Proc. Rome WMO–UNESCO Symposium.* Paris (UNESCO – *Arid Zone Research Series* **XX**).

NAMIAS, J. (1963*b*) Large-scale air-sea interactions over the North Pacific from summer 1962 through the subsequent winter, *J. Geophys. Res.*, **68** (22), 6171–86. Washington.

NAMIAS, J. (1965) Short-period climatic fluctuations, *Science*, **147** (3659), 696–706 (12 February 1965). New York.

NAMIAS, J. (1969) Seasonal interactions between the North Pacific Ocean and the atmosphere during the 1960s, *Monthly Weather Rev.*, **97** (3), 173–92. Washington.

References

NEY, E. P. (1959) Cosmic radiation and the weather, *Nature*, **183**, 451–2 (14 February 1959). London.

OUTI, M. (1961, 1962) Climatic variations in the North Pacific subtropical zone and solar activity during the past ten centuries, *Bull. Kyoto Gakugei Univ.*, **B**, No. 19, 41–61 (1961); No. 20, 25–48 (1962).

PALMER, C. E. (1953) The impulsive generation of certain changes in the tropospheric circulation, *J. Met.*, **10**, 1–9. Lancaster, Pa.

PIGGOTT, W. R. and SHAPLEY, A. H. (1962) The ionosphere over Antarctica. Pp. 111–26 in *Antarctic Research* (Ed. WEXLER, H., RUBIN and CASKEY). *Amer. Geophys. Union, Geophys. Monograph No. 7*, Washington.

PRESSMAN, J.; REIDY, W. and TANK, W. (1963) Rocket pollution of the atmosphere, *IAS Paper*, **63–83**. New York (Inst. of Aerospace Sciences).

RAGHAVAN, K. (1961) Solar influences on the monsoon storms in the Bay of Bengal, *Weather*, **16**, 59–64. London.

RATCLIFFE, R. A. S. and MURRAY, R. (1970) New lag associations between North Atlantic sea temperatures and European pressure applied to long-range weather forecasting, *Quart. J., R. Met. Soc.*, **96**, 226–46. London.

RENNELL, J. (1832) *An Investigation of the Currents of the Atlantic Ocean*. London (published by Rivington for Lady Rodd). (The sea temperatures characteristic for each month are entered as point values on Rennell's Current Charts, which are separate from the volume.)

ROBERTS, W. O. (1963) Does variable solar activity affect stratospheric circulation?, *Met. Abhandl.*, **36**, 341–52. Berlin (Inst. f. Met. Geophys. der Freien Universität).

RODEWALD, M. (1956, 1958*a*) Beiträge zur Klimaschwankung im Meere: 6. Die Temperaturen der pazifischen Küstengewässer Nordamerikas 1920–1951, *Deutsche Hydr. Zeit.*, **9**, 187–8; and *Ibid.*, 8. 1920–55, *Deutsche Hydr. Zeit.*, **10**, 147–51. Hamburg.

RODEWALD, M. (1958*b*, 1964 and 1968) Beiträge zur Klimaschwankung im Meere: 10. Die Anomalie der Wassertemperaturen . . . an der Küste Perus im Jahre 1955, *Deutsche Hydr. Zeit.*, **10**, 78–82; see also *Deutsche Hydr. Zeit.*, **17**, 105–14 and **21**, 21–2. Hamburg.

RUBASHEV, B. M. (1964) *Problems of Solar Activity*. Moscow–Leningrad (Main Astronom. Obsy). (English translation by NASA translation service NASA TT F-244, Washington.)

SAPPER, K. (1917) Beiträge zur Geographie der tätigen Vulkane, *Z. f. Vulk.*, **3**, 65–197. Berlin.

SAPPER, K. (1927) *Vulkankunde*. Stuttgart (Engelhorn).

SAWYER, J. S. (1965) Notes on the possible physical causes of long-term weather anomalies, *W.M.O. Tech. Note, No. 66*, 227–48. Geneva (W.M.O. No. 162, T.P. 79).

SAZONOV, B. I. (1964) *High-level Pressure Formations and Solar Activity*. Leningrad (Glav. Geofiz. Obs., Gidrometeoizdat, 131 pp. in Russian).

SAZONOV, B. I. (1966) On the possible role of cosmic ray particles in solar-tropospheric relationships, *Trudy*, **198**, 89–106. (In Russian.) Leningrad (Glav. Geofiz. Obs.).

SAZONOV, B. I. (1967) The latitude variation of the state of disturbance of the atmosphere, *Trudy*, **211**, 3–16. (In Russian.) Leningrad (Glav. Geofiz. Obs.).

SAZONOV, B. I. and KASOGLEDOVA, S. V. (1968) Conditions for the intensification of meridional circulation patterns in high latitudes, *Trudy*, **227**, 21–7. (In Russian.) Leningrad (Glav. Geofiz. Obs.).

SAZONOV, B. I. and ZORINA, V. P. (1967) On the nature of stratospheric warming, *Trudy*, **211**, 31–45. (In Russian.) Leningrad (Glav. Geofiz. Obs.).

SCHAEFER, V. J. (1966) Ice nuclei from automobile exhausts and iodine vapour, *Science*, **154**, 1555–7. Washington.

SCHELL, I. I. (1961) The ice off Iceland and the climates of the last 1200 years, *Geogr. Ann.*, **43**, 354–62. Stockholm.

SCHERHAG, R. (1950) Bestehen Zusammenhänge zwischen der elfjährigen Sonnenfleckenperiode und der allgemeinen Zirkulation?, *Deutsche Hydr. Zeit.*, **3**, 108–11. Hamburg.

SCHOVE, D. J. (1955) The sunspot cycle 649 B.C. to A.D. 2000, *J. Geophys. Res.*, **60** (2), 127–46. Washington.

SCHOVE, D. J. (1962) Auroral numbers since 500 B.C., *J. Brit. Astronom. Ass.*, **72** (1), 30–5. London.

SCHUURMANS, C. J. E. (1969) The influence of solar flares upon the tropospheric circulation, *Med. en Verhandl.*, **92**. De Bilt (K. Ned. Met. Inst.).

SCHWARZBACH, M. (1961) *Das Klima der Vorzeit.* Stuttgart (Enke), 2. Auflage.

SEMENOV, V. G. (1967) The undersurface's role in forming atmospheric circulation anomalies. Chapter VIII, pp. 210–28 (in Russian), in R. F. BURLUCKI and others, *Fluctuations in the General Circulation of the Atmosphere and Long-range Weather Forecasts.* Leningrad (Gidromet. Izdat.).

SEPTER, E. (1926) Sonnenflecken und Gewitter in Sibirien, *Met. Zeit.*, **43**, 229–31. Braunschweig.

SMED, J. (1946, 1948) Hydrography, Cons. perm. internat. pour l'exploration de la mer, *Annales Biologiques*, **3**, 10–24, and **5**, 10–15. Copenhagen.

SMED, J. (1965) Variation of the temperature of the surface water in areas of the northern North Atlantic, 1876–1961, *Internat. Comm. for the northwest Atlantic Fisheries, Special Pubn.* Nr. 6. Dartmouth, Nova Scotia.

SPEERSCHNEIDER, C. I. H. (1915) Om isforholdence i Danske farvande i aeldre og nyere tid: aarene 690–1860, *Meddelelser*, **2**. Copenhagen (Dansk. Met. Inst.).

SPEERSCHNEIDER, C. I. H. (1931) The state of the ice in the Davis Strait 1820–1930, *Meddelelser*, **8**. Copenhagen (Dansk. Met. Inst.).

STAGG, J. M. (1931) Atmospheric pressure and the state of the Earth's magnetism, *Nature*, **127**, 4020. London.

References

STAGG, J. M. (1964) Atmospheric contamination and climatic stability, *New Scientist*, **408**, 627–8 (10 September 1964). London.

STRANZ, D. (1959) Solar activity and the altitude of the tropopause near the equator, *J. Atm. & Terrest. Phys.*, **16**, 180–2. London.

STUBBS, P. (1963) May rockets disturb the weather?, *New Scientist*, **350**, 230–3 (1 August 1963). London.

SUDA, T. (1963) The effect of solar activity on some meteorological phenomena, *Geophys. Mag.*, **31**, 551–621. Tokyo (Japanese Met. Agency).

SUESS, H. E. (1965) Secular variations of the cosmic-ray-produced carbon-14 in the atmosphere and their interpretations, *J. Geophys. Res.*, **70**, 5937–52. Washington.

SUESS, H. E. (1968) Climatic changes, solar activity, and the cosmic-ray production rate of natural radiocarbon, *Met. Monographs*, **8**, No. 30, 146–50. Boston, Mass. (Amer. Met. Soc.).

SUMNER, E. J. (1951) Unusual deepening of a frontal depression over the British Isles, *Met. Mag.*, **80**, 130–9. London.

TAULIS, E. (1934) De la distribution des pluies au Chili, *Matériaux pour l'étude des calamités*, No. **33**, 3–20. Geneva (Soc. de Géogr.).

TETRODE, P. (1952) Sunspots and the occurrence of unprecedented excesses of mean temperature over a wide area, *Weather*, **7**, 14–15. London.

THORARINSSON, S. (1944) Tefrokronologiska studier på Island, *Geogr. Ann.*, **26**, 1–217. Stockholm.

THORARINSSON, S. *et al.* (1959) On the geology and geomorphology of Iceland: V. Tephra layers and Tephrochronology, *Geogr. Ann.*, **41**, 152–6. Stockholm.

TROUP, A. J. (1962) A secular change in the relation between the sunspot cycle and temperature in the tropics, *Geofisica pura e applicata*, **51**, 184–98. Milan.

TUCKER, G. B. (1964) Solar influence on the weather, *Weather*, **19**, 302–11. London.

VINOGRADOV, N. D. (1967) The relation between atmospheric circulation types and surface water temperature in the northern Atlantic, *Probl. Arkt. Antarkt.*, **25**, 44–53. (In Russian.) Leningrad (Arkt. i Antarkt. Inst., Nauč.-Issled).

VITELS, L. A. (1949) *Byull, K.I.S.O.*, No. 1. (In Russian.)

VITELS, L. A. (1952) Singularities in the variations of intensity of the Icelandic and north European depressions, *Bull. Acad. Sci., Sér. Geophys.*, 1952, No. **4**, 98–102. Moscow.

WAGNER, A. (1940) *Klima-änderungen und Klimaschwankungen.* Braunschweig (Vieweg).

WAHL, E. W. (1968) A comparison of the climate of the eastern United States during the 1830s with current normals, *Monthly Weather Rev.*, **96** (2), 73–82. Washington.

WALKER, G. T. (1915) Correlations in seasonal variations of weather: IV – sunspots and rainfall; V – sunspots and temperature; VI – sunspots and pressure, *Mem. of the Indian Met. Department*, **21**, Nos. 10–12. Poona.

WARD, F. W. (1960) Atmospheric temperatures and solar activity – I. 100 mb in the northern hemisphere auroral zone, *J. Met.*, **17**, 130–4. Lancaster, Pa.

WEXLER, H. (1936) A note on dust in the atmosphere, *Bull. Amer. Met. Soc.*, **17**, 303–5. Boston, Mass.

WEXLER, H. (1956) Variations in insolation, general circulation and climate, *Tellus*, **8**, 480–94. Stockholm.

WIESE, W. (1924) Polareis und atmosphärische Schwankungen, *Geogr. Ann.*, **6**, 273–99. Stockholm.

WILLETT, H. C. (1949*a*) Long-period fluctuations of the general circulation of the atmosphere, *J. Met.*, **6** (1), 34–50. Lancaster, Pa.

WILLETT, H. C. (1949*b*) Solar variability as a factor in the fluctuations of climate in geological time, *Geogr. Ann.*, **36**, 295–315. Stockholm.

WILLETT, H. C. (1964) Evidence of solar climatic relationships, in *Weather and our Food Supply*, 123–51. Ames, Iowa (Iowa State Univ., CAED-20).

YAMAMOTO, T. (1967) On the climatic change along the current of historical times in Japan and its surroundings, *Jap. Phil. Mag.*, **76** (3), 115–41. Tokyo.

Appendix I

BAUR, F. (1964) Neufestsetzung der Epochen der Minima und Maxima der Sonnenflecken, *Met. Abhandl.*, **L**, Heft 3. Berlin (Inst. f. Met. und Geophys., Free Univ. Berlin).

BRAY, J. R. (1965) Forest growth and glacier chronology in northwest North America in relation to solar activity, *Nature*, **205** (No. 4970) 440–3. London (30 January 1965).

GLEISSBERG, W. (1958) The 80-years sunspot cycle, *J. Brit. Astronom. Ass.*, **68**, No. 4, 142–52. London.

GNEVYSHEV, M. N. (1967) On the 11 years cycle of solar activity, *Solar Activity*, **1**, 107–20. Dordrecht (Reidel & Co.).

KOSSINNA, E. (1921) Die Tiefen des Weltmeeres, *Inst. f. Meereskunde, Veröff., N.F., A. Geogr.-naturwiss.*, Reihe, Heft 9. Berlin (Universität).

LINK, F. (1964) Manifestations de l'activité solaire dans le passé historique, *Planetary and Space Science*, **12**, 333–48. London (Pergamon).

SCHOVE, D. J. (1955) The sunspot cycle, 649 B.C. to A.D. 2000, *J. Geophys. Res.*, **60**, 127–46. London.

SUESS, H. E. (1965) Secular variations of the cosmic ray-produced carbon-14 in the atmosphere and their interpretations, *J. Geophys. Res.*, **70** (No. 23) 5937–52. Washington.

VITINSKII, YU. I. (1965) *Solar Activity Forecasting*. (Akad. Nauk. SSSR. Leningrad 1962). English edn, Jerusalem (Israel Program for Scientific Translations).

Additional sources referred to in connexion with Table App. I.10 (*b* iii)

FLOHN, H. (1963) Zur meteorologischen Interpretation der pleistozänen Klimaschwankungen, *Eiszeit und Gegenwart*, **14**, 153–60. Öhringen/Württ.

References

GIOVINETTO, M. R. and SCHWERDTFEGER, W. (1966) Analysis of a 200-year snow accumulation series from the South Pole, *Archiv. f. Met., Geophys. und Biokl.*, **A, 15**, 227–50. Vienna.

LAMB, H. H. (1967) Climatic variations affecting the Far South, *Proceedings of the I.C.P.M.–S.C.A.R.–W.M.O. Symposium on Polar Meteorology* (*Geneva, September 1966*). Geneva (W.M.O. Tech. Note No. 8).

LAMB, H. H.; PROBERT-JONES, J. R. and SHEARD, J. W. (1962) A new advance of the Jan Mayen glaciers and a remarkable increase of precipitation, *J. Glaciol.*, **4**, 355–65. Cambridge.

SCHWARZBACH, M. (1961) *Das Klima der Vorzeit.* Stuttgart (Enke Verlag) 2nd edn.

WOLDSTEDT, P. (1958) *Das Eiszeitalter: II. Europa, Vorderasien Nordafrika . . .* Stuttgart (Enke Verlag).

WOLDSTEDT, P. (1965) *Das Eiszeitalter: III. Afrika, Asien, Australien und Amerika.* Stuttgart (Enke Verlag).

Appendix III

BERNARD, E. A. (1962) Théorie astronomique des pluviaux et interpluviaux du Quaternaire africain, *Acad. Roy. des Sci. d'Outre-Mer, Classe des Sci. Nat. et Méd.* (*Nouvelle série*) **XII**, fasc. 1. Brussels.

MILANKOVITCH, M. (1930) Mathematische Klimalehre und astronomische Theorie der Klimaschwankungen, *Handbuch der Klimatologie* (KÖPPEN & GEIGER), **I**, Teil **A**. Berlin.

PETTERSSON, O. (1930) Flodkraften och vattenutbyttet mellan de tropiska och de polara haven, *Svenska Hydrografisk-Biologiska Kommissionens Skrifter. Ny Serie, Hydrografi*, **VIII**. Göteborg.

PETTERSSON, O. (1930) The tidal force, *Geogr. Ann.*, **12**, 261–322. Stockholm.

Appendix IV

LAMB, H. H. and JOHNSON, A. I. (1966) Secular variation of the atmospheric circulation since 1750, *Geophys. Mem.*, **110**, London (Met. Office).

Chapter 11

Bibliography on climatic classifications

ALISOV, B. P. (1963) Geographical types of climates, *Met. i Gidr.*, **6**, 16–25. Leningrad.

BLÜTHGEN, J. (1966) *Allgemeine Klimageographic.* Berlin (Walter de Gruyter – 2. Auflage: in the series *Lehrbuch der allgemeinen Geographie*).

BUDYKO, M. I. (1955) Klimatičeskie uslovija uvlažnenija na materikach, *Soobščenie* **1–2**, *Izvestija Akad. Nauk SSSR, Ser. geogr.* Moscow.

FLOHN, H. (1957) Zur Frage der Einteilung der Klimazonen, *Erdkunde*, **11**, 161–75. Bonn.

GENTILLI, J. (1958) *A Geography of Climate*. Perth (Univ. Western Australia) 2nd edn.
GIESE, E. (1969) Die Klimaklassifikation von BUDYKO und GRIGOR'EV, *Erdkunde*, **23**, 317–25. Bonn.
GORCZYNSKI, W. (1948) Decimal system of world climates, *Przeglad Met. Hydr.*, **1**, 30–48. Warsaw.
GRIGOR'EV, A. A. and BUDYKO, M. I. (1959) Klassifikacija klimatov SSSR, *Izvestija Akad. Nauk. SSSR, Ser. geogr.* Moscow.
HENDL, M. (1960) Entwurf einer genetischen Klimaklassifikation auf Zirkulationsbasis, *Z. f. Met.*, **14**, 46–50. Berlin.
HENDL, M. (1963) *Einführung in die physikalische Klimatologie*, **II**, *Systematische Klimatologie*. Berlin (VEB Deutscher Verlag der Wiss.).
KÖPPEN, W. (1900) Versuch einer Klassifikation der Klimate vorzugsweise nach ihren Beziehungen zur Pflanzenwelt, *Geogr. Z.*, **6**, 593–611, 657–79. Leipzig.
KÖPPEN, W. (1918) Klassifikation der Klimate nach Temperatur, Niederschlag und Jahreslauf, *Petermanns geogr. Mitteilungen*, **64**, 193–203, 243–8. Gotha.
KÖPPEN, W. (1923) *Die Klimate der Erde*. Berlin, Leipzig.
KÖPPEN, W. (1931) *Grundriss der Klimakunde*. Berlin, Leipzig.
KÖPPEN, W. (1936) Der geographische System der Klimate, *Handbuch der Klimatologie* (KÖPPEN & GEIGER), Band **I**, Teil C. Berlin (Borntraeger).
PENCK, A. (1910) Versuch einer Klimaklassifikation auf physiographischer Grundlage, *Sitz.-Ber. Kgl. Preuss. Akad. Wiss., Phys.-math. Klasse*, **12**, 236–46. Berlin.
SUPAN, A. (1879) Die Temperaturzonen der Erde, *A. Petermanns Mitteilungen aus J. Perthe's Geographischer Anstalt* (**25**. Jahrgang). Gotha.
THORNTHWAITE, C. W. (1933) The climates of the Earth, *Geogr. Rev.*, **23**, 433–40. New York.
THORNTHWAITE, C. W. (1943) Problems in the classification of climates, *Geogr. Rev.*, **33**, 233–55. New York.
THORNTHWAITE, C. W. (1948) An approach to a rational classification of climate, *Geogr. Rev.*, **38**, 55–94. New York.
TREWARTHA, G. T. (1954) *An Introduction to Climate*. New York (McGraw-Hill).
WARD, R. DE C. (1905) The climatic zones and their subdivisions, *Bull. Amer. Geogr. Soc.*, **5**. New York.

Bibliography on urbanization effects

CHANDLER, T. J. (1962) London's urban climate, *Georg. J.*, **128** (3), 279–302. London.
DRONIA, H. (1967) Der Stadteinfluss auf den weltweiten Temperaturtrend, *Met. Abhandl.*, **74** (4). Berlin (Inst. f. Met. Geophys. der Freien Univ. Berlin).
ERIKSEN, W. (1964) Beiträge zum Stadtklima von Kiel, *Schriften des Geogr. Inst. der Univ.*, **22** (1). Kiel.

References

PETERSON, J. T. (1969) *The Climate of Cities: a Survey of Recent Literature*. Raleigh, North Carolina (U.S. Dept. of Health, Education & Welfare: Public Health Service, Air Pollution Control Admin.).

STEINHAUSER, F.; ECKEL, O. and SAUBERER, F. (1955, 1957, 1959) *Klima und Bioklima von Wien*. Vienna (Stadtmagistrat & Gesell. f. Met. – in 3 parts).

World Meteorological Organization (1970) Urban climates, *Proceedings of the W.M.O.–W.H.O. Symposium on Urban Climates and Building Technology, Brussels 1969. W.M.O. Tech. Note No. 108*. Geneva (W.M.O., No. 254, T.P. 141).

Geographical Index

Those places mentioned only in the climatic table (pp. 522–44) are not included.

Geographical Index

Subject Index

Subject Index

Subject Index

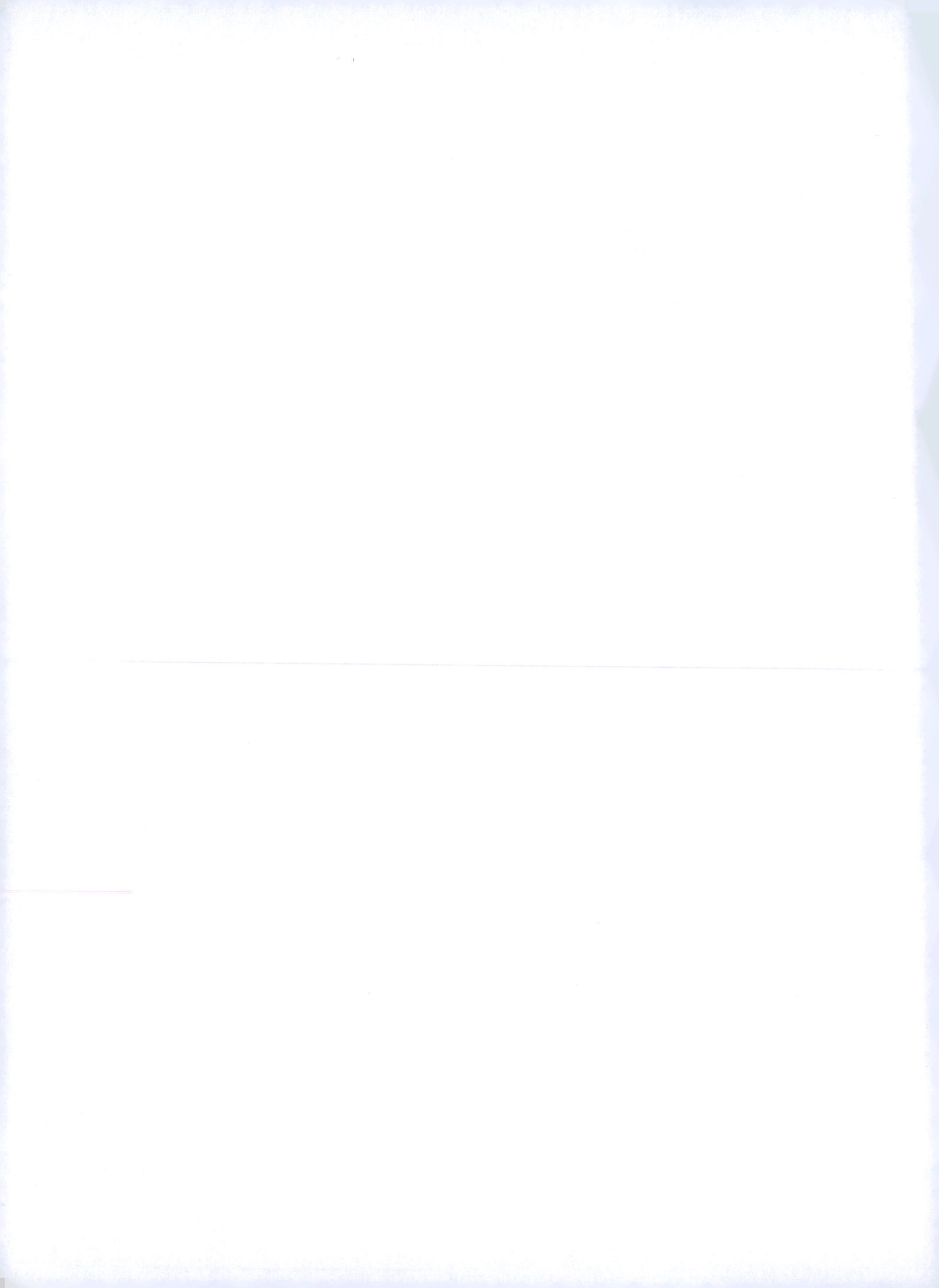

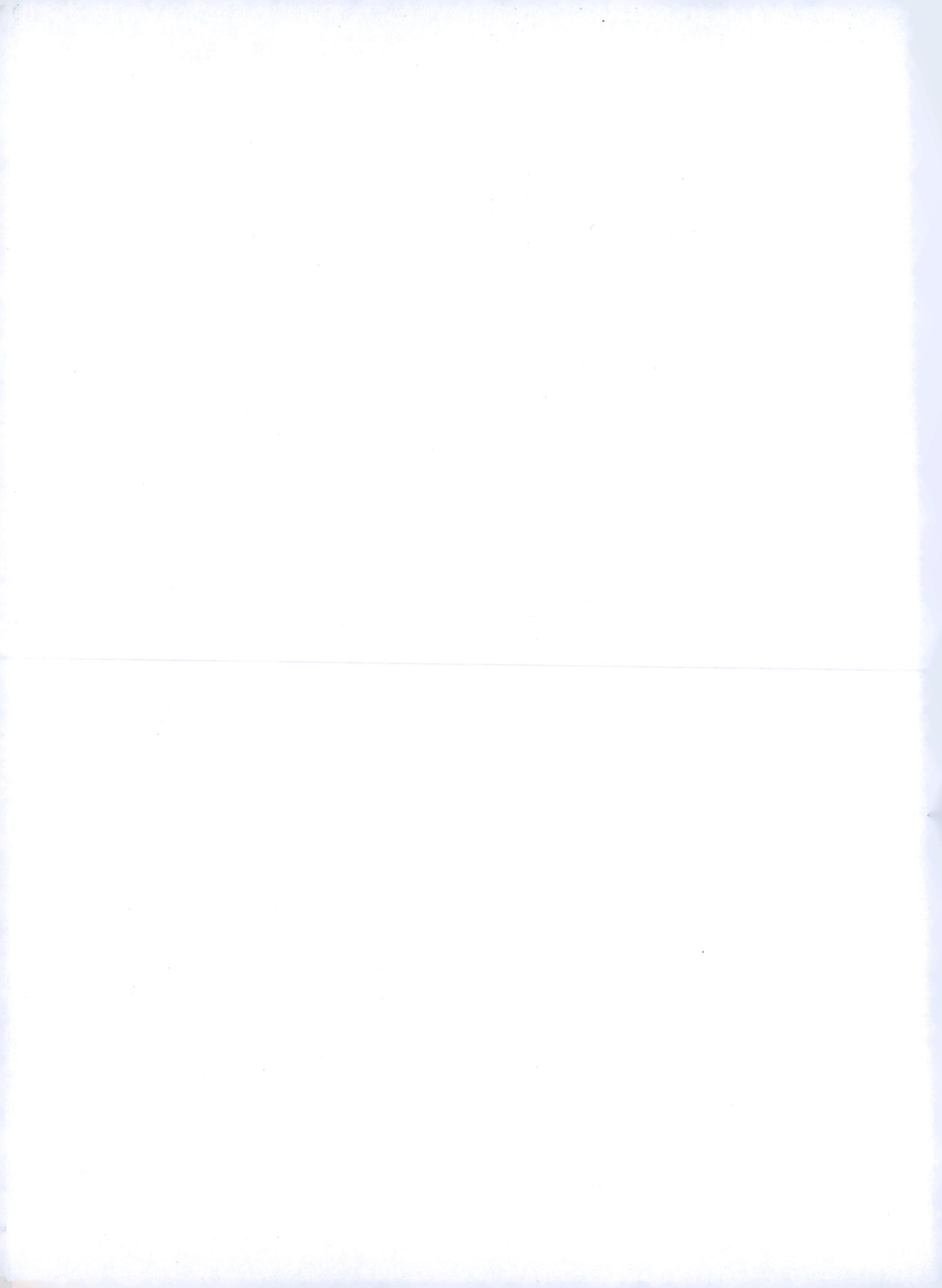